THE U.S. ATLAS
OF NUCLEAR FALLOUT
1951-1962

VOLUME I. TOTAL FALLOUT

THE U.S. ATLAS OF NUCLEAR FALLOUT 1951-1962

VOLUME I: TOTAL FALLOUT

RICHARD L. MILLER

WITH ANALYSIS BY
STELU DEACONU

LEGIS BOOKS
THE WOODLANDS, TEXAS

Copyright © 2000 by LEGIS BOOKS, a division of Legis Corp.

All rights reserved. No part of this book may be reproduced or transmitted in any form or by any means, electronic or mechanical, including photocopying, recording, or by any information storage and retrieval system, without written permission from the publisher.

LEGIS BOOKS
Division of Legis Corp.
P.O. Box 7888
The Woodlands, Texas 77387-7888

Printed in the United States of America

Printing number

1 2 3 4 5 6 7 8 9 10

Library of Congress Cataloging-in-Publication Data

Miller, Richard L. (Richard Lee)
The U.S. Atlas of Nuclear Fallout: Volume I. Total Fallout
1. Nuclear weapons—Testing.
2. Radioactive fallout—United States.
3. United States—History.
4. United States—Geography.
5. Science Reference.

ISBN: 1-881043-06-1 (Hardcover Edition)

ISBN 1-881043-11-8 (Softcover Edition)

For Janet Gordon

THE U.S. ATLAS OF NUCLEAR FALLOUT 1951-1962

VOLUMES IN THE SERIES

VOLUME I: **TOTAL FALLOUT**	ISBN 1881043-118	DEC 2000
VOLUME II: **RADIONUCLIDES**	ISBN 1881043-150	JUL 2001
VOLUME III: **COUNTY COMPARISONS**	ISBN 1881043-126	JUL 2001
VOLUME IV: **RAINOUTS**	ISBN 1881043-07X	DEC 2001
VOLUME V: **ANALYSIS**	ISBN 1881043 088	JUN 2002
ABRIDGED EDITION (VOLUMES I, II, III)	ISBN 1881043-096	AUG 2001

OVERVIEW

Radioactive fallout from the Nevada aboveground nuclear tests was dispersed in uneven patterns across the entire United States. During the aboveground test period the Atomic Energy Commission collected fallout data from aircraft tracking the nuclear clouds and also from approximately 300 fallout detection sites. On the rare occasions the information was made public, the fallout was discussed primarily in terms of the individual radionuclide strontium-90. The only other information regarding fallout involved the nuclear cloud trajectories, some of which were made public as the result of Congressional Hearings taking place in 1958 and 1959.[1]

In 1983-84 the Congress of the United States directed the Secretary of Health and Human Services to "conduct scientific research and prepare analyses necessary to develop valid and credible methods to estimate the thyroid doses of Iodine-131 that are received by individuals from nuclear bomb fallout (and) to develop valid and credible assessments of the exposure to Iodine-131 that the American people received from the Nevada atmospheric nuclear bomb tests."[2] [3]

In October, 1997 the 120,000-page report was published detailing county-by-county deposition amounts of I-131 as ranges described by geometric mean and geometric standard deviation. The data was categorized by nuclear test.[4] Due to the size of the document, the information was disseminated over the Internet as downloadable ASCII files.

During 1998 and 1999 I downloaded these files and reduced the range data to *arithmetic means of lognormal distributions*. Then, using ratios developed by Dr. Harry G. Hicks of Lawrence Livermore Laboratory[5], I

---

[1] "Fallout From Nuclear Weapons Tests": *Hearings Before the Special Subcommittee on Radiation of the Joint Committee on Atomic Energy, Congress of the United States, 86th Congress, First Session, On the Fallout From Nuclear Weapons Tests*. Washington, DC. 1959
[2] Public Law 97-414. (excerpt).
[3] Estimated Exposures and Thyroid Doses Received by the American People from Iodine-131 in Fallout Following Nevada Atmospheric Nuclear Bomb Tests. National Cancer Institute. October 1997. p. ES.1 Executive Summary.
[4] Page number estimate by Dr. Bruce W. Wachholz. Personal communication, Spring, 1998.
[5] Also known as the Hicks Tables. References for the full set is found in the Bibliography.

derived county-by-county deposition estimates for total fallout and approximately 50 radionuclides. This book consists of 2-D and 3-D maps of those estimates, displaying areas of highest fallout and radionuclide concentrations for the time period 1951-1962.

Section 12 of this book compares the fallout deposition rates against National Cancer Institute and Centers for Disease Control cancer rate values using summed R and T (n-2) values as well as Spearman rho, pooled standard error, Multivariate Adaptive Regression Spline™ and Poisson regression analysis. The test statistics and probability values included assume a single pairwise comparison. Populations included the entire United States for comparison with the NCI 2000 Atlas and a 513-county area in the Midwest for comparison against the NCI's 1983 and Centers for Disease Control WONDER data.[6]

The Bonferroni correction for multiple comparisons is made and suggests that many fallout components correlate positively with specific cancer rates at significance levels in excess of $p<0.05$ (see Tables G1-G5).

After using the Bonferroni correction for multiple comparisons, the Poisson regression analysis suggests that total fallout results in positive rate ratios (seven equal intervals) for white female lymphosarcoma 1979-1995; white female colon cancer 1979-1996; and white female brain cancer 1979-1996.

The findings suggest that exposure to nuclear fallout may have played a causal role in cancers, particularly in the states of Iowa, Illinois, Kansas, Missouri and Nebraska. This is supported by the strength of the association and by the graded deposition-response relationships. However, unidentified causal exposures cannot be ruled out.

Replication studies, further statistical analysis of the fallout data, particularly on a regional basis, and on-site evaluation of watersheds for long-lived radionuclides will help to elucidate the effects of the Nevada atmospheric nuclear testing on the American people.

---

[6] See Bibliography.

CONTENTS

Acknowledgements	iii
Preface	v
1. **RANGER SERIES**	1-1
RANGER ABLE	1-3
RANGER BAKER	1-4
RANGER EASY	1-8
RANGER BAKER-2	1-9
RANGER FOX	1-14
2. **BUSTER JANGLE SERIES**	2-1
BUSTER ABLE	2-3
BUSTER BAKER	2-5
BUSTER CHARLIE	2-9
BUSTER DOG	2-13
BUSTER EASY	2-17
JANGLE SUGAR	2-23
JANGLE UNCLE	2-26
3. **TUMBLER-SNAPPER SERIES**	3-1
TS ABLE	3-3
TS BAKER	3-7
TS CHARLIE	3-11
TS DOG	3-17
TS EASY	3-21

TS FOX	3-24
TS GEORGE	3-29
TS HOW	3-35
4. UPSHOT-KNOTHOLE SERIES	**4-1**
UK ANNIE	4-3
UK NANCY	4-7
UK RUTH	4-13
UK DIXIE	4-18
UK RAY	4-21
UK BADGER	4-24
UK SIMON	4-29
UK ENCORE	4-37
UK HARRY	4-41
UK GRABLE	4-46
UK CLIMAX	4-49
5. TEAPOT SERIES	**5-1**
TP WASP	5-3
TP MOTH	5-6
TP TESLA	5-9
TP TURK	5-13
TP HORNET	5-17
TP BEE	5-20
TP ESS	5-22
TP APPLE-1	5-27
TP WASP PRIME	5-31

TP HA	5-33
TP POST	5-35
TP MET	5-39
TP APPLE-2	5-44
TP ZUCCHINI	5-47
6. PLUMBBOB SERIES	**6-1**
PB BOLTZMANN	6-3
PB FRANKLIN	6-7
PB WILSON	6-9
PB PRISCILLA	6-14
PB HOOD	6-19
PB DIABLO	6-27
PB JOHN	6-31
PB KEPLER	6-33
PB OWENS	6-35
PB STOKES	6-38
PB SHASTA	6-43
PB DOPPLER	6-47
PB FRANKLIN PRIME	6-51
PB SMOKY	6-53
PB GALILEO	6-59
PB WHEELER	6-63
PB COULOMB B	6-65
PB LAPLACE	6-67
PB FIZEAU	6-71

PB NEWTON	6-75
PB WHITNEY	6-80
PB CHARLESTON	6-83
PB MORGAN	6-89
7. SHOT SEDAN	7-1
8. COUNTIES OF HIGHEST AVERAGE FALLOUT BY SHOT	8-1
9. COUNTY FALLOUT VALUES BY SHOT SERIES	9-1
10. COUNTIES RANKED BY RAINOUT POTENTIAL	10-1
11. RADIONUCLIDES	11-1
12. NUCLEAR FALLOUT AND CANCER	12-1
APPENDIX A: CALCULATION PROCEDURES	A-1
APPENDIX B I-131 ESTIMATE AND TOTAL FALLOUT CURVES	B-1
APPENDIX C: ESTIMATED VALUES FOR I-131	C-1
APPENDIX D: MULTIPLIERS : I-131 TO TOTAL FALLOUT	D-1
APPENDIX E: UNCERTAINTY AND ERROR PROPAGATION	E-1
APPENDIX F: ANALYSIS PROCEDURES	F-1
APPENDIX G: TABLES: NUCLEAR FALLOUT AND CANCER	G-1

ACKNOWLEDGEMENTS

This book would have been impossible without the dedication and commitment of the scientists of the National Cancer Institute, who produced the original I-131 deposition data. Theirs was the most difficult task of all, and they have my deepest appreciation and respect.

I would also like to thank the following software companies for donating or providing discounted software to support my research: Palisades Corporation (@Risk and Risk Optimizer); Attar Software (Xpert Rule); Decisioneering (Crystal Ball Predictor); Spotfire (Spotfire Pro); MapInfo (MapInfo Pro); Golden Software (Didger); Cytel Software (EGRET); Salford Systems (CART and MARS); and SPSS (Systat). And, of course, a tip of the hat to Microsoft, maker of the only software that should be classified as heavy equipment: Excel 97. Additionally, I would like to express thanks to the makers of Spreadsheet Assistant™ for Excel 97 and 2000. Without it, the work would have taken four times as long to complete.

Thanks to Dr. Lynn Anspaugh of the University of Utah for his initial help and suggestions regarding the Hicks Tables, and especially to Dr. Owen Hoffman of SENES, Oak Ridge, Tennessee, for his assistance and suggestions regarding geometric means, geometric standard deviations, arithmetic means of log normal distributions and risk simulations.

It should be noted that Cliff Honicker performed the original investigations on rainouts in both South Dakota and Minnesota, and also kindly provided me with the AEC reports of investigations. The device specs were provided by the indefatigable nuke researcher Chuck Hansen.

Special thanks to Leah Stratmann, for her excellent work transcribing the 1983-4 NCI Cancer Atlas for the 513 Midwestern counties, and to the fine technical writers Roy Constible and Mary Constible for the many hours spent transcribing the Hicks Tables to Excel. Thirty radionukes and not a single mistake. Also a tip of the hat to Brian Thorne and Walter Hirsch and Dr. H.W. Cummins for their unflagging encouragement and optimism.

And now the legal stuff: The map that appears on the cover of this book and the city lists of fallout trajectories are reprinted with the publisher's permission from

Under the Cloud: The Decades of Nuclear Testing by Richard L. Miller, copyright © 1991 by Richard L. Miller (Two-Sixty Press, 1999).

The maps that appear with the Estimated Sedan Trajectories are reproduced with the kind permission of the American Map Corporation.

Finally, I would like to thank my mother, Ruth E. Miller and my wife, Kim and daughter Trace for their continued interest and encouragement and support throughout the long days and nights working on this project. This book belongs to them.

PREFACE

The era of aboveground nuclear testing at the Nevada Test Site began on January 27, 1951 with the detonation of Ranger ABLE. For over ten years millions of Curies of radioactive isotopes were deposited across the country, making their way into the food chain and exposing several generations of Americans to radiation. While the Atomic Energy Commission was aware that a thunderstorm could theoretically scavenge an entire nuclear cloud within an hour, there was no effort to evaluate hot spots and rainouts downwind of the Test Site. Similarly, there were relatively few attempts to evaluate potential links between nuclear fallout and disease rates.

In 1981 Harry G. Hicks published tables that allowed researchers to theoretically determine deposition rates of over a hundred radionuclides. Using the Hicks Tables, and fallout data from the 1950s and early 1960s, any researcher with a computer could determine deposition values for any of over a hundred radioisotopes.. Despite this, no research project was ever launched.

In 1997, after 15 years of work, the National Cancer Institute published fallout data for every county in the continental United States. Yet, the values were for only one radionuclide, the relatively short-lived I-131, a radioisotope that accounts for approximately 2 percent of the total fallout deposited across the country.

At the time of the publication of this book (December, 2000), there are indications of further research by the NCI and other government agencies regarding analysis of possible links between radionuclides in fallout and cancer.

Hopefully this series of books will contribute to that process.

INTRODUCTION

In January 1983, the 97th Congress directed the National Institute of Health to evaluate the extent and effects of Iodine-131 (I-131) fallout resulting from the Nevada nuclear tests conducted in the 1950s and early 1960s.[1,2]

The task was finally completed and published in 1997. The report consisted of I-131 fallout deposition rates from nearly 60 aboveground and subsurface nuclear tests—for each of more than 3000 counties. The finished product included data for up to 20 fallout days subsequent to the nuclear test. All together, the report was equivalent to more than 120,000 pages of data.

Officials at the National Cancer Institute decided that publication of such a mass of data would be feasible only over the Internet. Interested citizens could download data for their own home counties and learn just how much I-131 was deposited from each test, and when. Not only that, they could enter their birth date and estimated milk consumption and read an estimate of the radiation exposure to their thyroid.

There were some problems, however. The uncertainty involved in the original sampling procedures and analysis resulted in data with a wide range. So, instead of presenting the information as minimums and maximums over a certain confidence interval, the information was given in terms of the Geometric mean (GM) and Geometric standard deviation (GSD). Statistically, the geometric mean represents the average estimate of the log normal distribution, and the geometric standard deviation refers to the uncertainty in the estimate. In practical terms, the geometric mean refers to the average estimated fallout while the geometric standard deviation refers to the uncertainty around that average.

While these statistical procedures were certainly appropriate for the subject, it was unclear how the data should be interpreted. Simply dividing the GM by the GSD results in a minimum value (lower bound), while multiplying the GM by this same GSD value will result in a *maximum* value (upper bound) with a confidence interval of 67%. Squaring the GSD before performing the same mathematical operations results in a wider range—lower minimums and higher maximums—but with a 97% confidence interval. In essence, each pair of NCI fallout data—the GM and GSD combination— represents *ranges* of values.

This presented an obvious problem: Should the reader assume that the county in question received the amount represented by the geometric mean? Even if the reader took the time to make the calculations for each particular shot-day, should she assume the maximum value, or the minimum value—and at which confidence interval?

The natural inclination would be to simply take the difference between the maximum and minimum and divide by two to obtain an average. But this procedure would work only if the distribution were normally distributed—as in a bell-shaped curve. And fallout is *not* normally distributed. Given the NCI's calculations, fallout is *log*normally distributed. To average the fallout using the standard procedures would result in an overestimation the geometric means of the fallout by 4.7 times. As a compromise I chose to evaluate the data based upon a special form of average: the **arithmetic mean of the lognormal distribution**. The formula is expressed as:

$$AM = GM\left(\exp\left(\frac{(\ln GSD)^2}{2}\right)\right)$$

Where AM = Arithmetic mean of the lognormal distribution;

GM = Geometric mean of the lognormal distribution

GSD = Geometric standard deviation of the lognormal distribution

exp = exponent e = 2.7183

ln = natural logarithm

Using the total fallout values from shot Tumbler-Snapper GEORGE (TS-7) as an example, the arithmetic means of the lognormal distribution values are typically equal to one-third (32.8%) of the simple means-between-bounds (97% confidence interval) values and twice (1.97 times) the raw geometric mean values.

I-131 AND TOTAL FALLOUT

The published NCI data included values for only a single radioactive component of fallout—one of the radioactive isotopes of the element Iodine. This particular radioactive isotope, I-131, makes up approximately 2 percent of the initial radionuclide output. In theory, one can determine total fallout simply by multiplying the I-131 activity values by 50 to arrive at the total fallout values. In practice, however, it is considerably more complicated.

For one thing, different nuclear tests produced different groups of radionuclides as well as varying amounts of I-131. To make matters even more complicated, each radionuclide has its own particular chemical and physical composition and decay scheme. Thus, the ratio of I-131 to total fallout for a particular test will vary depending on which day post-shot it is evaluated. The ratio between the activity level of I-131 and the activity level of total fallout on the day after the nuclear test will be significantly different than the ratio between the I-131 activity and total fallout activity on the second day after the test. In short, the ratio varies not only by the nuclear test, but also according to how many days have elapsed since detonation.

All these assumptions were theoretical. It would take experimentation and hard work to determine the actual ratios between I-131 and total fallout (as well as the other radionuclides).

Fortunately, that work had been done —and published— by Livermore physicist Harry G. Hicks in 1981. Through experimentation and long hours on a mainframe computer, Hicks calculated ratios for more than 100 radionuclides produced by every aboveground test ever detonated at the Nevada Test Site.

The work came to be called the Hicks Tables and, importantly, I-131 was among the radionuclides listed. After some interval analysis to recover the missing days (the Hicks Tables included only days 1, 2, 5 and 10), ratios were calculated that could be used to determine total fallout from I-131 activity values. From this point, relative total fallout activities could be determined, by nuclear test and by shot day, for every county in the United States.

The subject of total fallout forms the basis for this first volume. Subsequent volumes will characterize the fallout patterns by radionuclide and compare additional fallout values for each county.

Hopefully, it will provide a new perspective on an important part of our history and will stimulate further research in the field.

ORGANIZATION OF THE BOOK

Volume I of the **U.S. ATLAS OF NUCLEAR FALLOUT** is divided into the following main sections:

- Sections 1-7 include maps of total fallout organized by series and by specific nuclear test, arranged in chronological order.

- Section 8 includes tabulated data for counties with the highest average fallout (arithmetic mean of the log normal distribution).

- Section 9 includes fallout values for each US county, organized by nuclear test series.

- The Appendices contain technical data regarding procedures used to convert the NCI's I-131 data to Total Fallout data, uncertainty analyses, estimated I-131 values, Total Fallout multipliers, and equations used to estimate I-131 and total fallout values given the number of days post shot.

All information regarding Total Fallout is based upon analysis of the 1997 National Cancer Institute I-131 study.

SECTIONS 1-7:
MAPS AND CITY LISTS

Parts 1 through 7 include city lists and maps organized by nuclear test ("shot"). The information includes these components: city list with detonation data, U.S. Weather Service trajectory map, 2-D maps of total fallout, 3-D maps of relative average fallout; shot characteristics, and occasionally, photos of the nuclear detonation.

CITY LIST WITH DETONATION DATA

When a nuclear test occurs, radioactive debris, consisting of shot tower remnants, volatilized dirt and bomb casing material ascend into the air. Debris is then carried (generally) east with the wind. The debris will usually follow different paths depending upon the wind direction and speed. For that reason, there are often several cloud trajectories associated with a given nuclear test. Given the dispersion characteristics of the weather patterns, the nuclear clouds spread out as they travel further from the test site. During the nuclear tests, the

Atomic Energy Commission, in conjunction with the U.S. Department of Defense, conducted flights to follow and track the nuclear clouds as they cross the continent. The results are the maps included in this section.

Using the official Department of Defense maps in conjunction with modern highway maps, I attempted to determine which cities and towns were in the centerline path of the nuclear clouds. More than 10,000 sites were eventually identified. That list is included and typically begins the subsection for each nuclear test. In addition to a list of cities and towns in the path of the cloud is information regarding the shot date, time, area, sponsor (either Los Alamos or Livermore), yield in kilotons, radiation level at ground zero, height of the burst and the cloud top height.

U.S. WEATHER SERVICE TRAJECTORY MAPS

Following the City List is a copy of the Atomic Energy Commission-U.S. Weather Service Trajectory Map of the cloud trajectory. Some, such as those associated with the 1951 Ranger series, are very basic and display few details.

2-D MAPS

Two-dimensional maps are presented showing areas of fallout by county. The information is organized two different ways: first by **equal intervals** of data and then by **equal numbers** of counties. The equal interval maps and the equal number maps both represent the same total fallout values (arithmetic mean of lognormal distributions of fallout) but the data is categorized differently. Equal interval maps divide the fallout **values** into four equivalent categories, and then shade counties according to the fallout category.

Equal number maps divide the counties into four equally-numbered groups, then assign shading based upon relative amounts of fallout among the groups.

Maps organized by equal number of counties will typically show greater fallout coverage, while the maps organized by equal intervals will show only the counties with the *highest* fallout amounts.

3-D MAPS: RELATIVE AVERAGE TOTAL FALLOUT

Here, a three-dimensional prism map of the United States represents the relative total fallout for each NCI-evaluated nuclear test. County borders are projected into the third dimension (z axis) in proportion to their deposition of fallout *relative to the entire United States* and is based on the **equal interval map** (see above). Thus, only the counties with the highest relative amount of fallout will ascend above the plane of the map. Some shots may include several 3-D prism maps with close-ups of hot spots or fallout paths.

CHARACTERISTICS

This section lists characteristics of the particular nuclear test, including size and weight of the nuclear device, associated radionuclides produced and fallout patterns.

SECTION 8:
COUNTIES OF HIGHEST AVERAGE FALLOUT

This section ranks the top 120 counties with highest average (arithmetic mean of the lognormal distribution) calculated levels of fallout for a particular nuclear test. Activity is given in microCuries per square meter deposition.

SECTION 9:
COUNTY TOTAL FALLOUT DATA BY SERIES

In this table the arithmetic mean of the log normal distribution is calculated for each county and for each major aboveground shot series from RANGER BAKER through SEDAN.

In this section, activity is given in nanoCuries per square meter, a rate equivalent to 3.7 particles or rays produced per second over an area of one square meter.

Counties are ranked for fallout, with number 1 representing the highest estimated fallout levels to 3093 representing the lowest estimated fallout levels. Several counties near the Nevada Test Site were divided into sections labeled 1, 2, 3 etc. These represent different geological differences within counties that may have affected close-in fallout.

SECTION 10:
POTENTIAL RAINOUTS

One of the main concerns of the Atomic Energy Commission during the aboveground test series was that a nuclear cloud might come into contact with a storm system and bring high levels of radioactive fallout to earth with the rain. Termed variously rainouts (when the nuclear cloud is *within* the rain or snow cloud) or washouts (when the nuclear cloud is *below* the rain cloud), the potential for high levels of radiation exposure during these events was quite real. While there are a number of well-known events involving rainouts—for example, the Kodak incident in 1951 and the Utica, NY rainout in 1953—an examination of the NCI data suggests that rainouts occurred with some regularity.

While there have never been epidemiological studies regarding communities exposed to the potentially intense fallout during rainouts, the Department of Defense was aware of the problem. A 1977 DOD-Energy Department book, *The Effects of Nuclear Weapons*, concluded that a rainfall of only 1.1 hours could—theoretically—scavenge 99 percent of a nuclear cloud: Included on the next page is a table published in 1977 detailing information regarding rainouts and washouts.

Table 9.74a

ESTIMATED RAINFALL DURATION FOR RAINOUT

Percent of Cloud Scavenged	Duration of Rainfall (hours)
25	0.07
50	0.16
75	0.32
90	0.53
99	1.1

Table 9.74b

ESTIMATED RAINFALL DURATION FOR WASHOUT

Percent of Cloud Scavenged	Duration of Rainfall (hours)		
	Light	Moderate	Heavy
25	8	1.6	0.8
50	19	3.8	1.9
75	38	7.7	3.6
90	64	13	6.4
99	128	26	13

TABLE DISCUSSING RAINOUTS AND WASHOUTS[3]

Section 10 consists of a table titled Rainout Potential, which ranks counties for the potential exposure to not only rain but nuclear fallout. The counties were first ranked 1 through 3094, with 1 being the highest and 3094 being the lowest numbers, for exposure to nuclear fallout (total fallout). Then, after converting the NCI's precipitation indices to millimeters, the counties were ranked in terms of the amount of rainfall that had occurred within 10 days of the nuclear test. After normalizing the two arrays, I obtained a final ranking that included information for both fallout and precipitation. A county with a low rank number would have received a high combination of fallout and rainfall, whereas a county with a high rank number would have received a low combination of fallout and rainfall.

While absence of associated precipitation does not ensure low fallout—the Troy-Utica "rainout" in 1953 was probably due to dry downbursts preceding the storm—rainfall can significantly increase the amount of fallout deposited. Subsequent volumes will discuss rainouts in greater detail.

SECTION 11:
RADIONUCLIDES AND ANALYSIS

This section includes information on radionuclides found in fallout and includes maps and tables showing deposition patterns of such radionuclides as beryllium-7, cobalt-60 and cesium-137.

SECTION 12:
NUCLEAR FALLOUT AND CANCER

Section 12 discusses the various statistical techniques used analyze fallout and cancer rates. Included are results from statistical procedures such as Spearman correlation coefficient, pooled standard error, Multiple Adaptive Regression Splines™ and Poisson regression. Tables listing results of these statistical analyses are found in this section as well as in Appendix G.

TECHNICAL APPENDICES A-G

The appendices contain technical information regarding calculation methods and information regarding uncertainties associated with the values appearing in this book. Appendix G includes statistical analyses of fallout components and cancer.

NAMING CONVENTIONS

In this book, the nuclear tests are generally referred to first by the series, then by the specific name, as in Upshot-Knothole HARRY. Abbreviated, the shot may be listed as UK9 (shot HARRY was the ninth listed by the National Cancer Institute's I-131 survey in the Upshot-Knothole Series), followed by the number of the fallout day. For example, UK9-3 refers to the third day of fallout resulting from the ninth shot in the Upshot-Knothole Series, as listed by the National Cancer Institute.

Where space permits, shots are listed by the series name or abbreviation, followed by the shot name. Some nuclear tests, such as the Plumbbob underground shot, RAINIER, were not included in the NCI study. Others, such as Teapot BEE and Teapot ESS were combined. Also, some tests, such as Ranger-FOX (R-3) and Plumbbob STOKES (PB-7), which apparently were associated with minimal fallout, were excluded from some lists and maps. A complete list of abbreviations used in the book is found at the end of this section.

SHOT DAYS

The term "shot day" refers to a particular day after detonation. For example, TS1-1 refers to the day of detonation of the first shot in the Tumbler-Snapper series, that is, Tumbler-Snapper ABLE. In this case, that day would be April 1, 1952. The term TS1-2 refers to the first day after detonation of Tumbler-Snapper 1, or April 2, 1952. TS1-3 refers to the second day after the detonation. In this case, since TS1 refers to shot Tumbler-Snapper ABLE, and since that detonation occurred on April 1, 1952 (see above), TS1-3 refers to April 3, 1952.

CURIES AND NANOCURIES

Radioactive forms of elements are called radioactive isotopes or *radionuclides*. These radionuclides are found in great quantity in nuclear fallout. Depending upon their internal nuclear makeup, the individual atoms will change from a state of higher energy but less stability, to a state of lower energy but greater stability. As these transformations occur, radiation is given off from the nucleus of the atom. While the type of radiation depends on the radionuclide, the transformation itself can be thought of as the central radioactive "event." That is, a single unit of radiation from a radioactive material such as fallout is direct result of the transformation of a single atom from a state of higher energy to a state of lower energy. The energy difference is, in effect, the radiation itself.

Scientists term such transformation *disintegration*. A radionuclide is said to disintegrate or decay from a state of relative instability to a state of greater stability. All things being equal, the faster the rate of decay or transformation of these radioactive atoms the more radiation is emitted. A substance that decays rapidly, emitting more units of radiation per second is said to have a higher activity than a substance that decays slowly, emitting fewer units of radiation per second.

This rate of decay, or *activity*, can be measured in units called *Curies*. One Curie of activity is defined as producing 3.7×10^{10} disintegrations per second.

As an illustration, suppose someone somehow traveled to ground zero and scooped up a vial containing a mixture of very hot nuclear fallout. If a perfectly calibrated Geiger counter placed near the vial recorded 37,000,000,000 clicks per second, the vial could be said to have an activity equal to *one Curie.*

An aboveground nuclear test such as occurred in the 1950s, released fallout into the atmosphere that was equal to *millions* of Curies of activity.

Dispersion of a 1000-ft diameter nuclear cloud over thousands of cubic miles of air will decrease the overall activity considerably. As a consequence, the radioactivity of the fallout, when it finally fell to earth, was measured not in Curies, but in millionths and billionths of a Curie.

In Section 8, the fallout values for counties with the highest levels of fallout are expressed in terms of microCuries. In this book the term microCurie is represented by the symbol μCi. *One microCurie is defined as radiological decay equal to 37,000 disintegrations per second.* (37,000,000,000 divided by 1,000,000).

The fallout values used for the maps in Sections 1-7 and for the Table in Section 9 are given in *nanoCuries*. A nanoCurie represents an activity level equal to one thousandth of a microCurie. The term nanoCurie is represented in this book by the symbol nCi. A nanoCurie represents an activity equal to 37 disintegrations per second.

The 1997 National Cancer Institute study gave fallout levels in terms of nanocuries per square meter (nCi/sq meter). A square meter is equal to 10.76 square feet, or an area 3 ft. 3 inches on a side. Thus, if you read that fallout from a particular shot produced an average of 1 nCi/sq meter on a given county on a particular day, you may reasonably conclude that for each square plot of land 3.3 ft on a side, the fallout produced 37 units of radiation. Similarly, since there are 2,589,988.1 square meters in a square mile, you can conclude that, on average, 1 nCi/square meter is equivalent to 95,829,559.7 units of radiation produced per square mile. (37 x 2,589,988.1)

RELATIVE FALLOUT VALUES

A primary purpose of this book is to show *relative* fallout levels for each county from each nuclear test. To do that, arithmetic means were calculated from the geometric means and geometric standard deviations given in the NCI tables. These arithmetic means were then used as central values to evaluate relative fallout levels. Where tables or maps refer to Averages, the term specifically refers to the *Arithmetic mean of the log-normal distribution*.

UNCERTAINTY:
MORE ON GEOMETRIC STANDARD DEVIATION

In the book version of their I-131 fallout study the National Cancer Institute included this information about geometric mean and geometric standard deviation:

> "The uncertainties expressed in terms of geometric standard deviations, GSD, implying that 67% of the values in the distribution associated with a best estimate, GM, are expected to lie between GM/GSD and GM x GSD, while 97% are expected to range from GM/(GSD)2 and GM x (GSD)2. For example, if an average thyroid dose to a particular population group from a given test is listed with a best estimate, GM, of 0.4 rad and with an associated uncertainty, GSD, of 2.5, this means
>
> (a) That there is a 50% probability that the true value of the average thyroid dose is greater than 0.4 rad, and conversely, that there is a 50% probability that the average thyroid dose is lower than 0.4 rad; and,
>
> That the distribution of the expected values is such that there is a 67% probability that the true value of the average thyroid dose likes between:
>
> GM/GSD = 0.4/2.5 = 0.16 rad; and GM x GSD = 0.4 x 2.5 = 1 rad.
>
> And that there is a 97% probability that the true value of the average thyroid dose likes between:
>
> GM/(GSD)2 = 0.4/6.25 = 0.06 rad, and GM x (GSD)2 = 0.4 x 6.25 = 2.5 rad."
>
> The estimates provided in the Annexes and in the Subannexes for the average doses to the various population groups show that the associated GSDs range, in general between 2 and 10, the lowest GSDs being usually related to populations living in the vicinity in the NTS in areas for which County Data Base or Town Data Base data were available. The highest GSDs are associated with the dose estimates for which the depositions of I 131 were assessed with the meteorological approach."

It should be noted, however, that $GM/(GSD)^2$ and $GM \times (GSD)^2$ represents the 95th rather than the 97th confidence interval.

As discussed earlier, the geometric standard deviation (GSD). is the *primary measure of uncertainty* regarding the NCI fallout values. The GSD represents the range of possible values around the geometric mean. The greater the geometric standard deviation, the greater is the uncertainty surrounding the estimated mean (GM) and thus the greater the spread between the minimum and maximum estimated values. For example, just because a county is listed as receiving high average total fallout doesn't *necessarily* mean that the county was in a hot zone. It may have been in an area where the uncertainty regarding the fallout level was high. In the NCI study, counties located near the Nevada Test Site typically were associated with smaller geometric standard deviations than were counties located farther away.

The total fallout calculations in this book were derived from I-131 arithmetic means. These values were then used to calculate the arithmetic means of the Total Fallout. A more detailed discussion of the mathematical procedures is contained in the Appendices.

In sum, the data represents not only estimated fallout levels, but also uncertainties associated with measurement. Thus, the numbers found in the tables should be interpreted carefully. Hopefully, they are starting points for further research.

ABBREVIATION	SHOT/SHOT SERIES	DATES	INCLUDED IN NCI STUDY?
R	Ranger Series	Jan 27-Feb 6 1951	Yes
BJ	Buster-Jangle Series	22 Oct-29 Nov 1951	Yes
TS	Tumbler-Snapper Series	15 Apr Jun 1952	Yes
UK	Upshot-Knothole Series	17 Mar-4 Jun 1953	Yes
TP	Teapot Series	18 Feb-15 May 1955	Yes
PB	Plumbbob Series	28 May-7 Oct 1957	Yes
S	Shot SEDAN	6 Jul 1962	Yes
(Ranger Able)	Ranger ABLE	27 Jan 1951	No
R1	Ranger BAKER	28 Jan 1951	Yes
(Ranger Easy)	Ranger EASY	1 Feb 1951	No
R2	Ranger BAKER-2	2 Feb 1951	Yes
R3	Ranger-FOX	6 Feb 51	Yes
(Buster-Jangle Able)	Buster-Jangle ABLE	22 Oct 1951	No
BJ1	Buster-Jangle BAKER	28 Oct 1951	Yes
BJ2	Buster-Jangle CHARLIE	30 Oct 1951	Yes
BJ3	Buster-Jangle DOG	1 Nov 1951	Yes
BJ4	Buster-Jangle EASY	5 Nov 1951	Yes
BJ5	Buster-Jangle SUGAR	19 Nov 1951	Yes
BJ6	Buster-Jangle UNCLE	29 Nov 1951	Yes
TS1	Tumbler-Snapper ABLE	1 Apr 1952	Yes
TS2	Tumbler-Snapper BAKER	15 Apr 1952	Yes
TS3	Tumbler-Snapper CHARLIE	22 Apr 1952	Yes
TS4	Tumbler-Snapper DOG	1 May 1952	Yes

ABBREVIATION	SHOT/SHOT SERIES	DATES	INCLUDED IN NCI STUDY?
TS5	Tumbler-Snapper EASY	7 May 1952	Yes
TS6	Tumbler-Snapper FOX	25 May 1952	Yes
TS7	Tumbler-Snapper GEORGE	1 Jun 1952	Yes
TS8	Tumbler-Snapper HOW	5 Jun 1952	Yes
UK1	Upshot-Knothole ANNIE	17 Mar 1953	Yes
UK2	Upshot-Knothole NANCY	24 Mar 1953	Yes
UK3	Upshot-Knothole RUTH	31 Mar 1953	Yes
UK4	Upshot-Knothole DIXIE	6 Apr 1953	Yes
UK5	Upshot-Knothole RAY	11 Apr 1953	Yes
UK6	Upshot-Knothole BADGER	18 Apr 1953	Yes
UK7	Upshot-Knothole SIMON	25 Apr 1953	Yes
UK8	Upshot-Knothole ENCORE	8 May 1953	Yes
UK9	Upshot-Knothole HARRY	19 May 1953	Yes
UK10	Upshot-Knothole GRABLE	25 May 1953	Yes
UK11	Upshot-Knothole CLIMAX	4 Jun 1953	Yes
TP1	Teapot WASP	18 Feb 1955	Yes
TP2	Teapot MOTH	22 Feb 1955	Yes
TP3	Teapot TESLA	1 Mar 1955	Yes
TP4	Teapot TURK	7 Mar 1955	Yes
TP5	Teapot HORNET	12 Mar 1955	Yes
TP6	Teapot BEE	22 Mar 1955	Combined with Teapot ESS
TP6	Teapot ESS	23 Mar 1955	Combined with Teapot BEE
TP7	Teapot APPLE-1	29 Mar 1955	Yes

ABBREVIATION	SHOT/SHOT SERIES	DATES	INCLUDED IN NCI STUDY?
(Teapot Wasp Prime)	Teapot WASP PRIME	29 Mar 1955	No
(Teapot HA)	Teapot High Altitude	6 Apr 1955	No
TP8	Teapot POST	9 Apr 1955	Yes
TP9	Teapot MET	15 Apr 1955	Yes
TP10	Teapot APPLE-2	5 May 1955	Yes
TP11	Teapot ZUCCHINI	15 May 1955	Yes
PB1	Plumbbob BOLTZMANN	28 May 1957	Combined with Plumbbob FRANKLIN
PB1	Plumbbob FRANKLIN	2 Jun 1957	Combined with Plumbbob BOLTZMANN
PB2	Plumbbob WILSON	18 Jun 1957	Yes
PB3	Plumbbob PRISCILLA	24 Jun 1957	Yes
PB4	Plumbbob HOOD	5 Jul 1957	Yes
PB5	Plumbbob DIABLO	15 Jul 1957	Yes
(Plumbbob John)	Plumbbob JOHN	19 Jul 1957	No
PB6	Plumbbob KEPLER	24 Jul, 1957	Combined with Plumbbob OWENS
PB6	Plumbbob OWENS	25 Jul 1957	Combined with Plumbbob KEPLER
PB7	Plumbbob STOKES	7 Aug 1957	Yes (but not fallout listed)
PB8	Plumbbob SHASTA	18 Aug 1957	Yes
PB9	Plumbbob DOPPLER	23 Aug 1957	Yes
PB10	Plumbbob FRANKLIN PRIME	30 Aug 1957	Yes (but no fallout listed)
PB11	Plumbbob SMOKY	31 Aug 1957	Yes
PB12	Plumbbob GALILEO	2 Sep 1957	Yes

ABBREVIATION	SHOT/SHOT SERIES	DATES	INCLUDED IN NCI STUDY?
PB13	Plumbbob WHEELER	6 Sep 1957	Combined with Plumbbob COULOMB B and Plumbbob LAPLACE
PB13	Plumbbob COULOMB B	6 Sep 1957	Combined with Plumbbob WHEELER and Plumbbob LAPLACE
PB13	Plumbbob LA PLACE	8 Sep 1957	Combined with Plumbbob WHEELER and Plumbbob COULOMB B
PB14	Plumbbob FIZEAU	14 Sep 1957	Yes
PB15	Plumbbob NEWTON	16 Sep 1957	Yes
PB 16	Plumbbob WHITNEY	23 Sep 1957	Yes
PB 17	Plumbbob CHARLESTON	28 Sep 1957	Yes
PB18	Plumbbob MORGAN	7 Oct 1957	Yes
S or SEDAN	Storax SEDAN	6 Jul 1962	Yes

NOTES

1 Eisler, Peter and Steve Sternberg. "Fallout: Did It Harm? Study shows contaminants fell far from Nevada test site." USA Today. 26 Jul 1997.
2 Section 7(a) of Public Law 97-414 directs the Secretary of Health and Human Services to "(1) conduct scientific research and prepare analysis necessary to develop valid and credible assessments of the risks of thyroid cancer that are associated with thyroid doses of Iodine 131; (2) conduct scientific research and prepare analysis necessary to develop valid and credible methods to estimate the thyroid doses of Iodine 131 that are received by individuals from nuclear bomb fallout; and (3) conduct scientific research and prepare analysis necessary to develop valid and credible assessments of the exposure to Iodine 131 that the American people received from the Nevada atmospheric nuclear bomb tests"
3 Glasstone, Samuel and Philip KJ. Dolan (Editors) *The Effects of Nuclear Weapons.*, United States Department of Defense and the Energy Research and Development Administration, 1977. p. 419.

SECTION 1

RANGER SERIES

Jan 27–Feb 6, 1951

SHOT RANGER ABLE
(Not Included in the NCI Study)

RANGER: ABLE*

Detonation Date: 27 Jan 1951 • *Detonation Time:* 5:45 A.M. • *Area:* Frenchman's Flat • *Sponsor:* Los Alamos • *Yield:* 1 kt • *Radiation Level at Ground Zero:* 0.5 R/hr • *Height of Burst:* 1,060 ft • *Cloud Top Height:* 17,000 ft

10,000-Foot Trajectory: NEVADA: Lincoln County • UTAH: Orem, Provo, Vernal • COLORADO: Pueblo • KANSAS: (Crossed the border at Greeley, left Kansas near Wallace) • MISSOURI: St. Joseph, Cameron, Chillicothe, Brookfield, Atlanta, Bethel, West Quincy • ILLINOIS: Quincy, Jacksonville, Springfield • INDIANA: Lafayette, Kokomo, Marion, Berne • OHIO: Springfield, Dayton, Columbus, Canton • PENNSYLVANIA: Beaver Falls, Kittanning, Williamsport, Wilkes-Barre • NEW YORK: Port Jervis, Middletown, Poughkeepsie • CONNECTICUT: New Haven, New London • RHODE ISLAND: Newport • MASSACHUSETTS: Fall River.

30,000-Foot Trajectory: NEVADA: Caliente • UTAH: Cedar City, Panguitch, Moab • COLORADO: Crested Butte, Buena Vista, Colorado Springs • KANSAS: (The clouds crossed trajectories over Castle Rock at the border of Gove and Trego counties in Kansas), Topeka, Lawrence • MISSOURI: Kansas City, Harrisonville, Warrensburg, Sedalia, Columbia, St. Louis • ILLINOIS: Alton, Vandalia, Robinson • INDIANA: Bloomington, Columbus, Greensburg • OHIO: Hamilton, Athens, Marietta • PENNSYLVANIA: Uniontown, Lancaster, Reading • NEW JERSEY: New Brunswick.

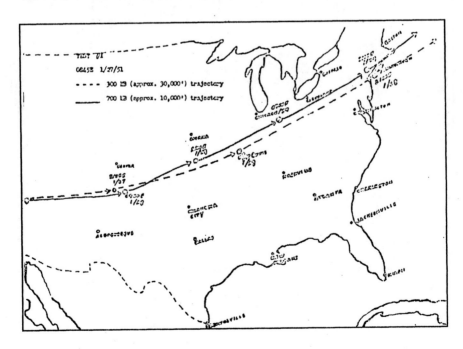

TRAJECTORY: RANGER ABLE

SHOT 1: RANGER BAKER-1 (R1)

RANGER: BAKER-1

Detonation Date: 28 Jan 1951 • *Detonation Time:* 5:52 A.M. • *Area:* Frenchman's Flat • *Sponsor:* Los Alamos • *Yield:* 8 kt • *Radiation Level at Ground Zero:* 15 R/hr • *Height of Burst:* 1,080 ft • *Cloud Top Height:* 35,000 ft

10,000-Foot Trajectory: COLORADO: Boulder • MISSOURI: St. Joseph, Clark • ILLINOIS: Alton, Terre Haute • OHIO: Montgomery, New Philadelphia • PENNSYLVANIA: Kittanning • NEW YORK: Middletown • CONNECTICUT: New Haven • MAINE: Nantucket.

30,000-Foot Trajectory: UTAH: St. George • COLORADO: Fort Carson, Arapahoe (This trajectory crossed over into Kansas above 4,039-foot Mount Sunflower, the highest point in Kansas) • KANSAS: Palmer • NEBRASKA: Falls City • MISSOURI: Coatsville • IOWA: Burlington • INDIANA: South Bend • MICHIGAN: Gibralter • NEW YORK: Auburn, Glens Falls • NEW HAMPSHIRE: Portsmouth.

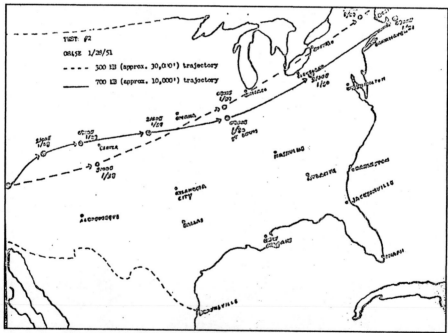

TRAJECTORY: RANGER BAKER (R-1)

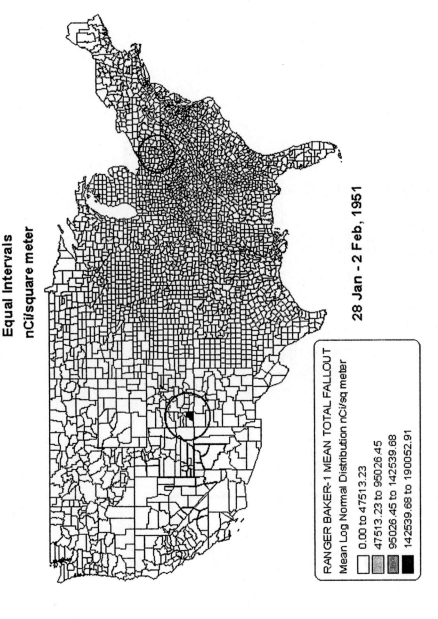

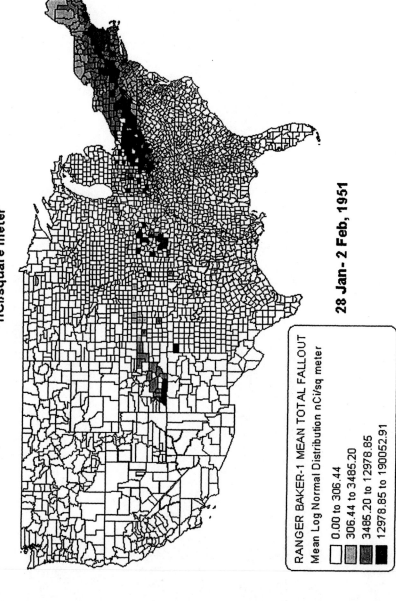

RELATIVE TOTAL FALLOUT: RANGER BAKER

JAN 28, 1951 8 KT

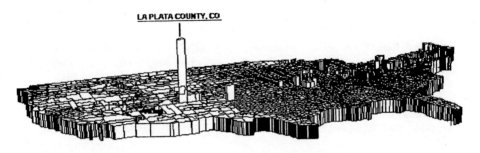

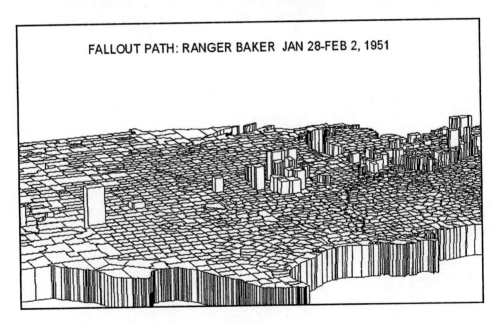

CHARACTERISTICS: RANGER BAKER-1

The test device weighed 10,800 lbs and resembled the "Fat Man" device dropped on Nagasaki. Ranger-BAKER-1 ranks fourth among the nuclear tests in the percentage of Np-239 contained in the fallout. The deposition pattern of Ranger BAKER-1 is most similar to that of shot Buster-Jangle CHARLIE (R value: 0.32, T value: 19.07).

SHOT RANGER EASY
(Not Included in the NCI Study)

RANGER: EASY

Detonation Date: 1 Feb 1951 • *Detonation Time:* 5:47 A.M. • *Area:* Frenchman's Flat • *Sponsor:* Los Alamos • *Yield:* 1 kt • *Radiation Level at Ground Zero:* 0.5 R/hr • *Height of Burst:* 1,080 ft • *Cloud Top Height:* 12,500 ft

10,000-Foot Trajectory: Crossed San Bernadino County, California, into Mexico over the Cabeza Prieta National Wildlife Refuge. The nuclear debris came back into the United States at Fort Meyers, Florida, and left the continental airspace over Sebastian, Florida.

30,000-Foot Trajectory: TEXAS: El Paso, Austin, Beaumont • LOUISIANA: Baton Rouge • ALABAMA: Montgomery • NORTH CAROLINA: Durham • VIRGINIA: Nassawadox.

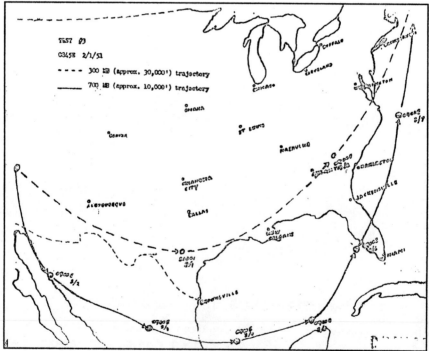

TRAJECTORY: RANGER EASY

SHOT 2: RANGER BAKER-2 (R2)

RANGER: BAKER-2

Detonation Date: 2 Feb 1951 • *Detonation Time:* 5:49 A.M. • *Area:* Frenchman's Flat • *Sponsor:* Los Alamos • *Yield:* 8 kt • *Radiation Level at Ground Zero:* 15 R/hr • *Height of Burst:* 1,100 ft • *Cloud Top Height:* 28,000 ft

10,000-Foot Trajectory: The nuclear debris crossed north of Las Vegas and into Arizona, passing north of Phoenix and leaving the state at Paradise. It entered Mexico airspace over the Alamo Hueco Mountains in New Mexico, then crossed back into the United States over Ruidosa, Texas. From there is passed out over Big Bend National Park, then to Carrizo Springs, Texas, and out over the Gulf at Aransas Pass, Texas. It entered Florida airspace at St. Petersburg and left near Melbourne.

30,000-Foot Trajectory: The nuclear debris at 30,000 feet passed over St. George, Utah, crossed the state and entered New Mexico airspace north of Crystal. From there it moved south of Los Alamos, then to Clovis, New Mexico, where it crossed into Texas near Bovina. The cloud passed over Petersburg and Waco and left the coast near Winnie, just north of Houston. Later it crossed the Florida peninsula, passing over Miami.

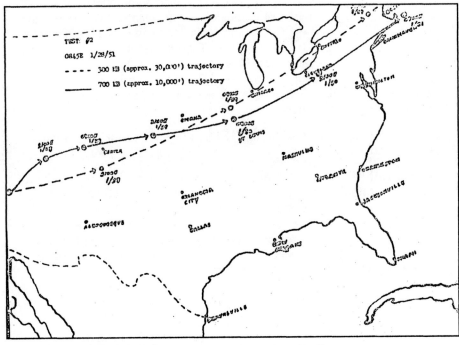

TRAJECTORY: RANGER BAKER-2 (R-2)

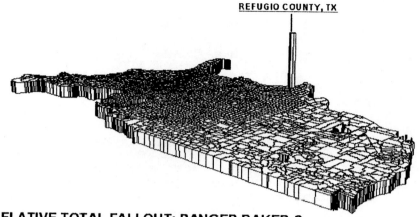

RELATIVE TOTAL FALLOUT: RANGER BAKER-2

CHARACTERISTICS: RANGER BAKER-2

The test device weighed 10,800 lbs and resembled the "Fat Man" device dropped on Nagasaki. Ranger BAKER-2 ranks fifth among nuclear tests in the percentage of Np-239 contained in the fallout. The deposition pattern for Ranger BAKER-2 is most similar to that of shot Tumbler-Snapper CHARLIE (R:0.20, T: 11.49). The county of highest deposition was Refugio, TX. Fallout from this nuclear test was deposited across 182 counties (6 percent of the U.S.). The average fallout per County amount was 1302 nCi/sq meter, ranking this shot 46th for average fallout among nuclear tests.

The day 1 ratio of total fallout to I-131 is 139.1 to 1, ranking 9th highest among all nuclear tests.

Ranger BAKER-2 deposited an estimated maximum of 7.5 milliCuries of fallout on Refugio County, TX on Feb 3, 1951. The Refugio, TX fallout accounted for the 3rd hottest shot day in the history of the nuclear test program. Other counties in the fallout path of Ranger BAKER-2 were Dade County, FL with a maximum of 20.45 microCuries total fallout and Broward County, FL with a maximum estimated value of 8.55 microCuries total fallout.

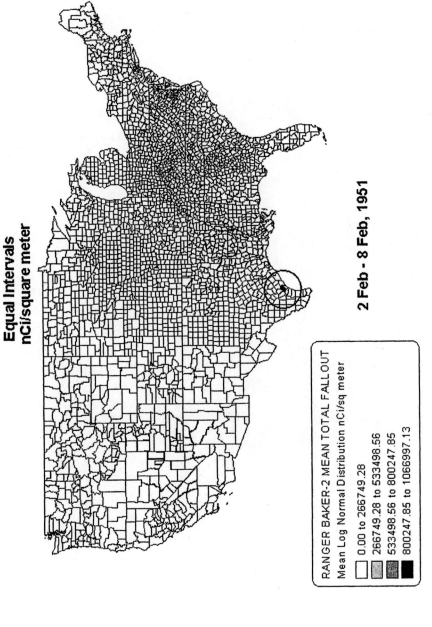

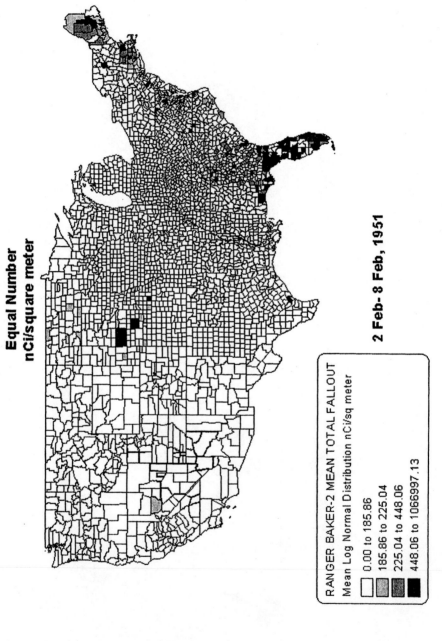

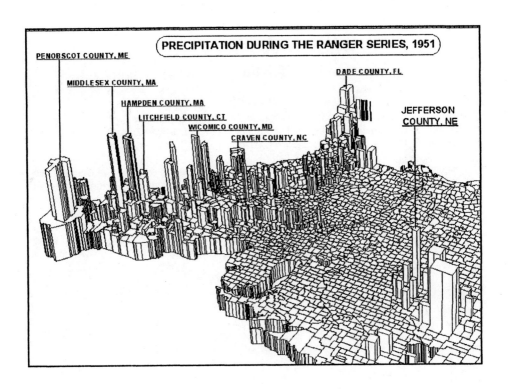

SHOT RANGER FOX
(Not Included in the NCI Study)

RANGER: FOX

Detonation Date: 6 Feb 1951 • *Detonation Time:* 5:47 A.M. • *Area:* Frenchman's Flat • *Sponsor:* Los Alamos • *Yield:* 22 kt • *Radiation Level at Ground Zero:* 15 R/hr • *Height of Burst:* 1,435 ft • *Cloud Top Height:* 43,000 ft

10,000-Foot Trajectory: This portion of the nuclear cloud passed over Mohave County, Lake Mead, and Prescott, and left United States airspace east of Pedgregosa, Arizona. It crossed back into the U.S. at Laredo, Texas, and left the state at Corpus Christi.

30,000-Foot Trajectory: This trajectory moved north of the 10,000-foot trajectory, passing over Flagstaff, Arizona. After briefly crossing Mexico airspace, the debris crossed the Texas cities of El Paso, Roma, McAllen, Pharr, Donna, Mercedes, Harlingen, San Benito, and Brownsville.

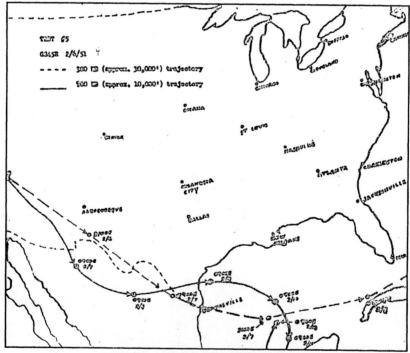

TRAJECTORY: RANGER FOX

SECTION 2

BUSTER-JANGLE SERIES

Oct 22-Nov29, 1951

SHOT BUSTER ABLE

(Not Included in the NCI Study)

BUSTER: ABLE

Detonation Date: 22 Oct 1951 • *Detonation Time:* 6:00 A.M. • *Area:* 7-5 • *Sponsor:* Los Alamos • *Yield:* Less than 0.1 kt • *Radiation Level at Ground Zero:* 300 counts per minute alpha radiation • *Height of Burst:* 100 ft • *Cloud Top Height:* 8,000 ft

8,000-Foot Trajectory: The cloud of debris passed north of Las Vegas and traveled on to Flagstaff, Arizona, circling down past the Colcord Mountains and Black Canyon, passing near Prescott then back over Flagstaff. From there, the debris traveled north, crossing into Utah over the Rainbow Plateau. The cloud followed Glen Canyon then drifted north over Roan Cliffs and East Tavaputs Plateau and into Colorado at Dinosaur. The cloud crossed into Wyoming at Power Wash and moved on to Rawlins. It left the state near Redbird and crossed into South Dakota near Parker Peak.

The cloud crossed the Badlands, then moved northeast, leaving South Dakota near Corona and entering Minnesota at Clinton. It then crossed over to Kingsdale, Minnesota, and into Wisconsin at Moose Junction. The BUSTER: ABLE cloud reentered Minnesota at Washburn and finally left United States airspace over Lake Superior.

CHARACTERISTICS: BUSTER ABLE

A first attempt to detonate the Buster-ABLE device met with failure. Due in part to faulty circuitry, nothing happened (a similar event would occur years later during the Plumbbob DIABLO event). When the device finally detonated, there was only a minimum yield. Official sources reported that the nuclear yield was "less than 0.1 kiloton" but unofficial sources suggest the yield was less than a pound.

The tower was damaged but still standing after the test.

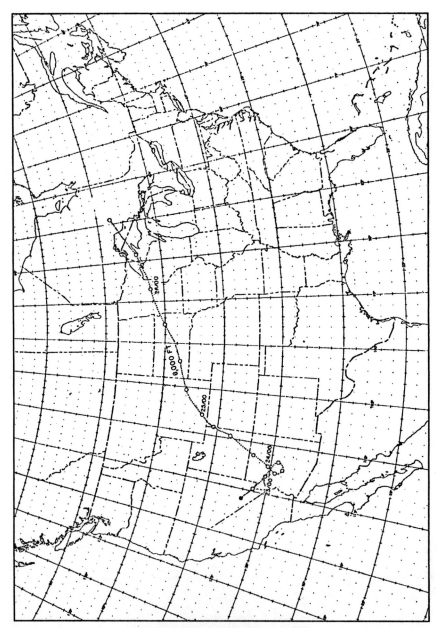

TRAJECTORY: BUSTER ABLE

SHOT 1: BUSTER BAKER (BJ1)

BUSTER: BAKER

Detonation Date: 28 Oct 1951 • *Detonation Time:* 7:20 A.M. • *Area:* 7-3 • *Sponsor:* Los Alamos • *Yield:* 3.5 kt • *Radiation Level at Ground Zero:* 6 R/hr • *Height of Burst:* 1,118 ft • *Cloud Top Height:* 31,700 ft

All trajectories traveled west over California in a band extending between Mount Whitney-Cape San Martin and Stove Pipe Wells-Guadalupe. The entire cloud crossed over San Luis Obispo County in California, circled, then moved east over the rest of the United States. One component of the 8,000-foot trajectory split off from the rest of the cloud and circled back over the Los Angeles area. The debris was tracked along four trajectories: the 8,000-, 10,000-, 24,000-, and 30,000-foot levels. Interestingly, three of the clouds—the 18,000-, 24,000-, and 30,000-foot levels—all crossed over the same point: Concordia, Kansas.

10,000-Foot Trajectory: This level moved north of Phoenix, Arizona, over Montrose, Colorado, then southeast toward Amarillo, Texas. It followed the Red River and crossed over the towns of Eldorado, Oklahoma, Wichita Falls and Denison, Texas (Dwight D. Eisenhower's boyhood home). It crossed over Hugo, Oklahoma, Ashtown, Texas, the state of Arkansas, and entered Mississippi near Tupelo. Other cities and areas under the 10,000-foot trajectory included: TENNESSEE: Chattanooga, Knoxville • VIRGINIA: Roanoke, Alexandria • Washington, D.C. • DELAWARE: New Castle • PENNSYLVANIA: Philadelphia • NEW YORK: New York • RHODE ISLAND: Providence • MASSACHUSETTS: Lexington, Boston • MAINE: Washington County.

24,000-Foot Trajectory: CALIFORNIA: Barrett • NEVADA: Laughlin, Lake Mead • UTAH: Hilldale, Cisco • COLORADO: Grand Junction, Denver, Hale • KANSAS: Goodland, Dresden, Blair, Concordia • MISSOURI: St. Joseph, Chillicothe, Brookfield, Macon, Hannibal • ILLINOIS: Jacksonville, Champaign, Danville • INDIANA: Logansport • OHIO: Celina, Greensburg, Youngstown • PENNSYLVANIA: Sharon, Tioga City (The 8,000- and 24,000-foot trajectories crossed over Elkland) • NEW YORK: Elmira, Glens Falls • VERMONT: Pawlet, West Hartford • NEW HAMPSHIRE: Hanover, Kearsarge • MAINE: (Near the White Mountain National Forest) Waterville, Bangor, Princeton.

30,000-Foot Trajectory: This followed the debris at the 24,000-foot level, the two crossing paths at Concordia, Kansas. The debris at the 30,000-foot level then crossed over Leavenworth, Kansas, and entered Missouri over Liberty. Traveling east, the cloud passed over Terre Haute, Indiana; Cincinnati, Ohio; and left the U.S. over Atlantic City, New Jersey.

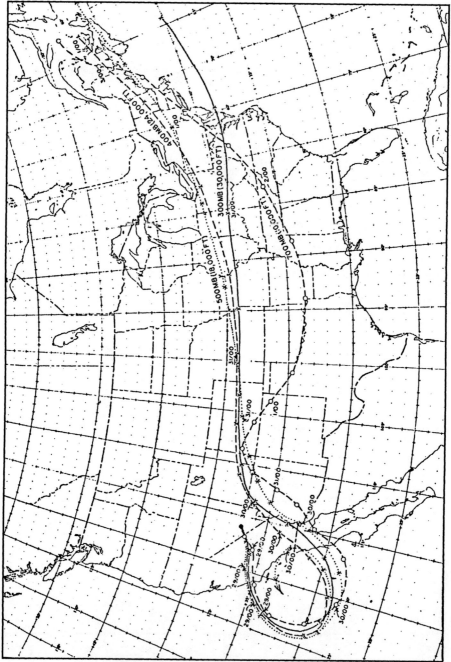

TRAJECTORY: BUSTER BAKER (BJ1)

TOTAL FALLOUT: BUSTER BAKER
Equal Numbers
nCi/square meter

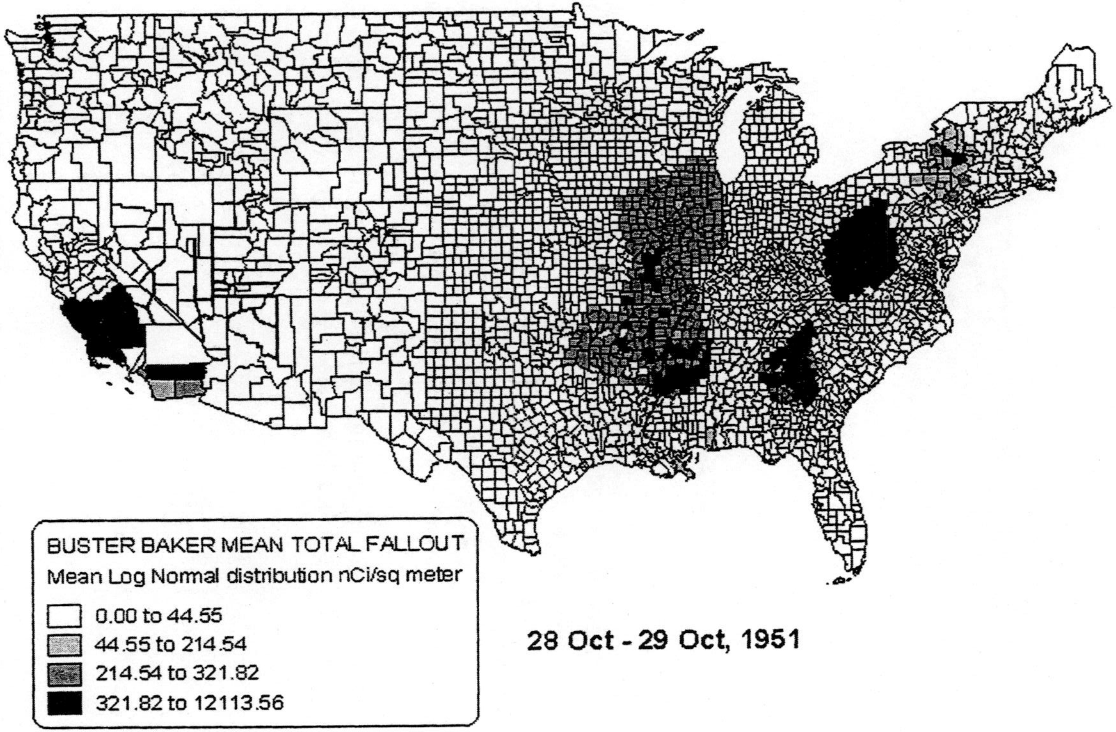

28 Oct - 29 Oct, 1951

BUSTER BAKER MEAN TOTAL FALLOUT
Mean Log Normal distribution nCi/sq meter

- 0.00 to 44.55
- 44.55 to 214.54
- 214.54 to 321.82
- 321.82 to 12113.56

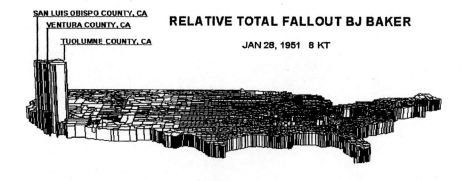

CHARACTERISTICS: BUSTER BAKER

Buster Baker ranks first among the nuclear tests in the percentage of Co-50, Eu-155, Pm-147 and Tc-99m contained in fallout, and 3rd in the percentage of Am-241 in fallout. The deposition pattern for this shot is most similar to that of Buster-Jangle CHARLIE (R:0.30, T: 17.64). The US county of highest deposition: Fresno, CA. Fallout from this nuclear test was deposited in 527 counties (17 percent of the U.S. counties.) The average fallout amount per county was 302.9 nCi/sq meter (0.3029 µCi/sq meter), ranking this shot 52nd for average fallout among nuclear tests.

The day 1 ratio of total fallout to I-131 is equal to 87.4 to 1, ranking 47th highest among all nuclear tests.

SHOT 2: BUSTER CHARLIE (BJ2)

BUSTER: CHARLIE

Detonation Date: 30 Oct 1951 • *Detonation Time:* 7:00 A.M. • *Area:* 7-3 • *Sponsor:* Los Alamos • *Yield:* 14 kt • *Radiation Level at Ground Zero:* 5 R/hr • *Height of Burst:* 1,132 ft • *Cloud Top Height:* 41,000 ft

10,000-Foot Trajectory: This part of the atomic cloud moved east, then south over Lake Mead. It crossed Phoenix, Arizona, then moved south into Mexico. It curved back to reenter the U.S. at Texas near the town of Lajitas. It then met up with the 18,000-foot trajectory over Houston, and crossed it again near Mobile, Alabama. From there, the radioactive cloud moved northeast over Columbus, Georgia and Charlotte and Greensboro, North Carolina. North of Richmond, Virginia, it began to curve east and crossed the East Coast over Atlantic City, New Jersey, but followed the coastline up to New York City. It then moved north crossing Danbury, Connecticut, and Pittsfield, Massachusetts, then followed the Vermont-New Hampshire border north to Canada. The cloud then followed the U.S.-Canada border until it left United States airspace.

18,000-Foot Trajectory: This trajectory crossed Death Valley, California, north of Bakersfield, continued west and left the coast over Point Conception, California. Once over the Pacific, however, it turned back east, crossing Mexico and then reentered United States airspace over Texas's Big Bend area. It crossed over San Antonio, Houston, and Beaumont-Port Arthur, Texas, and then entered Louisiana west of Lake Charles. The cloud crossed Louisiana and Mississippi and entered Alabama just north of Mobile. The debris continued on over Georgia and South Carolina, leaving the coast near Charleston.

24,000-Foot Trajectory: This trajectory crossed Fresno, California, and then, along with the rest of the atomic cloud, moved out over the Pacific. It then doubled back and entered Texas near Esperanza. It tracked just south of Lubbock, crossed Wichita Falls and Denison, Texas, then entered Arkansas airspace. Other cities it crossed included Little Rock, Arkansas; Memphis, Tennessee; Gilbert, West Virginia; Washington, D.C.; and Dover, Delaware.

30,000-Foot Trajectory: This trajectory moved west north of Los Angeles, crossed the coastline, then reentered the United States at Cowlic, Arizona. Later it crossed over Albuquerque and Santa Fe, New Mexico, then moved north, eventually crossing the 40,000-foot trajectory at Sterling, Nebraska. Other cities 30,000 feet below the atomic cloud: Indianola and Iowa City, Iowa; Rockford and Waukegan, Illinois; Lansing and Port Huron, Michigan; Toronto; Montreal; Presque Isle, Maine.

40,000-Foot Trajectory: This trajectory moved northeast over the northern tier of Colorado counties, crossing the cities of Fort Collins and Sterling. It then moved across the *second* southern tier of Nebraska counties, crossing the countryside just south of Lincoln. It followed the Missouri-Iowa border, then moved north over Burlington, Iowa, and Peoria, Illinois. Other cities and areas 40,000 feet under the atomic cloud: Fort Wayne, Indiana; Toledo, Ohio; the southern tier of New York counties from Jamestown to Kingston; and Worcester, Boston, and Lexington, Massachusetts.

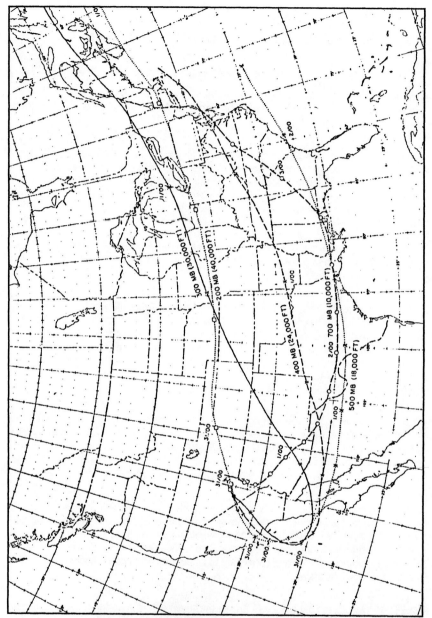

TRAJECTORY: BUSTER CHARLIE (BJ2)

CHARACTERISTICS: BUSTER CHARLIE

Like the devices in the Ranger series, the Buster CHARLIE test device resembled the "Fat Man" bomb detonated over Nagasaki. The deposition pattern for Buster CHARLIE is most similar to that of Buster DOG (R: 0.43; T: 26.85).

Fallout from this test was deposited across 95 percent of the U.S. or, 2937 counties. The average fallout amount per county was 12,266 nCi/sq meter, ranking this shot 8th for average fallout among the nuclear tests.

The day 1 ratio of total fallout to I-131 is 139.7 to 1, the 8th highest among all nuclear tests. Buster CHARLIE ranks 9th in deposition of Pr-144 and 21st in deposition of Te-127m across the U.S.

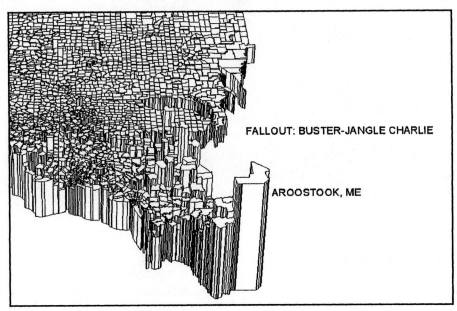

EAST COAST FALLOUT: BUSTER CHARLIE

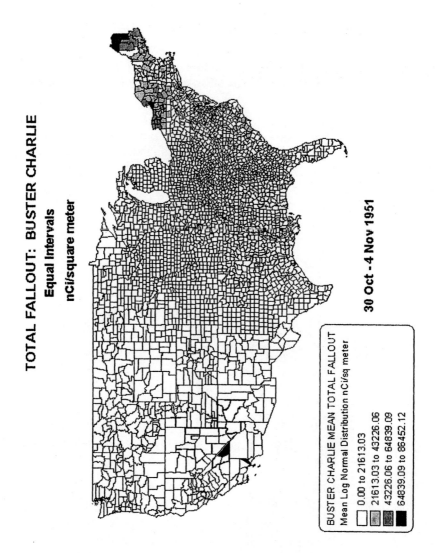

U.S. FALLOUT ATLAS : TOTAL FALLOUT

SHOT 3: BUSTER DOG (BJ3)

BUSTER: DOG

Detonation Date: 1 Nov 1951 • *Detonation Time:* 7:30 A.M. • *Area:* 7-3 • *Sponsor:* Los Alamos • *Yield:* 21 kt • *Radiation Level at Ground Zero:* 20 R/hr • *Height of Burst:* 1,416 ft • *Cloud Top Height:* 46,000 ft

The 18,000- 24,000- 30,000- and 40,000-foot levels all moved east in a wide arc, crossing over: Las Vegas, Nevada; Phoenix, Arizona; El Paso, San Angelo, Austin, Waco, Texas; and Shreveport, Louisiana. The band tightened over the north part of Madison County, Mississippi, at Camden (home of the Casey Jones State Museum), then spread in a wide arc over the entire Northeast, from a line from Knoxville, Tennessee–Ottawa, Canada, to a line from Asheville, North Carolina–Atlantic City, New Jersey.

The track of the atomic clouds passed over: NEVADA: Las Vegas, Henderson, Boulder City • ARIZONA: Dolan Springs, Hackberry, Dewey, Humboldt, Tonto Basin, San Carlos, Clifton • NEW MEXICO: Silver City, Tyrone, Bayard, Las Cruces, University Park, Mesquite • TEXAS: (Guadalupe Peak), Dell City, Pecos, Barstow, Royalty, Grandfalls, Big Lake, Sherwood, Vick, Eden, Millersview, Placid, Elm Grove, Bee House, Pearl, McGregor, Waco, Mexia, Red Lake, Palestine, Neches, Turney, New Salem, Gary, Carthage • LOUISIANA: Keatchie, Gloster, Castor, Lucky, Liberty Hill, Quitman, Vernon, Richwood, Raysville, Holly Ridge, Warden, Monticello • MISSISSIPPI: Onward, Valley, Yazoo City, Pickens, Goodman, Kosciusko, High Point, Sessums, Artesia, Columbus • ALABAMA: Fernbank, Belk, Corona, Jasper, Empire, Hayden, Oneonta, Rosa, Hendricks, Altoona, Mountainboro, Collinsville, Ringgold • GEORGIA: Summerville, Echota, Calhoun, Resaca, Ellijay, Cherrylog, Morganton • NORTH CAROLINA: Hayesville, Tusquitee, Kyle, Bryson City, Cherokee, Waterville • TENNESSEE: Johnson City, Elizabethton, Hunter, Carter, Shady Valley • VIRGINIA: Sugar Grove, Wytheville, Pulaski, Radford, McCoy, Newport, Blacksburg, Abbot, New Castle, Barbours Creek, Eagle Rock, Lexington, Greenville, Crimora, Lydia, Haywood, Boston, Rixeyville, New Baltimore, Gainesville, Arlington • Washington, D.C. • MARYLAND: Bethesda, Spencersville, Columbia, Baltimore, Sweet Air • PENNSYLVANIA: Oxford, West Chester, King of Prussia, Norristown, Philadelphia, Warminister, Newtown • NEW JERSEY: Trenton, Hopewell, Princeton, Piscataway, Wesfield, Cranford, Newark, Paterson, Hackensack, Paramus • NEW YORK: Millwood, Katona, Salem Center, White Plains • CONNECTICUT: Wilton, Naugatuck, Meridan, Middletown, Portland, Marlborough, Willimantic, Putnam, Harrisville • RHODE ISLAND: Woonsocket • MASSACHUSETTS: Millville, Bellingham, Holliston, Framingham, Needham, Wellesley, Newton, Brookline, Boston, Cambridge, Somerville, Malden, Lexington, Saugus, Lynn, Wakefield, Peabody, Salem, Gloucester, Pigeon Cove.

The 10,000-foot level moved south, crossing Mexico, then the southern tier of Texas counties: Zapata, Starr, Hidalgo, and Cameron.

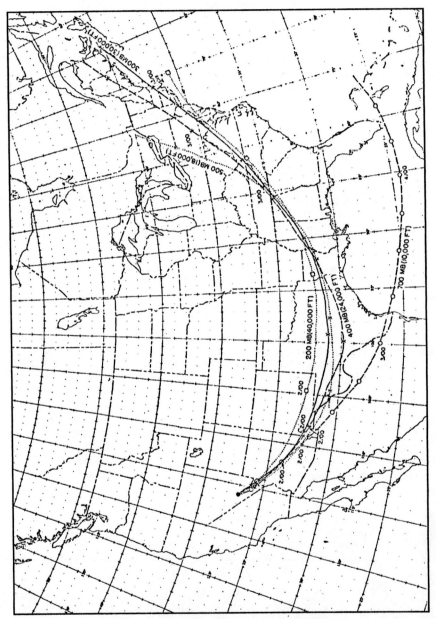

TRAJECTORY: BUSTER DOG (BJ3)

CHARACTERISTICS: BUSTER DOG

Among the nuclear tests, Buster DOG is 6TH for percentage of Co-58 and Co-60 in fallout. The deposition pattern for Buster DOG is most similar to that of Buster EASY. The counties of highest deposition include the New York counties of Chautauqua, Erie, Genesee, Niagara and Orleans. Estimated maximums for all: 210 µCi/sq meter of fallout on the same day, Nov 6, 1951.

Fallout from this nuclear test was deposited across 2163 counties (41 percent of the U.S.). The average fallout amount was 3.8 µCi/sq meter, ranking this shot 38th for average fallout among the nuclear tests.

The day one ratio of total fallout to I-131 for Buster DOG is 146.3 to 1, making it the 6th highest total fallout to I-131 ratio among all nuclear tests.

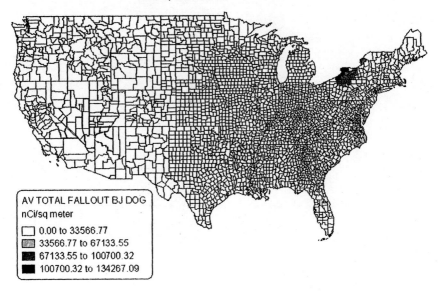

RELATIVE AVERAGE TOTAL FALLOUT BJ DOG
NOV 1, 1951 21 KT

AV TOTAL FALLOUT BJ DOG
nCi/sq meter
- 0.00 to 33566.77
- 33566.77 to 67133.55
- 67133.55 to 100700.32
- 100700.32 to 134267.09

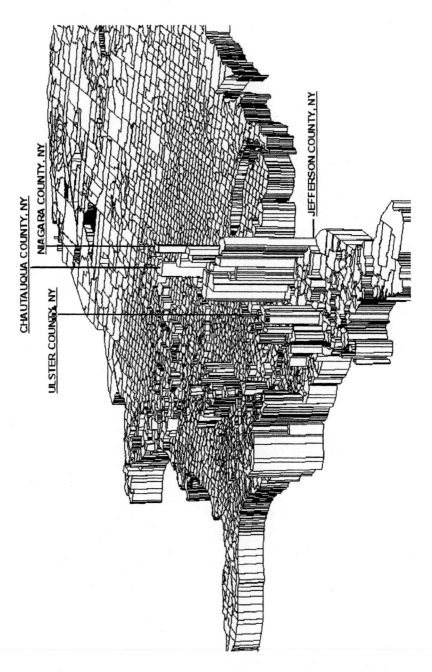

U.S. FALLOUT ATLAS : TOTAL FALLOUT

SHOT 4: BUSTER EASY (BJ4)

BUSTER: EASY

Detonation Date: 5 Nov 1951 • *Detonation Time:* 8:30 A.M. • *Area:* 7 • *Sponsor:* Los Alamos • *Yield:* 31 kt • *Radiation Level at Ground Zero:* 20 R/hr • *Height of Burst:* 1,314 ft • *Cloud Top Height:* 50,000 ft

The 24,000-, 30,000-, 40,000-, and 46,000-foot trajectories all crossed over Las Vegas, then moved southwest in a wide arc across the United States and Gulf. Among cities under the high-altitude nuclear clouds: Phoenix and Tucson, Arizona; Juarez, Mexico–El Paso, Laredo, San Antonio, Corpus Christi, Texas; Jacksonville, Florida.

10,000-Foot Trajectory: This part of the nuclear cloud moved southwest to Santa Barbara, Lompoc, and Santa Maria, California. It then followed the coastline to San Francisco where it crossed the 18,000-foot trajectory. It then moved back over Marysville, California, and into Nevada over the Smoke Creek Desert, where it continued Northeast. Over Dunphy, Nevada, it began to turn back to the south. It was last tracked over northern Milliard County in Utah.

18,000-Foot Trajectory: This part of the nuclear cloud followed essentially the same path as the 10,000-foot level, but after crossing that trajectory over San Francisco, it moved northeast, passing over Nevada, then Idaho. The cloud moved over the Teton Range into Wyoming, then drifted over Jackson and Thermopolis, finally leaving the state north of Kirtley. The cloud at 18,000-feet then moved across the northern tier of Nebraska counties before entering Iowa at Sioux City. Once in Iowa, it crossed over Storm Lake, Fort Dodge, Iowa Falls, and Cedar Falls. Soon after, it entered Illinois airspace at Elizabeth. From there it passed over Rockford, Woodstock, and Waukegan, then crossed into Michigan. Other cities 18,000 feet under the nuclear cloud included: MICHIGAN: Kalamazoo • OHIO: Toledo, Cleveland, and Youngstown • PENNSYLVANIA: Philadelphia • NEW JERSEY: Atlantic City.

RELATIVE TOTAL FALLOUT BJ EASY
NOV 5, 1951 31 KT

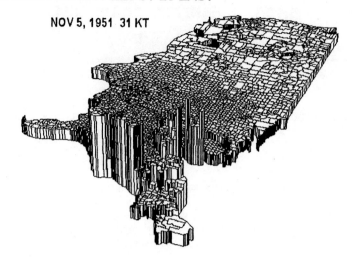

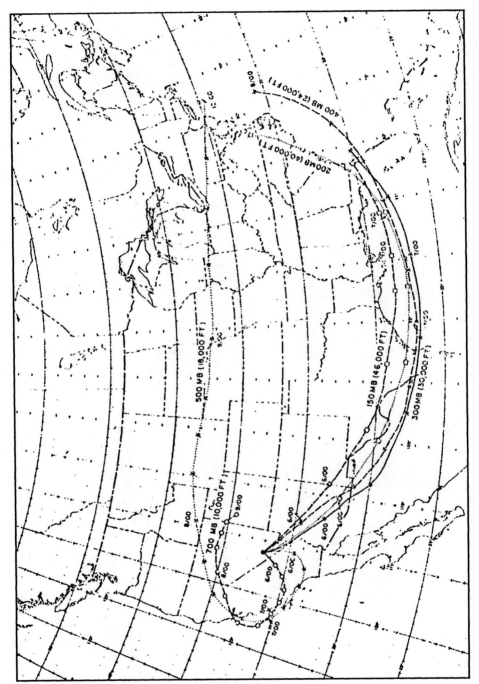

TRAJECTORY: BUSTER EASY (BJ4)

TOTAL FALLOUT: BUSTER EASY
Equal Intervals
nCi/square meter

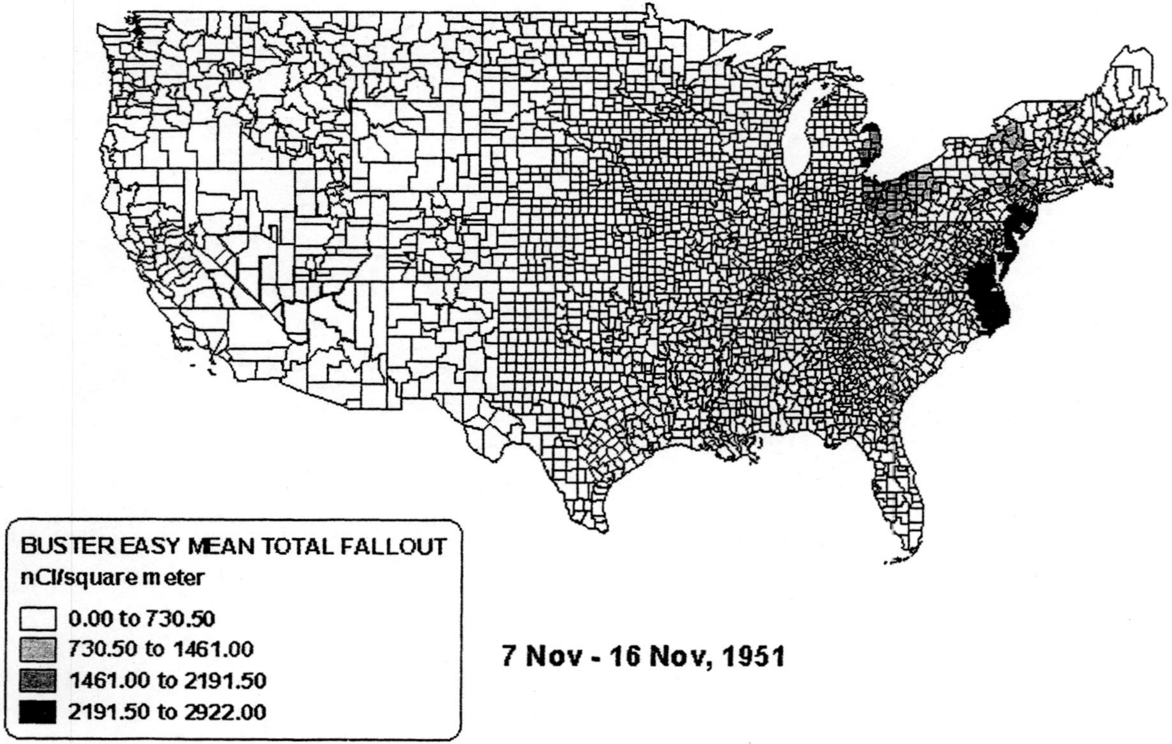

7 Nov - 16 Nov, 1951

BUSTER EASY MEAN TOTAL FALLOUT
nCi/square meter

- 0.00 to 730.50
- 730.50 to 1461.00
- 1461.00 to 2191.50
- 2191.50 to 2922.00

U.S. FALLOUT ATLAS : TOTAL FALLOUT

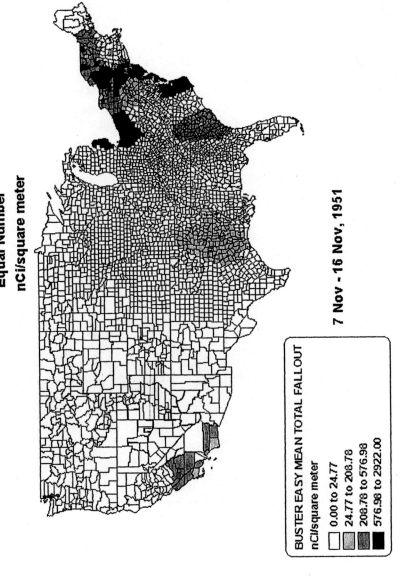

CHARACTERISTICS: BUSTER EASY

Buster EASY ranks 2nd among the nuclear tests in the percentage of Co-58 in fallout and 6th for percentage of Tc-99m in fallout. Among the aboveground nuclear tests this shot ranks 1st in deposition of the radionuclide Nb-95m.

The deposition pattern for Buster EASY is most similar to Buster-DOG (R: 0.51; T: 33.07). The counties of highest fallout deposition included: Kent DE, New Castle DE, Sussex DE, Worcester MA, Burlington NH, Camden NJ, Cape May NJ, Cumberland NJ, Gloucester NJ and Mercer NJ. Estimated maximum activity: 24.35 µCi/sq meter total fallout on Nov 10, 1951.

The day one ratio of total fallout to I-131 was 121 to 1, ranking it 17th highest among all nuclear tests for total fallout-to-I-131 ratio.

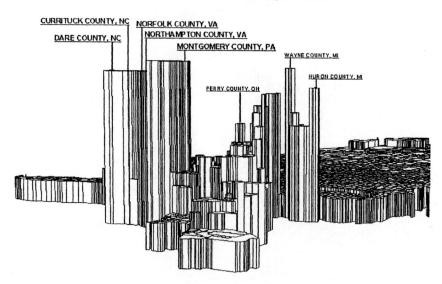

RELATIVE TOTAL FALLOUT BJ EASY
NOV 5, 1951 31 KT

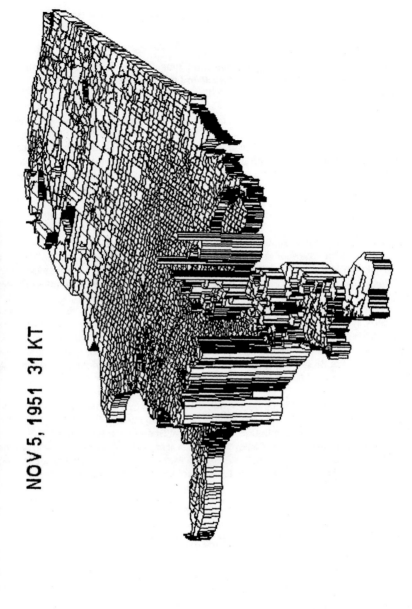

SHOT 5: JANGLE SUGAR (BJ5)

JANGLE: SUGAR

Detonation Date: 19 Nov 1951 • *Detonation Time:* 9:00 A.M. • *Area:* 9 • *Sponsor:* Department of Defense • *Yield:* 1.2 kt • *Radiation Level at Ground Zero:* 7,500 R/hr • *Height of Burst:* 3.5 ft • *Cloud Top Height:* 15,000 ft

Low-level debris from Sugar crossed north over Nye County, Nevada, passed along the border of Eureka and White Pine counties, and left Nevada through Elko County. It then crossed over the Duck Valley Indian Reservation and proceeded on toward Caldwell, Rayette, and Weiser, Indiana. From there, it curved west toward Huntington, Oregon, where it followed the track of now-U.S. 84 North across Baker, La Grande, and Pendleton. It then moved into the airspace of Washington State, where it passed over Richland, curled north around the Hanford Atomic Energy Works, and was then deflected south and west by the Saddle Mountains. The lower level was last detected over Yakima, Washington.

10,000-Foot Trajectory: Sugar's 10,000-foot section moved north over the Great Salt Lake, then crossed into the southeastern corner of Idaho at Malad City. It left the state near Irwin and entered Wyoming over the Teton Pass near Jackson. The cloud passed over Cody, Powell, and Lovell, then crossed into Montana airspace over the Crow Indian Reservation. From there it drifted over the villages of Bighorn, Sanders, Angela, and Union before it turned southeast toward Glendive. This section of the atomic cloud finally left the state at Wibaux. From there, the debris at 10,000 feet altitude entered North Dakota at Beach. It crossed Bullion Butte, New England, Elgin, Bismarck, Steele, Jamestown, Valley City, and finally left the state at Fargo. Upon entering Minnesota airspace, it crossed Georgetown, Ogema, (Itasca State Park), Deep River, Hibbing, and Virginia before leaving the state, and the United States, at Grand Portage, Minnesota.

14,000-Foot Trajectory: This trajectory moved northeast over Ogden, Utah and entered Wyoming airspace at Sage. It then crossed over to La Barge and the Wind River Indian Reservation. The cloud entered South Dakota airspace at Spearfish. It left South Dakota at Herreid and entered North Dakota at Ashley. From there, it moved over Jamestown and left the state at Grand Forks. The 14,000-foot cloud then moved into Minnesota airspace, crossing Thief Lake. It finally left the U.S. at Warroad, Minnesota, near Lake of the Woods.

CHARACTERISTICS: JANGLE SUGAR

Jangle SUGAR ranks 1st among the nuclear tests in the percentage of Nb-95m contained in fallout and 2nd for percentages of Sr-90, Y-90 and Y-91. The deposition pattern resulting from this test is most similar to that of Plumbbob SHASTA (R:0.37; T: 22.45). The county receiving the highest deposition of total fallout from Jangle SUGAR was Clearwater, ID. Other counties receiving high levels of fallout were Valley, ID and Boise, ID. Fallout from Jangle SUGAR was deposited across 615 counties (20% of the U.S.).

RELATIVE AVERAGE TOTAL FALLOUT: BJ SUGAR
NOV 19, 1951 1.2 KT

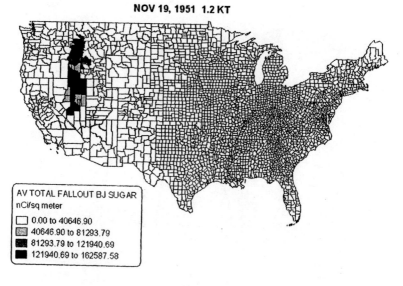

AV TOTAL FALLOUT BJ SUGAR
nCi/sq meter
- 0.00 to 40646.90
- 40646.90 to 81293.79
- 81293.79 to 121940.69
- 121940.69 to 162587.58

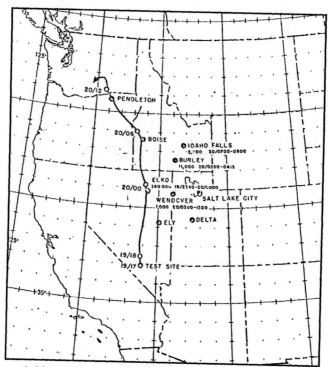

LOW-LEVEL DEBRIS: JANGLE SUGAR (BJ5)

U.S. FALLOUT ATLAS : TOTAL FALLOUT

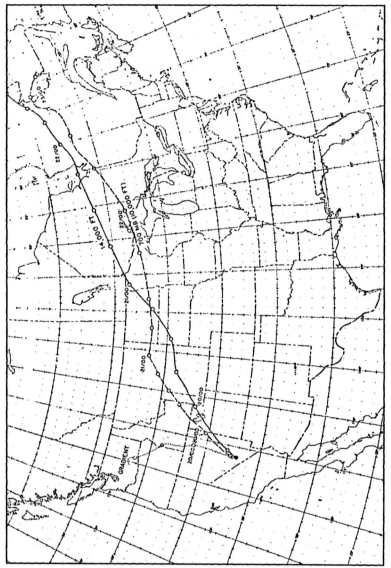

TRAJECTORY: JANGLE SUGAR (BJ5)

SHOT 6: JANGLE UNCLE (BJ6)

JANGLE: UNCLE

Detonation Date: 29 Nov 1951 • *Detonation Time:* 12:00 Noon • *Area:* 10 • *Sponsor:* Los Alamos, Department of Defense • *Yield:* 1.2 kt • *Radiation Level at Ground Zero:* 7,500 R/hr • *Height of Burst:* – 17 ft; underground burst: filled shaft in Nevada soil • *Cloud Top Height:* 11,500 ft

The low, surface-level portion of the Uncle cloud traveled due north over Nye County, Nevada, across the Diamond Mountains and over Huntington, Jiggs, the Te-Moak Indian Reservation, and Elko. It split up into two clouds over North Fork, Nevada. Shortly thereafter, tracking of this portion of the Uncle cloud was terminated.

10,000-Foot Trajectory: This part of the cloud crossed Nye County and moved toward the northeast over Railroad Valley and Lund. It crossed into Utah at Trout Creek, then passed over Salt Lake City. It left the state near the town of Upton. After entering Wyoming airspace, it crossed the towns of Mills, Church Buttes, Farson, Rawlins, Como, Medicine Bow, Garrett, Dwyer, and Guernsey. Other cities under the section of the nuclear cloud at 10,000-foot altitude: NEBRASKA: Agate, Crawford, Chadron, Whiteclay • SOUTH DAKOTA: (Slim Butte Mountain), Oglala, Potato Creek, Norris, White River, Kimball, Mitchell, Sioux Falls • MINNESOTA: Luverne, Wilmot, Windom, Odin, Truman, Matawan, Blooming Prairie, Rochester, Winona • WISCONSIN: Ettrick, Black River Falls, Pray, Arpin, Neopit, Beaver, Wausaukee • MICHIGAN: Banat, Carney, Steuben, Seney, Paradise • MAINE: (Kelly Brook Mountain), (Rocky Mountain), St. John, Fort Kent, Notre Dame.

CHARACTERISTICS: JANGLE UNCLE

Jangle UNCLE ranks 2nd among the nuclear tests for percentage of U-237 in fallout. The deposition pattern from this test is most similar to that of Tumbler-Snapper BAKER (TS2) (R:0.22; T: 12.31). The county of highest deposition was White Pine, NV where the estimated maximum (95% confidence interval) on Nov 29, 1951 was 184.08 µCi/sq meter. Other high-deposition counties were Lincoln(2) NV with an estimated maximum activity of 139 µCi/sq meter and Nye(2) NV with an estimated maximum activity of 112.98 µCi/sq meter. The day one ratio of total fallout to I-131 for this shot is moderate at 98.6 to 1.

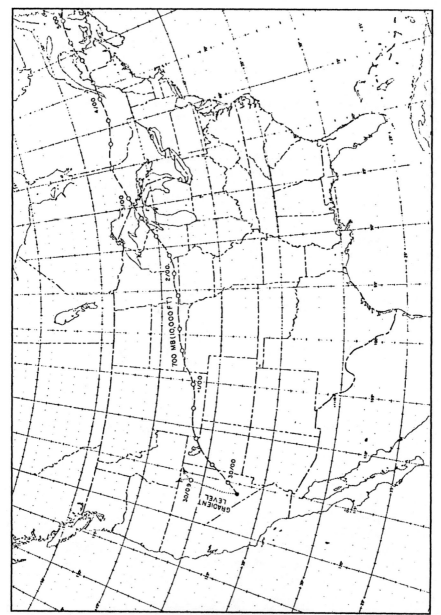

TRAJECTORY: JANGLE UNCLE (BJ6)

RELATIVE AVERAGE TOTAL FALLOUT: BJ UNCLE
NOV 29, 1951 1.2 KT

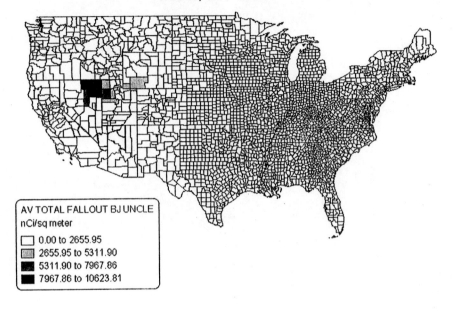

U.S. FALLOUT ATLAS : TOTAL FALLOUT

SECTION 3

TUMBLER-SNAPPER SERIES

April-Jun 5, 1952

SHOT 1: TUMBLER-SNAPPER ABLE (TS-1)

TUMBLER-SNAPPER: ABLE

Detonation Date: 1 Apr 1952 • *Detonation Time:* 9:00 A.M. • *Area:* Frenchman's Flat • *Sponsor:* Los Alamos, Department of Defense • *Yield:* 1 kt • *Radiation Level at Ground Zero:* 5 R/hr • *Height of Burst:* 793 ft • *Cloud Top Height:* 16,200 ft

10,000-Foot Trajectory: NEVADA: Alamo, Hiko, Pioche • UTAH: Wah Wah Springs, Oak City, Fountain Green, Milburn, Soldier Summit, Roosevelt, Bridgeport • WYOMING: Rawlins, Wheatland, Hawk Springs • NEBRASKA: Kimball • COLORADO: Sterling, Yuma, Kirk, Stratton, Cheyenne Wells • KANSAS: Syracuse, Ulysses, Arkalon, Hayne, Liberal • OKLAHOMA: Forgan, Beaver, Slapout, Harmon, Taloga, Geary, Seminole, McAlester, Page • ARKANSAS: Mena, Sulphur Springs, Hot Springs, Pine Bluff, Crumrod • MISSISSIPPI: Hillhouse, Batesville, Bonneville, Doskie • TENNESSEE: Lowryville, Topsy, Mount Pleasant, Columbia, Nashville, Portland • KENTUCKY: Glasgow, Campbellsville, Harrodsburg, Stamping Ground • OHIO: Georgetown, Hillsboro, London, Port Clinton • MICHIGAN: Detroit, Pontiac, Flint, Saginaw, Bay City, Clare, Big Rapids, Muskegon • INDIANA: Portage, Francisville, Sitka, Young America, Kokomo, Anderson, Rushville, New Trenton • OHIO: Cincinnati, Bethel, Georgetown, Portsmouth, Buckhorn, Gallipolis • WEST VIRGINIA: Ashton, Left Hand, Sutton, Harper, Martinsburg • MARYLAND: Baltimore • DELAWARE: Wilmington • PENNSYLVANIA: Philadelphia • NEW JERSEY: Trenton, Twin Rivers, Englishtown, Leonardo • NEW YORK: Lindenhurst, Mastic Beach, Southhampton • MASSACHUSETTS: Martha's Vinyard, Hyannis, Barnstable, E. Orleans.

16,000-Foot Trajectory: NEVADA: Rox, Carp • UTAH: St. George, Leeds, Mount Carmel Junction, Mexican Hat, Navajo Indian Reservation • COLORADO: Cortez, Durango,. Summitville, Alamosa, Fort Garland, Trinidad • OKLAHOMA: Black Mesa Summit (highest point in Oklahoma) near Kenton, Castaneda, Keyes, Eva, Guyman, Gray, Catesby, Woodward, Selling, Longdale, Bison, Stillwater, Tulsa, Tahlequa, Stilwell • ARKANSAS: Mountainburg, Mountaintop, Clarksville, Russellville, Beebe, Brinkley, Marianna • MISSISSIPPI: Prichard, Senatobia, Abbeville, Tupelo, Tremont • ALABAMA: Bexar, Hamilton, Jasper, Birmingham, Bemiston, Woodland • GEORGIA: Texas, Hogansville, Topeka Junction, Macon, Irwinton, Wrightsville, Rocky Ford, Springfield • SOUTH CAROLINA: Ridgeland, Burton, Beaufort, Edisto Island.

CHARACTERISTICS: TS ABLE

As with most of the shots in the previous series, the device was based on the Nagasaki Mk4 bomb, which weighed in at 10,500 lbs. The shot Tumbler-Snapper ABLE ranks 2nd among the nuclear tests for percentage of Np-239 contained in fallout. The deposition pattern from this shot was most similar to that of Ranger BAKER-1 (R: 0.22; T: 12.24). The counties of highest deposition included the Arkansas counties of Franklin, Johnson, Montgomery, Newton, Perry, Pope and Yell; the Missouri counties of Barry and Christian; and the Kansas county of Crawford. Estimated maximum activity: 254 µCi/sq meter total fallout on April

3,1952. Other counties receiving high fallout with an estimated maximum fallout activity of 241 µCi/sq meter each were the Nebraska counties of Banner, Dawes and Sioux.

The day one ratio of total fallout to I-131 for Tumbler-Snapper ABLE is 206.7 to 1, which is the highest ratio among all aboveground nuclear tests, including SEDAN. Fallout from Tumbler-Snapper ABLE was deposited in 2089 counties (68 percent of the U.S.).

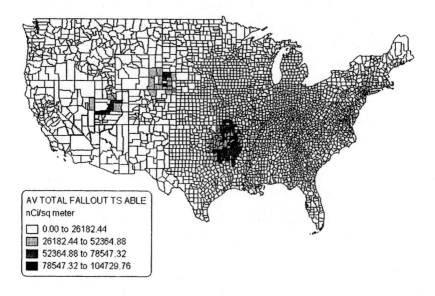

RELATIVE AVERAGE TOTAL FALLOUT: TS ABLE

APR 1, 1952 1 KT

AV TOTAL FALLOUT TS ABLE
nCi/sq meter
- 0.00 to 26182.44
- 26182.44 to 52364.88
- 52364.88 to 78547.32
- 78547.32 to 104729.76

RELATIVE TOTAL FALLOUT: TS ABLE

APR 1, 1952 1 KT

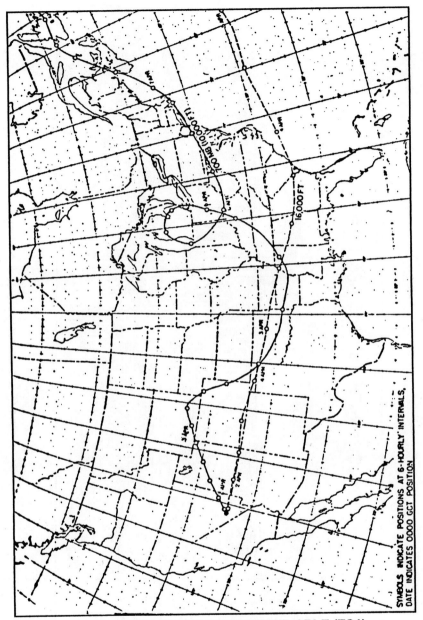

TRAJECTORY: TUMBLER-SNAPPER ABLE (TS1)

U.S. FALLOUT ATLAS : TOTAL FALLOUT

SHOT 2: TUMBLER-SNAPPER BAKER (TS2)

TUMBLER-SNAPPER: BAKER

Detonation Date: 15 Apr 1952 • *Detonation Time:* 9:30 A.M. • *Area:* 7-3 • *Sponsor:* Los Alamos, Department of Defense • *Yield:* 1 kt • *Radiation Level at Ground Zero:* 4.5 R/hr • *Height of Burst:* 1,109 ft • *Cloud Top Height:* 15,700 ft

10,000-Foot Trajectory: NEVADA: Goodsprings, Sandy • CALIFORNIA: Cima, Ludlow, San Bernadino, Redlands, Ontario, Pomona, Whittier, Buena Park, Long Beach, Redondo Beach, Inglewood • ARIZONA: San Luis, Somerton, Wellton, Growler, Quartzite, Bouse, Signal, Yucca, Kingman, Temple Bar • NEVADA: Overton, Glendale, Galt, Pioche • UTAH: Garrison, Provo, American Fork, Murray, Kerns, Salt Lake City, Syracuse • NEVADA: Shafter, Currie, Cherry Creek, Warm Springs, Alamo, Las Vegas • CALIFORNIA: Goffs, Vidal, Blythe, Palo Verde.

16,000-Foot Trajectory: NEVADA: Las Vegas • ARIZONA: Hualpi Indian Reservation, (Grand Canyon National Monument) • UTAH: Rockville, Springdale, Long Valley Junction, Hatch, Lund, Zane, Beryl, Hamlin Valley • NEVADA: Ursine, Pioche, Adaven, Warm Springs, Tonopah, Arlemont • CALIFORNIA: Benton Station, Benton, (Yosemite National Park), Bridgeport • NEVADA: Sweetwater, Babbitt, Hawthorne, Thorne, (Quartz Mountain), Austin, Winnemucca, Paradise Valley, Rebel Creek • IDAHO: Triangle, Silver City, Murphy, Nampa, Boise, Horsehoe Bend, Crouch, Warm Lake, Yellow Pine • MONTANA: Hamilton, Bearmouth, Helmville, Blackleaf, Conrad, Shelby, Sunburst, Sweetgrass.

RELATIVE AVERAGE TOTAL FALLOUT: TS BAKER
APR 15, 1952 1 KT

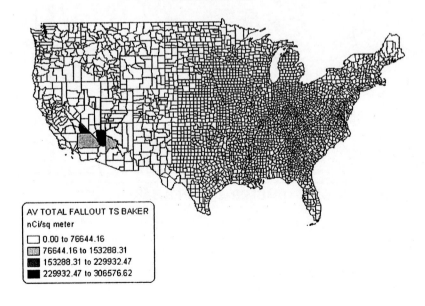

AV TOTAL FALLOUT TS BAKER
nCi/sq meter
☐ 0.00 to 76644.16
▨ 76644.16 to 153288.31
▩ 153288.31 to 229932.47
■ 229932.47 to 306576.62

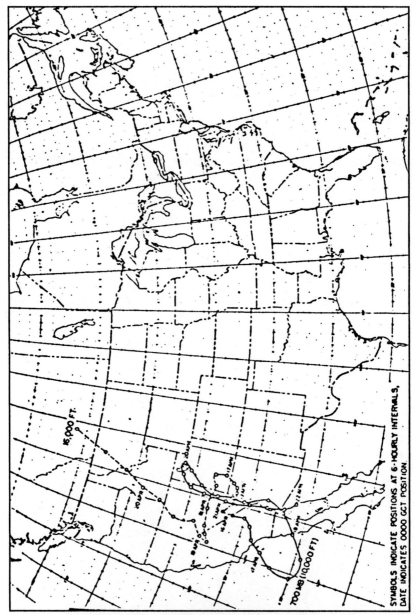

TRAJECTORY: TUMBLER-SNAPPER BAKER (TS2)

3-8 U.S. FALLOUT ATLAS : TOTAL FALLOUT

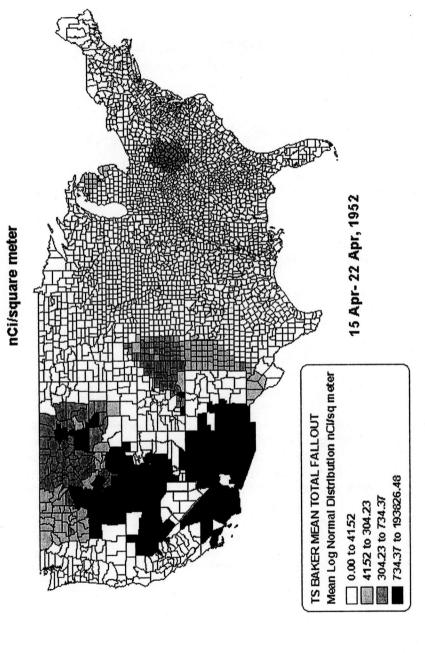

RELATIVE TOTAL FALLOUT: TS BAKER
APR 15, 1952 1 KT

CHARACTERISTICS: TS BAKER

The Tumbler-Snapper BAKER device was based on the Nagasaki bomb design (Mk4) and weighed approximately 10,800 lbs. Tumbler-Snapper BAKER ranks 3rd for fallout percentage of Np-239 and U-239 and 4th for U-240. The deposition pattern for Tumbler-Snapper BAKER is most similar to Teapot CLIMAX (TP11). The county of highest deposition was Inyo, CA with an estimated maximum activity of 569.46 µCi/sq meter total fallout.

Fallout from this nuclear test was deposited on 406 counties, comprising only 13 percent of the total U.S. counties.

The ratio of total fallout to I-131 for Tumbler-Snapper BAKER ranks second highest among nuclear tests at 201.5 to 1.

SHOT 3: TUMBLER-SNAPPER CHARLIE (TS3)

TUMBLER-SNAPPER: CHARLIE

Detonation Date: 22 Apr 1952 • *Detonation Time:* 9:30 A.M. • *Area:* 7-3 • *Sponsor:* Los Alamos, Department of Defense • *Yield:* 31 kt • *Radiation Level at Ground Zero:* 0.15 R/hr • *Height of Burst:* 3,447 ft • *Cloud Top Height:* 42,000 ft

10,000-Foot Trajectory: NEVADA: Pahrump • CALIFORNIA: Shoshone, Tecopa, Fort Irwin, Pioneer Point, Little Lake, Kings, (Canyon National Park), June Lake, (Mono Lake), Coleville, Markleeville • NEVADA: Gardnerville, Carson City, Crystal Bay • CALIFORNIA: Truckee, Portola, Susanville, Bieber, White Horse, Tule Lake • OREGON: Malin, Beatty, Silver Lake, Millican, (Stevenson Mountain), Mayville, Condon, Arlington • WASHINGTON: Whitcomb, Horse Heaven Hills, Prosser, Ephrata, Soap Lake, Mansfield, Disautel, Havillah, Chesaw • MICHIGAN: Sault Ste. Marie, Barbeau, Hessel, Freedon, Boyne Fals, Lansing, Battle Creek, Kinderhook • INDIANA: Orland, Fort Wayne, Muncie, Connersville, Batesville, Madison • KENTUCKY: Carrollton, Frankfort, Sunnybrook • TENNESSEE: Chanute, Boatland, Crossville, Chattanooga • GEORGIA: Dalton, Carrollton, LaGrange • ALABAMA: Lafayette, Midway, Dothan • FLORIDA: Campbellton, Marianna, Apalachicola, Tampa, Arcadia, Homestead, Florida City, Key Largo.

18,000-Foot Trajectory: CALIFORNIA: Beacon Station, San Bernadino, Anaheim, Santa Ana, Buena Park, Whittier, San Mateo, San Francisco, Santa Rosa, Ukia, Willits, Blue Lake, Gasquet • OREGON: Remote, Swisshome, Tidewater, Eddyville, Willamina, Blaine, Westport • WASHINGTON: Grays River, Doty, Olympia, Bremerton, Burlington, Mount Baker • MINNESOTA: Laurel, Silverdale, Aurora, Castle Danger • WISCONSIN: Ironwood, Pulaski, Green Bay, Two Creeks • INDIANA: Gary, Hammond • ILLINOIS: Kankakee, Rantoul, Centralia, Carbondale • MISSOURI: Cape Girardeau, Belmont • KENTUCKY: Hickman • TENNESSEE: Nankipoo, Collierville, Rossville • MISSISSIPPI: Byhalia, Eupora, Meridian • ALABAMA: Bogueloosa, St. Stephens, Perdido • FLORIDA: Pensacola, Fort Meyers, Hialeah, Miami.

24,000-Foot Trajectory: This part of the cloud followed the California-Nevada, California-Arizona border, then up through San Diego. The other cities are: CALIFORNIA: El Centro, Victorville, Fresno, Merced, Modesto, Sacramento, Burney, Dorris • OREGON: Klamath Falls, Belknap Springs, Sandy • WASHINGTON: (Mount St. Helens), Puyallup, Auburn, Kent, Bellevue, Everett, Sumas • MINNESOTA: Grand Marais • MICHIGAN: Beacon Hill, Houghton, Negaunee, Garden, Traverse City, Battle Creek • INDIANA: Kendallville, Marion, Indianapolis, Lamar • KENTUCKY: Owensboro, Hopkinsville • TENNESSEE: Clarksville, Fort Campbell, Wayland Springs • ALABAMA: Florence, Birmingham, Montgomery, Dothan • FLORIDA: Chattahoochee, Tallahassee, Williston, Daytona Beach.

30,000-Foot Trajectory: NEVADA: Las Vegas • ARIZONA: Signal, Phoenix, Casa Grande, S. Tucson, Bisbee, Douglas • TEXAS: Eagle Flat, Rankin, Barnhart, San Marcos, Cuero, Victoria, Port O'Connor • FLORIDA: Crystal River, Ocala, Jacksonville • NORTH CAROLINA: Wilmington, Jacksonville, New Bern, Elizabeth City.

40,000-Foot Trajectory: ARIZONA: Grand Canyon, Flagstaff • NEW MEXICO: Zuni Indian Reservation, Grants, Los Alamos, Amistad • TEXAS: Hartley, Dumas, Borger, Canadian • OKLAHOMA: Regdon, Oklahoma City, Muskogee • ARKANSAS: Fayetteville, Mountain Home • MISSOURI: Pontiac, Mountain View, Ellington, Minimum, St. Marys • ILLINOIS: Rockwood, Mount Vernon, Bridgeport • INDIANA: Vincinnes, Indianapolis, Anderson, Muncie, Portland • OHIO: Lima, Lorain • NEW YORK: Buffalo, Rochester, Mexico, Camden, Port Henry • VERMONT: Montpelier, Barre • NEW HAMPSHIRE: Lisbon • MAINE: Augusta, Bar Harbor.

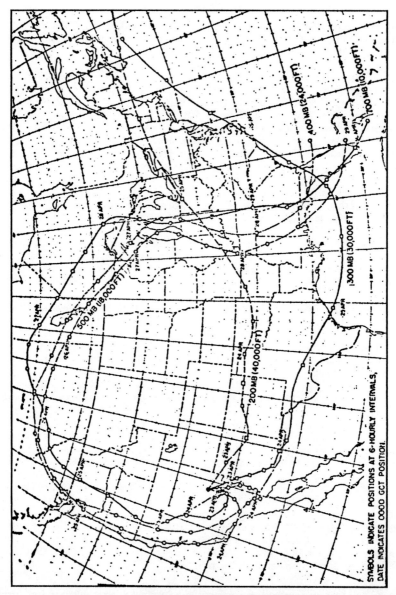

TRAJECTORY: TUMBLER-SNAPPER CHARLIE (TS3)

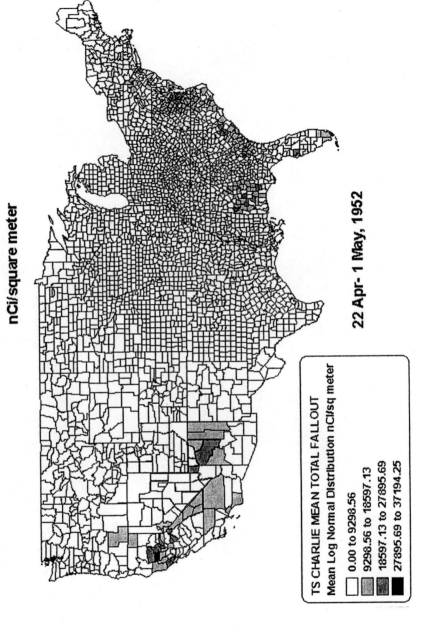

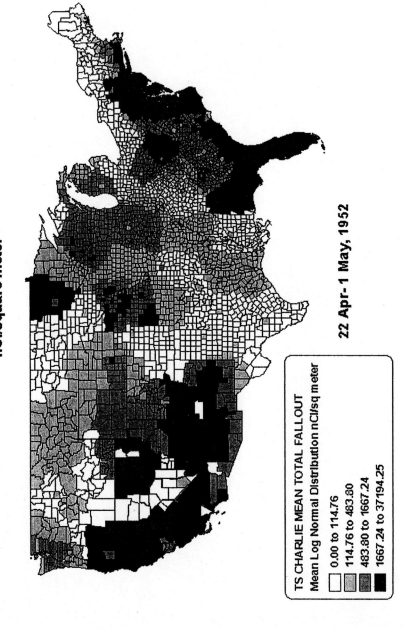

FALLOUT IN EASTERN US FROM: TS CHARLIE

APR 22, 1952 31 KT

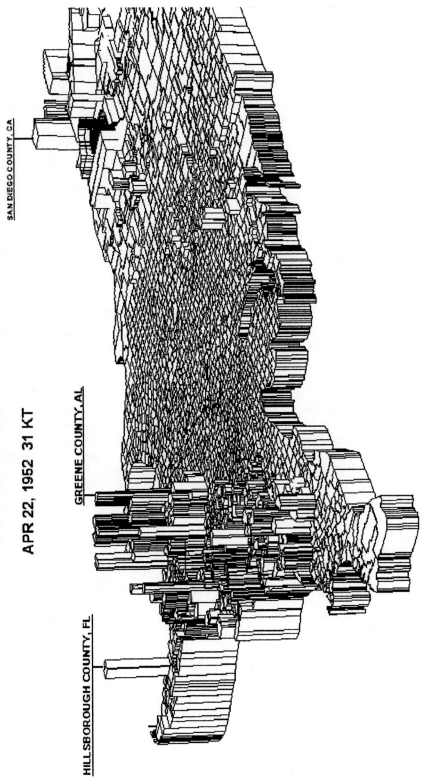

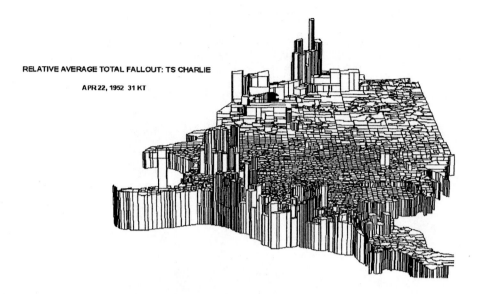

RELATIVE AVERAGE TOTAL FALLOUT: TS CHARLIE
APR 22, 1952 31 KT

CHARACTERISTICS: TS CHARLIE

The device was based on the Mk4 design, similar to the Nagasaki bomb, and weighed 10,440 lbs. Tumbler-Snapper CHARLIE is unique in that it ranks 1st among the nuclear tests, for percentage of Co-57 and Pr-144 in fallout and 2nd for percentage of U-240 in fallout. In addition, this test ranks 1st in the amount of the radionuclide Pr-144 deposited across the U.S. Fallout from this test was deposited on 3062 counties, (99 percent of the U.S.). The day one total fallout-to-I-131 for this nuclear test is 153.3 to 1, the 4th highest among all nuclear tests (after Tumbler-Snapper ABLE, Tumbler-Snapper BAKER, and Teapot ESS).

The deposition pattern for Tumbler-Snapper CHARLIE is most similar to Buster-Jangle CHARLIE (BJ2) (R:0.26; T: 15.19). The county with the highest amount of fallout from this test was Glenn CA: Estimated maximum activity: 225.29 µCi/sq meter of fallout on April 25, 1952. Other counties receiving high levels of fallout from this nuclear test (estimated maximum activity: 163.28 µCi/sq meter each) included the Alabama counties of Conecuh, Geneva, Greene, Lowndes and Wilcox; and the Florida county of Holmes.

SHOT 4: TUMBLER-SNAPPER DOG (TS4)

TUMBLER-SNAPPER: DOG

Detonation Date: 1 May 1952 • *Detonation Time:* 8:30 A.M. • *Area:* 7, Target 3 • *Sponsor:* Los Alamos • *Yield:* 19 kt • *Radiation Level at Ground Zero:* 10 R/hr • *Height of Burst:* 1,040 ft • *Cloud Top Height:* 44,000 ft

18,000-Foot Trajectory: UTAH: St. George • ARIZONA: Flagstaff, Prescott, Truxton, Crozier, Valentine • UTAH: St. George, (Zion National Park), Cedar City, Parowan, Thompsonville, Castle Dale, Vernal • COLORADO: Dinosaur, Sunbeam • WYOMING: Hanna, Medicine Bow, Casper, Moorcroft, (Devil's Tower National Monument), Hulett • MONTANA: Alzada, Albion, Capitol • SOUTH DAKOTA: Camp Hook, Ladner, Ludlow • NORTH DAKOTA: Bowman, New England, Antelope, Washburn, McClusky, Devil's Lake, Park River, Cavalier, Pembina • PENNSYLVANIA: Erie, Oil City, DuBois, Clearfield, Carlisle, York, Lancaster • MARYLAND: Taneytown, Eldersburg, Parksville, Pikesville, Baltimore, Edgemore, Centreville, Henderson • DELAWARE: Hartley, Dover, Kilts Hummock • NEW JERSEY: Villas, Rio Grande, Wildwood.

CHARACTERISTICS: TS DOG

The deposition pattern for Tumbler-Snapper DOG is most similar to shot Teapot TESLA (TP3) (R:0.41; T:25.10). The county of highest deposition was Iron, UT with an estimated maximum activity of 40.34 µCi/sq meter on May 1, 1952. Other counties receiving high fallout from this nuclear test were the Indiana counties of Carroll, Clinton and Tippecanoe, each receiving an estimated maximum activity level of 22.66 µCi/sq meter on May 5, 1952.

The day one ratio of total fallout to I-131 for this test is 124.2 to 1, ranking it 12th among all aboveground nuclear tests.

FALLOUT PATH: TS DOG

MAY1, 1952 19 KT

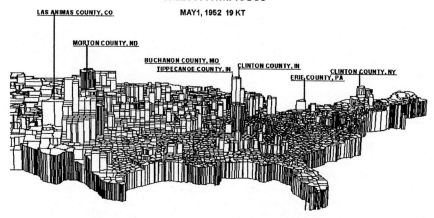

RELATIVE AVERAGE TOTAL FALLOUT: TS DOG

MAY1, 1952 19 KT

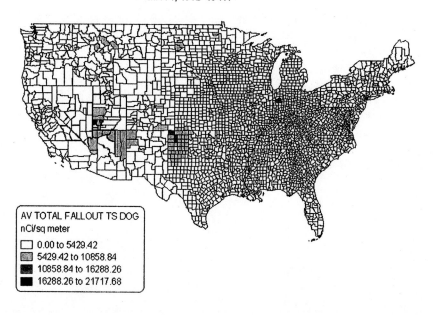

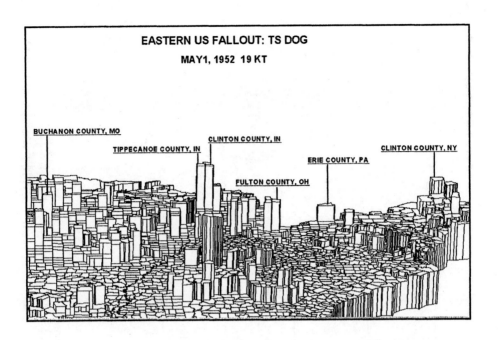

U.S. FALLOUT ATLAS : TOTAL FALLOUT   3-19

TRAJECTORY: TUMBLER-SNAPPER DOG (TS4)

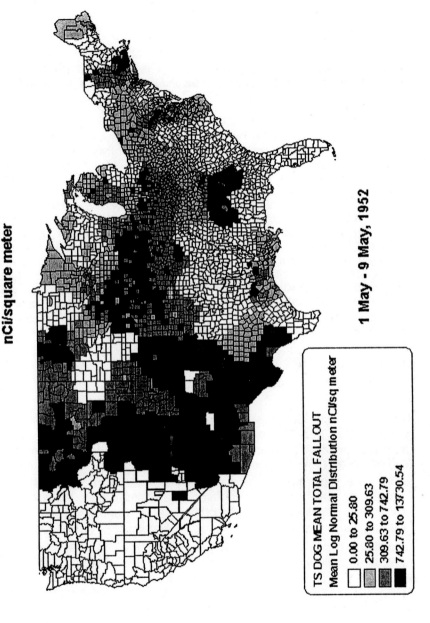

SHOT 5: TUMBLER-SNAPPER EASY (TS5)

TUMBLER-SNAPPER: EASY

Detonation Date: 7 May 1952 • *Detonation Time:* 4:15 A.M. • *Area:* T-1 • *Sponsor:* Los Alamos • *Yield:* 12 kt • *Radiation Level at Ground Zero:* 500 R/hr • *Height of Burst:* 300 ft • *Cloud Top Height:* 34,000 ft

10,000-Foot Trajectory: NEVADA: (Black Rock Summit), Hobson, Cave Creek, Metropolis, Contact • IDAHO: Amsterdam, Twin Falls, Jerome, Shoshone, Richfield, Arco, Mud Lake, St. Anthony • WYOMING: Bechler, (Yellowstone National Park: Lone Star Geyser, Shoshone Lake, Lewis Lake, West Thumb, Yellowstone Lake, Sylvan Pass, East Entrance, Pahaska, Powell, Elk Basin, Deaver, Frannie, Parkman, Sheridan, Rocky Point, Colony • SOUTH DAKOTA: Arpan, Newell, Creighton, Cottonwood, Long Valley, Tuthill • NEBRASKA: Nenzel, Kennedy, Brownlee, Thedford, Broken Bow, Miller, Kearney, Red Cloud • KANSAS: Webber, Concordia, Manhattan, Burlingame, Prescott • MISSOURI: Hume, Nevada, Hope, Manes, Rector, (Taum Sauk Mountain), Ironton, Lixville, New Wells • ILLINOIS: Aldridge, Carbondale, Carriers Mills, Ridgeway • INDIANA: Mount Vernon, Evansville, Boonville, Lama, Alton, Corydon, New Albany • KENTUCKY: Louisville, New Castle, Corinth, Mount Olivet, Maysville • OHIO: Ripley, Waverly, McArthur, Nelsonville, McConnelsville, Caldwell, Bellaire • WEST VIRGINIA: Benwood, Bethlehem, Wheeling, Valley Grove • PENNSYLVANIA: Washington, Pittsburgh, Punxsutawney, DuBois, Johnsonburg, Betula, Port Allegany, Myrtle • NEW YORK: Ceres, Bolivar, Friendship, Wiscoy, Genesco, Avon, West Rush, Henrietta, Rochester, Rondequolt, Sea Breeze.

CHARACTERISTICS: TS EASY

The device used for this test was small, only 22 inches in length and weighing 625 lbs. The yield was equivalent to 12 kilotons of TNT. The deposition pattern for Tumbler-Snapper EASY is most similar to shot SEDAN (R:0.37; T: 21.92). The county of highest deposition was Hot Springs, WY with an estimated maximum activity of 501.96 µCi/sq meter on May 7, 1952, followed by Natronia, WY with a estimated maximum activity of 425.66 µCi/sq meter. Other counties with high fallout from this nuclear test were Niobrara, WY (estimated maximum activity 326 µCi/sq meter); Fall River, SD (estimated maximum activity 322.32 µCi/sq meter) Teton, ID (estimated maximum activity 294.15 µCi/sq meter); and Dawes, NE (estimated maximum activity 270.69 µCi/sq meter), all on May 7, 1952. Fallout from this test was deposited across 2937 counties, or 95 percent of the United States.

The day one ratio of total fallout to I-131 for Tumbler-Snapper EASY is relatively low at 82 to 1.

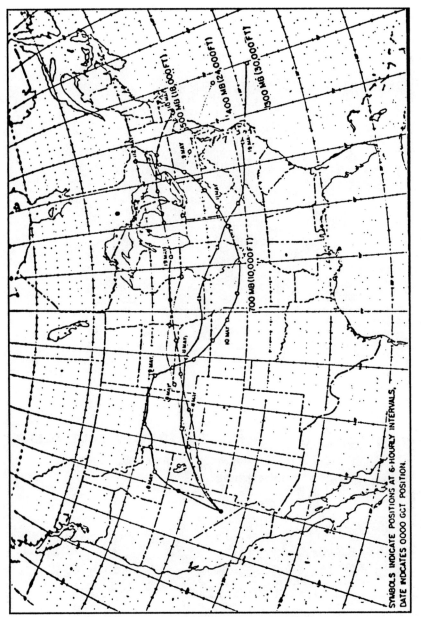

TRAJECTORY: TUMBLER-SNAPPER EASY (TS5)

RELATIVE AVERAGE TOTAL FALLOUT: TS EASY
MAY 7, 1952 12 KT

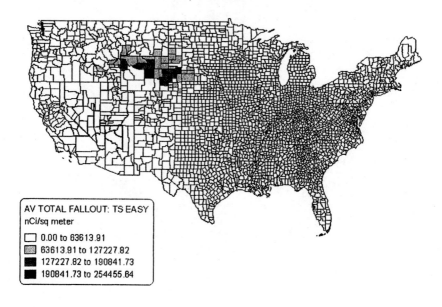

AV TOTAL FALLOUT: TS EASY
nCi/sq meter
- 0.00 to 63613.91
- 63613.91 to 127227.82
- 127227.82 to 190841.73
- 190841.73 to 254455.64

RELATIVE TOTAL FALLOUT: TS EASY
MAY 7, 1952 12 KT

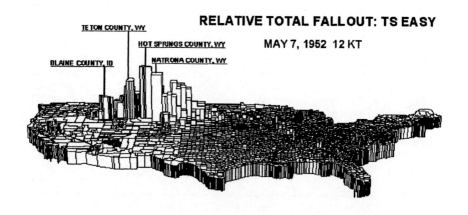

U.S. FALLOUT ATLAS : TOTAL FALLOUT

SHOT 6: TUMBLER-SNAPPER FOX (TS6)

TUMBLER-SNAPPER: FOX

Detonation Date: 25 May 1952 • *Detonation Time:* 4:00 A.M. • *Area:* 4 • *Sponsor:* Los Alamos • *Yield:* 11 kt • *Radiation Level at Ground Zero:* 2,000 R/hr • *Height of Burst:* 300 ft • *Cloud Top Height:* 41,000 ft

30,000-Foot Trajectory: NEVADA: Hilco, Ash Springs, Alamo, Caliente • UTAH: Minersville, Bicknell, Nanksville, Moab • COLORADO: Gateway, Orchard City, Bowie, Somerset, Oliver, Marble, Avon, Mount Powell, Heeney, Granby, Berthoud, Johnstown, Millikan, Greeley, Barnesville, Cornish, New Raymer • NEBRASKA: Chappell, Lomoyne, Tryon, Neligh, Hadar, Hoskins, Norfolk, Pender, Bancroft, Decatur • IOWA: Onawa, Dow City, Buck Grove, Manilla, Bagley, Perry, Gardiner, Moran, Elkhart, Baxter, Newburg, Belle Plaine, Cedar Rapids, Springville, Anamosa, Canton, Bellevue • ILLINOIS: Loran, Ridott, Rockford, Machesney Park • WISCONSIN: Beloit, Clinton, Delavan, Waterford, Wind Lake, Milwaukee • MICHIGAN: Muskegon, Ravenna, Sparta, Belding, Matherton, Maple Rapids, Eureka, Elsie, Oakley, Burt, Vassar, Caro, Gagetown, Bad Axe, Filion • NEW YORK: Madrid, Unionville, Potsdam, Parishville, Clear Lake, Saranac Lake, Lake Placid, Keene Valley, Moriah • VERMONT: Chimney Point, Cornwall, Salisbury, Sherburne Center, Windsor, Claremont, Deering, Nashua • MASSACHUSETTS: Lowell, Billerica, Woburn, Lexington, Medford, Arlington, Malden, Somerville, Cambridge, Boston, Hull, North Cohasset, Humarock, Kent Park, Rexhame, Ocean Bluff, Brant Rock, Green Harbor, Denis, South Yarmouth.

40,000-Foot Trajectory: NEVADA: Alamo, Ash Springs, Hiko, Caliente, Panaca, Carp, Elgin • UTAH: Milford, Cove Fort, Fillmore, Richfield, Mount Pleasant, Moroni, Price, East Carbon, Ouray, Bonanza • COLORADO: Massadonia, Blue Mountain, Elk Springs, Maybell, Sunbeam, Clark • WYOMING: Baggs, Dixon, Saven, Mountain Home, Woods Landing, Buford, Cheyenne, Carpenter • NEBRASKA: Kimball, Sidney, Lodgepole, Chappel, Ogallala, North Platte, Maxwell, Brady, Pleasanton, Grand Island, Aurora, York, Milford, Lincoln, Denton, Syracuse • MISSOURI: Tarkio, Maryville, Coffey, Chula, Brookfield, Marceline, College Mound, Duncan's Bridge, Holliday, Paris, Vandalia, Louisville, Auburn, New Hope • ILLINOIS: Alton, Edwardsille, Aviston, Bartelso, New Minden, Spring Garden, Broughton, Texas City, Ridgeway, Old Shawneetown • KENTUCKY: Spring Grove, Grove Center, Wheatcroft, Clay, Madisonville, Nortonville, Apex, Kirkmansville, Fearsville, Allegre, Claumour, Whippoorwill, Olmstead, Adairville • TENNESSEE: Orlinda, Portland, Bethpage, Hartsville, Carthage, Buffalo Valley, Bon Air, Roddy, Loudon, Glendale, Maryville, Townsend, Elkmont, Gatlinburg • NORTH CAROLINA: Asheville, Azalea, Vale, Reepville, Davidson, Concord, Albemarle, Candor, Manchester, Eastover, Clinton, Lyman, Fountain, Catherine Lake, Jacksonville, Camp Lejeune Marine Corps Base, Bear Inlet of Onslow Bay.

Note: Both the 30,000-foot and the 40,000-foot trajectory crossed over Chappel, Nebraska.

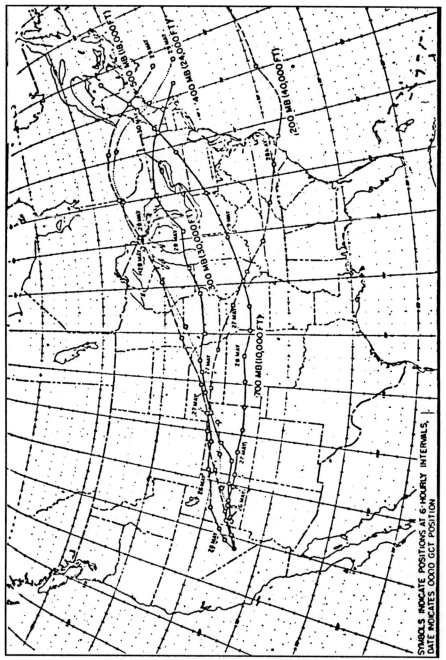

TRAJECTORY: TUMBLER-SNAPPER FOX (TS6)

CHARACTERISTICS: TS FOX

The FOX device, 40 inches in diameter and weighing 2700 lbs, was detonated atop a 300 foot tower at Nevada Test Site Area 4. The deposition pattern for Tumbler-Snapper FOX is most similar to that of Upshot-Knothole HARRY (UK9) (R: 0.29; T: 17.00). The county of highest deposition was Iron, UT with an estimated maximum. activity of 1056 µCi/sq meter total fallout on May 25, 1952. Other counties with high levels of fallout included Millard, UT (estimated maximum activity: 301.84 µCi/sq meter on May 25, 1952) and Chase, NE (estimated maximum activity: 356.15 Ci/sq meter on May 27, 1952).

Tumbler-Snapper FOX was unique in that fallout from this test was deposited over 3086 counties—nearly 100 percent of the entire United States land area. In that respect, the fallout from shot Tumbler-Snapper FOX was the most dispersed of any aboveground nuclear test.

The day one ratio of total fallout to I-131 for Tumbler-Snapper FOX is relatively low at 96 to 1.

Tumbler-Snapper FOX ranked 12th in the deposition of the radionuclide Sr-90, 13th in deposition of Ru-106 and 14th in deposition of Cs-137 across the United States. The fallout from this nuclear test contained *no* Co-60.

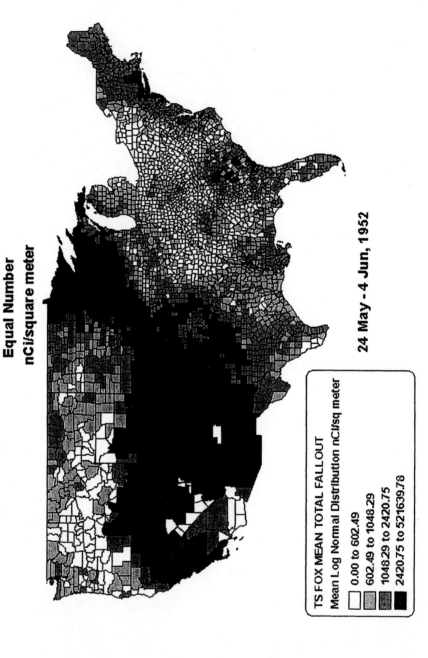

RELATIVE AVERAGE TOTAL FALLOUT: TS FOX
MAY 25, 1952 11 KT

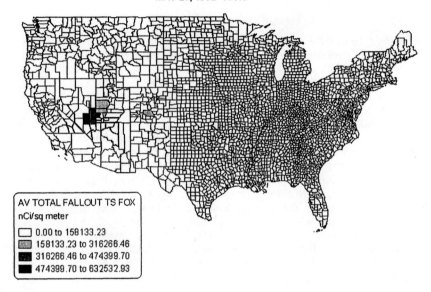

AV TOTAL FALLOUT TS FOX
nCi/sq meter
- 0.00 to 158133.23
- 158133.23 to 316266.46
- 316266.46 to 474399.70
- 474399.70 to 632532.93

RELATIVE TOTAL FALLOUT: TS FOX
MAY 25, 1952 11 KT

LINCOLN 2 COUNTY, NV
IRON 1 COUNTY, UT

SHOT 7: TUMBLER-SNAPPER GEORGE (TS7)

TUMBLER-SNAPPER: GEORGE

Detonation Date: 1 Jun 1952 • *Detonation Time:* 3:55 A.M. • *Area:* 3 • *Sponsor:* Los Alamos • *Yield:* 15 kt • *Radiation Level at Ground Zero:* 1,000 R/hr • *Height of Burst:* 300 ft • *Cloud Top Height:* 37,000 ft

35,000-Foot Trajectory: NEVADA: Caselton, Montello • UTAH: Artesian City, Carey, Antelope River, Darlington, Winsper, Humphrey • WYOMING: (Yellowstone National Park, Northern Campgrounds) • MONTANA: Dean, Joliet, Silesia, Castle Rock, Brandenburg, Ekalaka, Mill Iron • SOUTH DAKOTA: Ladner, Ludlow, Lodgepole, Glad Valley, Isabel, Ridgeview, Agar, Miller, Iroquois, Oldham, Elkton • MINNESOTA: Ruthton, Aroca, Dundee, Lakefield, Dunnell • IOWA: Buffalo Center, Forest City, Hanford, Marble Rock, Horton, Sumner, Osborne • WISCONSIN: Cornelia, Seymour Corners, Cadiz Springs • ILLINOIS: Durand, Belvidere, Chicago • INDIANA: Coburg, Hamlet, Argos, Luther, Roanoke, Zanesville, Ossian, Curryville, Salem • OHIO: Celina, New Brennen, Lena, Thackeray, Springfield, Brookwalter, Bloomingburg, Austin, Alma, Beaver, Bloom, Pedro Forrestdale • WEST VIRGINIA: Shoals, Eloise, Holden, Christian, Lincoln, Woolsey, Rock, Ingleside • VIRGINIA: Point Pleasant, Shorts Creek, Fancy Gap, Canal • NORTH CAROLINA: Mount Airy, Pinnacle, Cedar Lodge, Cid, Ophir, Norman • SOUTH CAROLINA: Cilo, Dillon, Mullins, Galivants Ferry, Toddville, Burgess, Brookgreen.

CHARACTERISTICS: TS GEORGE

The device was similar in size to the one used in the earlier FOX test: 2700 lbs and 40 inches in diameter. As with the FOX test, TS GEORGE was detonated from atop a 300-foot tower. The deposition pattern for Tumbler-Snapper GEORGE is most similar to that of the 1957 shot Plumbbob MORGAN (PB18) (R: 0.50; T: 31.78). The counties of highest fallout deposition included Knox, MO (estimated maximum activity: 1382 µCi/sq meter), Lewis, MO (estimated maximum activity: 1290 µCi/sq meter), Niobrara, WY (estimated maximum activity: 1199 µCi/sq meter), Appanoose, IA (estimated maximum activity: 1114 µCi/sq meter), and Scotland, MO (estimated maximum activity: 1111 µCi/sq meter)—all on June 3, 1952, and most associated with rain.

Tumbler-Snapper GEORGE deposited fallout on 2937 counties— 95 percent of the total counties in the United States, ranking it 3rd among the above ground tests for widespread fallout. Interestingly, this shot ranks 1st for deposition of the radionuclide Cs-137 (followed by Upshot-Knothole SIMON). It also ranks 1st in deposition of the radionuclides Nb-95 (again followed by Upshot-Knothole SIMON at 2nd place and Upshot-Knothole HARRY in 3rd place for this radionuclide.) From a radionuclide composition standpoint, Tumbler-Snapper GEORGE somewhat resembles the later shots UK SIMON and UK HARRY. Tumbler-Snapper GEORGE is also 2nd in the deposition of the radionuclide Nb-95m (with 1st place going to Buster-Jangle EASY and 3rd place held by Upshot-Knothole SIMON.)

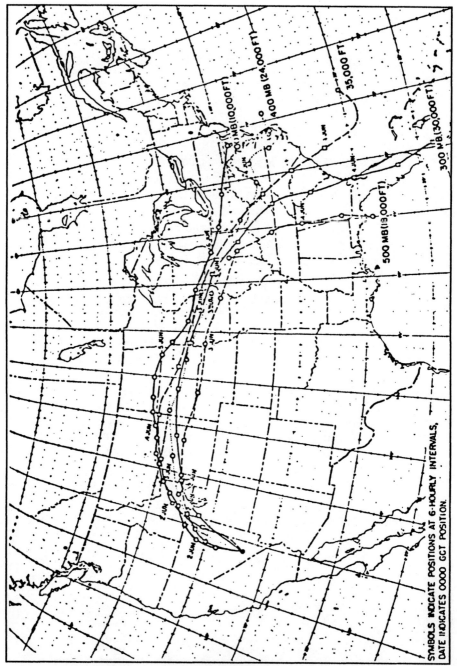

TRAJECTORY: TUMBLER-SNAPPER GEORGE (TS7)

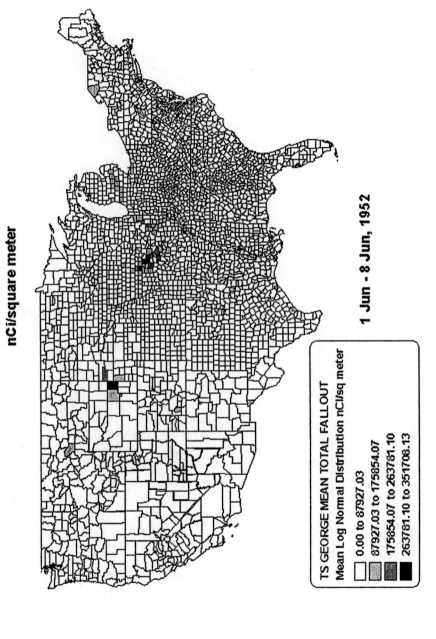

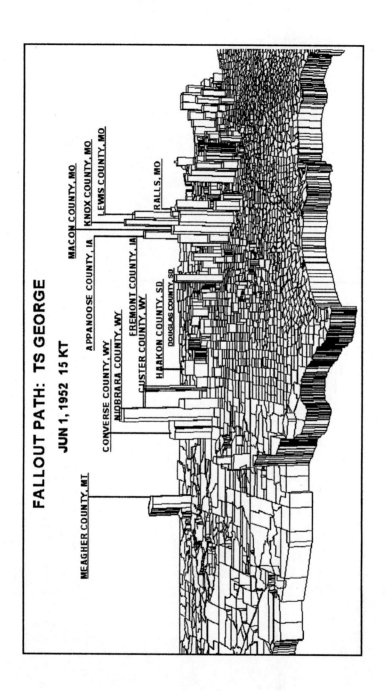

RELATIVE TOTAL FALLOUT: TS GEORGE

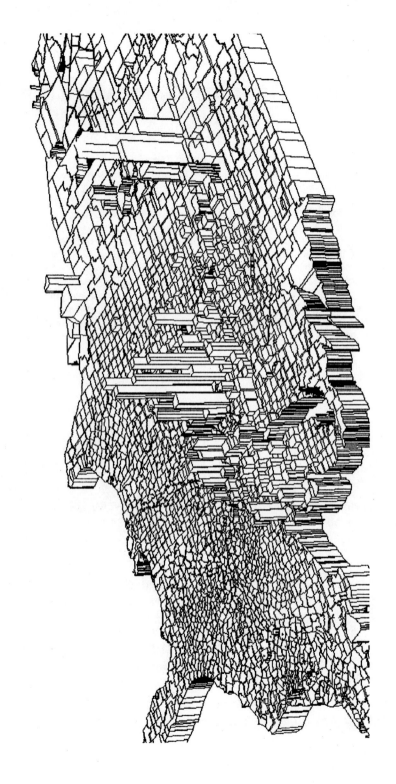

U.S. FALLOUT ATLAS : TOTAL FALLOUT

SHOT 8: TUMBLER-SNAPPER HOW (TS8)

TUMBLER-SNAPPER: How

Detonation Date: 5 Jun 1952 • *Detonation Time:* 3:55 A.M. • *Area:* 2 • *Sponsor:* Los Alamos • *Yield:* 14 kt • *Radiation Level at Ground Zero:* 1,500 R/hr • *Height of Burst:* 300 ft • *Cloud Top Height:* 41,800 ft

24,000-Foot Trajectory: NEVADA: Warm Springs, Round Mountain, Crescent Valley, Midas, Tuscarora • IDAHO: Riddle, Grasmere, Mayfield, Lowman • MONTANA: Darby, Bearmouth, Helmville, Greenfield, Power, Gildford, Kremlin • SASKATCHEWAN: Regina • NEW YORK: Youngstown, Model City, Niagara Falls, Tonawanda, Buffalo, Lancastor, Aurora, South Wales, Holland, Chaffee, Elton, North Cuba, Nile, Allentown • PENNSYLVANIA: Kinney, Gold, Carter Camp, Haneyville, Avis, Ranchtown, Cowan, New Berlin, Selinsgrove, Red Cross, Sharidan, Dayton, Jonestown, Bunker Hill, Avon, Iona, Clay, Ephrata, Groffdale, Gap, Atglen, Jennersville, Kelton, New London, Kemblesville • DELAWARE: Newark, Brookside, St. Georges, Taylors Bridge, Dupont Manor, Dover, Bowers, Slaughter Beach, Nassau, Cottonpatch Hill.

CHARACTERISTICS: TS HOW

This device was the first to use the beryllium neutron reflector surrounding the plutonium core. Tumbler-Snapper HOW is ranked 1st for percentage of both Am-241 and Sm-153 in nuclear fallout. In addition, perhaps due to rainouts soon after detonation, this shot resulted in some of the highest exposures of I-131 to the thyroid for Gem County, ID.

The deposition pattern for Tumbler-Snapper HOW is most similar to Plumbbob NEWTON (PB15) (R: 0.53; T: 34.85). Like Tumbler-Snapper GEORGE, the fallout was associated with rain. But unlike the earlier shot, the rainout occurred on the day of the shot, June 5, 1952. For that reason, the fallout level that day was among the highest calculated for any aboveground test. Counties with the highest total fallout deposition included Gem ID (estimated maximum. activity: 5278 µCi/sq meter), Custer ID (estimated maximum activity 5018 µCi/sq meter); Blaine ID (estimated maximum activity: 4756 µCi/sq meter), Lemhi ID (estimated maximum activity: 3,413 µCi/sq meter) and Deer Lodge MT (estimated maximum activity: 3021 µCi/sq meter).

This test deposited fallout over 3014 counties, or 97 percent of the U.S. The ratio of total fallout to I-131 for Tumbler-Snapper HOW is relatively low at 77.9 to 1. This shot is ranked 8th for deposition of Sr-90 across the United States and 9th for deposition of Cs-137.

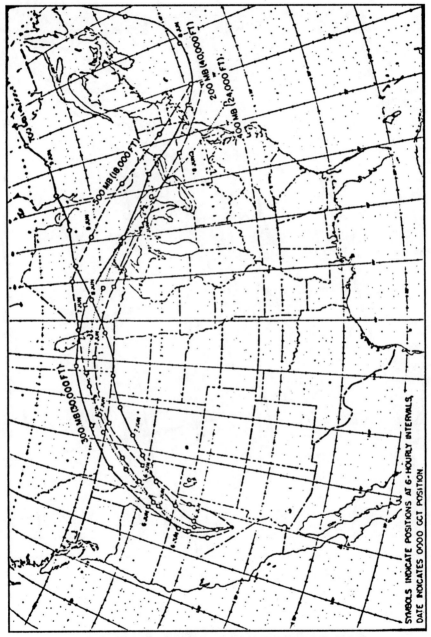

TRAJECTORY: TUMBLER-SNAPPER HOW (TS8)

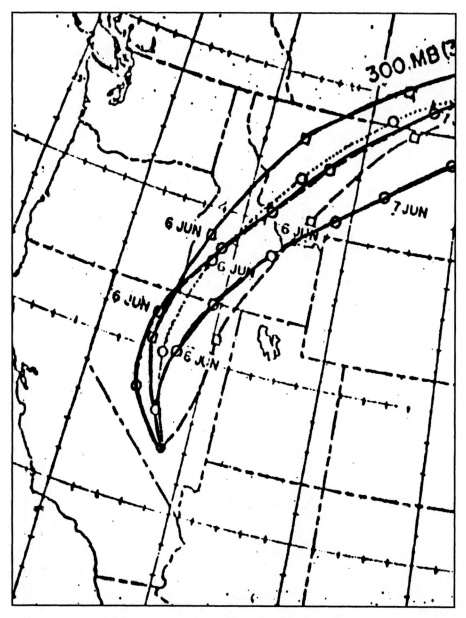

CLOSE-IN TRAJECTORY: TS HOW (TS8)

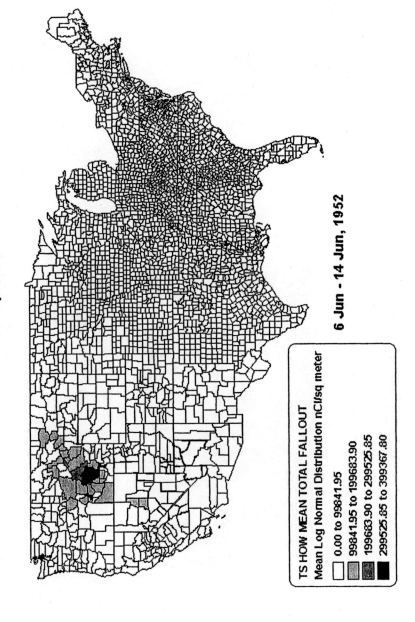

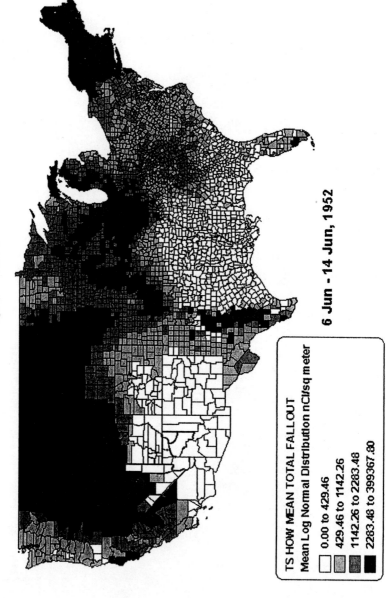

RELATIVE TOTAL FALLOUT: TS HOW
JUN 5, 1952 14 KT

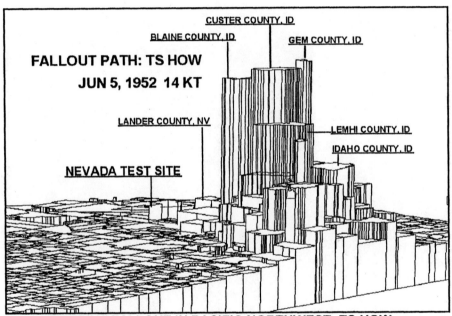

FALLOUT PATH: TS HOW
JUN 5, 1952 14 KT

RELATIVE FALLOUT IN PACIFIC NORTHWEST: TS HOW

SECTION 4

UPSHOT-KNOTHOLE SERIES

Mar 17–Jun 4, 1953

SHOT 1: UPSHOT-KNOTHOLE ANNIE (UK1)

UPSHOT-KNOTHOLE: ANNIE

Detonation Date: 17 Mar 1953 • *Detonation Time:* 5:20 A.M. • *Area:* 3 • *Sponsor:* Los Alamos • *Yield:* 16 kt • *Radiation Level at Ground Zero:* 2,000 R/hr • *Height of Burst:* 300 ft • *Cloud Top Height:* 41,000 ft

10,000-Foot Trajectory: NEVADA: Alamo, Elgin, Carp • UTAH: Modena, Beryl, Paragonah, Spry, Panguitch • COLORADO: Grand Junction, De Begue, Parachute, Glenwood Springs, Boulder, Longmont, Greeley, New Raymer, Sedgwick, Ovid • NEBRASKA: Julesburg, Ogallala, Keystone, Stapleton, Gandy, Anselmo, Taylor, Neligh, Pierce, Carroll, Concord, Jackson • IOWA: Sioux City, Lawton, Kingsley, Diamond Center, Hanover, Nemaha, Rockwell City, Somers, Slifer, Stratford, Story City, Roland, Marshalltown, Chelsea, Haven, South Amana, Tiffin, Iowa City, Downey, Moscow, Pleasant Prairie, Bettendorf, Davenport • ILLINOIS: Rock Island, Moline, Buda, Tiskilwa, McNabb, Kangley, Streator, Dwight, Kankakee • INDIANA: Lake Village, Thayer, Wilders, Brems, Grovertown, Hamlet, Koontz Lake, Teegarden, Lakeville, Woodland, Jimtown, Elkhart, Bristol • MICHIGAN: Vistula, Sturgis, Findley, Coldwater, Litchfield, Pulaski, Vandercook, Grass Lake, Ann Arbor, Plymouth, Livonia, Oak Park, Highland Park, Detroit • NEW YORK: Westfield, Point Chautauqua, Jamestown • PENNSYLVANIA: Westline, Guffey, Mount Jewett, Howard, Castle Garden, Spring Garden, Unityville, Elk Grove, Lopez, Sugar Run, Camptown, Little Meadows. • NEW YORK: Binghamton, Oxford, Norwich, Shelburne, Utica, Grant, Saranac Lake, Redford, Dannemora, Jericho, Champlain • MAINE: Dickey, St. John, Fort Kent, Frenchville.

40,000-Foot Trajectory: NEVADA: Alamo, Ash Springs, Hiko • UTAH: Modena, Beryl, Paragonah, Spry, Panguitch, Mount Ellen • COLORADO: Gateway, Delta, Hotchkiss, Paonia, Crested Butte, Granite, Tarryall, Palmer Lake, Elbert, Hale • KANSAS: St. Francis • NEBRASKA: Danbury, Oxford, Mascot, Ragan, Hildneth, Holstein, Roseland, Hastings, Harvard, Lushton, McCool Junction, Beaver Crossing, Malcolm, Ashland, Smithfield, Bellevue, Omaha • IOWA: Council Bluffs, Hancock, Marne, Lorah, Monteith, Minburn, Alleman, Maxwell, Melbourne, Marshalltown, Dysart, Walker, Troy Mills, Farley, Centralia, Asbury, Dubuque • WISCONSIN: Kieler, Cuba City, Avon, Lamont, Argyle, Monticello, Evansville, Edgerton, Fort Atkinson, Hebron, Palmyra, North Prairie, Genesee, West Allis, Milwaukee • MICHIGAN: Holland, Grand Rapids, Cedar Springs, Trufant, Six Lakes, Wyman, Mount Pleasant, Leaton, Edenville, Estey, Rhodes, Bentley, Omer, Twining, Tawas City, Au Sable • NEW YORK: Waddington, Chase Mills, Norfolk, Dickenson Center, Duane Center, Hawkeye, Clintonville • VERMONT: Huntington Center, Waitsfield, Northfield, Corinth, Bradford • NEW HAMPSHIRE: Wentworth, Stinson Lake, Melvin Village, Wawbeck, Tuttonboro, Wolfeboro Center, Wakefield, Sanbornville • MAINE: Emery Mills, Springvale, Sanford, Kennebunkport.

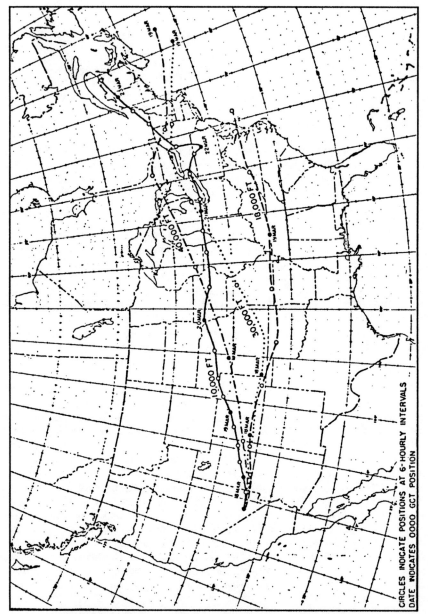

TRAJECTORY: UPSHOT-KNOTHOLE ANNIE (UK1)

CHARACTERISTICS: UK ANNIE

The device, designated as an "XR-3" design, weighed 2700 lbs and was detonated from atop a 300-foot tower. Upshot-Knothole ANNIE deposited fallout over 2937 counties, or approximately 95 percent of the total counties in the United States, thus ranking it 21st for widespread fallout. The deposition pattern for Upshot-Knothole ANNIE was most similar to Upshot-Knothole RUTH (Uk3) (R: 0.51; T: 32.89). The counties of highest fallout from this nuclear test included Washington, UT (maximum estimated activity: 2396 µCi/sq meter on Mar 17, 1953). Interestingly, on Mar 18, 1953 the fallout from this nuclear test also produced fallout hot spots on the East Coast, in the New Jersey counties of: Atlantic (estimated maximum activity: 306 µCi/sq meter); Monmouth (estimated maximum activity: 307 µCi/sq meter); Ocean (estimated maximum activity: 307 µCi/sq meter); Burlington (estimated maximum activity: *256 µCi/sq meter) and Cape May (estimated maximum activity: 230 µCi/sq meter).

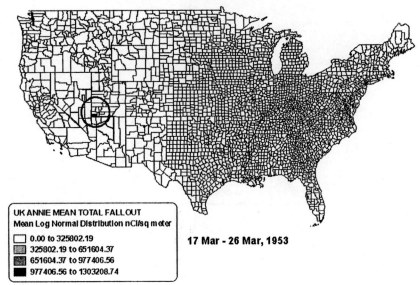

TOTAL FALLOUT: UPSHOT-KNOTHOLE ANNIE
Equal Intervals
nCi/square meter

UK ANNIE MEAN TOTAL FALLOUT
Mean Log Normal Distribution nCi/sq meter
- 0.00 to 325802.19
- 325802.19 to 651604.37
- 651604.37 to 977406.56
- 977406.56 to 1303208.74

17 Mar - 26 Mar, 1953

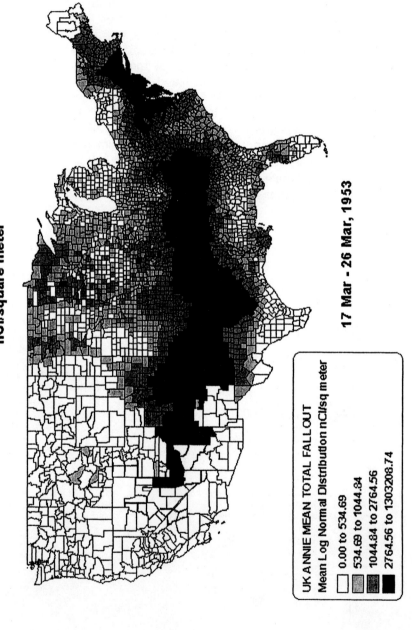

SHOT 2: UPSHOT-KNOTHOLE NANCY (UK2)

UPSHOT-KNOTHOLE: NANCY

Detonation Date: 24 Mar 1953 • *Detonation Time:* 5:10 A.M. • *Area:* 4 • *Sponsor:* Los Alamos • *Yield:* 24 kt • *Radiation Level at Ground Zero:* 2,000 R/hr • *Height of Burst:* 300 ft • *Cloud Top Height:* 41,500 ft

10,000-Foot Trajectory: NEVADA: (Summit Mountain), Dunphy, Tuscarora, Mountain City, Owyhee • IDAHO: Riddle, Grasmere, Mayfield, Lowman, Cobalt, Salmon • MONTANA: Butte, Boulder, Winston, Neihart, Windham, Benchland, Danvers, Hilger, Fergus, Roy, Mona, Andes • NORTH DAKOTA: Charleston, New Town, Plaza, Sawyer, Velva, Voltaire, Bergen, Balfour, Drake, Maddock, Oberon, Sheyenne, Warwick, Hamar, Tolna, Pekin, McVille, Kloten, Aneta, Northwood, Reynolds • MINNESOTA: Climax, Fertile, Rindal, Bejou, Naytahwaush, Akeley, Hackensack, Manhattan Beach, Fifty Lakes, Aitkin, Giese, Finlayson, Sandstone • WISCONSIN: Webster, Hertel, Barronett, Rice Lake, Canton, Cameron, Chetek, New Auburn, Cadott, Neillsville, Mather, Shennington, Cutler, New Lisbon, Lyndon Station, Baraboo, Prairie du Sac, Sauk City, Black Earth, Madison, Mount Horeb, Belleville, New Glarus, Monticello, Monroe, Juda • ILLINOIS: Rock Grove, Rock City, Ridott, Seward, Chana, Rockelle, Pawpaw, Earlville, Harding, Wedron, Marseilles, Seneca, Verona, Dwight, Reddick, Cabery, Stelle, Dunforth, Milford • INDIANA: Freeland Park, Pine Village, West Point, Romney, Kirkpatrick, Thorntown, Lebanon, Fayette, Royalton, Trader's Point, Indianapolis, Beech Grove, Acton, London, Fairland, Shelbyville, Waldron, St. Paul, Milford, Greensburg, Millhousen, Osgood, Farmers Retreat, East Enterprise, Quercus Grove, Florence • KENTUCKY: Warsaw, Glencoe, Poplar Grove, Mason, Corinth, Cynthiana, Millersburg, East Union, Sharpsburg, Owingsville, Sudith, Frenchburg, Mizel, Grassy Creek, Caney, Lakeville, Royalton, Blue River, Manton, Boldman, Pikeville, Chloe, Shelblana • VIRGINIA: Prater, Mount Heron, Lynn Spring, Red Ash, Cedar Bluff, Asberrys, Wytheville, Austinville, Hillsville, Laurel Fork, Vesta, Spencer, Glenwood, Harmony, Mayo, Cluster Springs, Clarksville, Alberta, Dophpin, Stony Creek, Savedge, Williamsburg, Mobjack, Mathews, Diggs, Birdsnest, Nassawadox.

40,000-Foot Trajectory: NEVADA: Alamo, Ash Springs, Hiko • UTAH: Molena, Beryl, Lund, Beaver, Koosharem, Ferron • COLORADO: Palisade, Cedaredge, Rogers Mesa, Lazear, Crawford, Maher, LaGareta, Center, Mosca, Fort Garland • NEW MEXICO: (Laughlin Peak), Sofia, Clapham, Stead • TEXAS: Amarillo (near the Pantex Plant), Pullman, Palo Duro, South Brice, Parnell, Swearingen, Crowell, Thulia, Mabelle, Markley, Wizard Wells, Boonsville, Cottondale, Keeter, Springtown, Forth Worth, Hurst, Arlington, Grand Prairie, Dallas, Duncanville, Bristol, Palmer, Telico, Gun Barrel City, Caney City, Malakoff, Bradford, Nechas, Rusk, Wells, Lufkin, Jasper, Newton, Bon Wier • LOUISIANA: Merryville, Reeves, LeBlanc, Indian Village, Hathaway, Evangeline, Rayne, Crowley, Maurice, Youngsville, New Iberia, Jeanerette, Baldwin, Franklin, Morgan City, Houma, Chauvin, Golden Meadow, Leeville, Grand Isle • FLORIDA: Venice, South Venice, North Port, Port Charlotte, Lakeport, Port Mayaca, Jupiter.

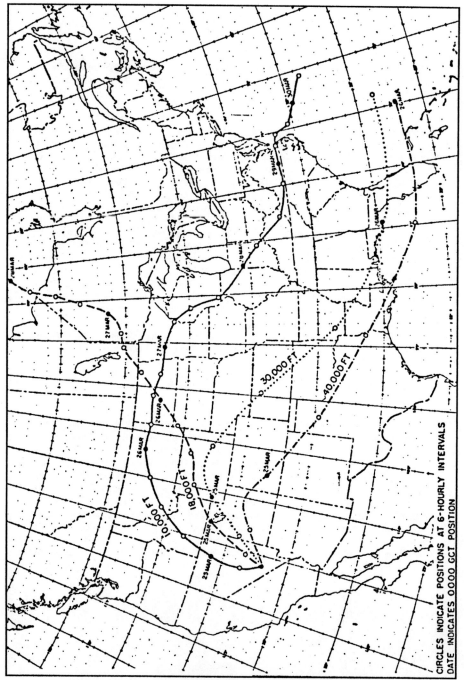

TRAJECTORY: UPSHOT-KNOTHOLE NANCY (UK2)

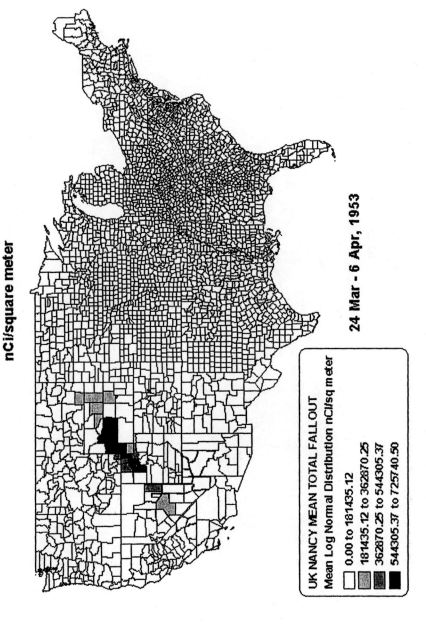

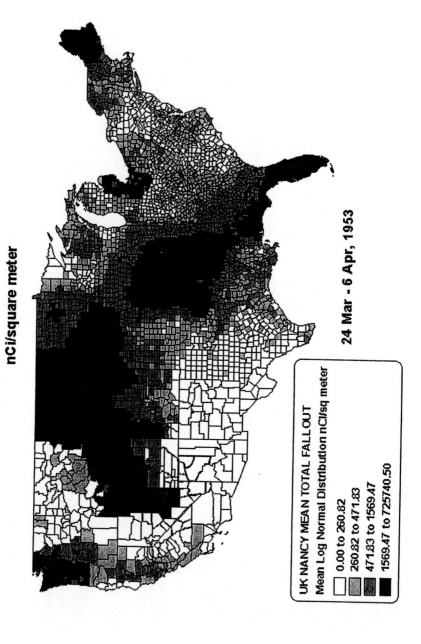

RELATIVE TOTAL FALLOUT: UK NANCY
MAR 24, 1953 24 KT

CHARACTERISTICS: UK NANCY

While the nuclear test was named "NANCY", the device that was the subject of the test was code-named "Nevada Zombie." A 3 ft diameter 10-foot long cylinder weighing 6.5-tons, Zombie was one of the biggest devices ever placed on top of a 300-foot tower. The shot was used to test a thermonuclear weapon design. Upshot-Knothole NANCY is ranked 3rd for percentage of Fe-59 in fallout, and 7th for percentages of Te-127, Te-127m and U-237. The deposition pattern for Upshot Knothole NANCY is most similar to Plumbbob SHASTA (PB8) (R: 0.49; T: 30.91). Fallout from this nuclear test was deposited on 97 percent of the U.S. counties (2997 counties) with an average per-county fallout deposition level of 14.6 µCi/sq meter.

Counties of highest fallout deposition include Morgan UT (estimated maximum activity: 1821 µCi/sq meter), Lincoln, WY (estimated maximum activity: 1670 µCi/Sq meter) and Sublette, WY (estimated maximum activity: 1658 µCi/sq meter).

The day one ratio of total fallout to I-131 is moderate at 91.6 to 1.

Upshot-Knothole NANCY also ranks 8th for deposition of the radionuclide Be-7, 9th for Co-60 and 10th among the nuclear tests for deposition of Cs-137

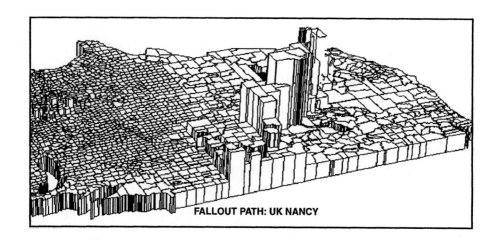

FALLOUT PATH: UK NANCY MAR 24, 1953

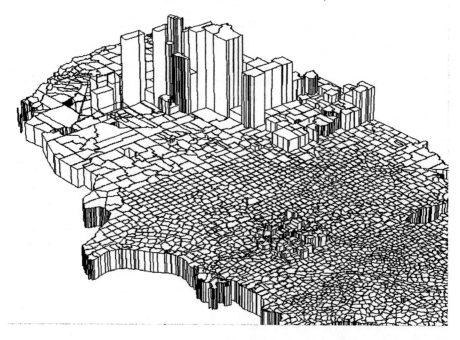

4-12 U.S. FALLOUT ATLAS : TOTAL FALLOUT

SHOT 3: UPSHOT-KNOTHOLE RUTH (UK3)

UPSHOT-KNOTHOLE: RUTH

Detonation Date: 31 Mar 1953 • Detonation Time: 5:00 A.M. • Area: 7–5a • Sponsor: Livermore Laboratory • Yield: 0.2 kt • Radiation Level at Ground Zero: 10 R/hr • Height of Burst: 304.69 ft • Cloud Top Height: 13,600 ft

10,000-Foot Trajectory: NEVADA: Indian Springs, Las Vegas, Winchester, Nelson, Cottonwood Cove, Riviera • CALIFORNIA: Needles, Vidal Junction, Rice, (Palen Mountains), (Chocolate Mountains), Glamis, (Sand hills) • ARIZONA: (Copper Mountains), (Sauceda Mountains), Casa Grande, Maricopa, Olberg, Palm Springs, Apache Junction, Tortilla Flat, Roosevelt, (Bear Mountain), Carrizo, Pinetop, Lakeside, (Greens Peak), Eagar, Alpine • NEW MEXICO: Luna, Reserve, (Tularosa Mountains), (Elk Mountain), Kingston, Las Cruces, La Mesa, Chamberino, La Union • TEXAS: El Paso, Horizon City, Clint, Andrews, Tarzan, Big Spring, Lomax, Otis Chalk, Silver, Robert Lee, Bronte, Maverick, Ballinger, Voss, Mozelle, Trickham, Brookesmith, Indian Creek, Mullin, Golthwaite, Center City, Evant, Pearl, Arnett, Fort Gates, Flat, Moffatt, Pendelton, Moody, Cego, Mooresville, Perry, Otto, Ben Hur, Groesbeck, Personville, Point Enterprise, Fairfield, Red Lake, Montalba, Jacksonville, Reese, Black Jack, St. Clair City, New London, Kilgore, Easton, Darco, Longview, Karnack, Smithland, Gray • LOUISIANA: Trees, Vivian, Gilliam, Hosston, Mira, Bolinger, Ida • ARKANSAS: Springhill, Welcome, Taylor, Emerson, Plainfield, Atlanta, Wasson, Urbana, Gerdner, Strong, North Crossett, West Crossett, Crossett, Parkdale, Empire, Eudora • MISSISSIPPI: Chaitham, Hollandale, Darloue, Belzoni, Cruger, Acona, West, Beatty, French Camp, Weir, Ackerman, Sturgis, Bradley, Longview, Starkville, Columbus, Mayhew, Caledonia • ALABAMA: Fernbank, Kingville, Covin, Howard, Carbon Hill, Townby, Saragossa, Manchester, Macedonia, Cold Springs, Hanceville, Center Hill, Blountsville, Liberty, Brooksville, Nixons Chapel, Douglas, Horton, Boaz, Sardis City, Keener, Sand Rock, Leesburg, Cedar Bluff, Gaylesville • GEORGIA: Coosa, Garden Lakes, Rome, Halls, Cassville, Pine Log, Rydal, White, Waleska, Nelson, Ball Ground, New Holland, Lula, Raoul, Hollingsworth, Avalon, Martin, Lavonia • SOUTH CAROLINA: Anderson, Williamston, Cheddar, Pelzer, Fork Shoals, Owings, Enoree, Cross Keys, Buffalo, Monarch, Union, Chester, Eureka, Richburg, Fort Lawn, Grace, Lancaster, Pageland, Mount Crogham, Rugby, Chesterfield • NORTH CAROLINA: Gibson, Laurel Hill, Old Hundred, Laurinburg, Wakulla, Red Springs, Shannon, Rex, St. Pauls, Jerome, Parkerburg, Ingold, Magnolia, Rose Hill, Beulaville, Richlands, Petersburg, New Bern, Arapahoe, Merritt, Pamlico, Whortonville, Oracoke, Hatteras, Frisco, Buxton, Cape Hatteras.

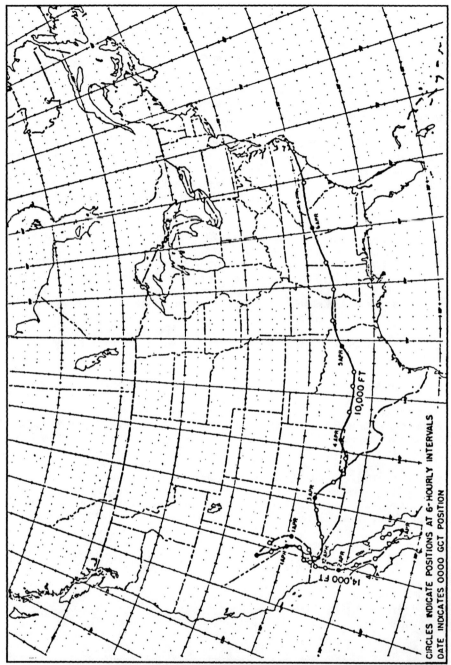

TRAJECTORY: UPSHOT-KNOTHOLE RUTH (UK3)

CHARACTERISTICS: UK RUTH

The device for the RUTH shot, code-named Hydride I, was the first University of California Radiation Lab bomb ever tested at the Nevada Test Site. Though huge (7400 lbs, 56 inches diameter, 66 inches long), the shot was a fizzle, vaporizing only the top 100 feet of the 300-foot shot tower. Among the nuclear tests, Upshot-Knothole RUTH is ranked 7th for Kr-85, Sr-90 and Y-90. The deposition pattern for Upshot-Knothole RUTH is most similar to Upshot-Knothole ANNIE (UK1) (R: 0.51; T: 32.89). Fallout from this nuclear test fell on 56 percent of the total U.S. counties (1733 counties out of 3094 designated by the NCI). Counties of highest deposition included Beckham OK, Greer OK, Harmon OK, Roger Mills OK, Armstrong TX, Carson TX Collingsworth TX, Donley TX and Parmer, TX. All received (estimated maximum activity) 83.58 µCi/sq meter on Apr 4, 1953.

Despite its low yield, low total fallout level, and the fact that it had been something of a fizzle, Upshot-Knothole RUTH has a relatively high day-one total fallout to I-131 ratio: 124 to 1.

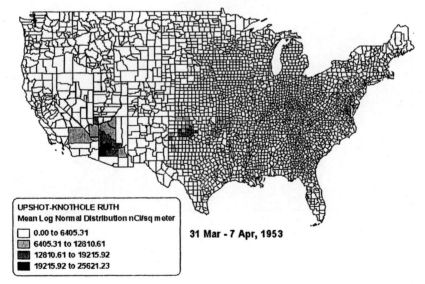

TOTAL FALLOUT: UPSHOT-KNOTHOLE RUTH
Equal Intervals
nCi/square meter

UPSHOT-KNOTHOLE RUTH
Mean Log Normal Distribution nCi/sq meter
- 0.00 to 6405.31
- 6405.31 to 12810.61
- 12810.61 to 19215.92
- 19215.92 to 25621.23

31 Mar - 7 Apr, 1953

TOTAL FALLOUT: UPSHOT-KNOTHOLE RUTH
Equal Number
nCi/square meter

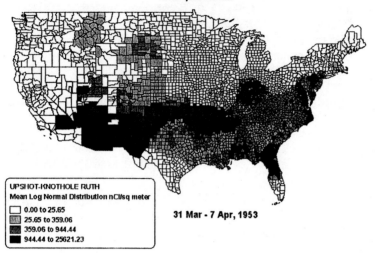

UPSHOT-KNOTHOLE RUTH
Mean Log Normal Distribution nCi/sq meter
- 0.00 to 25.65
- 25.65 to 359.06
- 359.06 to 944.44
- 944.44 to 25621.23

31 Mar - 7 Apr, 1953

FALLOUT PATH: UK RUTH

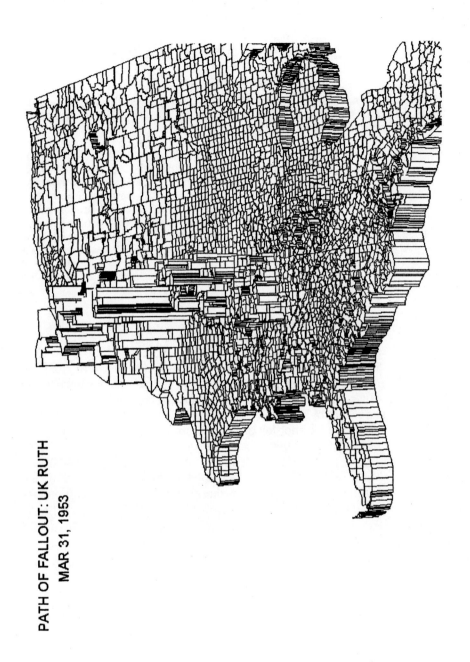

SHOT 4: UPSHOT-KNOTHOLE DIXIE (UK4)

UPSHOT-KNOTHOLE: DIXIE

Detonation Date: 6 Apr 1953 • *Detonation Time:* 7:30 A.M. • *Area:* 7-3 • *Sponsor:* Los Alamos • *Yield:* 11kt • *Radiation Level at Ground Zero:* 1.5 mR/hr • *Height of Burst:* 6,022 ft • *Cloud Top Height:* 45,000 ft

30,000-Foot Trajectory: NEVADA: Moapa, Logandale, Overton • ARIZONA: (Grand Canyon National Park), Flagstaff, Show Low, Nutrioso, Alpine • NEW MEXICO: Horse Springs, Negal, Hondo, Roswell, Caprock • TEXAS: Sundown, Lubbock, Roosevelt, Crosbytown, Croton, Truscott, Thalia, Electra, Wichita Falls, Terral, Illinois Bend, Dexter, Dennison, Ivanhoe, Paris, Reno, Detroit, Clarksville, Whaley, Texarkana • ARKANSAS: Homen, Patmos, Lamartine, Smackover, El Dorado, North Crossett, Crossett, Berlin, Wilmet, Indian, Readland • MISSISSIPPI: Glen Allan, Grace, Yazoo City, Vaughn, Carthage, McAffee, McDonald, Econdale • ALABAMA: Demopolis, Burnsville, Selma, Montgomery, Shorter, Tuskegee • GEORGIA: Fort Benning, Tazewell, Ideal, Montezuma, Elko, Hawkinsville, Gresston, Mount Vernon, Vidalia, Lyons, Santa Claus, Claxton, Pembroke, Bloomingdale • SOUTH CAROLINA: Port Wentworth, Forest Beach.

CHARACTERISTICS: UK DIXIE

The device for this shot (an airdrop) was a 10.5-ft x 5 ft cylinder weighing 3260 lbs. Upshot-Knothole DIXIE ranks 4th among the nuclear tests in the percentage of Co-58 in total fallout. The deposition pattern for UK DIXIE is most similar to that of Buster-Jangle CHARLIE (BJ2) (R: 0.25; T: 14.06). Nuclear debris from this shot fell on 50 percent of the total U.S. counties, or 1548 counties. Most of the nuclear fallout was deposited in the states of Massachusetts, Connecticut and Rhode Island and all deposition was associated with rain. Counties of highest deposition included the following: Plymouth Ma, (estimated maximum activity: 1260 µCi/sq meter), Washington RI (estimated maximum activity: 607 µCi/sq meter), Bristol MA (estimated maximum activity: 942 µCi/sq meter) and Middlesex CT (estimated maximum activity: 573 µCi/sq meter).

The ratio of total fallout to I-131 for DIXIE is relatively high at 125 to 1.

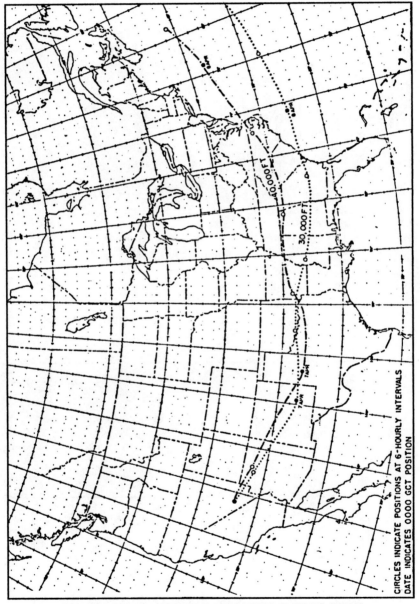

TRAJECTORY: UPSHOT-KNOTHOLE DIXIE (UK4)

TOTAL FALLOUT: UPSHOT-KNOTHOLE DIXIE
Equal Intervals
nCi/square meter

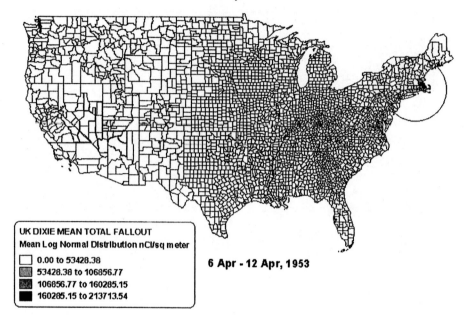

UK DIXIE MEAN TOTAL FALLOUT
Mean Log Normal Distribution nCi/sq meter
- 0.00 to 53428.38
- 53428.38 to 106856.77
- 106856.77 to 160285.15
- 160285.15 to 213713.54

6 Apr - 12 Apr, 1953

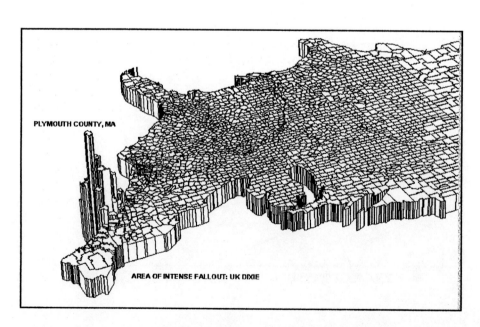

PLYMOUTH COUNTY, MA

AREA OF INTENSE FALLOUT: UK DIXIE

4-20 U.S. FALLOUT ATLAS : TOTAL FALLOUT

SHOT 5: UPSHOT-KNOTHOLE RAY (UK5)

UPSHOT-KNOTHOLE: RAY

Detonation Date: 11 Apr 1953 • *Detonation Time:* 4:45 A.M. • *Area:* 4a • *Sponsor:* Livermore Laboratory • *Yield:* 0.2 kt • *Radiation Level at Ground Zero:* 20 R/hr • *Height of Burst:* 100 ft • *Cloud Top Height:* 12,800 ft

10,000-Foot Trajectory: NEVADA: Lathrop Wells • CALIFORNIA: Death Valley Junction, (Greenwater Range), (Avawatz Mountains), (Bristol Mountains), Bagdad, Amboy, (Bristol Dry Lake), (Cadiz Dry Lake) (Palen Mountains), Blythe, East Blythe • ARIZONA: Ehrenburg, (Kofa Mountains), (Gila Bend Mountains), (Woolsey Peak), Maricopa Akchin Indian Reservation, Casa Grande, Stanfield, Eloy, Pichacho, Catalina, Mount Lemmon, (Lime Peak), Dragoon, (Chiricahua Peak) • NEW MEXICO: Rodeo, (Animas Mountains) • TEXAS: Socorro, Horizon City, (Hueco Mountains), (San Antonio Mountain) • NEW MEXICO: (Carlsbad Caverns National Park), (Sitting Bull Falls), Whites City, Loving, Carlsbad, Malaga, (Antelope Ridge), Bennett, Jal • TEXAS: Stanton, Greenwood, Sterling City, Water Valley, Carlsbad, Miles, Mereta, Millers View, Doole Fife, Placid, Hall, Algerita, San Saba, Lampasas, Watson, Prairie Dell, Bartlett, Granger, Thorndale, Navasota, Conroe, Moss Hill, Silsbee, Lumberton • LOUISIANA: Vinton, Edgerly, Sulphur, Westlake, Lake Charles, Iowa, Welsh, Jennings, Midland, Morse, Indian Bayou, Maurice, Youngsville, New Iberia, Jeanerette, Labadieville, Thibodaux, Raceland, Lockport, Larose, Lafite, Myrtle Grove, Magnolia, Port Sulphur, Empire, Buras, Triumph, Boothville, (North Pass) • FLORIDA: Marco, Goodland, Everglades City, Ochope, Monroe, Miccosukee Indian Reservation, Sweetwater, Miami, Miami Beach.

CHARACTERISTICS: UK RAY

Among the nuclear tests, Upshot-Knothole RAY is ranked 6th for percentages of Sr-89, Sr-90 and Y-90 in fallout. The deposition pattern for Upshot-Knothole RAY is most similar to that of Upshot-Knothole ANNIE (UK1) (R: 0.46; T: 28.46). Fallout from this shot was deposited on 2084 counties, or 67 percent of the total U.S. counties. The county of highest deposition was Yuma, AZ with an estimated maximum activity of 95.41 µCi/sq meter on Apr 11, 1953. The total fallout to I-131 ratio is relatively high at 121 to 1.

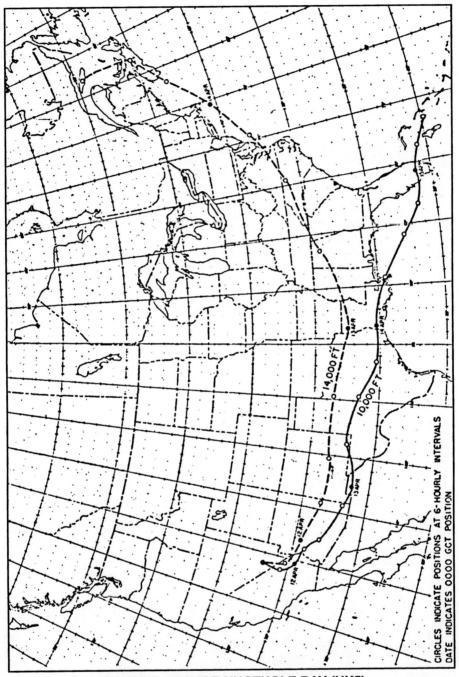

TRAJECTORY: UPSHOT-KNOTHOLE RAY (UK5)

TOTAL FALLOUT: UPSHOT-KNOTHOLE RAY
Equal Intervals
nCi/square meter

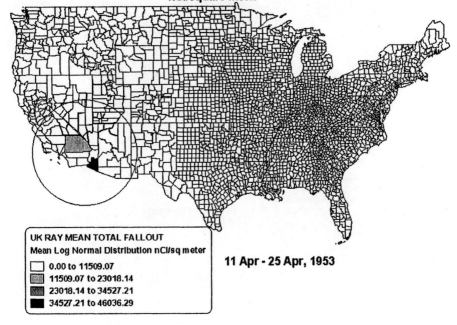

UK RAY MEAN TOTAL FALLOUT
Mean Log Normal Distribution nCi/sq meter
- 0.00 to 11509.07
- 11509.07 to 23018.14
- 23018.14 to 34527.21
- 34527.21 to 46036.29

11 Apr - 25 Apr, 1953

U.S. FALLOUT ATLAS : TOTAL FALLOUT   4-23

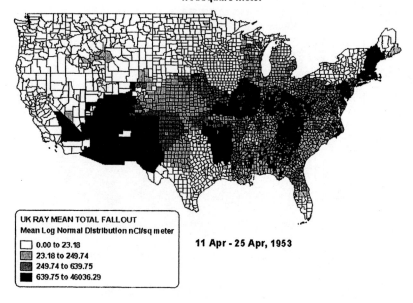

SHOT 6: UPSHOT-KNOTHOLE BADGER (UK6)

UPSHOT-KNOTHOLE: BADGER

Detonation Date: 18 Apr 1953 • *Detonation Time:* 4:35 A.M. • *Area:* 2 • *Sponsor:* Los Alamos • *Yield:* 23 kt • *Radiation Level at Ground Zero:* 1,200 R/hr • *Height of Burst:* 300 ft • *Cloud Top Height:* 36,000 ft

40,000-Foot Trajectory: NEVADA: Indian Springs, North Las Vegas, Henderson • ARIZONA: Hualpai Indian Reservation, Peach Springs, Nelson, Drake, Sedonia, Heber, Overgaard, Shumway, Vernon, Springerville • NEW MEXICO: Red Hill, Pie Town, Polvadera, Lemitar, Florida, Claunch, Ancho, Haystack Mountain, Elkins, Kenna, Milnesand, Pep, Lingo • TEXAS: Denver City, Higginbotham, Seminole, Three Leagues, Knott, Fairview, Big Spring, Otis Chalk, Edith, Tennyson, Orient, Rowena, Concho, Millersview, Lohn, Fife, Rochelle, Hall, San Saba, Lake Victor, Briggs, Youngsport, Little River, Heidenheimer, Temple, Westphalia, Durango, Barclay, Rosebud, Travis, Lott, Cedar Springs, Marlin, Oletha, Thronton, Donie, Buffalo, Oakwood, Elkhart, Palestine, Rusk, Gallatin, Ponta, New Salem, Laneville, Concord, Dotson, Gary, Carthage, Deadwood • LOUISIANA: Keatchie, Stonewall, Keithville, Gayles, McDude, Elm Grove, Janestown, Heflin, Arcadia, Bryceland, Grambling, Ruston, Vienna, Sibby, D'Arbonne, Rocky Branch, Monroe, Sterlington, Bastrop, Fairbanks, Collinston, Mer Rouge, Goodwill, Pioneer, Forest, Transylvania, Lake Providence • MISSISSIPPI: Cary, Fitler, Jonestown, Valley, Tinsley, Benton, Way, Vaughn, Camden, Ofahoma, Carthage, Edinburg, McAfee, Williamsville, Deemer, McDonald, Moscow, House, Daleville • ALABAMA: York, Livingston, Coatopa, Belmont, Demopolis, Prairieville, Suttle, Summerfield, Burnsville, Winslow, Independence, Booth, Prattville, Cobbs Ford, Ware, Milstead, Tuskegee, Society Hill, Uchee, Seale, Fort Mitchell • GEORGIA: Cusseta, Buena Vista, Murrays Crossroads, Oglethorpe, Montezuma, Unadilla, Hawkinsville, Hartford, Gresston, Plainfield, Mount Vernon, Alley, Higgston, Vidalia, Lyons, Choopee, Collins, Bellville, Claxton, Pembroke, Blitchton, Eden • SOUTH CAROLINA: Pritchardville, Bluffton, Hilton Head Island, Folly Field Beach.

CHARACTERISTICS: UK BADGER

The device for the BADGER test was code-named "Buzzard." Like many other thermonuclear-styled designs, Buzzard was quite large, weighing 7400 lbs with a diameter of 56 inches. Among the nuclear tests, Upshot-Knothole BADGER is ranked 7th for percentages of Sn-123, Sb-125 and Te-127m in fallout. The deposition pattern for Upshot-Knothole BADGER is most similar to Plumbbob WHEELER-COULOMB B and LAPLACE (PB13) (R:0.38; T: 22.74). Fallout from this shot was deposited on 3000 counties, or 97 percent of the total U.S. counties. Counties of highest deposition included: Archuleta, CO (estimated maximum activity: 1136 µCi/sq meter) and Clark, NV (estimated maximum activity: 1103 µCi/sq meter), both on April 18, 1953.

The day one ratio of total fallout to I-131 is moderately high at 111 to 1.

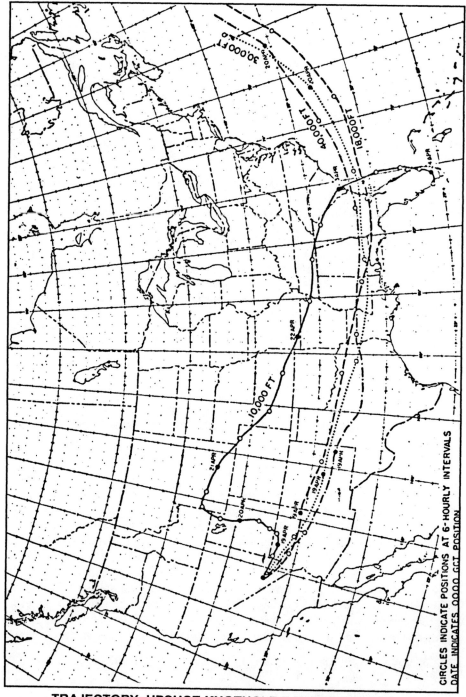

TRAJECTORY: UPSHOT-KNOTHOLE BADGER (UK6)

U.S. FALLOUT ATLAS : TOTAL FALLOUT

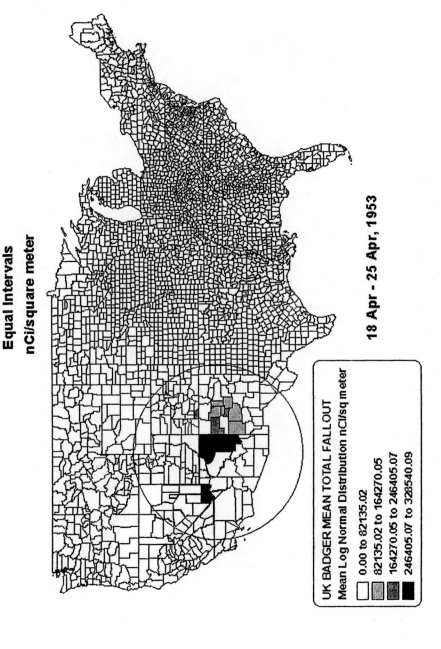

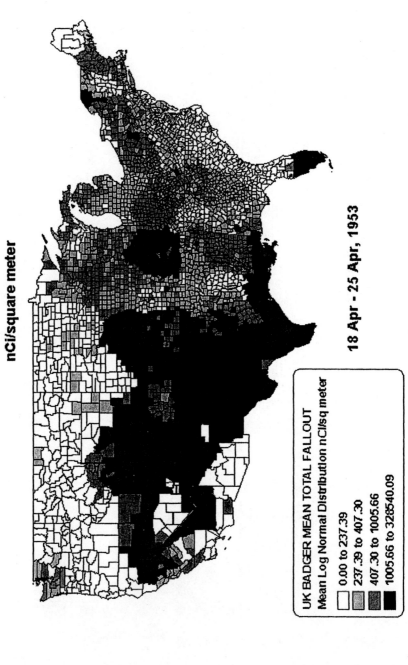

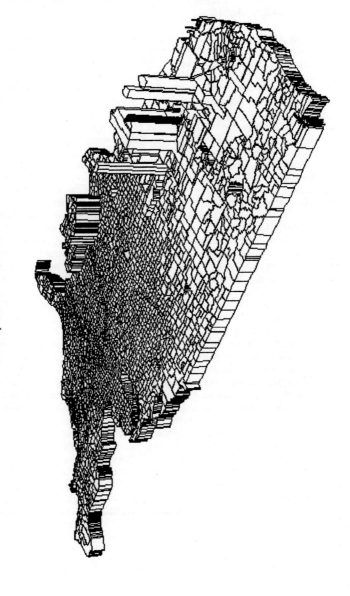

SHOT 7: UPSHOT-KNOTHOLE SIMON (UK7)

UPSHOT-KNOTHOLE: SIMON

Detonation Date: 25 Apr 1953 • *Detonation Time:* 4:30 A.M. • *Area:* 1 • *Sponsor:* Los Alamos • *Yield:* 43 kt • *Radiation Level at Ground Zero:* 300 R/hr • *Height of Burst:* 300 ft • *Cloud Top Height:* 44,000 ft

30,000-Foot Trajectory: NEVADA: Moapa Indian Reservation, Logandale, Overton • ARIZONA: Tusayan, Cameron, (Painted Desert), Navajo Indian Reservation, Hopi Indian Reservation, Hotevilla, Orabi, Second Mesa, Polacca, Keams Canyon, Nazlini, Fort Defiance, Window Rock • NEW MEXICO: Mexican Springs, Brimhall, Crown Point, Hospah, San Ysidro, Santo Domingo Pueblo, San Felipe Pueblo, Cerrillos, Madrid, Galisteo, Lamy, El Pueblo, Ribera, Serafina, Los Montoyas, Tucumcati, Porter • TEXAS: Adrian, Vega, Wildorado, Bushland, Amarillo, Pullman, Washburn, Conway, Claude, Goodnight, Howardwick, Samnorwood, Lutie, Aberdeen • OKLAHOMA: Madge, Vinson, Reed, Mangum, Hester, Blair, Warren, Snyder, Cache, Lawton, Hulen, Gas City, Duncan, Sunray, Velma, Countyline, Alma, Loco, Fox, Clemscot, Graham, Woodford, Newport, Springer, Gene Autry, Dickson, Baum, Ravia, Tishomingo, Nida, Armstrong, Caddo, Matoy, Cade, Bokchito, Boswell, Soper, Goodland, Hugo, Sawyer, Swink, Valient, Idabel, Haworth, Harris, Bokhoma, Tom • ARKANSAS: Arkinda, Fomby Homan, Stamps, Buckner, Magnolia, Calhoun, Macedonia, Emerson • LOUISIANA: Summerfield, Lillie, Spearsville, Farmerville, Linville, Spencer, Bastrop, Perryville, Fairbanks, Collinston, Oak Ridge, Epps, Monticello, Sondheimer • MISSISSIPPI: Blakely, Redwood, Bentonia, Ballard, Sharon, Offahoma, Carthage, Edinburg, Stallo, Plattsburg, McLeod, Prairie Point • ALABAMA: Pickensville, Beaver Town, Carrolton, Gordo, Echola, Port Birmingham, Bayview, Adamsville, Graysville, Fultondale, Gardendale, Mount Olive, Majestic, Pinson, Palmerdale, Remlap, Allgood, Whitney, Taits Gap, Gallant, Reece City, Gadsden, Keener, Leesburg, Sand Rock, Gaylesville, Ringgold • GEORGIA: Lylerly, Berryton, Holland, Crystal Springs, Summersville, Calhoun, Echota, Oakman, Whitestone, (Brasstown Bald, highest point in Georgia: 4,784 feet), Pillard • NORTH CAROLINA: Brasstown, Shooting Creek, Franklin, Iotla, Webster, Cullowhee, Sylva, Balsam, Hazelwood, Canton, Asheville, Leicester, Alexander, Jupiter, Barnardsville, Little Switzerland, Celo, Estatoe, Miraville, Spruce, Ingalls, Linville, Montezuma, Boone, Jefferson, Warrensville, Lansing, Grassy Creek • VIRGINIA: Independence, Fries, Sylvatus, Indian Valley, Alum Ridge, Copper Hill, Bent Mountain, Roanoke, Montvale, Big Island, Lowesville, Nellysford, White Hall, Free Union, Nortonsville, Ruckersville, Hood, Wolftown, Haywood, Boston, Viewtown, Amissville, Orlean, Old Tavern, Marshall, Rectorville, Halfway, Aldie, Oatlands, Hamilton, Waterford, Point of Rocks • MARYLAND: Tuscarora, Adamstown, Thurston, Buckeystown, Frederick, Mount Pleasant, Johnsville, Union Bridge, Ladiesburg, Uniontown, Middleburg, Tarrytown, Silver Run • PENNSYLVANIA: Hanover, Mount Royal, Newberrytown, Royalton, Hummelstown, Hershey, Shellsville, East Hanover, Bardnersville, Oak Grove, Ravine, Donaldson, Branch Dale, Buck Run, Gordon, Shenendoah, Fern Glen, Plymouth, Luzerne, Upper Exeter, Dalton, Lewisville, Royal, Winderdale • NEW YORK: Fish's Eddy, Shinhopple, Downsville, Lake Delaware, Bovina Center, Stamford, West Fulton, Schoharie, Schenectady, Rotterdam, Charlton, Factory Village, Saratoga Springs, Fort Howard, Hudson Falls, Kingsburg, Truthville, Poultney, Hampton • VERMONT: West Haven, Benson, Brandon, Forest Dale, Goshen, Alpine Village, Roxbury, Northfield, Berlin, Montpelier, Barre, North Montpelier, Plainfield, East Calais, Marshfield, Cabot, Walden, North Danville, Lyndon, Burke, East Haven, Lemington, Colebrook • NEW HAMPSHIRE: Stewartstown Hollow, Mount Pisgah, Magalloway Mountain • MAINE: Eustis, West Forks, Moosehead, Kokadjo, (Mount Katahdin, highest point in Maine), Patten, Crystal, Island Falls, Shin Pond, Knowles Corner, Bridgewater, Blaine, Mars Hill.

The 40,000-foot trajectory passed over the highest points in the states of Georgia and Maine.

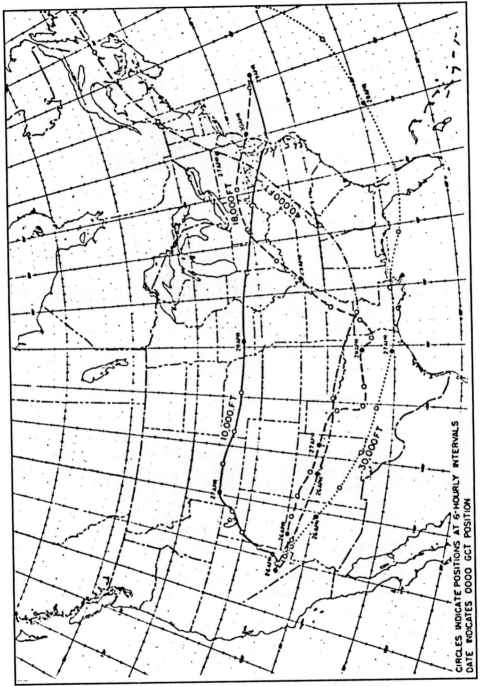

TRAJECTORY: UPSHOT-KNOTHOLE SIMON (UK7)

RELATIVE TOTAL FALLOUT: UK SIMON
APR 15, 1953 43 KT

CHARACTERISTICS: UK SIMON

The device for the SIMON shot, code-named "Simultaneity", was approximately 3 ft in diameter and 18.5 ft in length and weighed 5.5 tons. Among the nuclear tests, Upshot-Knothole SIMON is ranked 2nd for deposition of Cs-137 and Nb-95, 3rd for percentage of Be-7 and 4th for the percentage of U-237 in the total fallout. The deposition pattern for this test is most similar to Upshot-Knothole ENCORE (UK8) (R: 0.51; T: 32.81). Among all the tests, Upshot-Knothole

Counties of highest close-in fallout include: Clark, NV (estimated maximum activity: 3882 µCi/sq meter on Apr 25, 1953) and Coconino, AZ (estimated maximum activity: 2197 µCi/sq meter on Apr 25, 1953).

Interestingly, fallout from this nuclear test was involved in a significant hot spot in the Eastern United States, on April 26, 1953. On that day the following counties received significant fallout (in microCuries per square meter, estimated maximum activity): Albany NY—894; Columbia NY—1058; Fulton NY—1769; Rensselaer, NY—2694; Saratoga, NY—1808; Schenectady, NY—1260; Warren, NY—3068; Washington, NY—3226; Addison, VT—1043; and Bennington, VT—1519. The fallout was associated with a rainout. However, it has not been determined if the nuclear material was brought to earth by precipitation or by dry downdrafts that were part of the storm front. The ratio of total fallout to I-131 is moderate at 98 to 1.

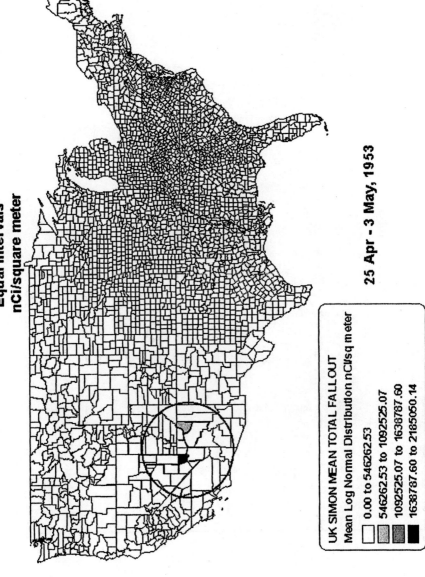

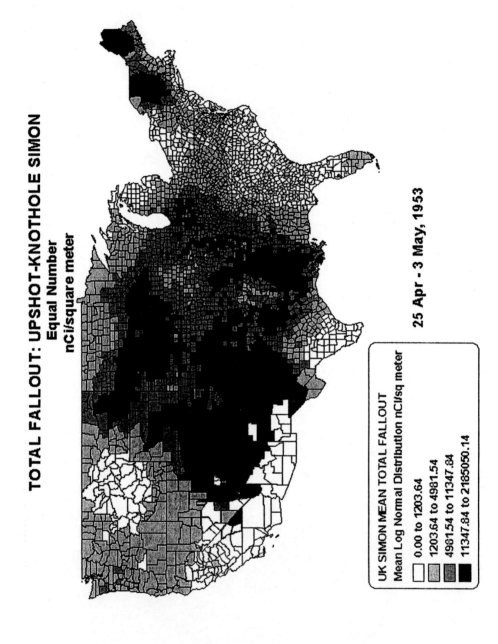

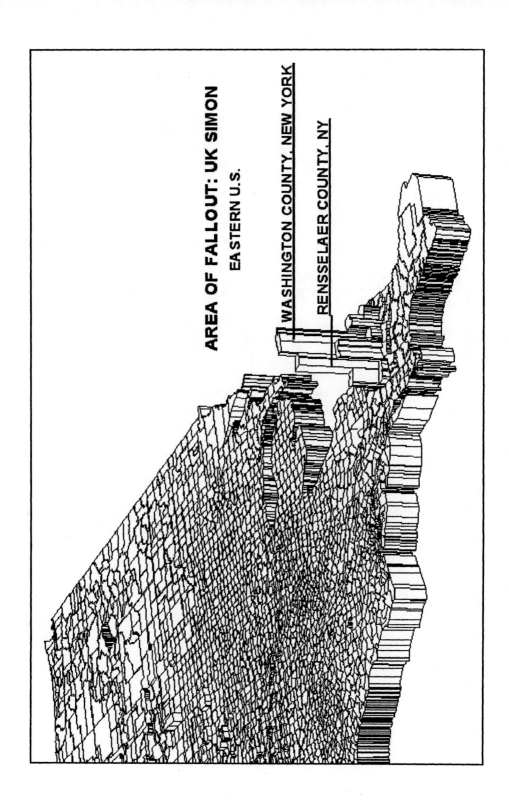

FALLOUT IN EASTERN U.S. FROM UPSHOT-KNOTHOLE SIMON

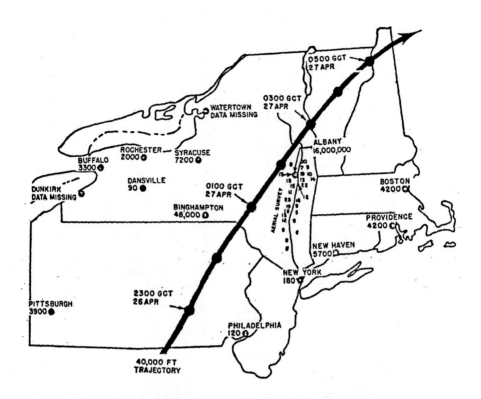

PATH OF THE SIMON CLOUD OVER NEW YORK STATE
APRIL 25, 1953

SHOT 8: UPSHOT-KNOTHOLE ENCORE (UK8)

UPSHOT-KNOTHOLE: ENCORE

Detonation Date: 8 May 1953 • *Detonation Time:* 7:30 A.M. • *Area:* Frenchman's Flat • *Sponsor:* Los Alamos, Department of Defense • *Yield:* 27 kt • *Radiation Level at Ground Zero:* 25 R/hr • *Height of Burst:* 2,423 ft • *Cloud Top Height:* 42,000 ft

10,000-Foot Trajectory: NEVADA: Moapa, Logandale, Overton • ARIZONA: (Mount Dellenbaugh), Havasupai Indian Reservation, North Rim, Cedar Ridge, Kaibito • UTAH: (Glen Canyon), (Cataract Canyon), (Canyonland National Park), (La Sal Mountains), Castleton, Mount Tomasaki • COLORADO: Gateway, Whitewater, Skyway, Grand Mesa, (Grand Mesa National Forest), Snowmass, Ruedi, Meredith, Norrie, (Tennessee Pass), Climax, Leadville, (Fremont Pass), Jefferson, Como, Deckers, Greenland, Palmer Lake, Peyton, Rush, Eads • KANSAS: Tennis, Wright, Windthorst, Brenham, Haviland, Wellsford, Coats, Sawyer, Isabel, Duquoin, Harper, Danville, Freeport, Argonia, Corbin, South Haven, Ashton, Arkansas City, Silverdale, Maple City, Cedar Vale, Wauneta, Elk City, Sycamore, Morehead, Galesburg, Erie, St. Paul, Walnut, Hepler, Farlington, Hiattville, Garland • MISSOURI: Deerfield, Richards, Horton, Prairie City, Appleton City, Ohio, Deepwater, Brownington, Tightwad, Coal, Leesville, Branden, Mora, Bahner, Otterville, Beaman, Pilot Grove, Bunceton, Bellair, Lone Elm, Lamine, Boonville, Wooldridge, Rocheport, New Franklin, Midway, Hinton, Harrisburg, Rucker, Sturgeon, Clark, Centralia, Paris, Stoutsville, Monroe City, Hassard, Hunnewell, Withers Mill, Palmyra, West Quincy • ILLINOIS: Quincy, Ursa, Mendon, Loraine, Bigneck, Stillwell, Bowen, Denver, Augusta, Plymouth, Colmar, Fondon, Macomb, Bardolph, New Philadelphia, Bushness, Marietta, Blyton, Ellisville, Fairview, London Mills, Rapatee, Farmington, Yates City, Elmwood, Oak Hill, Brimfield, Dunlap, Princeville, Edelstein, Speer, Camp Grove, Henry, Putnam, McNabb, Granville, Spring Valley, Oglesby, Cedar Point, Peru, LaSalle, Ottawa, Dayton, Wedrun, Norway, Serena, Sheridan, Newark, Sandwich, Plano, Yorkville, Bristol, Oswego, Montgomery, Autota, Batavia, Wheaton, Elmhurst, Bloomingdale, Roselle, Des Plaines, Highland Park, Glencoe, Winnettca, Wilmette • MICHIGAN: Graafschap, Beechwood, Holland, New Holland, Hudsonville, Jenison, Grandville, Wyoming, Walker, Grand Rapids, Cannonsburg, Bostwick Lake, Belding, Fenwick, Butternut, Vickeryville, Crystal, Sumner, Ithoca, Breckenridge, Wheeler, Merrill, Iva, Hemlock, Laporte, Freeland, Midland, Bay City, Essexville, Oakhurst, Bay Park, Sebewaing, Bay Port, Pigeon, Caseville, Pinnebog, Kinde, Port Austin, Grind Stone City • MAINE: Pittston Farm, Seboomook, North East Carry, (Caribou Lake), (Harrington Lake), (North Brother Mountain), (Mount Katahdin, highest point in Maine), Stacyville, Sherman Station, Sherman Mills, (Otter Lake Mountain), Orient.

CHARACTERISTICS: UK ENCORE

The Upshot-Knothole ENCORE device weighed 8330 lbs and was airdropped from a height of 19,000 feet. Among the nuclear tests, Upshot-Knothole ENCORE is ranked 7th for percentage of Co-57 in total fallout. The deposition pattern for Upshot-Knothole ENCORE is most similar to Upshot-Knothole SIMON (UK7) (R: 0.51; T: 32.81). Nuclear debris from this test was deposited on 2227 counties, or 72 percent of the total number of U.S. counties. County of highest deposition: Custer, MT (548 µCi/sq meter estimated maximum activity on May 8, 1953.) Other counties with high deposition amounts from this shot (all in µCi/sq meter, estimated maximum activity) included: Dawson MT—328; Fallon MT—328; McCone MT—328; Prairie MT—328; Richland MT—548; Roosevelt MT—328; Sheridan MT—328; Wibaux MT—328 and Billings ND—328

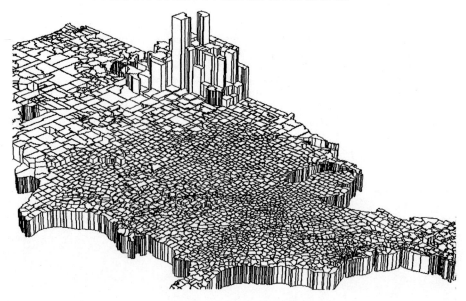

PRIMARY FALLOUT ZONE: UK ENCORE

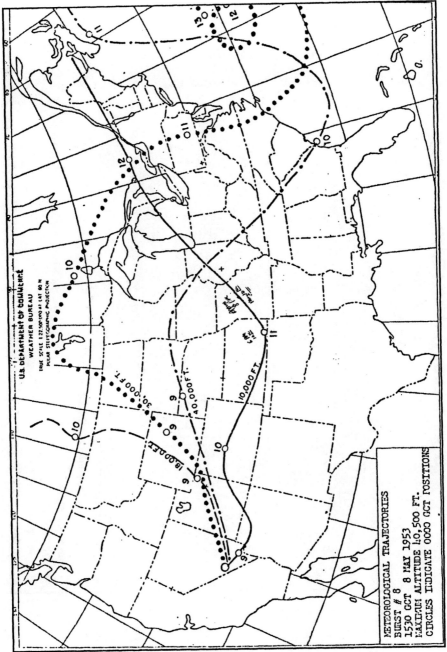

TRAJECTORY: UPSHOT-KNOTHOLE ENCORE (UK8)

U.S. FALLOUT ATLAS : TOTAL FALLOUT

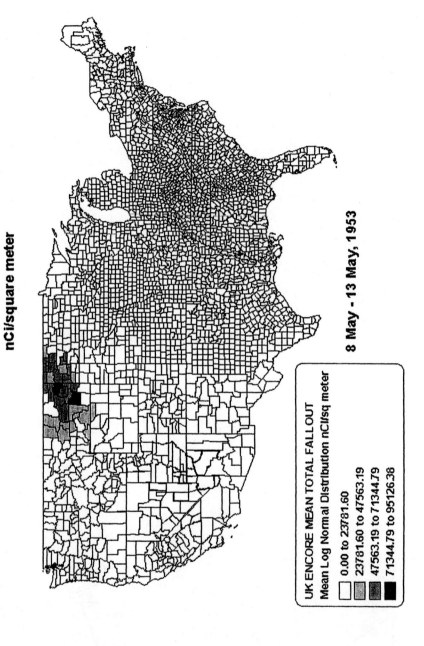

SHOT 9: UPSHOT-KNOTHOLE HARRY (UK9)

UPSHOT-KNOTHOLE: HARRY

Detonation Date: 19 May 1953 • *Detonation Time:* 4:05 A.M. • *Area:* 3a • *Sponsor:* Los Alamos • *Yield:* 32 kt • *Radiation Level at Ground Zero:* 10 R/hr • *Height of Burst:* 300 ft • *Cloud Top Height:* 42,500 ft

10,000-Foot Trajectory: NEVADA: Preston, Ely, Tippett • UTAH: Callao, Grantsville, Salt Lake City, Layton, Ogden, Woodruff • WYOMING: Frontier, Diamondville, Kemmerer, Fontenelle, Farson, Yoder • NEBRASKA: Scottsbluff, Gering, McGrew, Bridgeport, Broadwater, Lisco, Oshkosh, Lewellen, Ogallala, Roscoe, Farnam, Kearney, Lowell, Prosser, Hansen, Trumbull, Harvard, Lushton, Cordova, Pleasantdale, Denton, Lincoln, Avoca, Union • IOWA: Shenandoah, Bingam, Yorktown, Clarinda, Sharpsburg, Gravity, Conway, Kellerton, Decatur City, Leon, High Point, Corydon, Promise City, Plano, Centerville, West Grove, Bloomfield, Hillsboro, Salem, Danville • ILLINOIS: Monmouth, Galesburg, Wataga, Victoria, Fayette, Toulon, Wyoming, Putnam, Oglesby, LaSalle, Ottowa, Wedron, Lisle, Chicago • MICHIGAN: Allendale, Wyoming, Grand Rapids, Alpine, Sparta, Sherman City, Farwell, St. Helen, Luzerne, Metz, Hawks, Rogers City.

18,000-Foot Trajectory: NEVADA: Elgin, Carp • UTAH: Santa Clara, St. George, Hurricane, Virgin, Rockville, Springdale, Mt. Carmel, Blanding • COLORADO: Pleasant View, Stoner, Spar City, Masonic Park, Center, Cedarwood, Rocky Ford, Las Animas, McClave, Big Bend, Wiley • KANSAS: Lydia, Rush Center, Timkin, Shaffer, Olmitz, Hoisington, Redwing, Odin, Lorraine, Falun, Mentor, Kipp, Holland, Pearl, Abilene, Junction City, St. Marys, Mayetta, Half Mound, Atchison • MISSOURI: St. Joseph, Rochester, Fairport, McFall, Blue Ridge, Goshen, Princeton, Mercer • IOWA: Seymour, Centerville, Ottumwa, Bladensburg, Abington, Brighton, Cranston, Muscatine, Fairport, Montpelier, Davenport, Bettendorf • ILLINOIS: Rock Island, Moline, Hillsdale, Lyndon, Rock Falls, Sterling, Dixon, Rochelle, Hillcrest, Clare, Genoa, Hampshire, Huntley, Algonquin, Libertyville, Waukeegan, North Chicago • MICHIGAN: Douglas, Saugatuck, Hamilton, Bentheim, Green Lake, Caldeonia, Clarksville, Lake Odessa, Portland, Fowler, St. Johns, Eureka, Bannister, Brant, St. Charles, Saginaw.

RELATIVE TOTAL FALLOUT UK HARRY
MAY 19, 1953 32 KT

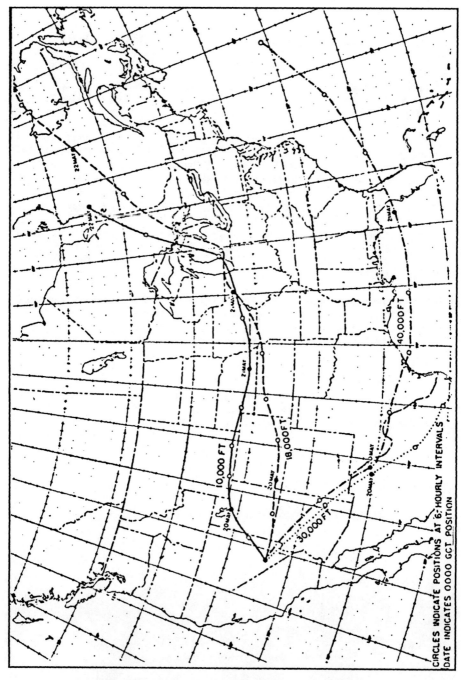

TRAJECTORY: UPSHOT-KNOTHOLE HARRY (UK9)

CHARACTERISTICS: UK HARRY

The device for shot HARRY was code-named "Hamlet." It was 56 inches in diameter, 66 inches in length, and weighed 4 tons. Among the nuclear tests, Upshot-Knothole HARRY is ranked 3rd for deposition of Cs-137; Nb-95; Sr-90; and Zr-95; 4th for deposition of Nb-95m and Pr-144; 5th for percentage of U-240 and Ru-106; 6th for percentage of I-130 and Te-127m, 8th for deposition of Co-60; 10th for deposition of Eu-155; 13th for percentage of Sr-89, Sr-90 and Y-90 and 16th for deposition of Be-7.

The deposition pattern for Upshot-Knothole HARRY is most similar to Upshot-Knothole CLIMAX (UK11) (R: 0.50; T: 32.26).

Fallout from this shot was deposited on 3046 counties, or 98 percent of the total counties in the U.S, ranking it the 6th for fallout dispersion.

The county with the highest deposition amount was also the county with one of the highest levels of fallout in the entire nuclear test series: Gunnison, CO. While this county received high amounts of fallout from a number of different nuclear tests, UK HARRY produced an exceptionally heavy fallout for this area: On May 19, 1953 Gunnison County, Colorado received an estimated maximum activity of 9609 µCi/square meter. Other counties in the southwest received slightly less (the following numbers in microCuries estimated maximum activity): Coconino, AZ—8037; Conejos CO—8036; Mineral CO—4224; Kane UT—6857; Washington(2) UT—7171.

RELATIVE TOTAL FALLOUT: UK HARRY

MAY 19, 1953 32 KT

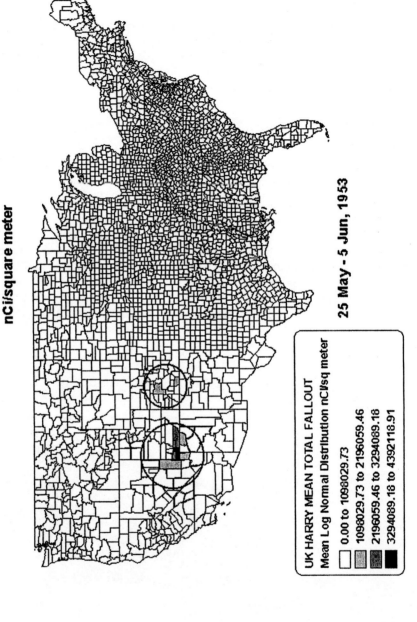

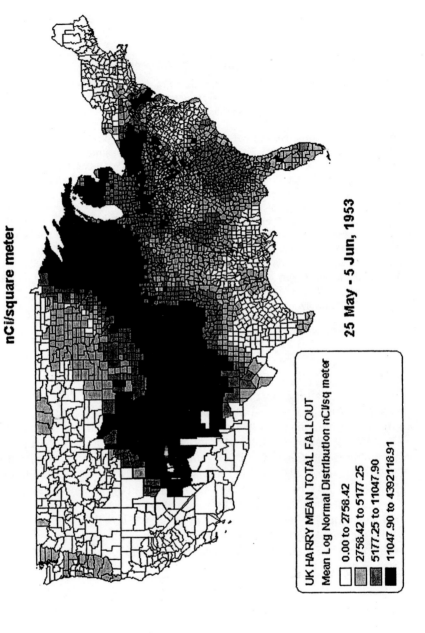

SHOT 10: UPSHOT-KNOTHOLE GRABLE (UK10)

UPSHOT-KNOTHOLE: GRABLE

Detonation Date: 25 May 1953 • *Detonation Time:* 7:30 A.M. • *Area:* Frenchman's Flat • *Sponsor:* Los Alamos • *Yield:* 15 kt • *Radiation Level at Ground Zero:* 10 R/hr • *Height of Burst:* 524 ft • *Cloud Top Height:* 35,000 ft

10,000-Foot Trajectory: NEVADA: (Grant Range), Ruth, Steptoe, Currie, (Snow Water Lake), Thousand Springs • IDAHO: Three Creek, Glenns Ferry, Mountain Home, Mayfield, Boise, Pearl, Montour, Payette, Weiser • OREGON: Pleasant Valley, Baker, Haines, North Powder, (Blue Mountains), Pilot Rock, Echo, Irrigon • WASHINGTON: Paterson, (Horse Heaven Hills), Yakima Indian Reservation, (Lincoln Plateau), (Mount Adams Wilderness Area) (Gifford Peak),(Gifford Pinchot National Forest), North Bonneville, Skamania • OREGON: Gresham, Boring, Eagle Creek, Estacada, (Goat Mountain), (Round Mountain), (Monument Peak), (Tombstone Pass), (Three Sisters Wilderness Area), Bend, (Bear Creek Butte), Post (Spanish Peak), Hamilton, (Meadow Creek Pass), (Blue Mountains), La Grande, Island City, Alicel, Imbler, Minam, Walowa • IDAHO: Waha, Lewiston, Lapwai, Spalding, Julietta, Southwick, Helmer, Boyill, Avery, Saltese, Mullan • MONTANA: Belknap, (Mount Headley), (Blue Mountain), Tray • IDAHO: Moyie Springs, Meadow Creek, (Moyie Falls), Nordman • WASHINGTON: (Molybdenite Mountain), Lost Creek, Arden, Addy, Cedonia, Hunters, Marlin, Moses Lake, Sunnyside, Grandview, Mabton, (Horse Heaven Hills) • OREGON: Arlington, Blalock, Rock Creek, Mikkalo, Kent, Shaniko, Antelope, Ashwood, (Grizzly Mountain).

CHARACTERISTICS: UK GRABLE

The device used for this shot was 54.4 inches long and weighed 803 lbs. It was fired from a cannon that weighed 85 tons and had a range of 20 miles. Among the nuclear tests, UK GRABLE is ranked 2nd for percentages of W-181, W-185 and W-188m 4th for percentages of Mn-54; and 6th for percentages of Ce-144, Nb-95 and Zr-95. The deposition pattern for Upshot-Knothole GRABLE is most similar to Tumbler-Snapper HOW (TS8) (R: 0.46; T: 28.61). Fallout from GRABLE was deposited on 98 percent of the U.S. counties (3025 counties). Counties of highest deposition included (in µCi/sq meter estimated maximum activity) Ferry WA—207; Bonner ID—109; Okanogan, WA—141; Pend Oreille WA—158; and Stevens WA—136. All of the aforementioned depositions of fallout occurred on a single day, June 5, 1953.

The ratio of total fallout to I-131 for this shot is relatively high at 124 to 1.

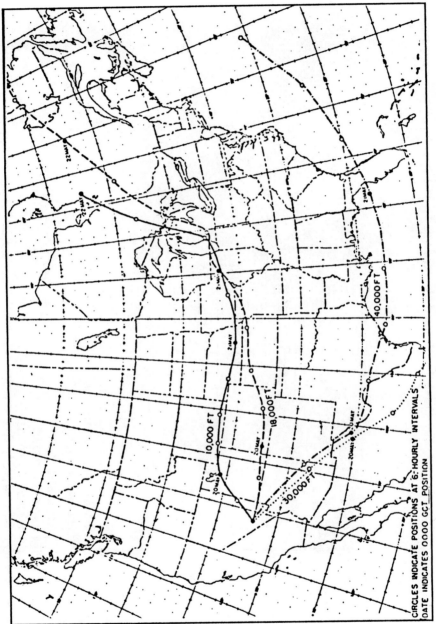

TRAJECTORY: UPSHOT-KNOTHOLE GRABLE (UK10)

U.S. FALLOUT ATLAS : TOTAL FALLOUT

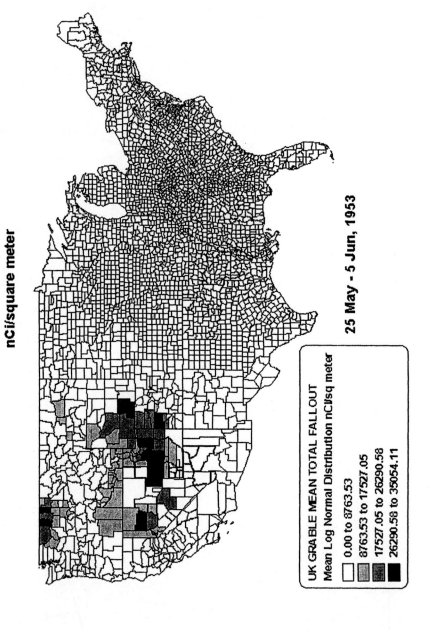

SHOT 11: UPSHOT-KNOTHOLE CLIMAX (UK11)

UPSHOT-KNOTHOLE: CLIMAX

Detonation Date: 4 Jun 1953 • *Detonation Time:* 3:15 A.M. • *Area:* 7-3 • *Sponsor:* Los Alamos • *Yield:* 61 kt • *Radiation Level at Ground Zero:* 10 R/hr • *Height of Burst:* 1,334 ft • *Cloud Top Height:* 42,700 ft

40,000-Foot Trajectory: NEVADA: (Pintwater Range), Carp • UTAH: (Snow Canyon), St. George, Washington, Hurricane, Rockville, Mount Carmel Junction, (Cottonwood Canyon), (White Canyon), (Cataract Canyon), (Angel Arch), La Sal Junction, Caselton • COLORADO: Gateway, Glade Park, Grand Junction, Orchard Mesa, Fruitvale, Clifton, De Beque, Parachute, Rulison, Rio Blanco, Buford, (Williams Fork Mountains), Haydon, Clark, Hahns Peak • WYOMING: (Blackhall Mountain), Riverside, Ryan Park, (Medicine Bow Peak), McFaddin, Rock River, Esterbrook, Douglas, Orin, Newcastle, Four Corners • SOUTH DAKOTA: Cheyenne Crossing, Savoy, Central City, Whitewood, Spearfish, St. Onge, Fruitdale, Nisland, Newell, Zeona, Bison, Meadow, Morristown, Watanga • NORTH DAKOTA: Shields, Breien, Fort Rice, Huff, Moffitt, Sterling, Steele, Dawson, Crystal Springs, Predina, Goldwin, Buchanan, Pingree, Spiritwood Lake, Courtenay, Wimbledon, Hannaford, Dazey, Sibley, Luverne, Blabon, Hope, Clifford, Galesburg, Blanchard, Hillsboro, Kelso • MINNESOTA: Climax, Shelly, Halstad, Lackhart, Gary, Rindal, Bejou, Ebro, Bagley, Shelvin, Solway, Bemidji, Turtle River, Tenstrike, Hines, Blackduck, Bergville, Effie, Craigville, Bois Fort, Nelt Lake, Ash Lake, Cusson, Crane Lake.

CHARACTERISTICS: UK CLIMAX

The device for the CLIMAX test was 30.6 inches wide, 15 ft long, and weighed 1840 lbs. Among the nuclear tests, Upshot-Knothole CLIMAX is ranked 1st in percentages of Ce-141 and Ce-144 in fallout and 2nd for percentages of Pr-144 in fallout. The deposition pattern for UK CLIMAX is most similar to Teapot TESLA (TP3) (R: 0.53; T: 35.09). Fallout from Upshot-Knothole CLIMAX was deposited on 2726 counties, or 88 percent of the total number of counties in the United States.

Counties of highest deposition included Coconino, AZ at 289 µCi/sq meter (estimated maximum activity) and Clark NV at 143 µCi/sq meter (estimated maximum activity)—both on May 25, 1953. The day one total fallout to I-131 ratio for CLIMAX is in the upper moderate range at 121 to 1.

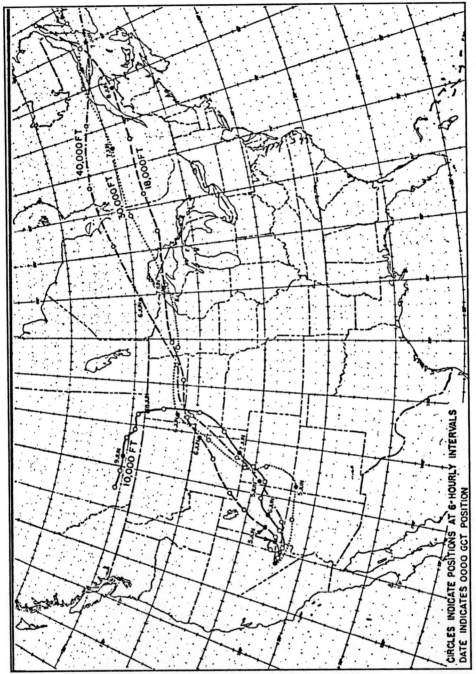

TRAJECTORY: UPSHOT-KNOTHOLE CLIMAX (UK11)

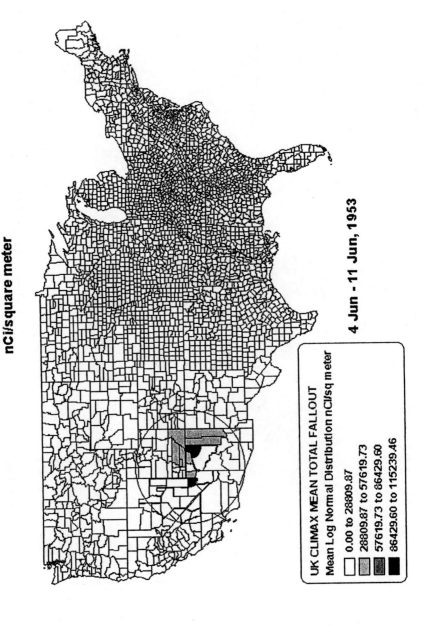

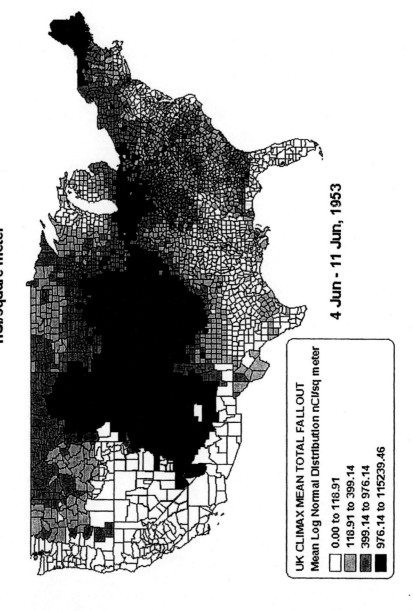

SECTION 5

TEAPOT SERIES

Feb 18–May 15, 1955

SHOT 1: TEAPOT WASP (TP1)

TEAPOT: WASP

Detonation Date: 18 Feb 1955 • *Detonation Time:* 12:00 noon • *Area:* T-7-4 • *Sponsor:* Los Alamos • *Yield:* 1 kt • *Radiation Level at Ground Zero:* 10 R/hr • *Height of Burst:* 762 ft • *Cloud Top Height:* 21,500 ft

18,000-Foot Trajectory: NEVADA: (Sheep Range), Moapa Indian Reservation, (Valley of Fire State Park), Overton • ARIZONA: (Grand Wash Cliffs), (Poverty Mountain), Havasupai Indian Reservation, (Grant Canyon), Tusayan, Cameron, Gray Mountain, Steamboat Canyon, Sunrise Springs, Cornfield, Ganado, Cross Canyon, Fort Defiance, Sawmill • NEW MEXICO: Mexican Springs, Tohatchi, Brimhall, Standing Rock, Pueblo Pintado, La Jara, Regina, Gallina, Coyote, Vallecitos, Las Tables, Tres Piedras, Cerro, Amalia • COLORADO: La Valley, Stonewall, Gulpare, Aguilar, Timpas, La Junta, Swink, North La Junta, Cheraw, Eads, Cheyenne Wells, Arapahoe • KANSAS: (Mount Sunflower, highest point in Kansas: 4,039 feet), Edson, Brewster, Achilles, Traer, Cedar Bluffs • NEBRASKA: Danbury, Cambridge, Holbrook, Elwood, Lexington, Berwyn, Almeria, Rose, Bassett, Burton.

At this point, the 16,000-foot trajectory split into two sections; one traveled east; the other west.

East Trajectory: SOUTH DAKOTA: Burke, Gregory, Oacoma, Reliance, Lower Brule, Harrold, Gettysburg, Lowry, Java, Artas • NORTH DAKOTA: Hague, Napoleon, Dawson, Lake Williams, Sykeston, Cathay, Sheyenne, Devils Lake, Webster, Hampden, Loma, Langden (into Manitoba).

West Trajectory: SOUTH DAKOTA: Wewela, Winner, Witten, Draper, Hajes, Cherry Creek, Howes, Marcus, Redowl, Stoneville, Fairpoint, (Mud Buttes) • WYOMING: Beulah, Aladdin, Sundance, Moorcroft, (Devils Tower), Rozet, Sussex, Powder River, Natrona, (Rattlesnake Hills), (Ferris Mountains), Rawlins, (Medicine Bow National Forest) • COLORADO: Columbine, Clark, Milner, Toponas, McCoy, Bond, Aron, Golman, Redcliff, Leadville, Texas Creek, Fairview, San Isabel, Green Towers, Walsonburg, Pryor, Aguilar, Ludlow, El Moro, Trinidad, (Raton Pass) • NEW MEXICO: Folsom, Des Moines, Capulin, Sofia, Mount Dora, Clapham, Stoad, Sedan • TEXAS: Perico, Dalhart, Cactus, Gruver, Spearman, Waka, Farnsworth, Perryton, Twitchell, Booker, Darrouzett • OKLAHOMA: Catesby, May, Camp Houston • KANSAS: Hardtner, Sharon, Attica, Harper, Duquoin, Milton Violla, Cionmel, Clearwater, Shulte, Wichita, Andover, Towanda, El Dorado, Thrall, Madison, Hartford, Strawn, Sharpe, Halls Summit, Waverly, Agricola, Williamsburg, Homewood, Ottawa, LeLoup, Edgerton, Wellsville, Gardner, Lenexa, Overland Park, Shawnee, Edwardsville, Kansas City • MISSOURI: Kansas City, Parkville, Gladstone, Kearney, Holt, Elmira, Mirabile, Kingston, Hamilton, Breckenridge, Lock Springs, Jamesport, Trenton, Dunlap, Galt, Osgood, Harris, Newton, Lucerne, Powersville • IOWA: Numa, Centerville, Brazil, Mystic, Rathbun, Moravia, Avery, Eddyville, Cedar, Wright, Keomah, Rose Hill, What Cheer, Thornburg, Millersburg, Williamsburg, West Ammana, Watkins, Norway, Atkins, Palo, Center Point, Waller, Quasqueton, Winthrop, Manchester, Lamart, Dundee, Strawberry Point, Osborie, Elkader, St. Olaf, Farmersburg, Freelich, Monona, Watson, Vallen, Waterville, Lansing. • WISCONSIN: De Soto, Red Mound, Victory, Genoa, Chaseburg, Coon Valley, La Crosse, Middle Ridge, Bangor, Rockland, West Salem, Cataract, Melrose, Shamrock, Black River Falls, Alma Center, Merrillan, Humbird, Willard, Thorp, Lublin, Kennan, Catawba, (Butternut Lake), Mailan, High Bridge, Marengo, Bad River Indian Reservation, Ashland, Washburn, La Pointe, Bayfield, Red Cliff, (York Island), (Bear Island) • MINNESOTA: Little Marais, (Nine Mile Lake), (Snowbank Lake), Orleans, Noyes, Humbolt, Northcote, Bowesmont • NORTH DAKOTA: Joliette, Bowesmont, Auburn, Oakwood, Minto, Warsaw, Ardoch, Johnstown, Manvel, Makinock, Holmes, Reynolds, Buxton, Mayville, Cummings, Kelso, Grandin, Gardiner • MINNESOTA: Georgetown, Moorhead, Dilworth, Lyndon, Rollag, Cormorant, Vergas, Perham, Sebeka, Nimrod, Poplar, Pequot Lakes, Breezy Point, Hassman, McGregor, Tamerack, Wright, Cromwell, Cloquet, Twig, Hermantown, Arnold, Lax Lake, Finland, Little Marais, Taconite Harbor, Schroeder, Tofte, Lutsen, Grand Marais, Grand Portage Indian Reservation.

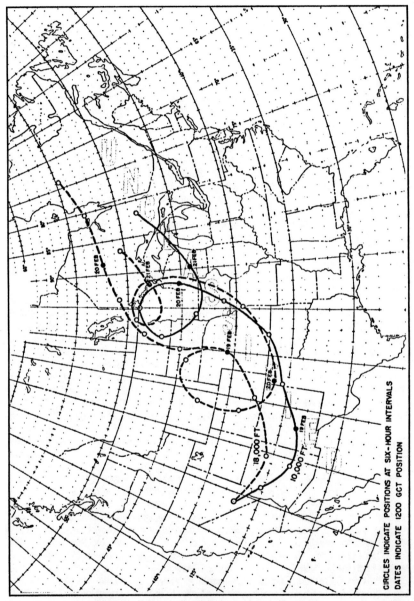

TRAJECTORY: TEAPOT WASP (TP1)

CHARACTERISTICS: TEAPOT WASP

The WASP device was relatively small, consisting of a 120 lb, 22-inch diameter sphere packed inside a bomb case. The total weight of the unit was approximately 1500 lbs. Among the nuclear tests, Teapot WASP is ranked 2nd for percentages of Pm-147 and 3rd for Tc-99m and Eu-155. The deposition pattern for Teapot WASP is most similar to Tumbler-Snapper BAKER (TS2) (R: 0.10; T: 5.86). County of highest deposition: San Bernadino, CA.

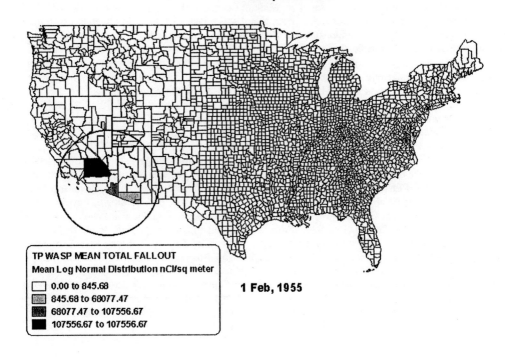

TOTAL FALLOUT: TEAPOT WASP
Equal Number
nCi/square meter

TP WASP MEAN TOTAL FALLOUT
Mean Log Normal Distribution nCi/sq meter
- ☐ 0.00 to 845.68
- 845.68 to 68077.47
- 68077.47 to 107556.67
- ■ 107556.67 to 107556.67

1 Feb, 1955

SHOT 2: TEAPOT MOTH (TP2)

TEAPOT: MOTH

Detonation Date: 22 Feb 1955 • *Detonation Time:* 5:45 A.M. • *Area:* 3 • *Sponsor:* Los Alamos • *Yield:* 2 kt • *Radiation Level at Ground Zero:* 500 R/hr • *Height of Burst:* 300 ft • *Cloud Top Height:* 24,200 ft

24,000-Foot Trajectory: NEVADA:(Spotted Range), Indian Springs, Las Vegas, Winchester, N. Las Vegas, Henderson, Boulder City • ARIZONA: Willow Beach, Dolan Springs, Chloride, Prescott, Mayer, (Mazatlan Mountains), Tonto Basin, (Rockinstraw Mountain), San Carlos Indian Reservation, Morenci, Clifton, (Macerick Hill) • NEW MEXICO: Mule Creek, Buckhorn, Kingston, Hillsboro, Caballo, (Caballo Mountains), Alamogordo, Sunspot, Sacramento, Weed, Dunken, Hope, Atoka, Loco Hills, Humble City, Hobbs • TEXAS: Seminole, Lamesa, Arvana, Mesquite (in Borden County), Gail, Fluvanna, Dermott, Rotan, Hamlin, Tuxedo, Stamford, Avoca, Woodson, South Bend, Eliasville, Graham, Bryson, Jacksboro, Wizard Wells, Chico, Alvord, Greenwood, Slidell, Pilot Point, Tioga, Dorchester, Sherman, Ambrose, Ivanhoe, Telephone, Elwood, Riverby, Direct, Belk, Chicota, Arthur City, Albion • OKLAHOMA: Ord, Frogville, Valiant, Wright City, Millerton, Garvin, Golden, Glover, Broken Bow, Eagleton, DeQueen • ARKANSAS: Gilham, New Hope, Langley, Salem, Glenwood, Bonnerdale, Hempwallace, Lake Hamilton, Magnet Cove, (Hot Springs National Park), Benton, Little Rock, Shannon, Wrightville, Scott, Lonoke, Carlisle, Des Arc, Cotton Plant, Hunter, Wynne, Colt, Gieseck, Parkin, Turrell, Frenchmans Bayou, Joiner, Pecan Point • TENNESSEE: Millington, Kerrville, Rosemark, Brownsville, Belle Eagle, Alamo, Fruitland, Gibson, Milan, Cades, Atwood, Trezevant, McKenzie, Mansfield, Maniyville, Big Sandy, Faxon, Tennessee Ridge, Erin, Shiloh, Marian, Cunningham, Fredonia, Port Royal, Cedar Hill, Springfield, Orlinda, Cross Plains, Mitchellville • KENTUCKY: Rapids, New Roe, Chapel Hill, Scottsville, Maynard, Cedar Springs, Austin, Roseville, Eighty Eight, Edmonton, Gradyville, Milltown, Casy Creek, Creston, Pricetown, Yosemite, Middleburg, King Mountain, Gum Sulphur, Conway, Morrill, Kerby Knob, Drip Rock, Old Landing, Zoe, Vortex, Malaga, Elsie, Leatha, Wheelersburg, Volga, Sitka, Nippa, Lowmansville, Clifford • WEST VIRGINIA: Radnor, Dunlow, Sias, Sumerco, Ruth, Rand, Charleston, Bomant, Maysel, Clay, Tesla, Sutton, Newville, Cleveland, Czar, Cassity, Beverly, Elkins, Red Creek, Maysville, Points, Paw Paw, Great Cacapon • PENNSYLVANIA: Lashley, Big Cove Tannery, Webster Mills, Cito, Cove Gap, LeMasters, Markes, Fort Loudon, St. Thomas, Chambersburg, Stoufferstown, Fayetteville, Wenksville, Laurel, Bendersville, Goodyear, Latimore, Clear Spring, Franklintown, Maytown, Lewisburg, Middletown, Deodate, Lawn, Mount Wilson, Mount Gretna, Iona, Richland, Mount Pleasant, Dauberville, Windsor Castle, Klinesville, Krumsville, Stony Run, Saegersville, Laurys Station, Pennsville, Bearsville, Wind Gap, Roseto, Bossandville, Slaterford, Delaware, East Stroudsburg • NEW JERSEY: Millbrook, Five Points, Myrtle Grove, Crandon Lakes, Augusta, Branchville, Pettettown, McAfee, Hamburg, Vernon • NEW YORK: Amity, Warwick, Edenville, Bellvale, Sugar Loaf, Central Valley, Arden, Harriman, West Point, Garrison Lake, Carmel, Lake Carmel, Putnam Lake • CONNECTICUT: New Fairfield, Candlewood Shores, Brookfield, Roxbury Falls, Hotchkissville, Minortown, Watertown, Plymouth, Pequabuck, Whigville, Unionville, Bloomfield, Windsor Locks, Kings Corner, Enfield, Hazardville • MASSACHUSETTS: Hampden, Monson, Brimfield, Spencer, Leicester, Worcester, Morningdale, South Berlin, Berlin, Marlborough, Hudson, Maynard, West Concord, Concord, Beford, Pinehurst, North Reading, Middleton, Topsfield, Ipswich, Little Neck, (Plum Island State Park).

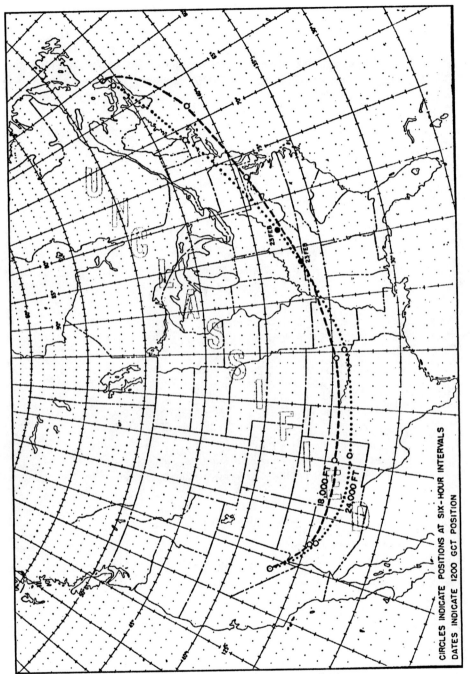

TRAJECTORY: TEAPOT MOTH (TP2)

CHARACTERISTICS: TEAPOT MOTH

The MOTH device weighed 445 lbs. Among the nuclear tests, Teapot MOTH is ranked 2nd for Cs-137 and Ba-137m and 5th for I-132, I-133, I-135, Rh-103m and Ru-103. The deposition pattern for Teapot MOTH is slightly similar to Teapot BEE and ESS (TP6), however, the Spearman R is only 0.05 with a T score of 2.53. County of highest (and only) recorded deposition: Clark, NV.

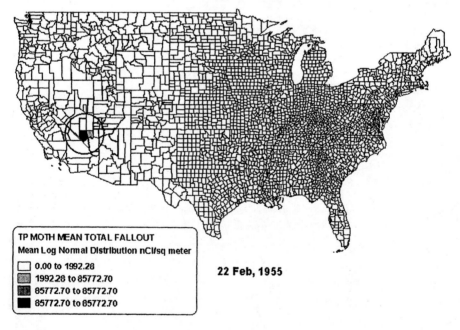

TOTAL FALLOUT: TEAPOT MOTH
Equal Number
nCi/square meter

TP MOTH MEAN TOTAL FALLOUT
Mean Log Normal Distribution nCi/sq meter
- 0.00 to 1992.28
- 1992.28 to 85772.70
- 85772.70 to 85772.70
- 85772.70 to 85772.70

22 Feb, 1955

SHOT 3: TEAPOT TESLA (TP3)

TEAPOT: TESLA

Detonation Date: 1 Mar 1955 • *Detonation Time:* 5:30 A.M. • *Area:* 9b • *Sponsor:* Livermore Laboratory • *Yield:* 7 kt • *Radiation Level at Ground Zero:* 750 R/hr • *Height of Burst:* 300 ft • *Cloud Top Height:* 30,000 ft

24,000-Foot Trajectory: NEVADA: (Spotted Range), (Sheep Range), (Morman Peak), Mesquite • ARIZONA: Littlefield, Colorado City, Cane Beds, Moccasin, Kaibab Indian Reservation, Kaibab, Fredonia, Page, (Tower Butte), (Monument Valley), Mexican Water • NEW MEXICO: Shiprock, (Shiprock Peak: 7,178 feet), Waterflow, Fruitland, Flora Vista, Aztec, Manero, Rutherton, Brazos, Eusenada, Tierra, Amarilla, (San Juan Mountains), Tres Piedras, Cerro, Questa, Red River, (Wheeler Peak, highest point in New Mexico: 13,161 feet), Maxwell, Grenville, Seneca • OKLAHOMA: Felt, Griggs, Guymon, Hardesty, Rosston, Laverne, Buffalo, Selman, Camp Houston, Capron, Burlington, Driftwood, Amorita, Byron, Wakita, Renfrow • KANSAS: Caldwell, Hunnewell, Ashton, Dexter, Oak Valley, Elk City, Sycamore, Morehead, Parsons, Strauss, McCune, Cherokee • MISSOURI: Mindenmines, Nashville, Kenoma, Golden City, Lockwood, Greenfield, Everton, Pennsboro, Ash Grove, Springfield, Williard, Walnut Grove, Glidewell, Fordland, Seymour, Mansfield, Norwood, Mountain Grove, Pine Crest, Mountain View, Birch Tree, Bartlett, Winna, Fremont, Van Buren, Chicopee, Williamsville, Hendrickson, Asherville, Wappapello, Puxico, Acornridge, Aguilla, Salcedo, Sikeston, Miner, Bertrand, Charleston, Anniston, Wyatt • KENTUCKY: Bardwell, Arlington, Fancy Farm, Mayfield, Brewers, Dexter, Fairdealing, Golden Pond, Cadiz, Caledonia, Hopkinsville, Mewstead, Pembroke, Elkton, Trenton, Daysville, Olmstead, Middleton, Franklin, Mount Aerial, Chapel Hill, Petroleum, Holland, Scottsville, Flippin, Mud Lick, Grandview, Thompkinsville, Judio, Littrell, Kettle, Mount Pisgah, Co-Operative, Pine Knot, Whitley City, Hollyhill, Mountain Ash, Gatliff, Middlesboro, Harrogate • TENNESSEE: Kyle's Ford, Okolona, Mount Carmel, Church Hill, Kingsport, Blountville, Avoca, Bluff City, Sadie, Buladeen, Shouns, Mountain City • NORTH CAROLINA: Creston, Clifton, Warrensville, Jefferson, Index, Laurel Springs, McGrady, Traphill, Austin, Elkin, Jonesville, Boonville, Smithtown, Flint Hill, Pfafftown, Enon, Lewsville, Winston-Salem, Union Cross, Jamestown, High Point, Archdale, Climax, Pleasant Gardens, Julian, Liberty, Pittsboro, Bynum, New Hill, Holly Springs, Fuquay-Varina, Willow Springs, Smithfield, Four Oaks, Goldsboro, Seymour Johnson Air Force Base, Seven Springs, Rivermont, Wise Forks, Trenton, Rhems, Pollocksville, Riverdale, Croatan, Hewelock, Newport, Morehead City, Otway, Beaufort, Parkers Island, (Shackleford Banks), (Cape Lookout).

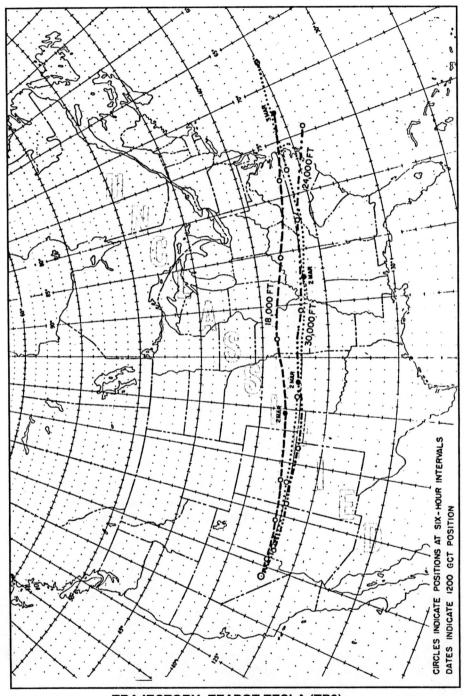

TRAJECTORY: TEAPOT TESLA (TP3)

CHARACTERISTICS: TEAPOT TESLA

The device for the TESLA shot was code-named "Cleo." Among the nuclear tests, Teapot TESLA is ranked 1st for percentages of Ba-137m, Cs-137, Rh-103m and Ru-103; 2nd for percentage of I-132 and 6th for percentage of Ag-109m. The deposition pattern for Teapot TESLA is most similar to that of Teapot TURK (TP4) (R:0.66; T: 48.46). The Spearman correlation between these two deposition patterns is among the highest in the entire series of aboveground tests.

Teapot TESLA deposited fallout on 1196 counties, equal to 39 percent of the total number of U.S. counties. Counties of highest estimated maximum activity included: Washington UT (415 µCi/sq meter estimated maximum activity) followed by the Colorado counties of Adams, Arapahoe, Boulder, Clear Creek, Douglas, Eagle, Gilpin, Grand, Jefferson and La Plata—all received a 246 µCi/sq meter fallout (estimated maximum activity) on March 2, 1955. The ratio of total fallout to I-131 for this shot is relatively low at 79 to 1.

RELATIVE TOTAL FALLOUT TP TESLA
MAR 1, 1955 7 KT

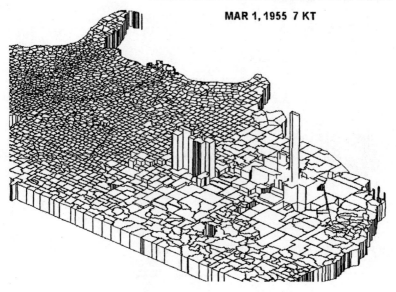

TOTAL FALLOUT: TEAPOT TESLA
Equal Intervals
nCi/square meter

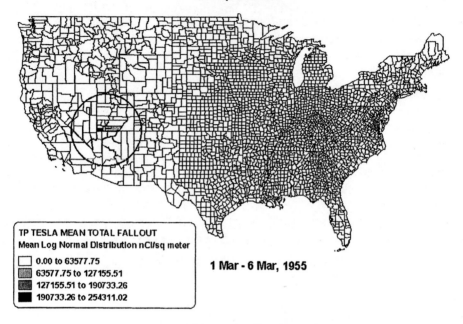

1 Mar - 6 Mar, 1955

TOTAL FALLOUT: TEAPOT TESLA
Equal Number
nCi/square meter

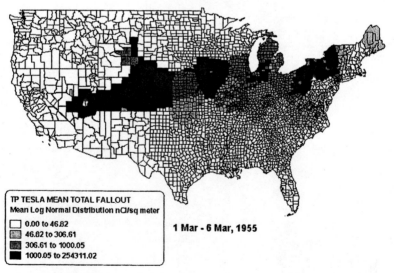

1 Mar - 6 Mar, 1955

SHOT 4: TEAPOT TURK (TP4)

TEAPOT: TURK

Detonation Date: 7 Mar 1955 • *Detonation Time:* 5:20 A.M. • *Area:* 2 • *Sponsor:* Livermore Laboratory • *Yield:* 43 kt • *Radiation Level at Ground Zero:* 1,000 R/hr • *Height of Burst:* 500 ft • *Cloud Top Height:* 44,700 ft

40,000-Foot Trajectory: NEVADA: (Mormon Peak) • UTAH: (Snow Canyon), St. George, Washington, Rockville, Kanab, (Vermillion Cliffs), Glen Canyon, Navajo Indian Reservation, (Monument Valley), Mexican Hat • COLORADO: Ute Mountain Indian Reservation, Red Mesa, Southern Ute Indian Reservation, Tiffany, Allison, Arboles, Juanita, Chrome, Ortiz • NEW MEXICO: Chama, San Miguel, Las Pinos, Amalia, Costilla, (Ortiz Peak), Capulin, Grenville, (Mount Dora), Clayton • TEXAS: Texline, Stratford, McKibben, Notla, Glazier, Canadian • OKLAHOMA: Durham, Crawford, Roll, Thomas, Geary, Calumet, El Reno, Yukon, Bethany, Oklahoma City, Midwest City, Woods, Dale, Shawnee, Little, Butner, Wetumka, Vernon, Indianola, Russellville, Quinton, Lewisville, Kinta, Wells, Wister, Poteau, Howe, Monroe • ARKANSAS: Midland, Hartford, Waldron, Harvey, Gravelly, Story, Buckville, Jessieville, Owensville, (Hot Springs National Park), Benton, Redfield, Wright, Sherrill, Althelmer, Lodge Corner, DeWitt, DeLuce, Crumrod, Snow Lake • MISSISSIPPI: Perthshire, Deeson, Shelby, Drew, Minter City, Philipp, La Flora, Duck Hill, Lodi, Tomnolen, Reform, Sturgis, Brooksville, Macon, Prairie Point • ALABAMA: Dancy, Clinton, Stewart, Akron, Havana, Helberger, Sprott, Plantersville, Winslow, Autaugaville, Hanter, Montgomery, Pile Road, Mathews, Fitzpatrick, Enon, Three Notch, Springhill • GEORGIA: Troutman, Westran, Bronwood, Oakfield, Ashburn, Sycamore, Mystic, Lax, Upton, Douglas, Bickley, Dixie Union, Blackshear, Nahonta, Atkinson, Brunswick, (St. Simons Island).

CHARACTERISTICS: TEAPOT TURK

The device for shot TURK, code-named "Linda," weighed 2325 lbs and was 30.5 inches by 61.3 inches. Among the nuclear tests, Teapot TURK is ranked 6th for percentage of Br-82 in fallout and 8th for percentage of Be-7 in fallout. The deposition pattern for Teapot TURK is most similar to Teapot TESLA (R: 0.66; T: 48.46). Fallout from TURK was deposited over 3008 counties. The county of highest deposition: Inyo, CA with an estimated maximum activity of 1043 µCi/sq meter total fallout. Other areas receiving high fallout were the Colorado counties of Adams, Arapahoe, Boulder, Clear Creek, Douglas, Eagle, Gilpin, Grand, Jefferson, and LaPlata. All received 922 µCi/sq meter (estimated maximum activity) on March 8, 1955.

The total fallout to I-131 ratio for Teapot TURK was relatively low at 79 to 1.

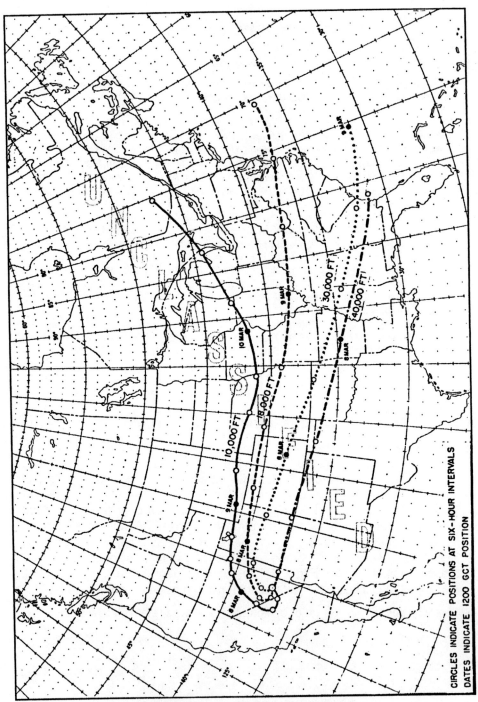

TRAJECTORY: TEAPOT TURK (TP4)

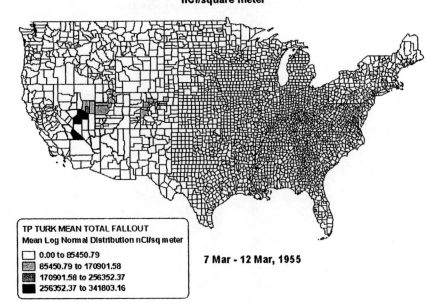

TOTAL FALLOUT: TEAPOT TURK
Equal Intervals
nCi/square meter

TP TURK MEAN TOTAL FALLOUT
Mean Log Normal Distribution nCi/sq meter
- 0.00 to 85450.79
- 85450.79 to 170901.58
- 170901.58 to 256352.37
- 256352.37 to 341803.16

7 Mar - 12 Mar, 1955

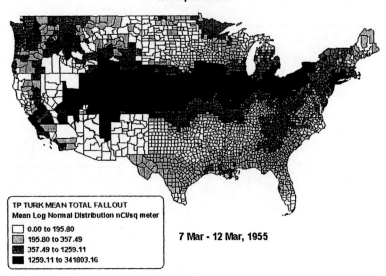

TOTAL FALLOUT: TEAPOT TURK
Equal Number
nCi/square meter

TP TURK MEAN TOTAL FALLOUT
Mean Log Normal Distribution nCi/sq meter
- 0.00 to 195.80
- 195.80 to 357.49
- 357.49 to 1259.11
- 1259.11 to 341803.16

7 Mar - 12 Mar, 1955

RELATIVE TOTAL FALLOUT TP TURK
MAR 7, 1955 43 KT

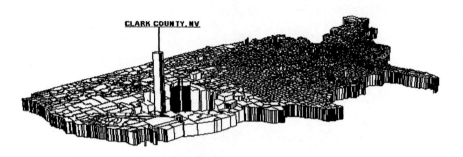

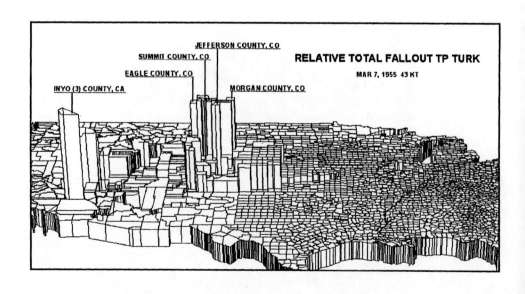

U.S. FALLOUT ATLAS : TOTAL FALLOUT

SHOT 5: TEAPOT HORNET (TP5)

TEAPOT: HORNET

Detonation Date: 12 Mar 1955 • *Detonation Time:* 5:20 A.M. • *Area:* 3a • *Sponsor:* Los Alamos • *Yield:* 4 kt • *Radiation Level at Ground Zero:* 2,500 R/hr • *Height of Burst:* 300 ft • *Cloud Top Height:* 37,000 ft

40,000-Foot Trajectory: NEVADA: (Papoose Range), Carp • UTAH: (Snow Canyon), Gunlock, Shivwits, Hurricane, La Verkin, Pintura, (Zion National Park), Glendale, Orderville, Mount Carmel Junction, (Valley of the Gods) • COLORADO: Squaw Point, Pleasant View, Silverton, Olney Springs, Numa, Ordway, Sugar City, Chivington, Sheridan Lake, Towner • KANSAS: Horace, Tribune, Selkirk, Lepti, Marienthal, Modor, Scott City, Manning, Grigston, Dighton, Shields, Utica, Ransom, Brownell, McCracken, Liebenthal, Loretta, Galatia, Susank, Beaver, Dubuque, Ellsworth, Carneiro, Brookville, Salina, Kipp, Abilene, Enterprise, Pear, Skiddy, Alta Vista, Hessdale, Keene, Topeka, Pauline, Big Spring, Williamstown, McLouth, Fairmount, Kansas City • MISSOURI: Kansas City, Raytown, Independence, Blue Springs, Wellington, Mayview, Higginsville, Alma, Mount Leonard, Marshall, Malta Bend, Slater, Forest Green, Roanoke, Mount Airy, Moberly, Middle Grove, Paris, Perry, Center, New London, Saverton • ILLINOIS: New Canton, New Salem, Valley City, Chapin, Literberry, Prentice, Pleasant Plains, Salisbury, Cantrall, Williamsville, Mount Pulaski, Chestnut, Kenny, Lane, Clinton, Weldon, De Laud, Mansfield, Fisher, Dewey, Rantoul, Ludlow, Clarence, East Lynn, Hoopeston • INDIANA: Ambra, Talbot, Fowler, Reynolds, Norway, Sitka, Royal Center, Lucerne, Metea, Perrysburg, Twelve Mile, Denver, Chili, Roann, Urbana, Andrews, Huntington, Zanesville, Ossian, Poe, Hoagland, Williams, Dixon, Monmouth • OHIO: Dixon, Cavett, Scott, Kalida, Pandora, Benton Ridge, Vanlue, Carey, Tymochtee, Belle Vernon, Sycamore, Benton, Chatfield, Tiro, Ganges, Epworth, Paradise Hill, Ashland, England Station, Rowsburg, New Pittsburg, Wooster, Orrville, Massillon, Perry Heights, Canton, Fairhope, Paris, Bayard, Hanoverton, West Point, Frederickstown • PENNSYLVANIA: Blackhawk, Fallston, East Rochester, Fernway, Wexford, Dorseyville, Oakmart, Plum, Sardis, Export, Delmont, Crabtree, New Derry, Derry, Wilpen, Tire Hill, Paint, Scalp Level, Ogletown, Weyant, Imber, New Enterprise, Riddlesburg, Kearney, Enid, Gracey, Clear Ridge, Fort Littleton, Fennettsburg, Metal, Creenvillage, Arendtsville, Table Rock, New Chester, New Oxford, Abbotstown, Spring Grove, Stoverstown, Rye, Felton, Pleasant Grove, Kyleville, Little Britain, Oxford, Strickersville • DELAWARE: Newark, Christiana, Bear, Red Lion, Delaware City • NEW JERSEY: New Castle, Harrisonville, Pointers, Salem, Renton, Alloway, Aldine, Centerton, Norma, Vineland, East Vineland, Milmay, Dorothy, Bargaintown, West Atlantic City, Atlantic City.

CHARACTERISTICS: TEAPOT HORNET

The HORNET device weighed 500 lbs and was detonated atop a 300-foot tower at the Nevada Test Site Area 2. Among the nuclear tests, Teapot HORNET is ranked 6th in percentages of Cd-115, In-115m and In-117 in total fallout. The deposition pattern for Teapot HORNET is most similar to that of Teapot APPLE-2 (TP10) (R: 0.43; T: 26.26). Fallout from Teapot HORNET was deposited on 2926 U.S. counties. County of highest deposition: Clark, NV.

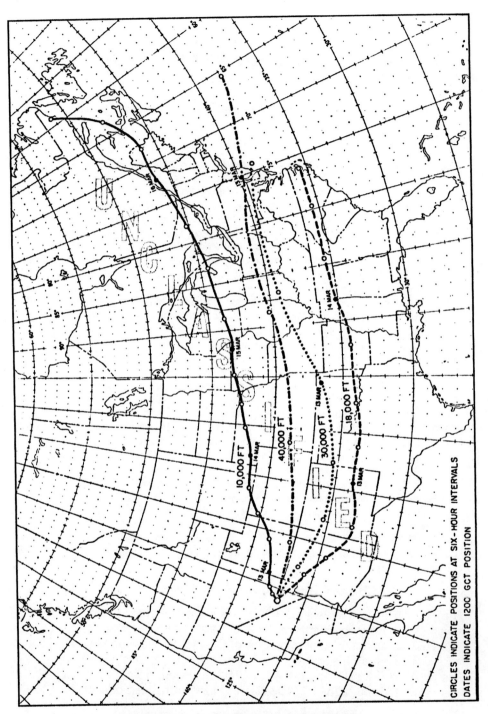

TRAJECTORY: TEAPOT HORNET (TP5)

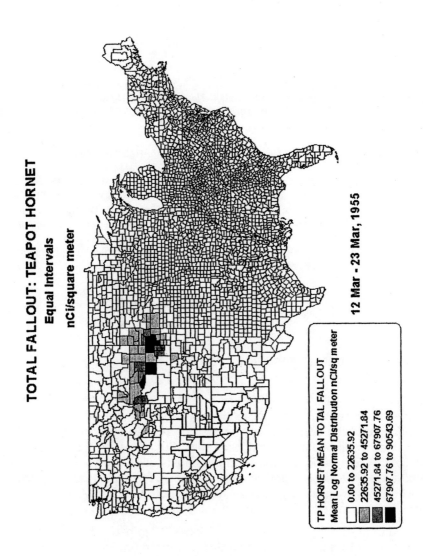

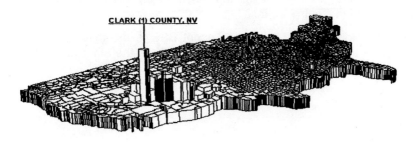

U.S. FALLOUT ATLAS : TOTAL FALLOUT

SHOT 6a: TEAPOT BEE (TP6A)

TEAPOT: BEE

Detonation Date: 22 Mar 1955 • *Detonation Time:* 5:05 A.M. • *Area:* 7-1a • *Sponsor:* Los Alamos • *Yield:* 8 kt • *Radiation Level at Ground Zero:* 2,000 R/hr • *Height of Burst:* 500 ft • *Cloud Top Height:* 39,700 ft

18,000-Foot Trajectory: NEVADA: Cottonwood Cove • ARIZONA: Bullhead City, Yucca, Congress, Wickenburg, Sun City, Phoenix, Scottsdale, Mesa, Mammoth, San Manuel, Cochise, Portal • NEW MEXICO: Rodeo, (Alamo Hueco Mountains) • TEXAS: Alpine, Sanderson, Dryden, Loma Alta, Campwood, Hondo, San Antonio, Union Valley, Shiner, Rock Island, Eagle Lake, Houston, Crosby, Shiloh, Devers, Beaumont, Orange • LOUISIANA: Vinton, Sulphur, Lake Charles, Church Point, Grand Coteau, Baton Rouge, Livingston, Hammond, Covington • MISSISSIPPI: Picayune, Caesar, Necaise, Wade, Nurley • ALABAMA: Mobile, Malbals, Robertsdale, Lillian • FLORIDA: Pensacola, Panama City, Highland Park, Chipola Park, Tallahassee, Waukeena, Lamont, Taylor, Jacksonville.

CHARACTERISTICS: TEAPOT BEE AND ESS

Though combined by the National Cancer Institute in the 1997 I-131 study, Teapot BEE and ESS are remarkably different shots, both in radionuclide content and in total fallout-to-I-131 ratios. The nuclear component of the BEE device consisted of a small sphere approximately 17 inches in diameter and weighing only 130 lbs. The ESS device, by contrast, was a 4-ton, six foot-long device placed at the bottom of a steel shaft drilled 70 feet beneath the desert floor. The BEE test undoubtedly resulted in significant dispersal of radioactive iron fume, while the fallout from ESS contained materials found in the desert alluvium.

Among the nuclear tests, Teapot BEE ranks 2nd in percentages of Pb-203 and 6th in I-133 and I-135. Teapot ESS is ranked first in Np-239, 3rd in U-237 and 7th in Na-24. The deposition pattern for the two combined shots is most similar to that of Plumbbob BOLTZMANN (PB1) ((R: 0.41; T: 25.16). County of highest total deposition: Clark, NV. Significant fallout also occurred in Central Texas.

The total fallout to I-131 ratio for Teapot BEE is the third lowest in the aboveground test series at 73 to 1. Shot ESS, on the other hand, has the third **highest** total fallout to I-131 ratio at 199 to 1.

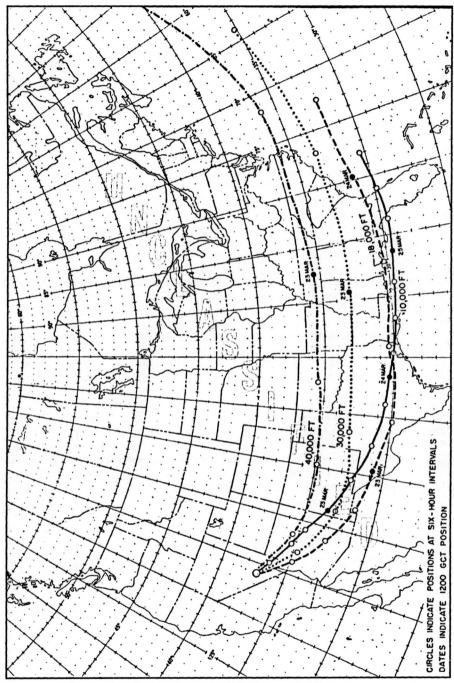

TRAJECTORY: TEAPOT BEE (TP6A)

SHOT 6B: TEAPOT ESS (TP6B)

TEAPOT: ESS

Detonation Date: 23 Mar 1955 • *Detonation Time:* 12:30 PM. • *Area:* T-10a • *Sponsor:* Los Alamos, Department of Defense • *Yield:* 1 kt • *Radiation Level at Ground Zero:* 6,000 R/hr • *Height of Burst:* −67 ft • *Cloud Top Height:* 12,000 ft

10,000-Foot Trajectory: NEVADA: (Pintwater Range), Mopa, Logandale, Overton, (Virgin Peak) • ARIZONA: (Virgin Mountains), (Mount Trumbull), Supai, (Grand Canyon), Tusayan, Cameron, Gray Mountain, (Painted Desert), (Stephen Butte), White Cone, Steamboat Canyon, Cornfields, Sunrise Springs, Ganada, (Cross Canyon), Fort Defiance, St. Michaels • NEW MEXICO: Fence Lake, Trechado, (Gallinas Mountains), Alamo Navajo Indian Reservation, Polvadera, Lemitar, Florida, (Los Pinos Mountains), (Chupadera Mesa), Claunch, Ancho, (Arroyo Del Macho), (Haystack Mountain), Elkins, Milnesand • TEXAS: Griffith, Lehman, Morton, Pettit, Levelland, Lubbock, Roosevelt, Posey, Cap Rock, Canyon Valley, Kalgary, Jayton, Peacock, Swenson, Old Glory, Aspermont, Rule, Haskell, Woodson, Eliasville, Graford, Mineral Wells, Weatherford, Fort Worth, Grand Prairie, Arlington, Irving, Dallas, Mesquite, Balch Springs, Lawrence, Wills Point, Edgewood, Alba, Quitman, Golden, Forest Hill, Hainesville, Rhonesboro, Grice, Thomas, LaFayett, Lone Star, Daingerfield, Red Hill, Atlanta, Queen City, Cass • ARKANSAS: Brightstar, Fort Lynn, Fouke, Lewisville, Falcon, Rosston, Waterloo, Childester, Amy, Holly Springs, Ramsey, Ivan, Staves, Randall, Ladd, Moscow, Bayou, Lodge Corner, DeWitt, Ethel, St. Charles, Lambrook, Lake View, Wabash, West Helena, Friars Point • MISSISSIPPI: Hill House, Rena Lara, Sherard Baugh, Farrell, Rudyard, Jonestown, Darling, Curtis Station, Pleasant Grove, Sardis, Harmontown, Looxahoma, Tyro, Thyatyra, Waterford, Lows Hill, Potts Camp, Winborn, Ashland, Peoples, Falkner, Chalybeate, Kossuth, Wenasoga, Corinth, Farmington, Kendrick • TENNESSEE: Michie, Acton, Childers Hill, Nixon, Burnt Church, Houston, Center, Crossroads, Ethridge, Webber City, Waco, Mooresville, Lewisburg, Shelbyville, Manchester, Ragsdale, Viola, Irving College, (Cumberland Plateau), Hitchcox, Mount Crest, Cold Spring, (Walden Ridge), Grandview, Roddy, Rockwood, Kingston, Solway, Ball Camp, Knoxville, Mascot, Blaine, Jefferson City, Shiloh, Talbott, Russellville, Whitesburg, St. Clair, Baileytown, McCloud, Fall Branch, Sullivan Gardens, Colonial Heights, Blountsville, Bristol • VIRGINIA: Bristol, Oscoala, Lodi, Adwolf, Attoway, Crockett, Wytheville, Draper, Graysontown, Christianburg, Alleghany Spring, Bent Mountain, Roanoke, Stewartsville, Bedford, New London, Lynchburg, Spout Spring, Appomattox, Guinea Mills, Macon, Richmond, Old Church, King William, Stevensville, Morattico, Burgess, Fair Port, Lilian, Reedville, (Smith Point).

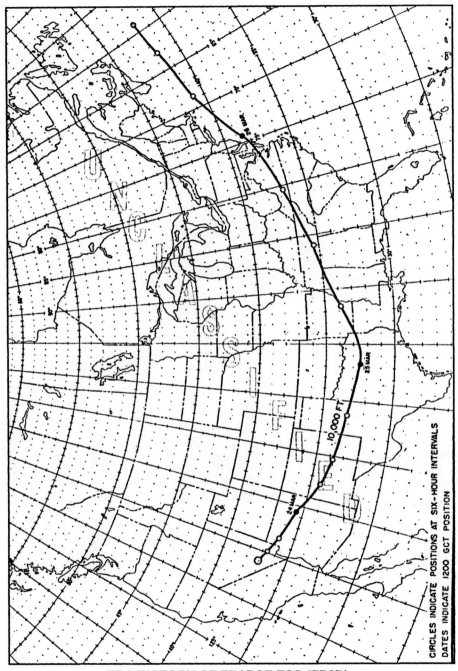

TRAJECTORY OF TEAPOT ESS (TP6B)

U.S. FALLOUT ATLAS : TOTAL FALLOUT

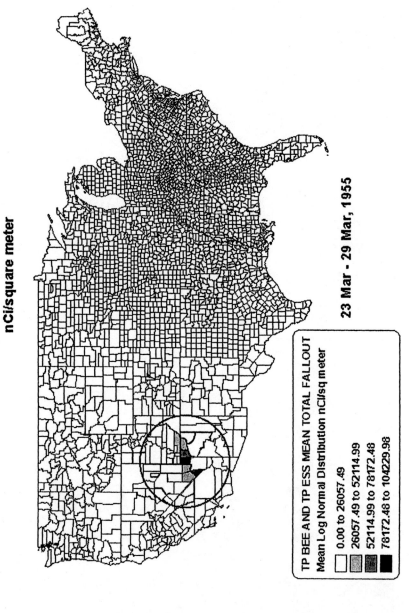

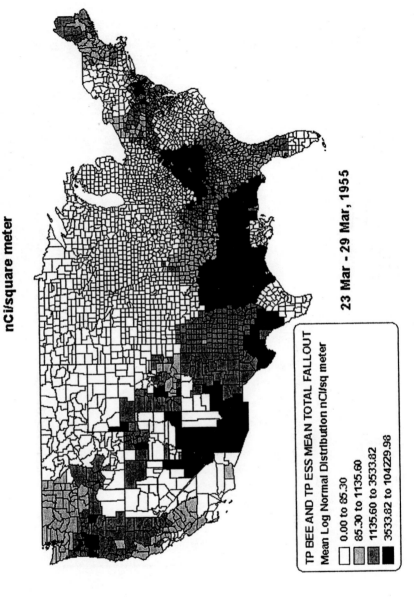

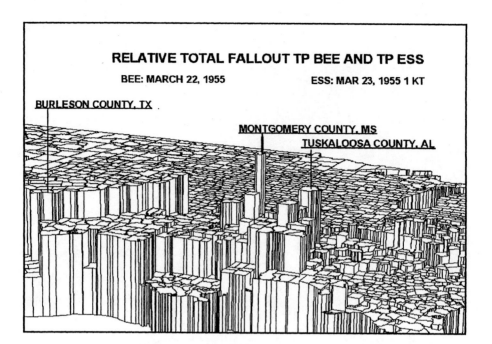

SHOT 7: TEAPOT APPLE-1 (TP7)

TEAPOT: APPLE-1

Detonation Date: 29 Mar 1955 • *Detonation Time:* 4:55 A.M. • *Area:* 4 • *Sponsor:* Los Alamos • *Yield:* 14 kt • *Radiation Level at Ground Zero:* 500 R/hr • *Height of Burst:* 500 ft • *Cloud Top Height:* 32,000 ft

10,000-Foot Trajectory: NEVADA: (Papoose Range), (Groom Lake), (Timpahute Range), Lake Valley, Baker • UTAH: Garrison, Grantsville, Salt Lake City, (Antelope Island), (Castle Rock), Tremonton, Newton, Plymouth, Cornish • IDAHO: Snowville, Woodruff, Samaria, Malad City, Virginia, McCammon, Swan Valley, Driggs, Felt, Tetonia • WYOMING: (Mount Moran), (Mount Gancock), (Yellowstone Park: Heart Lake, Yellowstone Lake), Pahaska, (Windy Mountain), (Bear Tooth Pass) • MONTANA: Red Lodge, Washoe, Bearcreek, Bridger, Pryor, Crow Indian Reservation, Hardin, Hathaway, Miles City, Locate, Ismay, Ollie • NORTH DAKOTA: Galva, (Tracy Mountain), Southeart, Dickinson, Marshall, Dodge, Halliday, Twin Buttes, White Shield, Roseglen, Ryder, Douglas, Sawyer, Logan, Surrey, Minot, Norwich, Granville, Towner, Berwick, Barton, Mylo, Agate, Porth, Rocklake, Calvin, Hannah.

CHARACTERISTICS: TEAPOT APPLE-1

The nuclear component of the APPLE-1 device was 29.5 inches wide, 74.6 inches long and weighed 2300 lbs. Among the nuclear tests, Teapot APPLE-1 is ranked 6th in percentage of U-240 in fallout. The deposition pattern for Teapot APPLE-1 is most similar to that of Teapot ZUCCHINI (TP11) (R: 0.43, T: 26.37). Fallout from Teapot APPLE 1 was deposited on 3062 counties.

Counties of highest deposition (all figures in µCi/sq meter, estimated maximum activity): Texas, OK—940.7; Stevens, KS—940.7; Oldam, TX, Potter, TX—940.7; Lincoln, NV—388.9; all on Mar 29, 1955.

With regard to radionuclide content, Teapot APPLE-1 appears to represent a "standard" atomic test. In deposition of radionuclides, Teapot APPLE 1's rank among all aboveground tests is as follows:

Co-60—26; Nb-95m—30; Nb-95—28; Cs-137—28; Be-7—25; Mn-54—28; Pr-144—29; Ru-106—30; Te-127m—29; Sr-90—24.

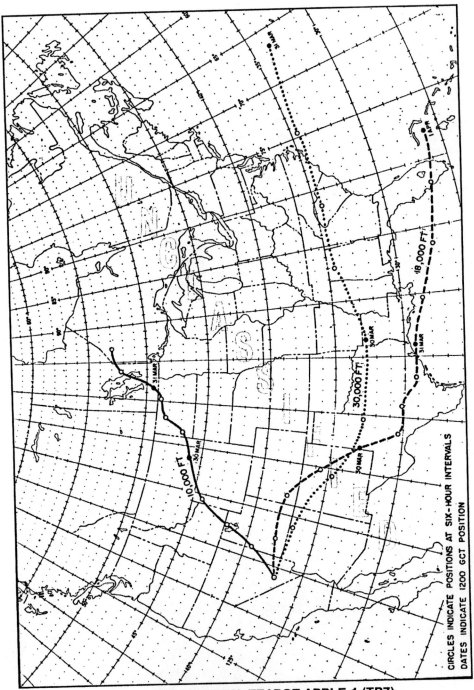

TRAJECTORY: TEAPOT APPLE-1 (TP7)

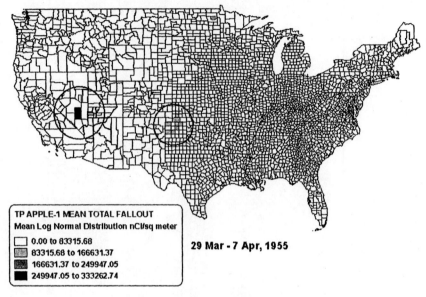

TOTAL FALLOUT: TEAPOT APPLE-1
Equal Intervals
nCi/square meter

TP APPLE-1 MEAN TOTAL FALLOUT
Mean Log Normal Distribution nCi/sq meter
- 0.00 to 83315.68
- 83315.68 to 166631.37
- 166631.37 to 249947.05
- 249947.05 to 333262.74

29 Mar - 7 Apr, 1955

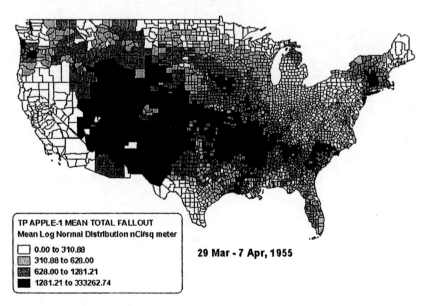

TOTAL FALLOUT: TEAPOT APPLE-1
Equal Number
nCi/square meter

TP APPLE-1 MEAN TOTAL FALLOUT
Mean Log Normal Distribution nCi/sq meter
- 0.00 to 310.88
- 310.88 to 628.00
- 628.00 to 1281.21
- 1281.21 to 333262.74

29 Mar - 7 Apr, 1955

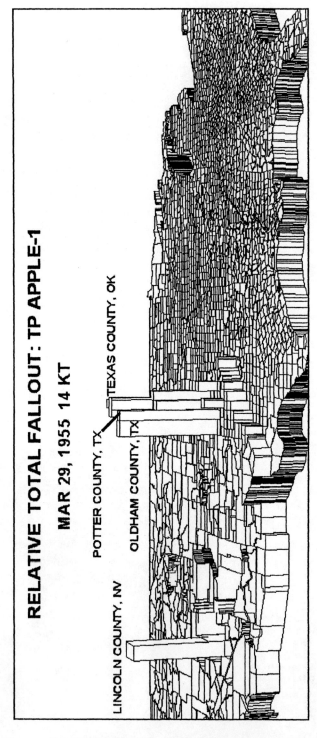

SHOT TEAPOT WASP PRIME
(Not Included in NCI Study)
TEAPOT: WASP PRIME

Detonation Date: 29 Mar 1955 • *Detonation Time:* 10:00 A.M. • *Area:* T-7-4 • *Sponsor:* Los Alamos • *Yield:* 3 kt • *Radiation Level at Ground Zero:* 10 R/hr • *Height of Burst:* 739 ft • *Cloud Top Height:* 32,000 ft

18,000-Foot Trajectory: NEVADA: (Spotted Range), (Pintwater Range), Elgin • UTAH: Enterprise, Newcastle, Cedar City, Enoch, Summit, Parowan, Paragonah, Circleville, Junction, Greenwich, Fremont, (Thousand Lake Mountain), (San Rafael Knob), Willowblind, Uintah and Quray Indian Reservation, Bonanza • COLORADO: Dinosaur, Blue Mountain • WYOMING: Baggs, (Medicine Bow National Forest), Saratoga, (Medicine Bow Peak), Bosler, Iron Mountain, Meriden, Albin • NEBRASKA: Potter, Sidney, Colton, Sunol, Lodgepole, Chappell, Big Springs, Brule, Ogallala, Keystone, Dunning, Almeria, Taylor, Elyria, Ord, North Loup, Scotia, Cotesfield, Elba, Cushing, Palmer, Archer, Central City, Hordville, Marquette, Polk, Arborville, Benedict, Bradshaw, York, Waco, Utica, Beaver Crossing, Cordova, Crete, Wilber, Kramer, Clatonia, Pickrell, Crab Orchard, Tecumseh, Elk Creek, Howe, Stella, Shubert, Barada • MISSOURI: Corning, Craig, Maitland, Graham, Barnard, Guilford, Skidmore, Arkoe, Conception Junction, Ravenwood, Parnell, Grant City, Sheridan • IOWA: Redding, Delphos, Benton, Mount Ayr, Tingley, Arispe, Talmadge, Lorimore, Barney, East Peru, Hanley, Winterset, Patterson, Des Moines, Bonderant, Clyde, Rhodes, Melbourne, Van Cleve, Haverhill, Marshalltown, Le Grand, Clutier, Elberon, Van Horne, Newhall, Atkins, Covington, Cedar Rapids, Mount Vernon, Tipton, Bennett, New Liberty, Sunbury, Maysville, Walcott, Davenport, Bettendorf • ILLINOIS: Moline, Oak Grove, Arion, Andove, Nekoma, Bishop Hill, Galva, LaFayette, Speer, Edelstein, Chillicothe, Rome, Metamora, Eureka, Hudson, Normal, Holder, Bellflower, Blue Ridge, Urbana, Champaign, Philo, Sidney, Allerton, Sidell, Ridge Farm, Scotland • INDIANA: Cayuga, Lodi, Newport, Annapolis, Marshall, Milligan, Morton, Clinton Falls, Brick Chapel, Fillmore, Stilesville, Hazelwood, Mooresville, Smith Valley, Stones Crossing, Greenwood, Whiteland, Rocklane, Boggstown, Needham, Shelbyville, Waldron, St. Paul, Sandusky, Oldenburg, St. Peters, South Gate, St. Leon, New Trenton, West Harrison, Harrison • OHIO: Harrison, New Baltimore, Miamitown, Groesbeck, Monfort Heights, Brentwood, Deer Park, Montgomery, Madeira, Cincinnati, Norwood, Milford, Owensville, Crosstown, Greenbush, Bulford, Gath, Mowrystown, May Hill, Locust Grove, Mount Joy, Sedan, Fire Brick, Eifort, Blackfork, Gallia, Rodney, Gallipolis • WEST VIRGINIA: Point Pleasant, Mount Alto, Cottageville, Liverpool, Reedy, Mount Zion, Sand Ridge, Copen, Cedarville, Napier, Arlington, Rock Cave, Alton, Cassity, Valley Bend, (Spruce Knob, highest point in West Virginia), Judy Gap, Fort Seybert • VIRGINIA: Broadway, Tenth Legion, Alma, Stanley, Brightwood, Leon, Burr Hill, Todds Tavern, Massaponax, Guinea, Woodford, Sparta, Beulaville, Mangohick, Aglett, Walkerton, King William, Port Richmond, West Point, Barhamsville, Williamsburg, Fort Eustis, Lawson, Rescue, Newport News, Norfolk, Chesapeake, Virginia Beach.

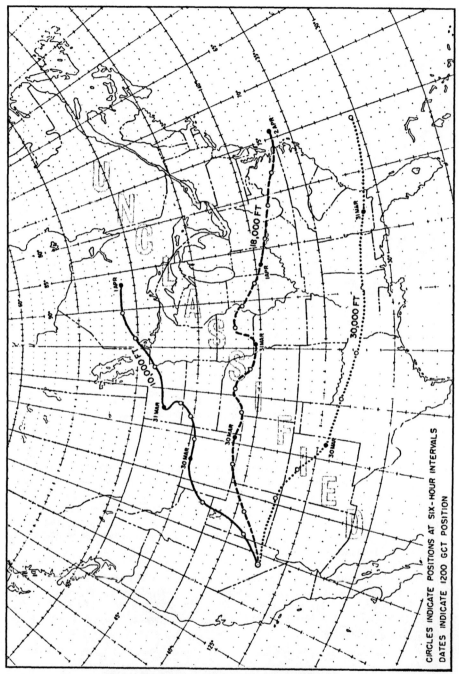

TRAJECTORY: TEAPOT WASP PRIME

SHOT TEAPOT HA (HIGH ALTITUDE)
(Not Included in the NCI Study)
Teapot: HA

Detonation Date: 6 Apr 1955 • *Detonation Time:* 10:00 A.M. • *Area:* T-5 • *Sponsor:* Department of Defense • *Yield:* 3 kt • *Radiation Level at Ground Zero:* None detected • *Height of Burst:* 32,582 +/- 100 ft • *Cloud Top Height:* 55,000 ft

45,000-Foot Trajectory: NEVADA: (Sheep Range), Mesquite, Bunkerville • ARIZONA: Littlefield, (Mount Bangs), (Kaibab Plateau), (Marble Canyon), Cedar Ridge, (Wildcat Peak), Tonalea, Cow Springs, Chilchinbito, Rough Rock, Many Farms, Tsaile • NEW MEXICO: Sanostee, Toadlena, Newcomb, Burnham, Blanco, Nageezi, Ojito, Gavilan, Llaves, El Rito, Vallecitos, Petaca, La Madera, Taos, Pueblo, Talpa, Arroyo Seca, Angel Fire, (Aqua Fria), (Christo Range), (Philmont Boy Scout Ranch), Cimarron, Miami, Springer, Maxwell, (Laughlin Peak), Grenville, Sierra Grande, Moses • OKLAHOMA: Wheeless, Boise City, Keyes, Tyrone, Lookout, Manchester • KANSAS: Elkard, Liberal, Hardtner, Kiowa, Caldwell, Arkansas City, Silverdale, Maple City, Hewins, Elgin, Chautauqua, Caney, Tyro, Coffeyville, Edna, Chetopa, Melrose, Baxter Springs, Treece, Lowell • MISSOURI: Joplin, Fidelity, Sarcoxie, Wentworth, Pierce City, Freistatt, Verona, Marionville, Aurora, Crane, Hurley, Jamesville, Union City, Highlandville, Chadwick, Garrison, Ava, Squires, Gentryville, Drury, Vanzant, (Cedar Knobs), Pomona, Olden, White Church, Peace Valley, Thomasville, Wilderness, Handy, Eastwood, Grandlin, Poplar Bluff, Ash Hill, Fisk, Dudley, Dexter, Essex, Matthews, Whiting, E. Prairie, Wolf Island • KENTUCKY: Hickman, Moscow, Oakton, Cayce, Crutchfield, Water Valley, Cuba, Tri City, Lynn Grove, Murray, Hamlin, La Fayette, Oak Grove, Henleytown, Tiny Town, Keysburg, Adairville, Prices Mill, Providence, Rapids, New Roe, Adolphus, Akersville, Gamaliel, Hestand, Vernan, Littrell, Albany, Static, Sunnybrook, Mount Pisgah, Co-Operative, Pine Knot, Strunk, Saxton, Pearl, Pruden, Fonde, Edgewood, Middleboro, (Cumberland Gap), Weber City, Morrison City • TENNESSEE: Kingsport, Lynn Garden, Bloomingdale, Orebank, Bristol, Sutherland, Laurel Bloomery • NORTH CAROLINA: Lansing, Grassy Creek, Piney Creek, Twin Oaks, Sparta, Lowgap, Bottom, Mount Airy, Bannertown, Francisco, Lawsonville, Sandy Ridge, Stoneville, Eden, Ruffin, Pelham, Yancyville, Concord, Leasburg, Gentry's Store, Allensville, Stovall, Williamsboro, Middleburg, Manson, Norlina, Macon, Warrenton, Vaughan, Littleton, Sunny Side, Roanoke Rapids, Weldon, Jackson, Conway, Petecasi, Murfreesboro, Winston, Union, Tunis, Cofield, Gatesville, Sandy Cross, Trotville, Hobbsville, Morgan's Corner, Shawboro, Gregory, Belcross, Camden, Currituck, Maple, Barco, Waterlily, Aydlett, Carolla.

CHARACTERISTICS: TEAPOT HA

The HA device was airdropped by a B36 from an altitude of over 32,000 ft and allowed to descend via parachute to the detonation height of 739 ft. The device, including bomb casing, weighted 1085 lbs.

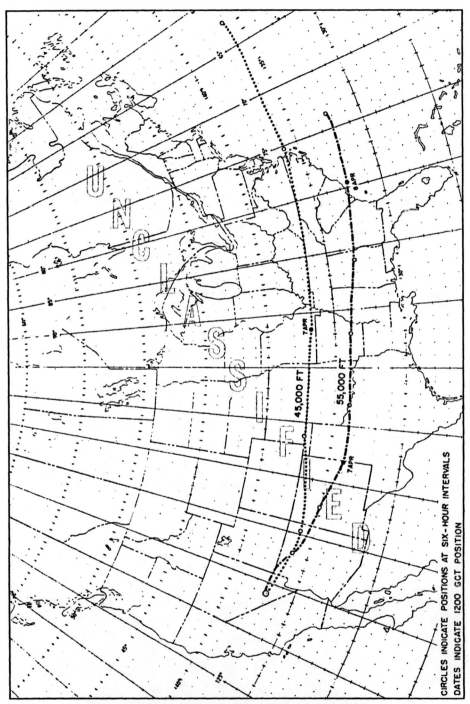

TRAJECTORY: TEAPOT HA (HIGH ALTITUDE)

SHOT 8: TEAPOT POST (TP8)

Teapot: Post

Detonation Date: 9 Apr 1955 • *Detonation Time:* 4:30 A.M. • *Area:* 9c • *Sponsor:* Livermore Laboratory • *Yield:* 2 kt • *Radiation Level at Ground Zero:* 1,000 R/hr • *Height of Burst:* 300 ft • *Cloud Top Height:* 15,500 ft

14,000-Foot Trajectory: NEVADA: Sandy, Goodsprings, Jean, Searchlight, Laughlin • ARIZONA: Kingman, Drake, Flagstaff, Winona, White Cone, Steamboat Canyon, Nazlini • NEW MEXICO: Crystal, Naschitti, Brazos, Ensenada, Tierra Amarilla, Cerro, Questa, Red River, Capulin, Moses • OKLAHOMA: Boise City, Keyes, Eva, Optima, Adams, Beaver, Laverne, Freedom, Cherokee, Vining, Medford • KANSAS: Arkansas City, Cambridge, Grenola, Moline, Fredonia, Thayer, St. Paul, Girard, Frontenac • MISSOURI: Lamar, Dadeville, Eudora, Elkland, Grovespring, Success, Oscar, Centerville, Lesterville, Cobalt City, Patton, Sedgewickville, Daisy, New Wells, Pocahontas • ILLINOIS: Alto Pass, Carbondale, Carterville, Energy, Johnston City, Akin, Diamond City, Springerton, Burnt Prairie, Bellmont, St. Francisville • INDIANA: Vincennes, Bicknell, Westphalia, Bloomfield, Bloomington, Unionville, Helmsburg, Spearsville, Marietta, Shelbyville, Rushville, Harrisburg, Richmond • OHIO: Gettysburg, Verona, Nashville, Tipp City, North Hampton, Springfield, Catawba, Tradersville, Columbus, Whitehall, Jacksontown, Gratiot, Hopewell, Zanesville, Opperman, Mount Ephraim, Calais, Ozark, Clarington • WEST VIRGINIA: Wileyville, Fairmont, Newburg, Arthur, Rig, Lost City • VIRGINIA: Edinburg, Mount Jackson, Sperryville, Culpepper, Fredericksburg, Woodford, Bowling Green, Millers Tavern, Dutton.

CHARACTERISTICS: TEAPOT POST

The nuclear device used for shot POST was 34.2 inches long and weighed 322 lbs. Among the nuclear tests, Teapot POST is ranked 1st in percentage of Rh-106, Ru-106, Te-127 and Te-127m and ranked 2nd for percentage of Ag-109m, Pd-111m and Sb-126 in fallout. The deposition pattern for Teapot POST is most similar to Plumbbob GALILEO (PB12) (R: 0.39; T: 23.35). Fallout from this shot was deposited over 2914 counties (94 percent of the total counties). Counties of highest deposition: Tooele, UT; Wasatch, UT; Morgan, UT; White Pine, NV and Salt Lake, UT. All received less than 100 µCi/sq meter of total fallout (estimated maximum activity).

The total fallout to I-131 ratio for this shot was one of the lowest of the aboveground test program at 72 to 1. Only Plumbbob KEPLER had a lower total fallout-to –I-131 ratio at 71 to 1.

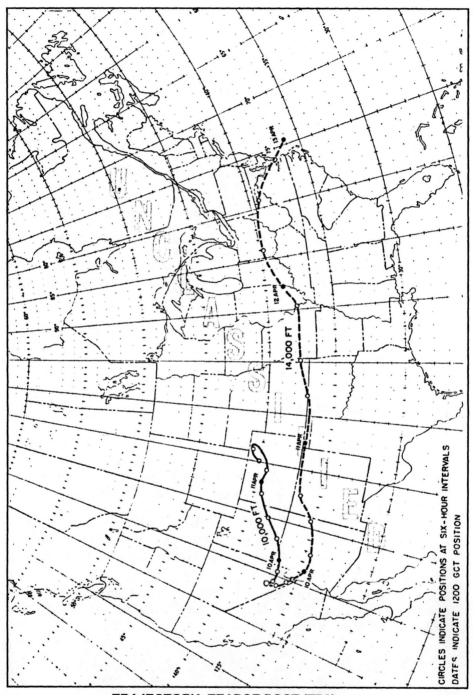

TRAJECTORY: TEAPOT POST (TP8)

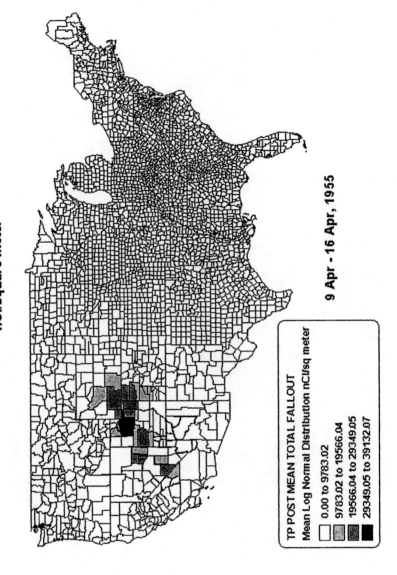

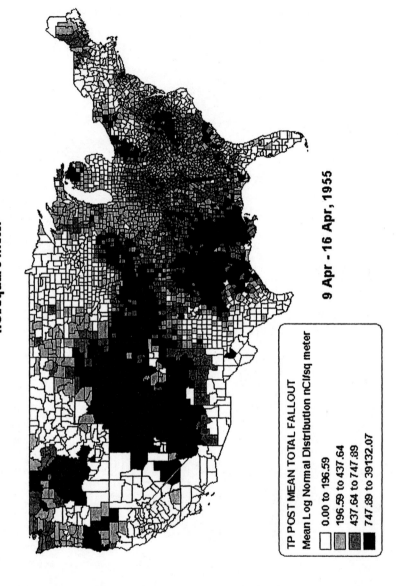

SHOT 9: TEAPOT MET (TP9)

TEAPOT: MET

Detonation Date: 15 Apr 1955 • *Detonation Time:* 11:15 A.M. • *Area:* Frenchman's Flat • *Sponsor:* Los Alamos, Department of Defense • *Yield:* 22 kt • *Radiation Level at Ground Zero:* 10 R/hr • *Height of Burst:* 400 ft • *Cloud Top Height:* 40,300 ft

40,000-Foot Trajectory: NEVADA: (Pintwater Range), (Pahranagat Range), (Delmar Mountains), Elgin • UTAH: Uvada, (Escalante Desert), Circleville, Junction, Greenwich, Fremont, (San Rafael Knob), (Windowblind Peak), (East Tavaputs Plateau), (Sweetwater Canyon) • COLORADO: (Douglas Pass), Buford, Phippsburg, Oak Creek, (Muddy Pass), Rand, Gould, (Cameron Pass), Poudre Park, Livermore, Wellington, Hereford • NEBRASKA: (Johnson Township, highest point in Nebraska), Kimball, Dix, Potter, Gurley, Oshkosh, Arthur, (Three Mile Lake), Halsey, Purdum, Emmet, O'Neill, Page, Winnetoon, Bazile Mills, Bloomfield, Bow Valley, Hartingen, Obert, Maskell • SOUTH DAKOTA: Vermillion, Burbank, Hub City, Junction City, Big Springs • IOWA: Elk Point, LeMars, Remsen, Peterson, Cornell, Gillett Grove, Ayrshire, Rodman, Whittemore, Hobarton, Algona, Sexton, Wesley, Hutchins, Britt, Duncan, Miller, Ventura, Clear Lake, Emery, Central Heights, Mason City, Nora Springs, Floyd, Colwell, Alata Vista, Jerico, Little Turkey, Fort Atkinson, Ossian, Frankville, Rossville, Volney, Waukon • WISCONSIN: Prairie du Chien, Wauzeka, Woodman, Ridgeway, Dodgeville, Daleyville, Paoli, Belleville, Edgerton, Indianford, Milton, Lima Center, Burlington, Union Grove, Racine, Kenosha • MICHIGAN: Hagar Shores, Riverside, Milburg, Watervliet, Keeler, Dowagiac, Volina, Wakelee, Vandalia, Jones, Constantine, Mottville • INDIANA: Bristol, Middlebury, Shipshewana, Lagrange, Plato, Stroh, Woodruff, Hudson, Ashley, Hamilton • OHIO: Edgerton, Farmer, Brunersburg, Defiance, Ayersville, New Bavaria, Holgate, Prentiss, Belmore, Townwood, McComb, Findlay, Upper Sandusky, Wyandott, Caledonia, Denmark, Mount Gilead, Pulaskiville, Waterford, Fredericktown, Howard, Millwood, Newcastle, New Guilford, West Bedford, Cooperdale, New Moscow, Adams Mills, Conesville, Bloomfield, Cambridge, Greenwood, Whigville, Calais, Mittonsburg, Woodsfield, Laings, Round Botton, Hannibal • WEST VIRGINIA: New Martinsville, Porters Falls, Reader, Folsom, Sedalia, Clarksburg, Rangoon, Belington, Elkins, Red Creek, Harmon, Onego, Milam • VIRGINIA: (Shenandoah Mountains), Bergton, Fulks Run, Timberville, Broadway, New Market, Tenth Legion, Alma, Haywood, Brightwood, Madison, Aroda, Rapidan, Paytes, Post Oak, Snell, Thornburg, Milford, Owenton, Stephens Church, Bruington, Truhart, Saluda, Glenns, Hartfield, Hudgins, Diggs, (Chesapeake Bay), Eastville, Cheriton, (Wreck Island), (Cobb Island).

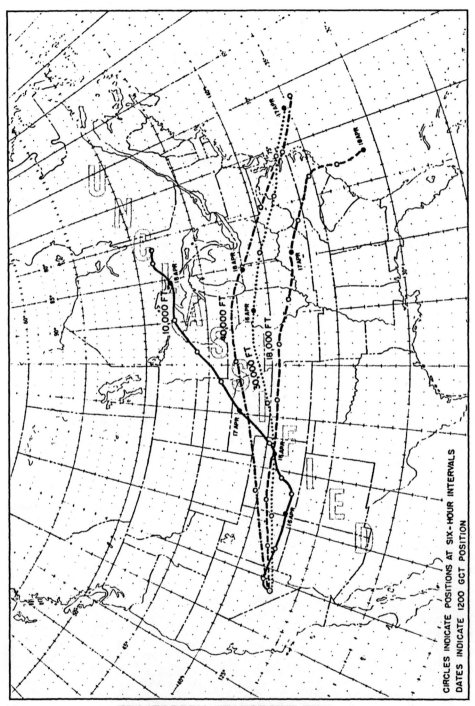

TRAJECTORY: TEAPOT MET (TP9)

CHARACTERISTICS: TEAPOT MET

The nuclear component for the MET ("Military Effects Test") shot was an 800 lb, 30-inch diameter sphere. The design was based on the earlier Buster Jangle EASY design. Teapot MET was a somewhat unique shot, ranking *first* among the nuclear tests for percentages of Br-82, Kr-85, Sr-89, Sr-90, Y-90 and Y-91 in fallout. The deposition pattern for Teapot MET is most similar to Plumbbob NEWTON (PB15) (R: 0.38; T: 22.92).

Fallout from this nuclear test was deposited over 3018 counties, 98 percent of the total counties in the U.S. The counties of highest estimated maximum activity included:

- Beaver UT (1057 µci/sq meter);
- Bowman ND (733 µCi/sq meter);
- Mesa CO (: 820 µCi/sq meter);
- Emery UT (: 846 µCi/sq meter); and
- Grand UT (: 846 µCi/sq meter), all on Apr 15, 1955.

The total fallout to I-131 ratio for shot Met is moderate at 88 to 1.

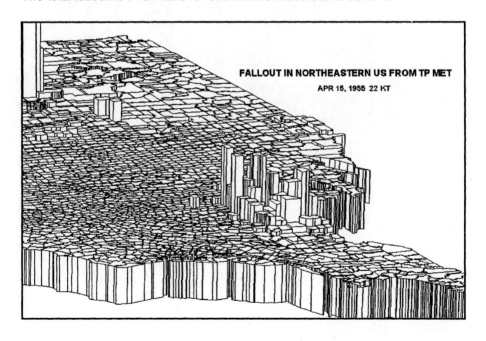

FALLOUT IN NORTHEASTERN US FROM TP MET
APR 15, 1955 22 KT

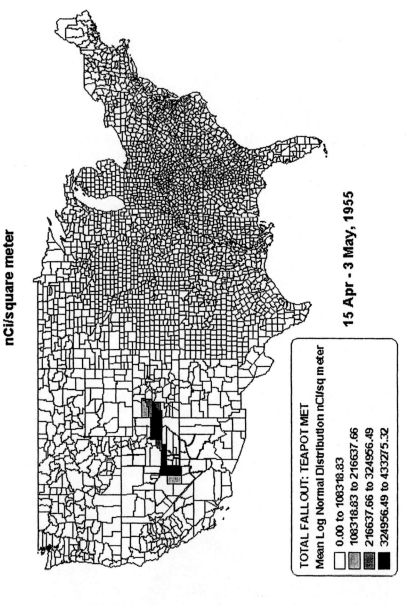

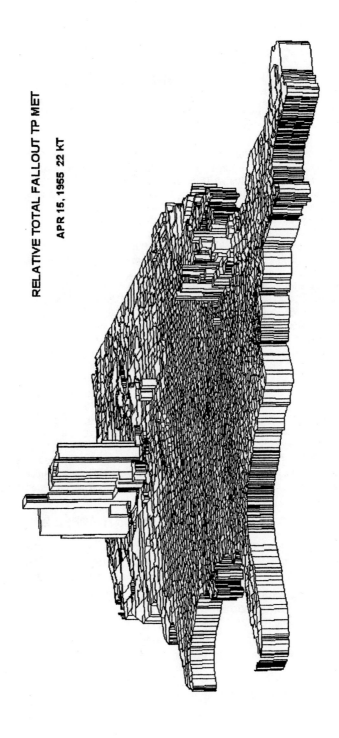

SHOT 10: TEAPOT APPLE-2 (TP10)

TEAPOT: APPLE-2

Detonation Date: 5 May 1955 • *Detonation Time:* 5:10 A.M. • *Area:* 1 • *Sponsor:* Los Alamos • *Yield:* 29 kt • *Radiation Level at Ground Zero:* 500 R/hr • *Height of Burst:* 500 ft • *Cloud Top Height:* 51,000 ft

30,000-Foot Trajectory: NEVADA: (Shoshone Peak), (Belted Range), (Egan Range), (Snake Range), (Humbolt National Forest) • UTAH: Trout Creek, Faust, Vernon, Fairfield, Oram, Provo, (Daniels Pass), Fruitland, Upalco, Myton, Randlett • COLORADO: Rangely, Rio Blanco, New Castle, Glenwood Springs, Cardiff, Cattle Creek, El Jezebel, Basalt, Snowmass, Woody Creek, Aspen, (Mount Elbert, highest point in Colorado), Twin Lakes, (Weston Pass), Fairplay, Estabrook, Pine, Foxton, South Platte, Leuviers, Littleton, Bennett, Strasburg, Woodrow, Akron, Platner • NEBRASKA: Champion, Wellfleet, Lexington, Overton, Elm Creek, Kearney, Minden, Norman, Bladen, Superior, Hardy • KANSAS: Republic, Rydal, Wayne, Agenda, Clyde, Vining, Clifton, Morganville, Fort Riley, Manhattan, Ogden, Miller, Osage City, Melvern, Agricola, Waverly, Harris, Garnett, Bush City, Blue Mound, Mantey, Harding, Fulton • MISSOURI: Richard, Deerfield, Nevada, Milo, Montevallo, Jerico Springs, Arcola, Dadeville, Walnut Grove, Ash Grove, Williard, Springfield, Nichols, Battlefield, Nixa, Linden, Ozark, Sparta, Chadwick, Bradleyville, Brownbranch, Longrun, Theodosia, Isabella, Howards Ridge • ARKANSAS: Gemalel, Viola, Mitchell, Wiseman, Sidney, Mount Pleasant, Cave City, Saffell, Dowdy, Egypt, Cash, Otwell, Bay, Caraway, Trumann, Keiser, Marle, Driver, Osceola • TENNESSEE: Garland, Covington, Stanton, Hillville, Uptonville, Henderson, Enville, Saltillo, Clifton, Center, Webber City, Campbellsville, Wales, Frankewing, McBurg, Fayetteville, Skinem, Coldwater, Bellview, Lincoln, Elora, Anderson, Richard City, New Hope • ALABAMA: Bryant • GEORGIA: New England, (Lookout Mountain), Chickamauga, Tunnel Hill, Elton, Crandell, Cherry Log, Robertstown, Helen, Batesville, Turnerville, Tallulah Falls • SOUTH CAROLINA: Long Creek, Westminster, Seneca, Clemson, Utica, Central, Pelzer, Woodville, Owings, Gray Court, Cross Anchor, Sedalia, Delta, Whitmire, Tuckertown, Shelton, Salemn Crossroads, White Oak, Winnsboro Mills, Langtown, DeKalb, Lucknow, Una, Lydia, Oats, Florence, Quinby, Peedee, Blue Brick, Marion, Zion, Nichols, Mullins • NORTH CAROLINA: Tabor City, Mollie, Nakina, Cruise Island, Winnabow, Bolivia, Boiling Spring Lakes, Seabreeze, Carolina Beach, Kure Beach.

CHARACTERISTICS: TEAPOT APPLE-2

The deposition pattern for Teapot APPLE 2 is most similar to that of Plumbbob shots WHEELER-COULOMB B and LAPLACE (PB13) (R: 0.58; T: 39.53). Fallout from this test was deposited over 3045 counties (98.4% of the total). Counties of highest deposition included (all in µCi/sq meter, estimated maximum activity:): White Pine NV— 533.6 on May 5, 1955; Costilla CO—734.29 on May 6, 1955 and Shelby MO—372 on May 7, 1955. Teapot APPLE-2 ranks 4th in the percentage of Sr-90 produced and 11th in the percentage of Be-7 produced.

The total fallout to I-131 ratio for the APPLE-2 shot is relatively low at 84 to 1. That is only slightly less than that for the APPLE-1 shot at 85 to 1.

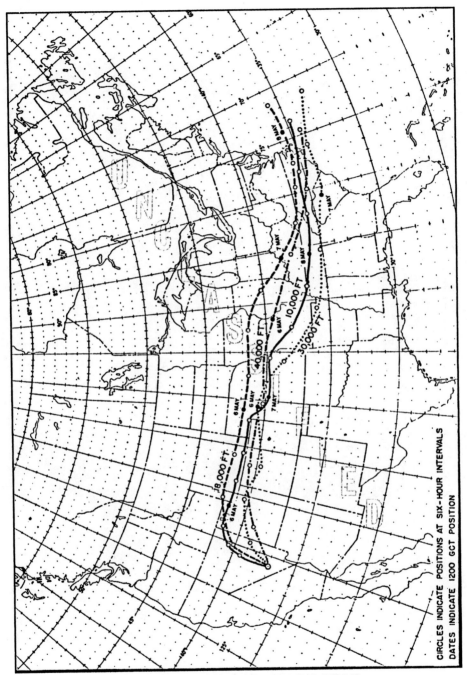

TRAJECTORY: TEAPOT APPLE-2 (TP10)

TOTAL FALLOUT: TEAPOT APPLE-2
Equal Intervals
nCi/square meter

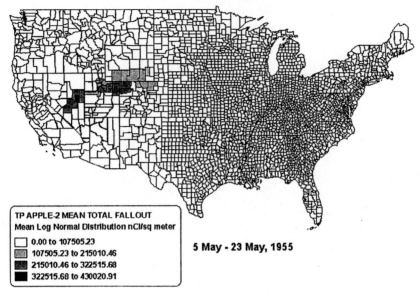

TP APPLE-2 MEAN TOTAL FALLOUT
Mean Log Normal Distribution nCi/sq meter
- 0.00 to 107505.23
- 107505.23 to 215010.46
- 215010.46 to 322515.68
- 322515.68 to 430020.91

5 May - 23 May, 1955

TOTAL FALLOUT: TEAPOT APPLE-2
Equal Number
nCi/square meter

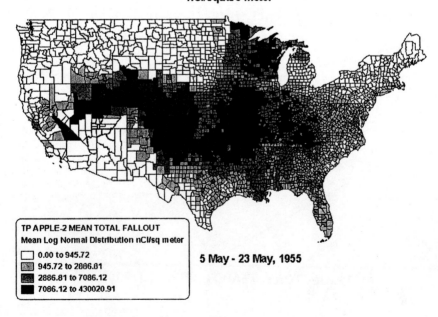

TP APPLE-2 MEAN TOTAL FALLOUT
Mean Log Normal Distribution nCi/sq meter
- 0.00 to 945.72
- 945.72 to 2886.81
- 2886.81 to 7086.12
- 7086.12 to 430020.91

5 May - 23 May, 1955

SHOT 11: TEAPOT ZUCCHINI (TP11)

TEAPOT: ZUCCHINI

Detonation Date: 15 May 1955 • *Detonation Time:* 5:00 A.M. • *Area:* 7-1a • *Sponsor:* Los Alamos • *Yield:* 28 kt • *Radiation Level at Ground Zero:* 500 R/hr • *Height of Burst:* 500 ft • *Cloud Top Height:* 40,000 ft

40,000-Foot Trajectory: NEVADA: (Delmar Mountains) • UTAH: (Morman Range), Central, New Harmony, Cedar City, Summit, Parowan, Panguitch, Teasdale, Grover, Torrey, Bicknell, (Temple Mountain), Green River, (East Tavaputs Plateau), Bonanza • COLORADO: Dinosaur, (Dinosaur National Monument), Greystone, (Vermillion Bluffs) • WYOMING: Table Rock, Wamsutter, Red Desert, (Great Divide Basin), (Green Mountains), Jeffrey City, Gas Hills, Moneta, Lyside, Lost Cabin, Hyattsville, Greybull, Shell, (Bald Mountain), Kane, Lovell, (Horseshoe Bend) • MONTANA: (Bighorn Lake), Crow Indian Reservation, Billings, Lockwood, Huntley, Shepherd, Warden, Klein, Roundup, Forestgrove, Hilger, Christina, Suffolk, Illad, Eagleton, Rocky Boy • MINNESOTA: Greenbush, Strathcona, (Mud Lake), Goodridge, High Landing, Trail, Gully, Lengby, Ebro, Roy Lake, Ponsford, Snellman, Osage, Wolf Lake, Midway, Evergreen, Hillview, Bluffton, Wrightstown, Eagle Bend, Clotho, Little Sauk, Ward Springs, Melrose, Greenwald, St. Martin, Roscoe, Paynesville, Crow River, Grove City, Litchfield, Greenleaf, Corvuso, Cedar Mills, Hutchinson, Stewart, Brownton, Ferando, New Auburn, Winthrop, Gaylord, Norseland, St. Peter, Kasota, Mankato, Eagle Lake, Smiths Mill, St. Clair, Alma City, Waldorf, Matawan, Minnesota Lake, Freeborn, Manchester, Albert Lea, Twin Lakes, Emmons, Glenville • IOWA: Silver Lake, Kensett, Mankey, Rock Falls, Mason City, Nora Springs, Rockford, Marble Rock, Greene, Packard, Clarksville, Shell Rock, Waverly, Finchford, Cedar Falls, Waterloo, Garrison, Van Horne, Newhall, Norway, Walford, East Amana, Amana, Tiffin, Iowa City, Coralville, Hills, Lone Tree, Conesville, Fredonia, Wapello, Toolesboro, Oakville, Northfield Kingston • ILLINOIS: Oquawka, Gulf Port, Media, Bushnell, New Philadelphia, Ipava, Summum, Bath, Chandlerville, Curran, Chatlam, Glenarm, Pawnee, Morrisonville, Wenonah, Nokomis, Ramsey, St. Elmo, Altamont, Edgewood, La Clede, Iola, Louisville, Clay City, Wynoose, Samsville, Bone Gap, Gards Point, Mount Carmel • INDIANA: East Mount Carmel, Johnson, Owensville, Haubstadt, Warrenton, Elberfeld, Daylight, Chandler, Boonville, Pelzer, Yankeetown, Hatfield, Eureka, Richland, Patronville • KENTUCKY: Owensboro, Philpot, Whitesville, Fordsville, Short Creek, Leitchfield, Grayson Springs, Millerstown, Jonesville, Pikes View, Campbellsville, Hatcher, Yuma, Casy Creek, Pellyton, Dunnville, Mintonville, (Daniel Boone National Forest), Lidu, Bush, Bluehole, Ogle, Brightshade, Roark, Chappell, Gilley, Benham • VIRGINIA: Keokee, Big Stone Gap, East Stone Gap, Fort Blackmore, Nickelsville, Collinwood, Benhams • NORTH CAROLINA: Baldwin, Fleetwood, Moravian Falls, Love Valley, Union Grove, Turnersburg, Harmony, Cooleemee, Churchland, Cid, Denton, Farmer, Utah, Seagrove, Highfalls, Parkwood, Spout Springs, Spring Lake, Fayetteville, Vanda, Stedman, Autreyville, Parkersburg, Ingold, Harrells, Penderlea, Watha, Holly Ridge.

CHARACTERISTICS: TEAPOT ZUCCHINI:

The device used for ZUCCHINI was 40 inches in diameter, 80.5 inches long, weighed 2935 lbs and was detonated atop a 500-ft steel tower. Teapot ZUCCHINI is ranked 2nd for percentages of Br-82, Kr-85 and Sr-89 in the total fallout. The deposition pattern for Teapot ZUCCHINI is most similar to Plumbbob BOLTZMANN (PB1) (R: 0.58; T: 39.44). Fallout from this shot was deposited on 3018 counties, or 98 percent of the total U.S. Counties of highest deposition included Iron UT (estimated maximum activity: 857 µCi/sq meter on May 15 1955) and Clark NV (estimated maximum activity: 406.4 µCi/sq meter on May 15, 1955).

There was some speculation that a rainout associated with debris cloud from ZUCCHINI occurred in Harney County, Oregon shortly after May 15, 1955. However, according to the NCI data, the fallout level at Harney County reached only 30 µCi/sq meter, estimated maximum activity.

The total fallout to I-131 ratio for Teapot ZUCCHINI is relatively low at 85 to 1.

TOTAL FALLOUT: TEAPOT ZUCCHINI

Equal Intervals
nCi/square meter

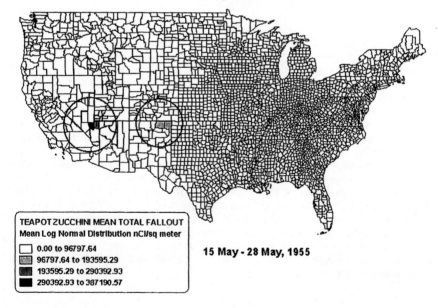

15 May - 28 May, 1955

TEAPOT ZUCCHINI MEAN TOTAL FALLOUT
Mean Log Normal Distribution nCi/sq meter
- 0.00 to 96797.64
- 96797.64 to 193595.29
- 193595.29 to 290392.93
- 290392.93 to 387190.57

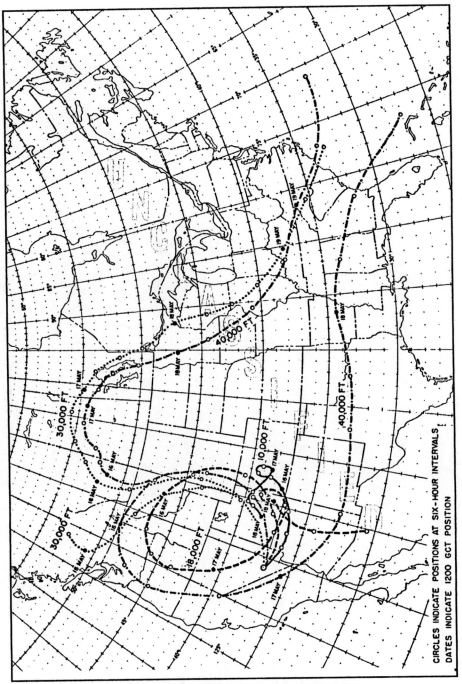

TRAJECTORY: TEAPOT ZUCCHINI (TP11)

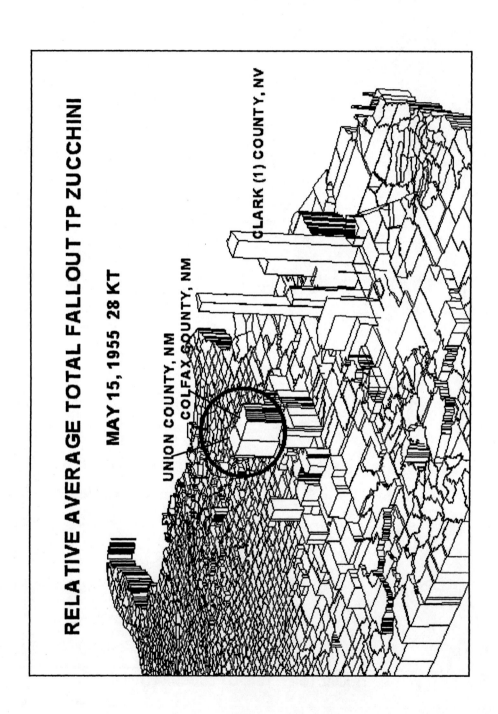

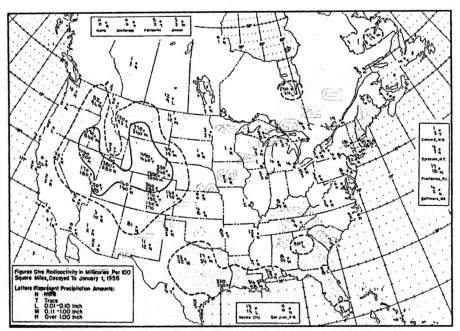

PRECIPITATION AMOUNTS DURING TEAPOT SERIES

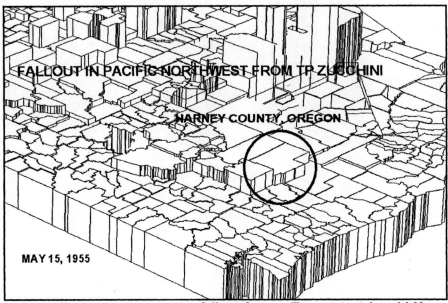

According to some reports, fallout from a Teapot test in mid-May-1955 contaminated areas of Harney County, Oregon. Based on fallout and precipitation data, it is likely the fallout was from Teapot ZUCCHINI.

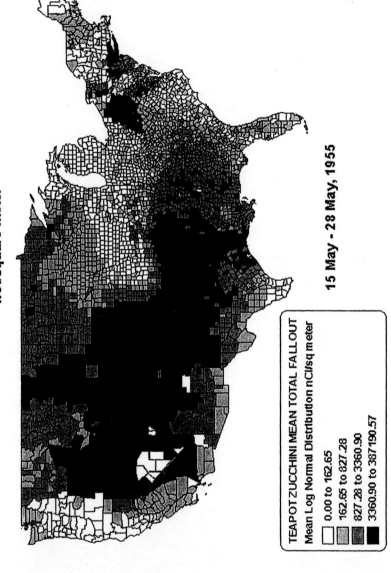

TOTAL FALLOUT: TEAPOT ZUCCHINI

Equal Number nCi/square meter

15 May - 28 May, 1955

TEAPOT ZUCCHINI MEAN TOTAL FALLOUT
Mean Log Normal Distribution nCi/sq meter

- 0.00 to 162.65
- 162.65 to 827.28
- 827.28 to 3360.90
- 3360.90 to 387190.57

SECTION 6

PLUMBBOB SERIES

May 28-Oct 7, 1957

SHOT 1: PLUMBBOB BOLTZMANN (PB1)

PLUMBBOB: BOLTZMANN

Detonation Date: 28 May 1957 • *Detonation Time:* 5:55 A.M. • *Area:* 7c • *Sponsor:* Los Alamos • *Yield:* 12 kt • *Radiation Level at Ground Zero:* 20 R/hr • *Height of Burst:* 500 ft • *Cloud Top Height:* 33,000 ft

10,000-Foot Trajectory: NEVADA: Warm Springs, Manhattan, Luning, Hawthorne, Wellington, Lake Tahoe • CALIFORNIA: Tahoe Pines, (Squaw Valley Ski Area), Placerville, Loomis, Lincoln, Trobridge, Marysville, Yuba City, Hamilton City, Richfield, Red Bluff, Cottonwood, Redding, Weed, Edgewood, Gazelle, Yreka, Hornbrook, Hilt • OREGON: Ashland, Medford, Shady Grove, Drew, Cottage Grove, Springfield, Albany, Salem, Brooks, Newburg, Beaverton, Hillsboro, Burlington, Columbia City, Rainier • WASHINGTON: Longview, Lexington, Ryderwood, Curtis, Chehalis, Fords Prairie, Littlerock, Shelton, Port Angeles.

RELATIVE TOTAL FALLOUT: PB BOLTZMANN

MAY 28, 1957 10 KT

CHARACTERISTICS: PLUMBBOB BOLTZMANN

The device used for the BOLTZMANN shot was 31.6 inches long, 18 inches wide and weighed 295 lbs. The nuclear component weighed only 144.6 lbs. Reportedly, the shot cab atop the 500-ft tower was weighted with 9 tons of sand and 15 tons of paraffin as a means to shield instruments. Among the nuclear tests, Plumbbob BOLTZMANN ranks 5th in percentage of Cs-137 in fallout The deposition pattern for Plumbbob BOLTZMANN is most similar to Teapot ZUCCHINI (TP11) (R: 0.58; T: 39.44). Fallout from this nuclear test was deposited over 2781 U.S. counties. The counties of highest deposition included Nye NV (estimated maximum activity: 1419 µCi/sq meter); Lander NV (estimated maximum activity: 598 µci/sq meter) and Mineral NV (estimated maximum activity: 585 µCi/sq meter) all on May 28, 1957. The day one total fallout to I-131 ratio for this shot was, at 74 to 1, the 4th lowest of all aboveground nuclear tests.

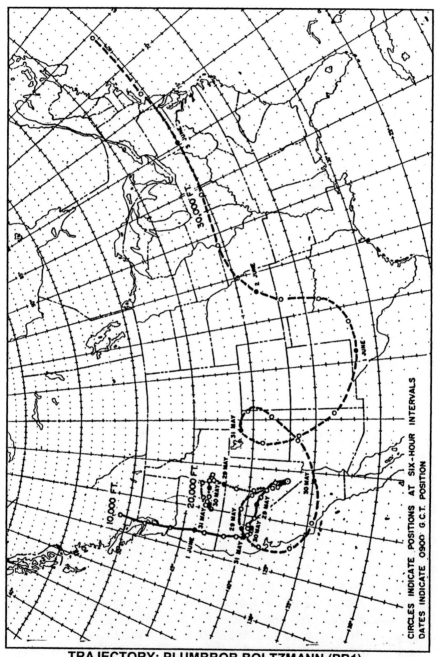

TRAJECTORY: PLUMBBOB BOLTZMANN (PB1)

TOTAL FALLOUT: PLUMBBOB BOLTZMANN
Equal Intervals
nCi/square meter

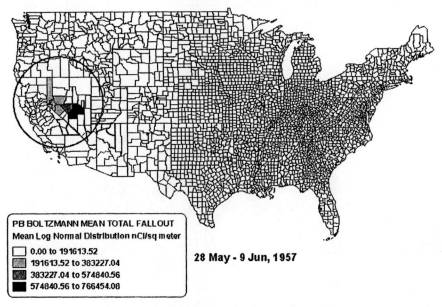

28 May - 9 Jun, 1957

PB BOLTZMANN MEAN TOTAL FALLOUT
Mean Log Normal Distribution nCi/sq meter
- 0.00 to 191613.52
- 191613.52 to 383227.04
- 383227.04 to 574840.56
- 574840.56 to 766454.08

TOTAL FALLOUT: PLUMBBOB BOLTZMANN
Equal Number
nCi/square meter

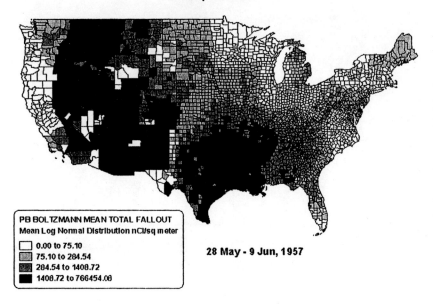

28 May - 9 Jun, 1957

PB BOLTZMANN MEAN TOTAL FALLOUT
Mean Log Normal Distribution nCi/sq meter
- 0.00 to 75.10
- 75.10 to 284.54
- 284.54 to 1408.72
- 1408.72 to 766454.08

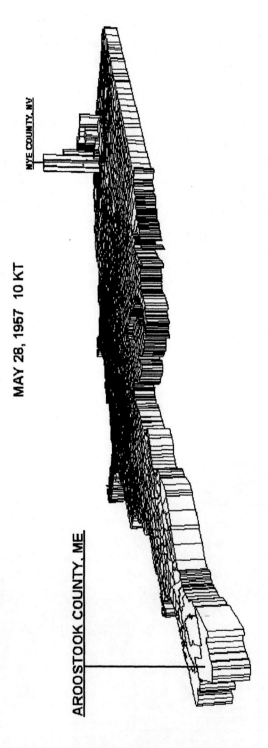

SHOT PLUMBBOB FRANKLIN
(Not included in NCI Study)

PLUMBBOB: FRANKLIN

Detonation Date: 2 Jun 1957 • *Detonation Time:* 4:55 A.M. • *Area:* 3 • *Sponsor:* Los Alamos • *Yield:* 140 tons • *Radiation Level at Ground Zero:* 35 R/hr • *Height of Burst:* 300 ft • *Cloud Top Height:* 16,700 ft

16,000-Foot Trajectory: NEVADA: (Belted Range), (Egan Range) • UTAH: (Frisco Peak), Minersville, Enoch, Cedar City, (Zion National Park), Rockville, Hilldale • ARIZONA: Colorado City, (Mount Trumbull), Hualapai Indian Reservation, Nelson, Yucca • CALIFORNIA: Needles, (Sada Mountains), (Tiefort Mountains), Argos, Borosolvay, Dunmovin, Olancha, Cartago, (Mount Whitney), (Mount Woodworth), (Florence Lake), Mono, Hot Springs, (Ritter Range), (Eagle Peak), (Sweetwater Mountains) • NEVADA: Wellington, Smith, Wabuska, Silver Springs, Fernley, Nixon, (Seven Troughs Range), (Jackson Mountains), (Disaster Peak) • OREGON: (Trout Creek Mountains), Rome, Arock, Danner • IDAHO: Melba, Kuna, Bowmont, Mora, Boise, Idaho City, Stanley, Sunbeam, Bonanza, Challis, May, Leadore, (Continental Divide) • MONTANA: Bell, Lima, (Beaverhead National Forest), Raynolds Pass, West Yellowstone • WYOMING: (Mount Holmes), Canyon, Pelican Cone, (Sleeping Giant Mountain), (Buffalo Bill Dam), Cody, (Elk Butte), Manderson, Ten Sleep, Sussex, (Thunder Basin National Grassland) • NEBRASKA: Chadron, Hay Springs, (Big Hill), Broken Bow, Litchfield, Poole, Wood River, Diniphan, Hansen, Trumbull, Harvard, Saronville, Sutton, Geneva, Milligan, Tobias, Swanton, Plymouth, Beatrice, Filley, Elk Creek, Howe, Stella, Shubert, Barada • MISSOURI: Craig, Corning, Maryville, Skidmore, Arkoe, Conception Junction, Gentry, Bethany, Modena, Spickard, Osgood, Milan, Green Castle, Novenger, Pure Air, Kirksville, Hardland, Edina, Knox City, La Belle, Monticello, Canton • ILLINOIS: Meyer, Mendon, Camp Point, Clayton, Mount Sterling, Ripley, Bluff Spring, Philadelphia, Ashland, Pleasant Plains, Springfield, Riverton, Mechanicsburg, Decatur, Elwin, Mt. Zion, Lake City, Lovington, Arthur, Arcola, Bushton, Kansas, Paris, Grandview, Elbridge, Vermillion, Dennison • INDIANA: St. Mary-of-the-Woods, North Terre Haute, Brazil, Harmony, Reelsville, Manhattan, Belle Union, Crown Center, Hall, Bethany, Waverly, Stones Crossing, New Whiteland, Whiteland, Needam, Shelbyville, Rays Crossing, Blue Ridge, New Salem, Milroy, Richland, Andersonville, Laurel, Blooming Grove, Bath, Mixerville • OHIO: Morning Sun, West Elkton, Greenbush, Franklin, Springboro, Lytle, Ridgeville, Corwin, Kingman, Roseville, Sabina, Staunton, Good Hope, Austin, Greenland, Andersonvile, Kinnickinnick, South Bloomingville, Hue, Mount Pleasant, Orland, New Plymouth, The Plains, Athens, Canaanville, Stewart, Little Hocking • WEST VIRGINIA: Parkersburg, Cedar Grove, Belmont, Wick Mountain, Alma, Lima, Alvy, Smithfield, Mannington, Arnettsville, Westover, Morgantown, Hazelton • PENNSYLVANIA: Markleysburg, Addison, Springs, West Salisbury, Boynton, Salisbury, Pocahontas, Pleasant Union, Wellersburg, Purcell, Inglesmith, Buck Valley, Plum Rush, Sylven, Claylick, Welsh Run, Shady Grove, Waynesboro, Fountain Dale, Green Mounds, Barlow • MARYLAND: Silver Run, Union Mills, Hampstead, Whitehouse, Butler, Sparks, Sunnybrook, Sweet Air, Perry Hall, White Marsh, Chase, Newtown, Melitota, Chestertown, Barclay, Roberts, Mount Zion, Henderson • DELAWARE: Petersburg, Sandtown, Felton, Masten Corner, Chestnut Knoll, Milford, Lincoln, Milton, Nassau, Harbeson, Angola, Seabreeze, Cottonpatch Hill.

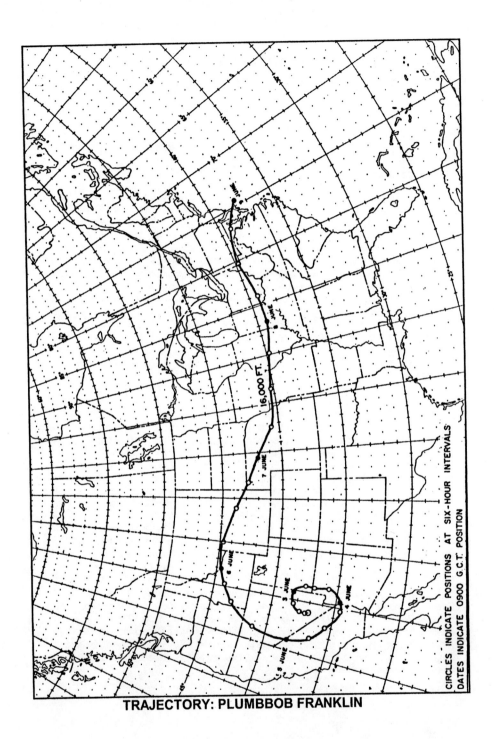

TRAJECTORY: PLUMBBOB FRANKLIN

U.S. FALLOUT ATLAS : TOTAL FALLOUT

SHOT 2: PLUMBBOB WILSON (PB2)

PLUMBBOB: WILSON

Detonation Date: 18 Jun 1957 • *Detonation Time:* 4:55 A.M. • *Area:* 9a • *Sponsor:* Livermore Laboratory • *Yield:* 10 kt • *Radiation Level at Ground Zero:* 20 R/hr • *Height of Burst:* 500 ft • *Cloud Top Height:* 35,000 ft

10,000-Foot Trajectory: NEVADA: Lathrop Wells, (Funeral Mountains) • CALIFORNIA: (Death Valley), (Last Chance Range) • NEVADA: (Magruder Mountain), Lida, (Montezuma Peak), Tonopah, (Pinion Peak), (Pancake Range), (Mount Hamilton), Steptoe, Tippett • UTAH: Ibapah, (Great Salt Lake Desert), Grantsville, Salt Lake City, Layton, Syracuse, Clearfield, Roy, Ogden, Laketown • WYOMING: Cokeville, Calpert, La Barge, (Continental Peak), (Great Divide Basin), Bairvil, Muddy Gap, (Emigrant Gap), Casper, Clareton, Newcastle, Osage, Four Corners • SOUTH DAKOTA: Cheyenne Crossing, Deadwood, Lead, Sturgis, Stoneville, Faith, Red Elm, Glencross, Trail City, Mobridge, Glenham, Eureka • NORTH DAKOTA: Venturia, Ashley, Fredonia, Jud, Nortonville, Millarton, Montpelier, Ypsilanti, Eckelson, Rogers, Dazey, Sibley, Cooperstown, Sharon, Aneta, Northwood, Kempton, Karimore, Arvilla, Honeyford, Gilby, McKinock, Manvel, Ardoch • MINNESOTA: Oslo, Aluorado, Argyle, Strandquist, Strathcona, Wannaska, Pencer, Roosevelt, Wheelers Point.

CHARACTERISTICS: PLUMBBOB WILSON

Among the nuclear tests, Plumbbob WILSON is ranked 2nd for percentage of Cm-242 in fallout. The deposition pattern for this shot is most similar to that for Upshot-Knothole CLIMAX (UK11) (R: 0.26, T: 15.21). Fallout from this test was deposited over 2689 counties (87% of the total). This shot also ranks 8th in amount of Mn-54 deposition across the U.S. The day one total fallout to I-131 ratio is moderately high at 101 to 1.

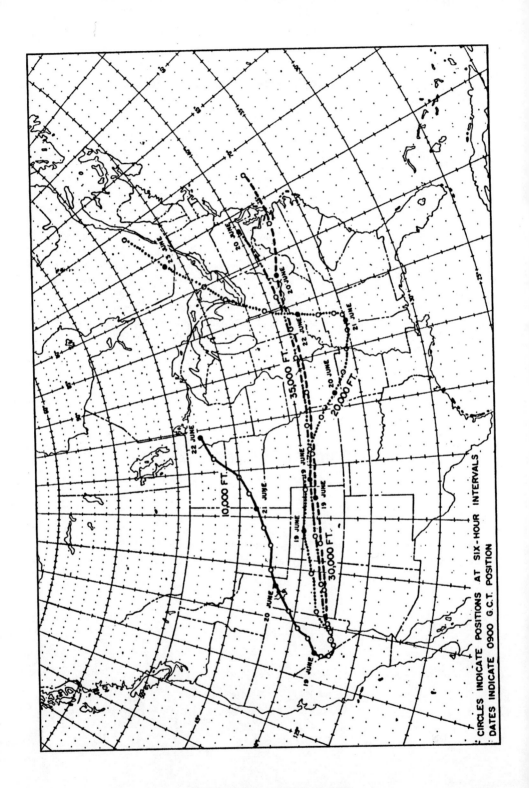

U.S. FALLOUT ATLAS : TOTAL FALLOUT

TRAJECTORY: PLUMBBOB WILSON (PB2)

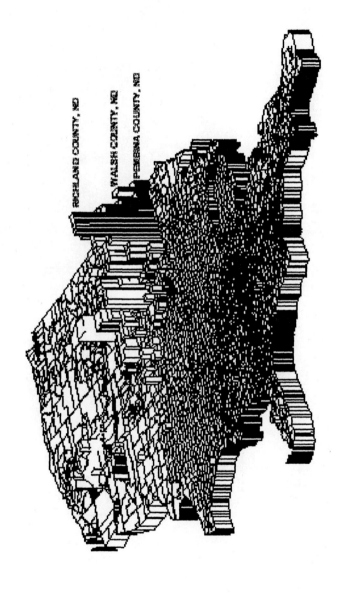

U.S. FALLOUT ATLAS : TOTAL FALLOUT

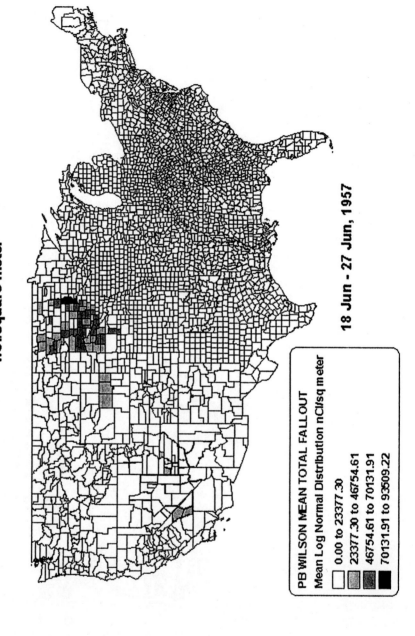

TOTAL FALLOUT: PLUMBBOB WILSON
Equal Number
nCi/square meter

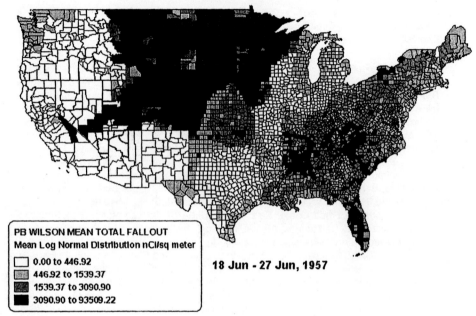

PB WILSON MEAN TOTAL FALLOUT
Mean Log Normal Distribution nCi/sq meter
- 0.00 to 446.92
- 446.92 to 1539.37
- 1539.37 to 3090.90
- 3090.90 to 93509.22

18 Jun - 27 Jun, 1957

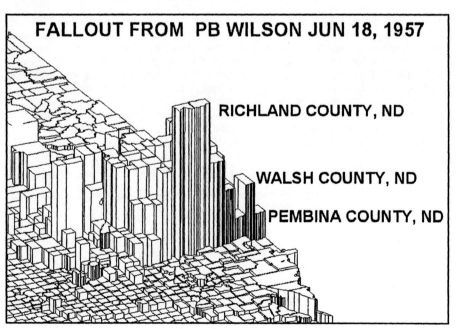

FALLOUT FROM PB WILSON JUN 18, 1957

RICHLAND COUNTY, ND

WALSH COUNTY, ND

PEMBINA COUNTY, ND

U.S. FALLOUT ATLAS : TOTAL FALLOUT

SHOT 3: PLUMBBOB PRISCILLA (PB3)

PLUMBBOB: PRISCILLA

Detonation Date: 24 Jun 1957 • *Detonation Time:* 6:30 A.M. • *Area:* Frenchman's Flat • *Sponsor:* Los Alamos, Department of Defense • *Yield:* 37 kt • *Radiation Level at Ground Zero:* 500 R/hr • *Height of Burst:* 700 ft • *Cloud Top Height:* 43,000 ft

40,000-Foot Trajectory: NEVADA: Alamo, Ash Springs, Caliente • UTAH: Uvada, Moderna, Beryl, Circleville, Junction, Loa, Lyman, Bicknell, (Canyons National Park), Dead Horse Point, La Sal • COLORADO: Bedrock, Mount Wilson, Lizard Head Pass, San Juan Mountains, Conejos Peak, (Rio Grande National Forest), Ortiz • NEW MEXICO: Los Pinos, Cerro, Valdez, Eagle Nest, Arroyo Ceso, Angel Fire, Black Lake, Ocate, Wagon Mound, Roy, Solano, Mosquero, Galegos • TEXAS: Glenrio, Hereford, Arney, Happy, Tulia, Rock Creek, Silverton, Gasoline, Flomot, White Flat, Paducah, Sneedville, Chalk, Benjamin, Rhineland, Munday, Throckmorton, Woodson, Crystal Falls, Caddo, Frankell, Huckaby, Stephenville, Selden, Duffau, Iredell, Meridian, Cranfill's Gap, Valley Mills, China Spring, Waco, Bellmead, Riesel, Otto, Ben Hur, Groesbeck, Thornton, Oletha, Jewett, Donie, Buffalo, Crockett, Latexo, Kennard, Ratcliff, Clawson, Central, Lufkin, Redland, Etoile, Chireno, Norwood, Penning, San Augustine, Fords, Corner, Sexton • LOUISIANA: Zwole, Natchitoches, Robaline, Natchez, Hazlewood, St. Maurice, Atlanta, Packton, Urania, Kelley, Olla, Clarks, Rosefield, Holum, Grayson, Fort Necessity, Extension, Jigger, Chase, Newlight, Somerset, Afton, Mound • MISSISSIPPI: Edwards, Oakley, Bolton, Raymound Forest, Jackson, Pelhatchie, Morton, Kalem, Forest, Conehatla, Lake, Decatur, Duffee, Collinsville, Susqualena, Martin, Obadiah, Daleville, Lauderdale • ALABAMA: Sumpterville, Epes, Forkland, Greensboro, Helberger, Stanton, Maplesville, Fairview, Coopers, Verbena, Mountain Creek, Equality, Nixburg, Cottage Grove, Jacksons Gap, Dadeville, Lafayette, Buffalo, Fredonia • GEORGIA: La Grange, Stovall, Greenville, Gay, Woodbury, Molena, Meansville, Zebulon, Aldora, Milner, Barnesville, Forsyth, Juliette, East Juliette, Round Oak, Wayside, Devereaux, Sparta, Calverton, Jewell, Dearing, Boneville, Harlem, Grovetown, Martinez • SOUTH CAROLINA: Augusta, Graniteville, Aiken, Vaucluse, Wagener, New Holland, Thor, Pelion, Gaston, Gadsden, Eastover, Statesburg, Horatio, Dalzell, Oswego, St. Charles, Lynchburg, Atkins, Cartersville, Timmonsville, Florence, Quinby, Peedee, Blue Brick, Sellers, Floyd Dale, Kemper • NORTH CAROLINA: Orrum, Bladenburg, Abbottsburg, Butters, Elizabethtown, Tomahawk, Harrells, Delway, Wallace, Teachey, Rose Hill, Greenovers, Chinquapin, Lyman, Beulaville, Fountain, Catherine Lake, Richlands, Petersburg, Belgrade, Pollocksville, Riverdale, Croatan, Rhems, New Bern, Minnesott Beach, Arapahoe, Olympia, Merritt, Stonewall, Bayboro, Mesic, Florence, Hoboken, Pamlico Point, Bluff Point, New Holland, Gulrock, Rodantha, Waves, Salvo.

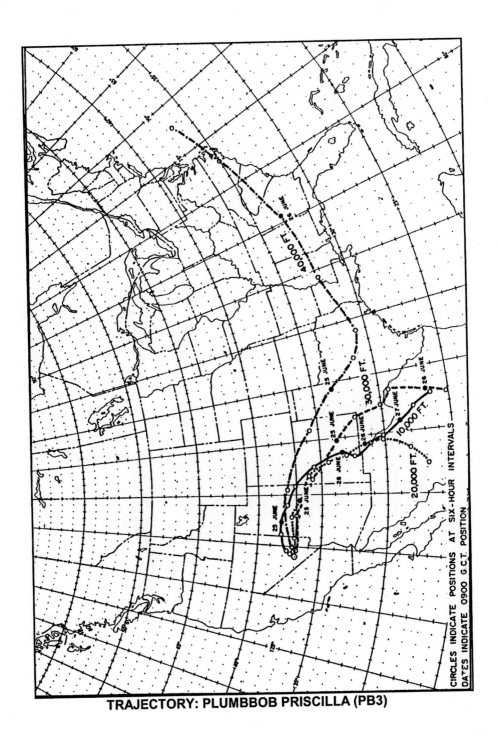

TRAJECTORY: PLUMBBOB PRISCILLA (PB3)

U.S. FALLOUT ATLAS : TOTAL FALLOUT

CHARACTERISTICS: PLUMBBOB PRISCILLA

The device used in the PRISCILLA test weighed 581.4 lbs. Among the nuclear tests, Plumbbob PRISCILLA ranks 2nd for percentage of Na-24 and Ce-144 in fallout. The deposition pattern for this shot is most similar to Teapot MET (TP9) (R: 0.09, T: 5.05). Fallout from this test was deposited on 2941 counties. Counties of highest deposition included Avoyelles LA, Acadia LA, Wilkinson MS, West Feliciana LA, Assumption, LA and Beauregard, LA .(all approximately 350 µCi/sq meter *average* total fallout). Hottest shot days included the following Louisiana counties with estimated *maximum* activity of 702 µCi/sq meter: Acadia, Assumption, Avoyelles, Beuregard, Calcasieu, Cameron, East Baton Rouge, Evangeline, Grant and Iberia. Fallout from this shot covered 2941 counties (94% of the total). The day one total fallout to I-131 ratio for this shot is moderately high at 115 to 1.

TOTAL FALLOUT: PLUMBBOB PRISCILLA
Equal Number
nCi/square meter

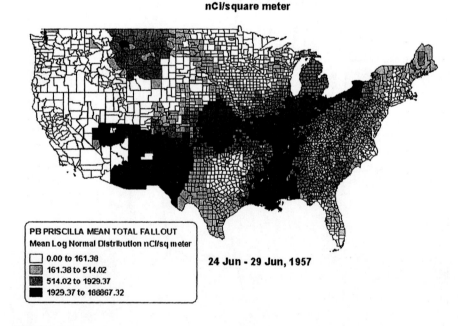

PB PRISCILLA MEAN TOTAL FALLOUT
Mean Log Normal Distribution nCi/sq meter
- 0.00 to 161.38
- 161.38 to 514.02
- 514.02 to 1929.37
- 1929.37 to 188867.32

24 Jun - 29 Jun, 1957

TOTAL FALLOUT: PLUMBBOB PRISCILLA
Equal Intervals
nCi/square meter

24 Jun - 29 Jun, 1957

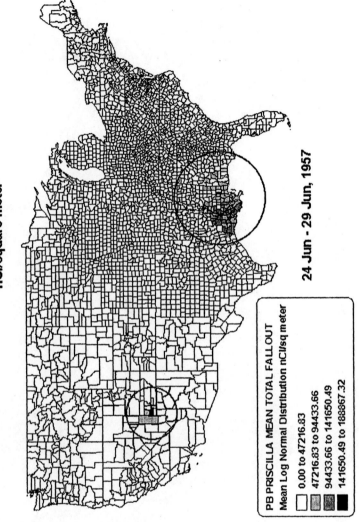

PB PRISCILLA MEAN TOTAL FALLOUT
Mean Log Normal Distribution nCi/sq meter

- 0.00 to 47216.83
- 47216.83 to 94433.66
- 94433.66 to 141650.49
- 141650.49 to 188867.32

RELATIVE TOTAL FALLOUT: PB PRISCILLA
JUN 24, 1957 37 KT

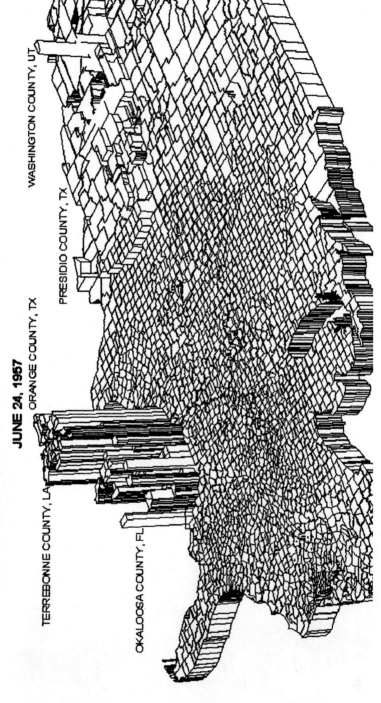

SHOT 4: PLUMBBOB HOOD (PB4)

PLUMBBOB: HOOD

Detonation Date: 5 Jul 1957 • *Detonation Time:* 4:40 A.M. • *Area:* 9a • *Sponsor:* Livermore Laboratory • *Yield:* 74 kt • *Radiation Level at Ground Zero:* 100 R/hr • *Height of Burst:* 1,500 ft • *Cloud Top Height:* 48,000 ft

10,000-Foot Trajectory: NEVADA: (Railroad Valley), Duckwater, Duckwater Indian Reservation, (Butte Mountains), Currie, Wendover • UTAH: (Great Salt Lake Desert), (Dolphin Island), Snowville • IDAHO: Woodruff, Samaria, Dayton, Clifton, Banida, Mink Creek, Bern, Bennington • WYOMING: Smoot, (Salt River Range), Halfway, Merna, (Gannett Peak, highest point in Wyoming: 13,806 feet), (Wind River Range), Burris, Hamilton Dome, Winchester, Worland, (Powder River Pass), Upton, Four Corners • SOUTH DAKOTA: Nemo, Piedmont, Blackhawk, Hereford, Milesville, Hayes, Pierre, Lower Brule Indian Reservation, Crow Creek Indian Reservation, Wessington Springs, Lane, Artesian, Carthage, Roswell, Ramona, Madison, Nunda, Rutland, Ward • MINNESOTA: Verdi, Ruthton, Garvin, Godahl, La Salle, Garden City, Good Thunder, Mapleton, Waldorf, New Richland, Blooming Prairie, Oslo, Hayfield, High Forest, Stewartville, Chatfield, Peterson, Whalan, Houston, Bratsberg, Brownsville • WISCONSIN: Stoddard, Genoa, Chaseburg, Newry, Westby, Bloomingdale, Rockton, Valley, Hillsboro, Union Center, Wonewoc, Lake Delton, Wisconsin, Dells, Briggsville, Friesland, Cambria, Fox Lake, Waupun, Brownsville, Knowles, Le Roy, Le Mira, Kewaskum, Random Lake, Belgium, Cedar Grove • MICHIGAN: Roosevelt Park, Norton Shores, Cloverville, Conklin, Alpine, Grand Rapids, Saranac, Portland, Eagle, Grand Lodge, Lansing, Okemos, Haslett, Williamston, Bell Oak, Oak Grove, Fowlerville, Howell, Hartland, Highland, Milford, Walled Lake, Commerce, Farmington Hills, Southfield, Ferndale, Oak Park, Highland Park, Detroit, Grosse Point Woods, (Lake St. Clair) • ONTARIO: Windsor, Tecumseh, Belle River, S. Woodslee, Comber, Wheatley • OHIO: (Hubbard Homestead), Ashtabula, North Kingsville, Bashnell, Kellogsville, Monroe Center • PENNSYLVANIA: Pennside, Springboro, Hickernell, Mosiertown, Saegertown, New Richmond, Blooming Valley, Townville, Troy Center, Gresham, Titusville, E. Titusville, Pleasantville, West Hickory, Starr, Guitonville, Marienville, Roses, Halfton, Portland Mills, Brandy Camp, Toby, Byrnedale, Force, Weedville, Caledonia, Karthaus, Pine Glen, Clarence, Howard, Jacksonville, Nittany, Lamar, Clintondale, Tylerville, Livonia, Hartleton, Millmont, White Springs, Dice, New Berlin, Kratzerville, Shamokin Dam, Sunbury, Selinsgrove, Edgewood, Shamokin, Gowen City, Ranshaw, Kulpmont, Locust Gap, Helfenstein, Gordon, Primrose, Minersvile, Jonestown, Marlin, Pottsville, Mount Carbon, Schuylkill Haven, Adamsdale, Orwigsburg, Deer Lake, Eckville, Port Clinton, Albany, Lenhartsville, Klinesville, Grimville, Maxatawney, Trexlertown, Breinigsville, Alburtis, Mertztown, Macungie, Shimerville, Old Zionsville, Zionsville, Spinnerstown, Milford Square, Rich Hill, Hargersville, Blooming Glen, Dublin, Plumsteadville, Gardenville, Danboro, Fountainville, Mechanicsville, Lahaska, Buckingham, Doylestown, Furlong, Pineville, Penns Park, Washington Crossing, Yardley, Marrisville • NEW JERSEY: Ewing, Trenton, White Horse, Groveville, Creamridge, Hornerstown, Lakehurst, Ridgeway, Pine Lake Park, Toms River, Gilford Park, Pine Beach, Bay Shore, Island Heights, (Pelican Island), Seaside Heights, Seaside Park, South Seaside Park.

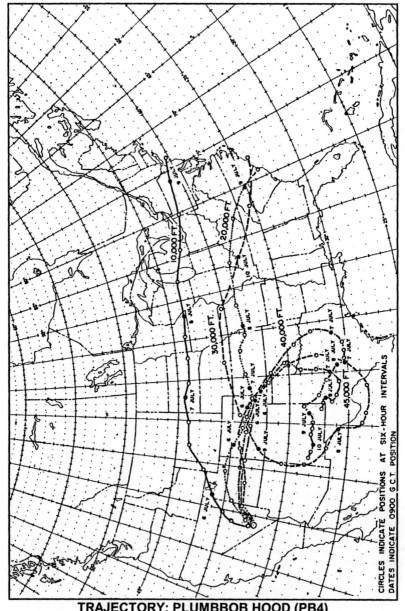

TRAJECTORY: PLUMBBOB HOOD (PB4)

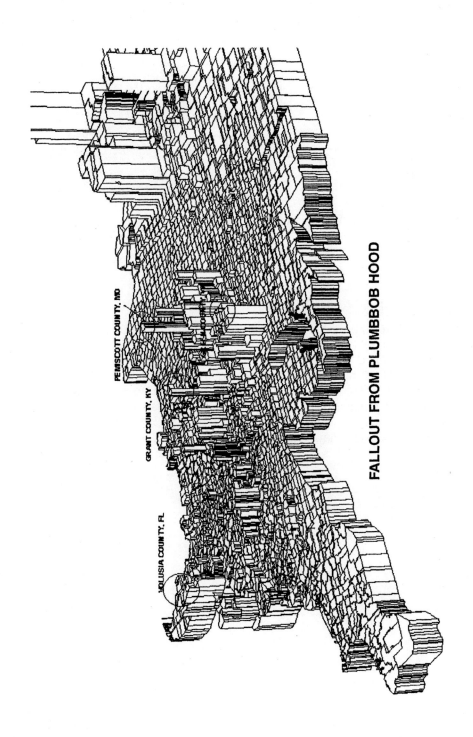

U.S. FALLOUT ATLAS : TOTAL FALLOUT   6-21

RELATIVE TOTAL FALLOUT: PB HOOD
JUL 5, 1957 74 KT

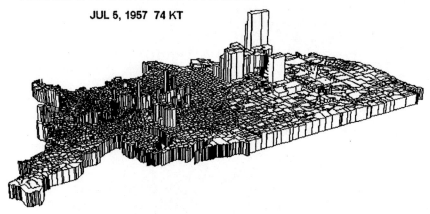

CHARACTERISTICS: PLUMBBOB HOOD

The device for the thermonuclear shot HOOD weighed 395 lbs, was 12.1 inches in diameter and 42.2 inches in length. Among the nuclear tests, Plumbbob HOOD was very similar in radionuclide composition to the later shot SEDAN. It ranks 2nd for percentages of Au-198, Au-199 and Be-7 (1st being SEDAN). No other shot, in fact, produced Au-198 or Au-199 in fallout. The deposition pattern for HOOD is most similar to Upshot-Knothole CLIMAX (UK11) (R: 0.58, T: 39.44). Counties of highest deposition included Pima AZ, Santa Cruz AZ, Osage MO, Colfax NM and Harding NM. HOOD's day one total fallout to I-131 ratio is moderately high at 110 to 1.

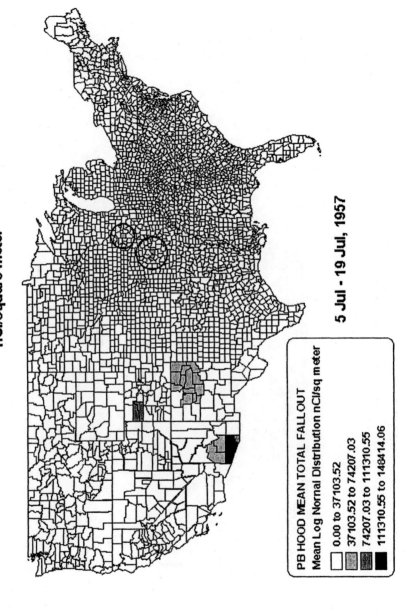

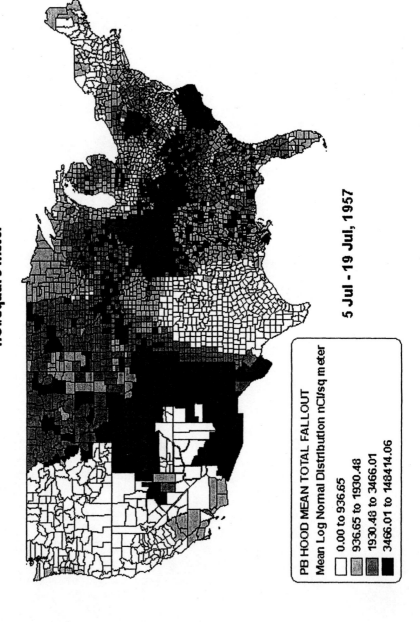

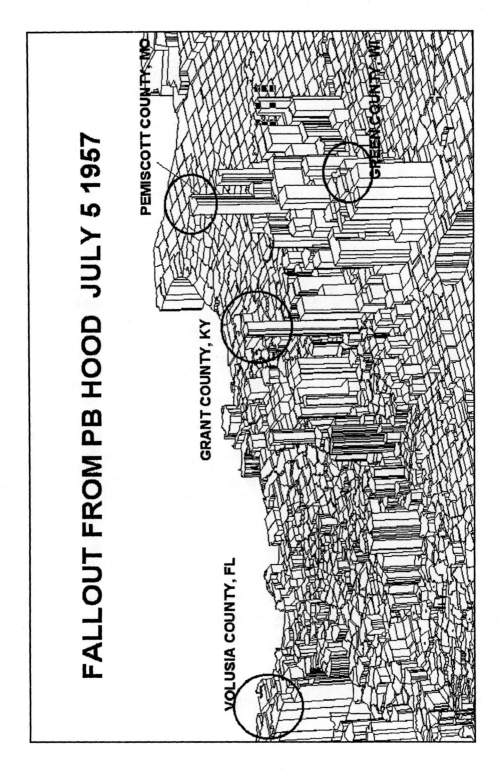

SHOT 5: PLUMBBOB DIABLO (PB5)

PLUMBBOB: DIABLO

Detonation Date: 15 Jul 1957 • *Detonation Time:* 4:30 A.M. • *Area:* 2b • *Sponsor:* Livermore Laboratory • *Yield:* 17 kt • *Radiation Level at Ground Zero:* 200 R/hr • *Height of Burst:* 500 ft • *Cloud Top Height:* 32,000 ft

30,000-Foot Trajectory: NEVADA: Indian Springs, Las Vegas, Overton, Logandale, Mesquite • UTAH: Santa Clara, Pine Valley, New Harmony, Kanarraville, Cedar City, Enoch, Summit, Parowan, Paragonah, Circleville, Junction, Marysvale, Monroe, Annabella, Glenwood, Sigurd, Salina, Price, Helper, Kenilworth, Bridgeland, Roosevelt, Cedarview, Whiterocks, Tridell, Dry Fork, (Mount Lena) • WYOMING: (Richard's Mountain), Red Desert, Creston, Shirley Basin, (Laramie Mountains), Douglas, Orpha, Lance Creek • SOUTH DAKOTA: Pringle, Fairburn, Hermosa, Owanka, Wall, Quinn, Ottumwa, Hayes, Fort Pierre, Pierre, Blunt, Harrold, Holabird, Highmore, Ree Heights, Miller, St. Lawrence, Veyland, Wessington, Bonilla, Hitchcock, Broadland, Carpenter, Willow Lake, Bryant, Vienna, Lake Norden, Hayti, Dempster, Estelline, Toronto, Astoria • MINNESOTA: Hendricks, Wilno, Ivanhoe, Marshall, Lynd, Revere, Lamberton, Sanborn, Comfrey, Sveadahl, Darfur, St. James, Grogan, Lewisvile, Winnebago, Huntley, Delavan, Frost, Bricelyn, Kiester • IOWA: Scarville, Lake Mills, Leland, Ventura, Clear Lake, Thornton, Latimer, Iowa Falls, New Providence, St. Anthony, Clemons, Melbourne, Newton, Rasnor, Otley, Flagler, Pershing, Attica, Maysville, Weller, Melrose, Iconium, Mystic, Brazil, Centerville, Numa, Cincinnati • MISSOURI: Hartford, Pure Air, Melrose, Ethel, New Cambria, Kaseyville, Prairie Hill, Salisbury, Forest Green, Glasgow, Lisbon, Boonesboro, Boonville, Pilot Grove, Bellair, Syracuse, Stover, Climax Springs, Urbana, Louisburg, Halfway, Brighton, Pleasant Hope, Glidewell, Springfield, Clever, Aurora, McDowell, Ridgley, Exeter • ARKANSAS: Pea Ridge, Bentonville, Rogers, Highfill, Robinson, Siloam Springs • OKLAHOMA: West Siloam Springs, Kansas, Daks, Leach, Peggs, Yonkers, Wagover, Red Bird, Coweta, Leonard, Mounds, Slick, Beggs, Nuyaka, Henryetta, Pharoah, Vernon, Hanna, Indianola, Krebs, Alderson, Dow, Haileyville, Hartshorne, Daisy, Jumbo, Miller, Farris, Boehler, Sunkist, Cade, Bokchito, Utika, Achille, Hendrix • TEXAS: Denison, Bells, Ida, Tom Bean, Sherman, Gunter, Tioga, Pilot Point, Sanger, Boliver, Krum, Ponder, Argyle, Justin, Roanoke, Bartonville, Grand Prairie, Arlington, Irving, Midlothian, Waxahachie, Maypearl, Italy, Frost, Milford, Martens, Emmett, Irene, Malone, Mount Calm, Axtell, Mart, Riesel, Ben Hur, Otto, Perry, Marlin, Cedar Springs, Rosebud, Baileyville, Burlington, Ben Arnold, Cameron, Minerva, Milano, Rockdale, Tanglewood, Old Dime Box, Lincoln, Dime Box, Hills, Giddings, Winchester, West Point, Plum, Muldoon, High Hill, Engle, Moravia, Breslau, Shiner, Sweet Home, Halletsville, Speaks, Cordele, Louise, Danevang, Bay City.

CHARACTERISTICS: PLUMBBOB DIABLO

The device for the DIABLO shot was 16.2 inches in diameter, 68.4 inches in length and weighed 1352 lb. Among the nuclear tests, Plumbbob DIABLO is ranked 6th for percentage of Pb-203 in fallout. The deposition pattern for this shot is most similar to Upshot-Knothole GRABLE (UK10) (R: 0.42; T: 25.85). Plumbbob DIABLO is ranked 3rd in the amount of Te-127m deposited nationwide and 6th in the amount of Ru-106. Counties of highest deposition included Nye NV,

White Pine NV, Lincoln NV, Stark ND, Teton ID, Utah UT, Dunn ND. The day one ratio of total fallout to I-131 is moderately low at 91 to 1.

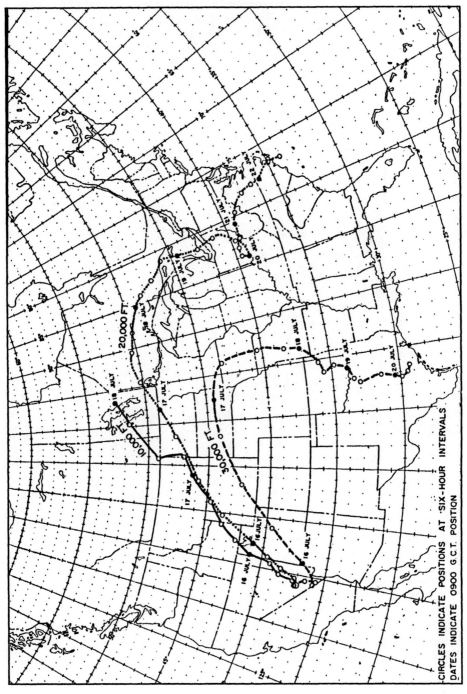

U.S. FALLOUT ATLAS : TOTAL FALLOUT

TRAJECTORY: PLUMBBOB DIABLO (PB5)

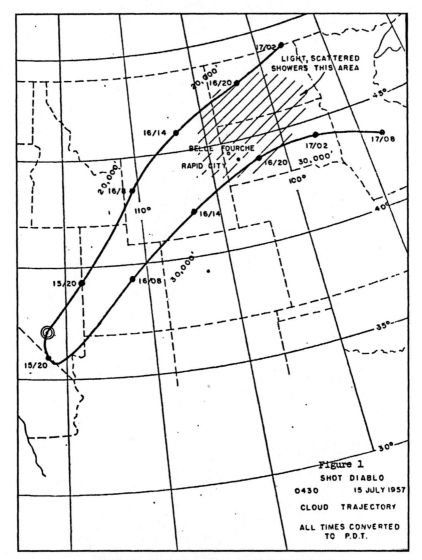

Figure 1
SHOT DIABLO
0430   15 JULY 1957
CLOUD TRAJECTORY
ALL TIMES CONVERTED TO P.D.T.

THE BELLE FOURCHE INVESTIGATION

Relatively high levels of radioactivity were reported in the Belle Fourche, SD area in late August, 1957. Investigators from the Atomic Energy Commission suggested that the increased levels of radioactivity may have come from either shot DIABLO or shot Stokes. This map was taken from the final report of the investigation.

TOTAL FALLOUT: PLUMBBOB DIABLO
Equal Intervals
nCi/square meter

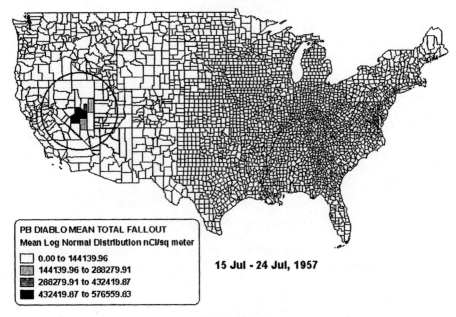

15 Jul - 24 Jul, 1957

TOTAL FALLOUT: PLUMBBOB DIABLO
Equal Number
nCi/square meter

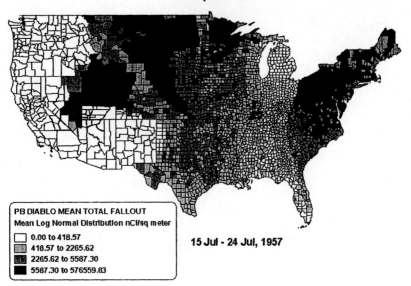

15 Jul - 24 Jul, 1957

PLUMBBOB JOHN
(Not Included in the NCI Study)

PLUMBBOB: JOHN

Detonation Date: 19 Jul 1957 • *Detonation Time:* 7:00 A.M. • *Area:* 10 • *Sponsor:* Department of Defense • *Yield:* Approximately 2 kt • *Radiation Level at Ground Zero:* None detected • *Height of Burst:* 20,000 ft • *Cloud Top Height:* 44,000 ft

40,000-Foot Trajectory: NEVADA: Currant, Ruth, Steptoe • UTAH: Wendover, Park Valley, Clear Creek • IDAHO: Naf, Bridge, Sublett, Rockland, American Falls, Pocatello, Chubbock, Blackfoot, Basalt, Shelly, Idaho Falls, Ammon, Lincoln, Ucon, Ririe, Heise, Archer, Rexburg, Teton, Drummond, Marysville, Warm River •. WYOMING: (Old Faithful in Yellowstone National Park), (Craig Pass), Canyon, (Dunraven Pass), Tower Junction • MONTANA: (Gallatin National Forest), Limestone, Greycliff, Reedpoint, Rapelje, Belmont, Lavina, Flatwillow, Cat Creek, Mosby, (Fort Peck Lake), Frazier, Fort Peck Indian Reservation, Plentywood, Westby.

CHARACTERISTICS: PLUMBBOB JOHN

The plutonium core for this air-to-air device was 17 inches in diameter, was 17.25 inches in length and weighed 221 lbs.

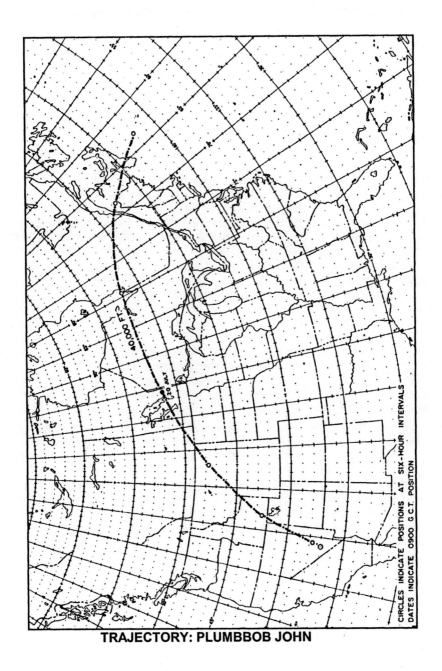

TRAJECTORY: PLUMBBOB JOHN

SHOTS 6A AND 6B

PLUMBBOB KEPLER AND PLUMBBOB OWENS
(PB6)

PLUMBBOB: KEPLER

Detonation Date: 24 Jul 1957 • *Detonation Time:* 4:50 A.M. • *Area:* 4 • *Sponsor:* Los Alamos • *Yield:* 10 kt • *Radiation Level at Ground Zero:* 100 R/hr • *Height of Burst:* 500 ft • *Cloud Top Height:* 28,000 ft

20,000-Foot Trajectory: NEVADA: (Papoose Range), (Timpahute Range), • UTAH: Garrison, (Cedar Mountains), (Great Salt Lake), Corinne, River City, Honeyville, Garland, Fielding, Newton, Clarkston, Lewiston • IDAHO: Franklin, St. Charles, Paris, Bloomington, Montpelier • WYOMING: Daniel, Pinedale, Crowheart, Wind River Indian Reservation, (Owl Creek Mountains), Thermopolis, East Thermopolis, Lucerne, Kirby, Ten Sheep, (Powder River Pass), Buffalo, Weston, Colony • SOUTH DAKOTA: Hoover, Zeona, Meadow, Bullhead, Walker • NORTH DAKOTA: Fort Yates, Strasburg, Wishek, Lehr, Merricourt, Monango, Crete, Gwinner, Milnor, De Lamere, Wyndmere, Barney, Wahpeton • MINNESOTA: Breckenridge, Everdell, Fergus Falls, Underwood, Urbank, Perkers Prairie, Rose City, Flensburg, Little Falls, Buckman, Ramey, Morrill, Milaca, Grandy, Stanchfield, Stark, Harris • WISCONSIN: Atlas, Trade Lake, Frederic, Clam Falls, Barranett, Haugen, Brill, Mikana, Ladysmith, Jump River, Little Chicago, Wausau, Schofield, Ringle, Hatley, Norrie, Wittenberg, Pella, Caroline, Embarrass, Navarino, Nichols, Briarton, Seymour, DePere, Morrison, Wayside, Reedsville, Whitelaw, Manitowoc • MICHIGAN: Whitehall, Dalton, Casnovia, Harrisburg, Sparta, Belmont, Rockford, Cannonsburg, Lowell, Saramac, Portland, Eagle, Grand Lodge, Lansing, Millett, Dimondale, Golt, Mison, Dansville, Millville, Plainfield, Gregory, Hell, Ann Arbor, Dixboro, Ypsilanti, Carleton, Rockwood, South Rockwood, Newport, Estral Beach • OHIO: Lorain, Sheffield Lake, North Ridgeville, Elyria, Columbia Station, Brunswick, Weymouth, Sharon Center, Norton, Barberton, Liberty, Greensburg, Marshand, North Canton, East Canton, Mapleton, Oneida, Pattersonville, Wattsville, North Canton, East Canton, Mapleton, Oneida, Pattersonville, Wattsville, Bergholz, Richmond, Steubenville, Colliers • WEST VIRGINIA: Follansbee, Wellsburg • PENNSYLVANIA: Penowa, Avella, West Middletown, Washington, Gabby Heights, Vestaburg, LaBelle, Millsboro, East Millsboro, Isabella, Filbert, Hibbs, New Salem, Uledi, Newcomer, Fairchance, Elliotsville • MARYLAND: Asher Glade, Selbysport, Friendville, Accident, Bitlinger, Luke, Bloomington, Westernport, McCoole • WEST VIRGINIA: Keyser, Burlington, Junction, Rio, Wardensville • VIRGINIA: Mount Olive, Maurertown, Detrick, Bentonville, Viewtown, Rixeyville, Brandy Station, Richardsville, Flat Run, Fredericksburg, Four Mile Fork, Guinea, Villboro, Bowling Green, Spark, Millers Tavern, Center Cross, Laneview, Church View, Urbanna, Locust Hill, Hartfield, Dixie, Amburg, Deltaville, Gwynn, Hudgins, Diggs, Eastville, Cheriton, Bay View, (Ship Shoal Island).

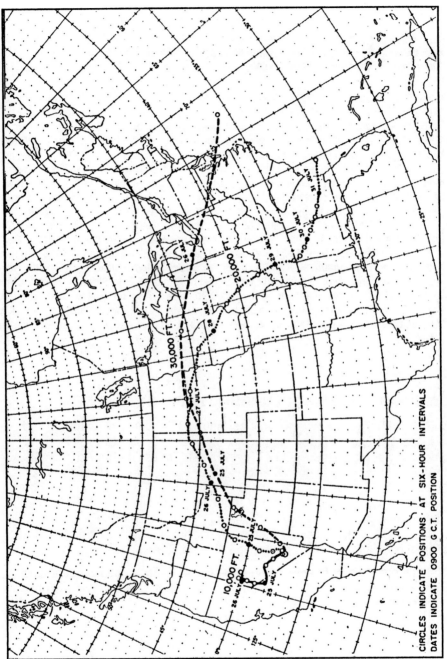

TRAJECTORY: PLUMBBOB KEPLER (PB6A)

PLUMBBOB: OWENS

Detonation Date: 25 Jul 1957 • *Detonation Time:* 6:30 A.M. • *Area:* 9b • *Sponsor:* Livermore Laboratory • *Yield:* 9.7 kt • *Radiation Level at Ground Zero:* 1,000 R/hr • *Height of Burst:* 500 ft • *Cloud Top Height:* 35,000 ft

10,000-Foot Trajectory: NEVADA: Currant, (Pancake Summit), Ruby Valley, Lamoille, (Secret Pass), (Humboldt Range), Wells, Metropolis, Thousand Springs, Contact, Jackpot • IDAHO: Rogerson, Kimberly, Eden, Dietrich, Picabo, Gannett, (Borah Peak, highest point in Idaho), Small • WYOMING: (Yellowstone National Park), West Thumb, Pahaska, (Windy Mountain), (Bear Tooth Pass) • MONTANA: Belfry, St. Xavier, Busby, Lame Deer, Ashland, San Labre Mission, (Custer National Forest) • SOUTH DAKOTA: Camp Crook, Zeona, Maurine, Isabel, Firested, Timber Lake, Wakpale, Mobridge, Herreid, Artas • NORTH DAKOTA: Zeeland, Wishek, Nortonville, Jud, Grand Rapids, La Moure, Verona, Stirum, Gwinner, Cayuga, Sisseton Indian Reservation • SOUTH DAKOTA: Velben, Hammer, Claire City, Sisseton, Peever, Big Stone City • MINNESOTA: Beardsley, (Hartford Beach State Park), Ortonville, Odessa, Louisburg, Milan, Watson, Montevideo, Wegdahl, Granite Falls, North Redwood, Evan, Cobden, Sleepy Eye, Leavenworth, Darfur, Butterfield, Mountain Lake, Bergen, Alpha, Jackson, Petersburg • IOWA: Huntington, Superior, Terril, Dickens, Greenville, Rossio, Peterson, Cherokee, Quimby, Washta, Pierson, Climbing Hill, Sloan • NEBRASKA: Walthill, Pender, Wisner, Clarkson, Schuyler, Columbus, Bellwood, Rising City, Surprise, Cresham, Waco, Lushton, Grafton, Sutton, Ong, Davenport, Ruskin, Hardy • KANSAS: Webber, Lovewell, Formoso, Montrose, Jewell, Randall, Hunter, Ash, Grove, Lucas, (Wilson Lake).

CHARACTERISTICS: PLUMBBOB KEPLER AND OWENS

The KEPLER device was 44 inches long, 28.5 inches in diameter and weighed 1517 lbs. Among the nuclear tests, Plumbbob KEPLER is 1st for production of Ag-109m, Ag-112, Cd-115, In-115m, Pd-111m, Pd-112, Sb-126 and I-130. Plumbbob OWENS is 1st for production of Tb-161 and 2nd for Ag-112, Cd-115, In-115m, In-117, Pd-112, Sm-153 and Tc-99m. The deposition pattern for these two shots (as combined by the National Cancer Institute) is most similar to that of Plumbbob DOPPLER (PB9) (R: 0.39: T: 23.55.) Counties of highest deposition include: Esmeralda NV, Mineral NV, Churchill NV, Beaverhead MT and Granite MT.

The day one ratio of total fallout to I-131 for KEPLER is the lowest among all the aboveground tests at 81 to 1, while shot OWENS was slightly higher at 86 to 1.

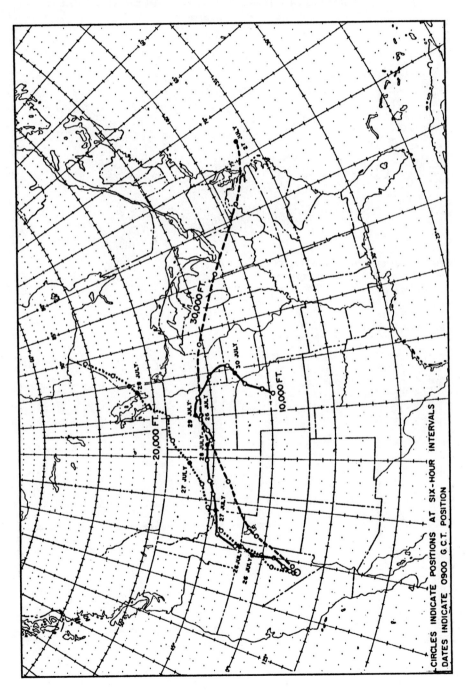

U.S. FALLOUT ATLAS : TOTAL FALLOUT

TRAJECTORY: PLUMBBOB OWENS (PB6B)

TOTAL FALLOUT PLUMBBOB SHOTS KEPLER AND OWENS
Equal Intervals
nCi/square meter

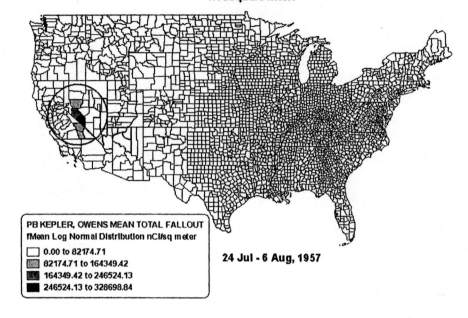

PB KEPLER, OWENS MEAN TOTAL FALLOUT
fMean Log Normal Distribution nCi/sq meter
- 0.00 to 82174.71
- 82174.71 to 164349.42
- 164349.42 to 246524.13
- 246524.13 to 328698.84

24 Jul - 6 Aug, 1957

TOTAL FALLOUT PLUMBBOB SHOTS KEPLER AND OWENS
Equal Number
nCi/square meter

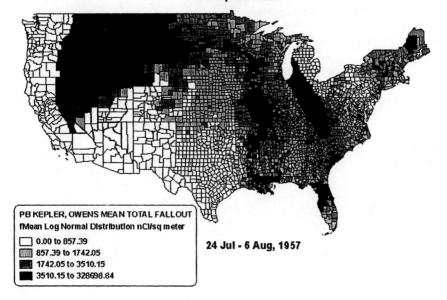

PB KEPLER, OWENS MEAN TOTAL FALLOUT
fMean Log Normal Distribution nCi/sq meter
- 0.00 to 857.39
- 857.39 to 1742.05
- 1742.05 to 3510.15
- 3510.15 to 328698.84

24 Jul - 6 Aug, 1957

SHOT 7: PLUMBBOB STOKES (PB7)

PLUMBBOB: STOKES

Detonation Date: 7 Aug 1957 • *Detonation Time:* 5:25 A.M. • *Area:* 7b • *Sponsor:* Los Alamos • *Yield:* 19 kt • *Radiation Level at Ground Zero:* 50 R/hr • *Height of Burst:* 1,500 ft • *Cloud Top Height:* 37,000 ft

30,000-Foot Trajectory: NEVADA: Ely, Tippett • UTAH: Ibapah, Brigham City, Logan, Pickleville • WYOMING: Geneva, Afton, Bondurant, Dubois, Meeteetse, Lovell, Kane • MONTANA: (Bighorn Canyon), Saint Xavier, (Custer Battlefield), Castle Rock, Hathaway, Fort Keough, Miles City, Wibaux • NORTH DAKOTA: Grassy Butte, Parshall, Makoti, Logan, Granville, Denbigh, Towner, Rugby, Wolford, Bisbee, Egeland, Munich, Langdon, Leroy, Pembina • CANADA: Montreal • NEW YORK: Moopers Forks, West Chazy, Beekmantown, Plattsburgh, Clintonville, Westport, Mineville, Crown Point, Ticonderoga, Putnam Station, Clemons, Whitehall, South Hartford, Cossayuna, Salem, Greenwich, Johnsonville, West Steventown, New Concord, Hillsdale, West Copake, Hope Farm, Wingdale, Holmes, Brewster, Pleasantville, White Plains, New Rochelle, New York City.

CHARACTERISTICS: PLUMBBOB STOKES

The device for the STOKES test was 44 inches in length with a diameter of 28.5 inches. It weighed 1517 lbs. The Stokes shot ranked 2nd for (theoretical) percentage of Zr-95 in fallout. However, the NCI Plumbbob STOKES data suggests this test deposited no fallout. But an investigation by researchers for the Atomic Energy Commission strongly suggested that either the STOKES or the earlier DIABLO shot deposited significant fallout near the town of Belle Fourche, SD.

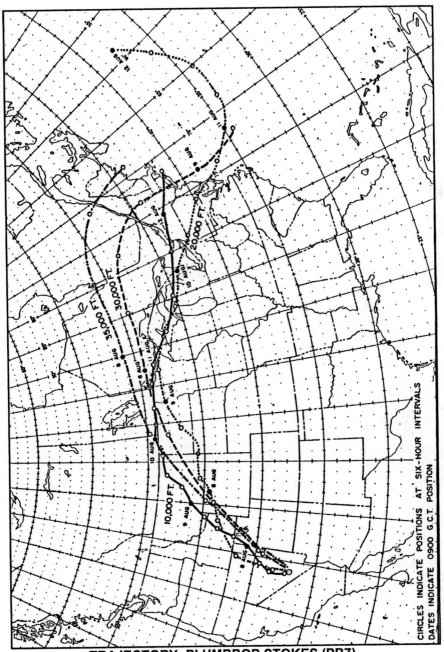

TRAJECTORY: PLUMBBOB STOKES (PB7)

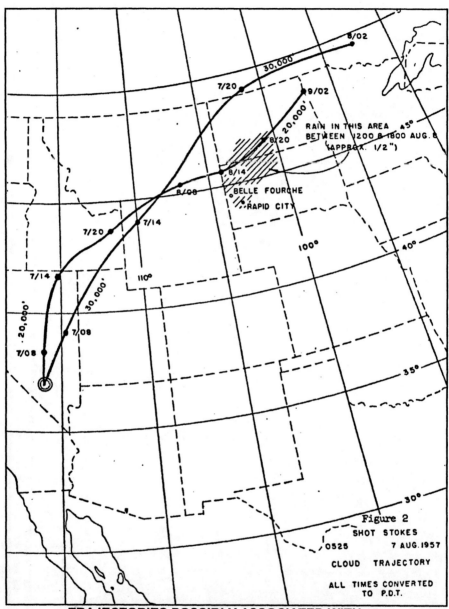

**TRAJECTORIES POSSIBLY ASSOCIATED WITH
THE BELLE FOURCHE RADIOACTIVITY**

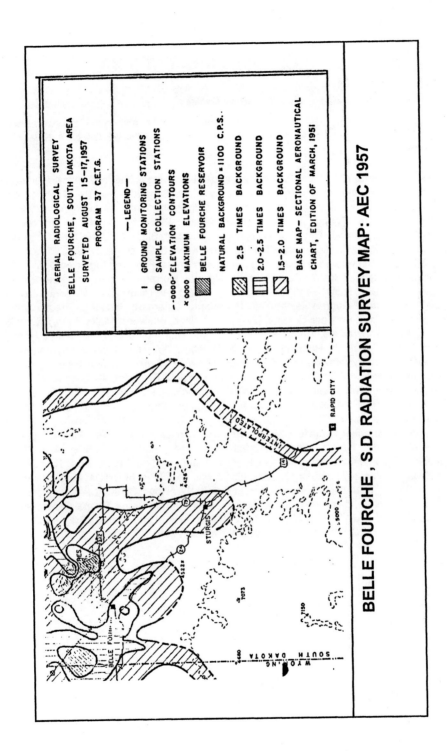

BELLE FOURCHE, S.D. RADIATION SURVEY MAP: AEC 1957

U.S. FALLOUT ATLAS : TOTAL FALLOUT

SHOT 8: PLUMBBOB SHASTA (PB8)

PLUMBBOB: SHASTA

Detonation Date: 18 Aug 1957 • *Detonation Time:* 5:00 A.M. • *Area:* 2a • *Sponsor:* Livermore Laboratory • *Yield:* 17 kt • *Radiation Level at Ground Zero:* 500 R/hr • *Height of Burst:* 500 ft • *Cloud Top Height:* 32,000 ft

20,000-Foot Trajectory: NEVADA: (Belted Range), (Reveille Range), (Railroad Valley), (Pancake Range), (Pancake Summit), Cherry Creek, (Becky Peak), (Sugarloaf Peak) • UTAH: (Great Salt Lake), (Promontory Mountains), Corinne, Bear, River City, Wellsville, Deweyville, Logan, Providence, Hyrum, Round Valley, Laketown, (Bear Lake) • WYOMING: (Commissary Ridge), La Barge, Atlantic City, Gas Hills, (Pine Mountain), Bill • SOUTH DAKOTA: Edgemont, Smithwick, Manderson, Porcupine, Wounded Knee, Batesland • NEBRASKA: Merriman, Eli, Rose, Raeville, Petersburg, Creston, Leigh Ames, Fremont, Inglewood, Elk City, Bennington, Omaha • IOWA: Council Bluffs, Mineola, Malvern, Strahan, Imogene, Essex, Norwich, Yorktown, Shambaugh, Braddyville • MISSOURI: Clearmont, Pickering, Ravenwood, Conception, Gentryville, McFall, Pattenburg, Altamont, Gallatin, Breckinridge, Mooresville, Ludlow, Braymer, Coloma, Bogard, Carrollton, Bosworth, Malta Bend, Marshall, Napton, Nelson, Syracuse, Tipton, Fortuna, Latham, High Point, Etterville, Marys Home, Capps, St. Anthony, Iberia, Hancock, Dixon, Devils Elbow, Duke, Flat, Licking, Raymondville, Yukon, Eunice, Summersville, Pine Crest, Mountain View, Teresita, Thomasville, Rover, Alton, Couch, Myrtle • ARKANSAS: Dalton, Noland, Walnut Ridge, Hoxie, Sedgwick, Bono, Harrisburg, Payneway, Twist, Marion, West Memphis • MISSISSIPPI: Glower, Barks, Eudora, Arkabutla, Coldwater, Seratobia, New Town, Looxahoma, Taylor, Springdale, Water Valley, Bunner, Shepherd, Pittsboro, Calhoun City, Slate Spring, Houhenlinden, Maben, Sherwood, Reform, Bradley, Longview, Sturgis, Brooksville, Macon, Shuqualak, Paulette, Binnsville • ALABAMA: Panola, Warsaw, Gainsville, Boligee, Forkland, Prairieville, Gallion, Founsdale, Uniontown, Safford, Alberta, Boykin, Darlington, Rosebud, Oakhill, Pine Apple, Red Level, River Falls, Andalusa, Carolina, Libertyville, Florala • FLORIDA: Paxton (highest point in Florida), Glendale, Ponce de Leon, Live Oak, New Hope, Crystal Lake, Bennett, Wewahitchka, Sumatra, Carrabelle, (Gulf of Mexico), Palm Harbor, Ozona, Tampa, Center, Balm, Fort Lonesome, Fort Green, Wauchula, Zolfo Springs, Lake Placid, Brighton, (Lake Okeechobee), Palm City, Stuart, Jensen Beach, Sewall's Point.

CHARACTERISTICS: PLUMBBOB SHASTA

The device used for the SHASTA shot resembled the one used in the DIABLO test. It weighed 1435 lbs, was 16.9 inches in diameter and 69.2 inches in length. Among the nuclear tests, Plumbbob SHASTA was ranked 5th in percentage of Pb-203 in fallout. The deposition pattern for this shot is similar to Upshot-Knothole NANCY (UK2) (R: 0.49, T: 30.91). SHASTA deposited fallout on 2301 U.S. counties (74 percent). SHASTA was ranked 6th in the deposition of Be-7 on U.S. counties and 15th in the deposition of Mn54. Counties of highest deposition included Eureka NV, White Pine NV, Billings ND, Brookings SD, Cherry NE.. On Aug 18 1957, Eureka County NV received an estimated maximum activity level of 1000 µCi/sq meter. Other counties receiving high estimated maximum activity levels included White Pine, NV (695 µCi/sq meter) and Park, WY (274 µCi/sq meter.)

T

RELATIVE TOTAL FALLOUT: PB SHASTA
AUG 18, 1957 17 KT

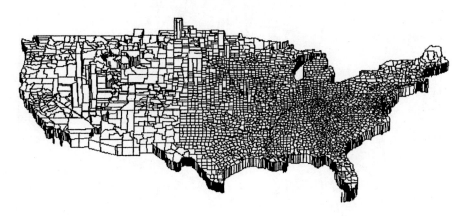

U.S. FALLOUT ATLAS : TOTAL FALLOUT

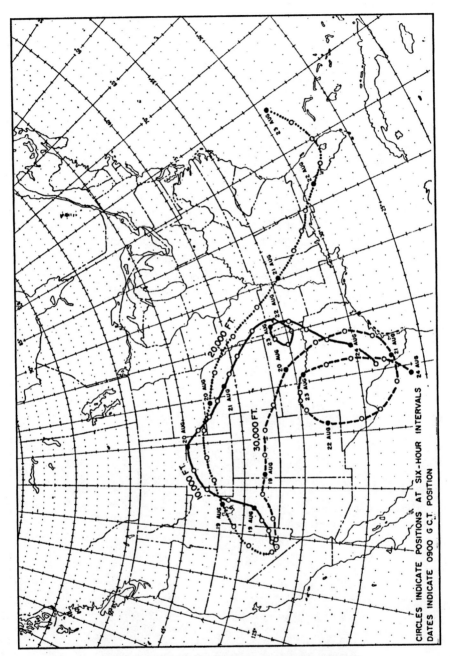

TRAJECTORY: PLUMBBOB SHASTA (PB8)

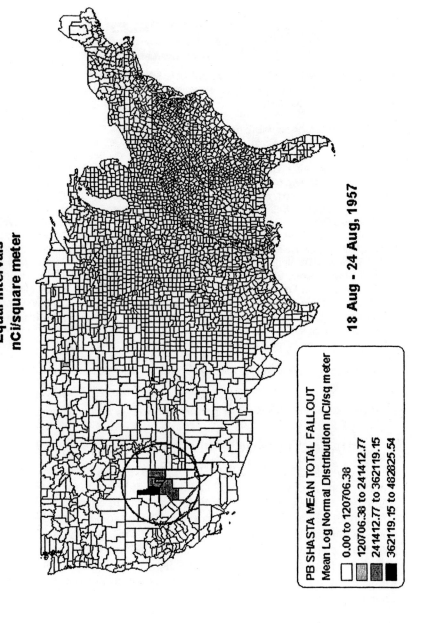

SHOT 9: PLUMBBOB DOPPLER (PB9)

PLUMBBOB: DOPPLER

Detonation Date: 23 Aug 1957 • *Detonation Time:* 5:30 A.M. • *Area:* 7a • *Sponsor:* Los Alamos • *Yield:* 11 kt • *Radiation Level at Ground Zero:* 50 R/hr • *Height of Burst:* 1,500 ft • *Cloud Top Height:* 38,000 ft

40,000-Foot Trajectory: NEVADA: (Papoose Range), (Pahranga Range), Hiko, (Wilson Creek Range) • UTAH: (Indian Peak), (Black Rock Desert), Flowell, Holden, Fayette, Ephraim, Manti, Hiawatha, Price, Wellington, Sunnyside, East Carbon, (Desolation Canyon), (Sweetwater Canyon) • COLORADO: (Douglas Pass), Rio Blanco, Burns, Bond, Green Mountain Camp, Geeny, (Arapahoe National Forest), Black Hawk, Central City, Golden, Arvada, Denver, Englewood, Aurora, Watkins, Bennett, Strasburg, Byers, Deer Trail, Shaw, Seibert, Vona, Stratton, Bethune, Burlington • KANSAS: Kanorado, Brownville, Monument, (Castle Rock), Brownell, McCracken, Hargrave, La Crosse, Bison, Shaffer, Otis, Olmitz, Albert, Helzer, Great Bend, Ellinwood, Silica, Raymond, Alden, Saxman, Sterling, Nickerson, Medora, Hutchinson, Halstead, Furley, Towanda, Haverhill, Leon, Latham, Elk Falls, Elk City, Bolton, Jefferson, Deating, Coffeyville • OKLAHOMA: South Coffeyville, Centralia, Vinita, Ketchum, Jay, Colcord, Siloam Springs, Watts, Ballard • ARKANSAS: Weddington, Farmington, Greenland, Prairie, Brentwood, Winslow, Ozark, Hunt, Atlas, Coal Hill, Midway, (Magazine Mountain, highest point in Arkansas: 2,753 feet), Belleville, Plainview, Rover, Mountain Pine, Royal, Lake Hamilton, Bismark, Caddo Valley, Arkadelphia, Curtis, Amy, Chidester, Camden, Buena Vista, Louann, Smackover, Shaler, Newell, Wesson, Junction City • LOUISIANA: Summerfield, Lisbon, Vienna, Ruston, Grambling, Clay, Ausley, Quitman, Hudson, Georgetown, Fishville, Kolin, Echo, Bunkie, Whiteville, Beggs, Washington, Grand Coteau, Carencro, Lafayette, Broussard, Youngsville, Abbeville, Delcambre, Avery Island, (Shell Keys).

The 40,000-foot trajectory passed over the highest point in Arkansas.

CHARACTERISTICS: PLUMBBOB DOPPLER

The device used in the DOPPLER test was 17 inches in diameter, 26 inches long and weighed only 275 lbs. Among the nuclear tests, Plumbbob DOPPLER is ranked 1st for percentage Na-24 produced and 9th for deposition of Mn-54 nationwide. The deposition pattern for this shot is similar to Plumbbob NEWTON (PB15) (R: 0.41; T: 25.29). Counties of highest deposition included the Iowa counties of Adair, Audubon, Boone, Carroll, Cass, Dallas, Davis, Franklin, Jasper and Madison. On Aug 26 1957, all received (estimated maximum activity) fallout levels of 509 µCi/sq meter.

The day one ratio of total fallout to I-131 for this shot is moderately high at 104:1.

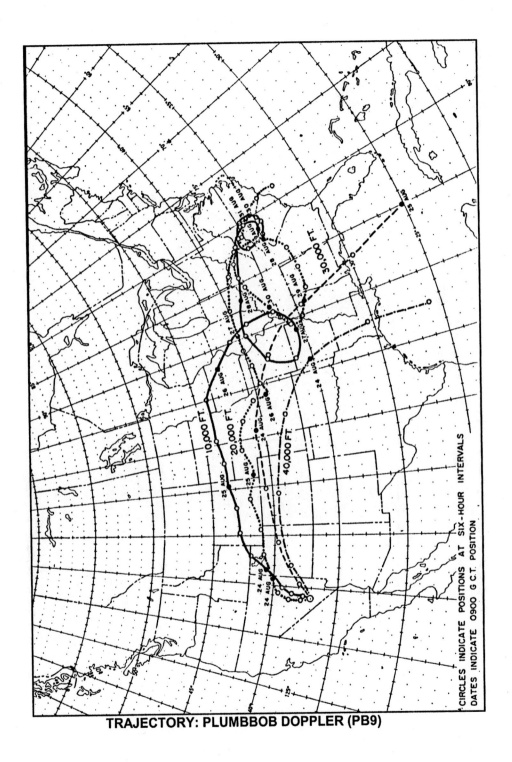
TRAJECTORY: PLUMBBOB DOPPLER (PB9)

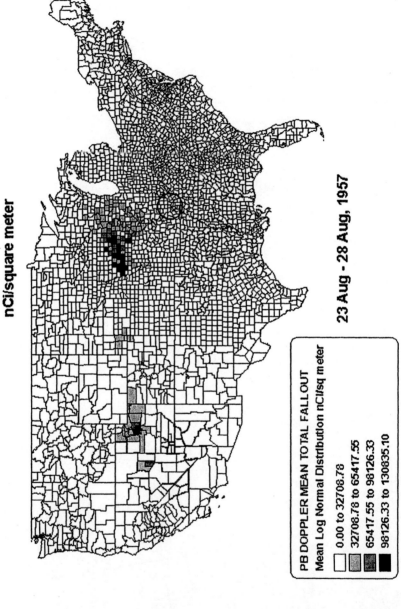

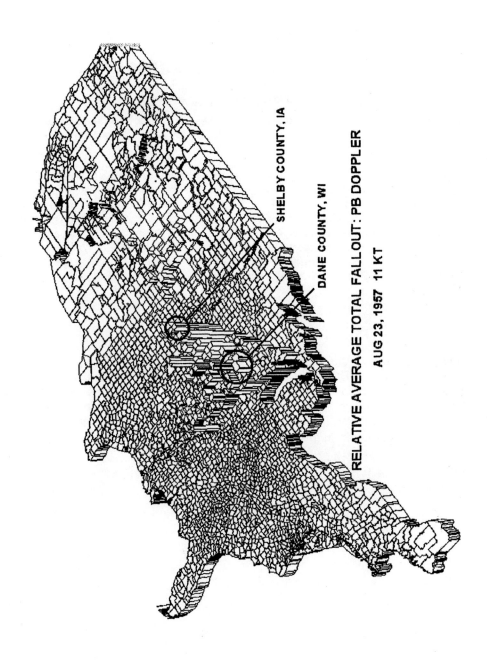

SHOT 10: PLUMBBOB FRANKLIN PRIME (PB-10)

PLUMBBOB: FRANKLIN PRIME

Detonation Date: 30 Aug 1957 • *Detonation Time:* 5:40 A.M. • *Area:* 7b • *Sponsor:* Los Alamos • *Yield:* 4.7 kt • *Radiation Level at Ground Zero:* 100 R/hr • *Height of Burst:* 750 ft • *Cloud Top Height:* 32,000 ft

30,000-Foot Trajectory: NEVADA: Ash Springs, Alamo, Hiko, Caliento • UTAH: Deseret, Hinckley, Nephi, Provo, Orem, Hailstone, Lake Town • WYOMING: Cokeville, Bondurant, Valley, (Buffalo Bill Dam), (Bear Tooth Pass) • MONTANA: Red Lodge, Bear Creek, Joliet, Laurel, Billings, Musselshell, Sand Springs, Glasgow, Geitana • NEW YORK: Potsdam, Parishville, Saranac Lake, Whitehall • VERMONT: Poultney, (Green Mountain National Forest), Houghtonville, Cambridgeport, Bellows Falls • NEW HAMPSHIRE: Walpole, Surry, West Peterborough, Noone, Greenville, Mason • MASSACHUSETTS: Woburn, Lexington, Arlington, Medford, Boston, Beechwood, Egypt, Green Garbor, Hyannis Port.

CHARACTERISTICS: PLUMBBOB FRANKLIN PRIME

This was a retest of the fizzled FRANKLIN shot that had taken place early in the test series. This device had a diameter of 21.4 inches and was 45.8 inches in length. Among the nuclear tests, Plumbbob FRANKLIN PRIME was theoretically 1st for percentages of Nb-95 and Zr-95 in fallout. According to the National Cancer Institute, however, there was *no* fallout deposition recorded from this test.

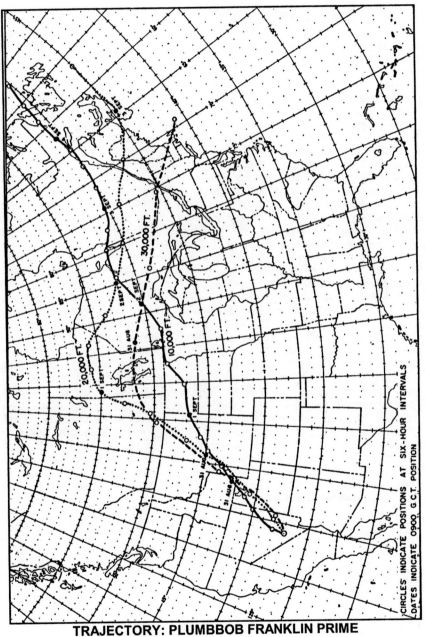

TRAJECTORY: PLUMBBOB FRANKLIN PRIME

SHOT 11: PLUMBBOB SMOKY (PB11)

PLUMBBOB: SMOKY

Detonation Date: 31 Aug 1957 • *Detonation Time:* 5:30 A.M. • *Area:* 2c • *Sponsor:* Livermore Laboratory • *Yield:* 44 kt • *Radiation Level at Ground Zero:* 300 R/hr • *Height of Burst:* 700 ft • *Cloud Top Height:* 38,000 ft

10,000-Foot Trajectory: NEVADA: (Shoshone Peak), Beatty, (Funeral Mountains) • CALIFORNIA: (Death Valley), Baker, Kelso, Essex • ARIZONA: Wikieup, (Snow Mountain), Valentine, Truxton, Peach Springs, Lake Mead, (Hoover Dam) • NEVADA: (Sheep Range), (Timpahute Range), (Grant Range), Preston, Ely, East Ely, (Shell Creek Range), Goshute Indian Reservation • UTAH: Goshute Indian Reservation, (Great Salt Lake Desert), (Cedar Mountains), Grantsville, (Great Salt Lake), (Antelope Island), Brigham City, Wellsville, Logan, Smithfield, Richmond • IDAHO: Weston, Franklin, Whitney, Preston, Benida, Mink Creek, Thatcher, Bench, Grace, Soda Springs, Conda, Freedom • WYOMING: Etna, Freedom, Merna, (Wind River Range), (Wolverine Peak), Fort Washakie, Ethete, Kinnear, Riverton, Moneta, Hiland, Waltman, Powder River, Natrona, Bill • SOUTH DAKOTA: Rockford, Nemo, Piedmont, Elm Springs, Milesville, Hayes, Fort Pierre, Pierre, Crow Creek Indian Reservation, Gann Valley, Crow Lake, Storla, Mitchell, Ethan, Dimrock, Milltown, Menno, Mayfields, Volin, Vermillion, Elk Point, Jefferson, North Sioux City, Sioux City, South Sioux City • IOWA: Bronson, Climbing Hill, Holly Springs, Smithland, Ticonic, Ute, Dunlap, Earling, Westphalia, Marne, Atlantic, Cumberland, Mount Etna, Prescott, Stringtown, Diagonal, Mount Ayr • MISSOURI: Hatfield, Blue Ridge, Gilmar City, Bedford, Avalon, Hale, (Swan Lake), Triplett, Brunswick, Glasgow, Fayette, New Franklin, Rocheport, Wooldridge, Huntsdale, Columbia, Ashland, Holts Summit, Wainwright, Loose Creek, Bland, Bele, Cuba, Steelville, Cherryville, Courtois, Edgehill, (Taum Sauk Mountain, highest point in Missouri), Glover, Sabula, Coldwater, Gipsy, Zalma, Sturdivant, Advance, Painton, Morley, Blodgett, Diehlstadt, Charleston, Wyatt, Wilson City • KENTUCKY: Bardwell, Kirbyton, Fancy Farm, Mayfield, Kirksey, Murray, New Concord • TENNESSEE: Faxon, Waverly, Hurricane Mills, Bold Spring, Spot, Shipps Bend, Centerville, Twomey, Chapel Hill, Mount Pleasant, Culleoka, Mooresville, Lewisburg, Belfast, Richmond, Booneville, Lynchburg, Winchester, Estill Springs, Decherd, Alto, Sewanee, Monteagle, Sequatchie, Victoria, (Signal Mountain), Ridgeside, McDonald, Waterville, Parksville, Benton, Reliance, Servilla, Cokercreek, (Great Smoky Mountains National Park) • NORTH CAROLINA: Webster, Waynesville, Asheville, Little Switzerland, Celo, Collettsville, Valmead, Kings Creek, Hamptonville, Flint Hill, Rural Hall, Dennis, Reidsville, Ruffin, Pelham, Providence, Blanch, Milton • VIRGINIA: Danville, Turbeville, South Boston, Red Oak, Fairview, Chase City, Dundas, McKenney, DeWitt, Burrowsville, Brandon, Williamsburg, Gloucester, Ware Neck, Gwynn, (Chesapeake Bay), Onancock, Tasley, Mappsville, Chincoteague.

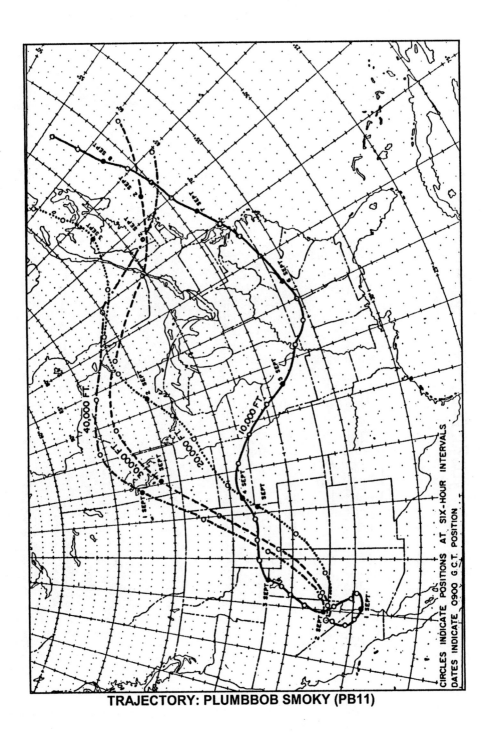

TRAJECTORY: PLUMBBOB SMOKY (PB11)

U.S. FALLOUT ATLAS : TOTAL FALLOUT

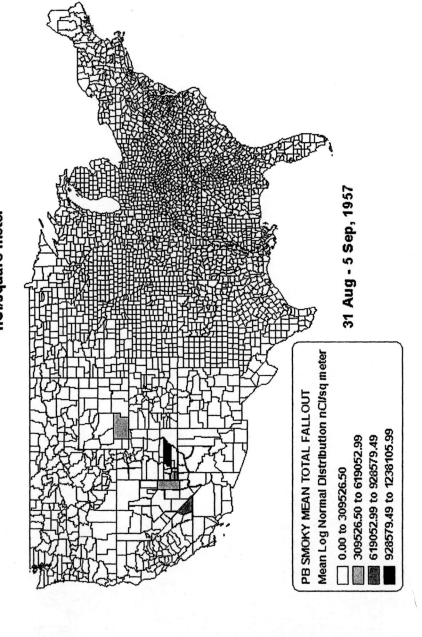

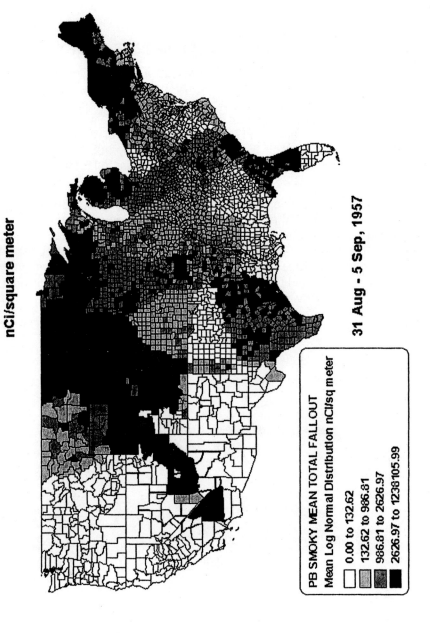

CHARACTERISTICS: PLUMBBOB SMOKY

The device for the SMOKY shot was 50 inches in diameter, 10.5 feet long, and weighed nearly five tons. Among the nuclear tests, Plumbbob SMOKY ranks 4th in the percentage of Be-7 in fallout and 5th for *deposition* of Be7 nationwide. It is also ranked 7th in the percentage of Cd-115, In-115m and In-117 in fallout and 11th for deposition of Mn-54, 15th for deposition of Sr-90 and 28th for deposition of Ru-106 nationwide. The deposition pattern for this shot is most similar to Plumbbob NEWTON (PB15) (R: 0.41; T: 25.29).

Counties of highest deposition included Garfield UT, Inyo CA, Washington UT, Sweetwater WY and Lincoln NV. Estimated maximum activities in µCi/sq meter resulting from this shot included the following: Inyo CA— 2638; Garfield UT—2566; Washington UT—1046; Sweetwater WY—965. The hottest average activity level (total fallout) resulting from this test occurred at Garfield County, UT: 1536 µCi/sq meter (1.536 milliCuries per square meter activity).

The day one ratio of total fallout to I-131 for shot SMOKY is relatively low at 87 to 1.

RELATIVE TOTAL FALLOUT: PB SMOKY
AUG 31, 1957 44 KT

SHOT 12: PLUMBBOB GALILEO (PB12)

PLUMBBOB: GALILEO

Detonation Date: 2 Sept 1957 • *Detonation Time:* 5:40 A.M. • *Area:* 1 • *Sponsor:* Los Alamos • *Yield:* 11 kt • *Radiation Level at Ground Zero:* 200 R/hr • *Height of Burst:* 500 ft • *Cloud Top Height:* 37,000 ft

40,000-Foot Trajectory: NEVADA: (Pahute Mesa), Ash Springs • UTAH: (Dixie National Forest), Central, New Harmony, Kanarraville, Hatch, Tropic, (Natural Bridges National Monument) • COLORADO: Northdale, Dove Creek, Ophir, Red Mountain, (Spring Creek Pass), (Mesa Peak), Timpas, Toonerville • KANSAS: Dodge City, Windthorst, Preston, Varner, Waterloo, Garden Plain, Wichita, Latham, Howard, Neodesha, Pittsburg • MISSOURI: Mindonmines, Lockwood, Springfield, Cabool, Pine Crest, Garwod, Bell City • KENTUCKY: Cairo, Paducah • ILLINOIS: Unionville, Hamletsburg, New Liberty • KENTUCKY: Smithland, Salem, Shady Grove, Manitou, Livermore, Bedo, Sulphur Springs, Hardin Springs, Stephensburg, Elizabethtown, Frederickstown, Mooresville, Willisburg, Duncan, McAffee, Mayo, Troy, Lexington, Mount Sterling, Morehead, Soldier, Jacobs, Lawton, Grahn, Grayson, Kilgore, Rush • WEST VIRGINIA: Huntington, Milton, Hometown, Goldtown, Kentuck, Speed, Milestone, Glenville, Troy, Westan, Lost Creek, Quiet Dell, Bridgeport, Grafton, McGee, Arthurdale, Morgantown, Hazelton • VIRGINIA: Markleysburg, Addison, Listonburg, St. Paul, Glencoe, Fairhope, Rainsburg, Clearville, Crystal Spring, Harrisonville, Metal, Shippenburg, Walnut Bottom, Mount Holly Springs, Lisburn, Harrisburg, High Springs, Lebanon, Mount Aetna, Trexler, Wanamakers, Jacksonville, Palmerton, Effort, Reeders, Henryville, Milford, Millrift • NEW YORK: Rio, Port Jervis, Slate Hill, Johnson, Ridgebury, New Hampton, Goshen, Blooming Grove, Salisbury Mills, Mountainville, Cornell-on-Hudson, Nelsonville, Kent Cliffs, Lake Carmel, Putnam Lake • CONNECTICUT: New Fairfield, Hawleyville, Southbury, Waterbury, Marion, Plantsville, Kensington, Berlin, Wethersfield, Rocky Hill, Glastonbury, Manchester, South Willington, Westford, North Ashford, Quinebaug • MASSACHUSETTS: Dudley Hill, Dudley, Webster, Manchaug, West Sutton, Sutton, Riverdale, Northbridge, Upton, Milford, Holliston, Sherborn, Harding, Dover, Needham, Dedham, Brookline, Boston, Winthrop.

CHARACTERISTICS: PLUMBBOB GALILEO

The device for the GALILEO test was 12.4 inches in diameter, 5 ft. long and weighed 848 lbs. Among the nuclear tests, Plumbbob GALILEO is ranked 1st for percentages of Co-60, Fe-59, Mn-54, Sn-123, Sb-125 and Te-125m in fallout. The deposition pattern for this shot is most similar to that of Plumbbob SHASTA (PB8) (R: 0.44; T: 27.28). Plumbbob GALILEO deposited fallout over 2975 counties, 96 percent of the total U.S. counties. Counties of highest deposition included LaFourche LA, St. Charles LA, Nye NV, Sumner KS, El Paso CO, McPherson KS, Sedgwick KS. Overall, activity levels for this shot were rather low:, On Sep 7, 1957 the highest estimated maximum estimated deposition was for LaFourche, LA for a level of 286 µCi/sq meter. The day one total fallout to I-131 ratio for this shot is moderately low at 92:1.

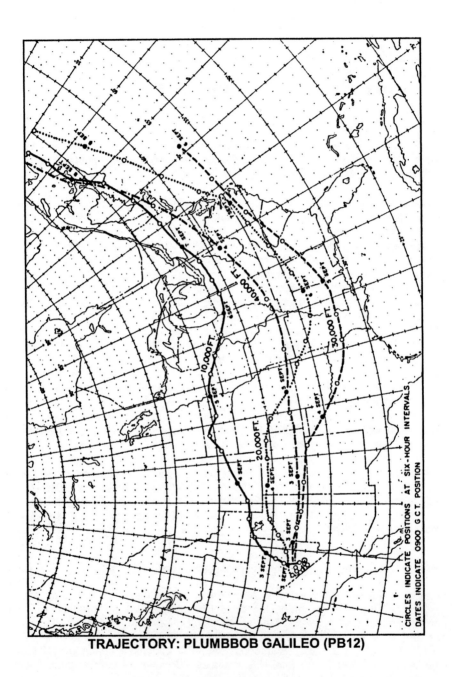
TRAJECTORY: PLUMBBOB GALILEO (PB12)

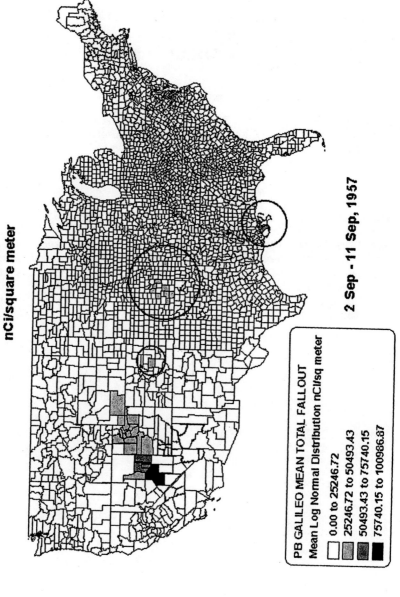

SHOT 13A,B,C: PLUMBBOB WCL (PB13)

SHOTS WHEELER, COULOMB B AND LAPLACE

PLUMBBOB: WHEELER

Detonation Date: 6 Sept 1957 • *Detonation Time:* 5:45 A.M. • *Area:* 9a • *Sponsor:* Livermore Laboratory • *Yield:* 197 tons • *Radiation Level at Ground Zero:* 10 R/hr • *Height of Burst:* 500 ft • *Cloud Top Height:* 17,000 ft

10,000-Foot Trajectory: NEVADA: Goldfield, Hawthorne, Smith, Weed Heights, Reno, Sparks • CALIFORNIA: Twin Bridges, Dardanelle, Mammoth Lakes, Death Valley, Needles • ARIZONA: Drake, Sedonia, Winslow, Holbrook • NEW MEXICO: Fence Lake, Grants, Seyboyeta, Los Alamos, Santa Fe, San Ignacio, Mosquero • TEXAS: Dalhart, Morse (30 miles north of Wheeler, Texas), Higgins • OKLAHOMA: Arnet, Cestos, Longdale, Lacey, Bison, Douglas, Covington, Red Rock, Bowring • KANSAS: Dearing, Mound Valley, Fort Scott • MISSOURI: Nevada, Walker, Schell City, Rockville, Deepwater, Clinton, Windsor, Green Ridge, Sedalia, Boonville, Clark, Paris, Hannibal • ILLINOIS: Hall, Beverly, Fishhook, Beardstown, Kilbourne, Delavan, Danvers, El Paso, Streator, Lisbon, Plattville, Plainfield, Cicero, Chicago • MICHIGAN: Beechwood, Lamont, Sparta, Pierson, Lakeview, Clare, Gladwin, Selkirk, Lincoln.

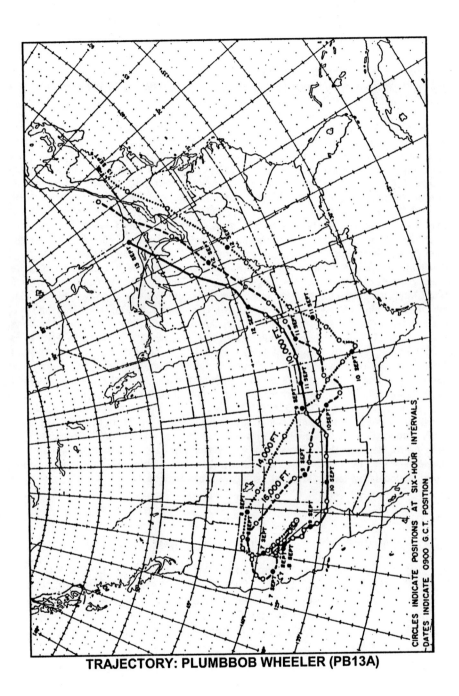

TRAJECTORY: PLUMBBOB WHEELER (PB13A)

Plumbbob: Coulomb B

Detonation Date: 6 Sept 1957 • *Detonation Time:* 1:05 P.M. • *Area:* 3g • *Sponsor:* Los Alamos • *Yield:* 0.3 kt • *Radiation Level at Ground Zero:* 100 R/hr • *Height of Burst:* 3 ft • *Cloud Top Height:* 18,000 ft

10,000-Foot Trajectory: NEVADA: (Shoshone Peak), (Sarcobatus Flat) • CALIFORNIA: (Scotty's Castle), Bishop Indian Reservation, Bishop, Chalfant, Benton • NEVADA: (Boundary Peak, highest point in Nevada), Mount Montgomery, Basalt, (Monte Christo Range), Tonopah, (Mud Lake), (Reveille Range), Templute, Caliente • ARIZONA: Littlefield, (Mount Delenbaugh), Red Lake, (Humphreys Peak, highest point in Arizona), Winona, Leupp Corner, Winslow, (Painted Desert), (Petrified Forest National Park), Zuni Indian Reservation • NEW MEXICO: Zuni Indian Reservation, Ramah Navajo Indian Reservation, Laguna Indian Reservation, Los Chavez, Bosque Farms, Isleta, Tijeras, Edgewood, Moriarty, Clines Corners, Santa Rosa, Puerto De Luna, McAlister, House, Field, Pleasant Hill • TEXAS: Bovina, Rhea, Friona, Black, Summerfield, Hereford, Umbarger, Canyon, Lake Tanglewood, Pullman, Washburn, Panhandle, White Deer, Kings Mill, Pampa, Hoover, Miami, Canadian, Glazier, Higgins • OKLAHOMA: Shattuck, Gage, Fargo, Fort Supply, Camp Houston • KANSAS: Aetna, Medicine Lodge, Zenda, Murdock, Midway, Andale, Bentley, Sedgwick, Elbing, Florence, Cedar Point, Clements, Elmdale, Dunlap, Bushong, Eskridge, Dover, Topeka, Kiro, Meriden, Rock Creek, Valley Falls, Nortonville, Cummings, Atchison, Doniphan, Wathena, Blair • MISSOURI: St. Joseph, Avenue City, Rochester, Union, Star, King City, Gentryville, New Hampton, Bethany, Ridgeway, Blythedale • IOWA: Allerton, Corydon, Bethlehem, Confidence, Melrose, Georgetown, Hiteman, Eddyville, Rose Hill, What Cheer, Keswick, North English, Parnell, Oxford, Homestead, Amana, Swisher, Shueyville, Ely, Martelle, Anamosa, Amber, Scotch Grove, Bernard, St. Donatus • ILLINOIS: Galena, Charles Mound (highest point in Illinois) • WISCONSIN: South Waynbe, Browntown, Monroe, Brodhead, Footville, Janesville, Lima Center, Whitewater, Mukwonago, Big Bend, Franklin, Muskego, Hales Corners, Greenfield, Cudahy, St. Francis, Milwaukee • MICHIGAN: Montague, Hesperia, Aetna, Ramona, White Cloud, Big Rapids, Rodney, Barryton, Sherman City, Brinton, Farwell, Vernon City, Clare, Beaverton, Rhodes, Bentley, Sterling, Maple Ridge, Twining, Turner, National City, Tawas City, East Tawas, Au Sable, Oscoda.

CHARACTERISTICS: PB WHEELER, COULOMB B AND LAPLACE

Among the nuclear tests, Plumbbob WHEELER is 1st for production of La-141, while COULOMB B is 1st for I-132, I-133 and I-135. Plumbbob LAPLACE is 2nd for La-141. The composite deposition pattern for these three shots is similar to Teapot APPLE-2 (TP10) (R: 0.58, T: 39.53). Counties of highest deposition: Buffalo NE, Hall NE, Mitchell KS, Kearney NE, Pawnee KS.

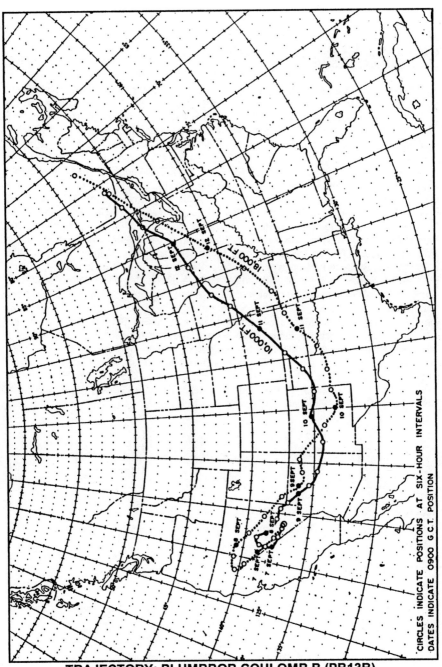

TRAJECTORY: PLUMBBOB COULOMB B (PB13B)

U.S. FALLOUT ATLAS : TOTAL FALLOUT

Plumbbob: La Place

Detonation Date: 8 Sept 1957 • *Detonation Time:* 6:00 A.M. • *Area:* 7b • *Sponsor:* Los Alamos • *Yield:* 1 kt • *Radiation Level at Ground Zero:* 50 R/hr • *Height of Burst:* 750 ft • *Cloud Top Height:* 20,000 ft

10,000-Foot Trajectory: NEVADA: (Spotted Range), (Sheep Range), (Valley of Fire State Park), Logandale, Overton • ARIZONA: (Virgin Mountains), (Grand Wash Cliffs), (Grand Canyon National Park), Hualiai Indian Reservation, (Trinity Mountain), Ash Fork, (Casner Mountain), Sedonia, (Mogolion Plateau), Holbrook, Woodruff, Zuni Indian Reservation • NEW MEXICO: Zuni Indian Reservation, Ramah Navajo Indian Reservation, Belen, Los Chavez, (Manzano Mountains), (Mosca Peak), Escabosa, Edgewood, Clines Corners, Pastura, Sumner Lake, House, McAlester, Forrest, Grady, Cameron • TEXAS: Vega, Ady, Exell, Masterson, Pringle, Morse, McKibben, Spearman, Waka, Farnsworth, Perryton, Twitchell, Booker • OKLAHOMA: Clear Lake, Knowles, Gate • KANSAS: Englewood, Sitka, Protection, Coldwater, Wilmore, Belvidere, Croft, Coats, Pratt, Natrona, Preston, Neola, Plevna, Abbyville, Willowbrook, Medora, Hutchinson, Buhler, Elyria, Lehigh, Durham, Pilsen, Lost Springs, Burdick, Delavan, Wilsey, Hessdale, Keene, Williard, Kiro, Topeka, Silver Lake, Elmont, Valley Falls, Half Mound, Nortonville, Cummings, Atchinson, Doniphan, Wathena, Elwood, Blair • MISSOURI: St. Joseph, Nodaway, Amazonia, Savannah, King City, Gentryville, Albany, Martinsville, Brooklyn, Eagleville, Blythdale • IOWA: Davis City, Pleasanton, Woodland, Corydon, Millerton, Melrose, Georgetown, Albia, Avery, Eddyville, Cedar, Wright, Fremont, Delta, Sigourney, South English, Iowa City, Morse, Cedar Bluff, Buchanan, Clarence, Massillon, Lost Nation, Elwood, Maquoketa, Springbrook • ILLINOIS: Savanna Army Depot, Loran, Kent, Pearl City, Freeport, Eleroy, Dakota, Davis, Shirland, Harrison, Rockton, South Beloit • WISCONSIN: Beloit, Clinton, Allen, Darien, Lyons, Springfield, Honey Creek, Rochester, Burlington, Franksville, Wind Point, North Bay • MICHIGAN: Muskegon, Cloverville, Ravenna, Baily, Kent City, Sand Lake, Trufant, Langston, Stanton, Elm Hall, Sumner, Alma, Ithaca, Nelson, Saginaw, Frankentrost, Reese, Richville, Gifford, Akron, Fairgrove, Colling, Cass City, New Greenleaf, Tyro, Parisville, Ruth.

RELATIVE TOTAL FALLOUT: PB WHEELER, COULOMB-B, LAPLACE

SEP 6, 1957 WHEELER: 197 TONS, COULOMB B 300 TONS, LAPLACE: 1 KT

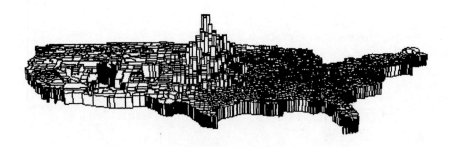

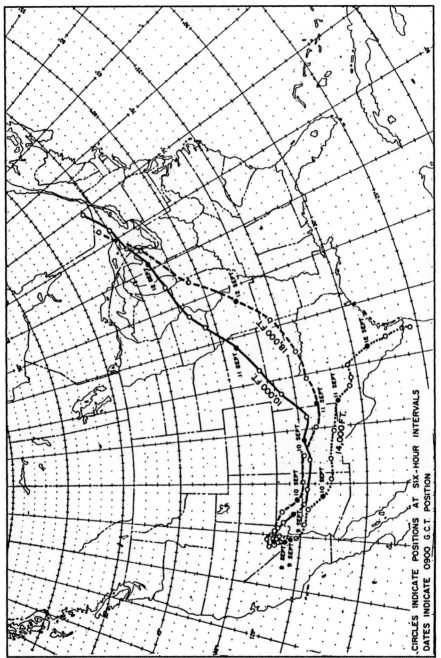

TRAJECTORY: PLUMBBOB LAPLACE (PB13C)

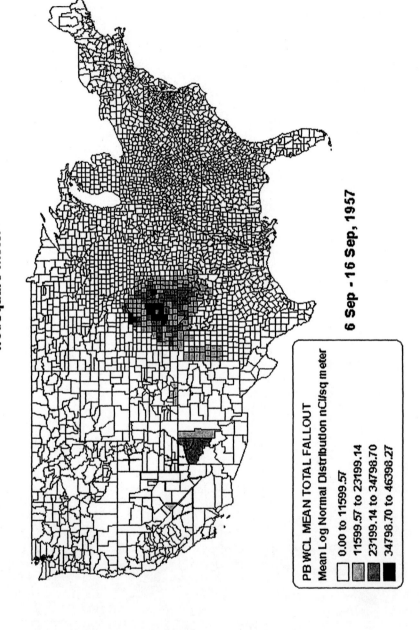

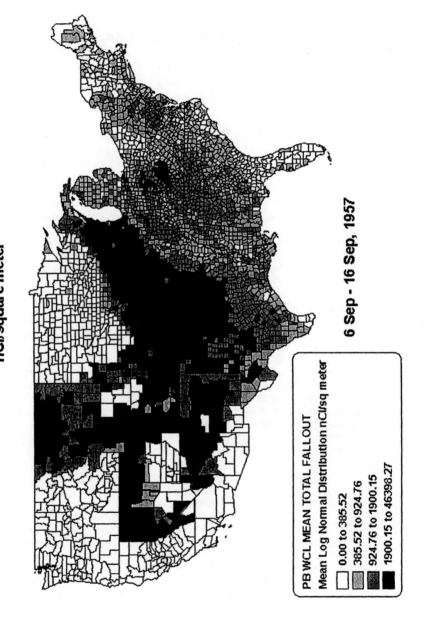

SHOT 14: PLUMBBOB FIZEAU (PB14)

PLUMBBOB: FIZEAU

Detonation Date: 14 Sept 1957 • *Detonation Time:* 9:45 A.M. • *Area:* 3b • *Sponsor:* Los Alamos • *Yield:* 11 kt • *Radiation Level at Ground Zero:* 100 R/hr • *Height of Burst:* 500 ft • *Cloud Top Height:* 40,000 ft

40,000-Foot Trajectory: NEVADA: (Pahute Mesa), Nellis Air Force Range, Tonopah, Luning, Shurz, Yerington, Wabuska, Carson City, Silver City, Glenbrook • CALIFORNIA: Tahoe City, Baxter, Emigrant Gap, Dutch Flat, Grass Valley, Nevada City, Loma Rica, Live Oak, Pennington, Colusa Indian Reservation, Maxwell, Ladoga, Ukiah, Calpella, Navarro, Elk, San Francisco, Alameda, Oakland, Berkeley, Orinda, Danville, Brentwood, Middle River, Byron, Stockton, Holt, French Camp, Farmington, Copperopolis, Tuttletown, Tuolumne, Sonora, (Yosemite National Park), Lee Vining • NEVADA: (Boundary Peak, highest point in Nevada), Montgomery, Coaldale, Tonopah, Warm Springs, Ursine, Ploche • UTAH: Beryl, Land, Parowan, Tropic, Cannonville, Henrieville, Mexican Hat • COLORADO: Ute Mountain Indian Reservation, Marvel, Oxford, Chimney Rock, Pagosa Springs, Capulin, La Jara, Bountiful, Sanford, Lasauses, San Acacio, Viejo San Acacio, San Luis, Gulnare, Ludlow, Hoehne, Villegreen, Kim, Stonington • KANSAS: Richfield, Moscow, Fowler, Minneola, Croft, Coats, St. Leo, Murdock, Cheney, Goddard, Schulte, Wichita, Andover, Augusta, Haverhill, Leon, Climax, Toronto, Batesville, Yates Center, Rose, Piqua, Iola, Gas, LaHarpe, Moran, Bayar, Xenia, Mapleton, Mantey, Harding, Fulton • MISSOURI: Hume, Rich Hill, Prairie City, Ohio, Mount Zion, Warsaw, Whitakerville, Gravois Mills, Eldon, Etterville, Henley, St. Thomas, Westphalia, Mount Sterling, Drake, Stony Hill, Washington, Clover Bottom, Labadie, Pond, Ballwin, Manchester, St. Louis • ILLINOIS: East St. Louis, Collinsville, Edwardsville, Marine, Old Ripley, Pocahontas, Vandalia, Haferstown, Bluff City, Browstown, Altamont, Effingham, Teutopolis, Montrose, Jewett, Hidalgo, Greenup, Casey, Martinsville, Marshall • INDIANA: Prairieton, Terre Haute, Cory, Prairie City, Poland, Cunot, Quincy, Lewisville, Centerton, Brooklyn, Bargersville, Whitelaud, Needham, Shelbyville, Rays Crossing, Manilla, Rushville, Glenwood, Connersville, Alquina, Brownsville, Liberty, Kitchell • OHIO: Sugar Valley, Glenwood, Little Richmond, Trotwood, Northridge, Dayton, Sulpher Grove, Fairborn, Medway, Snyderville, Springfield, Catawba, Tradersvle, Plumwood, Hayden, Linworth, Minerva Park, Croton, Lock, Homer, Millwood, Tiverton, Killbuck, Berlin, Trail, Winesburg, Beech City, Bolivar, Howenstein, North Industry, Mapleton, Robertsville, Bayard, New Alexander, New Garden, Guilford, East Fairfield, Signal, East Palestine, Negley • PENNSYLVANIA: New Galilee, Ellwood City, Mount Chestnut, Windward Heights, Chicora, Frogtown, Wattersonville, Tidal, Climax, New Bethlehem, South Bethlehem, Ringgold, Panic, Prescottville, Sandy, Oklahoma, Home Camp, (Moshannon State Forest), Keating, Cooks Run, West Renovo, Haneyville, Waterville, Quiggleville, Warrensville, Muncy Valley, Sonestown, Nordmont, Beaumont, Orange, Upper Exeter, Ransom, Scranton, Mount Cobb, Hamlin, Lakeville, Tafton, Greeley, Shohola, Pond Eddy, Millrift • NEW JERSEY: (High Point, highest point in New Jersey: 1,803 feet) • NEW YORK: Port Jervis, Huguenot, Mechanicstown, Campbell Hall, New Windsor, Beacon, Farmers Mills, Ludingtonville, Patterson, Towners • CONNECTICUT: New Fairfield, Brookfield, Brookfield Center, Roxbury Falls, Woodbury, Middleburg, Waterbury, Woodtick, Southington, Kensington, New Britain, Berlin, East Berlin, Rocky Hill, East Glastonbury, Diamond Lake, Willimantic, North Windham, South Killingly • RHODE ISLAND: Foster, North Scituate, Saundersville, Providence, East Providence • MASSACHUSETTS: Taunton, Raynham Center, North Middleboro, Warrentown, Silver Lake, Kingston, Tinkertown, South Duxbury.

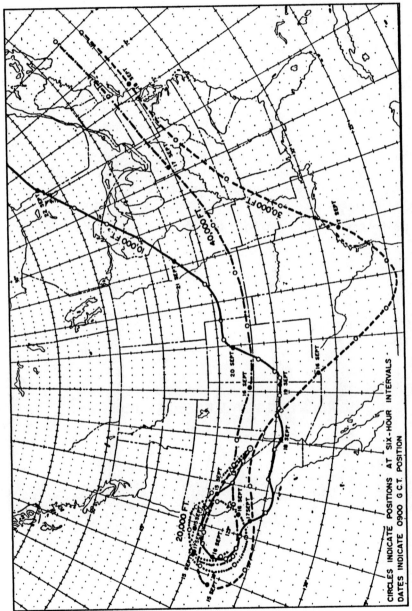

TRAJECTORY: PLUMBBOB FIZEAU (PB14)

CHARACTERISTICS: PLUMBBOB FIZEAU

The device used for the FIZEAU shot was 16 inches in diameter, 31.75 inches long, and weighed just 131.3 lbs. Among the nuclear tests, Plumbbob FIZEAU (PB14) is ranked 5th for percentages of Cd-115, In-115m and In-117 in fallout. The deposition pattern for this shot is similar to the composite pattern for the Plumbbob shots WHEELER, COULOMB B and LAPLACE (R: 0.29, T: 16.76). Fallout from this test was deposited on only 651 counties (21 percent of the total). Counties of highest deposition included Mineral NV, Carson City NV, Douglas NV, Lyon NV, Storey NV. Fallout levels were relatively light. The estimated maximum activity for the series, occurring at Mineral, NV on Sep 14, 1957 was only 119 Ci/sq meter.

The day one total fallout to I-131 ratio for this shot was relatively low at 75 to 1.

RELATIVE TOTAL FALLOUT: PB FIZEAU
SEP 14, 1957 11 KT

TRAJECTORY: PLUMBBOB FIZEAU (PB14)

TOTAL FALLOUT: PLUMBBOB FIZEAU
Equal Intervals
nCi/square meter

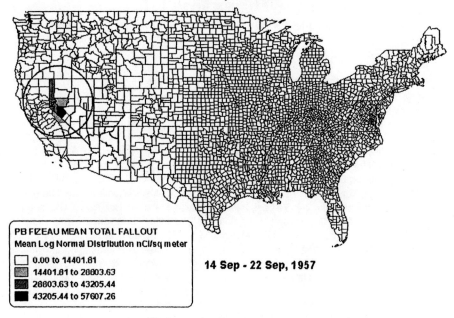

PB FIZEAU MEAN TOTAL FALLOUT
Mean Log Normal Distribution nCi/sq meter
- 0.00 to 14401.81
- 14401.81 to 28803.63
- 28803.63 to 43205.44
- 43205.44 to 57607.26

14 Sep - 22 Sep, 1957

TOTAL FALLOUT: PLUMBBOB FIZEAU
Equal Number
nCi/square meter

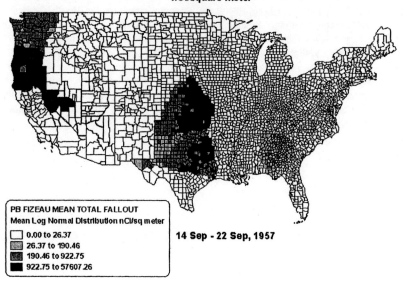

PB FIZEAU MEAN TOTAL FALLOUT
Mean Log Normal Distribution nCi/sq meter
- 0.00 to 26.37
- 26.37 to 190.46
- 190.46 to 922.75
- 922.75 to 57607.26

14 Sep - 22 Sep, 1957

SHOT 15: PLUMBBOB NEWTON (PB15)
PLUMBBOB: NEWTON

Detonation Date: 16 Sept 1957 • *Detonation Time:* 5:50 A.M. • *Area:* 7b • *Sponsor:* Los Alamos • *Yield:* 12 kt • *Radiation Level at Ground Zero:* 50 R/hr • *Height of Burst:* 1,500 ft • *Cloud Top Height:* 32,000 ft

30,000-Foot Trajectory: NEVADA: (Papoose Lake), (Papoose Range), (Pahranagat Range), Ash Springs, Caliente, Panaca • UTAH: Uvada, Modena, Lund, Minersville, Beaver, (Delano Peak), Loa, Bicknel, Teasdale, Torrey, Cainesville, Hanksville, (Canyonlands National Park), (Lisbon Valley) • COLORADO: Norwood, Ridgeway, (Sheep Mountain), Powderhorn, Doyleville, Sargents, (Ouray Peak), Poncha Pass, Wellsville, Howard, Parkdale, (Royal Gorge), Canon City, Fort Carson, Fountain, Widefield, Ellicott, Yoder, Rush, Hugo, Arriba, Flagler, Hale • KANSAS: St. Francis • NEBRASKA: Benkelman, Danbury, Naponee, Red Cloud, Superior, Hardy, Byron, Chester, Hubbell, Thompson, Steele City, Barneston, DuBois, Falls City, Preston, Rulo • KANSAS: Cedar Bluffs, Lyle, Long Island, Woodruff, Thornburg, Womer, Northbranch, Burr Oak, Webber, Republic, Narka, Mahaska, Hollenberg, Oketo, Summerfield, Bern, Berwick, Reserve, Iowa Sac and Fox Indian Reservation, White Cloud • MISSOURI: Fortescue, Mound City, Maitland, Barnard, Guilford, Stanberry, Conception Junction, Albany, Bethany, Mount Moriah, Modena, Mill Grove, Newtown, Pollock, Lemons, Graysville, Worthington, Downing, Memphis, Arbela, Farmington, Athens • IOWA: Argyle, Summitville, Montrose, New Boston • ILLINOIS: Nauvoo, Fort Madison, Niota, Raritan, Swan Creek, St. Augustine, Rapatee, Yates City, Elmwood, Kickapoo, Alta, Rome, Chillicothe, Washburn, Rutland, Long Point, Manville, Blackstone, Gardner, Braceville, Symerton, Wilton Center, Monee, Chicago Heights • INDIANA: Highland, Griffith, Gary, Lake Station, Ogden Dunes, Burns Harbor, Town of Pines, Trail Creek • MICHIGAN: Bertrand, Eagle Lake, Adamsville, Constantine, Centreville, Nottawa, Colon, Fairfax, Sherwood, Hodunk, Union City, Tekonsha, Homer, Concord, Parma, Jackson, Waterloo, Hamburg, New Hudson, Farmington Hills, Pontiac, Birmingham, Rochester Hills, New Baltimore, Marine City • NEW YORK: Sandy Pond, Ellisburg, Pierrepont Manor, Adams, Lorraine, Worth, Copenhagen, West Carthage, Carthage, Indian River, (Alder Bed Mountain), Sabattis, (Seymour Mountain), Keene Valley, Elizabethtown, Wadhams, Whallonsburg • VERMONT: Westport, Vergennes, Irasville, Waltsfield, Montpelier, Berlin, Barre, Groton, Ryegate • NEW HAMPSHIRE: Lyman, Lisbon, Franconia, Bethlehem, Twin Mountain, Bretton Woods, North Chatham • MAINE: Center, Louell, Waterford, South Paris, Hebron, North Monmouth, Monmouth, Winthrop, Hallowell, Gardinier, Randolph, Coopers Mills, Washington, Northport, Lincolnville, Harborside, Islesboro, Brooksville, Blue Falls, North Sedgwick, Bar Harbor, Corea, Winter Harbor, Petit Manan Pt.

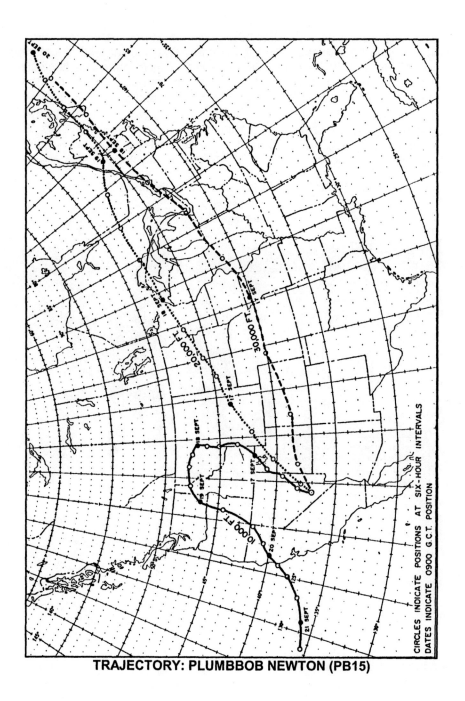

TRAJECTORY: PLUMBBOB NEWTON (PB15)

U.S. FALLOUT ATLAS : TOTAL FALLOUT

CHARACTERISTICS: PLUMBBOB NEWTON

The device used for the NEWTON shot was 28 inches in diameter, 39 inches long and weighed 1346 lbs. Among the nuclear tests, Plumbbob NEWTON was 1st for percentage of Cm-242 in fallout. The deposition pattern for this shot is similar to that of Tumbler-Snapper HOW (TS8) (R: 0.53, T: 34.85). Fallout from this test was deposited on 1531 counties (49 percent of the total.) Counties of highest deposition included Garfield NE, Codington SD, Kingsbury SD, Hamlin SD, Miner SD and Clark SD. Fallout levels were relatively light. The estimated maximum activity at Codington, SD, occurring on Sep 17 1957, was only 204 µCi/sq meter.

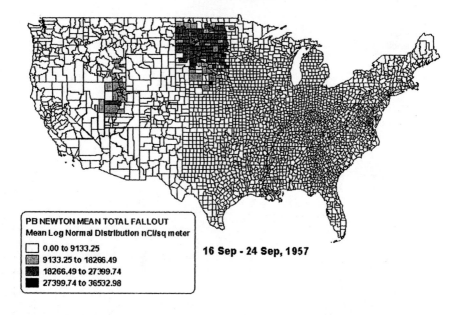

TOTAL FALLOUT: PLUMBBOB NEWTON
Equal Intervals
nCi/square meter

PB NEWTON MEAN TOTAL FALLOUT
Mean Log Normal Distribution nCi/sq meter
- 0.00 to 9133.25
- 9133.25 to 18266.49
- 18266.49 to 27399.74
- 27399.74 to 36532.98

16 Sep - 24 Sep, 1957

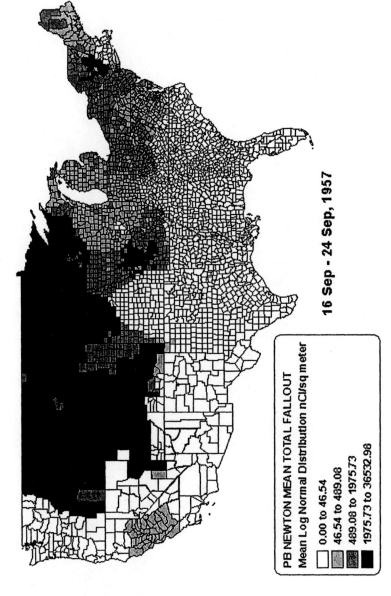

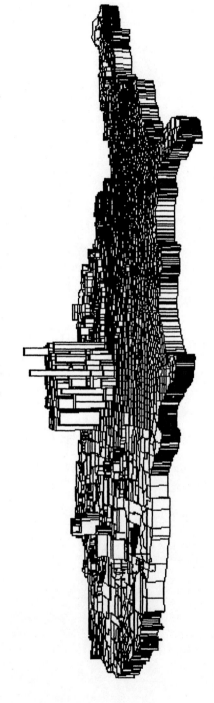

SHOT 16: PLUMBBOB WHITNEY (PB16)

PLUMBBOB: WHITNEY

Detonation Date: 23 Sept 1957 • *Detonation Time:* 5:30 A.M. • *Area:* 2 • *Sponsor:* Livermore Laboratory • *Yield:* 19 kt • *Radiation Level at Ground Zero:* 200 R/hr • *Height of Burst:* 500 ft • *Cloud Top Height:* 30,000 ft

10,000-Foot Trajectory: NEVADA: Warm Springs, Crescent Valley, Carlin, Metropolis, Pequop, Montello • UTAH: (Grouse Creek Mountains), Park Valley • IDAHO: Holbrook, Buist, Arbon, Pocatello, Chubbuck, Riverside, Terreton, Dubois, Spencer, Humphrey • MONTANA: Monida, Sheridan, Twin Bridges, Silver Star, Butte, McQueen, Basin, Rimini, Blossburg, Marysville, Canyon Creek, Choteau, Miller Colony, Bynum, Pendroy, Valier, Ethridge, Sweetgrass.

RELATIVE TOTAL FALLOUT: PB WHITNEY

SEP 23, 1957 19 KT

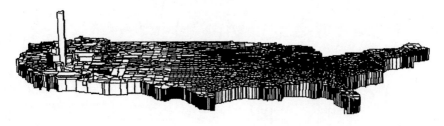

CHARACTERISTICS: PLUMBBOB WHITNEY

Among the nuclear tests, Plumbbob WHITNEY is ranked 2nd for percentages of Am-241 and Fe-59 and 3rd for percentage of Pb-203 in fallout. The deposition pattern for this shot is similar to that of Buster-Jangle CHARLIE (R: 0.30; T: 17.78). WHITNEY deposited fallout on 2982 counties (96 percent). Counties of highest deposition included Esmeralda NV, Churchill NV, Mineral NV, Lander NV, Mono CA. Fallout from this test had low to moderate activity. Esmeralda, NV received an estimated maximum activity of 569 µCi/sq meter, while Mineral, NV received an estimated maximum activity of only 179 µCi/sq meter, both on Sep 23, 1957.

The day one total fallout to I-131 ratio for Plumbbob WHITNEY was relatively low at 91 to 1.

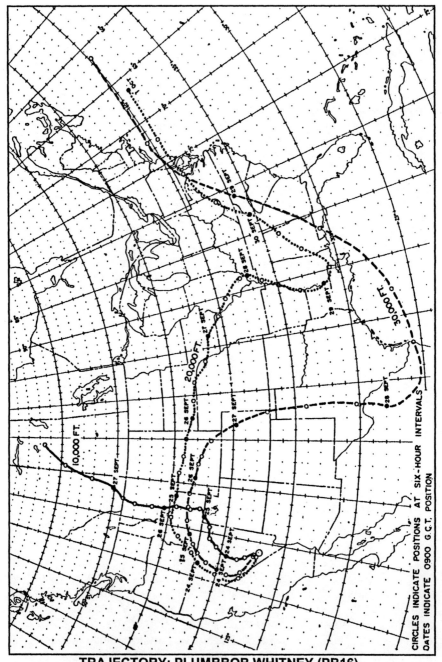

TRAJECTORY: PLUMBBOB WHITNEY (PB16)

TOTAL FALLOUT: PLUMBBOB WHITNEY
Equal Intervals
nCi/square meter

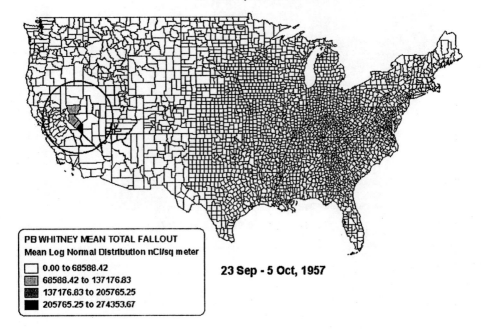

PB WHITNEY MEAN TOTAL FALLOUT
Mean Log Normal Distribution nCi/sq meter
- 0.00 to 68588.42
- 68588.42 to 137176.83
- 137176.83 to 205765.25
- 205765.25 to 274353.67

23 Sep - 5 Oct, 1957

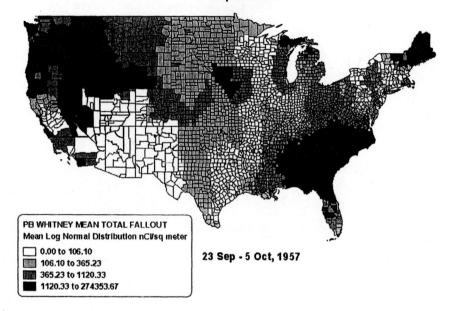

SHOT 17: PLUMBBOB CHARLESTON (PB17)

PLUMBBOB: CHARLESTON

Detonation Date: 28 Sept 1957 • *Detonation Time:* 6:00 A.M. • *Area:* 9a • *Sponsor:* Livermore Laboratory • *Yield:* 12 kt • *Radiation Level at Ground Zero:* 50 R/hr • *Height of Burst:* 1,500 ft • *Cloud Top Height:* 32,000 ft

20,000-Foot Trajectory: NEVADA: (Belted Range), (Revielle Range), (Pancake Range), (Diamond Mountains), Ruby Valley, Wells, (Delano Peak) • IDAHO: Oakley, Marion, Burley, Rupert, Minidoka, Atomic City, Taber, Roberts, Terneton, Hamer, St. Anthony, Island Park • WYOMING: (Yellowstone National Park: Old Faithful), Canyon, Silver Gate • MONTANA: (Granite Peak; highest point in Montana), Alpine, Red Lodge, Luther, Roberts, Franberg, Edgar, Pryor, Crow Indian Reservation, Hardin, Brandenberg, Valborg, Powderville, Ekalaka, Mill Iron • SOUTH DAKOTA: Buffalo, Reva, Sorum, Usta, Red Elm, Dupree, Eagle Butte, Mission Ridge, Blunt, Stephen, Wessington Springs, Letcher, Farmer, Emery, Monroe, Lennox, Worthing, Canton • IOWA: Beloit, Lebanon, Struble, West LeMars, LeMars, Moville, Climbing Hill, Holly Springs, Rodney, Castana, Moorhead, Logan, Neola, McClelland, Treynor, Hastings, Malvern, Randolph, Anderson, Riverton, Hamburg • MISSOURI: Watson • NEBRASKA: Brownville, Nemaha, Verdun, Dawson • KANSAS: Berwick, Sabetha, Kelley, Goff, Corning, Havensville, Onaga, Louisville, Wamego, Wabaunsee, Alta Vista, Council Grove, Strong City, Elmdale, Matfield Green, El Dorado, Haverhill, Douglass, Rock, Akron, Udall, Oxford, Geuda Springs, Ashton • OKLAHOMA: Braman, Billings, Covington, Hayward, Lovell, Crescent, Cashion, Piedmont, El Reno, Union City, Cogar, Dutton, Anadarko, Cyril, Apache, Fletcher, Geronimo, Devol • TEXAS: Burkburnett, Mankins, Dundee, Padgett, Crystal Falls, Breckenridge, Eastland, Rising Star, Brownwood, Trickham, Whon, Fife, Lohn, Pear Valley, London, Junction, Segovia, Barksdale, Camp Wood, (Military Mountain), (Kelly Peak), (Boiling Mountain), (Salmon Peak), (Turkey Mountain), Spofford, (Cline Mountain), Blewett, Dabney, Normandy, Eagle Pass.

CHARACTERISTICS: PLUMBBOB CHARLESTON

The device used for the CHARLESTON shot was 22.4 inches in diameter, 46.6 inches long and weighed 1225 lbs. Among the nuclear tests, Plumbbob CHARLESTON is ranked 2nd for percentage of Ce-141 in fallout. The deposition pattern for this shot is most similar to Plumbbob WILSON (PB2) (R: 0.35; T: 20.94). Fallout from this test covered 3067 counties, ranking its fallout pattern the 3rd most dispersed in the aboveground test series (after Tumbler-Snapper FOX (1) and Upshot-Knothole SIMON (2). Fallout activity, however, was light. Counties with highest fallout amounts included Coffee TN Oconee SC and Clay NC, each with an estimated maximum activity: 137 µCi/sq meter on Sep 30, 1957. The counties of Grundy TN, Sequatchie TN, Bedford TN, and Towns GA each received slightly lower estimated maximums on the same day. The day one total fallout to I-131 ratio for Plumbbob CHARLESTON was moderate at 106 to 1.

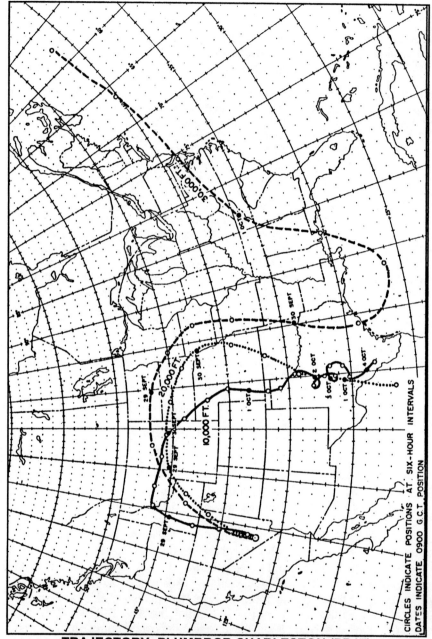

TRAJECTORY: PLUMBBOB CHARLESTON (PB17)

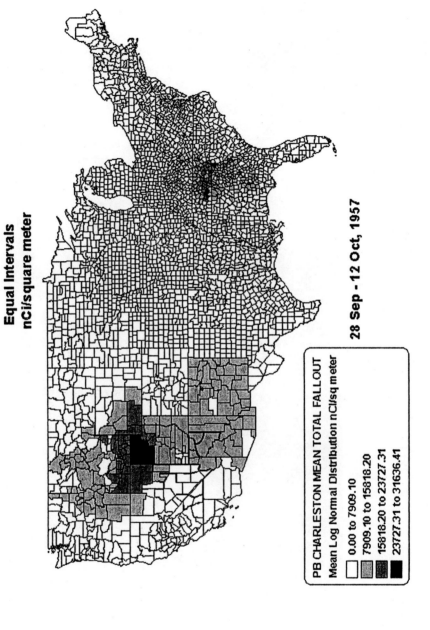

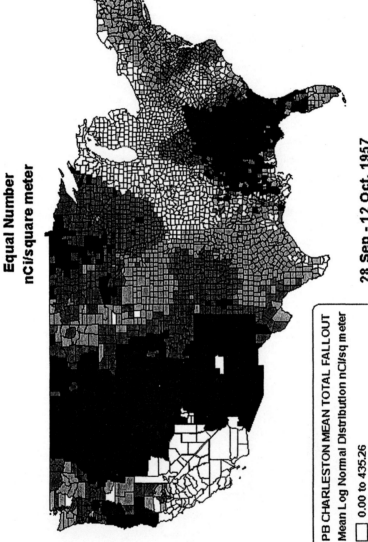

RELATIVE TOTAL FALLOUT: PB CHARLESTON
SEP 23, 1957 12 KT

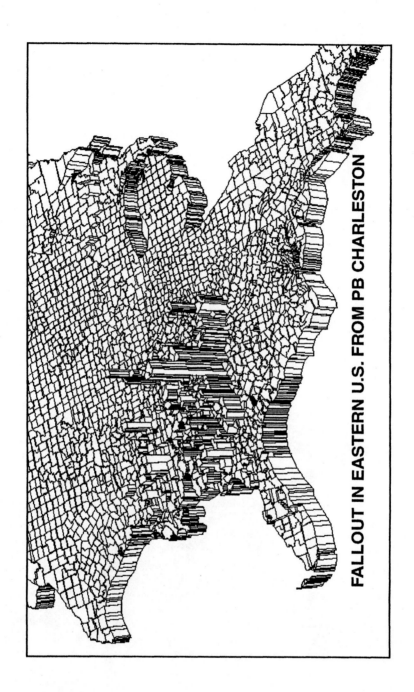

SHOT 18: PLUMBBOB MORGAN (PB18)

PLUMBBOB: MORGAN

Detonation Date: 7 Oct 1957 • *Detonation Time:* 5:00 A.M. • *Area:* 9a • *Sponsor:* Livermore Laboratory • *Yield:* 8 kt • *Radiation Level at Ground Zero:* 100 R/hr • *Height of Burst:* 500 ft • *Cloud Top Height:* 40,000 ft

20,000-Foot Trajectory: NEVADA: (Spotted Range), (Pintwater Range), (Sheep Range), (Mormon Peak) • ARIZONA: Colorado City, Cane Beds, Moccasin, Kaibab Indian Reservation, Fredonia, Page, (Monument Valley), Mexican Water, Teec Nos Pos • UTAH: St. George, Hildale, Kanab, Glen Canyon, (Navajo Mountain), (Rainbow Bridge National Monument), (Monument Valley), Mexican Hat, Ute Mountain Indian Reservation • COLORADO: Ute Mountain Indian Reservation, (Mesa Verde National Park), Hesperus, Durango, Trimble, Hermosa, (Sunlight Peak), Spar City, Wagon Wheel Gap, (Rio Grande National Forest), (Sangre De Cristo Mountains), Lincoln Park, Rockvale, Florence, Canon City, Penrose, Portland, Fort Carson, Fountain, Yoder, Rush, Hugo, Flagler, Seibert, Hale • KANSAS: St. Francis • NEBRASKA: Benkelman, Stratton, Trenton, Stockville, Eustis, Lexington, Miller, Amherst, Pleasanton, Poole, Ravenna, Boelus, Cairo, Chapman, Hordville, Marquette, Polk, Arborville, Gresham, Ulysses, Bee, Dwight, Valparaiso, Agnew, Ceresco, Ashland, Memphis, Gretna, Springfield, Cedar Creek, LaPlatte, Bellevue • IOWA: Mineola, Silver City, Henderson, Macedonia, Griswold, Lyman, Cumberland, Berea, Fontanelle, How, Arbor Hill, Cumming, Norwalk, West Des Moines, Colfax, Prairie City, Newton, Oakland Acres, Grinnell, Belle Plaine, Luzerne, Blairstown, Norway, Fairfax, Covington, Cedar Rapids, Puralta, Springville, Fairview, Anamosa, Amber, Center Junction, Onslow, Fulton, Andrew, Springbrook, Green Island • ILLINOIS: Savanna, Pearl City, German Valley, Seward, Winnebago, Rockford, Belvidere, Marengo, Union, Crystal Lake, Libertyville, Wauconda, Lake Bluff, Lake Forest, North Chicago • MICHIGAN: Paw Paw Lake, Glendale, Lawrence, Kalamazoo, Westwood, Galesburg, Springfield, Battle Creek, Brownlee Park, Mershall, Devereaux, Jackson, Munith, Stockbridge, Unadilla, Pinckney, Hell, Hamburg, Lakeland, New Hudson, Wixom, Walled Lake, Orchard Lake, Sylvan Lake, Pontiac, Birmingham, Troy, Rochester Hills, Sterling, New Baltimore, Anchorville, Fair Haven, Pearl Beach • NEW YORK: (Lake Ontario), Dexter, Brownville, Watertown, Evans Mills, Philadelphia, Antwerp, Oxbow, Wegatchie, Somerville, Natural Dam, Gouverneur, Richville, DeKalb, Rensselaer Falls, Canton, Morley, Madrid, Chamberlain Corners, Chase Mills, Louisville, Massena, Massena Center, Rooseveltown • VERMONT: Rouses Point, Alburg Springs, Highgate Springs, East Franklin, Rickford, Stevens Mills, Missisquoi, North Troy, Beebe Plain, Rock Island, Derby Line, Norton, Canaan, Beecher Falls • NEW HAMPSHIRE: Pittsburg, (Shatney Mountain), Happy Corner, The Glen, (Deer Mountain), (Salmon Mountain) • MAINE: Coburn Gore, (Kirby Mountain), (Spencer Mountain), (Coburn Mountain), (Indian Pond), (Big Squaw Mountain), (Whitecap Mountain), (Jo-Mary Mountain), Millinocket, East Millinocket, Reed, (Wytopilock Lake), Haynesville, Orient, North Amity.

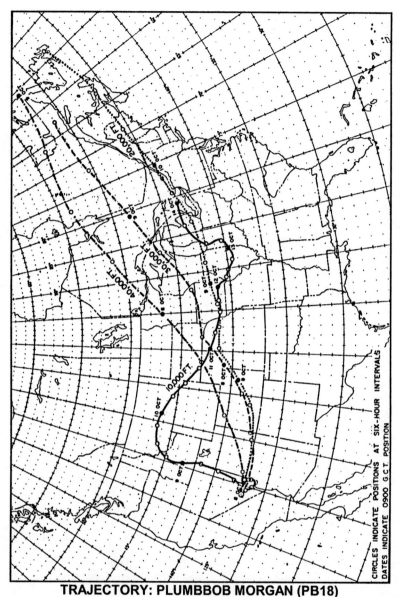

TRAJECTORY: PLUMBBOB MORGAN (PB18)

CHARACTERISTICS: PLUMBBOB MORGAN

Among the nuclear tests, Plumbbob MORGAN was 3rd in percentage of Cm-242 in fallout and 7th in the amount of Mn-54 deposited nationwide.. (Plumbbob WILSON was 2nd in Cm-242, while Plumbbob NEWTON was 1st for percentage of Cm-242 in fallout.). The deposition pattern for this shot is most similar to Upshot-Knothole SIMON (UK7) (R: 0.51; T: 32.67). Fallout from this test covered 2328 counties (75 percent of the total). Counties of highest deposition included El Paso CO (estimated maximum activity: 56.27 µCi/sq meter on Oct 8, 1957); Cheyenne CO, Kit Carson CO, Yuma CO, Boulder CO and Washington CO. One of the highest estimated maximum activities for fallout from shot MORGAN occurred on the eighth day post shot (Oct 14, 1957) for the following Missouri counties of Adair, Marion, Shelby and Monroe. The estimated maximum fallout activity for each of these counties equaled 56 µCi/sq meter. The day one total fallout to I-131 ratio for MORGAN is 102 to 1.

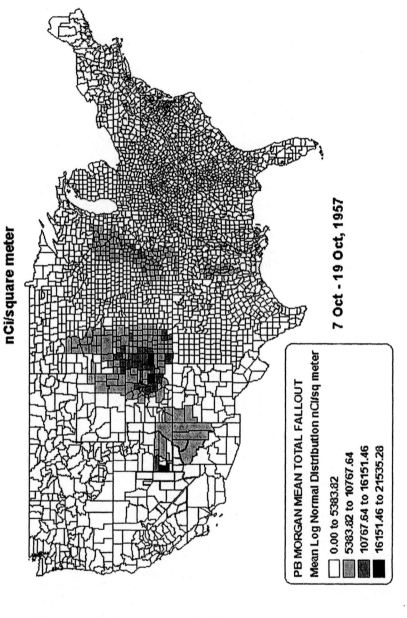

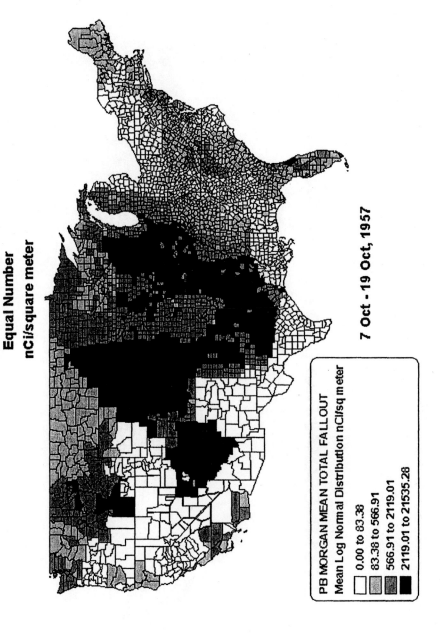

SECTION 7

SHOT SEDAN

July 6, 1962

STORAX SEDAN

STORAX: SEDAN

Detonation Date: 6 Jul 1962 • *Detonation Time:* 9:00 A.M. • *Area:* U10h • *Sponsor:* Livermore Laboratory • *Yield:* 110 kt • *Radiation Level at Ground Zero:* 500 R/hr • *Height of Burst:* −635 ft • *Cloud Top Height:* 16,000 ft

16,000-Foot Trajectory: NEVADA: (Belted Range), (Reveille Range), (Railroad Valley), Duckwater, (Mount Hamilton), (Pancake Summit), (Newark Lake), (Long Valley), (Ruby Lake), (Spruce Mountain), Shafter, (Pilot Peak) • UTAH: (Great Salt Lake Desert), (Hogup Mountains), (Hansels Mountains), (Dolphin Island In Great Salt), Snowville, Portage • IDAHO: Stone, Samaria, Woodruff, Malad City, Oxford, Clifton, Dayton, Banida, Mink Creek, Swan Lake, Liberty, Bern, Bennington, Ovid, Montpelier, Geneva • WYOMING: (Salt River Range), (Commissary Ridge), (Bridger-Teton National Forest), Marbleton, Boulder, Big Sandy, (Wind River Peak), Lander, Hudson, Hiland, Waltman, Powder River, Natrona, (Teapot Dome), Bill, Redbird • SOUTH DAKOTA: Provo, Ardmore, Pine Ridge • NEBRASKA: Eli, Cody, Kilgore, Crookston, Valentine, Sparks, Springview, Winnetoon, Creighton, Marquet, McLean, Randolph, Sholes, Wakefield, Emerso, Winnebago Indian Reservation, Homer, Winnebago • IOWA: Sloan, Rodney, Grant Center, Ticonic, Castana, Ute, Dow City, Earling, Defiance, Kirkman, Red Line, Jacksonville, Kimballton, Elk Horn, Brayton, Lorah, Wiota, Cumberland, Massena, Bridgewater, Williamson, Nevinville, Spaulding, Creston, Afton, Hopeville, Westerville, Leon, Woodland, Lineville • MISSOURI: South Lineville, Ravenna, Lucerne, Newtown, Harris, Milan, New Boston, Ethel, Callao, New Cambria, Bevier, Number Eight, College Mound, Cairo, Moberly, Middle Grove, Centralia, Thompson, Mexico, Auxvasse, Shamrock, Williamsburg, Readsville, Americus, Rhineland, Hermann, Gasconade, New Haven, Stony, Beaufort, Stanton, Richwoods, Fertile, Blackwell, Cadet, Bonne Terre, Desloge, Leadwood, Elvins, Flat River, Farmington, Junction City, Fredericktown, Patton, Scopus, Millersville, Burfordville, Whitewater, Dutchtown, Chaffee, New Hamburg, Benton, Diehlstadt, Charleston • KENTUCKY: Bardwell, Arlington, Milburn, Beulah, Fulgham, Wingo, Cuba, Lynnville, Fairbanks • TENNESSEE: Jones Mill, Paris, India, Springville, Manlyville, Big Sandy, Black Center, Eva, Denver, New Johnsonville, Hustburg, Bakerville, Sycamore Landing, (Tennessee National Wildlife Refuge), Only, Coble.

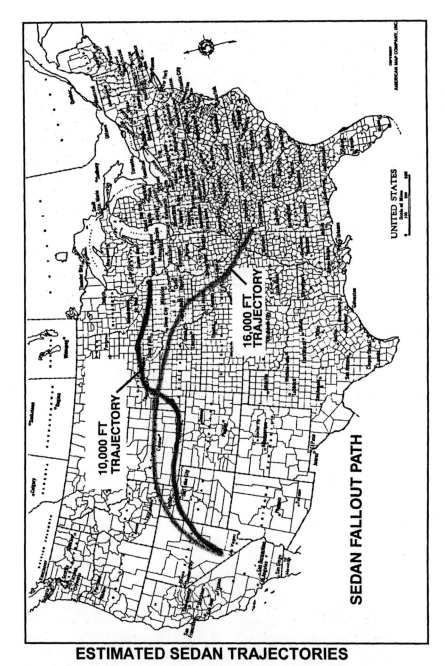

ESTIMATED SEDAN TRAJECTORIES

MAP NO. 761 REPRODUCED WITH PERMISSION FROM THE
AMERICAN MAP CORPORATION

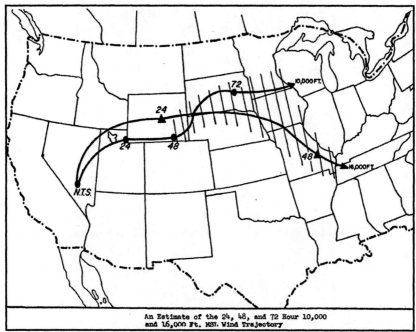

TRAJECTORY: SEDAN (SOURCE: DEFENSE DEPARTMENT)

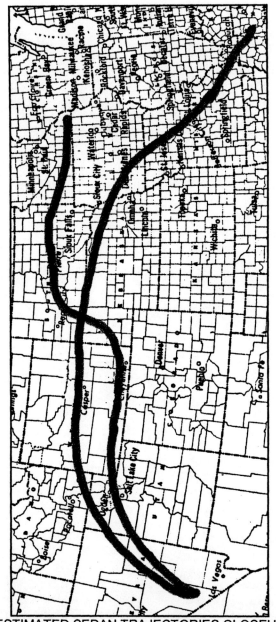

ESTIMATED SEDAN TRAJECTORIES CLOSEUP
MAP NO. 761 REPRODUCED WITH PERMISSION FROM THE
AMERICAN MAP CORPORATION

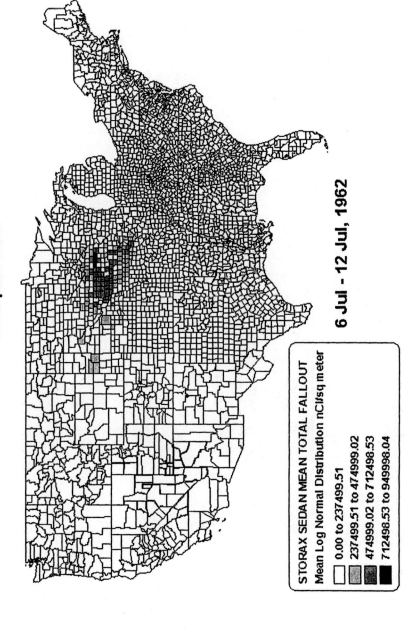

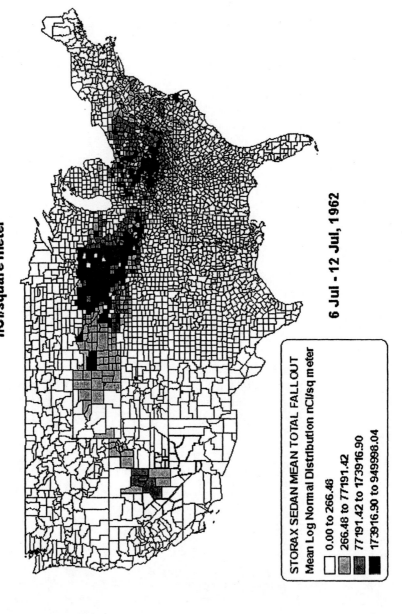

The Beryllium-7 deposition follows the SEDAN fallout pattern.
(Estimated Be7 relative deposition calculated at 1 yr postshot)

FALLOUT PATH: SHOT SEDAN JULY, 1962

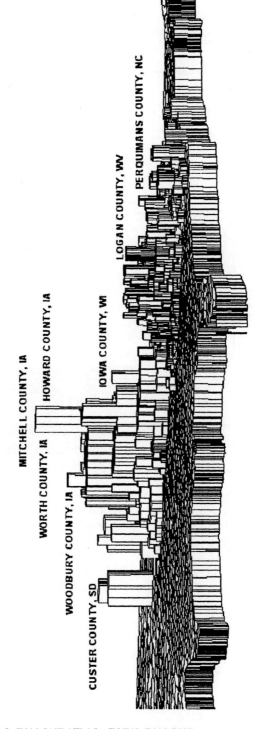

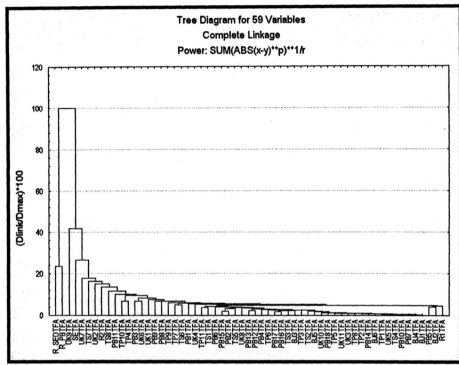

**Cluster Analysis of Deposition Patterns:
All Aboveground Nuclear Tests
1951-62**

This figure represents an analysis of deposition pattern similarities between nuclear tests. Here, the chart shows that the total fallout deposition pattern for SEDAN is relatively similar to that of UK9 and UK7. The terms R PBTFA and R-SEDTFA on the chart refer to the total fallout from 1951-57 and from 1951-62 respectively. The chart on the next page uses a different algorithm to evaluate differences and similarities between the SEDAN fallout and shots primarily from the Plumbbob series.

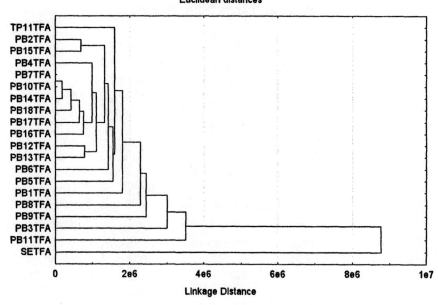

CHARACTERISTICS: SEDAN

The device used for the SEDAN shot was 17.1 inches in diameter, 38 inches long and weighed 467.9 lbs. Among the nuclear tests, shot Storax SEDAN was ranked 1ST for percentages of Au-198, Au-199, Be-7, Mo-99, Nd-147, Pb-203, W-181, W-185 and W-188 in fallout. It was ranked 2nd for percentages of Co-57, Co-60 and Mn-54 in fallout and 3rd for Na-24. The deposition pattern for SEDAN is similar to those of many earlier shots, including Teapot TESLA (R: 0.38, T: 22.67); Tumbler-Snapper Easy (R: 0.36); SEDAN deposited fallout on only 752 counties, approximately 24 percent of the total. While it ranked 45th in dispersion, SEDAN ranked first in activity per county. Shot SEDAN ranked first in overall deposition of the radionuclides Be-7, Mn-54, Ru-106 and Cm-242 across the country. It ranked 2nd in the deposition amount of Te-127m.

Counties of highest fallout deposition included: Howard IA, Mitchell IA, Worth IA, Washabaugh SD, Woodbury IA, Jefferson IA, Lee IA, Van Buren IA, Washington IA, Hancock IL and Dixon NE. Estimated maximum activities were calculated for a number of Midwestern and Eastern counties. Estimated *maximum* activities in µCi/sq meter were calculated for the following counties: Washabaugh SD: 6343 µCi/sq meter on July 7, 1962 (Washabaugh county, SD no longer exists*.); Howard, Mitchell and Worth counties in Iowa :6599 µCi/sq meter (each) on Jul 8, 1962; Hancock IL: 4314 µCi/sq meter on Jul 8, 1962; Faribault, MN: 4061 µCi/sq meter on Jul 8, 1962; Greenup, KY: 2497 µCi/sq meter on Jul 9, 1962; Boone WV: 2497 µCi/sq meter on Jul 9, 1962 and Logan WV: 2497 µCi/sq meter on July 9, 1962. The day one ratio of total fallout to I-131 for SEDAN was relatively low at 93 to 1.

*Since the time of the SEDAN shot, Washabaugh county, SD was absorbed into Jackson county, SD. Washabaugh was entirely within the present boundaries of the Pine Ridge Reservation.**

PHOTOS FROM THE
NEVADA TEST SITE: 1951-1962

BUSTER CHARLIE

BUSTER DOG

PHOTOS-2   U.S. FALLOUT ATLAS : TOTAL FALLOUT

TUMBLER-SNAPPER ABLE

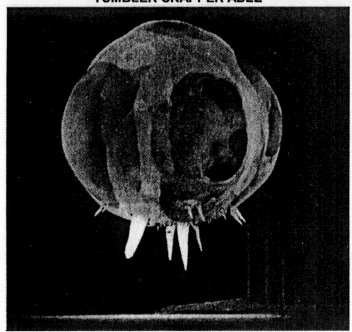
TUMBLER-SNAPPER GEORGE

UPSHOT-KNOTHOLE NANCY

UPSHOT-KNOTHOLE GRABLE

UPSHOT-KNOTHOLE HARRY

TEAPOT ESS

TEAPOT MET

PLUMBBOB HOOD

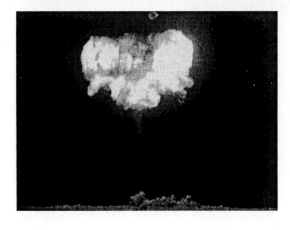
PLUMBBOB CHARLESTON

PLUMBBOB STOKES

PLUMBBOB PRISCILLA

PLUMBBOB FIZEAU

PLUMBBOB FRANKLIN

STORAX SEDAN

SEDAN CRATER

SECTION 8

COUNTIES WITH HIGHEST AVERAGE FALLOUT

1951-1962

COUNTIES OF HIGHEST AVERAGE FALLOUT FROM SHOT:

RANGER BAKER-1 (R-1) Jan 28, 1951 8 Kt

(microCuries per square meter)

Each Row Includes:
County Rank (most fallout = 1), State, County, Fallout Level in microCuries/square meter

Rank	State County	Fallout
1	CO LA PLATA	190.05
2	VT WINDHAM	48.70
3	OH BELMONT	48.61
4	OH GUERNSEY	48.19
5	WV OHIO	48.19
6	TX DALLAM	46.74
7	WV PLEASANTS	41.49
8	MO AUDRAIN	39.50
9	MO PIKE	39.50
10	OH ATHENS	38.66
11	MA ESSEX	38.35
12	MA MIDDLESEX	38.01
13	NH CARROLL	37.92
14	MA NORFOLK	37.60
15	MA SUFFOLK	37.60
16	NH ROCKINGHAM	37.60
17	NH STRAFFORD	37.60
18	MO GASCONADE	32.70
19	PA WAYNE	32.31
20	NY RENSSELAER	31.23
21	CT LITCHFIELD	30.19
22	CT TOLLAND	30.19
23	MA HAMPDEN	30.19
24	NY GREENE	30.19
25	NY SULLIVAN	30.19
26	NY ULSTER	30.19
27	MO BENTON	29.96
28	MO CALLAWAY	27.53
29	MO RALLS	26.72
30	MO DENT	25.01
31	MO PHELPS	25.01
32	CO ARCHULETA	24.57
33	CO CONEJOS	24.57
34	PA CAMERON	23.52
35	PA ELK	23.52
36	PA WASHINGTON	23.52
37	PA ALLEGHENY	23.29
38	PA INDIANA	23.29
39	PA WESTMORELAND	23.29
40	OH COLUMBIANA	23.23
41	OH GEAUGA	23.23
42	OH MORROW	23.23
43	OH RICHLAND	23.23
44	OH SENECA	23.23
45	PA BLAIR	23.18
46	PA CAMBRIA	23.18
47	OH MONROE	22.77
48	OH MORGAN	22.77
49	OH WASHINGTON	22.77
50	PA ARMSTRONG	22.77
51	PA BEAVER	22.77
52	PA CENTRE	22.77
53	PA CLEARFIELD	22.77
54	PA FAYETTE	22.77
55	PA GREENE	22.77
56	PA HUNTINGDON	22.77
57	PA MONTOUR	22.77
58	PA SNYDER	22.77
59	PA SULLIVAN	22.77
60	PA SUSQUEHANNA	22.77
61	PA TIOGA	22.77
62	PA UNION	22.77
63	WV MARSHALL	22.77
64	WV TYLER	22.77
65	OH ALLEN	22.15
66	OH MONTGOMERY	21.66
67	NY OTSEGO	21.62
68	VT BENNINGTON	21.62
69	NY DELAWARE	21.17
70	OH ASHLAND	20.92
71	OH CUYAHOGA	20.92
72	OH CLARK	20.91
73	OH DELAWARE	20.91
74	OH MARION	20.91
75	OH MIAMI	20.91
76	OH KNOX	20.40
77	WV WOOD	20.12
78	IN FRANKLIN	19.88
79	OH COSHOCTON	19.88
80	OH LICKING	19.88

COUNTIES OF HIGHEST AVERAGE FALLOUT FROM SHOT:

RANGER BAKER-2 (R-2) Feb 2, 1951 8 Kt

(microCuries per square meter)

Each Row Includes:
County Rank (most fallout = 1), State, County, Fallout Level in microCuries/square meter

Rank	State	County	Fallout
1	TX	REFUGIO	1067.00
2	FL	DADE	3.05
3	FL	COLLIER	1.63
4	FL	BROWARD	1.44
5	FL	HILLSBOROUGH	1.18
6	FL	LAKE	1.18
7	FL	PALM BEACH	1.11
8	FL	FRANKLIN	0.95
9	FL	LIBERTY	0.95
10	MD	PRINCE GEORG	0.90
11	CT	TOLLAND	0.78
12	ME	PENOBSCOT	0.78
13	FL	VOLUSIA	0.72
14	FL	ALACHUA	0.71
15	FL	BAKER	0.71
16	FL	BAY	0.71
17	GA	DOUGHERTY	0.71
18	GA	LUMPKIN	0.71
19	FL	CITRUS	0.57
20	FL	DE SOTO	0.57
21	FL	JEFFERSON	0.57
22	FL	LAFAYETTE	0.57
23	FL	MADISON	0.57
24	FL	POLK	0.57
25	FL	ST LUCIE	0.57
26	MA	MIDDLESEX	0.56
27	MD	WASHINGTON	0.56
28	NC	CASWELL	0.56
29	NC	CURRITUCK	0.56
30	VA	ESSEX	0.56
31	VA	HENRY	0.56
32	VA	LANCASTER	0.56
33	VT	ADDISON	0.56
34	FL	BREVARD	0.54
35	FL	DIXIE	0.54
36	FL	DUVAL	0.54
37	FL	GILCHRIST	0.54
38	FL	LEVY	0.54
39	FL	MANATEE	0.54
40	FL	PINELLAS	0.54
41	GA	EFFINGHAM	0.54
42	FL	COLUMBIA	0.52
43	FL	HAMILTON	0.52
44	FL	HERNANDO	0.52
45	FL	MARTIN	0.52
46	FL	ORANGE	0.52
47	FL	SANTA ROSA	0.52
48	FL	SUWANNEE	0.52
49	FL	TAYLOR	0.52
50	GA	CAMDEN	0.52
51	GA	CHARLTON	0.52
52	GA	CHATHAM	0.52
53	GA	FANNIN	0.52
54	GA	GORDON	0.52
55	AL	GENEVA	0.47
56	FL	OKALOOSA	0.47
57	GA	COLUMBIA	0.47
58	GA	JONES	0.47
59	NE	CHERRY	0.45
60	NE	CUSTER	0.45
61	NE	JEFFERSON	0.45
62	PA	BUCKS	0.45
63	MA	ESSEX	0.41
64	MA	FRANKLIN	0.41
65	ME	OXFORD	0.41
66	NC	BRUNSWICK	0.41
67	CT	NEW LONDON	0.37
68	CT	WINDHAM	0.37
69	MA	BARNS	0.37
70	MA	HAMPSHIRE	0.37
71	MA	NANTUCKET	0.37
72	MA	WORCESTER	0.37
73	ME	SOMERSET	0.37
74	RI	KENT	0.37
75	RI	NEWPORT	0.37
76	RI	PROVIDENCE	0.37
77	FL	HENDRY	0.36
78	FL	INDIAN RIVER	0.36
79	FL	NASSAU	0.36
80	FL	SEMINOLE	0.36

COUNTIES OF HIGHEST AVERAGE FALLOUT FROM SHOT:

BUSTER BAKER (BJ1) Oct 28, 1951 3.5 Kt

(microCuries per square meter)

Each row includes:
County Rank (most fallout = 1), State, County, Fallout Level in microCuries/square meter

Rank	State	County	Fallout
1	CA	FRESNO	12.11
2	CA	KERN	12.11
3	CA	KINGS	12.11
4	CA	MONTEREY	12.11
5	CA	SAN BENITO	12.11
6	CA	SAN LUIS OBI	12.11
7	CA	TULARE	12.11
8	CA	VENTURA	12.11
9	CA	SANTA BARBAR	5.03
10	GA	MERIWETHER	2.15
11	GA	PIKE	2.15
12	VA	GRAYSON	1.72
13	WV	PUTNAM	1.72
14	WV	RALEIGH	1.72
15	WV	SUMMERS	1.72
16	MS	MONROE	1.60
17	AL	RANDOLPH	1.29
18	GA	BALDWIN	1.29
19	GA	CLARKE	1.29
20	GA	DAWSON	1.29
21	GA	GREENE	1.29
22	GA	GWINNETT	1.29
23	GA	HANCOCK	1.29
24	GA	JACKSON	1.29
25	GA	MORGAN	1.29
26	GA	OCONEE	1.29
27	GA	PAULDING	1.29
28	GA	TALIAFERRO	1.29
29	GA	TROUP	1.29
30	GA	WARREN	1.29
31	GA	WILKINSON	1.29
32	NC	MACON	1.29
33	VA	BEDFORD	1.18
34	VA	PULASKI	1.18
35	VA	ROANOKE	1.18
36	VA	TAZEWELL	1.18
37	WV	BRAXTON	1.18
38	WV	KANAWHA	1.18
39	WV	MERCER	1.18
40	WV	TAYLOR	1.18
41	MS	ATTALA	0.91
42	MS	CARROLL	0.91
43	MS	HOLMES	0.91
44	MS	HUMPHREYS	0.91
45	MS	SHARKEY	0.91
46	MS	TATE	0.91
47	MS	WINSTON	0.91
48	GA	BARTOW	0.86
49	GA	BUTTS	0.86
50	GA	CHEROKEE	0.86
51	GA	DE KALB	0.86
52	GA	DODGE	0.86
53	GA	DOUGLAS	0.86
54	GA	FRANKLIN	0.86
55	GA	HALL	0.86
56	GA	HEARD	0.86
57	GA	HENRY	0.86
58	GA	JOHNSON	0.86
59	GA	LAMAR	0.86
60	GA	LAURENS	0.86
61	GA	MADISON	0.86
62	GA	NEWTON	0.86
63	GA	OGLETHORPE	0.86
64	GA	PICKENS	0.86
65	GA	RABUN	0.86
66	GA	ROCKDALE	0.86
67	GA	SPALDING	0.86
68	GA	STEPHENS	0.86
69	GA	UNION	0.86
70	GA	WALTON	0.86
71	GA	WASHINGTON	0.86
72	NC	HAYWOOD	0.86
73	SC	OCONEE	0.86
74	WV	RANDOLPH	0.71
75	AR	ARKANSAS	0.69
76	AR	GREENE	0.69
77	AR	JACKSON	0.69
78	AR	VAN BUREN	0.69
79	AR	WOODRUFF	0.69
80	CA	RIVERSIDE	0.69

COUNTIES OF HIGHEST AVERAGE FALLOUT FROM SHOT:

BUSTER CHARLIE (BJ-2) Oct 30, 1951 14 Kt

(microCuries per square meter)

Each Row Includes:
County Rank (most fallout = 1), State, County, Fallout Level in microCuries/square meter

Rank	State	County	Level
1	ME	AROOSTOOK	86.45
2	CA	INYO3	72.80
3	CA	LOS ANGELES	71.77
4	NY	OSWEGO	68.30
5	NY	ERIE	59.25
6	NY	GENESEE	56.61
7	NY	TOMPKINS	56.52
8	ME	PENOBSCOT	53.60
9	ME	PISCATAQUIS	50.86
10	NY	MONROE	50.14
11	NY	WAYNE	48.50
12	NY	JEFFERSON	48.48
13	NY	LIVINGSTON	48.31
14	NY	WYOMING	48.15
15	ME	WASHINGTON	47.73
16	NY	NIAGARA	46.00
17	NY	ORLEANS	45.42
18	NY	OTSEGO	44.89
19	NY	CHEMUNG	43.83
20	NY	ONTARIO	43.12
21	NY	HAMILTON	40.53
22	NY	DELAWARE	40.13
23	NY	SENECA	38.19
24	NY	CAYUGA	37.72
25	NY	LEWIS	37.57
26	ME	HANCOCK	37.54
27	NH	CARROLL	36.08
28	NY	ONONDAGA	35.72
29	NY	MADISON	35.24
30	NY	YATES	35.01
31	NY	CHENANGO	34.37
32	NY	CORTLAND	34.35
33	NY	FRANKLIN	34.31
34	NY	ONEIDA	33.25
35	NY	ALLEGANY	32.93
36	NC	CRAVEN	32.32
37	NY	SCHUYLER	31.72
38	ME	SAGADAHOC	31.50
39	NY	STEUBEN	31.13
40	ME	CUMBERLAND	30.70
41	ME	FRANKLIN	30.62
42	PA	WAYNE	30.49
43	NC	PITT	30.26
44	ME	OXFORD	30.03
45	ME	WALDO	30.02
46	NY	SULLIVAN	29.91
47	NC	ALAMANCE	29.84
48	NY	ESSEX	29.72
49	NY	ST LAWRENCE	29.54
50	NY	HERKIMER	29.49
51	VA	ARLINGTON	28.68
52	PA	WYOMING	28.05
53	NC	DUPLIN	27.72
54	DC	WASHINGTON	27.68
55	PA	LACKAWANNA	27.61
56	MD	PRINCE GEORG	27.08
57	NY	SARATOGA	26.98
58	ME	KENNEBEC	26.77
59	NC	HARNETT	26.75
60	NY	BROOME	26.33
61	VT	WINDSOR	26.24
62	NC	COLUMBUS	26.19
63	SC	MARION	26.06
64	ME	KNOX	25.84
65	NC	LEE	25.78
66	NC	CHATHAM	25.74
67	NY	TIOGA	25.63
68	VA	MECKLENBURG	25.51
69	NH	BELKNAP	25.46
70	NH	GRAFTON	25.33
71	NC	PENDER	25.22
72	NY	GREENE	25.03
73	SC	HORRY	24.98
74	NY	FULTON	24.93
75	ME	YORK	24.87
76	NY	CATTARAUGUS	24.73
77	NC	BLADEN	24.65
78	VA	FAIRFAX	24.57
79	MD	CHARLES	24.47
80	SC	FLORENCE	24.23

COUNTIES OF HIGHEST AVERAGE FALLOUT FROM SHOT:

BUSTER DOG (BJ3) Nov 1, 1951, 21 Kt

(microCuries per square meter)

Each Row Includes: County Rank (most fallout = 1), State, County, Fallout Level in microCuries/square meter

Rank	State	County	Level	Rank	State	County	Level	Rank	State	County	Level
1	NY	NIAGARA	43.70	33	VA	MIDDLESEX	12.94	65	NJ	WARREN	11.68
2	NY	CHAUTAUQUA	40.67	34	VA	NORTHUMBERLA	12.94	66	NY	BRONX	11.68
3	NY	ERIE	36.48	35	TN	ANDERSON	12.70	67	NY	NASSAU	11.68
4	NY	ORLEANS	35.31	36	TN	SEQUATCHIE	12.70	68	NY	WESTCHESTER	11.68
5	NY	GENESEE	34.26	37	NC	NORTHAMPTON	12.62	69	NC	PAMLICO	11.05
6	NY	JEFFERSON	30.38	38	VA	ISLE OF WIGH	12.62	70	OH	ERIE	10.53
7	NY	ALLEGANY	29.56	39	VA	SUFFOLK/NANS	12.62	71	GA	CATOOSA	9.63
8	NY	ONTARIO	29.56	40	VA	GREENSVILLE	12.45	72	GA	WALKER	9.63
9	PA	POTTER	29.56	41	MD	SOMERSET	12.24	73	GA	WHITFIELD	9.63
10	PA	WARREN	29.56	42	NC	BEAUFORT	12.24	74	NC	CHEROKEE	9.63
11	NY	STEUBEN	28.98	43	NC	BERTIE	12.24	75	TN	LOUDON	9.63
12	NY	WAYNE	28.39	44	NC	CARTERET	12.24	76	VA	CHESTERFIELD	9.52
13	NY	WYOMING	28.39	45	NC	CHOWAN	12.24	77	VA	HENRICO	9.52
14	NY	CATTARAUGUS	25.36	46	NC	GATES	12.24	78	GA	FLOYD	9.46
15	PA	ELK	25.36	47	NC	HERTFORD	12.24	79	TN	GRUNDY	9.46
16	PA	FOREST	25.36	48	NC	HYDE	12.24				
17	PA	MCKEAN	25.36	49	NC	MARTIN	12.24				
18	NY	YATES	24.20	50	NC	WASHINGTON	12.24				
19	NY	LIVINGSTON	23.15	51	VA	ACCOMACK	12.24				
20	NY	SENECA	22.84	52	VA	GLOUCESTER	12.24				
21	NJ	MORRIS	19.38	53	VA	HAMPTON	12.24				
22	NY	PUTNAM	19.38	54	VA	JAMES CITY	12.24				
23	NY	ULSTER	19.38	55	VA	MATHEWS	12.24				
24	VA	CHARLES CITY	14.15	56	VA	NEWPORT NEWS	12.24				
25	VA	PRINCE GEORG	14.15	57	VA	NORTHAMPTON	12.24				
26	NC	CAMDEN	13.38	58	VA	SOUTHAMPTON	12.24				
27	NC	CURRITUCK	13.38	59	VA	SURRY	12.24				
28	NC	DARE	13.38	60	VA	SUSSEX	12.24				
29	NC	PASQUOTANK	13.38	61	VA	YORK	12.24				
30	NC	PERQUIMANS	13.38	62	CT	FAIRFIELD	11.68				
31	VA	NORFOLK/CHES	13.38	63	NJ	ESSEX	11.68				
32	VA	LANCASTER	12.94	64	NJ	PASSAIC	11.68				

COUNTIES OF HIGHEST AVERAGE FALLOUT FROM SHOT:

BUSTER EASY (BJ4) Nov 5, 1951, 31 Kt

(microCuries per square meter)

Each Row Includes:
County Rank (most fallout = 1), State, County, Fallout Level in microCuries/square meter

Rank	State	County	Fallout
1	DE	KENT	2.92
2	DE	NEW CASTLE	2.92
3	DE	SUSSEX	2.92
4	MD	WORCESTER	2.92
5	NJ	BURLINGTON	2.92
6	NJ	CAMDEN	2.92
7	NJ	CAPE MAY	2.92
8	NJ	CUMBERLAND	2.92
9	NJ	GLOUCESTER	2.92
10	NJ	MERCER	2.92
11	NJ	OCEAN	2.92
12	NJ	SALEM	2.92
13	PA	BUCKS	2.92
14	PA	DELAWARE	2.92
15	PA	MONTGOMERY	2.92
16	PA	PHILADELPHIA	2.92
17	MI	WAYNE	2.75
18	MD	SOMERSET	2.70
19	NC	BEAUFORT	2.70
20	NC	BERTIE	2.70
21	NC	CAMDEN	2.70
22	NC	CARTERET	2.70
23	NC	CHOWAN	2.70
24	NC	CURRITUCK	2.70
25	NC	DARE	2.70
26	NC	GATES	2.70
27	NC	HERTFORD	2.70
28	NC	HYDE	2.70
29	NC	MARTIN	2.70
30	NC	NORTHAMPTON	2.70
31	NC	PAMLICO	2.70
32	NC	PASQUOTANK	2.70
33	NC	PERQUIMANS	2.70
34	NC	TYRRELL	2.70
35	NC	WASHINGTON	2.70
36	VA	ACCOMACK	2.70
37	VA	CHARLES CITY	2.70
38	VA	CHESTERFIELD	2.70
39	VA	DINWIDDIE	2.70
40	VA	GLOUCESTER	2.70
41	VA	GREENSVILLE	2.70
42	VA	HAMPTON	2.70
43	VA	HENRICO	2.70
44	VA	ISLE OF WIGH	2.70
45	VA	JAMES CITY	2.70
46	VA	KING AND QUE	2.70
47	VA	KING WILLIAM	2.70
48	VA	LANCASTER	2.70
49	VA	MATHEWS	2.70
50	VA	MIDDLESEX	2.70
51	VA	NEW KENT	2.70
52	VA	NEWPORT NEWS	2.70
53	VA	NORFOLK/CHES	2.70
54	VA	NORTHAMPTON	2.70
55	VA	NORTHUMBERLA	2.70
56	VA	PRINCE GEORG	2.70
57	VA	SOUTHAMPTON	2.70
58	VA	SUFFOLK/NANS	2.70
59	VA	SURRY	2.70
60	VA	SUSSEX	2.70
61	VA	YORK	2.70
62	MI	HURON	2.35
63	OH	ERIE	2.21
64	OH	LAKE	2.21
65	OH	OTTAWA	2.21
66	OH	HURON	1.94
67	OH	LORAIN	1.94
68	OH	RICHLAND	1.94
69	OH	SANDUSKY	1.94
70	OH	WYANDOT	1.94
71	MI	MACOMB	1.81
72	OH	SUMMIT	1.81
73	MI	OAKLAND	1.54
74	MI	SANILAC	1.54
75	MI	ST CLAIR	1.54
76	OH	ASHTABULA	1.54
77	OH	DELAWARE	1.54
78	OH	GEAUGA	1.54
79	OH	MARION	1.54
80	OH	MEDINA	1.54

COUNTIES OF HIGHEST AVERAGE FALLOUT FROM SHOT:

JANGLE SUGAR (BJ5) Nov 19, 1951 1.2 Kt

(microCuries per square meter)

Each Row Includes:
County Rank (most fallout = 1), State, County, Fallout Level in microCuries/square meter

Rank	State	County	Fallout
1	NV	NYE2	152.96
2	NV	EUREKA	106.82
3	NV	ELKO	105.23
4	ID	OWYHEE	91.04
5	ID	ADA	59.69
6	NV	LANDER1	56.41
7	ID	CLEARWATER	52.92
8	ID	ADAMS	50.30
9	ID	VALLEY	49.13
10	ID	IDAHO	48.01
11	ID	GEM	47.97
12	ID	ELMORE	45.35
13	ID	CANYON	45.24
14	ID	BOISE	44.79
15	NV	WHITE PINE2	40.22
16	NV	WHITE PINE1	40.01
17	NV	WHITE PINE3	37.79
18	ID	PAYETTE	36.45
19	ID	WASHINGTON	34.26
20	ID	CUSTER	33.20
21	ID	CAMAS	29.37
22	MT	MINERAL	28.25
23	NV	HUMBOLDT	27.53
24	UT	JUAB	18.73
25	NV	PERSHING	18.62
26	OR	HARNEY	18.28
27	OR	MALHEUR	18.28
28	ID	GOODING	18.16
29	ID	TWIN FALLS	18.16
30	ID	LINCOLN	9.14
31	ID	MINIDOKA	9.14
32	UT	BOX ELDER1	9.14
33	ID	JEROME	9.03
34	UT	TOOELE1	8.72
35	ID	CASSIA	8.62
36	NV	LANDER2	7.56
37	OK	GARFIELD	6.33
38	OK	MAJOR	6.23
39	UT	BOX ELDER2	4.63
40	KS	COWLEY	3.85
41	KS	SUMNER	3.85
42	OK	ALFALFA	3.85
43	OK	BLAINE	3.85
44	OK	BRYAN	3.85
45	OK	CANADIAN	3.85
46	OK	CHEROKEE	3.85
47	OK	CREEK	3.85
48	OK	DEWEY	3.85
49	OK	GRANT	3.85
50	OK	LINCOLN	3.85
51	OK	LOGAN	3.85
52	OK	MAYES	3.85
53	OK	MUSKOGEE	3.85
54	OK	NOBLE	3.85
55	OK	OKMULGEE	3.85
56	OK	OSAGE	3.85
57	OK	PAWNEE	3.85
58	OK	PAYNE	3.85
59	OK	ROGER MILLS	3.85
60	OK	ROGERS	3.85
61	OK	WAGONER	3.85
62	OK	WOODWARD	3.85
63	TX	GRAYSON	3.85
64	TX	WHEELER	3.85
65	NV	LINCOLN2	3.84
66	OK	GRADY	3.75
67	OK	KAY	3.75
68	OK	WASHINGTON	3.75
69	NE	ROCK	3.18
70	NV	CHURCHILL	2.82
71	UT	TOOELE2	2.82
72	OK	JACKSON	2.63
73	OK	WOODS	2.63
74	KS	CLARK	2.58
75	KS	EDWARDS	2.58
76	KS	HARPER	2.58
77	KS	KINGMAN	2.58
78	KS	KIOWA	2.58
79	KS	PRATT	2.58
80	KS	RUSH	2.58

COUNTIES OF HIGHEST AVERAGE FALLOUT FROM SHOT:

JANGLE UNCLE (BJ6) Nov 29, 1951 1.2 Kt

(microCuries per square meter)

Each Row Includes:
County Rank (most fallout = 1), State, County, Fallout Level in microCuries/square meter

Rank	State	County	Fallout
1	NV	WHITE PINE2	88.82
2	NV	LINCOLN2	85.19
3	NV	NYE2	79.78
4	NV	WHITE PINE1	46.98
5	NV	WHITE PINE3	28.14
6	NV	ELKO	8.56
7	UT	TOOELE1	5.89
8	NV	EUREKA	4.65
9	UT	JUAB	3.63
10	UT	BOX ELDER1	2.72
11	UT	TOOELE2	2.72
12	WY	UINTA	2.61
13	WY	SWEETWATER	2.46
14	ID	OWYHEE	1.82
15	ID	TWIN FALLS	1.82
16	UT	BOX ELDER2	1.82
17	WY	CARBON	1.82
18	UT	MILLARD	1.71
19	ID	CASSIA	1.70
20	ID	BEAR LAKE	0.91
21	ID	BINGHAM	0.91
22	ID	BONNEVILLE	0.91
23	ID	CANYON	0.91
24	ID	CARIBOU	0.91
25	ID	ELMORE	0.91
26	ID	FRANKLIN	0.91
27	ID	GOODING	0.91
28	ID	JEROME	0.91
29	ID	LINCOLN	0.91
30	ID	MINIDOKA	0.91
31	ID	ONEIDA	0.91
32	ID	POWER	0.91
33	OR	MALHEUR	0.91
34	UT	CACHE	0.91
35	UT	DAGGETT	0.91
36	UT	DUCHESNE	0.91
37	UT	RICH	0.91
38	UT	SUMMIT	0.91
39	UT	UINTAH	0.91
40	UT	WASATCH	0.91
41	WY	FREMONT	0.91
42	WY	LINCOLN	0.91
43	WY	SUBLETTE	0.91
44	ID	ADA	0.86
45	ID	BANNOCK	0.86
46	UT	DAVIS	0.86
47	UT	MORGAN	0.86
48	UT	SALT LAKE	0.86
49	UT	UTAH	0.86
50	UT	WEBER	0.86
51	NV	LINCOLN1	0.42

COUNTIES OF HIGHEST AVERAGE FALLOUT FROM SHOT:

TUMBLER-SNAPPER ABLE (TS1) Apr 1, 1952 1 Kt

(microCuries per square meter)

Each Row Includes:
County Rank (most fallout = 1), State, County, Fallout Level in microCuries/square meter

Rank	State	County	Fallout
1	NE	DAWES	140.77
2	NE	BANNER	140.57
3	NE	SIOUX	140.57
4	MO	CLAY	120.74
5	CO	PITKIN	98.21
6	KS	GRANT	80.31
7	KS	HODGEMAN	80.31
8	MO	MARIES	76.23
9	MO	VERNON	76.23
10	KS	WYANDOTTE	74.62
11	MO	BATES	74.62
12	MO	CEDAR	74.62
13	MO	GREENE	74.62
14	MO	HOWARD	74.62
15	MO	RAY	74.62
16	MO	SHELBY	74.62
17	MO	ST CLAIR	74.62
18	CO	ADAMS	71.13
19	CO	DENVER	71.13
20	CO	LARIMER	71.13
21	CO	LOGAN	71.13
22	CO	SEDGWICK	71.13
23	NE	CHEYENNE	71.13
24	NE	KIMBALL	71.13
25	NE	MORRILL	71.13
26	NE	SHERIDAN	71.13
27	WY	GOSHEN	71.13
28	WY	LARAMIE	71.13
29	CO	GILPIN	70.93
30	NE	BOX BUTTE	70.93
31	NE	HOOKER	70.93
32	WY	ALBANY	70.93
33	WY	CONVERSE	70.93
34	WY	NIOBRARA	70.93
35	CO	LAKE	65.97
36	KS	SHAWNEE	65.42
37	NE	RICHARDSON	65.42
38	NE	JOHNSON	63.82
39	KS	LYON	63.41
40	KS	NEOSHO	63.41
41	OK	GRANT	63.01
42	OK	LINCOLN	62.61
43	UT	BEAVER	56.63
44	MO	CAMDEN	51.73
45	MO	HOWELL	51.73
46	MO	MORGAN	51.73
47	MO	ST FRANCOIS	51.73
48	TX	RAINS	50.87
49	IL	CALHOUN	50.12
50	MO	ADAIR	50.12
51	MO	LAFAYETTE	50.12
52	MO	LINN	50.12
53	MO	POLK	50.12
54	MO	RANDOLPH	50.12
55	MO	SALINE	50.12
56	KS	LEAVENWORTH	46.39
57	KS	WABAUNSEE	46.39
58	KS	SALINE	44.79
59	NE	PAWNEE	44.79
60	KS	ALLEN	43.58
61	KS	BOURBON	43.58
62	KS	LINN	43.58
63	KS	WOODSON	43.58
64	AR	MONTGOMERY	43.07
65	OK	CHOCTAW	43.07
66	OK	MCCURTAIN	43.07
67	TX	CAMP	43.07
68	TX	CASS	43.07
69	TX	HOPKINS	43.07
70	TX	LAMAR	43.07
71	KS	CHASE	42.78
72	OK	KAY	42.78
73	MO	CHRISTIAN	42.76
74	MO	STONE	42.76
75	OK	ALFALFA	42.38
76	AR	NEWTON	42.13
77	KS	CRAWFORD	42.13
78	MO	BARRY	42.13
79	MO	DADE	42.13
80	MO	MCDONALD	42.13

COUNTIES OF HIGHEST AVERAGE FALLOUT FROM SHOT:

TUMBLER-SNAPPER BAKER (TS2) Apr 15, 1952 1 Kt

(microCuries per square meter)

Each Row Includes:
County Rank (most fallout = 1), State, County, Fallout Level in microCuries/square meter

Rank	State	County	Fallout
1	CA	INYO3	193.83
2	AZ	MOHAVE3	193.08
3	AZ	MOHAVE4	193.08
4	CA	SAN BERNADIN	91.14
5	AZ	YAVAPAI	75.75
6	NV	WHITE PINE2	33.67
7	CA	LOS ANGELES	24.28
8	NV	EUREKA	22.96
9	CA	ORANGE	18.93
10	CA	SAN DIEGO	18.82
11	CA	VENTURA	17.54
12	NV	WHITE PINE1	16.83
13	NV	WHITE PINE3	16.83
14	CA	KERN	15.89
15	UT	UTAH	15.35
16	AZ	SANTA CRUZ	15.31
17	AZ	PIMA	14.44
18	AZ	MARICOPA	14.38
19	AZ	PINAL	14.38
20	UT	IRON1	14.38
21	UT	JUAB	14.38
22	UT	MILLARD	14.38
23	CA	SANTA BARBAR	14.24
24	ID	OWYHEE	13.92
25	ID	TWIN FALLS	13.92
26	AZ	YUMA	13.56
27	ID	CASSIA	13.22
28	CA	IMPERIAL	12.09
29	ID	GOODING	10.90
30	ID	IDAHO	10.73
31	ID	JEROME	10.20
32	ID	ELMORE	9.97
33	AZ	COCONINO1	9.41
34	AZ	MOHAVE1	9.41
35	AZ	MOHAVE2	9.41
36	ID	ADA	9.41
37	ID	LINCOLN	9.28
38	ID	CANYON	8.81
39	NV	ELKO	8.31
40	NV	LANDER2	7.89
41	UT	BEAVER	7.19
42	UT	SANPETE	7.19
43	UT	BOX ELDER1	6.96
44	UT	TOOELE1	6.96
45	CA	RIVERSIDE	6.89
46	UT	KANE1	6.68
47	ID	MINIDOKA	6.49
48	AZ	COCHISE	6.26
49	AZ	GILA	6.26
50	AZ	GRAHAM	6.26
51	AZ	GREENLEE	6.26
52	UT	IRON2	6.26
53	UT	KANE2	6.26
54	AZ	COCONINO3	5.91
55	ID	ADAMS	5.63
56	ID	BOISE	5.63
57	ID	VALLEY	5.63
58	OR	GRANT	5.63
59	AZ	COCONINO2	5.33
60	CA	INYO1	5.10
61	CA	INYO2	5.10
62	NV	LANDER1	4.41
63	OR	BAKER	4.27
64	NV	CHURCHILL	4.17
65	UT	IRON3	4.17
66	UT	SEVIER	4.17
67	OR	MALHEUR	3.94
68	UT	TOOELE2	3.94
69	NV	DOUGLAS	3.25
70	NV	LYON	3.25
71	NV	PERSHING	3.25
72	NV	STOREY	3.25
73	NV	CARSON CITY	3.01
74	UT	BOX ELDER2	3.01
75	UT	PIUTE	3.01
76	AZ	APACHE	2.97
77	AZ	NAVAJO	2.97
78	ID	CAMAS	2.92
79	ID	CUSTER	2.92
80	ID	GEM	2.92

COUNTIES OF HIGHEST AVERAGE FALLOUT FROM SHOT:

TUMBLER-SNAPPER CHARLIE (TS3) Apr 22, 1952 31 Kt

(microCuries per square meter)

Each Row Includes:
County Rank (most fallout = 1), State, County, Fallout Level in microCuries/square meter

Rank	State County	Fallout	Rank	State County	Fallout	Rank	State County	Fallout
1	CA GLENN	37.19	30	AL CLARKE	17.97	59	SC CHARLESTON	14.55
2	AL LOWNDES	27.19	31	AL DALE	17.97	60	SC MARION	14.51
3	AL GENEVA	27.16	32	AL HALE	17.97	61	SC DILLON	14.44
4	AL WILCOX	27.03	33	AL SUMTER	17.97	62	CA SAN DIEGO	14.42
5	AL CONECUH	26.87	34	FL OKALOOSA	17.97	63	GA EFFINGHAM	14.19
6	AL GREENE	26.87	35	VA HANOVER	17.59	64	GA WAYNE	14.12
7	FL HOLMES	26.87	36	VA CHARLES CITY	17.50	65	SC JASPER	14.12
8	CA TEHAMA	25.20	37	VA PRINCE GEORG	17.29	66	AZ COCONINO2	14.11
9	SC LEE	25.04	38	VA GOOCHLAND	17.18	67	GA MCINTOSH	14.05
10	CA SONOMA	24.93	39	VA CULPEPER	17.17	68	CA PLACER	13.84
11	FL HILLSBOROUGH	24.93	40	VA HENRICO	17.06	69	CA COLUSA	13.80
12	AZ MOHAVE2	23.11	41	VA DINWIDDIE	17.06	70	PA MONROE	13.73
13	GA MONTGOMERY	22.64	42	VA BRUNSWICK	16.86	71	PA PIKE	13.72
14	GA WHEELER	22.64	43	VA KING GEORGE	16.46	72	CA SACRAMENTO	13.64
15	SC CLARENDON	22.08	44	VA FAUQUIER	16.35	73	PA LUZERNE	13.33
16	SC FLORENCE	20.66	45	VA KING WILLIAM	16.11	74	VA ORANGE	13.31
17	CA CALAVERAS	20.64	46	VA KING AND QUE	15.84	75	PA WAYNE	13.23
18	CA BUTTE	20.52	47	VA NEW KENT	15.70	76	VA RICHMOND	13.16
19	CA YUBA	20.52	48	VA GREENSVILLE	15.62	77	VA STAFFORD	13.12
20	CA NEVADA	20.51	49	GA LONG	15.49	78	VA ESSEX	13.05
21	CA AMADOR	20.47	50	GA LIBERTY	15.47	79	VA WESTMORELAND	13.05
22	CA ALPINE	20.43	51	GA BULLOCH	15.41	80	PA LACKAWANNA	12.98
23	AZ COCONINO3	19.47	52	GA SCREVEN	15.41			
24	SC SUMTER	18.92	53	GA BRYAN	15.39			
25	AL MARENGO	18.46	54	CA INYO1	14.64			
26	AL CHAMBERS	18.35	55	CA INYO2	14.64			
27	AL PIKE	18.30	56	CA INYO3	14.64			
28	AL COVINGTON	18.19	57	CA MONO	14.64			
29	AL DALLAS	18.19	58	SC DORCHESTER	14.63			

COUNTIES OF HIGHEST AVERAGE FALLOUT FROM SHOT:
TUMBLER-SNAPPER DOG (TS4) May 1, 1952 19 Kt

(microCuries per square meter)

Each Row Includes:
County Rank (most fallout = 1), State, County, Fallout Level in microCuries/square meter

Rank	State	County	Fallout		Rank	State	County	Fallout		Rank	State	County	Fallout
1	UT	IRON1	13.73		32	CO	DOLORES	3.43		63	IN	HOWARD	2.66
2	UT	IRON2	8.39		33	CO	MONTROSE	3.43		64	IN	TIPTON	2.66
3	AZ	MOHAVE3	6.41		34	CO	OURAY	3.43		65	CO	LAS ANIMAS	2.64
4	AZ	MOHAVE4	6.41		35	CO	SAN JUAN	3.43		66	OK	ELLIS	2.64
5	AZ	MOHAVE1	6.26		36	CO	SAN MIGUEL	3.43		67	AZ	GILA	2.57
6	AZ	MOHAVE2	6.26		37	NM	CATRON	3.43		68	CO	GARFIELD	2.57
7	AZ	COCONINO2	5.86		38	NM	SANDOVAL	3.43		69	CO	MOFFAT	2.57
8	UT	BEAVER	5.86		39	UT	EMERY	3.43		70	CO	RIO BLANCO	2.57
9	AZ	COCONINO1	4.86		40	UT	GRAND	3.43		71	NM	LOS ALAMOS	2.57
10	UT	IRON3	4.86		41	UT	WAYNE	3.43		72	NM	RIO ARRIBA	2.57
11	UT	MILLARD	4.86		42	AL	LIMESTONE	3.23		73	MO	BUCHANAN	2.54
12	OK	CIMARRON	4.81		43	AL	MADISON	3.23		74	NM	GUADALUPE	2.43
13	TX	MOORE	4.81		44	NM	DE BACA	3.14		75	NM	SANTA FE	2.43
14	TX	SHERMAN	4.81		45	NM	LINCOLN	3.14		76	NM	SOCORRO	2.43
15	AZ	NAVAJO	4.57		46	NM	ROOSEVELT	3.14		77	NM	TORRANCE	2.43
16	AZ	APACHE	4.43		47	KS	MEADE	3.07		78	CO	MESA	2.43
17	NM	MCKINLEY	4.43		48	KS	SEWARD	3.07		79	IN	RIPLEY	2.37
18	NM	SAN JUAN	4.43		49	OK	BEAVER	3.07		80	IA	MONONA	2.30
19	AZ	COCONINO3	4.31		50	OK	TEXAS	3.07					
20	NM	VALENCIA	4.29		51	TX	DEAF SMITH	3.07					
21	NV	LINCOLN2	4.24		52	TX	HANSFORD	3.07					
22	UT	SANPETE	3.86		53	TX	HARTLEY	3.07					
23	UT	SEVIER	3.86		54	TX	LIPSCOMB	3.07					
24	IN	CARROLL	3.70		55	TX	OCHILTREE	3.07					
25	IN	CLINTON	3.70		56	NM	CHAVES	2.97					
26	IN	TIPPECANOE	3.70		57	NM	EDDY	2.86					
27	AZ	YAVAPAI	3.57		58	NM	LEA	2.86					
28	CO	LA PLATA	3.57		59	UT	KANE2	2.86					
29	CO	MONTEZUMA	3.57		60	UT	GARFIELD	2.72					
30	UT	PIUTE	3.57		61	UT	KANE1	2.72					
31	CO	DELTA	3.43		62	UT	SAN JUAN	2.72					

U.S. FALLOUT ATLAS : TOTAL FALLOUT

COUNTIES OF HIGHEST AVERAGE FALLOUT FROM SHOT:

TUMBLER-SNAPPER EASY (TS5) May 7, 1952 12 Kt

(microCuries per square meter)

Each Row Includes:
County Rank (most fallout = 1), State, County, Fallout Level in microCuries/square meter

Rank	State	County	Fallout	Rank	State	County	Fallout	Rank	State	County	Fallout
1	WY	HOT SPRINGS	90.54	32	WY	CAMPBELL	26.89	63	IL	CALHOUN	17.15
2	WY	NATRONA	81.23	33	SD	BUTTE	26.86	64	GA	CARROLL	16.92
3	NE	SIOUX	79.14	34	WY	CROOK	26.76	65	CA	EL DORADO	16.49
4	WY	NIOBRARA	75.53	35	MT	PARK	26.72	66	SD	HUGHES	16.41
5	SD	FALL RIVER	71.98	36	SD	PENNINGTON	25.64	67	CA	TUOLUMNE	16.34
6	NE	DAWES	69.39	37	IL	SCOTT	25.42	68	SD	BON HOMME	16.24
7	WY	TETON	65.37	38	NE	CHERRY	24.69	69	CA	ALPINE	16.24
8	SD	SHANNON	63.14	39	WY	SHERIDAN	24.66	70	IL	JERSEY	16.18
9	SD	CUSTER	55.60	40	NE	SCOTTS BLUFF	24.50	71	NE	BROWN	16.13
10	ID	TETON	54.98	41	SD	TODD	24.48	72	IL	MACOUPIN	15.72
11	WY	GOSHEN	54.02	42	MT	SWEET GRASS	24.28	73	MT	WHEATLAND	15.64
12	WY	JOHNSON	52.35	43	CO	LARIMER	23.77	74	SD	JACKSON	15.53
13	WY	PLATTE	51.21	44	MT	POWDER RIVER	23.50	75	MT	CUSTER	15.46
14	WY	WASHAKIE	47.75	45	SD	HARDING	22.97	76	ID	CAMAS	15.40
15	WY	CONVERSE	44.61	46	SD	TRIPP	22.47	77	MT	STILLWATER	15.39
16	NE	SHERIDAN	44.50	47	SD	GREGORY	22.42	78	CO	LOGAN	15.28
17	SD	WASHABAUGH	44.36	48	MT	GALLATIN	21.24	79	SD	HAAKON	15.27
18	NE	BANNER	41.70	49	ID	MADISON	21.22	80	MT	GOLDEN VALLE	14.63
19	WY	PARK	40.90	50	NE	GRANT	20.90				
20	ID	BLAINE	38.01	51	NE	KEYA PAHA	20.74				
21	ID	FREMONT	36.31	52	SD	CHARLES MIX	20.21				
22	WY	WESTON	34.98	53	CA	AMADOR	20.10				
23	WY	BIG HORN	34.92	54	NC	DAVIDSON	20.01				
24	SD	BENNETT	33.37	55	WY	ALBANY	19.95				
25	SD	LAWRENCE	32.46	56	ID	LEMHI	19.80				
26	SD	MELLETTE	30.33	57	WY	LARAMIE	19.61				
27	ID	JEFFERSON	30.26	58	MT	CARTER	19.10				
28	NE	MORRILL	28.64	59	NE	BOYD	18.09				
29	ID	CLARK	27.66	60	NC	MOORE	17.99				
30	SD	MEADE	26.98	61	NE	ROCK	17.45				
31	NE	BOX BUTTE	26.97	62	MT	BEAVERHEAD	17.33				

COUNTIES OF HIGHEST AVERAGE FALLOUT FROM SHOT:

TUMBLER-SNAPPER FOX (TS6) May 25, 1952 11 Kt

(microCuries per square meter)

Each Row Includes:
County Rank (most fallout = 1), State, County, Fallout Level in microCuries/square meter

Rank	State	County	Fallout	Rank	State	County	Fallout	Rank	State	County	Fallout
1	NV	LINCOLN2	521.64	32	KS	CHEYENNE	52.74	63	NE	KEITH	38.84
2	UT	IRON1	509.73	33	UT	IRON3	52.32	64	CO	DOUGLAS	37.96
3	NV	LINCOLN1	302.37	34	NE	DEUEL	48.79	65	CO	SEDGWICK	36.86
4	UT	IRON2	130.90	35	NE	PERKINS	47.60	66	NE	DUNDY	36.37
5	UT	MILLARD	120.23	36	AZ	COCONINO1	46.41	67	UT	SUMMIT	36.36
6	UT	BEAVER	119.90	37	AZ	MOHAVE1	46.41	68	UT	WASATCH	36.36
7	CO	HINSDALE	99.67	38	AZ	MOHAVE2	46.41	69	UT	DAGGETT	36.14
8	CO	PHILLIPS	99.11	39	UT	UTAH	46.41	70	NM	BERNALILLO	35.64
9	UT	PIUTE	92.16	40	CO	DOLORES	45.42	71	NM	RIO ARRIBA	34.81
10	UT	SEVIER	92.16	41	CO	LA PLATA	45.42	72	CO	WELD	34.74
11	UT	EMERY	91.50	42	CO	MONTEZUMA	45.42	73	CO	ARCHULETA	34.49
12	UT	WAYNE	82.99	43	CO	OURAY	45.42	74	CO	GRAND	34.31
13	UT	GRAND	82.11	44	CO	SAN JUAN	45.42	75	NM	TAOS	34.04
14	CO	GARFIELD	72.49	45	CO	SAN MIGUEL	45.42	76	NM	SANTA FE	33.59
15	UT	CARBON	72.49	46	UT	DUCHESNE	45.31	77	MN	SHERBURNE	33.59
16	UT	GARFIELD	69.51	47	NM	SIERRA	45.20	78	CO	MINERAL	32.74
17	CO	MESA	68.58	48	AZ	COCONINO2	45.09	79	NE	RED WILLOW	32.64
18	UT	SANPETE	64.31	49	AZ	NAVAJO	44.76	80	AZ	GILA	32.38
19	CO	DELTA	63.65	50	AZ	APACHE	44.42				
20	CO	RIO BLANCO	63.43	51	NM	SAN JUAN	44.42				
21	UT	UINTAH	63.21	52	CO	YUMA	43.66				
22	CO	MOFFAT	62.88	53	NM	LOS ALAMOS	43.54				
23	WY	CARBON	62.22	54	NM	MCKINLEY	43.54				
24	NE	CHASE	57.34	55	NM	SANDOVAL	43.54				
25	CO	SAGUACHE	56.48	56	NM	VALENCIA	43.54				
26	UT	JUAB	55.70	57	CO	PUEBLO	43.13				
27	UT	KANE1	55.70	58	AZ	COCONINO3	42.52				
28	UT	KANE2	55.70	59	NE	HAYES	41.59				
29	CO	LOGAN	55.61	60	CO	HUERFANO	40.81				
30	CO	MONTROSE	54.48	61	CO	GUNNISON	40.49				
31	UT	SAN JUAN	54.48	62	NE	PHELPS	39.71				

COUNTIES OF HIGHEST AVERAGE FALLOUT FROM SHOT:

TUMBLER-SNAPPER GEORGE (TS7) Jun 1, 1952 15 Kt

(microCuries per square meter)

Each Row Includes:
County Rank (most fallout = 1), State, County, Fallout Level in microCuries/square meter

Rank	State	County	Fallout
1	MO	KNOX	351.71
2	MO	LEWIS	324.08
3	WY	NIOBRARA	310.25
4	IA	APPANOOSE	277.35
5	MO	SCOTLAND	274.93
6	SD	CUSTER	246.94
7	IA	MONROE	240.64
8	MO	RALLS	211.81
9	MO	AUDRAIN	208.57
10	MO	MACON	201.62
11	MO	MARION	198.74
12	MO	PIKE	198.74
13	IL	FORD	189.28
14	IL	PIATT	176.23
15	IL	CHRISTIAN	174.20
16	IL	MOULTRIE	174.17
17	IA	FREMONT	174.10
18	MO	SULLIVAN	173.28
19	IL	SHELBY	166.46
20	IL	MACON	164.91
21	IL	DOUGLAS	161.50
22	IL	ADAMS	161.46
23	MO	PUTNAM	155.34
24	IL	EFFINGHAM	151.62
25	IL	JASPER	150.31
26	MT	MEAGHER	149.80
27	WY	CONVERSE	141.80
28	MO	SCHUYLER	140.42
29	IA	WAYNE	132.25
30	MI	INGHAM	131.62
31	MI	LIVINGSTON	131.51
32	NE	SIOUX	131.27
33	IL	LAWRENCE	127.34
34	IL	RICHLAND	126.35
35	MI	CALHOUN	124.44
36	IN	KNOX	123.66
37	IA	RINGGOLD	117.91
38	IA	DAVIS	117.73
39	MI	KALAMAZOO	110.52
40	MI	GENESEE	110.29
41	OH	PUTNAM	109.19
42	MO	ADAIR	106.04
43	IA	TAMA	105.16
44	MI	VAN BUREN	104.90
45	MI	SHIAWASSEE	102.07
46	CO	MINERAL	102.04
47	IL	IROQUOIS	102.02
48	IN	WARREN	101.12
49	MI	EATON	100.47
50	MI	OTTAWA	98.95
51	TX	EL PASO	98.02
52	MI	CLINTON	96.75
53	IL	KANKAKEE	95.57
54	MO	LINCOLN	95.23
55	IA	MARION	94.54
56	NY	ST LAWRENCE	91.91
57	MI	LAPEER	91.85
58	OH	HANCOCK	91.44
59	MO	FRANKLIN	90.38
60	MO	LINN	89.49
61	MI	BARRY	87.74
62	IL	WILL	86.19
63	MO	SHELBY	85.60
64	IA	SHELBY	81.66
65	IN	DUBOIS	81.60
66	IN	CASS	81.30
67	IA	CASS	80.60
68	IA	ADAMS	80.22
69	SD	TURNER	79.84
70	SD	HAAKON	79.67
71	IL	CHAMPAIGN	78.41
72	MI	HILLSDALE	78.41
73	IA	MAHASKA	77.89
74	SD	HUTCHINSON	76.95
75	IA	UNION	76.47
76	MI	SAGINAW	76.26
77	IN	PULASKI	75.97
78	SD	DOUGLAS	75.74
79	OK	TULSA	75.66
80	IN	CARROLL	73.42

COUNTIES OF HIGHEST AVERAGE FALLOUT FROM SHOT:
TUMBLER-SNAPPER HOW (TS8) Jun 5, 1952 14 Kt

(microCuries per square meter)

Each Row Includes:
County Rank (most fallout = 1), State, County, Fallout Level in microCuries/square meter

Rank	State	County	Fallout	Rank	State	County	Fallout	Rank	State	County	Fallout
1	ID	GEM	399.37	32	NV	NYE2	93.95	63	MT	PONDERA	47.37
2	ID	CUSTER	385.26	33	NV	ELKO	92.93	64	ID	ADAMS	45.15
3	ID	BLAINE	362.15	34	MT	GALLATIN	92.79	65	MT	BLAINE	44.86
4	ID	LEMHI	265.64	35	NV	LANDER2	89.25	66	WA	ASOTIN	41.74
5	MT	DEER LODGE	239.28	36	ID	JEFFERSON	89.23	67	OR	UNION	39.44
6	ID	VALLEY	186.76	37	MT	LEWIS AND CL	87.85	68	ID	LATAH	38.01
7	ID	IDAHO	169.88	38	ID	CLEARWATER	86.46	69	MT	GLACIER	36.41
8	ID	CAMAS	167.53	39	OR	BAKER	86.17	70	ID	FREMONT	35.62
9	MT	CHOUTEAU	163.06	40	MT	MINERAL	85.28	71	ID	CANYON	35.06
10	MT	BEAVERHEAD	158.85	41	MT	TETON	77.04	72	MT	LINCOLN	33.98
11	MT	JUDITH BASIN	152.64	42	MT	HILL	75.80	73	NV	EUREKA	33.78
12	MT	POWELL	149.05	43	ID	SHOSHONE	73.81	74	OR	GRANT	31.24
13	ID	BOISE	145.78	44	MT	FLATHEAD	71.42	75	ID	CLARK	29.75
14	MT	SILVER BOW	145.45	45	ID	CASSIA	69.87	76	ID	BENEWAH	29.19
15	MT	JEFFERSON	145.20	46	ID	JEROME	69.87	77	MT	PHILLIPS	25.50
16	ID	GOODING	139.74	47	ID	MINIDOKA	69.87	78	ID	POWER	24.28
17	MT	BROADWATER	134.41	48	ID	TWIN FALLS	69.87	79	MT	PETROLEUM	23.33
18	ID	WASHINGTON	124.75	49	NV	HUMBOLDT	69.42	80	MT	CARBON	23.23
19	OR	WALLOWA	118.87	50	MT	PARK	62.67				
20	ID	LEWIS	118.64	51	MT	FERGUS	62.64				
21	MT	MEAGHER	116.25	52	WA	GARFIELD	61.37				
22	MT	MADISON	115.40	53	MT	CASCADE	61.33				
23	MT	MISSOULA	113.39	54	MT	RAVALLI	60.94				
24	ID	BUTTE	107.09	55	MT	TOOLE	59.19				
25	NV	LANDER1	105.93	56	MT	LIBERTY	58.91				
26	ID	ELMORE	104.81	57	MT	SANDERS	58.63				
27	ID	LINCOLN	104.81	58	ID	NEZ PERCE	54.33				
28	ID	OWYHEE	104.81	59	MT	LAKE	53.38				
29	ID	ADA	99.56	60	MT	GRANITE	53.10				
30	MT	WHEATLAND	95.52	61	UT	BOX ELDER1	52.58				
31	ID	PAYETTE	94.08	62	ID	MADISON	51.05				

COUNTIES OF HIGHEST AVERAGE FALLOUT FROM SHOT:
UPSHOT-KNOTHOLE ANNIE (UK1) Mar 17, 1953 16 Kt

(microCuries per square meter)

Each Row Includes:
County Rank (most fallout = 1), State, County, Fallout Level in microCuries/square meter

Rank	State	County	Fallout
1	UT	WASHINGTON2	1303.21
2	UT	WASHINGTON3	596.14
3	UT	KANE2	213.70
4	NV	LINCOLN1	120.42
5	UT	SAN JUAN	65.17
6	NJ	MONMOUTH	63.97
7	NJ	OCEAN	62.43
8	NY	KINGS	59.87
9	NY	QUEENS	55.75
10	NJ	ATLANTIC	54.88
11	NY	NASSAU	52.39
12	NJ	BURLINGTON	52.29
13	NV	CLARK1	51.94
14	NY	NEW YORK	50.45
15	NJ	MIDDLESEX	48.50
16	CO	MONTEZUMA	47.94
17	NY	RICHMOND	47.85
18	CO	LA PLATA	46.06
19	NJ	HUDSON	45.35
20	NJ	UNION	44.15
21	NJ	MERCER	42.87
22	NJ	ESSEX	41.47
23	NJ	CAPE MAY	41.42
24	NY	BRONX	41.27
25	NJ	BERGEN	40.02
26	NJ	CAMDEN	37.38
27	NY	WESTCHESTER	34.98
28	NJ	SOMERSET	33.76
29	NJ	CUMBERLAND	31.71
30	NJ	GLOUCESTER	30.93
31	NY	ROCKLAND	30.18
32	CT	FAIRFIELD	28.57
33	PA	PHILADELPHIA	28.44
34	AZ	APACHE	27.92
35	NY	SUFFOLK	27.80
36	NM	SAN JUAN	27.70
37	NM	RIO ARRIBA	27.48
38	NM	TAOS	27.48
39	NJ	PASSAIC	26.63
40	CO	CONEJOS	25.51
41	NJ	MORRIS	24.14
42	NJ	SALEM	23.63
43	NJ	HUNTERDON	22.81
44	NM	COLFAX	22.33
45	CO	LAS ANIMAS	22.22
46	TN	KNOX	21.73
47	PA	BUCKS	21.22
48	PA	DELAWARE	21.21
49	DE	SUSSEX	20.72
50	CO	COSTILLA	20.64
51	TN	ANDERSON	20.05
52	NY	PUTNAM	19.87
53	DE	KENT	18.91
54	AR	POLK	18.78
55	TX	DALLAS	18.76
56	TN	SEVIER	18.65
57	TX	GRAYSON	18.57
58	AR	LOGAN	18.44
59	NM	LOS ALAMOS	18.36
60	CO	BACA	18.28
61	NM	CHAVES	18.25
62	NM	DE BACA	18.25
63	NM	GUADALUPE	18.25
64	NM	MORA	18.25
65	NM	SAN MIGUEL	18.25
66	NM	CURRY	18.14
67	NM	HARDING	18.14
68	NM	QUAY	18.14
69	NM	ROOSEVELT	18.14
70	NM	UNION	18.14
71	OK	PUSHMATAHA	18.12
72	DE	NEW CASTLE	17.95
73	TX	ROCKWALL	17.90
74	AZ	MOHAVE2	17.83
75	TN	COCKE	17.77
76	TX	FANNIN	17.72
77	MO	TANEY	17.58
78	PA	MONTGOMERY	17.29
79	TN	CAMPBELL	17.29
80	AR	SCOTT	17.29

COUNTIES OF HIGHEST AVERAGE FALLOUT FROM SHOT:

UPSHOT-KNOTHOLE NANCY (UK2) Mar 24, 1953 24 Kt

(microCuries per square meter)

Each Row Includes:
County Rank (most fallout = 1), State, County, Fallout Level in microCuries/square meter

Rank	State	County	Fallout
1	UT	MORGAN	725.74
2	UT	DAVIS	723.10
3	UT	WEBER	679.47
4	UT	TOOELE2	635.02
5	UT	RICH	623.63
6	UT	SALT LAKE	598.89
7	WY	LINCOLN	568.59
8	WY	FREMONT	564.43
9	WY	SUBLETTE	564.43
10	UT	CACHE	536.50
11	NV	WHITE PINE3	492.93
12	UT	SUMMIT	451.90
13	ID	BEAR LAKE	443.88
14	UT	BOX ELDER2	443.88
15	WY	UINTA	286.32
16	ID	FRANKLIN	263.50
17	NV	NYE2	243.64
18	WY	CONVERSE	195.15
19	WY	CAMPBELL	194.99
20	WY	JOHNSON	194.72
21	MT	POWDER RIVER	194.67
22	WY	WASHAKIE	194.44
23	ID	CARIBOU	176.27
24	NV	LINCOLN2	168.92
25	MT	DAWSON	122.97
26	MT	CUSTER	122.96
27	ND	BOTTINEAU	122.86
28	ND	MCHENRY	122.86
29	NV	WHITE PINE2	122.36
30	UT	TOOELE1	119.78
31	UT	UTAH	96.49
32	ND	DUNN	92.54
33	ND	WARD	92.41
34	ND	BURKE	92.40
35	ND	MCKENZIE	92.36
36	UT	BOX ELDER1	88.82
37	ID	ONEIDA	87.55
38	NV	WHITE PINE1	87.15
39	WY	NATRONA	81.65
40	ND	MCLEAN	63.71
41	ND	SLOPE	62.14
42	MT	FALLON	62.09
43	ND	GOLDEN VALLE	62.09
44	ND	HETTINGER	62.07
45	MT	WIBAUX	62.04
46	ND	STARK	62.04
47	ND	BILLINGS	62.00
48	ND	MERCER	61.97
49	ND	GRANT	61.95
50	ND	MOUNTRAIL	61.91
51	ND	MORTON	61.90
52	ND	OLIVER	61.90
53	MT	SHERIDAN	61.87
54	MT	RICHLAND	61.85
55	MT	PRAIRIE	61.84
56	MT	ROOSEVELT	61.84
57	ND	RENVILLE	61.81
58	ND	DIVIDE	61.80
59	ND	BURLEIGH	61.79
60	ND	SHERIDAN	61.79
61	MT	DANIELS	61.78
62	MT	MCCONE	61.69
63	MT	VALLEY	61.57
64	AR	CRAIGHEAD	58.07
65	AR	FULTON	57.49
66	AR	PRAIRIE	49.98
67	UT	WASATCH	47.74
68	AR	CRITTENDEN	47.56
69	SD	LAWRENCE	47.17
70	SD	BUTTE	46.83
71	WY	CROOK	46.82
72	SD	HARDING	46.78
73	SD	MEADE	46.78
74	WY	WESTON	46.75
75	SD	SHANNON	46.72
76	MT	CARTER	46.71
77	SD	PERKINS	46.68
78	SD	CUSTER	46.68
79	ND	ADAMS	46.64
80	ND	BOWMAN	46.63

U.S. FALLOUT ATLAS : TOTAL FALLOUT

COUNTIES OF HIGHEST AVERAGE FALLOUT FROM SHOT:

UPSHOT-KNOTHOLE RUTH (UK3) Mar 31, 1953 0.2 Kt

(microCuries per square meter)

Each Row Includes:
County Rank (most fallout = 1), State, County, Fallout Level in microCuries/square meter

Rank	State	County	Fallout	Rank	State	County	Fallout	Rank	State	County	Fallout
1	AZ	MARICOPA	25.62	32	OK	TULSA	9.07	63	MO	MCDONALD	6.27
2	AZ	GILA	18.78	33	OK	BEAVER	8.66	64	AR	FRANKLIN	5.97
3	AZ	MOHAVE2	15.93	34	OK	ELLIS	8.66	65	KS	MEADE	5.93
4	AZ	YAVAPAI	14.51	35	OK	TEXAS	8.66	66	KS	SEWARD	5.93
5	OK	BECKHAM	14.12	36	TX	BRISCOE	8.66	67	TX	HALE	5.93
6	OK	GREER	14.12	37	TX	CASTRO	8.66	68	TX	HANSFORD	5.93
7	OK	HARMON	14.12	38	TX	CHILDRESS	8.66	69	TX	HUTCHINSON	5.93
8	OK	ROGER MILLS	14.12	39	TX	DEAF SMITH	8.66	70	TX	LUBBOCK	5.93
9	OK	WASHITA	14.12	40	TX	GRAY	8.66	71	TX	MOTLEY	5.93
10	TX	ARMSTRONG	14.12	41	TX	HEMPHILL	8.66	72	TX	OCHILTREE	5.93
11	TX	CARSON	14.12	42	TX	HOCKLEY	8.66	73	TX	OLDHAM	5.93
12	TX	COLLINGSWORT	14.12	43	TX	LIPSCOMB	8.66	74	TX	POTTER	5.93
13	TX	DONLEY	14.12	44	TX	LYNN	8.66	75	TX	SHERMAN	5.93
14	TX	PARMER	14.12	45	TX	ROBERTS	8.66	76	AR	BOONE	5.84
15	TX	WHEELER	14.12	46	TX	TERRY	8.66	77	MO	LAWRENCE	5.84
16	AZ	PINAL	14.08	47	OK	CRAIG	8.57	78	AR	LOGAN	5.61
17	AZ	MOHAVE3	11.23	48	OK	MCINTOSH	6.71	79	AZ	NAVAJO	5.40
18	AZ	MOHAVE4	11.23	49	AR	CRAWFORD	6.68	80	KS	CHEROKEE	5.31
19	CA	SAN BERNADIN	11.23	50	MO	BARRY	6.68				
20	AZ	COCONINO2	10.24	51	OK	ADAIR	6.68				
21	AZ	COCONINO3	9.52	52	OK	CREEK	6.68				
22	AR	MADISON	9.41	53	OK	HUGHES	6.68				
23	AR	WASHINGTON	9.41	54	OK	MUSKOGEE	6.68				
24	MO	NEWTON	9.41	55	OK	PITTSBURG	6.68				
25	OK	NOWATA	9.41	56	OK	CHEROKEE	6.54				
26	AR	BENTON	9.27	57	OK	OKFUSKEE	6.54				
27	OK	DELAWARE	9.27	58	OK	OKMULGEE	6.54				
28	OK	MAYES	9.27	59	OK	OTTAWA	6.54				
29	TX	HALL	9.13	60	OK	SEMINOLE	6.54				
30	AZ	GRAHAM	9.10	61	OK	WAGONER	6.34				
31	OK	ROGERS	9.07	62	AR	FAULKNER	6.31				

U.S. FALLOUT ATLAS : TOTAL FALLOUT

COUNTIES OF HIGHEST AVERAGE FALLOUT FROM SHOT:

UPSHOT-KNOTHOLE DIXIE (UK4) Apr 6, 1953 11 Kt

(microCuries per square meter)

Each Row Includes:
County Rank (most fallout = 1), State, County, Fallout Level in microCuries/square meter

Rank	State	County	Fallout		Rank	State	County	Fallout		Rank	State	County	Fallout
1	MA	PLYMOUTH	213.71		32	MA	HAMPSHIRE	14.87		63	FL	COLUMBIA	3.89
2	MA	NORFOLK	191.58		33	ME	YORK	13.96		64	FL	FLAGLER	3.89
3	MA	BRISTOL	156.46		34	CT	FAIRFIELD	11.47		65	FL	HERNANDO	3.89
4	MA	MIDDLESEX	136.12		35	MA	FRANKLIN	11.35		66	FL	HILLSBOROUGH	3.89
5	MA	SUFFOLK	126.99		36	ME	CUMBERLAND	10.77		67	FL	JEFFERSON	3.89
6	RI	BRISTOL	101.13		37	NH	CARROLL	10.30		68	FL	LAFAYETTE	3.89
7	RI	WASHINGTON	100.81		38	NH	GRAFTON	10.12		69	FL	LAKE	3.89
8	MA	DUKES	96.78		39	NH	COOS	9.48		70	FL	LEVY	3.89
9	MA	BARNS	95.69		40	CT	LITCHFIELD	9.36		71	FL	MARION	3.89
10	CT	MIDDLESEX	95.31		41	ME	SAGADAHOC	8.39		72	FL	NASSAU	3.89
11	MA	ESSEX	94.31		42	ME	KENNEBEC	7.91		73	FL	ORANGE	3.89
12	RI	KENT	84.09		43	ME	ANDROSCOGGIN	7.87		74	FL	OSCEOLA	3.89
13	RI	PROVIDENCE	74.30		44	FL	DIXIE	7.78		75	FL	PASCO	3.89
14	RI	NEWPORT	67.23		45	FL	GILCHRIST	7.78		76	FL	PINELLAS	3.89
15	CT	NEW LONDON	60.66		46	FL	HAMILTON	7.78		77	FL	POLK	3.89
16	CT	WINDHAM	56.64		47	FL	MADISON	7.78		78	FL	PUTNAM	3.89
17	CT	NEW HAVEN	55.07		48	GA	BACON	7.78		79	FL	SEMINOLE	3.89
18	MA	WORCESTER	53.19		49	GA	JEFF DAVIS	7.78		80	FL	ST JOHNS	3.89
19	NH	ROCKINGHAM	52.93		50	ME	OXFORD	7.73					
20	MA	NANTUCKET	49.57		51	VT	CALEDONIA	7.33					
21	CT	TOLLAND	44.32		52	VT	ESSEX	7.30					
22	NY	SUFFOLK	34.02		53	ME	LINCOLN	6.09					
23	NH	HILLSBOROUGH	33.09		54	ME	FRANKLIN	5.65					
24	CT	HARTFORD	29.69		55	VT	WINDHAM	5.18					
25	NH	CHESHIRE	23.57		56	NY	PUTNAM	4.84					
26	NH	MERRIMACK	23.40		57	FL	ALACHUA	3.89					
27	MA	HAMPDEN	21.48		58	FL	BAKER	3.89					
28	NH	BELKNAP	20.05		59	FL	BRADFORD	3.89					
29	NH	STRAFFORD	18.77		60	FL	BREVARD	3.89					
30	FL	LEON	15.57		61	FL	CITRUS	3.89					
31	NH	SULLIVAN	15.54		62	FL	CLAY	3.89					

COUNTIES OF HIGHEST AVERAGE FALLOUT FROM SHOT:

UPSHOT-KNOTHOLE RAY (UK5) Apr 11, 1953 0.2 Kt

(microCuries per square meter)

Each Row Includes:
County Rank (most fallout = 1), State, County, Fallout Level in microCuries/square meter

Rank	State	County	Fallout	Rank	State	County	Fallout	Rank	State	County	Fallout
1	AZ	YUMA	46.04	32	AR	CLARK	4.05	63	MA	ESSEX	2.84
2	CA	SAN BERNADIN	22.71	33	AR	COLUMBIA	4.05	64	MA	NANTUCKET	2.81
3	AZ	MOHAVE3	10.24	34	AR	HEMPSTEAD	4.05	65	ME	OXFORD	2.78
4	AZ	MOHAVE4	10.24	35	AR	HOWARD	4.05	66	NC	GRAHAM	2.78
5	UT	IRON1	6.53	36	AR	LAFAYETTE	4.05	67	NH	HILLSBOROUGH	2.78
6	UT	IRON3	6.23	37	AR	LITTLE RIVER	4.05	68	ME	ANDROSCOGGIN	2.73
7	AZ	YAVAPAI	5.56	38	AR	MONTGOMERY	4.05	69	AR	UNION	2.72
8	UT	BEAVER	5.51	39	AR	NEVADA	4.05	70	LA	CADDO	2.72
9	CA	INYO3	5.42	40	AR	PIKE	4.05	71	LA	NATCHITOCHES	2.72
10	UT	IRON2	5.20	41	AR	SEVIER	4.05	72	LA	RED RIVER	2.72
11	AZ	COCONINO3	5.12	42	LA	BOSSIER	4.05	73	TX	CAMP	2.72
12	AZ	COCONINO1	5.05	43	LA	LINCOLN	4.05	74	TX	CASS	2.72
13	AZ	MOHAVE2	5.05	44	LA	UNION	4.05	75	TX	FRANKLIN	2.72
14	UT	GARFIELD	5.05	45	LA	WEBSTER	4.05	76	TX	HARRISON	2.72
15	UT	KANE1	5.05	46	OK	CHOCTAW	4.05	77	TX	MARION	2.72
16	UT	KANE2	5.05	47	OK	MCCURTAIN	4.05	78	TX	MORRIS	2.72
17	UT	PIUTE	5.05	48	TX	BOWIE	4.05	79	TX	TITUS	2.72
18	UT	SEVIER	5.05	49	TX	RED RIVER	4.05	80	AZ	APACHE	2.64
19	UT	WAYNE	5.05	50	TX	UPSHUR	4.05				
20	AZ	COCONINO2	4.90	51	UT	EMERY	4.01				
21	AZ	MOHAVE1	4.73	52	CO	PITKIN	3.98				
22	CO	ROUTT	4.27	53	UT	GRAND	3.86				
23	AR	HOT SPRING	4.17	54	AZ	NAVAJO	3.62				
24	AR	PULASKI	4.17	55	TN	COCKE	3.03				
25	AR	BRADLEY	4.11	56	ME	CUMBERLAND	3.01				
26	AR	CLEVELAND	4.11	57	NH	ROCKINGHAM	2.99				
27	AR	DALLAS	4.11	58	MA	NORFOLK	2.95				
28	AR	GARLAND	4.11	59	NC	CLAY	2.92				
29	AR	GRANT	4.11	60	TN	SEVIER	2.92				
30	AR	SALINE	4.11	61	MA	MIDDLESEX	2.89				
31	AR	CALHOUN	4.05	62	ME	SAGADAHOC	2.85				

COUNTIES OF HIGHEST AVERAGE FALLOUT FROM SHOT:

UPSHOT-KNOTHOLE BADGER (UK6) Apr 18, 1953 23 Kt

(microCuries per square meter)

Each Row Includes:
County Rank (most fallout = 1), State, County, Fallout Level in microCuries/square meter

Rank	State	County	Fallout
1	AZ	COCONINO2	328.54
2	NV	CLARK1	324.13
3	AZ	NAVAJO	321.49
4	AZ	APACHE	319.19
5	NV	CLARK2	255.63
6	CO	ARCHULETA	217.12
7	NM	MCKINLEY	211.30
8	NM	VALENCIA	211.30
9	NM	SANDOVAL	156.71
10	NM	BERNALILLO	108.40
11	NM	CATRON	104.94
12	NM	SOCORRO	103.41
13	NM	SANTA FE	82.78
14	NM	TORRANCE	82.78
15	TX	HARRIS	82.03
16	TX	JASPER	81.94
17	TX	WALKER	81.89
18	TX	WALLER	81.80
19	LA	ACADIA	81.78
20	LA	ALLEN	81.78
21	LA	EVANGELINE	81.78
22	LA	RAPIDES	81.78
23	LA	VERMILION	81.78
24	LA	CAMERON	81.75
25	LA	CALCASIEU	81.73
26	TX	NEWTON	81.73
27	TX	FORT BEND	81.69
28	TX	CHAMBERS	81.66
29	LA	SABINE	81.65
30	TX	MONTGOMERY	81.64
31	TX	POLK	81.64
32	TX	SAN AUGUSTIN	81.62
33	LA	VERNON	81.59
34	TX	HARDIN	81.59
35	TX	ORANGE	81.59
36	TX	ANGELINA	81.55
37	TX	GRIMES	81.53
38	TX	LIBERTY	81.50
39	NM	LOS ALAMOS	81.49
40	TX	AUSTIN	81.46
41	TX	TYLER	81.40
42	TX	BRAZORIA	81.37
43	LA	LAFAYETTE	81.34
44	TX	GALVESTON	81.31
45	TX	TRINITY	81.30
46	TX	SABINE	81.30
47	TX	WASHINGTON	81.28
48	TX	SAN JACINTO	81.27
49	LA	GRANT	81.24
50	LA	BEAUREGARD	81.12
51	LA	JEFFERSON DA	81.08
52	NV	CLARK3	57.72
53	UT	IRON1	55.53
54	NM	GUADALUPE	40.75
55	NM	LINCOLN	40.75
56	NM	SIERRA	40.75
57	NM	SAN MIGUEL	40.49
58	NM	MORA	39.59
59	LA	ST HELENA	39.08
60	MS	PEARL RIVER	39.01
61	LA	ST BERNARD	38.99
62	MS	WALTHALL	38.97
63	AL	MOBILE	38.97
64	LA	IBERIA	38.94
65	LA	EAST FELICIA	38.87
66	LA	JEFFERSON	38.87
67	MS	JACKSON	38.82
68	MS	FORREST	38.80
69	MS	GEORGE	38.80
70	MS	GREENE	38.80
71	MS	PERRY	38.80
72	LA	PLAQUEMINES	38.77
73	LA	ST CHARLES	38.77
74	LA	ST TAMMANY	38.77
75	MS	HANCOCK	38.77
76	MS	HARRISON	38.77
77	LA	TERREBONNE	38.76
78	LA	LIVINGSTON	38.75
79	LA	ST JOHN THE	38.75
80	LA	WASHINGTON	38.75

COUNTIES OF HIGHEST AVERAGE FALLOUT FROM SHOT:

UPSHOT-KNOTHOLE SIMON Apr 25, 1953 43 Kt

(microCuries per square meter)

Each Row Includes:
County Rank (most fallout = 1), State, County, Fallout Level in microCuries/square meter

Rank	State	County	Fallout
1	NV	CLARK1	2185.05
2	AZ	COCONINO2	912.14
3	AZ	APACHE	537.54
4	NY	WASHINGTON	533.92
5	NY	WARREN	507.23
6	NY	RENSSELAER	488.57
7	NM	MCKINLEY	436.89
8	NM	CHAVES	393.20
9	AZ	NAVAJO	363.55
10	NM	SANDOVAL	338.39
11	NM	SANTA FE	335.90
12	NM	LINCOLN	333.53
13	NY	SARATOGA	302.04
14	NY	FULTON	294.53
15	NM	VALENCIA	262.11
16	VT	BENNINGTON	254.90
17	NM	TORRANCE	251.96
18	NM	CURRY	250.15
19	NM	DE BACA	250.15
20	NM	GUADALUPE	250.15
21	NM	ROOSEVELT	250.15
22	NY	SCHENECTADY	230.23
23	NM	QUAY	208.52
24	NY	ALBANY	186.57
25	NY	COLUMBIA	176.93
26	VT	ADDISON	176.61
27	AZ	COCONINO3	172.07
28	NM	LOS ALAMOS	166.77
29	NM	RIO ARRIBA	166.77
30	NM	SAN JUAN	166.77
31	NM	SAN MIGUEL	166.77
32	CO	DOLORES	165.53
33	AZ	MOHAVE1	156.16
34	AZ	COCONINO1	155.03
35	AZ	MOHAVE3	146.09
36	AZ	MOHAVE4	146.09
37	NM	BERNALILLO	145.15
38	AZ	MOHAVE2	139.35
39	UT	SAN JUAN	127.39
40	CO	MONTEZUMA	125.13
41	CO	LA PLATA	124.23
42	NV	LINCOLN1	117.10
43	VT	CHITTENDEN	102.45
44	CO	MESA	102.26
45	NY	MONTGOMERY	96.58
46	TX	COCHRAN	94.34
47	UT	BEAVER	93.64
48	TX	YOAKUM	90.45
49	UT	KANE1	90.27
50	UT	KANE2	90.27
51	CO	ARCHULETA	88.85
52	UT	IRON1	85.77
53	TX	BAILEY	84.16
54	VT	RUTLAND	83.89
55	CO	DELTA	83.38
56	CO	MONTROSE	83.38
57	CO	OURAY	83.38
58	CO	SAN JUAN	83.38
59	CO	SAN MIGUEL	83.38
60	NM	EDDY	83.38
61	NM	HARDING	83.38
62	NM	LEA	83.38
63	NM	MORA	83.38
64	NM	TAOS	83.38
65	NY	ESSEX	78.90
66	TX	GAINES	73.11
67	NY	SCHOHARIE	72.54
68	TX	HOCKLEY	71.59
69	CO	KIOWA	71.01
70	TX	TERRY	69.79
71	TX	PARMER	69.49
72	UT	WASHINGTON2	69.39
73	TX	LAMB	67.49
74	UT	GRAND	66.68
75	NY	GREENE	66.02
76	UT	WASHINGTON1	64.00
77	TX	GRAYSON	61.80
78	TX	FANNIN	60.25
79	NY	HAMILTON	58.38
80	CO	GARFIELD	58.33

COUNTIES OF HIGHEST AVERAGE FALLOUT FROM SHOT:

UPSHOT-KNOTHOLE ENCORE (UK8) May 8, 1953 27 Kt

(microCuries per square meter)

Each Row Includes:
County Rank (most fallout = 1), State, County, Fallout Level in microCuries/square meter

Rank	State	County	Fallout	Rank	State	County	Fallout	Rank	State	County	Fallout
1	MT	CUSTER	95.13	32	MT	BLAINE	36.47	63	MS	DE SOTO	8.82
2	MT	RICHLAND	95.02	33	ND	WILLIAMS	26.11	64	AR	LEE	8.50
3	MT	MCCONE	62.76	34	MT	TREASURE	25.66	65	OK	ROGER MILLS	8.46
4	MT	PRAIRIE	59.22	35	MT	CARBON	24.78	66	MS	COAHOMA	8.40
5	ND	BILLINGS	59.22	36	MT	DANIELS	23.21	67	AR	JACKSON	8.28
6	ND	BURKE	59.22	37	MT	GOLDEN VALLE	23.03	68	AR	POINSETT	8.28
7	MT	DAWSON	59.12	38	MT	STILLWATER	22.51	69	AR	WOODRUFF	8.28
8	MT	FALLON	59.12	39	MT	VALLEY	21.50	70	AR	INDEPENDENCE	8.21
9	MT	ROOSEVELT	59.12	40	ND	HETTINGER	21.39	71	TN	TIPTON	8.06
10	MT	SHERIDAN	59.12	41	ND	MCLEAN	21.39	72	MS	PANOLA	8.04
11	MT	WIBAUX	59.12	42	WY	BIG HORN	20.58	73	MS	TATE	8.04
12	ND	DIVIDE	59.12	43	MT	WHEATLAND	20.40	74	AR	CRAIGHEAD	8.01
13	ND	GOLDEN VALLE	59.12	44	ND	GRANT	19.57	75	AR	MISSISSIPPI	8.01
14	ND	MCKENZIE	59.12	45	ND	MCHENRY	19.57	76	AR	LAWRENCE	7.94
15	ND	RENVILLE	59.12	46	ND	MORTON	19.57	77	AR	LONOKE	7.94
16	ND	WARD	59.12	47	WY	HOT SPRINGS	19.09	78	MS	QUITMAN	7.79
17	ND	MOUNTRAIL	57.30	48	WY	PARK	16.54	79	OK	CIMARRON	7.77
18	ND	STARK	57.30	49	MT	SWEET GRASS	12.31	80	OK	WASHITA	7.77
19	ND	MERCER	54.67	50	ND	SHERIDAN	10.49				
20	ND	OLIVER	54.67	51	ND	BURLEIGH	10.19				
21	MT	GARFIELD	53.15	52	AR	CROSS	9.74				
22	MT	BIG HORN	42.04	53	MT	YELLOWSTONE	9.48				
23	ND	BOTTINEAU	39.65	54	AR	CRITTENDEN	9.11				
24	ND	DUNN	39.54	55	AR	MONROE	9.11				
25	ND	SLOPE	39.54	56	AR	ARKANSAS	8.89				
26	WY	SHERIDAN	39.50	57	AR	JEFFERSON	8.89				
27	MT	PETROLEUM	39.41	58	AR	PHILLIPS	8.89				
28	MT	ROSEBUD	38.89	59	AR	PRAIRIE	8.89				
29	MT	FERGUS	38.80	60	AR	ST FRANCIS	8.89				
30	MT	MUSSELSHELL	38.80	61	AR	WHITE	8.89				
31	MT	PHILLIPS	36.78	62	AR	CLEBURNE	8.82				

U.S. FALLOUT ATLAS : TOTAL FALLOUT

COUNTIES OF HIGHEST AVERAGE FALLOUT FROM SHOT:

UPSHOT-KNOTHOLE HARRY (UK9) May 19, 1953 32 Kt

(microCuries per square meter)

Each Row Includes:
County Rank (most fallout = 1), State, County, Fallout Level in microCuries/square meter

Rank	State	County	Fallout
1	UT	WASHINGTON2	4392.12
2	UT	WASHINGTON3	3642.40
3	UT	WASHINGTON1	2758.95
4	UT	KANE1	2334.06
5	UT	KANE2	2334.06
6	AZ	MOHAVE2	1986.84
7	NV	LINCOLN1	1716.20
8	CO	GUNNISON	1586.86
9	AZ	COCONINO1	1522.26
10	CO	CONEJOS	1452.03
11	AZ	MOHAVE1	1045.83
12	CO	ARCHULETA	926.47
13	CO	MINERAL	771.98
14	CO	HINSDALE	765.75
15	CO	RIO GRANDE	658.57
16	CO	PITKIN	627.41
17	UT	GARFIELD	568.41
18	UT	IRON3	502.13
19	UT	SAN JUAN	470.30
20	CO	DOLORES	466.90
21	CO	LA PLATA	466.90
22	CO	OURAY	466.90
23	CO	SAN JUAN	466.90
24	CO	SAN MIGUEL	466.90
25	CO	MONTROSE	463.51
26	CO	ALAMOSA	453.95
27	CO	SAGUACHE	406.29
28	UT	MILLARD	376.15
29	UT	SANPETE	376.15
30	CO	COSTILLA	376.07
31	CO	MONTEZUMA	374.79
32	UT	CARBON	374.79
33	CO	MESA	365.84
34	CO	LAKE	355.51
35	NM	BERNALILLO	351.01
36	CO	DELTA	323.30
37	NV	LINCOLN2	321.44
38	UT	IRON1	296.90
39	UT	WAYNE	282.11
40	AZ	APACHE	281.09
41	UT	EMERY	281.09
42	UT	JUAB	281.09
43	UT	GRAND	280.07
44	UT	UTAH	280.07
45	UT	WASATCH	280.07
46	NM	SANDOVAL	279.05
47	CO	GARFIELD	278.15
48	NM	MCKINLEY	278.15
49	NM	TORRANCE	275.09
50	UT	IRON2	266.15
51	NM	SAN JUAN	264.14
52	CO	EAGLE	238.74
53	UT	BEAVER	189.43
54	UT	PIUTE	188.75
55	UT	SEVIER	188.07
56	NM	LOS ALAMOS	186.72
57	NM	RIO ARRIBA	186.72
58	NM	SANTA FE	186.04
59	NM	VALENCIA	186.04
60	NM	TAOS	185.36
61	CO	CUSTER	164.71
62	CO	CHAFFEE	156.16
63	CO	SUMMIT	151.91
64	CO	HUERFANO	138.34
65	CO	FREMONT	118.59
66	NM	COLFAX	112.06
67	CO	LAS ANIMAS	107.60
68	CO	ROUTT	101.78
69	TX	HARTLEY	99.29
70	TX	DALLAM	99.03
71	NV	CLARK1	95.70
72	NM	MORA	92.00
73	NM	SAN MIGUEL	92.00
74	NM	GUADALUPE	91.66
75	NM	HARDING	91.32
76	NM	SOCORRO	91.32
77	TX	OLDHAM	90.89
78	NM	UNION	90.64
79	OK	CIMARRON	88.62
80	WI	SAWYER	76.50

U.S. FALLOUT ATLAS : TOTAL FALLOUT

COUNTIES OF HIGHEST AVERAGE FALLOUT FROM SHOT:

UPSHOT-KNOTHOLE GRABLE (UK10) May 25, 1953 15 Kt

(microCuries per square meter)

Each Row Includes:
County Rank (most fallout = 1), State, County, Fallout Level in microCuries/square meter

Rank	State	County	Fallout
1	NV	WHITE PINE3	35.05
2	WA	FERRY	35.04
3	UT	MILLARD	33.91
4	UT	SANPETE	33.63
5	UT	SEVIER	33.63
6	UT	JUAB	33.20
7	UT	UTAH	33.20
8	CO	DELTA	32.92
9	UT	DUCHESNE	32.92
10	UT	CARBON	32.77
11	WY	CARBON	31.06
12	CO	MESA	28.89
13	WA	PEND OREILLE	28.73
14	NV	PERSHING	27.36
15	NV	NYE2	26.05
16	WA	STEVENS	24.81
17	WY	SWEETWATER	24.06
18	WA	OKANOGAN	23.39
19	UT	EMERY	22.66
20	UT	GRAND	22.37
21	UT	WASATCH	22.37
22	CO	GARFIELD	22.23
23	CO	MONTROSE	22.23
24	CO	RIO BLANCO	22.09
25	CO	MOFFAT	21.94
26	UT	DAGGETT	21.94
27	UT	UINTAH	21.94
28	WY	UINTA	21.94
29	UT	SALT LAKE	20.96
30	WY	FREMONT	20.95
31	ID	BONNER	20.36
32	NV	LANDER2	20.09
33	NV	CARSON CITY	18.52
34	NV	CHURCHILL	18.52
35	NV	STOREY	18.52
36	NV	HUMBOLDT	17.47
37	NV	WASHOE	17.47
38	UT	BEAVER	17.07
39	UT	SUMMIT	16.39
40	MT	GARFIELD	14.08
41	WA	DOUGLAS	14.03
42	ID	KOOTENAI	13.69
43	NV	LYON	13.68
44	ID	LATAH	13.26
45	WA	GARFIELD	13.03
46	ID	BOUNDARY	12.71
47	MT	PETROLEUM	12.49
48	UT	GARFIELD	10.97
49	UT	PIUTE	10.97
50	UT	WAYNE	10.97
51	MT	WHEATLAND	10.94
52	UT	BOX ELDER2	10.69
53	UT	DAVIS	10.69
54	UT	MORGAN	10.69
55	UT	TOOELE2	10.69
56	UT	WEBER	10.69
57	ID	CASSIA	10.54
58	ID	ELMORE	10.54
59	ID	GOODING	10.54
60	ID	OWYHEE	10.54
61	ID	TWIN FALLS	10.54
62	UT	CACHE	10.54
63	UT	RICH	10.54
64	ID	JEROME	10.40
65	ID	LINCOLN	10.40
66	ID	MINIDOKA	10.40
67	ID	ONEIDA	10.40
68	ID	POWER	10.40
69	WA	CHELAN	10.39
70	ID	BEAR LAKE	10.26
71	ID	CARIBOU	10.26
72	ID	FRANKLIN	10.26
73	WY	LINCOLN	10.26
74	WY	SUBLETTE	10.26
75	ID	ADA	9.94
76	CO	PITKIN	9.91
77	VA	NEWPORT NEWS	9.90
78	WA	LINCOLN	9.85
79	VA	HAMPTON	9.80
80	NV	LANDER1	9.69

COUNTIES OF HIGHEST AVERAGE FALLOUT FROM SHOT:

UPSHOT-KNOTHOLE CLIMAX (UK11) Jun 4, 1953 61 Kt

(microCuries per square meter)

Each Row Includes:
County Rank (most fallout = 1), State, County, Fallout Level in microCuries/square meter

Rank	State County	Fallout
1	AZ COCONINO2	115.24
2	NV CLARK1	87.97
3	UT KANE1	47.69
4	AZ COCONINO1	44.82
5	AZ NAVAJO	38.55
6	UT KANE2	36.40
7	AZ MOHAVE2	34.93
8	UT GARFIELD	33.82
9	AZ APACHE	33.12
10	UT SAN JUAN	32.99
11	NM SAN JUAN	25.20
12	AZ MOHAVE1	22.83
13	AZ MOHAVE3	22.41
14	AZ MOHAVE4	22.41
15	UT EMERY	22.13
16	UT GRAND	21.85
17	CO DELTA	21.71
18	CO DOLORES	21.71
19	CO GARFIELD	21.71
20	CO MONTROSE	21.71
21	CO SAN MIGUEL	21.71
22	CO LA PLATA	21.57
23	CO OURAY	21.57
24	CO SAN JUAN	21.57
25	NM MCKINLEY	21.43
26	CO MONTEZUMA	21.13
27	WY CARBON	21.02
28	UT WAYNE	20.87
29	CO MESA	20.61
30	CO ROUTT	19.67
31	NV CLARK2	19.07
32	CO RIO BLANCO	16.42
33	NM LOS ALAMOS	16.28
34	NM RIO ARRIBA	16.28
35	NM SANDOVAL	16.28
36	WY SWEETWATER	14.44
37	AZ COCONINO3	13.39
38	UT IRON3	11.83
39	UT CARBON	11.55
40	CO JACKSON	11.23
41	NM BERNALILLO	11.16
42	CO MOFFAT	11.13
43	NM VALENCIA	11.13
44	WY FREMONT	10.86
45	KS WYANDOTTE	10.75
46	CO LARIMER	10.60
47	WY ALBANY	10.05
48	MO BUCHANAN	9.35
49	KS MIAMI	8.82
50	KS ANDERSON	8.79
51	CO GRAND	8.25
52	KS JEFFERSON	8.15
53	KS DOUGLAS	8.02
54	KS DONIPHAN	7.94
55	WY JOHNSON	7.45
56	KS ATCHISON	7.43
57	KS LEAVENWORTH	7.30
58	UT PIUTE	7.10
59	CO JEFFERSON	7.08
60	KS LINN	6.98
61	WY PLATTE	6.90
62	NM TAOS	6.82
63	KS BOURBON	6.72
64	CO PITKIN	6.59
65	KS ALLEN	6.49
66	CO PHILLIPS	6.45
67	WY LARAMIE	6.41
68	KS WOODSON	6.39
69	MO ANDREW	6.37
70	CO SUMMIT	6.36
71	CO GUNNISON	6.25
72	CO YUMA	6.16
73	WY SHERIDAN	6.03
74	WY NATRONA	6.00
75	CO SEDGWICK	5.95
76	MO CASS	5.94
77	CO LAKE	5.89
78	CO BOULDER	5.88
79	CO EAGLE	5.83
80	UT IRON2	5.79

COUNTIES OF HIGHEST AVERAGE FALLOUT FROM SHOT:

TEAPOT WASP (TP1) Feb 18, 1955 1 Kt

(microCuries per square meter)

Each Row Includes:
County Rank (most fallout = 1), State, County, Fallout Level in microCuries/square meter

Rank	State	County	Fallout
1	CA	SAN BERNADIN	107.56
2	AZ	YUMA	68.08
3	AZ	SANTA CRUZ	0.90
4	AZ	PIMA	0.85

COUNTIES OF HIGHEST AVERAGE FALLOUT FROM SHOT:

TEAPOT MOTH (TP2) Feb 22, 1955 2 Kt

(microCuries per square meter)

Each Row Includes:
County Rank (most fallout = 1), State, County, Fallout Level in microCuries/square meter

Rank	State	County	Fallout
1	NV	CLARK2	85.77
2	NV	CLARK1	1.99

COUNTIES OF HIGHEST AVERAGE FALLOUT FROM SHOT:
TEAPOT TESLA (TP3) Mar 1, 1955 7 Kt

(microCuries per square meter)

Each Row Includes:
County Rank (most fallout = 1), State, County, Fallout Level in microCuries/square meter

Rank	State	County	Fallout
1	UT	WASHINGTON2	254.31
2	UT	KANE2	91.68
3	UT	WASHINGTON3	68.79
4	UT	KANE1	64.42
5	NV	LINCOLN2	57.19
6	NV	LINCOLN1	49.78
7	CO	ADAMS	49.41
8	CO	BOULDER	45.13
9	CO	SUMMIT	45.13
10	CO	ARAPAHOE	44.96
11	CO	CLEAR CREEK	44.96
12	CO	DOUGLAS	44.96
13	CO	EAGLE	44.96
14	CO	GRAND	44.96
15	CO	JEFFERSON	44.96
16	CO	LAKE	44.96
17	CO	GILPIN	44.90
18	CO	MORGAN	44.90
19	UT	GARFIELD	38.02
20	UT	IRON3	36.62
21	AZ	MOHAVE2	30.06
22	AZ	MOHAVE1	29.87
23	AZ	COCONINO1	29.68
24	UT	IRON1	29.48
25	UT	IRON2	29.00
26	UT	SAN JUAN	28.13
27	UT	PIUTE	19.01
28	CO	DENVER	18.69
29	CO	SAN MIGUEL	18.52
30	CO	DOLORES	18.43
31	CO	OURAY	18.43
32	CO	SAN JUAN	18.43
33	UT	WASHINGTON1	16.54
34	UT	BEAVER	15.25
35	UT	SEVIER	14.26
36	CO	ROUTT	9.81
37	NE	LINCOLN	9.49
38	UT	EMERY	9.41
39	UT	WAYNE	9.41
40	UT	GRAND	9.31
41	CO	LA PLATA	9.21
42	CO	MONTEZUMA	9.21
43	CO	GUNNISON	9.00
44	CO	PITKIN	9.00
45	CO	BACA	8.92
46	CO	CHEYENNE	8.92
47	CO	KIOWA	8.92
48	CO	KIT CARSON	8.92
49	CO	PHILLIPS	8.92
50	CO	PROWERS	8.92
51	CO	WASHINGTON	8.92
52	CO	YUMA	8.92
53	KS	CHEYENNE	8.92
54	KS	DECATUR	8.92
55	KS	FINNEY	8.92
56	KS	FORD	8.92
57	KS	GOVE	8.92
58	KS	GRAHAM	8.92
59	KS	GRANT	8.92
60	KS	GRAY	8.92
61	KS	GREELEY	8.92
62	KS	HAMILTON	8.92
63	KS	HASKELL	8.92
64	KS	HODGEMAN	8.92
65	KS	KEARNY	8.92
66	KS	LANE	8.92
67	KS	LOGAN	8.92
68	KS	NESS	8.92
69	KS	NORTON	8.92
70	KS	RAWLINS	8.92
71	KS	SCOTT	8.92
72	KS	SHERIDAN	8.92
73	KS	STANTON	8.92
74	KS	THOMAS	8.92
75	KS	TREGO	8.92
76	KS	WALLACE	8.92
77	KS	WICHITA	8.92
78	NE	CHASE	8.92
79	NE	DAWSON	8.92
80	NE	DUNDY	8.92

COUNTIES OF HIGHEST AVERAGE FALLOUT FROM SHOT:

TEAPOT TURK (TP4) Mar 7, 1955 43 Kt

(microCuries per square meter)

Each Row Includes:
County Rank (most fallout = 1), State, County, Fallout Level in microCuries/square meter

Rank	State	County	Fallout	Rank	State	County	Fallout	Rank	State	County	Fallout
1	CA	INYO3	341.80	32	UT	DAGGETT	73.70	63	KS	GREELEY	53.05
2	NV	NYE2	322.23	33	WY	UINTA	73.42	64	KS	LANE	53.05
3	NV	WHITE PINE2	205.41	34	CO	MESA	69.50	65	CO	YUMA	53.03
4	CO	LAKE	168.12	35	CO	DENVER	67.84	66	NE	FURNAS	53.01
5	CO	SUMMIT	164.80	36	CO	GARFIELD	66.12	67	KS	HODGEMAN	52.94
6	CO	GRAND	163.68	37	UT	TOOELE1	60.65	68	NE	DUNDY	52.93
7	CO	GILPIN	160.53	38	UT	TOOELE2	60.65	69	KS	FINNEY	52.88
8	CO	CLEAR CREEK	160.48	39	UT	WEBER	60.29	70	NE	GOSPER	52.79
9	CO	BOULDER	159.56	40	UT	GRAND	59.56	71	KS	HAMILTON	52.74
10	CO	EAGLE	159.07	41	CO	RIO BLANCO	58.74	72	NE	CHASE	52.74
11	CO	JEFFERSON	158.98	42	WY	SWEETWATER	58.37	73	NE	FRONTIER	52.70
12	CO	ARAPAHOE	158.85	43	KS	THOMAS	53.49	74	KS	KEARNY	52.69
13	CO	DOUGLAS	158.80	44	KS	SHERIDAN	53.49	75	NE	HAYES	52.69
14	CO	ADAMS	158.56	45	CO	KIT CARSON	53.48	76	CO	PROWERS	52.65
15	CO	MORGAN	157.69	46	KS	GOVE	53.42	77	KS	FORD	52.62
16	UT	UTAH	151.68	47	KS	LOGAN	53.39	78	CO	PHILLIPS	52.57
17	NV	WHITE PINE3	138.40	48	KS	NORTON	53.34	79	KS	GRAY	52.50
18	UT	MILLARD	115.01	49	KS	DECATUR	53.30	80	NE	DAWSON	52.46
19	UT	JUAB	114.19	50	CO	CHEYENNE	53.29				
20	UT	SALT LAKE	114.06	51	KS	GRAHAM	53.27				
21	UT	WASATCH	113.01	52	CO	WASHINGTON	53.26				
22	UT	SUMMIT	112.55	53	KS	WICHITA	53.26				
23	UT	MORGAN	97.59	54	KS	WALLACE	53.26				
24	NV	WHITE PINE1	95.05	55	KS	RAWLINS	53.25				
25	CO	PITKIN	91.50	56	CO	KIOWA	53.25				
26	CO	GUNNISON	90.05	57	KS	SCOTT	53.20				
27	CO	ROUTT	83.98	58	KS	NESS	53.19				
28	CO	HINSDALE	82.97	59	KS	TREGO	53.18				
29	CO	MINERAL	82.84	60	NE	RED WILLOW	53.13				
30	UT	DAVIS	75.34	61	NE	HITCHCOCK	53.11				
31	UT	UINTAH	73.97	62	KS	CHEYENNE	53.11				

COUNTIES OF HIGHEST AVERAGE FALLOUT FROM SHOT:

TEAPOT HORNET (TP5) Mar 12, 1955 4 Kt

(microCuries per square meter)

Each Row Includes:
County Rank (most fallout = 1), State, County, Fallout Level in microCuries/square meter

Rank	State	County	Fallout
1	NV	CLARK1	126.37
2	AZ	COCONINO2	43.96
3	AZ	MOHAVE1	43.35
4	AZ	COCONINO1	42.85
5	AZ	COCONINO3	41.27
6	AZ	NAVAJO	34.78
7	UT	KANE1	34.28
8	NV	LINCOLN1	30.46
9	UT	IRON1	26.11
10	AZ	APACHE	26.01
11	AZ	MOHAVE2	23.87
12	UT	IRON3	17.34
13	NM	MCKINLEY	17.24
14	UT	IRON2	17.24
15	UT	KANE2	17.14
16	UT	GARFIELD	16.94
17	NM	VALENCIA	16.84
18	NM	SOCORRO	16.53
19	NV	LINCOLN2	15.69
20	NM	SAN JUAN	12.91
21	IN	LAWRENCE	12.81
22	NM	SANDOVAL	12.40
23	NM	LINCOLN	12.10
24	IN	MARTIN	11.91
25	IN	DUBOIS	11.42
26	IN	DAVIESS	11.38
27	IN	WASHINGTON	10.69
28	IN	ORANGE	10.65
29	IN	PIKE	10.51
30	NM	CHAVES	10.46
31	IN	JACKSON	10.10
32	IN	CLARK	9.92
33	IN	SCOTT	9.79
34	IN	GIBSON	9.22
35	KY	OLDHAM	8.94
36	IN	JEFFERSON	8.94
37	CO	DOLORES	8.67
38	CO	MONTEZUMA	8.67
39	IN	JENNINGS	8.57
40	UT	SAN JUAN	8.57
41	KY	TRIMBLE	8.39
42	NM	CATRON	8.37
43	NM	LOS ALAMOS	8.27
44	NM	RIO ARRIBA	8.27
45	NM	SIERRA	8.27
46	NM	TAOS	8.27
47	IL	WABASH	8.10
48	KY	JEFFERSON	8.07
49	NM	COLFAX	8.07
50	NM	MORA	8.07
51	NM	SAN MIGUEL	8.07
52	NM	SANTA FE	8.07
53	NM	TORRANCE	8.07
54	WY	LARAMIE	8.00
55	NM	CURRY	7.86
56	NM	DE BACA	7.86
57	NM	GUADALUPE	7.86
58	NM	HARDING	7.86
59	NM	QUAY	7.86
60	NM	ROOSEVELT	7.86
61	NM	UNION	7.86
62	KY	SHELBY	7.73
63	IL	EDWARDS	7.69
64	IN	MONROE	7.52
65	IN	PUTNAM	7.22
66	KY	WASHINGTON	7.20
67	IN	BROWN	7.12
68	IN	HENDRICKS	7.05
69	IN	GREENE	7.04
70	KY	GRANT	7.02
71	IN	OWEN	6.99
72	IN	CLAY	6.71
73	IL	LAKE	6.54
74	IN	HARRISON	6.48
75	IN	CRAWFORD	6.43
76	KY	HANCOCK	6.43
77	IN	SPENCER	6.36
78	IL	DU PAGE	6.32
79	IL	DE KALB	6.31
80	KY	PENDLETON	6.26

COUNTIES OF HIGHEST AVERAGE FALLOUT FROM SHOTS:

TEAPOT BEE AND ESS (TP6)
BEE: Mar 22, 1955 8 Kt
ESS: Mar 23, 1955 1 Kt
(microCuries per square meter)

Each Row Includes:
County Rank (most fallout = 1), State, County, Fallout Level in microCuries/square meter

Rank	State County	Fallout	Rank	State County	Fallout	Rank	State County	Fallout
1	NV CLARK3	104.23	31	OK COMANCHE	15.74	61	TX CHEROKEE	15.74
2	AZ MOHAVE2	102.51	32	OK COTTON	15.74	62	TX CLAY	15.74
3	AZ MOHAVE1	56.78	33	OK GARVIN	15.74	63	TX COLLIN	15.74
4	NV CLARK2	49.19	34	OK GRADY	15.74	64	TX COOKE	15.74
5	AZ COCONINO1	42.35	35	OK HASKELL	15.74	65	TX CORYELL	15.74
6	UT KANE1	28.08	36	OK HUGHES	15.74	66	TX DALLAS	15.74
7	NV CLARK1	26.74	37	OK JEFFERSON	15.74	67	TX DELTA	15.74
8	MS MONTGOMERY	24.74	38	OK JOHNSTON	15.74	68	TX DENTON	15.74
9	AZ YAVAPAI	21.33	39	OK LATIMER	15.74	69	TX ELLIS	15.74
10	AZ MOHAVE3	19.08	40	OK LE FLORE	15.74	70	TX ERATH	15.74
11	AZ MOHAVE4	19.08	41	OK LOVE	15.74	71	TX FALLS	15.74
12	AL TUSCALOOSA	18.18	42	OK MARSHALL	15.74	72	TX FANNIN	15.74
13	MS LEAKE	16.64	43	OK MCCLAIN	15.74	73	TX FRANKLIN	15.74
14	AL MARENGO	16.62	44	OK MCCURTAIN	15.74	74	TX FREESTONE	15.74
15	AR HEMPSTEAD	15.74	45	OK MCINTOSH	15.74	75	TX GRAYSON	15.74
16	AR HOWARD	15.74	46	OK MURRAY	15.74	76	TX GREGG	15.74
17	AR LAFAYETTE	15.74	47	OK PITTSBURG	15.74	77	TX HAMILTON	15.74
18	AR LITTLE RIVER	15.74	48	OK PONTOTOC	15.74	78	TX HARRISON	15.74
19	AR MILLER	15.74	49	OK POTTAWATOMIE	15.74	79	TX HENDERSON	15.74
20	AR MONTGOMERY	15.74	50	OK PUSHMATAHA	15.74	80	TX HILL	15.74
21	AR PIKE	15.74	51	OK SEMINOLE	15.74			
22	AR POLK	15.74	52	OK STEPHENS	15.74			
23	AR SCOTT	15.74	53	TX ANDERSON	15.74			
24	AR SEVIER	15.74	54	TX BELL	15.74			
25	OK ATOKA	15.74	55	TX BOSQUE	15.74			
26	OK BRYAN	15.74	56	TX BOWIE	15.74			
27	OK CARTER	15.74	57	TX BRAZOS	15.74			
28	OK CHOCTAW	15.74	58	TX BURLESON	15.74			
29	OK CLEVELAND	15.74	59	TX CAMP	15.74			
30	OK COAL	15.74	60	TX CASS	15.74			

U.S. FALLOUT ATLAS : TOTAL FALLOUT

COUNTIES OF HIGHEST AVERAGE FALLOUT FROM SHOT:

TEAPOT APPLE-1   Mar 29, 1955   14 Kt

(microCuries per square meter)

Each Row Includes:
County Rank (most fallout = 1), State, County, Fallout Level in microCuries/square meter

Rank	State	County	Fallout
1	NV	LINCOLN2	333.26
2	OK	TEXAS	160.85
3	KS	STEVENS	160.49
4	TX	OLDHAM	159.97
5	TX	POTTER	159.85
6	TX	HANSFORD	107.99
7	TX	HUTCHINSON	107.39
8	TX	OCHILTREE	106.75
9	TX	DALLAM	106.44
10	UT	IRON3	92.64
11	UT	WASHINGTON1	89.86
12	UT	GARFIELD	72.85
13	NV	LINCOLN1	71.59
14	UT	IRON1	69.71
15	UT	IRON2	65.52
16	OK	CIMARRON	53.91
17	TX	BRISCOE	53.90
18	TX	LIPSCOMB	53.88
19	TX	HARTLEY	53.81
20	TX	WHEELER	53.57
21	UT	KANE2	46.14
22	UT	PIUTE	40.24
23	AZ	COCONINO1	32.62
24	AZ	MOHAVE1	32.62
25	UT	KANE1	32.62
26	KS	MORTON	29.88
27	OK	BECKHAM	29.33
28	TX	ROBERTS	28.90
29	OK	ELLIS	28.72
30	OK	GREER	28.63
31	TX	SHERMAN	28.51
32	OK	ROGER MILLS	28.48
33	KS	SEWARD	28.47
34	OK	BEAVER	28.41
35	TX	MOORE	28.39
36	TX	ARMSTRONG	28.39
37	TX	DEAF SMITH	28.35
38	TX	CARSON	28.35
39	TX	SWISHER	28.12
40	KS	MEADE	27.79
41	TX	DONLEY	27.77
42	TX	CASTRO	27.56
43	TX	HALL	27.55
44	OK	WASHITA	27.48
45	TX	COLLINGSWORT	27.20
46	TX	MOTLEY	27.19
47	TX	GRAY	27.19
48	OK	HARMON	27.17
49	TX	FLOYD	27.15
50	TX	HEMPHILL	27.15
51	TX	CHILDRESS	27.14
52	TX	PARMER	27.14
53	TX	CROSBY	27.12
54	TX	BAILEY	27.12
55	TX	HALE	27.12
56	TX	LAMB	27.12
57	TX	LUBBOCK	27.09
58	TX	LYNN	27.06
59	TX	HOCKLEY	27.05
60	TX	TERRY	27.02
61	TX	COCHRAN	24.80
62	TX	YOAKUM	24.78
63	TX	GAINES	24.71
64	TX	ANDREWS	24.68
65	TX	LOVING	24.62
66	AZ	MOHAVE2	24.61
67	TX	ECTOR	24.57
68	TX	WINKLER	24.57
69	TX	CULBERSON	24.56
70	TX	REEVES	24.53
71	TX	WARD	24.53
72	TX	HUDSPETH	24.49
73	TX	EL PASO	24.47
74	TX	JEFF DAVIS	24.44
75	UT	WAYNE	23.93
76	UT	BEAVER	23.12
77	NM	HARDING	22.68
78	NM	UNION	22.68
79	CO	ARCHULETA	20.86
80	NM	COLFAX	19.20

COUNTIES OF HIGHEST AVERAGE FALLOUT FROM SHOT:

TEAPOT POST (TP8) Apr 9, 1955 2 Kt

(microCuries per square meter)

Each Row Includes:
County Rank (most fallout = 1), State, County, Fallout Level in microCuries/square meter

Rank	State	County	Fallout	Rank	State	County	Fallout	Rank	State	County	Fallout
1	UT	TOOELE2	39.13	32	UT	BEAVER	11.44	63	UT	SAN JUAN	4.26
2	UT	SALT LAKE	36.91	33	NV	LINCOLN2	10.08	64	UT	BOX ELDER1	4.17
3	UT	WASATCH	33.43	34	UT	IRON1	8.96	65	KS	STANTON	4.14
4	UT	MORGAN	33.17	35	UT	IRON2	8.96	66	ID	ADAMS	3.97
5	UT	SUMMIT	31.24	36	UT	IRON3	8.96	67	OR	GRANT	3.97
6	UT	DAVIS	31.07	37	UT	GARFIELD	8.69	68	ID	BOISE	3.92
7	UT	UTAH	29.44	38	UT	PIUTE	8.69	69	ID	WASHINGTON	3.89
8	NV	WHITE PINE2	29.00	39	UT	SEVIER	8.69	70	ID	PAYETTE	3.88
9	WY	UINTA	28.82	40	UT	EMERY	8.60	71	ID	GEM	3.85
10	UT	MILLARD	26.16	41	UT	WAYNE	8.60	72	OR	BAKER	3.85
11	NV	WHITE PINE1	25.62	42	CO	DOLORES	8.51	73	WY	CONVERSE	3.68
12	CO	RIO BLANCO	25.10	43	CO	SAN JUAN	8.51	74	KS	CHEYENNE	3.51
13	CO	MOFFAT	24.92	44	CO	SAN MIGUEL	8.51	75	CO	ROUTT	3.47
14	UT	DUCHESNE	24.92	45	UT	GRAND	8.51	76	UT	CACHE	3.28
15	UT	WEBER	24.92	46	UT	SANPETE	8.51	77	UT	RICH	3.28
16	UT	UINTAH	24.74	47	UT	TOOELE1	8.51	78	CO	SUMMIT	3.27
17	WY	SWEETWATER	24.74	48	WY	LINCOLN	8.34	79	NE	DUNDY	3.06
18	NV	NYE3	21.46	49	WY	SUBLETTE	8.25	80	WY	LARAMIE	3.01
19	CA	INYO3	18.27	50	CO	PITKIN	6.53				
20	NV	WHITE PINE3	18.09	51	ID	IDAHO	5.97				
21	CO	MONTROSE	16.94	52	WY	NATRONA	5.96				
22	UT	CARBON	16.76	53	ID	VALLEY	5.89				
23	UT	BOX ELDER2	16.58	54	CO	JACKSON	5.83				
24	UT	DAGGETT	16.49	55	AZ	COCONINO1	4.43				
25	WY	CARBON	16.49	56	AZ	MOHAVE1	4.43				
26	WY	FREMONT	16.49	57	AZ	MOHAVE3	4.43				
27	CO	GARFIELD	15.79	58	AZ	MOHAVE4	4.43				
28	CO	MESA	14.97	59	UT	KANE1	4.35				
29	UT	JUAB	12.95	60	UT	KANE2	4.35				
30	CO	OURAY	12.77	61	CO	LA PLATA	4.26				
31	CO	DELTA	11.96	62	CO	MONTEZUMA	4.26				

U.S. FALLOUT ATLAS : TOTAL FALLOUT

COUNTIES OF HIGHEST AVERAGE FALLOUT FROM SHOT:

TEAPOT MET (TP9) Apr 15, 1955 22 Kt

(microCuries per square meter)

Each Row Includes:
County Rank (most fallout = 1), State, County, Fallout Level in microCuries/square meter

Rank	State	County	Fallout	Rank	State	County	Fallout	Rank	State	County	Fallout
1	NV	LINCOLN1	433.28	32	OH	LAKE	29.68	63	KS	OSBORNE	15.20
2	UT	BEAVER	406.51	33	PA	ERIE	28.96	64	KS	PAWNEE	15.06
3	CO	MESA	382.05	34	NE	GRANT	28.92	65	PA	WARREN	15.04
4	UT	EMERY	325.27	35	OH	TRUMBULL	27.92	66	KS	RUSH	14.87
5	UT	GRAND	325.27	36	NE	MORRILL	27.48	67	OH	CHAMPAIGN	14.64
6	UT	PIUTE	325.27	37	PA	MERCER	26.63	68	KS	ELLIS	14.57
7	CO	DELTA	322.85	38	CO	GUNNISON	24.73	69	KS	LINCOLN	14.35
8	CO	GARFIELD	322.85	39	UT	JUAB	24.19	70	KS	PRATT	14.35
9	UT	SEVIER	243.92	40	NY	ERIE	23.94	71	UT	UTAH	14.11
10	NV	LINCOLN2	209.20	41	MI	ST CLAIR	21.48	72	UT	DAVIS	13.91
11	CO	RIO BLANCO	159.66	42	PA	CRAWFORD	21.25	73	UT	MORGAN	13.91
12	CO	MONTROSE	85.85	43	CO	DENVER	21.23	74	UT	TOOELE2	13.91
13	MI	WAYNE	85.84	44	CO	ROUTT	19.81	75	OH	HANCOCK	13.36
14	UT	IRON1	84.89	45	PA	VENANGO	19.39	76	ND	PEMBINA	13.32
15	OH	MEDINA	84.43	46	MI	CLINTON	19.37	77	UT	SALT LAKE	13.12
16	UT	WAYNE	80.94	47	KS	BARTON	19.22	78	KS	ROOKS	12.98
17	OH	CUYAHOGA	78.35	48	KS	RUSSELL	18.48	79	OH	WOOD	12.98
18	OH	SENECA	70.70	49	PA	FOREST	18.28	80	KS	KINGMAN	12.96
19	MI	MACOMB	57.98	50	CO	BOULDER	18.03				
20	OH	ERIE	49.64	51	KS	STAFFORD	17.77				
21	OH	GEAUGA	45.67	52	NY	NIAGARA	17.50				
22	MI	HURON	44.33	53	KS	RICE	16.95				
23	OH	SANDUSKY	41.06	54	MI	LAPEER	16.55				
24	MI	EATON	38.03	55	KS	NORTON	16.35				
25	OH	HURON	38.00	56	UT	IRON2	16.23				
26	MI	SANILAC	37.66	57	KS	ELLSWORTH	16.05				
27	NY	CHAUTAUQUA	37.51	58	MI	OAKLAND	16.03				
28	OH	LORAIN	36.27	59	KS	RENO	15.88				
29	UT	MILLARD	32.46	60	WY	FREMONT	15.62				
30	OH	PORTAGE	32.34	61	WY	CARBON	15.52				
31	MI	GENESEE	31.25	62	CO	GILPIN	15.24				

COUNTIES OF HIGHEST AVERAGE FALLOUT FROM SHOT:

TEAPOT APPLE-2 (TP10) May 5, 1955 29 Kt

(microCuries per square meter)

Each Row Includes:
County Rank (most fallout = 1), State, County, Fallout Level in microCuries/square meter

Rank	State	County	Fallout
1	NV	WHITE PINE2	430.02
2	NV	NYE2	316.28
3	UT	UTAH	238.18
4	NV	WHITE PINE1	238.09
5	UT	WASATCH	237.40
6	UT	DUCHESNE	235.65
7	CO	MOFFAT	233.13
8	UT	DAGGETT	233.13
9	UT	UINTAH	233.13
10	CO	RIO BLANCO	232.25
11	NV	WHITE PINE3	221.15
12	UT	JUAB	158.85
13	WY	CARBON	154.87
14	CO	COSTILLA	127.67
15	UT	SALT LAKE	119.41
16	UT	DAVIS	118.26
17	UT	MORGAN	117.39
18	UT	WEBER	117.39
19	WY	SWEETWATER	116.61
20	CO	LARIMER	114.28
21	CO	WELD	113.91
22	CO	JACKSON	113.06
23	WY	ALBANY	112.25
24	WY	PLATTE	111.47
25	UT	SUMMIT	110.40
26	CO	CONEJOS	96.93
27	KS	STEVENS	80.60
28	UT	MILLARD	80.30
29	UT	TOOELE2	78.84
30	CO	GARFIELD	77.19
31	CO	CLEAR CREEK	70.00
32	CO	GILPIN	68.93
33	CO	BENT	68.82
34	CO	OTERO	68.53
35	CO	LINCOLN	67.84
36	CO	CROWLEY	67.13
37	CO	LAS ANIMAS	66.82
38	CO	CUSTER	66.28
39	CO	RIO GRANDE	66.10
40	CO	ALAMOSA	66.09
41	CO	GRAND	65.49
42	MO	CAMDEN	65.41
43	CO	HUERFANO	64.94
44	CO	LAKE	63.53
45	MO	MORGAN	63.33
46	MO	LACLEDE	63.18
47	MO	WEBSTER	62.65
48	CO	DOUGLAS	62.45
49	CO	EAGLE	62.12
50	MO	OSAGE	61.87
51	CO	JEFFERSON	61.51
52	CO	ADAMS	61.47
53	CO	SUMMIT	61.18
54	MO	PETTIS	60.99
55	AR	PERRY	59.87
56	AR	SALINE	59.59
57	CO	BOULDER	59.57
58	CO	ARAPAHOE	58.86
59	MO	WRIGHT	58.82
60	NM	COLFAX	57.87
61	NM	UNION	57.87
62	CO	MORGAN	57.65
63	KS	FINNEY	57.26
64	MO	DOUGLAS	57.01
65	MO	HENRY	55.89
66	MO	OZARK	54.80
67	TX	HARTLEY	54.24
68	OK	TEXAS	53.91
69	CO	TELLER	52.55
70	CO	ELBERT	52.27
71	IN	PERRY	52.12
72	CO	CHAFFEE	51.53
73	CO	PARK	51.38
74	CO	SAGUACHE	51.38
75	CO	FREMONT	50.90
76	TX	LIPSCOMB	50.72
77	NM	HARDING	50.59
78	AR	POPE	50.49
79	IN	SPENCER	48.53
80	AR	VAN BUREN	48.50

COUNTIES OF HIGHEST AVERAGE FALLOUT FROM SHOT:

TEAPOT ZUCCHINI (TP11) 15 May, 1955 28 Kt

(microCuries per square meter)

Each Row Includes:
County Rank (most fallout = 1), State, County, Fallout Level in microCuries/square meter

Rank	State	County	Fallout		Rank	State	County	Fallout		Rank	State	County	Fallout
1	UT	IRON2	413.43		32	CO	BOULDER	36.05		63	CO	MESA	22.75
2	NV	CLARK1	387.19		33	UT	UTAH	35.45		64	UT	DUCHESNE	22.56
3	AZ	MOHAVE2	207.52		34	CO	LARIMER	34.55		65	ID	OWYHEE	22.36
4	UT	IRON3	180.05		35	CO	CLEAR CREEK	30.89		66	ID	CANYON	22.17
5	NM	COLFAX	130.08		36	CO	GILPIN	30.42		67	ID	ELMORE	22.17
6	NM	UNION	130.08		37	OK	PONTOTOC	29.90		68	OR	HARNEY	21.87
7	UT	WASHINGTON2	104.21		38	CO	DENVER	29.75		69	OR	MALHEUR	21.87
8	AZ	MOHAVE3	88.60		39	WY	CAMPBELL	29.42		70	UT	WASATCH	21.87
9	AZ	MOHAVE4	88.60		40	CO	JEFFERSON	29.02		71	MT	POWDER RIVER	21.84
10	UT	MILLARD	87.66		41	NV	CLARK2	27.29		72	CO	ARAPAHOE	21.58
11	UT	PIUTE	82.13		42	NM	CURRY	25.98		73	TX	FOARD	21.38
12	NM	HARDING	78.03		43	NM	GUADALUPE	25.98		74	UT	DAGGETT	21.29
13	NM	MORA	78.03		44	CO	DOUGLAS	24.59		75	UT	SUMMIT	21.29
14	UT	BEAVER	66.21		45	UT	EMERY	24.51		76	UT	DAVIS	21.09
15	UT	WASHINGTON3	62.19		46	CO	CUSTER	24.47		77	UT	MORGAN	21.09
16	UT	WASHINGTON1	60.67		47	UT	CARBON	24.41		78	UT	WEBER	21.00
17	UT	SEVIER	57.32		48	UT	GRAND	24.32		79	WY	UINTA	21.00
18	CO	PITKIN	54.24		49	CO	DELTA	24.12		80	TX	HALE	20.94
19	UT	SANPETE	53.79		50	CO	DOLORES	24.12					
20	NM	QUAY	52.05		51	CO	GARFIELD	24.12					
21	NM	SAN MIGUEL	52.05		52	CO	MONTEZUMA	24.12					
22	CO	MORGAN	51.43		53	CO	MONTROSE	24.12					
23	CO	HINSDALE	49.80		54	CO	SAN MIGUEL	24.12					
24	WY	CARBON	43.75		55	CO	OURAY	24.02					
25	CO	GUNNISON	43.63		56	CO	SAN JUAN	24.02					
26	UT	JUAB	43.26		57	CO	LA PLATA	23.93					
27	UT	GARFIELD	41.11		58	CO	RIO BLANCO	23.73					
28	CO	MINERAL	40.27		59	UT	UINTAH	23.73					
29	CO	ROUTT	38.07		60	UT	TOOELE2	23.66					
30	CO	GRAND	37.29		61	UT	BOX ELDER2	23.24					
31	WY	PLATTE	36.53		62	CO	MOFFAT	23.14					

COUNTIES OF HIGHEST AVERAGE FALLOUT FROM SHOT:

PLUMBBOB BOLTZMANN (PB12) May 28, 1957 12 Kt

(microCuries per square meter)

Each Row Includes:
County Rank (most fallout = 1), State, County, Fallout Level in microCuries/square meter

Rank	State	County	Fallout
1	NV	NYE2	766.45
2	NV	NYE1	519.96
3	NV	LANDER2	283.24
4	NV	MINERAL	282.27
5	NV	CHURCHILL	279.25
6	NV	WASHOE	253.86
7	NV	CARSON CITY	238.37
8	NV	DOUGLAS	238.37
9	NV	STOREY	237.50
10	NV	LYON	233.11
11	NV	PERSHING	160.00
12	NV	EUREKA	71.95
13	NV	LANDER1	40.48
14	UT	UTAH	24.87
15	UT	DAVIS	23.30
16	UT	MORGAN	23.30
17	UT	WEBER	23.30
18	UT	SALT LAKE	21.98
19	UT	DUCHESNE	18.70
20	UT	SUMMIT	18.70
21	UT	WASATCH	18.70
22	NV	ELKO	15.97
23	NV	HUMBOLDT	15.79
24	CO	LA PLATA	12.44
25	CO	MONTEZUMA	12.44
26	UT	CARBON	12.44
27	UT	EMERY	12.44
28	UT	GARFIELD	12.44
29	UT	KANE1	12.44
30	UT	KANE2	12.44
31	UT	SAN JUAN	12.44
32	UT	UINTAH	12.44
33	UT	WAYNE	12.44
34	AZ	MARICOPA	10.70
35	AZ	APACHE	10.41
36	AZ	COCHISE	10.41
37	AZ	COCONINO1	10.41
38	AZ	COCONINO2	10.41
39	AZ	COCONINO3	10.41
40	AZ	GILA	10.41
41	AZ	GRAHAM	10.41
42	AZ	GREENLEE	10.41
43	AZ	MOHAVE1	10.41
44	AZ	NAVAJO	10.41
45	AZ	PINAL	10.41
46	AZ	SANTA CRUZ	10.41
47	TX	WHARTON	9.49
48	TX	WALLER	9.36
49	TX	JACKSON	9.09
50	TX	COLORADO	9.05
51	TX	CALHOUN	8.73
52	NM	CATRON	8.66
53	NM	CHAVES	8.66
54	NM	COLFAX	8.66
55	NM	CURRY	8.66
56	NM	DE BACA	8.66
57	NM	DONA ANA	8.66
58	NM	EDDY	8.66
59	NM	GRANT	8.66
60	NM	GUADALUPE	8.66
61	NM	HARDING	8.66
62	NM	HIDALGO	8.66
63	NM	LEA	8.66
64	NM	LINCOLN	8.66
65	NM	LOS ALAMOS	8.66
66	NM	LUNA	8.66
67	NM	MCKINLEY	8.66
68	NM	MORA	8.66
69	NM	OTERO	8.66
70	NM	QUAY	8.66
71	NM	RIO ARRIBA	8.66
72	NM	ROOSEVELT	8.66
73	NM	SAN JUAN	8.66
74	NM	SAN MIGUEL	8.66
75	NM	SANDOVAL	8.66
76	NM	SANTA FE	8.66
77	NM	SIERRA	8.66
78	NM	SOCORRO	8.66
79	NM	TAOS	8.66
80	NM	TORRANCE	8.66

COUNTIES OF HIGHEST AVERAGE FALLOUT FROM SHOT:

PLUMBBOB WILSON (PB2) Jun 18, 1957 10 Kt

(microCuries per square meter)

Each Row Includes:
County Rank (most fallout = 1), State, County, Fallout Level in microCuries/square meter

Rank	State	County	Fallout
1	ND	RICHLAND	93.51
2	SD	ROBERTS	93.30
3	ND	CASS	56.62
4	ND	MERCER	56.52
5	SD	WALWORTH	56.43
6	ND	OLIVER	56.22
7	ND	FOSTER	56.18
8	SD	HYDE	56.12
9	SD	ZIEBACH	56.12
10	ND	SARGENT	55.98
11	SD	DAY	55.98
12	SD	FAULK	55.98
13	SD	DEWEY	55.93
14	SD	BEADLE	55.87
15	SD	CAMPBELL	55.87
16	SD	POTTER	55.87
17	SD	SULLY	55.87
18	ND	EMMONS	55.71
19	SD	BRULE	55.71
20	SD	CODINGTON	55.71
21	SD	LYMAN	55.71
22	SD	MELLETTE	55.71
23	SD	SANBORN	55.71
24	SD	SPINK	55.71
25	NE	BLAINE	55.65
26	NE	BROWN	55.65
27	SD	AURORA	55.55
28	SD	BUFFALO	55.55
29	SD	TODD	55.55
30	UT	IRON1	43.12
31	WY	NATRONA	39.86
32	ND	TRAILL	38.14
33	SD	CORSON	37.73
34	SD	MARSHALL	37.73
35	SD	HAAKON	37.68
36	SD	STANLEY	37.68
37	ND	LOGAN	37.63
38	ND	MORTON	37.63
39	ND	BURLEIGH	37.49
40	SD	BROWN	37.49
41	ND	MCINTOSH	37.38
42	ND	MCLEAN	37.38
43	ND	WALSH	37.38
44	SD	CLARK	37.38
45	SD	HAND	37.38
46	SD	TRIPP	37.38
47	ND	MCHENRY	37.22
48	ND	WARD	37.22
49	SD	BENNETT	37.22
50	SD	HAMLIN	37.22
51	SD	JERAULD	37.22
52	SD	JONES	37.22
53	SD	KINGSBURY	37.22
54	SD	MINER	37.22
55	SD	JACKSON	37.11
56	UT	IRON3	32.43
57	CA	INYO1	31.32
58	CA	INYO2	31.32
59	WY	CONVERSE	30.61
60	WY	NIOBRARA	26.21
61	UT	PIUTE	24.45
62	NV	LINCOLN2	21.88
63	UT	IRON2	21.00
64	UT	BEAVER	20.88
65	NV	NYE3	20.33
66	ND	GRANT	19.75
67	ND	BARNES	19.70
68	UT	WAYNE	19.44
69	ND	LA MOURE	19.34
70	SD	EDMUNDS	19.34
71	ND	STUTSMAN	19.29
72	ND	GRAND FORKS	19.24
73	NE	MADISON	19.24
74	ND	DICKEY	19.10
75	ND	KIDDER	19.00
76	KS	STANTON	18.85
77	ND	BOTTINEAU	18.84
78	ND	PEMBINA	18.84
79	ND	SHERIDAN	18.84
80	NE	GARFIELD	18.84

COUNTIES OF HIGHEST AVERAGE FALLOUT FROM SHOT:

PLUMBBOB PRISCILLA (PB3) Jun 24, 1957 37 Kt

(microCuries per square meter)

Each Row Includes:
County Rank (most fallout = 1), State, County, Fallout Level in microCuries/square meter

Rank	State	County	Fallout	Rank	State	County	Fallout	Rank	State	County	Fallout
1	UT	WASHINGTON1	188.87	32	MS	PIKE	85.11	63	AL	CONECUH	55.80
2	MS	FRANKLIN	139.24	33	LA	ST LANDRY	84.94	64	MS	STONE	55.70
3	LA	AVOYELLES	139.24	34	LA	ALLEN	84.51	65	AL	MONROE	55.67
4	MS	WILKINSON	139.03	35	LA	JEFFERSON DA	84.43	66	LA	ST JOHN THE	55.51
5	LA	WEST FELICIA	138.91	36	MS	HINDS	84.10	67	LA	ST JAMES	55.45
6	LA	POINTE COUPE	138.78	37	MS	LAUDERDALE	84.01	68	MS	GREENE	55.22
7	LA	ACADIA	138.75	38	MS	MARION	83.85	69	MS	JACKSON	54.66
8	LA	LA SALLE	138.04	39	LA	EAST FELICIA	83.66	70	MS	HARRISON	54.60
9	TX	ORANGE	138.01	40	AL	MARENGO	83.52	71	LA	LAFOURCHE	54.49
10	LA	TENSAS	137.94	41	LA	LIVINGSTON	82.72	72	LA	ST CHARLES	54.49
11	LA	GRANT	137.82	42	LA	WASHINGTON	82.70	73	UT	WASHINGTON3	52.16
12	MS	ADAMS	137.81	43	LA	ST TAMMANY	82.69	74	NM	MCKINLEY	33.80
13	LA	RAPIDES	137.74	44	LA	ASCENSION	82.66	75	AZ	APACHE	32.07
14	LA	BEAUREGARD	137.45	45	AL	MOBILE	82.59	76	MS	JASPER	31.42
15	LA	ST HELENA	137.04	46	MS	GEORGE	82.35	77	MS	SMITH	31.37
16	LA	EAST BATON R	136.99	47	MS	HANCOCK	82.27	78	MS	AMITE	30.83
17	LA	EVANGELINE	136.99	48	MS	PEARL RIVER	82.23	79	MS	WALTHALL	30.20
18	LA	IBERVILLE	136.93	49	FL	OKALOOSA	81.80	80	AL	WILCOX	30.18
19	LA	WEST BATON R	136.93	50	LA	PLAQUEMINES	81.63				
20	LA	IBERIA	136.82	51	LA	ST BERNARD	81.63				
21	LA	TANGIPAHOA	136.82	52	LA	VERMILION	81.59				
22	LA	ST MARTIN	136.42	53	UT	KANE2	67.77				
23	LA	ST MARY	136.42	54	UT	IRON1	59.51				
24	LA	ASSUMPTION	136.37	55	MS	COPIAH	58.17				
25	LA	TERREBONNE	135.83	56	MS	CLAIBORNE	58.10				
26	LA	LAFAYETTE	135.79	57	MS	JEFFERSON	57.98				
27	MS	WARREN	135.72	58	MS	CLARKE	56.81				
28	LA	CALCASIEU	135.70	59	NV	LINCOLN1	56.77				
29	LA	CAMERON	135.70	60	MS	WAYNE	56.74				
30	LA	CATAHOULA	85.39	61	MS	LAMAR	56.65				
31	LA	CONCORDIA	85.39	62	MS	FORREST	55.84				

COUNTIES OF HIGHEST AVERAGE FALLOUT FROM SHOT:

PLUMBBOB HOOD (PB4) Jul 5, 1957 74 Kt

(microCuries per square meter)

Each Row Includes:
County Rank (most fallout = 1), State, County, Fallout Level in microCuries/square meter

Rank	State	County	Fallout
1	AZ	PIMA	148.41
2	AZ	SANTA CRUZ	98.28
3	CO	GARFIELD	81.66
4	CO	RIO BLANCO	81.28
5	NM	COLFAX	73.63
6	NM	HARDING	73.63
7	NM	UNION	73.63
8	NM	MORA	73.09
9	NM	SAN MIGUEL	73.09
10	NM	SANTA FE	73.09
11	NM	TORRANCE	72.55
12	NM	BERNALILLO	67.97
13	NM	QUAY	55.22
14	NM	TAOS	55.22
15	NM	GUADALUPE	54.82
16	NM	LOS ALAMOS	54.41
17	NM	SANDOVAL	54.41
18	MO	OSAGE	53.81
19	NM	VALENCIA	53.60
20	MO	COLE	47.36
21	MO	MONITEAU	44.77
22	AZ	MARICOPA	41.34
23	AZ	PINAL	41.34
24	IL	GALLATIN	40.07
25	IL	KNOX	38.23
26	MO	MILLER	33.81
27	AZ	COCHISE	32.72
28	AZ	GILA	32.72
29	AZ	GRAHAM	32.72
30	MO	ST CHARLES	32.02
31	MO	MORGAN	29.85
32	MO	AUDRAIN	29.05
33	KY	CUMBERLAND	28.87
34	MO	CALLAWAY	28.57
35	IL	GREENE	27.68
36	NM	CURRY	27.61
37	NM	DE BACA	27.61
38	MO	CAMDEN	27.54
39	NM	ROOSEVELT	27.34
40	MO	PULASKI	27.12
41	IL	CHRISTIAN	27.10
42	MO	MARIES	26.77
43	MO	COOPER	26.23
44	MO	DOUGLAS	25.78
45	IL	STEPHENSON	25.72
46	IL	MACON	25.60
47	WI	GREEN	25.41
48	IL	WAYNE	25.37
49	IL	HENRY	25.34
50	MO	BOONE	25.12
51	AZ	GREENLEE	25.11
52	IL	WHITESIDE	25.09
53	MO	DALLAS	24.98
54	IL	PEORIA	24.78
55	MO	LINCOLN	24.64
56	IL	WINNEBAGO	24.49
57	IL	MONTGOMERY	23.82
58	IL	LEE	23.70
59	IL	OGLE	23.68
60	MO	PERRY	23.64
61	IL	MORGAN	23.39
62	IL	MCLEAN	23.38
63	IL	PUTNAM	23.36
64	IL	STARK	23.36
65	IL	JACKSON	23.30
66	IL	MARSHALL	23.26
67	MO	HENRY	23.18
68	KS	JOHNSON	23.02
69	MO	JOHNSON	23.00
70	IL	PERRY	22.98
71	MO	POLK	22.48
72	NC	HYDE	22.22
73	MO	GASCONADE	21.58
74	MO	SHELBY	21.02
75	MO	BATES	21.01
76	UT	GRAND	20.80
77	MO	BENTON	20.79
78	CO	DELTA	20.16
79	MO	CRAWFORD	20.16
80	MO	MONROE	20.03

COUNTIES OF HIGHEST AVERAGE FALLOUT FROM SHOT:

PLUMBBOB DIABLO (PB5) Jul 15, 1957 11 Kt

(microCuries per square meter)

Each Row Includes:
County Rank (most fallout = 1), State, County, Fallout Level in microCuries/square meter

Rank	State	County	Fallout		Rank	State	County	Fallout		Rank	State	County	Fallout
1	NV	NYE2	576.56		32	SD	BUTTE	28.90		63	VA	POWHATAN	18.89
2	NV	WHITE PINE2	299.97		33	ND	BILLINGS	27.39		64	CO	LOGAN	18.87
3	NV	LINCOLN2	222.21		34	CO	MOFFAT	25.78		65	NY	TOMPKINS	18.45
4	NV	WHITE PINE3	218.27		35	UT	UINTAH	25.78		66	NE	BANNER	18.34
5	NV	WHITE PINE1	111.52		36	WY	SUBLETTE	25.78		67	OH	WASHINGTON	18.16
6	UT	UTAH	89.61		37	VT	ESSEX	25.41		68	NE	GRANT	18.08
7	UT	JUAB	73.24		38	MT	DAWSON	24.74		69	WY	CAMPBELL	18.02
8	UT	WASATCH	70.14		39	MT	RICHLAND	24.59		70	UT	DAVIS	17.92
9	ND	STARK	48.88		40	ND	WILLIAMS	24.59		71	SD	PERKINS	17.82
10	ID	TETON	47.59		41	SD	LAWRENCE	23.70		72	NE	MORRILL	17.64
11	UT	MILLARD	45.80		42	SD	HARDING	23.39		73	UT	CARBON	17.59
12	UT	SALT LAKE	44.81		43	WY	CROOK	23.14		74	UT	MORGAN	17.59
13	UT	SANPETE	44.81		44	ND	HETTINGER	22.87		75	UT	DAGGETT	17.37
14	UT	TOOELE1	44.81		45	NE	CHEYENNE	22.73		76	NC	CHOWAN	17.34
15	UT	TOOELE2	44.81		46	MT	FALLON	22.53		77	OH	ATHENS	17.31
16	UT	SUMMIT	43.81		47	ND	BOWMAN	22.31		78	WV	TYLER	17.31
17	WY	UINTA	43.81		48	NE	KIMBALL	21.97		79	OH	COSHOCTON	17.21
18	WY	CARBON	42.92		49	NY	CORTLAND	21.47		80	OH	NOBLE	17.21
19	WY	FREMONT	42.92		50	NY	STEUBEN	21.47					
20	WY	SWEETWATER	42.92		51	NE	HOOKER	21.00					
21	ND	DUNN	37.91		52	WY	GOSHEN	20.99					
22	ND	GOLDEN VALLE	36.51		53	WY	NIOBRARA	20.53					
23	MT	WIBAUX	35.11		54	WV	MARION	20.35					
24	ND	MOUNTRAIL	35.11		55	NE	BUTLER	19.57					
25	ID	CLARK	34.66		56	ND	BARNES	19.43					
26	ID	JEFFERSON	34.66		57	ND	LOGAN	19.43					
27	ID	MADISON	34.66		58	MD	WORCESTER	19.36					
28	UT	DUCHESNE	34.30		59	NY	GENESEE	19.25					
29	ND	ADAMS	33.39		60	NY	ALLEGANY	18.94					
30	NV	LINCOLN1	31.66		61	SD	MELLETTE	18.93					
31	WY	WESTON	29.52		62	NE	HITCHCOCK	18.89					

COUNTIES OF HIGHEST AVERAGE FALLOUT FROM SHOTS:

PLUMBBOB KEPLER AND OWENS (PB6)
KEPLER: Jul 24, 1957 10 Kt
OWENS: Jul 25, 1957 9.7 Kt
(microCuries per square meter)

Each Row Includes:
County Rank (most fallout = 1), State, County, Fallout Level in microCuries/square meter

Rank	State	County	Fallout
1	NV	ESMERALDA2	328.70
2	NV	ESMERALDA1	256.26
3	NV	MINERAL	184.78
4	NV	CHURCHILL	121.94
5	CA	INYO1	88.07
6	CA	INYO2	88.07
7	NV	LYON	69.69
8	NV	DOUGLAS	55.22
9	MT	BEAVERHEAD	52.07
10	MT	GRANITE	51.14
11	MT	POWELL	47.98
12	KY	MENIFEE	47.69
13	MT	CASCADE	47.67
14	IN	DEARBORN	47.15
15	IN	RIPLEY	46.92
16	CA	MONO	46.65
17	MT	LEWIS AND CL	43.88
18	NV	CARSON CITY	43.57
19	MT	BROADWATER	41.52
20	NV	PERSHING	37.35
21	NV	WASHOE	37.35
22	MT	CHOUTEAU	37.21
23	NV	STOREY	36.99
24	UT	TOOELE1	36.99
25	MT	DEER LODGE	36.57
26	ID	LEMHI	36.31
27	KY	KENTON	36.23
28	ID	LATAH	36.23
29	ID	CUSTER	36.11
30	OR	UNION	35.99
31	ID	VALLEY	35.98
32	OR	BAKER	35.73
33	OR	GRANT	35.58
34	MT	SILVER BOW	35.13
35	OH	CLERMONT	34.91
36	MT	TOOLE	34.88
37	MT	JEFFERSON	34.60
38	MT	LAKE	33.65
39	OH	LAWRENCE	32.30
40	MT	GOLDEN VALLE	32.18
41	MT	MISSOULA	32.06
42	KY	BREATHITT	32.01
43	WV	MASON	31.91
44	OH	GALLIA	31.83
45	KY	WOLFE	31.65
46	IN	FRANKLIN	31.58
47	ID	IDAHO	31.32
48	MT	RAVALLI	30.91
49	KY	POWELL	30.50
50	ID	BONNER	30.46
51	MT	HILL	29.92
52	KY	ESTILL	29.85
53	OH	HIGHLAND	29.29
54	KY	BOYD	29.23
55	WV	CABELL	29.18
56	MT	STILLWATER	28.88
57	IN	OHIO	28.82
58	IN	DECATUR	28.67
59	MT	FERGUS	28.39
60	MT	FLATHEAD	28.26
61	MT	LIBERTY	28.07
62	MT	MINERAL	27.91
63	MT	SWEET GRASS	27.45
64	MT	TETON	27.36
65	MT	GLACIER	27.22
66	MT	JUDITH BASIN	27.10
67	ID	CLEARWATER	26.77
68	MT	SANDERS	26.67
69	NV	LANDER2	26.63
70	MT	PONDERA	26.60
71	ID	SHOSHONE	26.37
72	MT	GALLATIN	26.34
73	MT	PARK	26.13
74	MT	LINCOLN	26.12
75	ID	LEWIS	25.80
76	MT	MEAGHER	25.76
77	ID	CAMAS	25.70
78	OR	WALLOWA	25.68
79	WA	ASOTIN	25.60
80	ID	BUTTE	25.58

COUNTIES OF HIGHEST AVERAGE FALLOUT FROM SHOT:

PLUMBBOB SHASTA (PB8) 18 Aug, 1957 17 Kt

(microCuries per square meter)

Each Row Includes:
County Rank (most fallout = 1), State, County, Fallout Level in microCuries/square meter

Rank	State	County	Fallout
1	NV	EUREKA	482.83
2	NV	WHITE PINE2	335.38
3	NV	WHITE PINE3	263.85
4	NV	NYE2	254.89
5	NV	WHITE PINE1	249.74
6	ND	BILLINGS	136.54
7	SD	BROOKINGS	98.45
8	NE	CHERRY	97.60
9	ND	RANSOM	94.67
10	ND	DIVIDE	89.80
11	UT	TOOELE1	84.89
12	SD	KINGSBURY	64.67
13	ND	EMMONS	57.38
14	NV	ELKO	55.23
15	ND	STARK	54.45
16	ND	SLOPE	54.08
17	ND	MCKENZIE	53.87
18	SD	HARDING	53.25
19	MT	CARTER	53.09
20	ND	DICKEY	51.87
21	ND	RICHLAND	51.87
22	NE	PHELPS	51.34
23	NE	HOOKER	50.53
24	MT	FALLON	49.63
25	ND	STUTSMAN	49.56
26	NE	MADISON	49.56
27	SD	FAULK	49.56
28	ND	BURKE	49.29
29	ND	RENVILLE	48.47
30	NE	LINCOLN	48.25
31	MT	JUDITH BASIN	46.42
32	ND	BENSON	46.23
33	ND	FOSTER	46.23
34	NE	SALINE	46.21
35	SD	MCCOOK	46.20
36	WY	PARK	45.86
37	NE	NUCKOLLS	44.85
38	KS	REPUBLIC	44.58
39	NE	PLATTE	44.48
40	MO	PETTIS	44.34
41	MO	LAFAYETTE	44.01
42	IA	DECATUR	43.91
43	MO	HARRISON	43.84
44	MO	DE KALB	43.68
45	MO	GENTRY	43.68
46	MO	GRUNDY	43.68
47	MO	CALDWELL	43.68
48	MO	LIVINGSTON	43.68
49	MO	HENRY	43.66
50	SD	MINNEHAHA	43.64
51	MO	WORTH	43.64
52	IA	TAYLOR	43.53
53	MO	DAVIESS	43.52
54	MO	JOHNSON	43.52
55	MO	MERCER	43.52
56	MO	SALINE	43.52
57	MO	POLK	43.50
58	IA	RINGGOLD	43.37
59	SD	DAVISON	43.27
60	SD	HANSON	43.27
61	SD	LAKE	43.27
62	SD	MINER	43.27
63	SD	PENNINGTON	42.39
64	UT	TOOELE2	42.39
65	SD	SANBORN	41.81
66	SD	BENNETT	40.63
67	SD	JACKSON	40.63
68	UT	BOX ELDER1	40.59
69	UT	BOX ELDER2	40.59
70	UT	UTAH	40.59
71	SD	TODD	39.41
72	SD	HAAKON	39.03
73	NE	BLAINE	37.50
74	SD	STANLEY	37.31
75	SD	MELLETTE	36.49
76	ND	DUNN	35.98
77	SD	LYMAN	35.85
78	SD	AURORA	35.71
79	SD	CORSON	35.63
80	SD	TRIPP	35.02

COUNTIES OF HIGHEST AVERAGE FALLOUT FROM SHOT:

PLUMBBOB DOPPLER (PB9) Aug 23, 1957 11 Kt

(microCuries per square meter)

Each Row Includes:
County Rank (most fallout = 1), State, County, Fallout Level in microCuries/square meter

Rank	State	County	Fallout
1	IA	WAPELLO	130.84
2	IA	DAVIS	121.23
3	IA	TAMA	111.83
4	IA	JASPER	110.55
5	IA	POTTAWATTAMI	107.95
6	IA	CASS	107.70
7	IA	FRANKLIN	107.10
8	IA	SHELBY	106.60
9	IA	MONTGOMERY	106.36
10	IA	MADISON	106.12
11	IA	AUDUBON	105.90
12	IA	BOONE	105.85
13	IA	POLK	105.84
14	IA	DALLAS	105.41
15	IA	ADAIR	105.28
16	IA	CARROLL	105.26
17	UT	UTAH	103.16
18	IA	BLACK HAWK	99.05
19	IA	BENTON	98.80
20	IA	BREMER	97.95
21	IA	POWESHIEK	91.49
22	IA	LINN	82.36
23	IA	HENRY	81.99
24	IA	FAYETTE	81.01
25	IA	JACKSON	81.01
26	IA	DELAWARE	80.77
27	IA	BUCHANAN	80.64
28	IA	KEOKUK	80.64
29	NV	WHITE PINE2	79.88
30	MO	CLARK	78.57
31	IA	VAN BUREN	76.74
32	UT	WASATCH	74.47
33	IA	MARION	70.46
34	IA	MARSHALL	67.10
35	IA	STORY	67.10
36	IA	HAMILTON	65.76
37	IA	GREENE	65.41
38	IA	GUTHRIE	65.17
39	IA	HARDIN	65.17
40	IA	ADAMS	64.92
41	IA	WARREN	64.92
42	IA	CLARKE	64.79
43	IA	UNION	64.79
44	IA	LUCAS	64.09
45	IA	CEDAR	62.01
46	IA	IOWA	61.76
47	IA	CLAYTON	60.67
48	WI	DANE	60.46
49	IA	DUBUQUE	60.42
50	IL	OGLE	60.42
51	IL	DE KALB	60.28
52	IL	LOGAN	60.28
53	IA	DES MOINES	60.17
54	WI	GRANT	60.17
55	IL	MCDONOUGH	60.16
56	IL	JOHNSON	60.05
57	IL	WARREN	60.05
58	WI	SAUK	59.97
59	IL	HANCOCK	59.47
60	IL	BROWN	59.40
61	IL	SCOTT	59.40
62	IL	MORGAN	58.70
63	IL	PIKE	54.80
64	UT	JUAB	54.07
65	UT	TOOELE2	53.43
66	UT	DUCHESNE	52.28
67	UT	UINTAH	51.90
68	UT	DAVIS	49.94
69	IA	MAHASKA	48.77
70	UT	SALT LAKE	47.10
71	IA	GRUNDY	45.92
72	CO	PHILLIPS	44.49
73	IA	DECATUR	44.34
74	IA	TAYLOR	44.20
75	MO	NODAWAY	44.20
76	CO	SEDGWICK	43.50
77	IA	RINGGOLD	43.50
78	MO	WORTH	43.50
79	UT	BOX ELDER2	43.10
80	UT	SANPETE	43.10

COUNTIES OF HIGHEST AVERAGE FALLOUT FROM SHOT:

PLUMBBOB SMOKY (PB11)   Aug 31, 1957   11 Kt

(microCuries per square meter)

Each Row Includes:
County Rank (most fallout = 1), State, County, Fallout Level in microCuries/square meter

Rank	State	County	Fallout
1	UT	GARFIELD	1238.11
2	CA	INYO3	898.04
3	UT	WASHINGTON2	641.04
4	UT	WASHINGTON3	500.51
5	NV	LINCOLN1	393.34
6	WY	SWEETWATER	384.53
7	UT	WASHINGTON1	318.50
8	UT	UINTAH	235.72
9	UT	DAGGETT	232.42
10	AZ	MOHAVE2	230.98
11	WY	FREMONT	228.21
12	UT	IRON3	162.62
13	WY	CARBON	151.11
14	NV	CLARK1	109.14
15	UT	WAYNE	85.40
16	UT	EMERY	79.41
17	WY	SUBLETTE	76.10
18	WY	LINCOLN	60.68
19	ND	GRIGGS	60.47
20	ND	RANSOM	55.29
21	ND	RICHLAND	55.29
22	ND	LA MOURE	53.12
23	ND	BARNES	52.16
24	ND	PEMBINA	52.16
25	ND	CASS	51.07
26	ND	EDDY	51.07
27	ND	GRAND FORKS	51.07
28	ND	NELSON	51.07
29	ND	RAMSEY	51.07
30	ND	TRAILL	51.07
31	ND	WALSH	51.07
32	ND	DICKEY	49.99
33	ND	STEELE	49.99
34	ND	SARGENT	49.44
35	SD	MARSHALL	49.44
36	UT	IRON1	45.43
37	CA	SAN BERNADIN	44.25
38	UT	CARBON	41.90
39	UT	PIUTE	40.06
40	UT	DUCHESNE	38.85
41	WY	UINTA	38.15
42	ND	FOSTER	34.81
43	SD	ROBERTS	34.81
44	ND	CAVALIER	34.33
45	ND	MCINTOSH	34.27
46	ND	TOWNER	33.85
47	ND	BENSON	32.76
48	ND	ROLETTE	32.76
49	SD	EDMUNDS	32.70
50	MT	WIBAUX	32.62
51	ND	GOLDEN VALLE	31.93
52	ND	STUTSMAN	31.68
53	ND	STARK	31.65
54	ND	HETTINGER	31.38
55	ND	DUNN	31.14
56	ND	EMMONS	31.14
57	ND	KIDDER	31.14
58	ND	LOGAN	31.14
59	NE	ANTELOPE	31.14
60	NE	BOONE	31.14
61	NE	MADISON	31.14
62	SD	AURORA	31.14
63	SD	BROWN	31.14
64	SD	BRULE	31.14
65	SD	DAY	31.14
66	SD	FAULK	31.14
67	SD	MCPHERSON	31.14
68	SD	POTTER	31.14
69	SD	SPINK	31.14
70	SD	WALWORTH	31.14
71	UT	IRON2	31.07
72	UT	KANE2	30.32
73	ND	ADAMS	25.12
74	ND	BOWMAN	25.12
75	ND	SLOPE	25.12
76	NE	CEDAR	25.12
77	NE	PIERCE	25.12
78	SD	HARDING	25.12
79	SC	FLORENCE	24.42
80	UT	GRAND	23.83

COUNTIES OF HIGHEST AVERAGE FALLOUT FROM SHOT:

PLUMBBOB GALILEO (PB12) Sep 2, 1957 11 Kt

(microCuries per square meter)

Each Row Includes:
County Rank (most fallout = 1), State, County, Fallout Level in microCuries/square meter

Rank	State	County	Fallout
1	NV	NYE2	100.99
2	LA	LAFOURCHE	58.17
3	LA	ST CHARLES	55.11
4	NV	WHITE PINE1	52.61
5	NV	WHITE PINE2	52.61
6	NV	WHITE PINE3	52.61
7	UT	UTAH	46.17
8	UT	MILLARD	45.16
9	UT	MORGAN	36.92
10	UT	TOOELE1	36.17
11	UT	WASATCH	36.13
12	KS	SEDGWICK	35.54
13	UT	DAVIS	34.59
14	UT	TOOELE2	34.59
15	KS	SUMNER	34.04
16	UT	SUMMIT	33.85
17	UT	SALT LAKE	32.62
18	CO	EL PASO	31.83
19	NV	EUREKA	31.50
20	KS	MCPHERSON	29.10
21	UT	WEBER	27.66
22	UT	JUAB	27.08
23	KS	KINGMAN	26.96
24	UT	DAGGETT	26.53
25	UT	DUCHESNE	26.53
26	WY	SWEETWATER	26.53
27	WY	UINTA	26.53
28	KS	MORRIS	26.30
29	CO	DENVER	26.12
30	UT	BOX ELDER2	25.91
31	KS	BUTLER	25.85
32	KS	SHAWNEE	25.65
33	MO	VERNON	25.62
34	KS	DICKINSON	25.46
35	KS	LYON	25.27
36	KS	RENO	25.18
37	KS	SALINE	25.16
38	KS	WILSON	25.01
39	KS	COWLEY	24.68
40	KS	MARION	24.58
41	KS	HARVEY	24.26
42	KS	CRAWFORD	24.21
43	OK	ROGERS	24.20
44	KS	LABETTE	24.02
45	KS	GREENWOOD	23.94
46	MO	BARRY	23.62
47	MO	CEDAR	23.36
48	CO	ADAMS	23.25
49	CO	DOUGLAS	23.25
50	MS	COVINGTON	22.95
51	KS	COFFEY	22.78
52	CO	JEFFERSON	22.61
53	KS	JEFFERSON	22.46
54	KS	CHASE	22.44
55	KS	PRATT	22.03
56	KS	LEAVENWORTH	21.94
57	KS	STAFFORD	21.84
58	CO	CUSTER	21.81
59	OK	WAGONER	21.69
60	KS	LINCOLN	21.59
61	MO	CHRISTIAN	21.36
62	KS	OSAGE	21.32
63	KS	HARPER	21.06
64	KS	OTTAWA	21.05
65	KS	ANDERSON	21.00
66	CO	ARAPAHOE	20.94
67	KS	PAWNEE	20.61
68	CO	BOULDER	20.48
69	AR	CARROLL	20.26
70	KS	ATCHISON	20.26
71	KS	JOHNSON	20.12
72	MO	LAWRENCE	20.05
73	AR	BENTON	20.02
74	KS	EDWARDS	19.96
75	CO	PUEBLO	19.88
76	MO	GREENE	19.74
77	CO	LARIMER	19.55
78	OK	TULSA	19.52
79	KS	LINN	19.49
80	KS	RUSH	19.49

COUNTIES OF HIGHEST AVERAGE FALLOUT FROM SHOTS
PLUMBBOB WHEELER, COULOMB B AND LaPLACE
(PB13)
WHEELER: Sep 6, 1957, 0.197 Kt
COULOMB-B: Sep 6, 1957, 0.3 Kt
LaPLACE: Sep 8, 1957, 1 Kt
(microCuries per square meter)
Each Row Includes:
County Rank (most fallout = 1), State, County, Fallout Level in microCuries/square meter

Rank	State/County	Fallout	Rank	State/County	Fallout	Rank	State/County	Fallout
1	NE BUFFALO	46.40	31	AZ COCONINO2	30.09	61	NE PAWNEE	24.89
2	NE HALL	45.83	32	AZ COCONINO3	30.09	62	NE GAGE	24.87
3	NE KEARNEY	44.53	33	KS POTTAWATOMIE	29.91	63	KS CHASE	24.81
4	KS MITCHELL	43.62	34	KS OSBORNE	29.63	64	OK NOBLE	24.72
5	KS MCPHERSON	40.15	35	KS TREGO	29.60	65	KS GREENWOOD	24.56
6	KS PAWNEE	39.66	36	KS ELLIS	29.39	66	OK ALFALFA	24.54
7	KS BARTON	38.16	37	KS MEADE	28.72	67	OK KAY	24.21
8	KS OTTAWA	37.65	38	KS RENO	28.49	68	KS ELK	24.01
9	KS SALINE	37.49	39	KS HARVEY	28.45	69	OK ROGERS	23.81
10	KS RUSSELL	37.38	40	KS LANE	28.34	70	OK PAWNEE	23.78
11	KS CLOUD	37.29	41	KS SHAWNEE	28.33	71	KS CHAUTAUQUA	23.71
12	KS STAFFORD	36.46	42	KS MARSHALL	28.28	72	OK PAYNE	23.69
13	KS LINCOLN	36.23	43	KS WASHINGTON	27.93	73	OK MAJOR	23.48
14	KS RICE	35.44	44	KS LABETTE	27.79	74	NE FRANKLIN	23.25
15	KS WABAUNSEE	34.88	45	KS SCOTT	27.74	75	OK WOODS	23.12
16	KS FORD	34.77	46	OK HARPER	27.51	76	KS MORRIS	23.05
17	NE PHELPS	34.52	47	NE HARLAN	27.44	77	OK MAYES	22.82
18	KS ELLSWORTH	34.48	48	KS SUMNER	27.38	78	OK GARFIELD	22.79
19	KS DICKINSON	34.29	49	NE POLK	26.99	79	KS LYON	22.79
20	KS CLAY	33.76	50	KS COWLEY	26.67	80	OK LINCOLN	22.72
21	KS GEARY	33.24	51	KS JACKSON	26.25			
22	KS EDWARDS	32.61	52	KS JEFFERSON	26.12			
23	OK GRANT	32.57	53	OK WASHINGTON	26.05			
24	KS CLARK	32.18	54	NE JOHNSON	25.93			
25	KS RILEY	32.11	55	KS HASKELL	25.88			
26	KS RUSH	31.86	56	OK TULSA	25.79			
27	KS NESS	31.68	57	KS GRANT	25.78			
28	NE NANCE	31.28	58	OK OSAGE	25.62			
29	NE MERRICK	30.81	59	KS BUTLER	25.49			
30	KS MARION	30.34	60	OK CREEK	25.18			

COUNTIES OF HIGHEST AVERAGE FALLOUT FROM SHOT:

PLUMBBOB FIZEAU (PB14) Sep 14, 1957 11 Kt

(microCuries per square meter)

Each Row Includes:
County Rank (most fallout = 1), State, County, Fallout Level in microCuries/square meter

Rank	State	County	Fallout
1	NV	MINERAL	57.61
2	NV	CARSON CITY	39.80
3	NV	DOUGLAS	39.80
4	NV	LYON	39.80
5	NV	STOREY	39.80
6	NV	WASHOE	39.80
7	CA	MONO	23.62
8	NV	CHURCHILL	15.92
9	NV	NYE2	12.09
10	NV	NYE1	7.89
11	KS	BUTLER	6.01
12	KS	CHASE	6.01
13	KS	FRANKLIN	6.01
14	KS	LYON	6.01
15	KS	MORRIS	6.01
16	KS	OSAGE	6.01
17	KS	WOODSON	6.01
18	KS	COFFEY	5.68
19	KS	DOUGLAS	5.68
20	KS	GREENWOOD	5.68
21	KS	JOHNSON	5.68
22	KS	ANDERSON	5.53
23	KS	LINN	5.53
24	KS	MIAMI	5.53
25	KS	COWLEY	5.34
26	OK	OKLAHOMA	5.34
27	KS	ALLEN	5.30
28	KS	BOURBON	5.30
29	KS	ELK	5.30
30	KS	WILSON	5.30
31	MO	VERNON	5.30
32	OK	CADDO	5.30
33	OK	CANADIAN	5.30
34	OK	CHEROKEE	5.30
35	OK	GRANT	5.30
36	OK	LINCOLN	5.30
37	OK	MUSKOGEE	5.30
38	OK	OKFUSKEE	5.30
39	OK	WASHITA	5.30
40	KS	HARVEY	3.93
41	KS	JEFFERSON	3.93
42	KS	LEAVENWORTH	3.93
43	KS	NEOSHO	3.93
44	KS	WABAUNSEE	3.93
45	KS	HARPER	3.60
46	KS	JACKSON	3.60
47	KS	MARION	3.60
48	KS	MCPHERSON	3.60
49	KS	SUMNER	3.60
50	MO	BATES	3.60
51	OK	CUSTER	3.60
52	CA	LASSEN	3.55
53	CA	MODOC	3.55
54	CA	SISKIYOU	3.55
55	CA	TEHAMA	3.55
56	OR	CROOK	3.55
57	OR	LAKE	3.55
58	KS	CRAWFORD	3.45
59	OK	KAY	3.45
60	OK	TULSA	3.45
61	OK	ALFALFA	3.26
62	KS	CHEROKEE	3.22
63	MO	BARTON	3.22
64	MO	JASPER	3.22
65	NE	SEWARD	3.22
66	OK	BECKHAM	3.22
67	OK	BLAINE	3.22
68	OK	CRAIG	3.22
69	OK	CREEK	3.22
70	OK	DELAWARE	3.22
71	OK	KINGFISHER	3.22
72	OK	MAYES	3.22
73	OK	NOWATA	3.22
74	OK	OKMULGEE	3.22
75	OK	OTTAWA	3.22
76	OK	PAYNE	3.22
77	OK	WAGONER	3.22
78	KS	KINGMAN	2.89
79	KS	NEMAHA	2.89
80	CA	SHASTA	2.66

COUNTIES OF HIGHEST AVERAGE FALLOUT FROM SHOT:

PLUMBBOB NEWTON (PB15) Sep 14, 1957 12 Kt

(microCuries per square meter)

Each Row Includes:
County Rank (most fallout = 1), State, County, Fallout Level in microCuries/square meter

Rank	State	County	Fallout
1	NE	GARFIELD	36.53
2	SD	CODINGTON	35.71
3	SD	KINGSBURY	25.12
4	SD	HAMLIN	25.06
5	SD	MINER	24.82
6	SD	CLARK	23.94
7	SD	ROBERTS	23.76
8	ND	LA MOURE	23.62
9	ND	RICHLAND	23.62
10	SD	BEADLE	23.51
11	SD	MINNEHAHA	23.51
12	ND	BARNES	23.50
13	ND	TRAILL	23.50
14	ND	LOGAN	23.42
15	SD	WALWORTH	23.40
16	NE	LOGAN	23.17
17	SD	HUTCHINSON	23.17
18	SD	LAKE	23.17
19	ND	CASS	22.96
20	ND	DICKEY	22.96
21	ND	KIDDER	22.96
22	SD	MCPHERSON	22.84
23	ND	MORTON	22.75
24	ND	STEELE	22.75
25	ND	STUTSMAN	22.75
26	SD	DAY	22.75
27	SD	SPINK	22.70
28	ND	SARGENT	22.69
29	ND	MERCER	22.65
30	SD	AURORA	22.65
31	SD	FAULK	22.65
32	SD	EDMUNDS	22.62
33	ND	GRIGGS	22.41
34	ND	EMMONS	22.31
35	SD	HYDE	22.19
36	SD	MARSHALL	22.12
37	ND	BURLEIGH	21.97
38	ND	GRAND FORKS	21.97
39	ND	SHERIDAN	21.97
40	SD	DEWEY	21.97
41	SD	HAND	21.97
42	SD	JACKSON	21.97
43	SD	POTTER	21.97
44	SD	HAAKON	21.87
45	UT	JUAB	21.81
46	ND	MCLEAN	21.71
47	SD	ZIEBACH	21.67
48	ND	NELSON	21.63
49	SD	CORSON	21.57
50	ND	EDDY	21.54
51	ND	GRANT	21.52
52	ND	BENSON	21.48
53	SD	STANLEY	21.47
54	ND	OLIVER	21.04
55	SD	CAMPBELL	21.04
56	SD	SULLY	21.04
57	ND	BOWMAN	20.54
58	ND	GOLDEN VALLE	20.54
59	ND	HETTINGER	20.54
60	ND	STARK	20.54
61	SD	HARDING	20.54
62	ND	ADAMS	20.37
63	ND	BILLINGS	20.37
64	ND	DUNN	20.37
65	ND	MCKENZIE	20.37
66	SD	MEADE	20.37
67	SD	PERKINS	20.37
68	SD	GREGORY	17.63
69	SD	CHARLES MIX	16.89
70	SD	BRULE	16.83
71	SD	JERAULD	16.83
72	SD	SANBORN	16.83
73	SD	TRIPP	16.83
74	SD	BUFFALO	16.78
75	NE	BOYD	16.49
76	SD	MCCOOK	16.49
77	ND	FOSTER	16.44
78	ND	MCINTOSH	16.07
79	NE	CHERRY	16.02
80	SD	LYMAN	15.97

COUNTIES OF HIGHEST AVERAGE FALLOUT FROM SHOT:

PLUMBBOB WHITNEY (PB16) Sep 23, 1957 19 Kt

(microCuries per square meter)

Each Row Includes:
County Rank (most fallout = 1), State, County, Fallout Level in microCuries/square meter

Rank	State	County	Fallout
1	NV	ESMERALDA2	274.35
2	NV	ESMERALDA1	107.32
3	NV	CHURCHILL	91.75
4	NV	MINERAL	86.53
5	NV	LANDER2	78.71
6	NV	PERSHING	54.49
7	CA	MONO	48.97
8	NV	CARSON CITY	45.88
9	NV	DOUGLAS	45.88
10	NV	LYON	45.88
11	NV	STOREY	45.88
12	NV	WASHOE	45.88
13	NV	NYE1	39.08
14	NV	LANDER1	36.74
15	GA	FAYETTE	29.86
16	CA	INYO1	29.60
17	CA	INYO2	29.60
18	NV	HUMBOLDT	28.46
19	GA	CLAYTON	27.52
20	GA	FULTON	26.65
21	GA	DE KALB	26.61
22	GA	MONROE	24.93
23	GA	DOUGLAS	24.59
24	GA	GWINNETT	23.14
25	GA	CRAWFORD	23.13
26	GA	BIBB	22.06
27	GA	SCHLEY	22.01
28	GA	JONES	21.58
29	GA	DOOLY	21.57
30	GA	SPALDING	21.39
31	GA	PEACH	21.31
32	GA	PULASKI	21.24
33	GA	COLUMBUS	20.72
34	GA	COBB	20.31
35	GA	SUMTER	20.10
36	GA	PIKE	19.81
37	GA	HOUSTON	19.71
38	GA	LAMAR	19.70
39	GA	BUTTS	19.69
40	GA	TWIGGS	19.50
41	GA	CRISP	19.43
42	GA	MACON	19.14
43	GA	PAULDING	19.12
44	NV	ELKO	18.94
45	GA	HARRIS	18.76
46	GA	NEWTON	18.66
47	GA	HENRY	18.59
48	GA	WEBSTER	18.58
49	GA	CARROLL	18.47
50	GA	UPSON	18.24
51	GA	COWETA	18.21
52	GA	DODGE	18.19
53	GA	LEE	17.68
54	GA	MERIWETHER	17.52
55	GA	ROCKDALE	17.38
56	OR	HARNEY	17.37
57	OR	MALHEUR	17.37
58	GA	JASPER	17.27
59	GA	BLECKLEY	17.22
60	GA	TERRELL	16.88
61	ID	OWYHEE	16.70
62	WA	SPOKANE	16.50
63	WA	WALLA WALLA	16.41
64	GA	TALBOT	16.09
65	GA	WALTON	15.92
66	GA	TAYLOR	15.84
67	WA	COLUMBIA	15.81
68	ID	CANYON	15.64
69	GA	DOUGHERTY	15.56
70	AL	CLAY	15.32
71	GA	CHATTAHOOCHE	15.05
72	GA	EMANUEL	14.86
73	GA	RANDOLPH	14.86
74	GA	MORGAN	14.68
75	GA	PUTNAM	14.53
76	GA	HARALSON	14.51
77	GA	WILCOX	14.38
78	GA	BERRIEN	14.21
79	GA	TROUP	14.11
80	NV	NYE2	14.06

COUNTIES OF HIGHEST AVERAGE FALLOUT FROM SHOT:

PLUMBBOB CHARLESTON (PB17) Sep 28, 1957 12 Kt

(microCuries per square meter)

Each Row Includes:
County Rank (most fallout = 1), State, County, Fallout Level in microCuries/square meter

Rank	State	County	Fallout
1	SC	OCONEE	31.64
2	NC	CLAY	31.09
3	TN	COFFEE	31.06
4	UT	SALT LAKE	24.38
5	TN	GRUNDY	24.15
6	UT	BOX ELDER1	23.90
7	UT	BOX ELDER2	23.90
8	UT	CACHE	23.90
9	UT	DAVIS	23.90
10	UT	MORGAN	23.90
11	UT	RICH	23.90
12	UT	TOOELE1	23.90
13	UT	TOOELE2	23.90
14	UT	WEBER	23.90
15	GA	TOWNS	22.94
16	WY	SWEETWATER	22.80
17	WY	UINTA	22.80
18	NC	MACON	22.76
19	GA	RABUN	22.60
20	TN	SEQUATCHIE	22.42
21	TN	BEDFORD	22.22
22	TN	WARREN	22.08
23	TN	HAMILTON	21.85
24	NC	CHEROKEE	21.82
25	TN	CUMBERLAND	21.72
26	TN	ROANE	21.67
27	NC	SWAIN	21.59
28	TN	BLOUNT	21.53
29	TN	MARSHALL	21.39
30	NC	HAYWOOD	21.24
31	TN	BLEDSOE	21.23
32	NC	TRANSYLVANIA	21.21
33	NC	JACKSON	21.20
34	TN	MCMINN	21.02
35	TN	RHEA	20.93
36	TN	POLK	20.89
37	TN	BRADLEY	20.88
38	TN	VAN BUREN	20.85
39	TN	SEVIER	20.83
40	NV	ELKO	20.12
41	ID	FREMONT	20.12
42	WY	LINCOLN	19.02
43	GA	CRAWFORD	18.99
44	GA	PEACH	18.32
45	KY	FLOYD	18.21
46	GA	TALIAFERRO	18.13
47	GA	CLARKE	18.00
48	GA	WARREN	17.53
49	ID	BANNOCK	17.32
50	ID	BEAR LAKE	17.32
51	ID	BINGHAM	17.32
52	ID	BONNEVILLE	17.32
53	ID	CARIBOU	17.32
54	ID	CASSIA	17.32
55	ID	ELMORE	17.32
56	ID	FRANKLIN	17.32
57	ID	GOODING	17.32
58	ID	JEROME	17.32
59	ID	LINCOLN	17.32
60	ID	MINIDOKA	17.32
61	ID	ONEIDA	17.32
62	ID	OWYHEE	17.32
63	ID	POWER	17.32
64	ID	TWIN FALLS	17.32
65	GA	MCDUFFIE	17.23
66	GA	CLAY	17.18
67	GA	OGLETHORPE	16.79
68	GA	QUITMAN	16.73
69	MS	CHICKASAW	16.00
70	UT	SUMMIT	15.85
71	UT	UTAH	15.85
72	UT	WASATCH	15.85
73	ID	BLAINE	15.71
74	GA	DOUGLAS	15.58
75	GA	FAYETTE	15.55
76	GA	PAULDING	15.52
77	GA	GLASCOCK	15.39
78	WY	CARBON	15.24
79	WY	FREMONT	15.24
80	WY	SUBLETTE	15.24

U.S. FALLOUT ATLAS : TOTAL FALLOUT

COUNTIES OF HIGHEST AVERAGE FALLOUT FROM SHOT:

PLUMBBOB MORGAN (PB18) Oct 7m 1957 8 Kt

(microCuries per square meter)

Each Row Includes:
County Rank (most fallout = 1), State, County, Fallout Level in microCuries/square meter

Rank	State	County	Fallout
1	UT	IRON1	21.54
2	UT	WASHINGTON1	20.12
3	CO	EL PASO	19.98
4	CO	CHEYENNE	16.28
5	UT	IRON2	15.50
6	UT	IRON3	15.50
7	CO	KIT CARSON	15.37
8	UT	WASHINGTON3	15.36
9	CO	YUMA	15.10
10	UT	BEAVER	14.62
11	CO	BOULDER	13.25
12	CO	WASHINGTON	13.25
13	CO	BACA	13.10
14	CO	LINCOLN	12.87
15	KS	FORD	12.79
16	KS	GRAY	12.79
17	KS	MEADE	12.79
18	MO	SHELBY	12.71
19	MO	MARION	12.41
20	MO	MONROE	12.37
21	NE	KIMBALL	12.23
22	CO	LOGAN	12.17
23	CO	FREMONT	12.14
24	MO	ADAIR	12.11
25	MO	CARROLL	12.07
26	NE	GARDEN	12.07
27	CO	PUEBLO	11.95
28	CO	PHILLIPS	11.59
29	NE	CHEYENNE	11.51
30	CO	CLEAR CREEK	11.27
31	CO	PARK	11.27
32	CO	TELLER	11.27
33	KS	WALLACE	11.21
34	IL	HANCOCK	11.13
35	NE	KEITH	11.11
36	CO	ARAPAHOE	11.09
37	KS	CHEYENNE	11.09
38	KS	SHERMAN	11.09
39	CO	GILPIN	10.96
40	CO	JEFFERSON	10.94
41	NE	FRONTIER	10.81
42	NE	LINCOLN	10.72
43	KS	HASKELL	10.49
44	KS	SEWARD	10.49
45	CO	CUSTER	10.48
46	CO	HUERFANO	10.48
47	CO	SEDGWICK	10.48
48	KS	STEVENS	10.37
49	CO	ADAMS	10.32
50	NE	CUSTER	10.32
51	KS	ELLIS	10.30
52	KS	ROOKS	10.30
53	UT	GARFIELD	10.18
54	UT	PIUTE	10.18
55	UT	WAYNE	10.18
56	KS	MORTON	10.17
57	IA	CEDAR	10.11
58	CO	LA PLATA	10.06
59	CO	MONTEZUMA	10.06
60	AZ	APACHE	9.94
61	AZ	NAVAJO	9.94
62	UT	SAN JUAN	9.94
63	NE	DUNDY	9.91
64	IA	LOUISA	9.84
65	IA	VAN BUREN	9.84
66	NM	MCKINLEY	9.82
67	NM	SAN JUAN	9.82
68	IA	LEE	9.78
69	MO	CLARK	9.78
70	KS	LOGAN	9.75
71	KS	THOMAS	9.75
72	IA	DES MOINES	9.52
73	UT	KANE2	9.49
74	KS	GRAHAM	9.44
75	KS	NESS	9.44
76	KS	NORTON	9.44
77	KS	TREGO	9.44
78	NE	HAYES	9.43
79	AR	MILLER	9.35
80	KS	WICHITA	9.33

COUNTIES OF HIGHEST AVERAGE FALLOUT FROM SHOT:

STORAX SEDAN Jul 6, 1962 110 Kt

(microCuries per square meter)

Each Row Includes:
County Rank (most fallout = 1), State, County, Fallout Level in microCuries/square meter

Rank	State	County	Fallout
1	IA	HOWARD	950.00
2	IA	MITCHELL	950.00
3	IA	WORTH	950.00
4	SD	WASHABAUGH	860.52
5	IA	WOODBURY	665.68
6	IA	JEFFERSON	621.15
7	IA	LEE	621.15
8	IA	VAN BUREN	621.15
9	IA	WASHINGTON	621.15
10	IL	HANCOCK	621.15
11	MN	FARIBAULT	584.61
12	MN	MOWER	584.61
13	MN	WATONWAN	584.61
14	NE	DIXON	557.97
15	IA	PLYMOUTH	553.32
16	IA	OSCEOLA	551.97
17	IA	IDA	548.72
18	IA	BENTON	527.44
19	IA	BOONE	527.44
20	IA	BUENA VISTA	527.44
21	IA	CALHOUN	527.44
22	IA	CARROLL	527.44
23	IA	CHEROKEE	527.44
24	IA	CLAY	527.44
25	IA	CRAWFORD	527.44
26	IA	DICKINSON	527.44
27	IA	EMMET	527.44
28	IA	FLOYD	527.44
29	IA	GREENE	527.44
30	IA	HAMILTON	527.44
31	IA	HANCOCK	527.44
32	IA	HUMBOLDT	527.44
33	IA	IOWA	527.44
34	IA	JASPER	527.44
35	IA	KOSSUTH	527.44
36	IA	LYON	527.44
37	IA	MARSHALL	527.44
38	IA	MONONA	527.44
39	IA	O BRIEN	527.44
40	IA	PALO ALTO	527.44
41	IA	POCAHONTAS	527.44
42	IA	SAC	527.44
43	IA	SIOUX	527.44
44	IA	STORY	527.44
45	IA	WINNEBAGO	527.44
46	IA	WINNESHIEK	527.44
47	MN	ROCK	527.44
48	NE	DAKOTA	527.44
49	NE	STANTON	527.44
50	NE	THURSTON	527.44
51	SD	TURNER	527.44
52	SD	HAMLIN	451.85
53	NE	LOUP	436.77
54	SD	CUSTER	434.17
55	NE	HOLT	398.55
56	SD	KINGSBURY	361.42
57	SD	MOODY	361.42
58	KY	GREENUP	359.52
59	KY	LAWRENCE	359.52
60	KY	LEWIS	359.52
61	KY	TRIMBLE	359.52
62	OH	SCIOTO	359.52
63	WV	BOONE	359.52
64	WV	LOGAN	359.52
65	MO	CLARK	358.08
66	MO	KNOX	358.08
67	MO	SCOTLAND	358.08
68	KY	POWELL	354.59
69	SD	HUTCHINSON	346.94
70	SD	MCCOOK	346.94
71	SD	FALL RIVER	345.60
72	SD	SHANNON	345.60
73	SD	STANLEY	345.60
74	IA	ADAIR	345.56
75	IA	ADAMS	345.56
76	IA	APPANOOSE	345.56
77	IA	AUDUBON	345.56
78	IA	CASS	345.56
79	IA	CLARKE	345.56
80	IA	DAVIS	345.56

COUNTIES OF HIGHEST AVERAGE FALLOUT FROM NUCLEAR TESTS:

RANGER BAKER-1 THROUGH STORAX SEDAN
1951-1962

(microCuries per square meter)

Each Row Includes:
County Rank (most fallout = 1), State, County, Fallout Level in microCuries/square meter

Rank	State	County	Fallout	Rank	State	County	Fallout	Rank	State	County	Fallout
1	UT	WASHINGTON2	6800.7	30	UT	IRON2	1197.0	59	UT	RICH	917.5
2	UT	WASHINGTON3	4957.2	31	UT	WEBER	1193.9	60	CO	RIO GRANDE	917.1
3	UT	WASHINGTON1	3536.0	32	NV	WHITE PINE1	1193.8	61	CO	MONTROSE	913.1
4	NV	CLARK1	3414.6	33	CO	HINSDALE	1182.5	62	WY	SWEETWATER	913.0
5	NV	LINCOLN1	3378.4	34	UT	BEAVER	1180.5	63	UT	SEVIER	909.3
6	NV	NYE2	3119.2	35	CO	MESA	1176.6	64	NV	EUREKA	891.8
7	UT	KANE2	3103.0	36	UT	MILLARD	1166.7	65	NM	VALENCIA	887.2
8	AZ	MOHAVE2	2965.0	37	NM	MCKINLEY	1148.9	66	CO	LAKE	887.2
9	UT	KANE1	2812.3	38	CO	GARFIELD	1138.4	67	SD	CUSTER	882.2
10	UT	GARFIELD	2285.3	39	IA	HOWARD	1132.9	68	MO	KNOX	880.0
11	AZ	COCONINO1	2034.9	40	UT	WASATCH	1129.6	69	UT	DAGGETT	865.7
12	NV	LINCOLN2	2029.3	41	IA	MITCHELL	1114.2	70	UT	SANPETE	865.1
13	NV	WHITE PINE2	2021.2	42	IA	WORTH	1107.4	71	IA	WOODBURY	858.7
14	CO	GUNNISON	1990.4	43	TX	REFUGIO	1107.1	72	CO	DOLORES	850.7
15	CO	CONEJOS	1842.6	44	CO	PITKIN	1105.6	73	IA	VAN BUREN	848.6
16	NV	WHITE PINE3	1793.7	45	SD	WASHABAUGH	1100.0	74	WY	LINCOLN	839.7
17	AZ	COCONINO2	1624.4	46	UT	EMERY	1080.1	75	UT	CACHE	836.7
18	CA	INYO3	1618.4	47	WY	FREMONT	1079.1	76	WY	SUBLETTE	835.4
19	AZ	MOHAVE1	1565.2	48	UT	SUMMIT	1062.9	77	IA	LEE	829.1
20	CO	ARCHULETA	1463.2	49	UT	PIUTE	1051.6	78	IL	HANCOCK	822.2
21	AZ	APACHE	1413.3	50	UT	JUAB	1050.5	79	UT	CARBON	821.6
22	UT	IRON1	1407.2	51	UT	GRAND	1038.1	80	IA	CARROLL	820.3
23	UT	UTAH	1364.9	52	CO	LA PLATA	1026.6	81	UT	WAYNE	819.2
24	UT	MORGAN	1307.4	53	CO	DELTA	1024.7	82	NM	SANTA FE	816.6
25	UT	DAVIS	1297.5	54	NM	SANDOVAL	998.0	83	NM	BERNALILLO	812.9
26	UT	IRON3	1278.0	55	UT	SAN JUAN	946.5	84	UT	BOX ELDER2	805.5
27	UT	SALT LAKE	1253.8	56	UT	UINTAH	937.9	85	NM	TORRANCE	805.2
28	UT	TOOELE2	1220.4	57	AZ	NAVAJO	933.5	86	IA	JASPER	800.4
29	CO	MINERAL	1203.3	58	CO	RIO BLANCO	927.1	87	IA	BENTON	795.2

COUNTIES OF HIGHEST AVERAGE FALLOUT FROM NUCLEAR TESTS:

RANGER BAKER-1 THROUGH STORAX SEDAN 1951-1962

(microCuries per square meter)

Each Row Includes:
County Rank (most fallout = 1), State, County, Fallout Level in microCuries/square meter

Rank	State	County	Fallout
88	IA	BOONE	794.0
89	IA	MARSHALL	791.4
90	MO	SCOTLAND	785.6
91	IA	STORY	781.7
92	WY	CARBON	772.6
93	IA	WASHINGTON	772.3
94	IA	JEFFERSON	767.2
95	NE	DIXON	763.9
96	CO	MONTEZUMA	760.6
97	CO	OURAY	760.6
98	WY	UINTA	760.0
99	CO	EAGLE	759.8
100	CO	SAN MIGUEL	756.7
101	IA	APPANOOSE	754.3
102	CO	SAN JUAN	750.9
103	SD	TURNER	748.8
104	CO	SAGUACHE	747.7
105	MN	MOWER	747.6
106	IA	HAMILTON	745.5
107	IA	MONROE	744.7
108	MN	FARIBAULT	734.8
109	IA	GREENE	732.4
110	IA	IDA	727.7
111	UT	DUCHESNE	721.6
112	MN	WATONWAN	715.2
113	IA	O BRIEN	715.2
114	IA	MONONA	712.4
115	NE	THURSTON	711.6
116	IA	HUMBOLDT	711.1
117	IA	BUENA VISTA	709.9
118	NE	STANTON	708.3
119	IA	IOWA	706.9
120	CO	COSTILLA	706.8
121	IA	POCAHONTAS	704.8
122	IA	OSCEOLA	702.9
123	IA	PLYMOUTH	701.3
124	IA	SIOUX	701.0
125	IA	DAVIS	700.6
126	IA	CALHOUN	699.5
127	MO	LEWIS	699.0
128	IA	SAC	697.0
129	IA	WINNESHIEK	697.0
130	IA	CHEROKEE	693.0
131	IA	LYON	691.9
132	NE	DAKOTA	691.6
133	CO	ALAMOSA	689.4
134	IA	FLOYD	687.7
135	NM	SAN JUAN	686.2
136	IA	CLAY	684.6
137	IA	PALO ALTO	683.3
138	IA	KOSSUTH	679.6
139	IA	HANCOCK	679.5
140	MN	ROCK	677.6
141	IA	CRAWFORD	675.9
142	IA	WINNEBAGO	672.3
143	IA	DICKINSON	672.0
144	CO	MOFFAT	671.8
145	NV	ESMERALDA2	666.0
146	CO	SUMMIT	664.5
147	IA	CASS	663.9
148	IA	EMMET	663.6
149	IA	SHELBY	660.4
150	SD	HAMLIN	658.2
151	NV	MINERAL	656.2
152	IA	RINGGOLD	650.3
153	IA	WAPELLO	650.3
154	NM	LOS ALAMOS	646.5
155	SD	STANLEY	644.5
156	IA	MARION	636.6
157	NY	WASHINGTON	635.9
158	IA	MADISON	634.5
159	IA	WAYNE	629.3
160	IA	ADAIR	626.8
161	IA	ADAMS	625.3
162	MO	CLARK	624.8
163	IA	UNION	622.2
164	ID	BEAR LAKE	621.4
165	WY	NIOBRARA	617.6
166	NY	WARREN	615.8
167	IA	AUDUBON	613.1
168	CO	GRAND	609.7
169	UT	TOOELE1	607.3
170	IA	MAHASKA	606.6
171	MO	RALLS	605.4
172	NE	LOUP	605.0
173	NE	HOLT	604.9
174	SD	KINGSBURY	603.5

U.S. FALLOUT ATLAS : TOTAL FALLOUT

COUNTIES OF HIGHEST AVERAGE FALLOUT FROM NUCLEAR TESTS:

RANGER BAKER-1 THROUGH STORAX SEDAN
1951-1962

(microCuries per square meter)

Each Row Includes:
County Rank (most fallout = 1), State, County, Fallout Level in microCuries/square meter

Rank	State	County	Fallout		Rank	State	County	Fallout		Rank	State	County	Fallout
175	NY	RENSSELAER	603.2		204	SD	MINER	535.3		233	NE	WASHINGTON	482.7
176	SD	FALL RIVER	602.3		205	MO	SHELBY	533.4		234	CO	ROUTT	482.4
177	NV	LANDER2	597.0		206	NE	SHERMAN	533.2		235	SD	MOODY	481.4
178	NV	NYE1	595.6		207	NM	COLFAX	531.0		236	NM	CURRY	475.0
179	IA	DECATUR	594.1		208	NM	QUAY	527.6		237	MN	WABASHA	474.4
180	IA	CLARKE	593.3		209	IL	ADAMS	527.1		238	CO	ARAPAHOE	472.0
181	NV	CHURCHILL	591.7		210	IA	HENRY	524.3		239	IA	POWESHIEK	469.4
182	WY	CONVERSE	589.5		211	NM	MORA	520.0		240	NE	MADISON	468.8
183	SD	HUTCHINSON	589.4		212	NM	UNION	518.3		241	IA	FRANKLIN	468.7
184	NM	GUADALUPE	588.9		213	MO	SCHUYLER	517.4		242	WI	BUFFALO	468.5
185	AZ	MOHAVE3	577.7		214	CO	JEFFERSON	514.6		243	IL	PIKE	466.8
186	AZ	MOHAVE4	577.7		215	CO	ADAMS	512.4		244	MN	RICE	465.4
187	IA	DELAWARE	567.1		216	IL	MCDONOUGH	511.5		245	MN	LE SUEUR	465.0
188	SD	MCCOOK	566.0		217	CO	BOULDER	511.5		246	MN	PIPESTONE	464.0
189	NM	CHAVES	564.3		218	CO	DOUGLAS	511.0		247	IA	BREMER	462.9
190	SD	LAKE	563.9		219	IL	WARREN	508.8		248	IL	MOULTRIE	462.9
191	IA	KEOKUK	562.4		220	NM	HARDING	507.2		249	MO	MARION	459.6
192	SD	SHANNON	560.2		221	CO	MORGAN	504.7		250	MN	WASECA	459.5
193	NE	HOWARD	556.2		222	WI	IOWA	504.3		251	NV	LYON	459.2
194	NM	LINCOLN	555.4		223	IA	LOUISA	501.5		252	IL	SHELBY	457.1
195	NM	SAN MIGUEL	554.8		224	ID	BLAINE	499.6		253	MN	MARTIN	456.1
196	IA	TAMA	549.4		225	IL	FULTON	493.9		254	NM	ROOSEVELT	455.3
197	NM	RIO ARRIBA	545.7		226	NE	SIOUX	489.9		255	MO	AUDRAIN	454.9
198	CO	CLEAR CREEK	544.6		227	MO	PIKE	487.3		256	KY	GREENUP	451.9
199	CO	GILPIN	541.2		228	IA	BLACK HAWK	486.6		257	WV	LOGAN	451.5
200	IA	GUTHRIE	541.1		229	IA	JOHNSON	486.4		258	MN	MURRAY	450.8
201	ID	CUSTER	541.0		230	NM	DE BACA	485.8		259	MN	NICOLLET	448.3
202	ID	GEM	538.5		231	NM	TAOS	485.3		260	IA	MONTGOMERY	447.6
203	MO	PUTNAM	537.8		232	MN	FILLMORE	485.3		261	NV	CARSON CITY	447.3

COUNTIES OF HIGHEST AVERAGE FALLOUT FROM NUCLEAR TESTS:

RANGER BAKER-1 THROUGH STORAX SEDAN 1951-1962

(microCuries per square meter)

Each Row Includes:
County Rank (most fallout = 1), State, County, Fallout Level in microCuries/square meter

Rank	State	County	Fallout
262	NV	CLARK2	446.1
283	NV	WASHOE	445.0
264	KY	POWELL	443.9
265	MO	WORTH	442.0
266	MN	BROWN	440.8
267	TX	OLDHAM	440.7
268	NV	DOUGLAS	440.5
269	NV	STOREY	439.3
270	MN	BLUE EARTH	438.4
271	CO	CHAFFEE	436.7
272	IA	TAYLOR	436.6
273	TX	DALLAM	436.6
274	OH	SCIOTO	435.2
275	KY	LAWRENCE	434.9
276	IA	LINN	434.0
277	ID	FRANKLIN	433.0
278	NE	GARFIELD	431.8
279	NE	ANTELOPE	431.6
280	KY	TRIMBLE	431.4
281	CO	CUSTER	430.7
282	IA	FAYETTE	427.4
283	IA	BUCHANAN	426.8
284	KY	LEWIS	426.6
285	NE	CUMING	425.6
286	NV	ELKO	424.5
287	SD	HANSON	424.5
288	MO	MACON	423.0
289	IA	HARDIN	421.8
290	CO	LARIMER	421.1
291	SD	YANKTON	420.4
292	SD	BROOKINGS	419.7
293	SD	DAVISON	419.2
294	WV	BOONE	419.1
295	OH	BUTLER	411.4
296	NY	SARATOGA	409.9
297	MO	LINN	407.8
298	KS	STEVENS	407.1
299	IA	DALLAS	406.1
300	NY	FULTON	405.2
301	SD	UNION	404.1
302	MO	SULLIVAN	404.0
303	IA	POTTAWATTAMI	401.0
304	SD	LINCOLN	398.1
305	CO	FREMONT	397.6
306	ND	BILLINGS	395.8
307	OH	MONTGOMERY	395.5
308	CO	PHILLIPS	393.9
309	IL	CHRISTIAN	393.2
310	IN	ALLEN	392.7
311	WY	ALBANY	392.2
312	IA	BUTLER	391.9
313	NE	CEDAR	391.6
314	MO	ADAIR	391.6
315	NE	WAYNE	390.9
316	AZ	COCONINO3	390.7
317	KY	CAMPBELL	390.4
318	IN	WAYNE	388.4
319	NM	SOCORRO	387.8
320	OH	MEDINA	386.5
321	VT	BENNINGTON	386.0
322	KY	ROWAN	385.5
323	MT	MEAGHER	384.5
324	CO	LAS ANIMAS	383.3
325	KY	OWEN	382.3
326	ID	LEMHI	382.2
327	UT	BOX ELDER1	381.5
328	CO	HUERFANO	380.7
329	WY	CAMPBELL	380.3
330	WY	JOHNSON	380.0
331	TX	POTTER	377.6
332	MO	DAVIESS	377.0
333	IL	CASS	375.3
334	IA	WARREN	373.2
335	OH	SENECA	372.4
336	WI	GRANT	371.1
337	IL	COLES	370.4
338	SD	MINNEHAHA	369.0
339	OK	TEXAS	369.0
340	KY	CARROLL	368.7
341	SD	CLAY	367.3
342	NV	ESMERALDA1	366.6
343	IA	POLK	366.2
344	MT	POWDER RIVER	366.1
345	ID	VALLEY	365.8
346	IA	DUBUQUE	365.8
347	WY	PLATTE	365.1
348	SD	JERAULD	364.7

COUNTIES OF HIGHEST AVERAGE FALLOUT FROM NUCLEAR TESTS:

RANGER BAKER-1 THROUGH STORAX SEDAN
1951-1962

(microCuries per square meter)

Each Row Includes:
County Rank (most fallout = 1), State, County, Fallout Level in microCuries/square meter

Rank	State	County	Fallout
349	CO	BACA	364.5
350	IA	JACKSON	364.1
351	IA	CERRO GORDO	363.1
352	IL	BROWN	362.6
353	IL	JASPER	361.4
354	IA	PAGE	361.1
355	IA	FREMONT	360.4
356	IA	CLINTON	359.8
357	NE	NANCE	359.8
358	IL	SANGAMON	359.0
359	CO	KIOWA	358.9
360	IN	LAWRENCE	356.5
361	IA	CEDAR	356.0
362	MT	CUSTER	354.4
363	NE	MERRICK	354.1
364	IL	GREENE	353.1
365	NE	GREELEY	351.8
366	NE	DAWES	351.7
367	IA	CLAYTON	351.1
368	IA	WEBSTER	350.4
369	IL	MACON	350.3
370	MN	NOBLES	350.2
371	CO	TELLER	349.7
372	KY	KENTON	349.4
373	CO	DENVER	349.3
374	TX	HARTLEY	349.2
375	IA	MILLS	347.8
376	OH	CLARK	346.6
377	NE	PLATTE	345.9
378	KS	GRANT	345.6
379	IA	WRIGHT	345.5
380	IL	MONTGOMERY	345.2
381	CO	JACKSON	344.5
382	NE	COLFAX	344.3
383	OH	WARREN	343.9
384	SD	HAAKON	343.0
385	NV	PERSHING	341.9
386	IL	MACOUPIN	341.4
387	CO	WELD	341.3
388	KY	HARRISON	340.8
389	WY	WASHAKIE	340.5
390	ND	STARK	340.3
391	OH	CLINTON	340.2
392	KS	HODGEMAN	338.5
393	NE	VALLEY	338.0
394	MT	DAWSON	337.7
395	MT	DEER LODGE	337.5
396	IL	PIATT	336.2
397	CO	PARK	334.9
398	NE	POLK	334.9
399	NE	LINCOLN	334.4
400	NE	CHERRY	334.1
401	MN	WINONA	333.3
402	NE	LOGAN	333.0
403	WY	NATRONA	332.5
404	ID	IDAHO	332.4
405	CO	SEDGWICK	332.2
406	KY	NICHOLAS	332.2
407	IL	DOUGLAS	331.5
408	MN	DODGE	331.2
409	KS	NESS	330.6
410	IL	EFFINGHAM	330.6
411	IA	HARRISON	330.3
412	PA	VENANGO	330.1
413	NE	BUTLER	329.9
414	MN	GOODHUE	329.3
415	MN	STEELE	328.9
416	NE	CUSTER	327.2
417	IL	SCHUYLER	326.6
418	WI	VERNON	326.3
419	MN	HOUSTON	325.9
420	NE	BANNER	325.7
421	KS	FORD	325.3
422	NV	LANDER1	325.2
423	IL	HENDERSON	322.6
424	IA	ALLAMAKEE	322.2
425	KY	SCOTT	321.0
426	ID	OWYHEE	320.7
427	LA	GRANT	319.6
428	IL	MARSHALL	319.3
429	OH	HANCOCK	318.7
430	OH	CHAMPAIGN	318.4
431	NE	CHASE	317.6
432	CO	LINCOLN	317.3
433	KY	WOODFORD	316.9
434	ND	DUNN	316.8
435	IL	PEORIA	316.3

COUNTIES OF HIGHEST AVERAGE FALLOUT FROM NUCLEAR TESTS:

RANGER BAKER-1 THROUGH STORAX SEDAN 1951-1962

(microCuries per square meter)

Each Row Includes:
County Rank (most fallout = 1), State, County, Fallout Level in microCuries/square meter

Rank	State	County	Fallout		Rank	State	County	Fallout		Rank	State	County	Fallout
436	KY	CLARK	315.9		465	NE	MORRILL	305.7		494	NY	COLUMBIA	300.2
437	WV	MASON	315.5		466	NE	FURNAS	305.2		495	IN	GRANT	300.1
438	OH	HIGHLAND	314.9		467	NE	DOUGLAS	305.0		496	WV	WAYNE	299.7
439	OH	MIAMI	314.3		468	IN	HAMILTON	304.8		497	NE	SALINE	299.3
440	NE	KEITH	313.9		469	WV	LINCOLN	304.7		498	NE	RED WILLOW	299.1
441	ND	GOLDEN VALLE	313.3		470	KS	TREGO	304.6		499	TX	HUTCHINSON	298.4
442	LA	RAPIDES	313.3		471	KS	NORTON	304.4		500	NE	CLAY	298.4
443	NY	SCHENECTADY	313.1		472	ID	CAMAS	304.1		501	IL	WINNEBAGO	298.3
444	IA	DES MOINES	313.1		473	OH	KNOX	303.8		502	MN	LYON	298.2
445	OH	LICKING	312.6		474	KY	BATH	303.5		503	TX	BAILEY	298.1
446	KY	MENIFEE	312.6		475	NE	PHELPS	303.5		504	OH	JACKSON	297.7
447	KS	KEARNY	312.5		476	CO	BENT	303.1		505	VT	ADDISON	297.6
448	OK	CIMARRON	312.2		477	ID	CARIBOU	303.0		506	KY	BOYD	297.5
449	CO	LOGAN	311.7		478	KY	HENRY	303.0		507	IL	STARK	297.1
450	NM	CATRON	311.3		479	NE	PERKINS	302.7		508	ID	BOISE	297.1
451	IA	JONES	310.3		480	NE	BUFFALO	302.6		509	MT	WIBAUX	297.1
452	TX	PARMER	310.3		481	ND	MCKENZIE	302.5		510	ND	WARD	296.9
453	KY	FRANKLIN	309.7		482	MO	RANDOLPH	302.5		511	MA	NORFOLK	296.8
454	CA	SAN BERNADIN	309.0		483	CO	YUMA	302.4		512	IN	TIPTON	296.6
455	KS	STANTON	307.9		484	TX	HANSFORD	302.3		513	CO	PROWERS	296.3
456	TX	DEAF SMITH	307.9		485	KS	GRAHAM	302.1		514	NE	DAWSON	296.3
457	OH	ALLEN	307.2		486	OH	PERRY	302.0		515	NE	JOHNSON	295.8
458	MT	RICHLAND	307.1		487	SD	LYMAN	301.8		516	NE	FRONTIER	295.8
459	IL	WOODFORD	306.9		488	CO	WASHINGTON	301.8		517	IN	HOWARD	295.6
460	OH	RICHLAND	306.8		489	IL	KNOX	301.6		518	LA	BEAUREGARD	295.5
461	NE	GOSPER	306.1		490	WI	WALWORTH	301.3		519	MN	SIBLEY	295.4
462	CO	OTERO	306.1		491	KS	CHEYENNE	300.9		520	CO	KIT CARSON	295.1
463	MO	CHARITON	306.0		492	CO	EL PASO	300.6		521	MN	LINCOLN	295.0
464	NE	SAUNDERS	305.8		493	SD	JACKSON	300.5		522	OH	MORGAN	295.0

COUNTIES OF HIGHEST AVERAGE FALLOUT FROM NUCLEAR TESTS:

RANGER BAKER-1 THROUGH STORAX SEDAN
1951-1962

(microCuries per square meter)

Each Row Includes:
County Rank (most fallout = 1), State, County, Fallout Level in microCuries/square meter

Rank	State	County	Fallout	Rank	State	County	Fallout	Rank	State	County	Fallout
523	ND	MERCER	294.7	552	ID	JEFFERSON	288.4	581	SD	BRULE	281.0
524	IL	ROCK ISLAND	294.6	553	KS	FINNEY	288.1	582	MO	GRUNDY	280.9
525	ND	OLIVER	294.2	554	CO	CHEYENNE	288.1	583	IL	MERCER	280.0
526	CO	ELBERT	293.9	555	KS	GOVE	287.9	584	TX	CASTRO	279.7
527	CO	CROWLEY	293.9	556	KS	SCOTT	287.9	585	NC	PERQUIMANS	279.5
528	MN	JACKSON	293.7	557	WV	SUMMERS	287.9	586	KY	FLEMING	279.2
529	KS	WALLACE	293.6	558	MT	BEAVERHEAD	287.8	587	LA	EVANGELINE	279.2
530	MO	CAMDEN	293.6	559	NE	DUNDY	287.7	588	MT	JEFFERSON	279.1
531	OH	PIKE	293.4	560	KS	SALINE	287.7	589	NE	SHERIDAN	279.1
532	NE	CHEYENNE	293.3	561	WY	GOSHEN	287.2	590	NC	CURRITUCK	279.1
533	MO	MORGAN	293.0	562	KS	LOGAN	287.1	591	IL	MCLEAN	278.9
534	IN	WELLS	292.7	563	OH	FULTON	286.4	592	ND	BOTTINEAU	278.8
535	WI	LAFAYETTE	292.6	564	OH	PREBLE	286.0	593	NE	HITCHCOCK	278.7
536	IL	MASON	292.5	565	OK	TULSA	285.9	594	NE	HOOKER	278.7
537	MO	MONROE	292.3	566	IL	DE WITT	285.8	595	KS	WICHITA	278.4
538	NE	CASS	292.1	567	MO	CALDWELL	285.8	596	KY	ELLIOTT	277.9
539	IA	MUSCATINE	292.0	568	OH	ADAMS	285.7	597	NC	PASQUOTANK	277.7
540	LA	LAFAYETTE	291.5	569	MT	JUDITH BASIN	285.0	598	NC	CAMDEN	277.6
541	SD	MELLETTE	291.4	570	ND	RICHLAND	284.9	599	IN	MARION	277.4
542	KS	LANE	290.6	571	NE	HAYES	284.7	600	WV	KANAWHA	277.3
543	TX	OCHILTREE	290.3	572	KS	RUSSELL	284.5	601	KS	RAWLINS	276.8
544	KY	WOLFE	290.3	573	ND	BURKE	284.0	602	MT	BROADWATER	276.5
545	IL	HENRY	290.3	574	KY	GALLATIN	283.9	603	OH	LAWRENCE	276.3
546	KY	BREATHITT	289.8	575	MT	FALLON	282.5	604	SD	DEWEY	276.2
547	MO	RAY	289.5	576	ND	MCHENRY	282.5	605	LA	ACADIA	275.9
548	NE	THOMAS	289.1	577	OH	WYANDOT	282.5	606	KY	CARTER	275.8
549	ND	DIVIDE	288.6	578	KS	DECATUR	282.1	607	KY	MASON	275.7
550	MA	PLYMOUTH	288.5	579	KS	HAMILTON	282.0	608	ND	SLOPE	274.8
551	KY	ROBERTSON	288.4	580	KY	MAGOFFIN	281.1	609	MO	FRANKLIN	274.8

U.S. FALLOUT ATLAS : TOTAL FALLOUT

COUNTIES OF HIGHEST AVERAGE FALLOUT FROM NUCLEAR TESTS:

RANGER BAKER-1 THROUGH STORAX SEDAN
1951-1962

(microCuries per square meter)

Each Row Includes:
County Rank (most fallout = 1), State, County, Fallout Level in microCuries/square meter

Rank	State	County	Fallout
610	KS	GRAY	274.8
611	IL	LOGAN	274.4
612	LA	CALCASIEU	274.0
613	TX	LAMB	273.1
614	IL	TAZEWELL	273.1
615	KS	THOMAS	273.0
616	SD	ZIEBACH	272.8
617	IL	CALHOUN	272.7
618	TX	LIPSCOMB	272.7
619	ID	ADA	272.7
620	KS	HASKELL	271.9
621	TX	ORANGE	271.7
622	LA	CAMERON	271.5
623	IA	SCOTT	271.3
624	NY	ALBANY	271.3
625	WI	GREEN	270.7
626	SD	AURORA	270.6
627	NE	GAGE	270.6
628	IN	JACKSON	269.7
629	KS	GREELEY	269.5
630	MT	CHOUTEAU	269.4
631	MO	HARRISON	269.1
632	IL	BOONE	269.1
633	MT	MCCONE	268.9
634	NE	ADAMS	268.2
635	WV	MONROE	268.0
636	WI	ROCK	268.0
637	KS	SHERIDAN	267.6
638	IN	CASS	267.0
639	ID	TETON	266.9
640	IL	CLAY	266.7
641	MO	CALLAWAY	266.7
642	ID	ELMORE	266.6
643	IL	SCOTT	266.5
644	KY	JOHNSON	266.4
645	SD	HARDING	266.3
646	CO	PUEBLO	266.2
647	SD	MEADE	265.5
648	MO	PETTIS	265.5
649	WV	MCDOWELL	265.3
650	MT	POWELL	265.2
651	NE	SEWARD	265.2
652	KY	FLOYD	265.1
653	OK	CRAIG	265.0
654	KY	GRANT	264.7
655	WV	RALEIGH	264.6
656	MO	OSAGE	263.3
657	IN	GREENE	263.3
658	MT	WHEATLAND	263.2
659	MO	CLAY	262.7
660	ND	MOUNTRAIL	262.7
661	KS	SHAWNEE	262.6
662	VA	BATH	262.3
663	ND	MCLEAN	261.6
664	PA	ELK	261.6
665	MT	MADISON	261.5
666	KS	WABAUNSEE	261.3
667	WV	PENDLETON	261.0
668	NE	FILLMORE	260.8
669	IL	FORD	260.8
670	KS	LYON	260.5
671	IL	MORGAN	260.4
672	SD	BENNETT	260.2
673	WY	LARAMIE	259.6
674	SD	FAULK	259.2
675	VA	BOTETOURT	259.2
676	MO	MARIES	259.1
677	ID	CLARK	258.9
678	SD	TODD	258.4
679	MI	INGHAM	258.4
680	NE	ARTHUR	258.4
681	NE	DEUEL	258.3
682	ID	GOODING	257.8
683	ND	RENVILLE	257.5
684	SD	SULLY	257.5
685	IL	CHAMPAIGN	257.1
686	LA	ST HELENA	257.0
687	TX	HOCKLEY	256.9
688	MO	LIVINGSTON	256.4
689	OK	CREEK	256.4
690	NE	YORK	256.4
691	MT	ROOSEVELT	255.6
692	MT	SILVER BOW	255.0
693	MO	LAFAYETTE	254.6
694	KY	SHELBY	254.5
695	IL	JERSEY	254.5
696	MT	PRAIRIE	254.3

COUNTIES OF HIGHEST AVERAGE FALLOUT FROM NUCLEAR TESTS:

RANGER BAKER-1 THROUGH STORAX SEDAN 1951-1962

(microCuries per square meter)

Each Row Includes:
County Rank (most fallout = 1), State, County, Fallout Level in microCuries/square meter

Rank	State	County	Fallout
697	IN	DELAWARE	254.2
698	KY	MORGAN	254.1
699	TX	MOORE	253.9
700	MI	EATON	253.9
701	NE	NUCKOLLS	253.5
702	IN	HENDRICKS	252.9
703	MO	GREENE	252.6
704	MN	COTTONWOOD	252.2
705	LA	TANGIPAHOA	250.9
706	MO	VERNON	250.6
707	OK	KAY	250.6
708	TX	HALE	250.4
709	ND	MORTON	250.2
710	NE	KIMBALL	250.0
711	TX	BRISCOE	249.9
712	OK	CHEROKEE	249.7
713	ND	HETTINGER	249.7
714	MT	FERGUS	249.5
715	NE	HAMILTON	249.3
716	KY	PIKE	248.9
717	ID	WASHINGTON	248.8
718	WI	SAUK	248.7
719	WI	CRAWFORD	248.6
720	TX	COCHRAN	248.4
721	MO	POLK	248.3
722	ID	ONEIDA	248.3
723	MO	HOWARD	247.8
724	IL	RICHLAND	247.2
725	MO	MILLER	246.8
726	KY	JACKSON	246.7
727	OK	OTTAWA	246.7
728	KS	MORTON	246.2
729	IN	SWITZERLAND	246.0
730	TX	SWISHER	245.8
731	NE	MCPHERSON	245.6
732	IN	OWEN	245.5
733	NE	GARDEN	245.5
734	KY	OWSLEY	245.4
735	SD	CORSON	245.4
736	TX	EL PASO	245.0
737	OH	ATHENS	245.0
738	LA	LA SALLE	244.8
739	MT	GALLATIN	244.7
740	WV	MINGO	244.4
741	WY	CROOK	244.4
742	LA	TENSAS	244.1
743	ID	MADISON	243.9
744	KS	MARION	243.8
745	SD	POTTER	243.6
746	KS	BARTON	243.5
747	KS	CHASE	243.4
748	NE	BOX BUTTE	243.2
749	IN	CRAWFORD	243.2
750	NE	GRANT	243.1
751	OH	MADISON	243.0
752	SD	HYDE	242.9
753	MS	WARREN	242.9
754	SD	ROBERTS	242.6
755	MO	SALINE	242.0
756	MO	BENTON	241.7
757	MO	DE KALB	241.6
758	ND	EMMONS	241.5
759	MO	HENRY	241.4
760	MA	MIDDLESEX	241.3
761	MO	CEDAR	241.2
762	SD	JONES	241.2
763	KS	DICKINSON	240.9
764	KS	NEOSHO	240.5
765	WI	RICHLAND	240.5
766	MA	BRISTOL	240.3
767	KS	WYANDOTTE	240.1
768	OK	GRANT	240.0
769	MO	PULASKI	239.8
770	SD	SANBORN	239.7
771	MO	LINCOLN	239.6
772	LA	IBERIA	239.6
773	CA	INYO1	239.5
774	CA	INYO2	239.5
775	IN	MARTIN	239.4
776	MO	ATCHISON	239.0
777	IN	HENRY	238.4
778	MT	CARTER	238.2
779	MS	ADAMS	237.9
780	OK	WASHINGTON	237.8
781	KS	LEAVENWORTH	237.4
782	OH	GALLIA	237.4
783	PA	SNYDER	237.4

COUNTIES OF HIGHEST AVERAGE FALLOUT FROM NUCLEAR TESTS:

RANGER BAKER-1 THROUGH STORAX SEDAN 1951-1962

(microCuries per square meter)

Each Row Includes:
County Rank (most fallout = 1), State, County, Fallout Level in microCuries/square meter

Rank	State	County	Fallout
784	OH	PORTAGE	237.4
785	TX	TERRY	237.0
786	SD	SPINK	236.7
787	TX	WHEELER	236.7
788	ND	GRANT	236.4
789	KS	MEADE	236.4
790	OK	DELAWARE	236.2
791	NE	OTOE	235.9
792	LA	POINTE COUPE	235.9
793	ND	BURLEIGH	235.7
794	WV	MERCER	235.7
795	KS	MCPHERSON	235.6
796	KS	LINCOLN	235.5
797	NE	NEMAHA	235.3
798	LA	ALLEN	235.2
799	LA	WEST FELICIA	235.1
800	KS	OSBORNE	234.9
801	KS	ELLSWORTH	234.7
802	MT	SHERIDAN	234.4
803	OK	OSAGE	233.6
804	ND	ADAMS	233.5
805	ND	RANSOM	233.1
806	KS	OTTAWA	233.0
807	VA	POWHATAN	232.8
808	AZ	PIMA	232.8
809	NE	LANCASTER	232.6
810	MO	COLE	232.5
811	LA	WEST BATON R	232.5
812	TX	SHERMAN	232.3
813	IN	CARROLL	232.1
814	IN	JEFFERSON	232.0
815	VA	BUCHANAN	231.9
816	OH	CLERMONT	231.9
817	SD	PENNINGTON	231.6
818	LA	JEFFERSON DA	231.5
819	KY	ANDERSON	231.3
820	LA	TERREBONNE	231.2
821	SD	BUTTE	230.8
822	MO	WRIGHT	230.7
823	LA	VERMILION	230.6
824	OK	MAYES	230.5
825	IN	MONROE	230.5
826	LA	IBERVILLE	230.5
827	KS	COWLEY	230.5
828	IL	VERMILION	230.3
829	LA	EAST BATON R	230.1
830	NE	HALL	229.9
831	KS	KINGMAN	229.8
832	SD	WALWORTH	229.7
833	OK	ATOKA	229.7
834	OH	LOGAN	229.6
835	SD	TRIPP	229.4
836	KS	CHAUTAUQUA	229.4
837	SD	BUFFALO	229.1
838	MO	HOWELL	228.9
839	OH	PAULDING	228.9
840	OH	COSHOCTON	228.7
841	TX	LUBBOCK	228.7
842	KS	BUTLER	228.6
843	IN	KNOX	228.5
844	ND	SARGENT	228.5
845	KS	RENO	228.4
846	VA	HANOVER	228.3
847	KS	REPUBLIC	228.2
848	OH	BROWN	228.1
849	MS	WILKINSON	228.0
850	OK	PITTSBURG	227.9
851	MO	NEWTON	227.6
852	MO	LAWRENCE	227.5
853	NE	BROWN	227.5
854	OK	ROGER MILLS	227.5
855	KS	HARVEY	227.4
856	AR	FULTON	227.3
857	NE	WEBSTER	227.2
858	KS	ELLIS	227.2
859	VA	AMELIA	227.1
860	VA	GOOCHLAND	226.9
861	OK	ELLIS	226.6
862	KS	GREENWOOD	226.5
863	OK	WAGONER	226.4
864	OK	MUSKOGEE	226.3
865	KS	RUSH	226.1
866	SD	DOUGLAS	225.2
867	OH	UNION	225.0
868	MO	LACLEDE	224.8
869	ID	CASSIA	224.7
870	WV	WYOMING	224.6

COUNTIES OF HIGHEST AVERAGE FALLOUT FROM NUCLEAR TESTS:

RANGER BAKER-1 THROUGH STORAX SEDAN
1951-1962

(microCuries per square meter)

Each Row Includes:
County Rank (most fallout = 1), State, County, Fallout Level in microCuries/square meter

871 PA FULTON 224.4	900 NE THAYER 220.5	929 IN FRANKLIN 217.4
872 OK NOWATA 224.1	901 TX ARMSTRONG 220.5	930 ID BUTTE 217.1
873 SD PERKINS 223.7	902 LA ST MARTIN 220.3	931 VT CHITTENDEN 217.0
874 IN ORANGE 223.4	903 LA ASSUMPTION 220.2	932 PA FAYETTE 216.3
875 KS SEWARD 223.4	904 KY OLDHAM 220.2	933 ND CASS 215.9
876 ND BOWMAN 223.3	905 MA SUFFOLK 220.2	934 TX FANNIN 215.9
877 IN DUBOIS 223.3	906 OK SEQUOYAH 219.9	935 MO COOPER 215.3
878 KS ROOKS 223.2	907 ID PAYETTE 219.6	936 ID CLEARWATER 215.1
879 MO BATES 223.1	908 KY PENDLETON 219.6	937 WY TETON 214.9
880 OK ROGERS 222.9	909 MO TEXAS 219.4	938 ND FOSTER 214.6
881 TX DONLEY 222.7	910 MO DOUGLAS 219.4	939 SD CAMPBELL 214.3
882 SD LAWRENCE 222.6	911 IA LUCAS 219.3	940 KS MORRIS 214.3
883 KS MITCHELL 222.6	912 KS SUMNER 219.2	941 CA MONO 214.2
884 SD CHARLES MIX 222.5	913 MS FRANKLIN 219.1	942 AR LAWRENCE 214.1
885 OK LINCOLN 222.4	914 OH GUERNSEY 218.7	943 SD CODINGTON 213.7
886 SD GREGORY 222.3	915 WI DANE 218.7	944 NJ BURLINGTON 213.7
887 NV HUMBOLDT 222.2	916 AR BENTON 218.5	945 KS CRAWFORD 213.7
888 ID FREMONT 221.8	917 AR SHARP 218.5	946 WY HOT SPRINGS 213.7
889 LA ST MARY 221.8	918 VA CHESTERFIELD 218.5	947 TX FLOYD 213.2
890 SD BEADLE 221.8	919 OH MUSKINGUM 218.5	948 OK HUGHES 213.0
891 OK ALFALFA 221.3	920 MO WEBSTER 218.5	949 NE RICHARDSON 213.0
892 PA FOREST 221.2	921 ID LINCOLN 218.1	950 OH DARKE 213.0
893 TX CARSON 221.1	922 KY LESLIE 218.0	951 KS EDWARDS 212.9
894 MT PARK 221.0	923 IL MENARD 218.0	952 AR MADISON 212.8
895 IL CLARK 221.0	924 TX YOAKUM 217.9	953 KS WOODSON 212.5
896 MO MCDONALD 221.0	925 AR FAULKNER 217.9	954 OK PAWNEE 212.5
897 KY MONTGOMERY 220.8	926 KS RILEY 217.8	955 MT BLAINE 212.1
898 ND DICKEY 220.7	927 KS RICE 217.6	956 WY PARK 212.0
899 ND SHERIDAN 220.6	928 KS GEARY 217.5	957 MO OZARK 211.6

COUNTIES OF HIGHEST AVERAGE FALLOUT FROM NUCLEAR TESTS:

RANGER BAKER-1 THROUGH STORAX SEDAN 1951-1962

(microCuries per square meter)

Each Row Includes:
County Rank (most fallout = 1), State, County, Fallout Level in microCuries/square meter

Rank	State	County	Fallout		Rank	State	County	Fallout		Rank	State	County	Fallout
958	WY	WESTON	211.5		987	NE	JEFFERSON	205.9		1016	AR	VAN BUREN	201.9
959	MO	BARRY	211.2		988	TX	GRAYSON	205.7		1017	KS	WILSON	201.9
960	AR	PRAIRIE	211.2		989	MO	CHRISTIAN	205.7		1018	IN	UNION	201.9
961	AR	JACKSON	211.0		990	KS	HARPER	205.5		1019	KS	POTTAWATOMIE	201.8
962	MO	PHELPS	210.9		991	MO	CARROLL	205.5		1020	MT	DANIELS	201.7
963	SD	HAND	210.7		992	MO	JASPER	205.2		1021	PA	CRAWFORD	201.7
964	OK	NOBLE	210.7		993	OK	BEAVER	205.2		1022	IL	LAWRENCE	201.4
965	OK	MCINTOSH	210.6		994	IL	BOND	205.1		1023	TX	LAMAR	201.1
966	VA	FRANKLIN	210.5		995	TX	RANDALL	204.8		1024	KS	PAWNEE	200.9
967	PA	CLEARFIELD	210.5		996	SD	CLARK	204.6		1025	KY	BRACKEN	200.8
968	NE	BLAINE	210.3		997	PA	BEAVER	204.4		1026	OH	TUSCARAWAS	200.7
969	SD	DAY	210.3		998	KS	ALLEN	204.4		1027	IN	MADISON	200.6
970	NE	KEARNEY	210.3		999	NE	BOONE	204.3		1028	MO	JEFFERSON	200.3
971	AR	WASHINGTON	209.7		1000	IN	CLINTON	204.2		1029	IA	CHICKASAW	200.3
972	NE	BOYD	209.3		1001	ID	ADAMS	203.8		1030	MO	MERCER	200.3
973	NM	LEA	209.2		1002	AR	CRAIGHEAD	203.7		1031	NY	ESSEX	200.1
974	MO	DENT	208.7		1003	PA	MERCER	203.4		1032	OH	HAMILTON	200.1
975	MO	OREGON	208.3		1004	AR	IZARD	203.2		1033	OH	BELMONT	199.9
976	LA	AVOYELLES	208.2		1005	MO	GASCONADE	203.2		1034	IL	EDGAR	199.9
977	ID	TWIN FALLS	208.0		1006	IN	VERMILLION	203.1		1035	PA	CLARION	199.5
978	AR	PERRY	207.9		1007	AR	NEWTON	203.0		1036	OK	SEMINOLE	199.5
979	KS	ELK	207.9		1008	OH	TRUMBULL	203.0		1037	OK	MCCURTAIN	199.4
980	CT	MIDDLESEX	207.6		1009	LA	EAST FELICIA	202.7		1038	KS	STAFFORD	199.4
981	MO	CASS	207.5		1010	AR	POPE	202.7		1039	OH	LAKE	199.0
982	OH	ASHLAND	207.2		1011	AR	CLEBURNE	202.4		1040	ND	MCINTOSH	198.8
983	NE	HARLAN	207.2		1012	MO	ST CLAIR	202.4		1041	MT	ROSEBUD	198.8
984	TX	LYNN	206.8		1013	NM	EDDY	202.4		1042	TX	GAINES	198.7
985	KS	PRATT	206.5		1014	AR	BAXTER	202.2		1043	AR	FRANKLIN	198.5
986	OK	CHOCTAW	206.0		1015	TX	HALL	202.1		1044	OR	WALLOWA	198.5

COUNTIES OF HIGHEST AVERAGE FALLOUT FROM NUCLEAR TESTS:

RANGER BAKER-1 THROUGH STORAX SEDAN
1951-1962

(microCuries per square meter)

Each Row Includes:
County Rank (most fallout = 1), State, County, Fallout Level in microCuries/square meter

Rank	State	County	Fallout	Rank	State	County	Fallout	Rank	State	County	Fallout
1045	MO	JOHNSON	198.3	1074	OK	PUSHMATAHA	195.0	1103	MT	BIG HORN	191.6
1046	MO	MONITEAU	198.2	1075	AZ	SANTA CRUZ	195.0	1104	WV	MARSHALL	191.6
1047	PA	INDIANA	198.1	1076	OH	COLUMBIANA	194.8	1105	VT	WINDHAM	191.6
1048	KS	WASHINGTON	198.1	1077	OK	LATIMER	194.6	1106	AZ	YAVAPAI	191.5
1049	MO	NODAWAY	198.0	1078	IL	WHITESIDE	194.6	1107	IN	SHELBY	191.4
1050	TX	ROBERTS	197.6	1079	NJ	OCEAN	194.5	1108	ND	WALSH	191.4
1051	NE	FRANKLIN	197.5	1080	MT	MISSOULA	194.5	1109	IN	ST JOSEPH	191.3
1052	KS	MONTGOMERY	197.5	1081	OK	OKMULGEE	194.4	1110	KS	CHEROKEE	191.3
1053	MT	GARFIELD	197.5	1082	KY	PERRY	194.3	1111	ND	TRAILL	191.3
1054	LA	LIVINGSTON	197.4	1083	ND	LA MOURE	193.9	1112	MA	ESSEX	191.1
1055	TX	COLLINGSWORT	197.3	1084	TX	MOTLEY	193.9	1113	NE	KNOX	191.1
1056	PA	ARMSTRONG	197.3	1085	MT	CARBON	193.7	1114	MT	LEWIS AND CL	191.1
1057	AR	JOHNSON	197.3	1086	MO	DALLAS	193.4	1115	OH	WASHINGTON	190.9
1058	ID	LEWIS	197.3	1087	WY	SHERIDAN	193.4	1116	ND	BARNES	190.9
1059	NY	GREENE	197.3	1088	TX	GRAY	193.1	1117	NE	PIERCE	190.9
1060	SD	MARSHALL	197.2	1089	OK	ADAIR	192.9	1118	OK	HASKELL	190.9
1061	KS	PHILLIPS	197.0	1090	AR	MARION	192.9	1119	LA	ASCENSION	190.5
1062	MT	VALLEY	196.9	1091	PA	CAMBRIA	192.8	1120	KY	KNOTT	190.4
1063	KS	LABETTE	196.4	1092	MI	LIVINGSTON	192.7	1121	ND	LOGAN	190.2
1064	AR	INDEPENDENCE	196.1	1093	AR	WHITE	192.5	1122	KS	MARSHALL	190.1
1065	OH	WOOD	196.1	1094	OH	SANDUSKY	192.4	1123	OH	MAHONING	190.1
1066	IL	FAYETTE	196.0	1095	KS	COFFEY	192.3	1124	OK	BRYAN	189.9
1067	OR	BAKER	196.0	1096	OH	ERIE	192.3	1125	ID	JEROME	189.6
1068	KS	SMITH	195.9	1097	MT	SWEET GRASS	192.3	1126	IA	GRUNDY	189.6
1069	NE	PAWNEE	195.6	1098	NE	WHEELER	192.2	1127	LA	ST CHARLES	189.4
1070	IN	ADAMS	195.6	1099	IN	FULTON	191.9	1128	AR	CRAWFORD	189.3
1071	MO	BOONE	195.6	1100	PA	WESTMORELAND	191.9	1129	AR	WOODRUFF	189.1
1072	NE	KEYA PAHA	195.5	1101	OH	STARK	191.8	1130	VT	RUTLAND	189.0
1073	TX	HUDSPETH	195.1	1102	RI	WASHINGTON	191.8	1131	PA	WASHINGTON	188.9

COUNTIES OF HIGHEST AVERAGE FALLOUT FROM NUCLEAR TESTS:

RANGER BAKER-1 THROUGH STORAX SEDAN 1951-1962

(microCuries per square meter)

Each Row Includes:
County Rank (most fallout = 1), State, County, Fallout Level in microCuries/square meter

Rank	State	County	Fallout
1132	KY	SPENCER	188.8
1133	AR	POLK	188.8
1134	WV	TYLER	188.8
1135	NY	ERIE	188.7
1136	LA	LAFOURCHE	188.7
1137	SD	EDMUNDS	188.5
1138	OK	GARFIELD	188.5
1139	MO	CRAWFORD	188.4
1140	MI	GENESEE	188.4
1141	OK	BECKHAM	188.2
1142	MO	CLINTON	187.7
1143	PA	JEFFERSON	187.7
1144	MT	PETROLEUM	187.6
1145	KS	JOHNSON	187.1
1146	AR	SALINE	186.8
1147	TX	CHILDRESS	186.5
1148	NY	MONTGOMERY	186.5
1149	ME	AROOSTOOK	186.3
1150	OH	MEIGS	186.2
1151	KS	SHERMAN	186.2
1152	MT	GOLDEN VALLE	186.1
1153	OH	DELAWARE	186.0
1154	IN	WARREN	185.8
1155	KY	BOONE	185.6
1156	OK	PONTOTOC	185.5
1157	OH	MARION	185.4
1158	KS	CLAY	185.4
1159	TX	HEMPHILL	185.3
1160	IN	BARTHOLOMEW	185.3
1161	ND	WILLIAMS	185.2
1162	KS	LINN	185.0
1163	ND	STUTSMAN	184.7
1164	PA	LAWRENCE	184.7
1165	OH	HARRISON	184.6
1166	OH	MONROE	184.2
1167	MO	DADE	184.0
1168	MT	STILLWATER	184.0
1169	WV	UPSHUR	183.9
1170	IN	CLARK	183.8
1171	ND	SIOUX	183.8
1172	MO	ST FRANCOIS	183.7
1173	AR	RANDOLPH	183.6
1174	WI	JEFFERSON	183.6
1175	MS	HINDS	183.6
1176	TX	COLLIN	183.4
1177	NE	BURT	183.2
1178	KY	JESSAMINE	182.9
1179	KY	ESTILL	182.8
1180	OH	LUCAS	182.5
1181	ID	MINIDOKA	182.5
1182	IL	CUMBERLAND	182.5
1183	MT	MINERAL	182.5
1184	ND	GRAND FORKS	182.4
1185	NY	WYOMING	182.2
1186	AR	ARKANSAS	181.9
1187	LA	CONCORDIA	181.9
1188	AR	GREENE	181.9
1189	KS	FRANKLIN	181.9
1190	OK	WASHITA	181.7
1191	MO	BARTON	181.7
1192	IN	FLOYD	181.7
1193	OK	GREER	181.5
1194	KS	BARBER	181.5
1195	MI	OAKLAND	181.3
1196	NY	GENESEE	181.1
1197	KS	OSAGE	181.0
1198	KS	ATCHISON	180.9
1199	NY	HAMILTON	180.8
1200	KS	BOURBON	180.8
1201	OH	AUGLAIZE	180.7
1202	VA	ACCOMACK	180.7
1203	OH	MORROW	180.7
1204	OH	HOCKING	180.6
1205	ND	GRIGGS	180.5
1206	NM	SIERRA	180.4
1207	RI	BRISTOL	180.2
1208	OK	OKFUSKEE	179.9
1209	TX	COOKE	179.9
1210	WV	WOOD	179.9
1211	TX	CROSBY	179.8
1212	OH	CUYAHOGA	179.6
1213	KS	JACKSON	179.3
1214	AR	JEFFERSON	179.3
1215	AR	SEARCY	179.2
1216	MO	GENTRY	179.0
1217	KY	LETCHER	178.9
1218	AR	YELL	178.8

COUNTIES OF HIGHEST AVERAGE FALLOUT FROM NUCLEAR TESTS:

RANGER BAKER-1 THROUGH STORAX SEDAN
1951-1962

(microCuries per square meter)

Each Row Includes:
County Rank (most fallout = 1), State, County, Fallout Level in microCuries/square meter

Rank	State	County	Fallout	Rank	State	County	Fallout	Rank	State	County	Fallout
1219	SD	BROWN	178.7	1248	TX	DENTON	175.6	1277	LA	PLAQUEMINES	173.2
1220	LA	CATAHOULA	178.5	1249	NY	CHAUTAUQUA	175.6	1278	MT	MUSSELSHELL	173.2
1221	IL	LA SALLE	178.2	1250	LA	SABINE	175.6	1279	OH	FAYETTE	173.2
1222	MN	REDWOOD	178.1	1251	CT	NEW HAVEN	175.4	1280	KY	MADISON	173.2
1223	ND	NELSON	178.1	1252	KS	JEFFERSON	175.4	1281	PA	BUTLER	173.0
1224	VA	DICKENSON	178.0	1253	AR	CONWAY	175.3	1282	CT	TOLLAND	172.9
1225	AR	SEVIER	177.9	1254	WI	PEPIN	175.3	1283	IL	CRAWFORD	172.8
1226	OK	PAYNE	177.8	1255	AR	LOGAN	175.3	1284	NC	CHOWAN	172.8
1227	OH	ROSS	177.5	1256	MD	GARRETT	175.0	1285	IN	RANDOLPH	172.7
1228	AR	SEBASTIAN	177.3	1257	IL	IROQUOIS	174.7	1286	MI	CALHOUN	172.6
1229	KS	CLARK	177.1	1258	NJ	MONMOUTH	174.7	1287	MI	SHIAWASSEE	172.4
1230	SD	MCPHERSON	177.1	1259	AR	LONOKE	174.7	1288	MS	HANCOCK	172.3
1231	MO	MONTGOMERY	177.0	1260	AR	PULASKI	174.6	1289	MO	SHANNON	172.3
1232	KS	SEDGWICK	177.0	1261	LA	ST LANDRY	174.5	1290	IN	TIPPECANOE	172.2
1233	NJ	ATLANTIC	176.9	1262	LA	WASHINGTON	174.2	1291	WV	CALHOUN	172.2
1234	AR	MONTGOMERY	176.8	1263	OK	KINGFISHER	174.1	1292	OH	CRAWFORD	172.0
1235	ID	CANYON	176.7	1264	OK	LOGAN	174.1	1293	MO	RIPLEY	171.8
1236	NE	DODGE	176.7	1265	AL	MOBILE	174.1	1294	IL	LEE	171.8
1237	KS	JEWELL	176.6	1266	MO	STONE	174.0	1295	WV	GILMER	171.8
1238	NY	ST LAWRENCE	176.5	1267	RI	KENT	173.9	1296	WV	GREENBRIER	171.8
1239	KS	DOUGLAS	176.4	1268	OK	LE FLORE	173.9	1297	NY	NASSAU	171.7
1240	MS	PEARL RIVER	176.4	1269	WV	HARRISON	173.9	1298	OH	PUTNAM	171.6
1241	VT	GRAND ISLE	176.1	1270	LA	VERNON	173.7	1299	MS	PIKE	171.4
1242	KS	KIOWA	176.1	1271	AR	CLAY	173.6	1300	KS	BROWN	171.3
1243	NY	SCHOHARIE	175.9	1272	ND	EDDY	173.5	1301	MO	PLATTE	171.0
1244	NC	WASHINGTON	175.9	1273	LA	ST TAMMANY	173.4	1302	IN	HARRISON	170.8
1245	IN	WASHINGTON	175.8	1274	PA	CHESTER	173.4	1303	MT	PHILLIPS	170.5
1246	SD	HUGHES	175.7	1275	ND	BENSON	173.3	1304	KY	ROCKCASTLE	170.5
1247	KS	MIAMI	175.7	1276	PA	GREENE	173.3	1305	MI	LAPEER	170.3

U.S. FALLOUT ATLAS : TOTAL FALLOUT

COUNTIES OF HIGHEST AVERAGE FALLOUT FROM NUCLEAR TESTS:

RANGER BAKER-1 THROUGH STORAX SEDAN
1951-1962

(microCuries per square meter)

Each Row Includes:
County Rank (most fallout = 1), State, County, Fallout Level in microCuries/square meter

Rank	State/County	Level	Rank	State/County	Level	Rank	State/County	Level
1306	MO ST CHARLES	170.3	1335	NY ULSTER	167.2	1364	TX JASPER	162.8
1307	AR CARROLL	170.1	1336	OK HARMON	167.2	1365	CT FAIRFIELD	162.7
1308	VA CARROLL	170.0	1337	OH GREENE	167.2	1366	TX NEWTON	162.5
1309	TX ANDREWS	169.9	1338	KY HARLAN	167.1	1367	IN WABASH	162.4
1310	IL KANKAKEE	169.8	1339	WI LA CROSSE	166.9	1368	IN WARRICK	162.1
1311	AL MARENGO	169.7	1340	IL PUTNAM	166.4	1369	NV CLARK3	161.9
1312	MI HILLSDALE	169.6	1341	WV WETZEL	166.1	1370	KS ANDERSON	161.9
1313	NY LIVINGSTON	169.6	1342	AR LITTLE RIVER	166.1	1371	MI VAN BUREN	161.9
1314	AR BOONE	169.6	1343	MO HICKORY	166.0	1372	NY NIAGARA	161.8
1315	OH FAIRFIELD	169.5	1344	MI KALAMAZOO	165.6	1373	MS JEFFERSON	161.6
1316	ND PEMBINA	169.5	1345	PA SOMERSET	165.6	1374	LA ST JOHN THE	161.6
1317	IL OGLE	169.5	1346	MO NEW MADRID	165.5	1375	OK WOODS	161.6
1318	MO WAYNE	169.4	1347	IN PULASKI	165.5	1376	WV TUCKER	161.6
1319	ND STEELE	169.2	1348	WV LEWIS	165.3	1377	AR MISSISSIPPI	161.5
1320	KS NEMAHA	169.1	1349	AR ST FRANCIS	165.1	1378	MO WARREN	161.5
1321	AR SCOTT	169.1	1350	NE SARPY	164.9	1379	LA ST BERNARD	161.4
1322	AR STONE	169.0	1351	KY BOURBON	164.8	1380	OK LOVE	161.3
1323	MT GRANITE	168.9	1352	MS CLAIBORNE	164.6	1381	ND RAMSEY	161.3
1324	ID BANNOCK	168.8	1353	WI MONROE	164.4	1382	OH VAN WERT	161.1
1325	WV WIRT	168.8	1354	MS MARION	164.3	1383	WI SAWYER	161.0
1326	MT TREASURE	168.5	1355	IL STEPHENSON	164.3	1384	OH VINTON	160.8
1327	MT CASCADE	168.4	1356	MA BERKSHIRE	164.2	1385	AR HOWARD	160.7
1328	MS GEORGE	168.2	1357	NE SCOTTS BLUFF	163.8	1386	OK JOHNSTON	160.5
1329	TX LOVING	168.1	1358	TN GIBSON	163.7	1387	WV BRAXTON	160.1
1330	IN SCOTT	167.9	1359	MO CARTER	163.5	1388	SD BON HOMME	159.8
1331	WI JUNEAU	167.9	1360	TN WASHINGTON	163.3	1389	OH NOBLE	159.8
1332	MA BARNS	167.8	1361	TX DALLAS	163.2	1390	AR MONROE	159.4
1333	MO REYNOLDS	167.5	1362	MT HILL	163.0	1391	WY BIG HORN	159.3
1334	MA DUKES	167.4	1363	KS CLOUD	162.8	1392	MI LENAWEE	159.3

COUNTIES OF HIGHEST AVERAGE FALLOUT FROM NUCLEAR TESTS:

RANGER BAKER-1 THROUGH STORAX SEDAN 1951-1962

(microCuries per square meter)

Each Row Includes:
County Rank (most fallout = 1), State, County, Fallout Level in microCuries/square meter

Rank	State	County	Fallout
1393	TX	CULBERSON	159.2
1394	NY	ONTARIO	159.1
1395	OK	POTTAWATOMIE	159.0
1396	IN	DAVIESS	158.9
1397	KY	FAYETTE	158.8
1398	VA	KING WILLIAM	158.8
1399	MT	TETON	158.5
1400	IL	WILL	158.5
1401	MI	CLINTON	158.5
1402	PA	ERIE	158.4
1403	TN	LAKE	158.2
1404	NY	SUFFOLK	158.2
1405	PA	DELAWARE	158.1
1406	TN	WEAKLEY	158.1
1407	AR	GRANT	158.1
1408	TX	ANGELINA	158.0
1409	WI	TREMPEALEAU	158.0
1410	MO	DUNKLIN	157.9
1411	ND	KIDDER	157.8
1412	NY	QUEENS	157.8
1413	PA	WARREN	157.7
1414	NY	WAYNE	157.6
1415	NY	CLINTON	157.5
1416	IN	SULLIVAN	157.5
1417	TN	TIPTON	157.5
1418	LA	ST JAMES	157.4
1419	MO	JACKSON	157.4
1420	VA	TAZEWELL	157.3
1421	PA	CLINTON	157.3
1422	VT	FRANKLIN	157.1
1423	CT	NEW LONDON	157.1
1424	MS	LAUDERDALE	157.0
1425	WV	BROOKE	156.9
1426	WV	PUTNAM	156.9
1427	WV	FAYETTE	156.8
1428	NY	ALLEGANY	156.6
1429	MA	HAMPDEN	156.4
1430	CT	LITCHFIELD	156.3
1431	TX	SABINE	156.3
1432	PA	ALLEGHENY	156.3
1433	NE	ROCK	156.3
1434	WI	DUNN	156.1
1435	TX	SAN AUGUSTIN	156.1
1436	IN	JASPER	155.8
1437	NY	KINGS	155.5
1438	MA	WORCESTER	155.3
1439	TN	LAUDERDALE	155.1
1440	IN	HUNTINGTON	154.8
1441	NY	WESTCHESTER	154.6
1442	IL	GRUNDY	154.5
1443	WV	RANDOLPH	154.5
1444	TX	HOPKINS	154.4
1445	WI	WASHBURN	154.4
1446	AR	CLARK	154.4
1447	IN	VIGO	154.3
1448	MO	BUTLER	154.2
1449	WV	MARION	154.1
1450	AR	PIKE	154.1
1451	IN	WHITE	154.1
1452	IN	VANDERBURGH	153.7
1453	IL	RANDOLPH	153.6
1454	TX	JEFF DAVIS	153.5
1455	WV	NICHOLS	153.4
1456	TN	OBION	153.3
1457	TX	WALKER	153.2
1458	MS	FORREST	153.1
1459	OK	COAL	153.0
1460	TX	RAINS	152.7
1461	NJ	HUDSON	152.6
1462	MO	IRON	152.6
1463	IL	JOHNSON	152.6
1464	WV	MONONGALIA	152.5
1465	IN	MIAMI	152.5
1466	WI	WOOD	152.3
1467	MO	ST LOUIS	152.2
1468	MO	WASHINGTON	152.2
1469	NH	ROCKINGHAM	152.0
1470	MI	KENT	151.9
1471	TX	POLK	151.9
1472	CT	WINDHAM	151.8
1473	WV	GRANT	151.8
1474	VA	KING AND QUE	151.6
1475	TX	BOWIE	151.5
1476	OH	GEAUGA	151.4
1477	IN	SPENCER	150.9
1478	IL	DE KALB	150.8
1479	IN	PERRY	150.7

COUNTIES OF HIGHEST AVERAGE FALLOUT FROM NUCLEAR TESTS:

RANGER BAKER-1 THROUGH STORAX SEDAN 1951-1962

(microCuries per square meter)

Each Row Includes:
County Rank (most fallout = 1), State, County, Fallout Level in microCuries/square meter

Rank	State	County	Fallout
1480	PA	HUNTINGDON	150.7
1481	MO	PEMISCOT	150.6
1482	TX	TARRANT	150.6
1483	OK	CANADIAN	150.5
1484	AR	CRITTENDEN	150.3
1485	MT	TOOLE	150.0
1486	OH	LORAIN	150.0
1487	AZ	PINAL	150.0
1488	FL	OKALOOSA	150.0
1489	NJ	ESSEX	149.8
1490	NY	CATTARAUGUS	149.6
1491	MD	HARFORD	149.6
1492	WV	CLAY	149.5
1493	MS	STONE	149.4
1494	AZ	GILA	149.4
1495	VA	SMYTH	149.3
1496	NY	STEUBEN	149.0
1497	AZ	YUMA	149.0
1498	KY	MARTIN	148.9
1499	TX	HUNT	148.9
1500	WI	COLUMBIA	148.9
1501	OK	HARPER	148.8
1502	PA	YORK	148.7
1503	AR	CROSS	148.6
1504	AR	LAFAYETTE	148.6
1505	TN	MADISON	148.6
1506	IL	JEFFERSON	148.5
1507	ID	SHOSHONE	148.5
1508	NY	SULLIVAN	148.3
1509	MI	MONROE	148.3
1510	OK	CARTER	148.3
1511	WV	POCAHONTAS	148.1
1512	MT	RAVALLI	148.1
1513	TX	RED RIVER	148.1
1514	AR	MILLER	148.0
1515	TX	CHAMBERS	147.9
1516	OK	MAJOR	147.9
1517	MI	OTTAWA	147.8
1518	PA	CENTRE	147.4
1519	TX	VAN ZANDT	147.2
1520	NC	NORTHAMPTON	147.2
1521	NY	OTSEGO	147.1
1522	TN	DYER	147.1
1523	OH	HURON	146.8
1524	TN	HAYWOOD	146.7
1525	ME	PENOBSCOT	146.6
1526	MI	WAYNE	146.4
1527	WI	ADAMS	146.3
1528	OK	MARSHALL	146.2
1529	MD	BALTIMORE	146.1
1530	NY	PUTNAM	145.8
1531	WI	RUSK	145.7
1532	ID	POWER	145.6
1533	PA	CAMERON	145.5
1534	KY	HANCOCK	145.5
1535	TX	GRIMES	145.5
1536	RI	PROVIDENCE	145.5
1537	MO	MISSISSIPPI	145.5
1538	WI	JACKSON	145.4
1539	AZ	MARICOPA	145.4
1540	AR	DESHA	145.4
1541	WV	WEBSTER	145.3
1542	TX	WOOD	145.1
1543	KY	HICKMAN	145.1
1544	TX	WINKLER	145.0
1545	AR	LEE	144.9
1546	AR	NEVADA	144.9
1547	AR	HOT SPRING	144.9
1548	AR	GARLAND	144.7
1549	TX	TRINITY	144.6
1550	TX	CAMP	144.6
1551	NY	JEFFERSON	144.5
1552	AR	POINSETT	144.4
1553	KS	COMANCHE	144.3
1554	ND	WELLS	144.1
1555	VA	NEW KENT	144.1
1556	NC	BEAUFORT	144.0
1557	ND	CAVALIER	143.9
1558	MA	HAMPSHIRE	143.8
1559	IL	LIVINGSTON	143.8
1560	TX	LIBERTY	143.8
1561	IN	MONTGOMERY	143.7
1562	NY	FRANKLIN	143.6
1563	VA	WARREN	143.4
1564	MS	HARRISON	143.0
1565	NJ	UNION	143.0
1566	VA	FAIRFAX	143.0

COUNTIES OF HIGHEST AVERAGE FALLOUT FROM NUCLEAR TESTS:

RANGER BAKER-1 THROUGH STORAX SEDAN
1951-1962

(microCuries per square meter)

Each Row Includes:
County Rank (most fallout = 1), State, County, Fallout Level in microCuries/square meter

Rank	State	County	Fallout
1567	IL	WABASH	143.0
1568	WI	DOUGLAS	143.0
1569	OR	GRANT	142.9
1570	OK	COMANCHE	142.8
1571	NY	DELAWARE	142.8
1572	MS	JACKSON	142.8
1573	WI	EAU CLAIRE	142.8
1574	MS	DE SOTO	142.7
1575	NY	NEW YORK	142.7
1576	PA	NORTHUMBERLA	142.6
1577	TX	CASS	142.4
1578	TX	HARRIS	142.4
1579	VA	CHARLES CITY	142.2
1580	TX	TYLER	142.1
1581	TX	BRAZORIA	142.0
1582	KY	FULTON	141.8
1583	WI	MARQUETTE	141.7
1584	WI	ASHLAND	141.7
1585	IN	NEWTON	141.7
1586	MI	BARRY	141.5
1587	MS	QUITMAN	141.5
1588	PA	LACKAWANNA	141.3
1589	WI	BARRON	141.3
1590	KS	DONIPHAN	141.2
1591	VA	LOUDOUN	141.1
1592	OK	DEWEY	140.8
1593	OK	BLAINE	140.6
1594	MT	FLATHEAD	140.6
1595	MS	MARSHALL	140.3
1596	NJ	BERGEN	140.3
1597	ID	BONNEVILLE	140.3
1598	TX	MONTGOMERY	140.2
1599	AR	HEMPSTEAD	140.0
1600	TX	HARDIN	139.7
1601	WI	ST CROIX	139.7
1602	NJ	MIDDLESEX	139.4
1603	VA	FLOYD	139.4
1604	TX	ROCKWALL	139.3
1605	CT	HARTFORD	139.2
1606	WI	POLK	139.2
1607	MO	STODDARD	139.0
1608	AR	LINCOLN	138.9
1609	WI	PIERCE	138.8
1610	VA	CRAIG	138.6
1611	NY	RICHMOND	138.5
1612	TX	REEVES	138.5
1613	VA	HIGHLAND	138.4
1614	TX	DELTA	138.2
1615	NC	PERSON	138.2
1616	WI	BURNETT	138.1
1617	PA	WAYNE	138.1
1618	NY	ORLEANS	138.1
1619	TN	CROCKETT	138.0
1620	MN	DAKOTA	137.9
1621	NY	HERKIMER	137.8
1622	PA	LANCASTER	137.8
1623	MT	LIBERTY	137.8
1624	MN	CARLTON	137.8
1625	AR	PHILLIPS	137.7
1626	PA	ADAMS	137.5
1627	ME	PISCATAQUIS	137.5
1628	NH	CARROLL	137.4
1629	IN	POSEY	137.4
1630	AL	CONECUH	137.3
1631	VA	ROCKINGHAM	137.2
1632	WI	CHIPPEWA	137.2
1633	NY	OSWEGO	137.2
1634	MS	GREENE	137.2
1635	WI	CLARK	137.1
1636	IN	DECATUR	137.1
1637	MS	TATE	137.1
1638	RI	NEWPORT	137.1
1639	IN	BOONE	137.1
1640	AZ	GRAHAM	136.7
1641	IN	PORTER	136.7
1642	OK	CADDO	136.4
1643	PA	BEDFORD	136.3
1644	TX	WISE	136.2
1645	NJ	PASSAIC	136.1
1646	KY	DAVIESS	136.1
1647	TX	WALLER	135.9
1648	WI	IRON	135.9
1649	VA	FAUQUIER	135.7
1650	VT	WASHINGTON	135.6
1651	ME	WASHINGTON	135.6
1652	ID	LATAH	135.5
1653	VT	LAMOILLE	135.2

COUNTIES OF HIGHEST AVERAGE FALLOUT FROM NUCLEAR TESTS:

RANGER BAKER-1 THROUGH STORAX SEDAN 1951-1962

(microCuries per square meter)

Each Row Includes:
County Rank (most fallout = 1), State, County, Fallout Level in microCuries/square meter

Rank	State	County	Fallout
1654	NJ	MORRIS	135.2
1655	OK	WOODWARD	135.2
1656	TX	SAN JACINTO	135.0
1657	VA	ORANGE	135.0
1658	WA	GARFIELD	134.9
1659	MS	CHICKASAW	134.8
1660	MO	ANDREW	134.6
1661	VT	WINDSOR	134.4
1662	IL	WAYNE	134.3
1663	IN	LAKE	134.3
1664	IL	MONROE	134.2
1665	TX	MONTAGUE	134.1
1666	OK	MCCLAIN	134.1
1667	KY	GRAVES	134.0
1668	NY	BRONX	133.9
1669	MS	TIPPAH	133.8
1670	TN	FAYETTE	133.6
1671	WI	WAUSHARA	133.6
1672	IL	CARROLL	133.4
1673	OR	UNION	133.4
1674	NY	SENECA	133.4
1675	WI	BAYFIELD	133.3
1676	TX	FORT BEND	133.2
1677	VA	BEDFORD	133.2
1678	MS	WALTHALL	133.1
1679	VA	ROANOKE	133.0
1680	MS	PANOLA	132.9
1681	TX	AUSTIN	132.7
1682	VA	PITTSYLVANIA	132.6
1683	MS	COAHOMA	132.6
1684	NJ	CAPE MAY	132.6
1685	IN	FOUNTAIN	132.4
1686	VA	ALLEGHANY	132.4
1687	MO	BUCHANAN	132.4
1688	MS	LAMAR	132.3
1689	VA	MECKLENBURG	132.3
1690	NC	VANCE	132.3
1691	OK	CLEVELAND	132.2
1692	MO	TANEY	132.2
1693	MN	OLMSTED	132.1
1694	IL	ST CLAIR	132.1
1695	OK	JEFFERSON	131.9
1696	OK	STEPHENS	131.7
1697	IL	EDWARDS	131.7
1698	MN	WASHINGTON	131.6
1699	ND	TOWNER	131.6
1700	NJ	MERCER	131.6
1701	MS	TALLAHATCHIE	131.4
1702	TX	TITUS	131.3
1703	IN	CLAY	131.2
1704	TN	CLAIBORNE	131.2
1705	IL	KANE	131.2
1706	MT	LAKE	131.2
1707	VA	SPOTSYLVANIA	131.2
1708	AR	COLUMBIA	131.0
1709	TX	ELLIS	130.9
1710	OK	OKLAHOMA	130.7
1711	AR	DALLAS	130.7
1712	TN	CAMPBELL	130.5
1713	TX	WASHINGTON	130.3
1714	KY	CUMBERLAND	130.3
1715	NH	CHESHIRE	130.3
1716	OK	MURRAY	130.3
1717	NY	DUTCHESS	130.2
1718	MO	HOLT	130.1
1719	WI	PRICE	130.0
1720	MI	BRANCH	129.8
1721	MD	ALLEGANY	129.8
1722	PA	POTTER	129.8
1723	NY	ORANGE	129.7
1724	TN	HANCOCK	129.7
1725	MI	MACOMB	129.6
1726	OK	GRADY	129.5
1727	OK	GARVIN	129.3
1728	ID	NEZ PERCE	129.0
1729	IN	BENTON	129.0
1730	KY	BELL	128.9
1731	VA	PRINCE EDWAR	128.9
1732	WI	TAYLOR	128.9
1733	MS	YALOBUSHA	128.8
1734	TN	HARDEMAN	128.6
1735	VA	CAROLINE	128.4
1736	VA	RAPPAHANNOCK	128.4
1737	MS	AMITE	128.3
1738	VA	MADISON	128.3
1739	TX	GALVESTON	128.2
1740	IN	PARKE	128.1

COUNTIES OF HIGHEST AVERAGE FALLOUT FROM NUCLEAR TESTS:

RANGER BAKER-1 THROUGH STORAX SEDAN 1951-1962

(microCuries per square meter)

Each Row Includes:
County Rank (most fallout = 1), State, County, Fallout Level in microCuries/square meter

Rank	State	County	Fallout
1741	MS	TUNICA	128.0
1742	ND	ROLETTE	127.6
1743	TX	SMITH	127.5
1744	MI	SAGINAW	127.3
1745	VA	LUNENBURG	127.3
1746	NC	GRANVILLE	127.2
1747	VT	ORLEANS	127.0
1748	MS	BOLIVAR	126.8
1749	MA	FRANKLIN	126.6
1750	IL	MADISON	126.6
1751	MS	COPIAH	126.5
1752	MN	RAMSEY	126.4
1753	MS	MONTGOMERY	126.3
1754	NY	TOMPKINS	126.2
1755	AR	OUACHITA	126.0
1756	MS	BENTON	126.0
1757	TN	CARTER	125.7
1758	VT	ORANGE	125.6
1759	IL	JACKSON	125.6
1760	PA	PIKE	125.2
1761	MN	SCOTT	124.8
1762	PA	MCKEAN	124.7
1763	NH	STRAFFORD	124.6
1764	WI	PORTAGE	124.5
1765	VA	LEE	124.4
1766	NH	HILLSBOROUGH	124.3
1767	MT	PONDERA	124.2
1768	OH	SUMMIT	124.1
1769	MI	SANILAC	124.1
1770	IN	MORGAN	123.8
1771	VA	SUFFOLK/NANS	123.7
1772	ND	PIERCE	123.7
1773	NJ	WARREN	123.6
1774	AR	CLEVELAND	123.5
1775	MT	SANDERS	123.5
1776	MN	PINE	123.3
1777	NC	CHATHAM	123.3
1778	MO	CAPE GIRARDE	123.2
1779	AZ	GREENLEE	123.0
1780	VA	AMHERST	123.0
1781	WI	WAUPACA	123.0
1782	TX	WARD	122.9
1783	MN	FREEBORN	122.9
1784	IN	BROWN	122.9
1785	OK	COTTON	122.9
1786	NC	WILKES	122.8
1787	KY	CLAY	122.7
1788	MO	PERRY	122.6
1789	TX	FRANKLIN	122.3
1790	VA	RUSSELL	122.3
1791	IL	GALLATIN	122.2
1792	OK	KIOWA	122.2
1793	MS	GRENADA	121.9
1794	VA	SCOTT	121.9
1795	TX	CLAY	121.8
1796	VA	AUGUSTA	121.8
1797	MI	GRATIOT	121.8
1798	IL	BUREAU	121.7
1799	VA	PULASKI	121.6
1800	NC	MARTIN	121.4
1801	LA	JEFFERSON	121.3
1802	MS	CARROLL	121.3
1803	WI	ONEIDA	121.2
1804	AL	MONROE	121.1
1805	NJ	SOMERSET	121.1
1806	AR	UNION	121.1
1807	MS	PONTOTOC	121.0
1808	OK	CUSTER	120.9
1809	MN	CARVER	120.8
1810	KY	UNION	120.8
1811	VA	APPOMATTOX	120.5
1812	ME	HANCOCK	120.5
1813	TN	SEVIER	120.2
1814	AZ	COCHISE	120.1
1815	IN	RIPLEY	120.1
1816	TN	SULLIVAN	120.0
1817	OK	JACKSON	119.9
1818	NC	ASHE	119.9
1819	VA	MONTGOMERY	119.7
1820	PA	LUZERNE	119.5
1821	KY	PULASKI	119.4
1822	KY	KNOX	119.4
1823	WV	OHIO	119.3
1824	WV	HARDY	119.2
1825	MS	CLARKE	119.2
1826	WA	ASOTIN	119.1
1827	KY	BOYLE	119.0

COUNTIES OF HIGHEST AVERAGE FALLOUT FROM NUCLEAR TESTS:

RANGER BAKER-1 THROUGH STORAX SEDAN 1951-1962

(microCuries per square meter)

Each Row Includes:
County Rank (most fallout = 1), State, County, Fallout Level in microCuries/square meter

Rank	State	County	Fallout
1828	WI	VILAS	119.0
1829	VT	ESSEX	119.0
1830	MO	STE GENEVIEV	118.9
1831	VA	CAMPBELL	118.9
1832	NC	MACON	118.8
1833	MA	NANTUCKET	118.7
1834	AL	WILCOX	118.7
1835	MN	SHERBURNE	118.7
1836	TX	HARDEMAN	118.4
1837	KY	TODD	118.2
1838	IL	ALEXANDER	118.2
1839	PA	WYOMING	118.2
1840	VA	PATRICK	118.2
1841	VA	NELSON	118.2
1842	NH	GRAFTON	117.9
1843	WI	LINCOLN	117.9
1844	SC	FLORENCE	117.8
1845	KY	MARSHALL	117.8
1846	MI	GOGEBIC	117.8
1847	MI	HURON	117.7
1848	IL	PULASKI	117.6
1849	IN	RUSH	117.6
1850	NJ	SUSSEX	117.5
1851	TN	SHELBY	117.5
1852	LA	MADISON	117.5
1853	NY	ROCKLAND	117.5
1854	TN	CLAY	117.5
1855	OR	MALHEUR	117.1
1856	MS	WAYNE	117.1
1857	MS	YAZOO	117.1
1858	NH	SULLIVAN	116.9
1859	VA	FREDERICK	116.9
1860	NC	JONES	116.7
1861	MS	PERRY	116.6
1862	KY	RUSSELL	116.5
1863	KY	MERCER	116.4
1864	MS	TISHOMINGO	116.3
1865	IL	UNION	116.3
1866	MI	JACKSON	116.1
1867	MN	ST LOUIS	116.1
1868	MO	MADISON	116.1
1869	IN	PUTNAM	115.9
1870	TX	ECTOR	115.8
1871	NC	CLAY	115.7
1872	MS	LEE	115.3
1873	IN	DEARBORN	115.2
1874	TX	UPSHUR	115.1
1875	LA	WEBSTER	115.1
1876	MI	HOUGHTON	114.9
1877	MT	GLACIER	114.7
1878	NM	GRANT	114.7
1879	IL	MCHENRY	114.4
1880	VA	ROCKBRIDGE	114.3
1881	OH	OTTAWA	114.1
1882	TX	PARKER	114.1
1883	MN	ANOKA	114.0
1884	IN	GIBSON	114.0
1885	NH	BELKNAP	113.9
1886	MS	UNION	113.9
1887	TN	SCOTT	113.9
1888	TN	MCNAIRY	113.8
1889	NC	JACKSON	113.7
1890	KY	MCCRACKEN	113.7
1891	TX	KAUFMAN	113.6
1892	LA	RICHLAND	113.6
1893	OH	HARDIN	113.4
1894	VA	GILES	113.3
1895	NC	SWAIN	113.2
1896	TX	RUSK	113.1
1897	MO	SCOTT	113.1
1898	MI	KEWEENAW	113.0
1899	GA	RABUN	113.0
1900	TN	PICKETT	112.8
1901	NM	OTERO	112.7
1902	TX	MORRIS	112.7
1903	SC	OCONEE	112.5
1904	MS	ISSAQUENA	112.4
1905	PA	MONROE	112.2
1906	NH	MERRIMACK	112.2
1907	MS	PRENTISS	112.2
1908	IL	JO DAVIESS	112.2
1909	CA	LOS ANGELES	112.1
1910	MS	NEWTON	112.0
1911	NJ	GLOUCESTER	111.9
1912	MS	LAFAYETTE	111.9
1913	LA	OUACHITA	111.9
1914	TN	ANDERSON	111.8

COUNTIES OF HIGHEST AVERAGE FALLOUT FROM NUCLEAR TESTS:

RANGER BAKER-1 THROUGH STORAX SEDAN
1951-1962

(microCuries per square meter)

Each Row Includes:
County Rank (most fallout = 1), State, County, Fallout Level in microCuries/square meter

Rank	State	County	Fallout
1915	TX	MADISON	111.7
1916	NJ	CAMDEN	111.7
1917	MN	MCLEOD	111.6
1918	VA	GREENE	111.5
1919	GA	LIBERTY	111.5
1920	KY	BALLARD	111.5
1921	NC	TRANSYLVANIA	111.4
1922	NC	HAYWOOD	111.2
1923	AR	BRADLEY	111.2
1924	MN	CHISAGO	111.2
1925	LA	CADDO	111.2
1926	IN	JOHNSON	111.1
1927	ME	KNOX	111.0
1928	MS	SHARKEY	110.9
1929	IL	PERRY	110.9
1930	KY	GARRARD	110.8
1931	OH	ASHTABULA	110.8
1932	KY	OHIO	110.8
1933	VA	GRAYSON	110.7
1934	MI	ONTONAGON	110.7
1935	VT	CALEDONIA	110.7
1936	MS	ALCORN	110.5
1937	WI	MARATHON	110.4
1938	KY	HENDERSON	110.2
1939	VA	SHENANDOAH	110.2
1940	WI	WAUKESHA	110.2
1941	LA	BOSSIER	110.2
1942	KY	BULLITT	110.1
1943	VA	WASHINGTON	110.0
1944	IL	CLINTON	109.9
1945	TX	JOHNSON	109.8
1946	WV	PLEASANTS	109.8
1947	NY	LEWIS	109.5
1948	MN	MEEKER	109.4
1949	NJ	CUMBERLAND	109.4
1950	OH	FRANKLIN	109.3
1951	MN	AITKIN	109.2
1952	MN	BENTON	109.2
1953	MS	HUMPHREYS	109.1
1954	MN	WRIGHT	109.1
1955	WA	PEND OREILLE	108.7
1956	OH	WAYNE	108.7
1957	MT	YELLOWSTONE	108.5
1958	IN	PIKE	108.5
1959	NH	COOS	108.5
1960	GA	LONG	108.4
1961	MO	BOLLINGER	108.3
1962	ID	BINGHAM	108.0
1963	ME	OXFORD	107.9
1964	ME	WALDO	107.9
1965	TX	MARION	107.8
1966	MS	WASHINGTON	107.8
1967	NC	CHEROKEE	107.7
1968	NY	ONEIDA	107.7
1969	MS	LEFLORE	107.6
1970	AR	DREW	107.6
1971	WI	MARINETTE	107.5
1972	NY	CORTLAND	107.4
1973	TX	DAWSON	107.4
1974	OH	MERCER	107.4
1975	NY	YATES	107.3
1976	MI	ALLEGAN	107.2
1977	TX	DICKENS	107.2
1978	VA	WISE	106.8
1979	IL	POPE	106.7
1980	TN	ROANE	106.5
1981	IN	KOSCIUSKO	106.4
1982	NC	HENDERSON	106.4
1983	PA	BLAIR	106.3
1984	TN	CHESTER	106.2
1985	MN	ISANTI	106.0
1986	ME	LINCOLN	105.9
1987	NC	HYDE	105.9
1988	MN	RENVILLE	105.7
1989	VA	VIRGINIA BEA	105.7
1990	OK	TILLMAN	105.6
1991	TN	COCKE	105.4
1992	MS	CALHOUN	105.3
1993	CA	ALPINE	105.3
1994	NC	ORANGE	105.2
1995	ID	BENEWAH	105.1
1996	WI	GREEN LAKE	104.8
1997	NY	MADISON	104.8
1998	TX	GARZA	104.8
1999	IL	WASHINGTON	104.7
2000	TX	GREGG	104.7
2001	TN	BLOUNT	104.6

COUNTIES OF HIGHEST AVERAGE FALLOUT FROM NUCLEAR TESTS:

RANGER BAKER-1 THROUGH STORAX SEDAN
1951-1962

(microCuries per square meter)

Each Row Includes:
County Rank (most fallout = 1), State, County, Fallout Level in microCuries/square meter

Rank	State	County	Fallout
2002	WI	OCONTO	104.5
2003	TX	CHEROKEE	104.5
2004	LA	WINN	104.4
2005	ME	CUMBERLAND	104.1
2006	MN	KANABEC	104.0
2007	MS	SUNFLOWER	103.8
2008	LA	BIENVILLE	103.8
2009	OH	HOLMES	103.7
2010	TN	COFFEE	103.7
2011	TX	HENDERSON	103.6
2012	NM	LUNA	103.6
2013	TX	HARRISON	103.4
2014	AR	ASHLEY	103.2
2015	AL	GREENE	103.1
2016	MS	LINCOLN	103.1
2017	MS	JASPER	103.0
2018	KY	LIVINGSTON	102.9
2019	NJ	SALEM	102.8
2020	LA	UNION	102.8
2021	TN	BLEDSOE	102.7
2022	NY	CHEMUNG	102.5
2023	NJ	HUNTERDON	102.4
2024	WV	CABELL	102.4
2025	MN	LAKE	102.3
2026	TX	JACK	102.3
2027	NC	MITCHELL	102.2
2028	LA	JACKSON	102.2
2029	TN	GRUNDY	102.1
2030	ME	KENNEBEC	101.8
2031	ME	YORK	101.8
2032	ID	BONNER	101.7
2033	AL	GENEVA	101.7
2034	AL	BALDWIN	101.5
2035	WI	SHAWANO	101.4
2036	MI	MIDLAND	101.3
2037	IN	LA PORTE	101.2
2038	TN	MONROE	101.2
2039	MS	MADISON	101.0
2040	FL	LAKE	100.9
2041	TN	MCMINN	100.9
2042	ME	SAGADAHOC	100.9
2043	MS	WEBSTER	100.8
2044	KY	LINCOLN	100.7
2045	MS	OKTIBBEHA	100.6
2046	PA	TIOGA	100.6
2047	WI	MENOMINEE	100.6
2048	LA	MOREHOUSE	100.5
2049	IL	MARION	100.5
2050	TN	MONTGOMERY	100.3
2051	NC	YANCEY	100.3
2052	DE	NEW CASTLE	100.3
2053	MN	CROW WING	100.3
2054	ME	ANDROSCOGGIN	100.2
2055	NC	WATAUGA	100.2
2056	MN	KANDIYOHI	100.1
2057	LA	LINCOLN	99.9
2058	WI	DODGE	99.8
2059	MN	MILLE LACS	99.8
2060	GA	TOWNS	99.6
2061	TN	CUMBERLAND	99.5
2062	IN	HANCOCK	99.5
2063	NC	BURKE	99.5
2064	KY	JEFFERSON	99.5
2065	GA	WAYNE	99.5
2066	NY	CHENANGO	99.5
2067	NY	CAYUGA	99.3
2068	MI	TUSCOLA	99.3
2069	KY	WARREN	99.3
2070	DE	SUSSEX	99.2
2071	TX	WICHITA	99.2
2072	FL	HOLMES	99.2
2073	WI	FOND DU LAC	99.1
2074	PA	NORTHAMPTON	99.0
2075	LA	EAST CARROLL	99.0
2076	TX	FOARD	98.9
2077	TN	HENRY	98.7
2078	ME	FRANKLIN	98.7
2079	KY	WASHINGTON	98.7
2080	OH	HENRY	98.7
2081	MS	SCOTT	98.5
2082	MD	SOMERSET	98.4
2083	NY	MONROE	98.4
2084	KY	CHRISTIAN	98.3
2085	TN	DICKSON	98.3
2086	NC	GRAHAM	98.2
2087	KY	BUTLER	98.1
2088	IN	MARSHALL	97.8

COUNTIES OF HIGHEST AVERAGE FALLOUT FROM NUCLEAR TESTS:

RANGER BAKER-1 THROUGH STORAX SEDAN
1951-1962

(microCuries per square meter)

Each Row Includes:
County Rank (most fallout = 1), State, County, Fallout Level in microCuries/square meter

Rank	State	County	Fallout
2089	DE	KENT	97.7
2090	IN	STARKE	97.7
2091	MI	MENOMINEE	97.7
2092	TN	BRADLEY	97.6
2093	NC	BUNCOMBE	97.4
2094	TN	HAWKINS	97.3
2095	TN	CARROLL	97.3
2096	AL	CLARKE	97.3
2097	TN	FENTRESS	97.3
2098	PA	LEHIGH	97.2
2099	FL	SANTA ROSA	97.1
2100	TN	RHEA	97.1
2101	NM	HIDALGO	97.1
2102	IL	WHITE	97.0
2103	MN	TODD	96.9
2104	PA	MONTGOMERY	96.8
2105	AR	CHICOT	96.8
2106	LA	NATCHITOCHES	96.7
2107	LA	WEST CARROLL	96.7
2108	SC	PICKENS	96.6
2109	NC	MADISON	96.6
2110	TN	SEQUATCHIE	96.5
2111	IL	LAKE	96.5
2112	OR	HARNEY	96.4
2113	PA	SUSQUEHANNA	96.4
2114	VA	PRINCE GEORG	96.4
2115	MT	LINCOLN	96.3
2116	TX	FISHER	96.3
2117	KY	MONROE	96.2
2118	TN	POLK	96.1
2119	SC	CLARENDON	96.1
2120	WA	COLUMBIA	95.9
2121	TN	GREENE	95.9
2122	MI	IRON	95.8
2123	PA	BUCKS	95.8
2124	MI	ST CLAIR	95.6
2125	WI	CALUMET	95.5
2126	NC	AVERY	95.4
2127	MS	CLAY	95.2
2128	MN	COOK	95.1
2129	SC	LEE	95.0
2130	MN	STEARNS	95.0
2131	WI	WINNEBAGO	94.9
2132	TN	MORGAN	94.9
2133	OR	UMATILLA	94.8
2134	LA	DE SOTO	94.8
2135	TX	NAVARRO	94.8
2136	NC	DURHAM	94.6
2137	LA	FRANKLIN	94.6
2138	IL	WILLIAMSON	94.5
2139	TX	BORDEN	94.5
2140	PA	CARBON	94.5
2141	GA	BRYAN	94.5
2142	IN	JAY	94.4
2143	TX	PANOLA	94.3
2144	NC	WAKE	94.3
2145	WI	WASHINGTON	94.3
2146	VA	HENRICO	94.2
2147	TX	KNOX	94.2
2148	AL	WASHINGTON	94.0
2149	TX	COTTLE	94.0
2150	TX	KENT	94.0
2151	NC	PAMLICO	93.9
2152	KY	CARLISLE	93.9
2153	IL	FRANKLIN	93.8
2154	KY	WAYNE	93.8
2155	TX	HOOD	93.7
2156	MN	MORRISON	93.7
2157	MD	WORCESTER	93.6
2158	MN	HENNEPIN	93.6
2159	GA	WHEELER	93.5
2160	GA	TERRELL	93.4
2161	TX	PALO PINTO	93.3
2162	WI	SHEBOYGAN	93.3
2163	NC	CALDWELL	93.3
2164	AL	ESCAMBIA	93.2
2165	SC	JASPER	93.2
2166	PA	BERKS	93.1
2167	TX	STONEWALL	93.0
2168	IN	STEUBEN	93.0
2169	LA	CALDWELL	92.9
2170	WV	HANCOCK	92.8
2171	SC	CALHOUN	92.7
2172	NM	DONA ANA	92.7
2173	MN	CASS	92.7
2174	IL	DU PAGE	92.6
2175	MS	COVINGTON	92.5

U.S. FALLOUT ATLAS : TOTAL FALLOUT

COUNTIES OF HIGHEST AVERAGE FALLOUT FROM NUCLEAR TESTS:

RANGER BAKER-1 THROUGH STORAX SEDAN
1951-1962

(microCuries per square meter)

Each Row Includes:
County Rank (most fallout = 1), State, County, Fallout Level in microCuries/square meter

Rank	State	County	Fallout
2176	TN	LOUDON	92.5
2177	NC	YADKIN	92.5
2178	MN	KOOCHICHING	92.4
2179	TX	ARCHER	92.4
2180	AL	TUSCALOOSA	92.4
2181	GA	MONTGOMERY	92.3
2182	NC	NEW HANOVER	92.3
2183	NC	MOORE	92.3
2184	KY	CASEY	92.3
2185	NC	ONSLOW	92.3
2186	MI	SCHOOLCRAFT	92.3
2187	NC	ALEXANDER	92.3
2188	KY	WHITLEY	92.3
2189	KY	TRIGG	92.2
2190	NC	ALLEGHANY	92.1
2191	TX	LIMESTONE	92.0
2192	IL	KENDALL	92.0
2193	GA	EFFINGHAM	91.9
2194	MS	SMITH	91.9
2195	TX	HILL	91.9
2196	MN	CHIPPEWA	91.9
2197	TN	HUMPHREYS	91.7
2198	SC	MARION	91.7
2199	IN	ELKHART	91.6
2200	TN	UNICOI	91.4
2201	GA	FANNIN	91.4
2202	WI	LANGLADE	91.3
2203	MI	BARAGA	91.3
2204	GA	QUITMAN	91.3
2205	TN	CHEATHAM	91.3
2206	TN	DECATUR	91.2
2207	IL	SALINE	91.2
2208	VA	MIDDLESEX	91.1
2209	MI	MONTCALM	91.1
2210	IL	MASSAC	91.1
2211	GA	CARROLL	91.1
2212	SC	GREENVILLE	91.0
2213	NC	SURRY	90.9
2214	TN	JEFFERSON	90.8
2215	TN	HAMILTON	90.8
2216	ME	SOMERSET	90.7
2217	NC	DARE	90.7
2218	MN	WADENA	90.6
2219	NC	MCDOWELL	90.6
2220	KY	LAUREL	90.5
2221	TN	VAN BUREN	90.5
2222	TX	MARTIN	90.3
2223	MN	LAC QUI PARL	90.2
2224	TN	WARREN	90.2
2225	SC	ORANGEBURG	90.2
2226	GA	SCREVEN	90.1
2227	OH	CARROLL	90.1
2228	MS	MONROE	90.0
2229	IN	LAGRANGE	89.9
2230	GA	CLAY	89.9
2231	AL	COLBERT	89.9
2232	KY	MCCREARY	89.8
2233	MI	BERRIEN	89.8
2234	TN	PUTNAM	89.7
2235	SC	SUMTER	89.7
2236	GA	EVANS	89.6
2237	NC	POLK	89.6
2238	MN	ITASCA	89.4
2239	WI	FOREST	89.4
2240	SC	HORRY	89.3
2241	LA	RED RIVER	89.3
2242	GA	RANDOLPH	89.3
2243	TN	MARSHALL	89.3
2244	TX	WILBARGER	89.3
2245	GA	PULASKI	89.2
2246	MD	ST MARYS	89.2
2247	FL	ESCAMBIA	89.2
2248	IL	HAMILTON	89.1
2249	AL	HALE	89.0
2250	NC	CARTERET	89.0
2251	TN	JOHNSON	89.0
2252	LA	CLAIBORNE	88.9
2253	ID	KOOTENAI	88.8
2254	FL	HILLSBOROUGH	88.8
2255	MD	PRINCE GEORG	88.8
2256	NC	GUILFORD	88.8
2257	WA	WALLA WALLA	88.7
2258	TN	HENDERSON	88.7
2259	NC	CRAVEN	88.5
2260	VA	ARLINGTON	88.5
2261	SC	CHARLESTON	88.4
2262	GA	FAYETTE	88.2

COUNTIES OF HIGHEST AVERAGE FALLOUT FROM NUCLEAR TESTS:

RANGER BAKER-1 THROUGH STORAX SEDAN
1951-1962

(microCuries per square meter)

Each Row Includes:
County Rank (most fallout = 1), State, County, Fallout Level in microCuries/square meter

Rank	State	County	Level
2263	TX	MCLENNAN	88.2
2264	AR	CALHOUN	88.1
2265	TN	BEDFORD	88.1
2266	VA	NORTHUMBERLA	88.1
2267	GA	WHITE	88.0
2268	GA	SCHLEY	87.9
2269	MD	MONTGOMERY	87.9
2270	TX	ANDERSON	87.9
2271	GA	WEBSTER	87.9
2272	PA	SULLIVAN	87.7
2273	GA	CALHOUN	87.7
2274	KY	CALDWELL	87.7
2275	KY	MEADE	87.6
2276	TN	DAVIDSON	87.5
2277	GA	TATTNALL	87.4
2278	GA	CLAYTON	87.3
2279	SC	COLLETON	87.1
2280	WI	OZAUKEE	87.0
2281	MI	MUSKEGON	87.0
2282	TX	JEFFERSON	87.0
2283	MD	CALVERT	86.9
2284	NC	IREDELL	86.9
2285	TN	HARDIN	86.9
2286	NY	ONONDAGA	86.7
2287	VA	CULPEPER	86.6
2288	SC	DORCHESTER	86.6
2289	GA	CRAWFORD	86.6
2290	TN	HICKMAN	86.6
2291	AL	MADISON	86.5
2292	NC	RANDOLPH	86.5
2293	PA	BRADFORD	86.5
2294	WI	KENOSHA	86.4
2295	NC	PITT	86.4
2296	GA	CHATHAM	86.4
2297	NC	ALAMANCE	86.4
2298	GA	JENKINS	86.4
2299	VA	LANCASTER	86.3
2300	NC	PENDER	86.2
2301	GA	BULLOCH	86.2
2302	TN	HAMBLEN	86.1
2303	PA	SCHUYLKILL	86.1
2304	KY	EDMONSON	86.0
2305	TX	BOSQUE	86.0
2306	MN	YELLOW MEDIC	85.7
2307	MD	WICOMICO	85.7
2308	MD	CECIL	85.7
2309	NC	BERTIE	85.7
2310	TX	CALLAHAN	85.7
2311	MS	ITAWAMBA	85.6
2312	MS	LEAKE	85.6
2313	MI	WASHTENAW	85.6
2314	GA	GLYNN	85.5
2315	GA	PEACH	85.5
2316	MS	ATTALA	85.4
2317	GA	UNION	85.4
2318	SC	BERKELEY	85.3
2319	WI	KEWAUNEE	85.3
2320	GA	COLQUITT	85.2
2321	KY	CALLOWAY	85.2
2322	AL	PICKENS	85.2
2323	TN	KNOX	85.0
2324	AL	PIKE	85.0
2325	TN	MEIGS	85.0
2326	PA	MONTOUR	84.9
2327	AL	PERRY	84.9
2328	NC	DAVIDSON	84.9
2329	KY	METCALFE	84.8
2330	KY	WEBSTER	84.8
2331	VA	GREENSVILLE	84.7
2332	FL	VOLUSIA	84.7
2333	TX	NACOGDOCHES	84.5
2334	AL	LOWNDES	84.5
2335	NC	HERTFORD	84.5
2336	NC	CATAWBA	84.5
2337	AL	ETOWAH	84.4
2338	ID	BOUNDARY	84.3
2339	GA	DOUGHERTY	84.3
2340	AL	BARBOUR	84.3
2341	MS	CHOCTAW	84.2
2342	AL	DALE	84.2
2343	VA	DINWIDDIE	84.2
2344	VA	STAFFORD	84.2
2345	WA	WHITMAN	84.1
2346	GA	BAKER	84.0
2347	KY	MCLEAN	84.0
2348	VA	NEWPORT NEWS	84.0
2349	TN	MAURY	84.0

COUNTIES OF HIGHEST AVERAGE FALLOUT FROM NUCLEAR TESTS:

RANGER BAKER-1 THROUGH STORAX SEDAN
1951-1962

(microCuries per square meter)

Each Row Includes:
County Rank (most fallout = 1), State, County, Fallout Level in microCuries/square meter

Rank	State	County	Level		Rank	State	County	Level		Rank	State	County	Level
2350	CA	NEVADA	84.0		2379	AL	TALLADEGA	82.1		2408	GA	MITCHELL	81.1
2351	IN	OHIO	83.9		2380	MD	HOWARD	82.1		2409	GA	FLOYD	81.1
2352	TX	HASKELL	83.9		2381	AL	CULLMAN	82.1		2410	GA	MCINTOSH	81.1
2353	MN	SWIFT	83.7		2382	NC	CUMBERLAND	82.0		2411	GA	DOUGLAS	81.0
2354	SC	MARLBORO	83.7		2383	AL	SUMTER	82.0		2412	KY	MUHLENBERG	81.0
2355	TX	BAYLOR	83.6		2384	VA	NORFOLK/CHES	82.0		2413	GA	TOOMBS	81.0
2356	DC	WASHINGTON	83.5		2385	MS	RANKIN	82.0		2414	MI	NEWAYGO	81.0
2357	TX	SCURRY	83.5		2386	NY	TIOGA	82.0		2415	NC	WAYNE	81.0
2358	GA	GORDON	83.4		2387	NC	GATES	82.0		2416	GA	MONROE	80.9
2359	GA	BURKE	83.4		2388	IL	HARDIN	81.9		2417	SC	WILLIAMSBURG	80.9
2360	LA	ORLEANS	83.4		2389	TX	HOUSTON	81.9		2418	GA	PAULDING	80.9
2361	FL	HERNANDO	83.3		2390	MD	CHARLES	81.9		2419	GA	BACON	80.9
2362	AL	DE KALB	83.3		2391	GA	BLECKLEY	81.9		2420	TN	DE KALB	80.9
2363	KY	BRECKINRIDGE	83.3		2392	WI	MANITOWOC	81.9		2421	PA	UNION	80.8
2364	MD	ANNE ARUNDEL	83.2		2393	AL	COVINGTON	81.9		2422	TN	SUMNER	80.8
2365	TX	SOMERVELL	83.1		2394	NC	CLEVELAND	81.8		2423	NC	WARREN	80.8
2366	FL	LEON	83.1		2395	SC	BEAUFORT	81.8		2424	VA	KING GEORGE	80.7
2367	TN	WHITE	83.1		2396	AL	LAUDERDALE	81.8		2425	GA	MORGAN	80.7
2368	SC	SPARTANBURG	83.0		2397	MI	ALGER	81.7		2426	NC	RUTHERFORD	80.6
2369	GA	JEFF DAVIS	83.0		2398	AL	MORGAN	81.6		2427	TN	STEWART	80.6
2370	SD	DEUEL	82.9		2399	WV	JACKSON	81.6		2428	IN	JENNINGS	80.6
2371	NC	LENOIR	82.9		2400	VA	PRINCE WILLI	81.5		2429	MI	CASS	80.5
2372	FL	ST JOHNS	82.8		2401	GA	TWIGGS	81.5		2430	AL	CHAMBERS	80.5
2373	MD	KENT	82.8		2402	TN	PERRY	81.5		2431	VA	BRUNSWICK	80.4
2374	SC	DARLINGTON	82.7		2403	GA	JEFFERSON	81.5		2432	TN	ROBERTSON	80.4
2375	MI	ST JOSEPH	82.6		2404	TN	OVERTON	81.3		2433	GA	JOHNSON	80.4
2376	GA	GREENE	82.5		2405	NC	DUPLIN	81.3		2434	MI	DICKINSON	80.4
2377	VA	ISLE OF WIGH	82.4		2406	GA	LEE	81.2		2435	FL	BREVARD	80.4
2378	MD	TALBOT	82.2		2407	OH	JEFFERSON	81.2		2436	FL	SUMTER	80.4

COUNTIES OF HIGHEST AVERAGE FALLOUT FROM NUCLEAR TESTS:

RANGER BAKER-1 THROUGH STORAX SEDAN
1951-1962

(microCuries per square meter)

Each Row Includes:
County Rank (most fallout = 1), State, County, Fallout Level in microCuries/square meter

Rank	State	County	Fallout	Rank	State	County	Fallout	Rank	State	County	Fallout
2437	VA	NOTTOWAY	80.3	2466	GA	LANIER	79.1	2495	AL	CLEBURNE	77.8
2438	TN	UNION	80.3	2467	MS	LAWRENCE	79.1	2496	AL	DALLAS	77.8
2439	TX	ERATH	80.3	2468	FL	NASSAU	79.0	2497	FL	CALHOUN	77.6
2440	VA	MATHEWS	80.3	2469	GA	COLUMBUS	79.0	2498	MS	HOLMES	77.5
2441	PA	PHILADELPHIA	80.3	2470	SC	BAMBERG	78.9	2499	FL	POLK	77.5
2442	VA	SURRY	80.2	2471	VA	WESTMORELAND	78.9	2500	FL	ALACHUA	77.5
2443	WA	SPOKANE	80.2	2472	TX	EASTLAND	78.8	2501	FL	MADISON	77.3
2444	TX	KING	80.1	2473	TX	YOUNG	78.8	2502	TN	GRAINGER	77.3
2445	AL	BIBB	80.1	2474	AL	CLAY	78.8	2503	KY	CRITTENDEN	77.3
2446	GA	TROUP	80.0	2475	WI	OUTAGAMIE	78.5	2504	GA	BERRIEN	77.3
2447	MS	JONES	80.0	2476	GA	WILCOX	78.5	2505	IN	FAYETTE	77.2
2448	NY	SCHUYLER	80.0	2477	OH	SHELBY	78.4	2506	MS	KEMPER	77.2
2449	SC	DILLON	79.9	2478	GA	EMANUEL	78.4	2507	TN	WILLIAMSON	77.1
2450	CA	SIERRA	79.9	2479	TX	FREESTONE	78.4	2508	AL	ELMORE	77.0
2451	AL	JACKSON	79.9	2480	TN	LAWRENCE	78.3	2509	MN	POPE	77.0
2452	GA	APPLING	79.9	2481	AL	LEE	78.3	2510	KY	LYON	77.0
2453	MI	OCEANA	79.8	2482	OH	PICKAWAY	78.3	2511	SC	ANDERSON	76.9
2454	GA	HOUSTON	79.8	2483	VA	HENRY	78.3	2512	VA	NORTHAMPTON	76.9
2455	NC	FORSYTH	79.6	2484	KY	LEE	78.2	2513	GA	BALDWIN	76.8
2456	MN	BELTRAMI	79.5	2485	GA	CRISP	78.2	2514	GA	BIBB	76.8
2457	MD	DORCHESTER	79.5	2486	MD	FREDERICK	78.1	2515	GA	CLARKE	76.8
2458	NC	WILSON	79.4	2487	FL	FLAGLER	78.1	2516	FL	GILCHRIST	76.7
2459	GA	CAMDEN	79.4	2488	GA	CHATTAHOOCHE	78.1	2517	NC	ROBESON	76.6
2460	AL	MARSHALL	79.3	2489	MD	WASHINGTON	78.0	2518	AL	HOUSTON	76.6
2461	KY	BARREN	79.3	2490	TN	HOUSTON	78.0	2519	PA	FRANKLIN	76.5
2462	MI	LUCE	79.2	2491	WI	DOOR	77.9	2520	TX	LEON	76.5
2463	GA	SUMTER	79.2	2492	TX	STEPHENS	77.9	2521	VA	HAMPTON	76.4
2464	TX	HAMILTON	79.2	2493	AL	RANDOLPH	77.9	2522	FL	MARION	76.4
2465	GA	STEWART	79.1	2494	AL	JEFFERSON	77.8	2523	GA	GILMER	76.4

COUNTIES OF HIGHEST AVERAGE FALLOUT FROM NUCLEAR TESTS:

RANGER BAKER-1 THROUGH STORAX SEDAN 1951-1962

(microCuries per square meter)

Each Row Includes:
County Rank (most fallout = 1), State, County, Fallout Level in microCuries/square meter

Rank	State	County	Level
2524	TN	WAYNE	76.4
2525	GA	WARREN	76.3
2526	SC	KERSHAW	76.2
2527	MS	WINSTON	76.2
2528	GA	WALKER	76.2
2529	WI	FLORENCE	76.2
2530	GA	MCDUFFIE	76.1
2531	GA	MADISON	76.0
2532	TN	LINCOLN	76.0
2533	PA	COLUMBIA	76.0
2534	IN	NOBLE	76.0
2535	MS	SIMPSON	75.9
2536	MN	OTTER TAIL	75.9
2537	WV	PRESTON	75.9
2538	AL	TALLAPOOSA	75.9
2539	GA	TALIAFERRO	75.8
2540	GA	CLINCH	75.7
2541	SC	CHESTERFIELD	75.7
2542	MD	CAROLINE	75.7
2543	GA	WORTH	75.6
2544	KY	LOGAN	75.6
2545	MI	DELTA	75.5
2546	TN	JACKSON	75.5
2547	GA	WILKINSON	75.5
2548	GA	BANKS	75.5
2549	GA	TURNER	75.5
2550	GA	GWINNETT	75.5
2551	GA	MURRAY	75.4
2552	VA	SOUTHAMPTON	75.4
2553	MN	BECKER	75.3
2554	SC	CHEROKEE	75.2
2555	GA	TIFT	75.2
2556	GA	JACKSON	75.2
2557	VA	SUSSEX	75.1
2558	TN	GILES	75.1
2559	GA	CHATTOOGA	75.1
2560	TX	HOWARD	75.1
2561	VA	YORK	75.0
2562	VA	JAMES CITY	75.0
2563	VA	GLOUCESTER	75.0
2564	MD	QUEEN ANNES	74.9
2565	TX	SHELBY	74.9
2566	NC	ROWAN	74.9
2567	NC	COLUMBUS	74.8
2568	GA	ATKINSON	74.8
2569	GA	GLASCOCK	74.8
2570	SC	YORK	74.7
2571	GA	WASHINGTON	74.7
2572	MI	IOSCO	74.6
2573	NC	HARNETT	74.5
2574	GA	SPALDING	74.4
2575	FL	PINELLAS	74.3
2576	GA	COOK	74.3
2577	SC	HAMPTON	74.2
2578	TX	BELL	74.2
2579	GA	TELFAIR	74.2
2580	GA	HARALSON	74.1
2581	TN	BENTON	73.9
2582	SC	LAURENS	73.9
2583	FL	LIBERTY	73.8
2584	KY	HOPKINS	73.8
2585	NC	LEE	73.8
2586	GA	PIKE	73.8
2587	NC	BRUNSWICK	73.8
2588	GA	COWETA	73.8
2589	WV	MORGAN	73.8
2590	MI	BAY	73.7
2591	TN	MARION	73.7
2592	GA	BEN HILL	73.7
2593	TX	BROWN	73.7
2594	TX	WILLIAMSON	73.7
2595	NY	BROOME	73.5
2596	AL	CALHOUN	73.5
2597	TX	THROCKMORTON	73.5
2598	VA	ESSEX	73.5
2599	MI	MARQUETTE	73.4
2600	CA	AMADOR	73.4
2601	KY	SIMPSON	73.4
2602	KY	LARUE	73.3
2603	FL	DIXIE	73.2
2604	GA	HANCOCK	73.2
2605	VA	RICHMOND	73.2
2606	GA	DOOLY	73.1
2607	TX	SHACKELFORD	73.1
2608	TN	WILSON	73.1
2609	KY	CLINTON	73.0
2610	NC	MONTGOMERY	73.0

COUNTIES OF HIGHEST AVERAGE FALLOUT FROM NUCLEAR TESTS:

RANGER BAKER-1 THROUGH STORAX SEDAN
1951-1962

(microCuries per square meter)

Each Row Includes:
County Rank (most fallout = 1), State, County, Fallout Level in microCuries/square meter

Rank	State	County	Fallout	Rank	State	County	Fallout	Rank	State	County	Fallout
2611	FL	ORANGE	73.0	2640	AL	BULLOCK	72.1	2669	AL	ST CLAIR	70.9
2612	NC	LINCOLN	73.0	2641	IN	BLACKFORD	72.1	2670	GA	BUTTS	70.8
2613	WI	RACINE	73.0	2642	MN	BIG STONE	72.1	2671	GA	LUMPKIN	70.8
2614	AL	FRANKLIN	72.9	2643	NC	GREENE	72.0	2672	FL	FRANKLIN	70.8
2615	AL	RUSSELL	72.8	2644	AL	BLOUNT	72.0	2673	FL	PUTNAM	70.8
2616	SD	GRANT	72.8	2645	TX	FALLS	71.9	2674	GA	JASPER	70.7
2617	SC	GEORGETOWN	72.7	2646	NC	UNION	71.9	2675	GA	HARRIS	70.7
2618	GA	IRWIN	72.7	2647	GA	HABERSHAM	71.9	2676	SC	CHESTER	70.7
2619	MS	JEFFERSON DA	72.7	2648	AL	LIMESTONE	71.9	2677	OH	DEFIANCE	70.6
2620	TN	FRANKLIN	72.7	2649	MN	STEVENS	71.9	2678	GA	DE KALB	70.6
2621	GA	WHITFIELD	72.7	2650	MN	HUBBARD	71.9	2679	MS	NESHOBA	70.6
2622	FL	GULF	72.7	2651	NC	BLADEN	71.7	2680	GA	DECATUR	70.6
2623	TN	SMITH	72.6	2652	GA	STEPHENS	71.6	2681	SC	GREENWOOD	70.5
2624	GA	HALL	72.6	2653	SC	MCCORMICK	71.6	2682	NC	EDGECOMBE	70.5
2625	GA	DODGE	72.6	2654	MS	NOXUBEE	71.6	2683	GA	JONES	70.4
2626	VA	HALIFAX	72.6	2655	NC	NASH	71.5	2684	TX	MCCULLOCH	70.4
2627	FL	CITRUS	72.5	2656	PA	MIFFLIN	71.4	2685	IN	WHITLEY	70.4
2628	FL	SEMINOLE	72.5	2657	NC	SAMPSON	71.3	2686	MN	POLK	70.3
2629	GA	THOMAS	72.4	2658	WI	BROWN	71.3	2687	AL	BUTLER	70.2
2630	GA	MERIWETHER	72.3	2659	AL	AUTAUGA	71.3	2688	NC	STANLY	70.1
2631	GA	CANDLER	72.3	2660	SC	NEWBERRY	71.2	2689	GA	TAYLOR	70.1
2632	TN	MOORE	72.3	2661	FL	WAKULLA	71.2	2690	MI	ALCONA	70.1
2633	NC	HALIFAX	72.3	2662	WV	JEFFERSON	71.2	2691	VA	LOUISA	70.0
2634	AL	COFFEE	72.3	2663	MI	ARENAC	71.1	2692	SC	BARNWELL	70.0
2635	GA	CHEROKEE	72.3	2664	GA	PUTNAM	71.1	2693	NC	STOKES	70.0
2636	GA	OCONEE	72.2	2665	TX	ROBERTSON	71.0	2694	MI	ISABELLA	70.0
2637	GA	OGLETHORPE	72.2	2666	VA	CUMBERLAND	71.0	2695	AL	CHEROKEE	70.0
2638	GA	CATOOSA	72.1	2667	OH	WILLIAMS	71.0	2696	KY	HARDIN	69.9
2639	AL	CHOCTAW	72.1	2668	GA	LOWNDES	70.9	2697	PA	LYCOMING	69.9

COUNTIES OF HIGHEST AVERAGE FALLOUT FROM NUCLEAR TESTS:

RANGER BAKER-1 THROUGH STORAX SEDAN 1951-1962

(microCuries per square meter)

Each Row Includes:
County Rank (most fallout = 1), State, County, Fallout Level in microCuries/square meter

Rank	State	County	Fallout		Rank	State	County	Fallout		Rank	State	County	Fallout
2698	GA	LAMAR	69.9		2727	SC	LANCASTER	68.6		2756	GA	TALBOT	67.1
2699	GA	HENRY	69.9		2728	GA	PICKENS	68.6		2757	WV	DODDRIDGE	67.1
2700	GA	WILKES	69.7		2729	WV	RITCHIE	68.6		2758	VA	FLUVANNA	67.0
2701	TX	COMANCHE	69.7		2730	GA	EARLY	68.5		2759	GA	BARROW	67.0
2702	NC	GASTON	69.7		2731	GA	CHARLTON	68.5		2760	MI	ROSCOMMON	66.9
2703	GA	HART	69.7		2732	NC	RICHMOND	68.5		2761	GA	ELBERT	66.9
2704	TN	CANNON	69.6		2733	TX	MILAM	68.3		2762	SC	FAIRFIELD	66.8
2705	TX	CONCHO	69.5		2734	TX	JONES	68.3		2763	GA	COFFEE	66.7
2706	GA	WALTON	69.5		2735	GA	DAWSON	68.3		2764	SC	SALUDA	66.6
2707	NC	SCOTLAND	69.5		2736	GA	NEWTON	68.2		2765	TX	CORYELL	66.5
2708	KY	ADAIR	69.4		2737	GA	FRANKLIN	68.2		2766	VA	CLARKE	66.5
2709	MN	TRAVERSE	69.4		2738	FL	BAY	68.1		2767	MI	CLARE	66.3
2710	GA	PIERCE	69.3		2739	GA	RICHMOND	68.1		2768	NC	DAVIE	66.3
2711	AL	CHILTON	69.3		2740	AL	COOSA	68.0		2769	WV	TAYLOR	66.2
2712	GA	BARTOW	69.3		2741	PA	JUNIATA	67.9		2770	VA	WYTHE	66.2
2713	NC	HOKE	69.2		2742	GA	FORSYTH	67.9		2771	PA	CUMBERLAND	66.2
2714	SC	UNION	69.1		2743	FL	GADSDEN	67.8		2772	AL	CRENSHAW	66.1
2715	FL	SUWANNEE	69.1		2744	KY	GREEN	67.8		2773	GA	COBB	66.1
2716	AL	WALKER	69.0		2745	GA	BROOKS	67.7		2774	FL	JEFFERSON	66.1
2717	IL	COOK	69.0		2746	SC	ABBEVILLE	67.7		2775	AL	MACON	66.0
2718	TX	BRAZOS	69.0		2747	MS	LOWNDES	67.6		2776	SC	EDGEFIELD	65.9
2719	MN	PENNINGTON	69.0		2748	GA	BRANTLEY	67.6		2777	AL	SHELBY	65.9
2720	MN	KITTSON	68.9		2749	WV	BARBOUR	67.5		2778	MD	CARROLL	65.9
2721	TX	NOLAN	68.8		2750	AL	FAYETTE	67.5		2779	GA	LINCOLN	65.9
2722	SC	ALLENDALE	68.8		2751	KY	TAYLOR	67.4		2780	MI	ANTRIM	65.8
2723	FL	HAMILTON	68.8		2752	MN	CLEARWATER	67.4		2781	FL	JACKSON	65.7
2724	MI	CHIPPEWA	68.8		2753	KY	ALLEN	67.4		2782	FL	WASHINGTON	65.6
2725	MI	CHARLEVOIX	68.7		2754	GA	DADE	67.3		2783	FL	WALTON	65.6
2726	MN	DOUGLAS	68.7		2755	AL	LAWRENCE	67.1		2784	FL	BAKER	65.6

COUNTIES OF HIGHEST AVERAGE FALLOUT FROM NUCLEAR TESTS:

RANGER BAKER-1 THROUGH STORAX SEDAN
1951-1962

(microCuries per square meter)

Each Row Includes:
County Rank (most fallout = 1), State, County, Fallout Level in microCuries/square meter

Rank	State	County	Fallout
2785	NC	ANSON	65.3
2786	TX	BURLESON	65.3
2787	TX	MIDLAND	65.2
2788	TX	MITCHELL	65.2
2789	GA	UPSON	65.2
2790	KY	MARION	65.2
2791	WV	BERKELEY	65.2
2792	NC	MECKLENBURG	65.1
2793	GA	ECHOLS	65.0
2794	MI	MACKINAC	65.0
2795	AL	MARION	64.9
2796	NC	CASWELL	64.9
2797	NC	JOHNSTON	64.8
2798	WV	HAMPSHIRE	64.7
2799	GA	FULTON	64.7
2800	TX	STERLING	64.7
2801	AL	LAMAR	64.5
2802	NV	NYE3	64.4
2803	TX	COLEMAN	64.2
2804	CA	GLENN	64.1
2805	GA	ROCKDALE	64.1
2806	TX	RUNNELS	64.0
2807	GA	COLUMBIA	64.0
2808	SC	LEXINGTON	63.9
2809	AL	WINSTON	63.9
2810	MN	ROSEAU	63.9
2811	TX	LAMPASAS	63.9
2812	TN	LEWIS	63.8
2813	GA	MARION	63.7
2814	MI	CRAWFORD	63.7
2815	CA	KERN	63.7
2816	MI	CHEBOYGAN	63.5
2817	GA	GRADY	63.4
2818	NC	TYRRELL	63.4
2819	PA	DAUPHIN	63.2
2820	MN	MAHNOMEN	63.2
2821	FL	LAFAYETTE	63.1
2822	NC	CABARRUS	63.0
2823	WV	ROANE	62.8
2824	KY	HART	62.8
2825	CA	PLACER	62.7
2826	IN	DE KALB	62.7
2827	MN	GRANT	62.7
2828	TX	SAN SABA	62.7
2829	VA	PAGE	62.6
2830	GA	POLK	62.6
2831	PA	PERRY	62.4
2832	TN	RUTHERFORD	62.3
2833	TX	LEE	62.3
2834	PA	LEBANON	62.1
2835	VA	BUCKINGHAM	62.0
2836	GA	WARE	62.0
2837	GA	LAURENS	62.0
2838	FL	COLUMBIA	62.0
2839	KY	GRAYSON	61.8
2840	WV	MINERAL	61.7
2841	MN	RED LAKE	61.7
2842	VA	CHARLOTTE	61.4
2843	FL	TAYLOR	61.2
2844	MI	GLADWIN	61.2
2845	NC	ROCKINGHAM	61.2
2846	GA	TREUTLEN	61.1
2847	GA	MACON	61.0
2848	FL	CLAY	61.0
2849	AL	MONTGOMERY	61.0
2850	NC	FRANKLIN	61.0
2851	FL	OSCEOLA	60.9
2852	FL	PASCO	60.8
2853	SC	AIKEN	60.8
2854	SC	RICHLAND	60.7
2855	TX	MASON	60.7
2856	WI	MILWAUKEE	60.5
2857	FL	UNION	60.4
2858	CA	VENTURA	60.2
2859	GA	HEARD	60.1
2860	MN	LAKE OF THE	59.9
2861	TN	MACON	59.7
2862	TX	PRESIDIO	59.4
2863	KY	NELSON	59.4
2864	MI	MISSAUKEE	59.2
2865	MI	MECOSTA	59.2
2866	MN	WILKIN	58.8
2867	CA	EL DORADO	58.7
2868	TX	LLANO	58.6
2869	TN	TROUSDALE	58.5
2870	VA	BLAND	58.4
2871	AL	HENRY	58.2

U.S. FALLOUT ATLAS : TOTAL FALLOUT

COUNTIES OF HIGHEST AVERAGE FALLOUT FROM NUCLEAR TESTS:

RANGER BAKER-1 THROUGH STORAX SEDAN 1951-1962

(microCuries per square meter)

Each Row Includes:
County Rank (most fallout = 1), State, County, Fallout Level in microCuries/square meter

Rank	State	County	Level		Rank	State	County	Level		Rank	State	County	Level
2872	FL	COLLIER	58.0		2901	TX	IRION	54.4		2930	FL	GLADES	49.8
2873	TX	BLANCO	57.9		2902	MI	WEXFORD	54.4		2931	CA	SANTA CRUZ	49.1
2874	CA	CALAVERAS	57.9		2903	MI	OSCEOLA	54.3		2932	FL	MARTIN	49.0
2875	CA	YUBA	57.8		2904	MI	GRAND TRAVER	54.3		2933	CA	SAN LUIS OBI	48.9
2876	CA	TUOLUMNE	57.6		2905	FL	MONROE	54.3		2934	WA	STEVENS	48.9
2877	GA	MILLER	57.4		2906	TX	MENARD	54.0		2935	TX	KERR	48.4
2878	TX	BURNET	57.3		2907	FL	ST LUCIE	54.0		2936	FL	OKEECHOBEE	48.2
2879	MI	LAKE	57.3		2908	MI	MANISTEE	53.9		2937	TX	SCHLEICHER	48.2
2880	FL	SARASOTA	57.3		2909	TX	GILLESPIE	53.7		2938	TX	TRAVIS	48.0
2881	TX	TOM GREEN	57.3		2910	CA	LASSEN	53.7		2939	MI	MONTMORENCY	48.0
2882	WA	FERRY	57.2		2911	CA	TEHAMA	53.7		2940	TX	KENDALL	47.9
2883	TX	MILLS	57.1		2912	VA	ALBEMARLE	53.6		2941	CA	TULARE	47.7
2884	TX	COKE	57.1		2913	FL	MANATEE	53.1		2942	FL	INDIAN RIVER	47.5
2885	MN	MARSHALL	57.0		2914	CA	MONTEREY	52.9		2943	MI	OSCODA	47.4
2886	CA	PLUMAS	56.9		2915	TX	PECOS	52.6		2944	TX	BREWSTER	47.3
2887	MI	EMMET	56.9		2916	MI	OGEMAW	52.4		2945	MI	KALKASKA	47.1
2888	GA	SEMINOLE	56.4		2917	CA	SONOMA	51.5		2946	MI	LEELANAU	47.0
2889	MI	MASON	56.2		2918	FL	BROWARD	51.3		2947	TX	FRIO	46.8
2890	FL	HENDRY	55.9		2919	TX	COLORADO	51.2		2948	TX	MATAGORDA	46.7
2891	FL	PALM BEACH	55.8		2920	TX	WHARTON	51.1		2949	TX	CROCKETT	46.6
2892	MN	NORMAN	55.5		2921	MI	BENZIE	51.1		2950	CA	SAN DIEGO	46.5
2893	MI	IONIA	55.4		2922	TX	BANDERA	51.1		2951	CA	KINGS	46.3
2894	TX	REAGAN	55.2		2923	CA	BUTTE	50.6		2952	TX	TAYLOR	46.2
2895	TX	CRANE	55.0		2924	FL	DADE	50.4		2953	TX	COMAL	45.8
2896	TX	FAYETTE	55.0		2925	FL	LEE	50.3		2954	TX	UPTON	45.8
2897	MI	OTSEGO	54.9		2926	MI	PRESQUE ISLE	50.3		2955	FL	CHARLOTTE	45.5
2898	MN	CLAY	54.7		2927	CA	SANTA BARBAR	50.2		2956	CA	MODOC	45.2
2899	TX	GLASSCOCK	54.7		2928	TX	MEDINA	50.1		2957	FL	LEVY	44.9
2900	FL	BRADFORD	54.4		2929	FL	HIGHLANDS	49.9		2958	CA	SAN BENITO	44.8

COUNTIES OF HIGHEST AVERAGE FALLOUT FROM NUCLEAR TESTS:

RANGER BAKER-1 THROUGH STORAX SEDAN
1951-1962

(microCuries per square meter)

Each Row Includes:
County Rank (most fallout = 1), State, County, Fallout Level in microCuries/square meter

Rank	State	County	Fallout
2959	CA	MARIPOSA	44.7
2960	FL	DUVAL	44.6
2961	TX	KIMBLE	44.6
2962	CA	FRESNO	44.5
2963	WA	OKANOGAN	44.1
2964	MI	ALPENA	44.0
2965	TX	GONZALES	43.9
2966	TX	CALDWELL	43.9
2967	FL	DE SOTO	43.7
2968	TX	BASTROP	43.6
2969	TX	JIM WELLS	43.3
2970	CA	ORANGE	43.1
2971	TX	JACKSON	42.9
2972	TX	LAVACA	42.6
2973	TX	DIMMIT	42.3
2974	TX	BEXAR	42.0
2975	TX	DE WITT	41.9
2976	TX	HAYS	41.6
2977	FL	HARDEE	40.5
2978	TX	CALHOUN	40.2
2979	TX	JIM HOGG	40.0
2980	TX	BEE	40.0
2981	TX	UVALDE	39.8
2982	CA	MARIN	39.6
2983	TX	ZAVALA	39.0
2984	CA	LAKE	38.8
2985	TX	CAMERON	38.8
2986	CA	MADERA	38.6
2987	TX	SUTTON	38.2
2988	TX	WILLACY	38.1
2989	TX	STARR	37.9
2990	TX	ARANSAS	37.5
2991	CA	SACRAMENTO	37.3
2992	TX	MCMULLEN	37.1
2993	TX	KLEBERG	36.9
2994	CA	COLUSA	36.9
2995	OR	LAKE	36.5
2996	TX	KARNES	36.1
2997	TX	TERRELL	35.9
2998	TX	REAL	35.9
2999	TX	BROOKS	35.9
3000	TX	LIVE OAK	35.3
3001	TX	MAVERICK	35.3
3002	TX	EDWARDS	35.2
3003	TX	SAN PATRICIO	34.8
3004	CA	SANTA CLARA	34.4
3005	TX	ZAPATA	33.8
3006	TX	LA SALLE	33.6
3007	TX	VICTORIA	33.5
3008	OR	CLACKAMAS	33.4
3009	TX	ATASCOSA	33.4
3010	TX	KENEDY	33.4
3011	CA	SAN MATEO	33.3
3012	TX	HIDALGO	33.2
3013	TX	WILSON	33.0
3014	TX	GOLIAD	32.6
3015	TX	KINNEY	32.6
3016	TX	DUVAL	32.6
3017	WA	SKAMANIA	32.4
3018	CA	NAPA	32.0
3019	CA	SISKIYOU	32.0
3020	TX	GUADALUPE	31.9
3021	OR	LINCOLN	31.8
3022	TX	NUECES	31.7
3023	TX	WEBB	31.6
3024	OR	CROOK	31.5
3025	CA	TRINITY	31.5
3026	WA	DOUGLAS	31.5
3027	CA	DEL NORTE	31.3
3028	OR	MARION	31.1
3029	CA	MERCED	31.1
3030	OR	CURRY	31.1
3031	CA	STANISLAUS	31.1
3032	CA	SUTTER	31.1
3033	CA	RIVERSIDE	31.1
3034	OR	LANE	30.9
3035	CA	YOLO	30.9
3036	CA	ALAMEDA	30.9
3037	OR	KLAMATH	30.8
3038	CA	CONTRA COSTA	30.7
3039	OR	MORROW	30.5
3040	CA	IMPERIAL	30.5
3041	OR	CLATSOP	30.4
3042	OR	TILLAMOOK	30.1
3043	OR	DOUGLAS	29.8
3044	WA	PIERCE	29.8
3045	WA	SNOHOMISH	29.8

U.S. FALLOUT ATLAS : TOTAL FALLOUT

COUNTIES OF HIGHEST AVERAGE FALLOUT FROM NUCLEAR TESTS:

RANGER BAKER-1 THROUGH STORAX SEDAN
1951-1962

(microCuries per square meter)

Each Row Includes:
County Rank (most fallout = 1), State, County, Fallout Level in microCuries/square meter

Rank	State	County	Fallout
3046	CA	SAN JOAQUIN	29.4
3047	CA	HUMBOLDT	29.0
3048	OR	LINN	28.9
3049	OR	JOSEPHINE	28.6
3050	WA	LINCOLN	28.3
3051	CA	SHASTA	27.7
3052	WA	COWLITZ	27.4
3053	CA	SOLANO	27.4
3054	CA	MENDOCINO	27.3
3055	WA	CHELAN	27.3
3056	OR	COOS	27.3
3057	OR	WHEELER	26.9
3058	WA	KITTITAS	26.5
3059	WA	WAHKIAKUM	26.5
3060	OR	COLUMBIA	25.7
3061	OR	JEFFERSON	25.6
3062	TX	VAL VERDE	25.6
3063	WA	KING	25.1
3064	WA	ADAMS	24.6
3065	WA	CLARK	24.6
3066	OR	HOOD RIVER	24.6
3067	CA	SAN FRANCISC	24.3
3068	WA	GRAYS HARBOR	24.2
3069	WA	WHATCOM	24.1
3070	WA	BENTON	23.8
3071	WA	PACIFIC	23.8
3072	OR	GILLIAM	23.8
3073	WA	GRANT	23.5
3074	WA	LEWIS	23.4
3075	OR	DESCHUTES	23.2
3076	WA	JEFFERSON	23.0
3077	OR	MULTNOMAH	23.0
3078	WA	SKAGIT	22.9

COUNTIES OF HIGHEST AVERAGE FALLOUT FROM NUCLEAR TESTS:

RANGER BAKER-1 THROUGH STORAX SEDAN
1951-1962

(microCuries per square meter)

Each Row Includes:
County Rank (most fallout = 1), State, County, Fallout Level in microCuries/square meter

SIXTEEN COUNTIES WITH THE LOWEST ESTIMATED AVERAGE TOTAL FALLOUT 1951-1962

3079	WA	YAKIMA	22.6
3080	OR	SHERMAN	22.6
3081	OR	WASHINGTON	22.3
3082	OR	WASCO	22.3
3083	OR	POLK	22.1
3084	OR	JACKSON	21.8
3085	OR	BENTON	21.7
3086	WA	CLALLAM	21.2
3087	WA	FRANKLIN	21.2
3088	WA	KLICKITAT	21.2
3089	OR	YAMHILL	20.5
3090	WA	MASON	20.3
3091	WA	KITSAP	19.4
3092	WA	THURSTON	18.2
3093	WA	SAN JUAN	14.3
3094	WA	ISLAND	13.0

SECTION 9

TOTAL FALLOUT ACTIVITY BY SHOT SERIES

1951-1962

SECTION 9

TOTAL FALLOUT ACTIVITY BY SHOT SERIES

1951-1962

TOTAL FALLOUT ACTIVITY (µCi/sq meter) by SHOT SERIES WITH COUNTY RANK

COUNTY AND STATE	RANGER SERIES	RANGER RANK	B-J SERIES	B-J RANK	T-S SERIES	T-S RANK	U-K SERIES	UK RANK	TEAPOT SERIES	TP RANK	PLUMBBOB SERIES	PB RANK	SHOT SEDAN	SEDAN RANK
	1951		1951		1952		1953		1955		1957		1962	
AUTAUGA AL	0.00	0	3.25	1353	17.69	1897	11.18	2596	15.36	1396	19.75	2278	0.00	0
BALDWIN AL	0.00	0	4.79	1044	11.87	2374	17.57	2018	10.18	1897	49.13	830	0.00	0
BARBOUR AL	0.00	0	7.14	833	19.59	1765	9.52	2794	8.60	2045	33.32	1248	0.00	0
BIBB AL	0.00	0	2.68	1452	21.88	1667	11.98	2513	17.55	1201	22.23	2029	0.00	0
BLOUNT AL	0.00	0	2.77	1432	18.98	1808	12.32	2481	15.21	1417	17.64	2488	0.00	0
BULLOCK AL	0.00	0	9.99	659	12.48	2310	9.86	2762	8.96	2018	26.74	1616	0.00	0
BUTLER AL	0.00	0	6.01	918	17.78	1891	10.25	2712	12.92	1624	19.14	2338	0.00	0
CALHOUN AL	0.00	0	12.16	526	12.05	2352	10.46	2691	9.46	1971	24.11	1831	0.00	0
CHAMBERS AL	0.00	0	5.11	1000	26.40	1502	9.35	2821	10.54	1866	23.50	1889	0.00	0
CHEROKEE AL	0.00	0	12.53	507	5.96	2952	10.56	2680	10.68	1850	25.66	1709	0.00	0
CHILTON AL	0.00	0	3.62	1282	17.47	1907	10.78	2652	11.02	1818	21.63	2087	0.00	0
CHOCTAW AL	0.00	0	2.34	1525	2.23	3080	12.12	2498	21.63	880	25.92	1684	0.00	0
CLARKE AL	0.00	0	4.14	1184	22.11	1656	10.43	2694	23.33	786	30.78	1382	0.00	0
CLAY AL	0.00	0	3.61	1285	19.48	1777	10.19	2726	11.19	1802	29.97	1425	0.00	0
CLEBURNE AL	0.00	0	5.96	928	19.89	1752	10.79	2650	9.52	1967	26.30	1649	0.00	0
COFFEE AL	0.00	0	12.26	519	17.45	1910	10.08	2740	10.10	1903	19.13	2339	0.00	0
COLBERT AL	0.00	0	1.86	1679	13.52	2206	28.72	1349	16.60	1293	22.99	1940	0.00	0
CONECUH AL	0.00	0	4.42	1117	31.25	1364	11.57	2546	18.26	1123	59.72	648	0.00	0
COOSA AL	0.00	0	3.11	1375	17.33	1919	10.15	2731	9.97	1920	23.27	1910	0.00	0
COVINGTON AL	0.00	0	13.32	475	22.64	1627	10.88	2639	10.52	1868	19.53	2305	0.00	0
CRENSHAW AL	0.00	0	5.20	986	13.60	2196	9.39	2817	14.78	1453	19.17	2333	0.00	0
CULLMAN AL	0.00	0	9.03	710	22.40	1636	14.32	2302	10.92	1830	19.41	2318	0.00	0
DALE AL	0.00	0	11.20	586	22.07	1660	9.39	2815	9.85	1935	26.45	1639	0.00	0
DALLAS AL	0.00	0	3.63	1280	21.46	1693	13.93	2338	13.71	1558	20.46	2205	0.00	0
DE KALB AL	0.00	0	10.65	615	11.98	2361	12.91	2434	14.28	1508	26.13	1663	0.00	0
ELMORE AL	0.00	0	4.54	1096	22.31	1643	10.43	2693	8.44	2064	26.13	1665	0.00	0

TOTAL FALLOUT ACTIVITY (µCi/sq meter) by SHOT SERIES WITH COUNTY RANK

COUNTY AND STATE	RANGER SERIES	RANGER RANK	B-J SERIES	B-J RANK	T-S SERIES	T-S RANK	U-K SERIES	UK RANK	TEAPOT SERIES	TP RANK	PLUMBBOB SERIES	PB RANK	SHOT SEDAN	SEDAN RANK
	1951		1951		1952		1953		1955		1957		1962	
ESCAMBIA AL	0.00	0	5.99	921	13.26	2231	10.62	2670	13.36	1581	42.07	991	0.00	0
ETOWAH AL	0.00	0	11.89	543	12.75	2281	12.13	2497	15.70	1370	25.43	1729	0.00	0
FAYETTE AL	0.00	0	2.46	1498	11.75	2388	15.39	2193	12.15	1702	22.85	1955	0.00	0
FRANKLIN AL	0.00	0	1.72	1745	11.44	2426	18.86	1919	14.26	1509	22.29	2026	0.00	0
GENEVA AL	0.47	475	19.55	276	32.69	1322	9.76	2772	9.84	1937	23.60	1885	0.00	0
GREENE AL	0.00	0	8.66	734	31.45	1359	13.58	2373	24.24	744	21.90	2061	0.00	0
HALE AL	0.00	0	2.99	1393	26.43	1501	10.84	2644	25.49	677	19.50	2309	0.00	0
HENRY AL	0.27	506	3.10	1378	7.15	2838	7.95	2912	6.66	2325	27.64	1550	0.00	0
HOUSTON AL	0.27	507	8.86	722	12.56	2303	10.05	2744	6.85	2295	31.19	1349	0.00	0
JACKSON AL	0.00	0	10.23	642	12.15	2341	13.48	2383	13.74	1553	22.99	1938	0.00	0
JEFFERSON AL	0.00	0	3.13	1372	17.81	1887	14.93	2232	12.82	1634	23.02	1937	0.00	0
LAMAR AL	0.00	0	1.85	1683	9.68	2595	14.50	2283	5.99	2454	26.63	1624	0.00	0
LAUDERDALE AL	0.00	0	1.34	1921	8.50	2718	25.26	1507	15.71	1369	26.51	1632	0.00	0
LAWRENCE AL	0.00	0	1.97	1627	10.49	2527	18.10	1973	13.89	1539	16.15	2610	0.00	0
LEE AL	0.29	494	5.54	961	18.86	1817	10.01	2747	7.40	2202	29.86	1433	0.00	0
LIMESTONE AL	0.00	0	1.51	1838	12.83	2275	16.74	2082	11.34	1781	21.89	2062	0.00	0
LOWNDES AL	0.00	0	4.36	1128	33.19	1309	10.62	2671	12.75	1644	21.12	2128	0.00	0
MACON AL	0.00	0	5.22	983	17.36	1916	10.06	2742	8.06	2116	22.18	2035	0.00	0
MADISON AL	0.00	0	8.52	743	19.35	1784	17.83	1997	12.77	1642	20.10	2240	0.00	0
MARENGO AL	0.00	0	3.27	1346	22.43	1631	15.96	2138	26.03	654	86.83	337	0.00	0
MARION AL	0.00	0	1.57	1812	6.31	2914	16.73	2083	9.69	1956	26.88	1605	0.00	0
MARSHALL AL	0.00	0	11.66	555	9.79	2587	14.18	2315	14.61	1475	22.22	2030	0.00	0
MOBILE AL	0.00	0	4.12	1192	3.47	3053	52.48	613	7.55	2176	88.99	327	0.00	0
MONROE AL	0.00	0	4.85	1037	7.90	2765	11.26	2584	20.79	931	64.16	579	0.00	0
MONTGOMERY AL	0.00	0	3.38	1327	8.73	2684	10.74	2657	10.08	1908	23.24	1913	0.00	0
MORGAN AL	0.00	0	8.48	745	15.21	2054	16.24	2117	10.08	1909	24.35	1819	0.00	0

TOTAL FALLOUT ACTIVITY (µCi/sq meter) by SHOT SERIES WITH COUNTY RANK

COUNTY AND STATE	RANGER SERIES	RANGER RANK	B-J SERIES	B-J RANK	T-S SERIES	T-S RANK	U-K SERIES	UK RANK	TEAPOT SERIES	TP RANK	PLUMBBOB SERIES	PB RANK	SHOT SEDAN	SEDAN RANK
	1951		1951		1952		1953		1955		1957		1962	
PERRY AL	0.00	0	2.96	1395	19.24	1790	11.48	2560	23.34	785	23.65	1881	0.00	0
PICKENS AL	0.00	0	1.93	1652	10.62	2509	29.44	1311	20.21	967	19.54	2304	0.00	0
PIKE AL	0.00	0	7.00	837	24.98	1536	9.85	2763	12.44	1668	25.21	1746	0.00	0
RANDOLPH AL	0.00	0	7.40	817	13.15	2245	11.20	2595	7.11	2257	33.35	1244	0.00	0
RUSSELL AL	0.29	495	4.06	1202	17.42	1913	8.86	2863	8.52	2052	27.54	1558	0.00	0
SHELBY AL	0.00	0	4.91	1022	11.64	2402	11.41	2567	11.00	1820	22.43	2009	0.00	0
ST CLAIR AL	0.00	0	3.01	1389	17.89	1879	11.44	2565	14.66	1464	18.95	2356	0.00	0
SUMTER AL	0.00	0	2.66	1454	21.84	1670	11.49	2556	19.46	1023	23.11	1930	0.00	0
TALLADEGA AL	0.00	0	3.68	1272	26.80	1488	11.00	2622	11.47	1769	23.95	1849	0.00	0
TALLAPOOSA AL	0.00	0	4.61	1081	16.63	1958	10.20	2722	10.05	1911	28.78	1492	0.00	0
TUSCALOOSA AL	0.00	0	2.64	1456	11.12	2469	20.94	1775	27.96	598	26.04	1677	0.00	0
WALKER AL	0.00	0	2.51	1485	12.95	2265	15.47	2184	11.10	1812	21.85	2066	0.00	0
WASHINGTON AL	0.00	0	2.58	1468	11.62	2404	11.05	2611	18.85	1061	41.28	1005	0.00	0
WILCOX AL	0.00	0	3.64	1278	31.19	1367	11.51	2555	20.29	961	43.36	964	0.00	0
WINSTON AL	0.00	0	2.09	1586	11.56	2410	17.00	2056	11.66	1748	18.59	2384	0.00	0
ARKANSAS AR	0.00	0	4.92	1020	14.24	2137	76.80	323	45.59	349	32.33	1287	0.00	0
ASHLEY AR	0.00	0	4.44	1111	13.52	2204	40.38	881	21.98	863	26.09	1672	0.00	0
BAXTER AR	0.00	0	4.32	1143	35.30	1245	65.57	445	52.92	292	39.04	1046	0.00	0
BENTON AR	0.00	0	4.24	1155	68.10	650	35.40	1058	47.92	331	55.90	694	0.00	0
BOONE AR	0.00	0	4.35	1130	38.71	1156	37.36	998	56.35	275	33.37	1242	0.00	0
BRADLEY AR	0.00	0	6.61	872	21.50	1688	40.63	872	23.45	776	27.88	1537	0.00	0
CALHOUN AR	0.00	0	1.56	1820	8.23	2745	34.96	1078	18.30	1119	21.20	2117	0.00	0
CARROLL AR	0.00	0	5.10	1002	37.38	1191	39.94	898	40.32	410	41.72	999	0.00	0
CHICOT AR	0.00	0	3.79	1250	5.97	2951	45.01	747	21.54	888	17.14	2529	0.00	0
CLARK AR	0.00	0	5.79	937	45.00	1020	27.02	1429	45.21	352	38.76	1049	0.00	0
CLAY AR	0.00	0	4.71	1064	18.14	1856	53.23	601	45.75	348	41.17	1007	0.00	0

TOTAL FALLOUT ACTIVITY (µCi/sq meter) by SHOT SERIES WITH COUNTY RANK

COUNTY AND STATE	RANGER SERIES	RANGER RANK	B-J SERIES	B-J RANK	T-S SERIES	T-S RANK	U-K SERIES	UK RANK	TEAPOT SERIES	TP RANK	PLUMBBOB SERIES	PB RANK	SHOT SEDAN	SEDAN RANK
	1951		1951		1952		1953		1955		1957		1962	
CLEBURNE AR	0.00	0	4.43	1115	17.94	1876	78.00	310	61.29	252	35.36	1144	0.00	0
CLEVELAND AR	0.00	0	3.92	1227	14.80	2086	38.46	960	35.50	463	27.01	1597	0.00	0
COLUMBIA AR	0.00	0	4.76	1054	43.03	1056	29.18	1325	36.67	450	24.06	1834	0.00	0
CONWAY AR	0.00	0	5.84	934	30.95	1377	44.80	752	60.95	256	27.82	1539	0.00	0
CRAIGHEAD AR	0.00	0	4.29	1151	19.37	1782	94.38	225	32.70	504	42.37	985	0.00	0
CRAWFORD AR	0.00	0	3.23	1356	62.01	740	30.29	1270	41.47	398	46.80	882	0.00	0
CRITTENDEN AR	0.00	0	3.69	1271	12.35	2325	91.20	242	19.84	993	19.50	2308	0.00	0
CROSS AR	0.00	0	4.09	1199	17.01	1939	90.87	244	17.43	1210	18.07	2436	0.00	0
DALLAS AR	0.00	0	5.55	959	54.86	858	32.44	1185	20.20	969	35.14	1149	0.00	0
DESHA AR	0.00	0	3.97	1218	22.11	1658	73.85	352	23.63	769	25.08	1765	0.00	0
DREW AR	0.00	0	2.80	1428	20.59	1722	51.95	624	23.50	774	18.49	2402	0.00	0
FAULKNER AR	0.00	0	4.39	1122	35.08	1254	76.10	326	61.71	249	29.80	1438	0.00	0
FRANKLIN AR	0.00	0	4.23	1160	64.75	696	39.33	922	53.70	288	30.14	1416	0.00	0
FULTON AR	0.00	0	3.13	1370	24.99	1534	93.85	228	47.48	335	47.09	874	0.00	0
GARLAND AR	0.00	0	6.63	870	39.37	1135	27.91	1392	41.85	392	30.95	1368	0.00	0
GRANT AR	0.00	0	5.20	989	44.72	1026	35.95	1036	47.40	337	32.20	1292	0.00	0
GREENE AR	0.00	0	5.33	977	17.62	1901	71.60	375	43.27	379	36.52	1109	0.00	0
HEMPSTEAD AR	0.00	0	3.32	1341	31.99	1341	21.60	1724	51.86	302	25.98	1682	0.00	0
HOT SPRING AR	0.00	0	5.50	964	38.87	1149	28.41	1365	36.84	449	37.77	1073	0.00	0
HOWARD AR	0.00	0	5.13	996	43.48	1047	24.24	1567	44.34	364	43.47	960	0.00	0
INDEPENDENCE AR	0.00	0	4.88	1028	18.27	1847	74.75	340	44.92	358	44.21	939	0.00	0
IZARD AR	0.00	0	4.74	1056	19.09	1801	64.37	459	63.63	243	42.87	977	0.00	0
JACKSON AR	0.00	0	4.87	1034	23.12	1609	88.11	260	42.14	388	42.71	979	0.00	0
JEFFERSON AR	0.00	0	3.78	1255	21.84	1671	86.05	265	40.36	409	29.57	1451	0.00	0
JOHNSON AR	0.00	0	4.88	1029	53.58	875	33.35	1144	61.53	250	37.72	1075	0.00	0
LAFAYETTE AR	0.00	0	5.06	1004	44.16	1035	22.45	1668	56.39	274	26.15	1662	0.00	0

TOTAL FALLOUT ACTIVITY (µCi/sq meter) by SHOT SERIES WITH COUNTY RANK

COUNTY AND STATE	RANGER SERIES	RANGER RANK	B-J SERIES	B-J RANK	T-S SERIES	T-S RANK	U-K SERIES	UK RANK	TEAPOT SERIES	TP RANK	PLUMBBOB SERIES	PB RANK	SHOT SEDAN	SEDAN RANK
	1951		1951		1952		1953		1955		1957		1962	
LAWRENCE AR	0.00	0	5.35	976	19.61	1762	83.15	277	54.16	285	42.13	987	0.00	0
LEE AR	0.00	0	3.49	1306	13.50	2210	85.89	266	16.26	1322	22.29	2023	0.00	0
LINCOLN AR	0.00	0	2.91	1407	21.35	1698	60.50	495	36.59	451	20.91	2157	0.00	0
LITTLE RIVER AR	0.00	0	3.08	1380	48.16	963	21.10	1757	61.53	251	35.57	1135	0.00	0
LOGAN AR	0.00	0	4.52	1097	38.13	1170	41.76	838	48.78	323	37.81	1072	0.00	0
LONOKE AR	0.00	0	3.46	1314	20.84	1715	73.52	355	48.26	329	27.08	1593	0.00	0
MADISON AR	0.00	0	4.88	1030	61.77	745	48.87	665	48.74	324	41.87	994	0.00	0
MARION AR	0.00	0	4.45	1108	33.30	1304	61.19	488	57.60	269	32.59	1278	0.00	0
MILLER AR	0.00	0	2.39	1508	48.47	953	21.56	1727	47.14	339	30.46	1399	0.00	0
MISSISSIPPI AR	0.00	0	3.82	1243	25.09	1526	80.80	288	19.33	1033	25.69	1706	0.00	0
MONROE AR	0.00	0	3.48	1308	16.00	1996	92.09	240	21.02	917	22.45	2005	0.00	0
MONTGOMERY AR	0.00	0	5.69	947	60.53	765	25.71	1486	48.93	322	36.62	1108	0.00	0
NEVADA AR	0.00	0	5.18	990	38.52	1161	24.95	1522	44.09	369	32.78	1273	0.00	0
NEWTON AR	0.00	0	6.89	849	53.93	871	42.17	824	60.96	255	33.13	1259	0.00	0
OUACHITA AR	0.00	0	5.64	951	43.49	1045	26.55	1448	23.45	775	33.73	1221	0.00	0
PERRY AR	0.00	0	4.56	1086	47.34	978	41.35	846	75.57	215	33.73	1220	0.00	0
PHILLIPS AR	0.00	0	3.22	1359	24.98	1537	75.54	332	18.36	1111	26.06	1676	0.00	0
PIKE AR	0.00	0	5.15	992	34.31	1279	25.08	1518	48.39	328	36.41	1112	0.00	0
POINSETT AR	0.00	0	4.28	1153	13.26	2230	79.82	297	22.65	815	20.83	2163	0.00	0
POLK AR	0.00	0	5.73	942	35.90	1233	38.69	950	58.93	261	42.89	976	0.00	0
POPE AR	0.00	0	5.96	927	48.18	962	39.50	915	70.19	231	33.96	1212	0.00	0
PRAIRIE AR	0.00	0	4.79	1045	13.88	2172	93.23	232	49.51	317	43.20	967	0.00	0
PULASKI AR	0.00	0	5.65	949	30.38	1400	56.16	552	54.24	284	25.86	1691	0.00	0
RANDOLPH AR	0.00	0	5.10	1001	23.55	1588	78.53	307	20.61	944	46.91	877	0.00	0
SALINE AR	0.00	0	7.33	823	31.53	1356	33.67	1124	72.82	219	34.34	1188	0.00	0
SCOTT AR	0.00	0	4.37	1126	33.98	1285	35.64	1046	43.97	372	44.95	924	0.00	0

TOTAL FALLOUT ACTIVITY (µCi/sq meter) by SHOT SERIES WITH COUNTY RANK

COUNTY AND STATE	RANGER SERIES	RANGER RANK	B-J SERIES	B-J RANK	T-S SERIES	T-S RANK	U-K SERIES	UK RANK	TEAPOT SERIES	TP RANK	PLUMBBOB SERIES	PB RANK	SHOT SEDAN	SEDAN RANK
	1951		1951		1952		1953		1955		1957		1962	
SEARCY AR	0.00	0	6.64	869	31.60	1354	43.43	787	56.68	273	34.01	1210	0.00	0
SEBASTIAN AR	0.00	0	4.12	1189	44.29	1032	25.98	1475	57.16	271	39.72	1035	0.00	0
SEVIER AR	0.00	0	5.13	997	39.14	1143	24.13	1576	70.61	229	34.71	1165	0.00	0
SHARP AR	0.00	0	5.73	941	21.29	1700	77.41	317	63.36	244	42.38	984	0.00	0
ST FRANCIS AR	0.00	0	3.59	1288	13.27	2228	95.64	218	23.24	790	25.72	1703	0.00	0
STONE AR	0.00	0	4.78	1050	27.19	1481	66.01	440	37.78	436	27.27	1574	0.00	0
UNION AR	0.00	0	4.03	1208	20.22	1734	52.05	620	22.88	809	29.13	1474	0.00	0
VAN BUREN AR	0.00	0	6.12	910	35.62	1239	54.79	574	67.05	235	36.40	1113	0.00	0
WASHINGTON AR	0.00	0	4.88	1031	62.05	739	44.32	766	45.29	351	46.10	908	0.00	0
WHITE AR	0.00	0	5.27	980	15.54	2029	77.95	311	57.71	268	30.36	1403	0.00	0
WOODRUFF AR	0.00	0	5.02	1008	14.69	2100	73.07	360	52.19	299	34.19	1199	0.00	0
YELL AR	0.00	0	4.31	1146	47.69	973	38.94	936	46.51	344	36.22	1118	0.00	0
APACHE AZ	0.00	0	0.00	0	108.68	342	1205.05	15	72.47	222	69.85	494	0.00	0
COCHISE AZ	0.00	0	0.00	0	72.52	594	6.22	2979	8.20	2099	68.63	514	0.00	0
COCONINO1 AZ	0.00	0	0.00	0	86.60	451	1758.60	9	160.23	79	47.16	873	0.00	0
COCONINO2 AZ	0.00	0	0.00	0	88.20	434	1418.59	12	81.30	201	49.52	824	0.00	0
COCONINO3 AZ	0.00	0	0.00	0	90.01	415	202.15	111	62.24	247	49.52	825	0.00	0
GILA AZ	0.00	0	0.00	0	74.62	563	20.58	1803	20.62	943	60.06	642	0.00	0
GRAHAM AZ	0.00	0	0.00	0	79.35	512	10.91	2636	13.25	1590	68.63	513	0.00	0
GREENLEE AZ	0.00	0	0.00	0	80.23	505	6.22	2978	10.43	1878	61.59	621	0.00	0
MARICOPA AZ	0.00	0	0.00	0	61.56	746	27.33	1417	18.95	1054	64.03	582	0.00	0
MOHAVE1 AZ	0.00	0	0.00	0	96.89	375	1266.30	13	175.35	66	52.91	757	0.00	0
MOHAVE2 AZ	0.00	0	0.00	0	111.89	337	2221.97	7	391.23	18	265.46	31	0.00	0
MOHAVE3 AZ	0.00	0	0.00	0	252.87	230	215.95	105	116.44	136	17.95	2448	0.00	0
MOHAVE4 AZ	0.00	0	0.00	0	252.87	231	215.95	106	116.44	137	17.95	2449	0.00	0
NAVAJO AZ	0.00	0	0.00	0	91.53	404	741.96	22	61.10	254	63.29	593	0.00	0

TOTAL FALLOUT ACTIVITY (µCi/sq meter) by SHOT SERIES WITH COUNTY RANK

COUNTY AND STATE	RANGER SERIES	RANGER RANK	B-J SERIES	B-J RANK	T-S SERIES	T-S RANK	U-K SERIES	UK RANK	TEAPOT SERIES	TP RANK	PLUMBBOB SERIES	PB RANK	SHOT SEDAN	SEDAN RANK
	1951		1951		1952		1953		1955		1957		1962	
PIMA AZ	0.00	0	0.00	0	73.45	580	4.46	3032	11.39	1778	174.83	54	0.00	0
PINAL AZ	0.00	0	0.00	0	75.89	548	15.89	2143	16.00	1337	68.68	511	0.00	0
SANTA CRUZ AZ	0.00	0	0.00	0	72.67	591	2.80	3073	11.92	1725	134.19	118	0.00	0
YAVAPAI AZ	0.00	0	0.00	0	144.81	295	20.07	1837	29.68	563	23.31	1906	0.00	0
YUMA AZ	0.00	0	0.00	0	18.02	1870	46.04	721	70.71	228	16.57	2581	0.00	0
ALAMEDA CA	0.00	0	0.21	2857	37.81	1178	3.36	3056	4.85	2704	1.80	3084	0.00	0
ALPINE CA	0.00	0	1.27	1959	1250.27	114	18.14	1968	7.36	2211	2.44	3068	0.00	0
AMADOR CA	0.00	0	0.21	2858	303.51	210	4.48	3031	5.38	2574	2.20	3071	0.00	0
BUTTE CA	0.00	0	0.32	2768	96.34	378	4.00	3042	3.73	2921	3.15	3058	0.00	0
CALAVERAS CA	0.00	0	0.21	2859	115.46	329	4.32	3036	5.65	2519	2.00	3075	0.00	0
COLUSA CA	0.00	0	0.32	2769	99.75	362	3.34	3057	2.04	3074	1.82	3082	0.00	0
CONTRA COSTA CA	0.00	0	0.21	2860	53.76	874	3.53	3051	3.56	2941	1.78	3087	0.00	0
DEL NORTE CA	0.00	0	0.43	2607	9.59	2608	6.73	2964	3.15	2991	9.72	2937	0.00	0
EL DORADO CA	0.00	0	0.21	2861	194.37	260	4.21	3039	6.74	2310	2.19	3072	0.00	0
FRESNO CA	0.00	0	15.59	392	34.80	1264	2.82	3071	3.50	2948	2.00	3076	0.00	0
GLENN CA	0.00	0	0.32	2770	248.95	233	3.73	3047	2.22	3061	2.42	3069	0.00	0
HUMBOLDT CA	0.00	0	0.43	2608	20.42	1729	5.87	2992	1.39	3086	9.11	2953	0.00	0
IMPERIAL CA	0.00	0	1.62	1782	30.79	1382	1.53	3090	2.05	3072	3.98	3040	0.00	0
INYO1 CA	0.00	0	0.00	0	1931.18	90	2.98	3067	44.10	367	159.50	65	0.00	0
INYO2 CA	0.00	0	0.00	0	1931.18	91	2.98	3068	44.10	368	159.50	66	0.00	0
INYO3 CA	0.00	0	72.80	18	2200.48	88	38.98	934	379.90	19	905.12	3	0.00	0
KERN CA	0.00	0	15.59	393	54.43	863	2.01	3086	2.19	3065	3.59	3054	0.00	0
KINGS CA	0.00	0	15.59	394	30.77	1383	2.59	3078	4.30	2816	3.09	3059	0.00	0
LAKE CA	0.00	0	0.32	2771	74.32	567	3.38	3055	2.78	3029	1.93	3078	0.00	0
LASSEN CA	0.00	0	0.94	2128	356.35	195	7.52	2928	6.51	2358	5.92	2995	0.00	0
LOS ANGELES CA	0.00	0	71.77	20	31.18	1369	0.90	3091	4.51	2773	4.52	3023	0.00	0

TOTAL FALLOUT ACTIVITY (µCi/sq meter) by SHOT SERIES WITH COUNTY RANK

COUNTY AND STATE	RANGER SERIES	RANGER RANK	B-J SERIES	B-J RANK	T-S SERIES	T-S RANK	U-K SERIES	UK RANK	TEAPOT SERIES	TP RANK	PLUMBBOB SERIES	PB RANK	SHOT SEDAN	SEDAN RANK
	1951		1951		1952		1953		1955		1957		1962	
MADERA CA	0.00	0	0.11	2963	93.04	393	2.56	3079	6.57	2343	1.76	3089	0.00	0
MARIN CA	0.00	0	0.21	2862	45.86	1005	3.46	3053	2.73	3033	4.54	3021	0.00	0
MARIPOSA CA	0.00	0	0.11	2964	165.06	271	3.06	3065	6.39	2390	1.91	3079	0.00	0
MENDOCINO CA	0.00	0	0.32	2772	45.83	1006	2.75	3074	2.05	3073	2.17	3073	0.00	0
MERCED CA	0.00	0	0.11	2965	34.16	1282	2.72	3075	4.96	2671	1.72	3092	0.00	0
MODOC CA	0.00	0	0.43	2609	850.11	131	4.33	3035	5.58	2534	5.97	2994	0.00	0
MONO CA	0.00	0	0.00	0	3786.34	76	6.03	2986	27.69	606	140.69	99	0.00	0
MONTEREY CA	0.00	0	15.59	395	39.21	1140	2.71	3076	4.87	2699	3.90	3041	0.00	0
NAPA CA	0.00	0	0.21	2863	120.39	324	3.79	3046	2.11	3070	1.85	3080	0.00	0
NEVADA CA	0.00	0	1.98	1624	220.90	246	13.46	2384	8.42	2067	2.81	3063	0.00	0
ORANGE CA	0.00	0	1.52	1829	36.09	1226	1.88	3088	1.67	3082	3.65	3050	0.00	0
PLACER CA	0.00	0	0.21	2864	195.88	257	5.33	3012	7.45	2193	2.87	3062	0.00	0
PLUMAS CA	0.00	0	1.12	2028	480.72	189	7.13	2949	9.00	2011	2.63	3065	0.00	0
RIVERSIDE CA	0.00	0	1.97	1625	44.45	1031	1.67	3089	1.73	3081	4.37	3030	0.00	0
SACRAMENTO CA	0.00	0	0.21	2865	70.68	613	2.88	3069	2.21	3063	1.81	3083	0.00	0
SAN BENITO CA	0.00	0	15.59	396	32.85	1318	2.55	3080	4.63	2756	1.78	3086	0.00	0
SAN BERNADIN CA	0.00	0	0.00	0	104.50	345	45.61	725	111.07	154	47.79	865	0.00	0
SAN DIEGO CA	0.00	0	0.66	2324	64.97	691	1.95	3087	1.88	3078	4.33	3031	0.00	0
SAN FRANCISC CA	0.00	0	0.21	2866	48.30	959	2.54	3081	1.45	3085	2.20	3070	0.00	0
SAN JOAQUIN CA	0.00	0	0.21	2867	51.08	905	3.60	3049	2.76	3030	1.76	3088	0.00	0
SAN LUIS OBI CA	0.00	0	15.56	399	47.84	970	2.63	3077	4.91	2684	3.66	3049	0.00	0
SAN MATEO CA	0.00	0	0.21	2868	63.98	706	3.16	3062	4.80	2713	4.37	3029	0.00	0
SANTA BARBAR CA	0.00	0	6.43	882	65.66	688	2.25	3084	2.91	3015	3.61	3051	0.00	0
SANTA CLARA CA	0.00	0	0.21	2869	39.12	1144	3.30	3059	6.98	2273	1.98	3077	0.00	0
SANTA CRUZ CA	0.00	0	0.21	2870	78.85	517	3.22	3061	5.26	2600	4.58	3017	0.00	0
SHASTA CA	0.00	0	0.32	2773	90.79	408	4.89	3022	4.35	2809	6.06	2993	0.00	0

TOTAL FALLOUT ACTIVITY (µCi/sq meter) by SHOT SERIES WITH COUNTY RANK

COUNTY AND STATE	RANGER SERIES	RANGER RANK	B-J SERIES	B-J RANK	T-S SERIES	T-S RANK	U-K SERIES	UK RANK	TEAPOT SERIES	TP RANK	PLUMBBOB SERIES	PB RANK	SHOT SEDAN	SEDAN RANK
	1951		1951		1952		1953		1955		1957		1962	
SIERRA CA	0.00	0	1.38	1891	330.01	202	18.78	1922	9.59	1964	3.68	3048	0.00	0
SISKIYOU CA	0.00	0	0.43	2610	52.49	886	5.03	3021	2.87	3018	8.92	2958	0.00	0
SOLANO CA	0.00	0	0.21	2871	50.64	911	3.12	3064	1.93	3076	1.75	3090	0.00	0
SONOMA CA	0.00	0	0.32	2774	61.11	752	3.33	3058	2.12	3069	2.01	3074	0.00	0
STANISLAUS CA	0.00	0	0.21	2872	53.28	879	3.45	3054	2.58	3043	1.75	3091	0.00	0
SUTTER CA	0.00	0	0.21	2873	66.21	676	3.26	3060	3.00	3006	1.83	3081	0.00	0
TEHAMA CA	0.00	0	0.32	2775	97.18	373	4.34	3034	4.26	2827	4.10	3038	0.00	0
TRINITY CA	0.00	0	0.32	2776	31.49	1358	6.22	2977	2.21	3062	8.09	2970	0.00	0
TULARE CA	0.00	0	15.59	397	50.36	919	2.44	3082	5.52	2547	3.72	3046	0.00	0
TUOLUMNE CA	0.00	0	1.72	1744	200.72	252	3.84	3045	6.69	2320	2.74	3064	0.00	0
VENTURA CA	0.00	0	15.49	403	67.57	659	2.03	3085	2.45	3050	3.61	3052	0.00	0
YOLO CA	0.00	0	0.21	2874	53.88	872	3.15	3063	2.20	3064	1.79	3085	0.00	0
YUBA CA	0.00	0	0.32	2777	114.33	333	3.90	3044	5.88	2484	3.19	3057	0.00	0
ADAMS CO	0.00	0	1.33	1924	45.02	1019	31.24	1232	312.29	38	66.52	549	0.00	0
ALAMOSA CO	0.00	0	0.59	2391	90.78	409	486.09	55	120.67	120	30.50	1395	0.00	0
ARAPAHOE CO	0.00	0	1.16	2009	42.57	1068	31.23	1233	298.38	42	68.86	508	0.00	0
ARCHULETA CO	24.40	31	0.59	2388	312.42	207	1246.22	14	44.61	361	42.12	990	0.00	0
BACA CO	0.00	0	1.43	1875	49.74	930	145.72	141	125.86	103	51.06	792	0.00	0
BENT CO	0.00	0	1.28	1949	41.78	1081	92.10	239	109.55	157	54.96	715	0.00	0
BOULDER CO	0.00	0	1.53	1826	54.40	866	30.63	1257	324.02	34	67.72	527	0.00	0
CHAFFEE CO	0.00	0	0.59	2392	58.28	794	185.19	117	132.14	90	45.06	923	0.00	0
CHEYENNE CO	0.00	0	1.44	1870	51.91	892	57.01	539	112.99	148	59.26	652	0.00	0
CLEAR CREEK CO	0.00	0	1.07	2054	46.54	997	63.33	465	324.99	33	74.48	436	0.00	0
CONEJOS CO	24.40	32	0.59	2389	102.45	352	1548.85	11	165.58	74	44.74	929	0.00	0
COSTILLA CO	5.57	253	0.98	2105	57.53	811	438.22	66	179.99	62	40.31	1023	0.00	0
CROWLEY CO	0.00	0	0.91	2138	52.54	885	74.12	350	107.29	162	55.57	702	0.00	0

TOTAL FALLOUT ACTIVITY (µCi/sq meter) by SHOT SERIES WITH COUNTY RANK

COUNTY AND STATE	RANGER SERIES	RANGER RANK	B-J SERIES	B-J RANK	T-S SERIES	T-S RANK	U-K SERIES	UK RANK	TEAPOT SERIES	TP RANK	PLUMBBOB SERIES	PB RANK	SHOT SEDAN	SEDAN RANK
	1951		1951		1952		1953		1955		1957		1962	
CUSTER CO	0.00	0	1.18	2001	57.68	806	190.12	116	126.13	101	69.07	504	0.00	0
DELTA CO	0.00	0	0.00	0	139.78	304	468.36	60	418.21	15	64.58	572	0.00	0
DENVER CO	0.00	0	1.36	1909	41.53	1088	28.53	1359	170.35	69	53.79	740	0.00	0
DOLORES CO	0.00	0	0.00	0	94.23	389	673.91	29	92.48	171	34.45	1181	0.00	0
DOUGLAS CO	0.00	0	1.34	1917	72.19	600	43.67	780	306.94	40	67.23	533	0.00	0
EAGLE CO	0.00	0	0.59	2393	69.88	625	293.14	89	304.51	41	54.43	727	0.00	0
EL PASO CO	0.00	0	0.47	2564	77.20	533	62.26	474	66.82	237	96.51	270	0.00	0
ELBERT CO	3.34	295	1.39	1890	56.18	836	39.04	929	120.46	122	67.68	529	0.00	0
FREMONT CO	5.57	254	1.16	2008	67.58	658	137.62	151	126.09	102	57.26	676	0.00	0
GARFIELD CO	0.00	0	0.00	0	235.66	237	386.63	73	517.03	8	152.09	76	0.00	0
GILPIN CO	0.00	0	0.89	2150	46.23	1001	46.91	705	326.27	31	63.09	596	0.00	0
GRAND CO	0.00	0	1.32	1931	89.97	416	90.80	246	333.82	30	67.09	538	0.00	0
GUNNISON CO	0.00	0	0.59	2394	93.52	390	1632.97	10	187.18	61	40.56	1020	0.00	0
HINSDALE CO	4.08	259	1.28	1947	154.45	284	814.42	18	160.51	78	36.75	1104	0.00	0
HUERFANO CO	4.08	260	1.18	2003	118.44	327	158.85	131	111.28	152	44.04	945	0.00	0
JACKSON CO	0.00	0	1.37	1901	50.06	928	67.73	419	147.23	86	51.06	791	0.00	0
JEFFERSON CO	0.00	0	1.36	1910	62.94	726	41.19	852	317.01	35	69.86	493	0.00	0
KIOWA CO	0.00	0	0.89	2151	49.49	934	125.26	165	111.21	153	66.17	557	0.00	0
KIT CARSON CO	3.34	296	2.18	1562	55.95	841	42.62	808	121.56	115	62.03	616	0.00	0
LA PLATA CO	189.61	2	0.00	0	140.58	301	665.57	31	74.85	217	44.46	937	0.00	0
LAKE CO	0.00	0	1.53	1825	66.84	670	395.89	72	316.44	36	52.49	769	0.00	0
LARIMER CO	0.00	0	1.37	1902	71.25	608	40.92	861	175.70	64	70.81	475	0.00	0
LAS ANIMAS CO	0.00	0	1.48	1855	51.44	900	175.07	123	125.10	105	39.94	1028	0.00	0
LINCOLN CO	3.34	297	1.79	1720	60.54	764	54.54	577	120.59	121	71.60	465	0.00	0
LOGAN CO	0.00	0	0.89	2152	108.82	341	35.96	1034	45.04	356	72.01	460	0.00	0
MESA CO	0.00	0	0.00	0	254.43	226	526.72	49	505.75	11	66.25	554	0.00	0

TOTAL FALLOUT ACTIVITY (µCi/sq meter) by SHOT SERIES WITH COUNTY RANK

COUNTY AND STATE	RANGER SERIES	RANGER RANK	B-J SERIES	B-J RANK	T-S SERIES	T-S RANK	U-K SERIES	UK RANK	TEAPOT SERIES	TP RANK	PLUMBBOB SERIES	PB RANK	SHOT SEDAN	SEDAN RANK
	1951		1951		1952		1953		1955		1957		1962	
MINERAL CO	10.15	165	0.59	2390	144.77	296	848.22	17	153.15	82	33.46	1236	0.00	0
MOFFAT CO	0.00	0	0.00	0	320.40	204	90.62	247	345.58	29	161.38	64	0.00	0
MONTEZUMA CO	0.00	0	0.00	0	95.96	380	576.83	45	87.54	184	44.59	933	0.00	0
MONTROSE CO	0.00	0	0.00	0	196.22	256	599.80	40	193.06	59	57.09	679	0.00	0
MORGAN CO	0.00	0	0.89	2153	48.52	951	28.24	1374	325.55	32	69.93	491	0.00	0
OTERO CO	0.00	0	1.28	1950	45.79	1007	97.57	211	112.88	149	46.17	904	0.00	0
OURAY CO	4.30	257	0.00	0	94.23	387	583.87	42	82.31	198	40.09	1027	0.00	0
PARK CO	0.00	0	1.16	2010	68.79	641	81.59	284	127.23	98	63.66	590	0.00	0
PHILLIPS CO	0.00	0	0.89	2154	139.36	307	29.14	1327	115.11	142	80.29	388	0.00	0
PITKIN CO	0.00	0	0.59	2395	89.95	417	704.13	25	198.56	58	28.57	1505	0.00	0
PROWERS CO	0.00	0	1.28	1951	50.30	922	82.19	281	118.39	133	53.22	748	0.00	0
PUEBLO CO	5.57	255	1.25	1965	105.95	344	59.00	514	53.43	289	65.79	561	0.00	0
RIO BLANCO CO	0.00	0	0.00	0	267.12	219	147.08	139	507.40	10	198.75	45	0.00	0
RIO GRANDE CO	0.00	0	0.59	2396	79.17	513	692.59	27	115.90	140	42.13	988	0.00	0
ROUTT CO	0.00	0	0.59	2397	46.98	983	179.88	120	176.10	63	43.93	948	0.00	0
SAGUACHE CO	4.30	258	0.98	2106	112.31	336	440.02	65	125.08	106	54.86	718	0.00	0
SAN JUAN CO	4.08	261	0.00	0	94.23	388	581.74	44	80.68	204	34.45	1182	0.00	0
SAN MIGUEL CO	0.00	0	0.00	0	103.25	350	582.61	43	83.31	195	40.91	1014	0.00	0
SEDGWICK CO	0.00	0	0.89	2155	86.19	455	56.24	549	44.61	360	89.27	323	0.00	0
SUMMIT CO	0.00	0	1.32	1932	50.57	913	204.87	109	314.94	37	61.94	618	0.00	0
TELLER CO	0.00	0	1.14	2018	86.92	446	80.30	292	121.52	116	67.81	525	0.00	0
WASHINGTON CO	0.00	0	1.28	1952	62.73	728	35.34	1060	123.56	109	57.29	675	0.00	0
WELD CO	0.00	0	0.89	2156	60.30	768	22.71	1656	154.01	81	70.89	472	0.00	0
YUMA CO	0.00	0	1.05	2056	77.51	530	33.64	1126	118.45	131	62.88	602	0.00	0
FAIRFIELD CT	9.14	188	27.51	133	50.42	918	45.28	740	12.43	1671	13.57	2789	0.00	0
HARTFORD CT	11.84	153	15.93	382	36.82	1203	44.49	761	7.25	2227	18.58	2386	0.00	0

TOTAL FALLOUT ACTIVITY (µCi/sq meter) by SHOT SERIES WITH COUNTY RANK

COUNTY AND STATE	RANGER SERIES	RANGER RANK	B-J SERIES	B-J RANK	T-S SERIES	T-S RANK	U-K SERIES	UK RANK	TEAPOT SERIES	TP RANK	PLUMBBOB SERIES	PB RANK	SHOT SEDAN	SEDAN RANK
	1951		1951		1952		1953		1955		1957		1962	
LITCHFIELD CT	29.12	24	20.76	240	41.93	1078	34.01	1116	6.69	2319	18.58	2387	0.00	0
MIDDLESEX CT	9.36	175	18.43	311	38.62	1158	112.31	178	7.96	2127	16.71	2566	0.00	0
NEW HAVEN CT	9.32	176	18.37	313	44.26	1034	74.32	345	12.69	1648	12.95	2827	0.00	0
NEW LONDON CT	3.87	267	22.30	209	33.00	1315	74.49	343	7.01	2269	12.73	2832	0.00	0
TOLLAND CT	29.72	22	19.79	270	45.69	1009	55.37	561	6.97	2274	10.73	2908	0.00	0
WINDHAM CT	3.87	268	26.07	161	36.00	1228	67.86	418	6.97	2276	8.88	2959	0.00	0
WASHINGTON DC	0.19	543	27.98	125	14.86	2083	17.99	1981	5.48	2552	14.70	2721	0.00	0
KENT DE	1.17	383	19.99	263	13.77	2181	36.55	1017	7.42	2199	16.89	2550	0.00	0
NEW CASTLE DE	1.35	363	17.33	346	12.40	2321	35.78	1041	9.43	1974	21.19	2122	0.00	0
SUSSEX DE	0.00	0	16.97	354	15.09	2063	34.20	1110	9.06	2008	21.38	2101	0.00	0
ALACHUA FL	0.71	420	1.75	1737	12.41	2319	16.11	2127	5.97	2458	33.07	1262	0.00	0
BAKER FL	0.71	421	2.36	1516	11.79	2381	12.18	2492	5.51	2550	27.88	1538	0.00	0
BAY FL	0.71	422	8.40	750	11.14	2463	9.27	2831	8.73	2035	25.13	1761	0.00	0
BRADFORD FL	0.00	0	1.91	1665	11.15	2461	12.94	2429	3.36	2960	20.75	2171	0.00	0
BREVARD FL	0.54	450	0.38	2729	13.06	2253	14.01	2333	5.08	2641	38.23	1059	0.00	0
BROWARD FL	1.44	354	0.18	2911	13.18	2240	17.55	2021	2.67	3037	12.72	2833	0.00	0
CALHOUN FL	0.00	0	11.04	593	11.97	2364	8.98	2856	9.23	1993	32.18	1293	0.00	0
CHARLOTTE FL	0.00	0	0.18	2942	11.49	2420	15.88	2144	2.60	3040	12.98	2825	0.00	0
CITRUS FL	0.57	437	0.55	2440	12.96	2264	17.31	2040	3.35	2962	30.53	1392	0.00	0
CLAY FL	0.00	0	1.95	1632	14.22	2139	11.91	2521	5.10	2635	22.88	1951	0.00	0
COLLIER FL	1.63	351	0.28	2834	15.73	2015	19.40	1884	2.71	3034	14.80	2710	0.00	0
COLUMBIA FL	0.52	459	1.98	1623	11.45	2425	12.66	2452	3.42	2958	27.10	1591	0.00	0
DADE FL	3.05	302	0.68	2306	15.40	2042	16.98	2059	2.30	3056	9.75	2935	0.00	0
DE SOTO FL	0.57	438	0.06	2990	11.71	2392	14.74	2259	3.09	2997	10.86	2903	0.00	0
DIXIE FL	0.54	451	0.98	2110	12.60	2297	16.92	2063	3.84	2907	32.07	1303	0.00	0
DUVAL FL	0.54	452	3.50	1304	5.81	2963	8.08	2906	4.29	2817	19.51	2307	0.00	0

TOTAL FALLOUT ACTIVITY (µCi/sq meter) by SHOT SERIES WITH COUNTY RANK

COUNTY AND STATE	RANGER SERIES	RANGER RANK	B-J SERIES	B-J RANK	T-S SERIES	T-S RANK	U-K SERIES	UK RANK	TEAPOT SERIES	TP RANK	PLUMBBOB SERIES	PB RANK	SHOT SEDAN	SEDAN RANK
	1951		1951		1952		1953		1955		1957		1962	
ESCAMBIA FL	0.00	0	3.78	1254	11.89	2371	10.82	2646	10.36	1882	44.68	930	0.00	0
FLAGLER FL	0.00	0	1.11	2034	12.32	2330	13.71	2362	3.48	2950	38.10	1061	0.00	0
FRANKLIN FL	0.95	412	0.89	2159	11.28	2445	8.72	2871	4.42	2788	37.87	1070	0.00	0
GADSDEN FL	0.00	0	6.49	878	13.05	2255	9.96	2752	7.04	2267	25.86	1692	0.00	0
GILCHRIST FL	0.54	453	1.15	2011	12.00	2358	17.21	2048	3.99	2878	35.22	1147	0.00	0
GLADES FL	0.00	0	0.18	2914	12.54	2304	17.01	2055	3.20	2982	14.02	2764	0.00	0
GULF FL	0.00	0	2.22	1556	10.91	2486	9.08	2845	7.80	2140	37.56	1080	0.00	0
HAMILTON FL	0.52	460	2.40	1507	10.17	2562	17.74	2005	5.95	2466	26.42	1641	0.00	0
HARDEE FL	0.00	0	0.25	2851	12.36	2324	14.24	2313	2.80	3028	8.28	2966	0.00	0
HENDRY FL	0.36	486	0.38	2730	15.14	2060	18.46	1952	4.06	2867	14.43	2740	0.00	0
HERNANDO FL	0.52	461	0.28	2829	13.21	2236	17.43	2029	4.34	2810	36.96	1096	0.00	0
HIGHLANDS FL	0.00	0	0.18	2915	11.71	2391	15.86	2149	3.86	2900	14.79	2713	0.00	0
HILLSBOROUGH FL	1.18	377	0.16	2950	30.15	1405	15.73	2157	6.07	2444	28.26	1520	0.00	0
HOLMES FL	0.00	0	18.41	312	30.10	1406	9.44	2807	8.95	2019	27.10	1592	0.00	0
INDIAN RIVER FL	0.36	487	0.27	2838	11.98	2363	14.51	2282	3.85	2906	13.77	2779	0.00	0
JACKSON FL	0.29	496	4.37	1125	11.82	2376	8.56	2881	8.78	2029	26.91	1604	0.00	0
JEFFERSON FL	0.57	439	0.83	2196	10.90	2488	12.91	2433	4.73	2732	31.14	1352	0.00	0
LAFAYETTE FL	0.57	440	1.72	1746	12.57	2299	14.10	2325	3.98	2879	25.84	1694	0.00	0
LAKE FL	1.18	378	0.48	2556	11.92	2369	26.87	1436	6.87	2290	43.96	946	0.00	0
LEE FL	0.00	0	0.08	2984	14.04	2157	17.20	2049	2.54	3046	13.88	2772	0.00	0
LEON FL	0.29	497	0.47	2563	5.19	2997	24.35	1559	4.09	2860	41.82	996	0.00	0
LEVY FL	0.54	454	0.66	2326	8.40	2729	12.82	2442	3.45	2953	15.49	2660	0.00	0
LIBERTY FL	0.95	413	7.25	828	11.98	2362	9.12	2840	10.01	1915	29.83	1435	0.00	0
MADISON FL	0.57	441	2.25	1543	11.09	2471	18.15	1967	4.81	2712	33.61	1226	0.00	0
MANATEE FL	0.54	455	0.12	2961	14.93	2076	13.01	2423	2.15	3067	17.90	2453	0.00	0
MARION FL	0.00	0	0.70	2290	13.73	2183	15.17	2208	8.17	2101	31.30	1345	0.00	0

U.S. FALLOUT ATLAS : TOTAL FALLOUT

TOTAL FALLOUT ACTIVITY (µCi/sq meter) by SHOT SERIES WITH COUNTY RANK

COUNTY AND STATE	RANGER SERIES	RANGER RANK	B-J SERIES	B-J RANK	T-S SERIES	T-S RANK	U-K SERIES	UK RANK	TEAPOT SERIES	TP RANK	PLUMBBOB SERIES	PB RANK	SHOT SEDAN	SEDAN RANK
	1951		1951		1952		1953		1955		1957		1962	
MARTIN FL	0.52	462	0.18	2912	12.70	2287	17.70	2010	3.97	2882	11.66	2876	0.00	0
MONROE FL	0.27	508	0.96	2117	15.24	2050	20.97	1773	2.09	3071	12.00	2860	0.00	0
NASSAU FL	0.36	488	3.10	1376	10.84	2491	12.43	2472	11.20	1800	32.18	1294	0.00	0
OKALOOSA FL	0.47	476	3.22	1358	20.45	1727	10.32	2705	8.58	2047	89.42	322	0.00	0
OKEECHOBEE FL	0.00	0	0.18	2916	14.37	2127	14.50	2284	2.27	3058	13.97	2768	0.00	0
ORANGE FL	0.52	463	0.39	2725	12.06	2350	15.88	2145	5.21	2619	31.95	1310	0.00	0
OSCEOLA FL	0.00	0	0.48	2551	11.63	2403	15.86	2148	4.73	2734	22.68	1974	0.00	0
PALM BEACH FL	1.11	408	0.26	2847	13.78	2179	20.97	1774	3.31	2969	13.42	2796	0.00	0
PASCO FL	0.00	0	0.26	2848	11.60	2407	14.83	2239	5.47	2557	20.96	2145	0.00	0
PINELLAS FL	0.54	456	0.15	2956	12.94	2266	14.50	2285	4.47	2780	33.78	1218	0.00	0
POLK FL	0.57	442	0.25	2850	13.95	2169	16.15	2123	5.83	2488	32.54	1282	0.00	0
PUTNAM FL	0.00	0	2.03	1613	12.75	2283	12.91	2431	5.54	2545	30.97	1366	0.00	0
SANTA ROSA FL	0.52	464	5.70	944	11.99	2360	15.44	2187	9.55	1966	46.43	891	0.00	0
SARASOTA FL	0.00	0	0.15	2959	15.17	2058	17.85	1992	3.30	2972	16.98	2545	0.00	0
SEMINOLE FL	0.36	489	0.59	2399	11.25	2447	13.77	2354	10.43	1877	29.71	1443	0.00	0
ST JOHNS FL	0.00	0	2.43	1500	10.91	2484	12.39	2473	4.87	2697	43.50	958	0.00	0
ST LUCIE FL	0.57	443	0.49	2540	12.47	2311	23.76	1591	2.80	3027	11.30	2889	0.00	0
SUMTER FL	0.00	0	0.41	2700	10.94	2481	33.02	1156	7.69	2155	21.98	2055	0.00	0
SUWANNEE FL	0.52	465	2.64	1457	12.57	2300	13.91	2340	7.26	2222	26.68	1618	0.00	0
TAYLOR FL	0.52	466	0.80	2216	11.48	2422	13.54	2378	4.75	2727	26.12	1668	0.00	0
UNION FL	0.00	0	1.88	1672	11.14	2464	12.85	2440	3.48	2951	26.33	1646	0.00	0
VOLUSIA FL	0.72	419	0.80	2219	12.49	2308	14.72	2262	5.02	2655	40.74	1016	0.00	0
WAKULLA FL	0.00	0	0.56	2427	11.13	2465	13.69	2364	3.49	2949	38.61	1053	0.00	0
WALTON FL	0.29	498	4.78	1046	9.61	2603	10.03	2745	9.09	2005	25.16	1754	0.00	0
WASHINGTON FL	0.00	0	10.05	652	11.42	2427	9.11	2841	8.75	2033	21.67	2082	0.00	0
APPLING GA	0.27	509	6.23	899	16.14	1992	14.53	2279	5.70	2511	31.34	1341	0.00	0

TOTAL FALLOUT ACTIVITY (µCi/sq meter) by SHOT SERIES WITH COUNTY RANK

COUNTY AND STATE	RANGER SERIES	RANGER RANK	B-J SERIES	B-J RANK	T-S SERIES	T-S RANK	U-K SERIES	UK RANK	TEAPOT SERIES	TP RANK	PLUMBBOB SERIES	PB RANK	SHOT SEDAN	SEDAN RANK
	1951		1951		1952		1953		1955		1957		1962	
ATKINSON GA	0.00	0	7.57	801	13.16	2244	15.21	2204	6.24	2412	28.35	1517	0.00	0
BACON GA	0.27	510	4.41	1120	13.56	2200	18.73	1929	4.64	2751	31.48	1334	0.00	0
BAKER GA	0.00	0	11.48	563	15.93	1998	8.96	2858	5.39	2572	33.15	1256	0.00	0
BALDWIN GA	0.27	511	9.32	692	16.50	1965	12.33	2480	7.26	2221	25.67	1707	0.00	0
BANKS GA	0.00	0	8.52	742	10.41	2537	14.82	2240	8.79	2026	26.25	1650	0.00	0
BARROW GA	0.00	0	6.15	906	6.31	2915	11.57	2547	8.45	2061	29.57	1452	0.00	0
BARTOW GA	0.00	0	5.90	930	7.89	2767	12.45	2470	9.86	1934	27.73	1546	0.00	0
BEN HILL GA	0.27	512	6.85	852	12.82	2277	14.27	2309	4.71	2738	30.20	1412	0.00	0
BERRIEN GA	0.00	0	8.35	753	11.18	2457	14.69	2264	6.02	2450	31.71	1318	0.00	0
BIBB GA	0.27	513	7.92	784	10.52	2522	9.76	2771	7.22	2234	35.46	1141	0.00	0
BLECKLEY GA	0.27	514	11.21	584	10.97	2479	11.48	2559	8.04	2120	33.98	1211	0.00	0
BRANTLEY GA	0.00	0	6.92	846	13.80	2178	13.65	2368	4.29	2818	22.20	2032	0.00	0
BROOKS GA	0.00	0	6.41	883	10.78	2494	14.47	2288	6.95	2278	26.38	1642	0.00	0
BRYAN GA	0.27	515	13.12	481	29.55	1423	10.68	2661	5.96	2464	30.28	1407	0.00	0
BULLOCH GA	0.00	0	13.57	465	21.25	1702	11.46	2561	5.61	2526	30.15	1415	0.00	0
BURKE GA	0.00	0	16.84	357	15.19	2056	11.95	2516	6.39	2389	28.20	1522	0.00	0
BUTTS GA	0.00	0	7.26	827	7.63	2790	10.66	2662	5.88	2482	34.64	1168	0.00	0
CALHOUN GA	0.00	0	13.03	487	16.82	1950	9.14	2837	5.54	2544	34.32	1189	0.00	0
CAMDEN GA	0.52	467	3.53	1298	12.21	2337	14.60	2272	7.26	2224	33.50	1234	0.00	0
CANDLER GA	0.00	0	11.30	577	8.15	2751	15.20	2206	4.35	2806	28.80	1491	0.00	0
CARROLL GA	0.00	0	7.75	788	24.10	1569	10.97	2628	9.99	1918	33.57	1232	0.00	0
CATOOSA GA	0.00	0	11.96	537	9.17	2645	13.59	2372	11.62	1753	20.78	2167	0.00	0
CHARLTON GA	0.52	468	3.84	1239	11.59	2408	13.45	2385	6.98	2271	26.19	1656	0.00	0
CHATHAM GA	0.52	469	13.83	461	22.24	1649	9.75	2773	5.89	2479	29.48	1458	0.00	0
CHATTAHOOCHE GA	0.00	0	8.71	732	12.12	2344	8.88	2862	7.14	2254	35.24	1145	0.00	0
CHATTOOGA GA	0.00	0	11.90	542	9.67	2596	13.13	2406	13.85	1542	21.20	2118	0.00	0

TOTAL FALLOUT ACTIVITY (µCi/sq meter) by SHOT SERIES WITH COUNTY RANK

COUNTY AND STATE	RANGER SERIES	RANGER RANK	B-J SERIES	B-J RANK	T-S SERIES	T-S RANK	U-K SERIES	UK RANK	TEAPOT SERIES	TP RANK	PLUMBBOB SERIES	PB RANK	SHOT SEDAN	SEDAN RANK
	1951		1951		1952		1953		1955		1957		1962	
CHEROKEE GA	0.00	0	5.59	957	8.62	2702	12.19	2490	11.13	1808	29.77	1440	0.00	0
CLARKE GA	0.00	0	8.56	738	7.01	2853	12.22	2487	8.67	2039	33.43	1238	0.00	0
CLAY GA	0.00	0	14.96	425	13.54	2201	9.02	2852	6.43	2373	38.85	1048	0.00	0
CLAYTON GA	0.00	0	5.97	926	11.08	2472	10.91	2635	6.82	2297	46.80	884	0.00	0
CLINCH GA	0.27	516	6.18	904	10.23	2552	14.60	2273	6.23	2414	33.61	1227	0.00	0
COBB GA	0.00	0	4.76	1053	6.63	2891	11.00	2620	6.86	2292	33.33	1247	0.00	0
COFFEE GA	0.00	0	4.33	1139	14.98	2073	15.62	2167	4.90	2689	22.36	2017	0.00	0
COLQUITT GA	0.00	0	4.99	1010	13.70	2187	15.32	2198	6.28	2404	39.52	1037	0.00	0
COLUMBIA GA	0.47	477	9.97	661	3.42	3054	12.58	2460	6.74	2309	26.93	1601	0.00	0
COLUMBUS GA	0.27	517	9.05	706	12.50	2307	10.16	2728	7.48	2186	34.64	1169	0.00	0
COOK GA	0.00	0	6.97	840	10.95	2480	14.65	2266	5.96	2461	31.20	1348	0.00	0
COWETA GA	0.00	0	5.75	940	11.01	2475	9.10	2843	6.42	2377	33.79	1217	0.00	0
CRAWFORD GA	0.00	0	9.03	711	8.06	2760	9.92	2756	5.22	2614	48.20	854	0.00	0
CRISP GA	0.00	0	12.21	523	16.54	1963	9.14	2838	3.78	2916	33.06	1264	0.00	0
DADE GA	0.00	0	7.98	778	8.89	2674	12.17	2494	14.22	1515	18.86	2362	0.00	0
DAWSON GA	0.00	0	5.70	945	11.22	2450	13.18	2403	12.76	1643	20.88	2160	0.00	0
DE KALB GA	0.00	0	4.72	1062	4.37	3028	9.67	2781	8.81	2024	38.90	1047	0.00	0
DECATUR GA	0.00	0	6.85	853	12.62	2295	13.56	2376	4.74	2728	27.26	1575	0.00	0
DODGE GA	0.29	499	10.75	609	5.90	2954	10.17	2727	5.97	2460	34.94	1156	0.00	0
DOOLY GA	0.00	0	14.71	433	5.16	2999	9.60	2788	4.00	2875	35.81	1129	0.00	0
DOUGHERTY GA	0.71	435	11.11	588	13.66	2192	9.77	2770	6.01	2452	36.72	1106	0.00	0
DOUGLAS GA	0.27	518	5.69	948	5.73	2968	10.27	2709	8.78	2027	44.82	926	0.00	0
EARLY GA	0.00	0	4.96	1012	12.41	2320	10.26	2710	6.61	2332	28.95	1484	0.00	0
ECHOLS GA	0.00	0	2.29	1532	9.13	2650	14.63	2269	5.97	2459	27.62	1552	0.00	0
EFFINGHAM GA	0.54	457	15.70	389	24.43	1561	10.48	2690	5.41	2567	29.05	1481	0.00	0
ELBERT GA	0.00	0	7.49	810	4.29	3029	14.08	2326	9.80	1944	26.22	1654	0.00	0

U.S. FALLOUT ATLAS : TOTAL FALLOUT

TOTAL FALLOUT ACTIVITY (µCi/sq meter) by SHOT SERIES WITH COUNTY RANK

COUNTY AND STATE	RANGER SERIES	RANGER RANK	B-J SERIES	B-J RANK	T-S SERIES	T-S RANK	U-K SERIES	U-K RANK	TEAPOT SERIES	TP RANK	PLUMBBOB SERIES	PB RANK	SHOT SEDAN	SEDAN RANK
	1951		1951		1952		1953		1955		1957		1962	
EMANUEL GA	0.00	0	11.59	559	8.24	2743	11.02	2618	5.01	2656	36.28	1115	0.00	0
EVANS GA	0.00	0	11.05	592	17.93	1877	15.42	2189	5.47	2556	32.56	1280	0.00	0
FANNIN GA	0.52	470	9.48	686	17.61	1902	18.95	1913	14.21	1516	23.03	1936	0.00	0
FAYETTE GA	0.00	0	6.60	873	11.55	2412	9.77	2768	6.85	2294	48.31	848	0.00	0
FLOYD GA	0.00	0	12.61	503	12.13	2342	11.83	2526	11.31	1787	26.81	1610	0.00	0
FORSYTH GA	0.00	0	5.17	991	7.71	2784	12.05	2504	10.41	1880	28.48	1511	0.00	0
FRANKLIN GA	0.00	0	7.94	781	6.77	2871	13.03	2419	7.81	2138	27.37	1567	0.00	0
FULTON GA	0.27	519	3.41	1325	4.29	3031	9.46	2801	7.54	2179	37.20	1090	0.00	0
GILMER GA	0.00	0	6.76	857	10.85	2490	15.81	2152	14.23	1514	23.17	1922	0.00	0
GLASCOCK GA	0.00	0	11.82	548	6.67	2884	12.47	2468	6.47	2364	31.96	1308	0.00	0
GLYNN GA	0.00	0	7.56	804	12.92	2270	14.72	2260	7.26	2223	34.27	1193	0.00	0
GORDON GA	0.52	471	11.63	557	13.66	2191	13.79	2349	10.97	1823	27.50	1559	0.00	0
GRADY GA	0.00	0	4.28	1152	14.72	2097	10.75	2655	4.42	2790	25.34	1735	0.00	0
GREENE GA	0.29	500	13.82	462	6.30	2919	13.08	2411	8.35	2080	31.91	1313	0.00	0
GWINNETT GA	0.00	0	6.80	855	5.95	2953	11.55	2550	9.28	1991	37.57	1078	0.00	0
HABERSHAM GA	0.00	0	7.08	835	8.67	2693	16.70	2086	9.58	1965	23.20	1920	0.00	0
HALL GA	0.00	0	6.68	865	9.60	2605	14.24	2314	9.84	1936	27.29	1572	0.00	0
HANCOCK GA	0.00	0	10.20	646	8.70	2689	12.51	2466	5.82	2491	31.09	1355	0.00	0
HARALSON GA	0.00	0	5.61	954	6.02	2944	13.73	2357	11.11	1810	31.68	1321	0.00	0
HARRIS GA	0.00	0	5.99	922	12.44	2316	9.28	2829	7.33	2216	31.22	1347	0.00	0
HART GA	0.00	0	8.21	762	7.07	2848	16.02	2135	9.34	1982	23.61	1884	0.00	0
HEARD GA	0.27	520	6.71	861	10.53	2519	10.63	2667	7.46	2192	17.70	2477	0.00	0
HENRY GA	0.00	0	6.91	847	6.73	2876	10.48	2689	6.45	2367	34.25	1196	0.00	0
HOUSTON GA	0.00	0	11.23	581	10.93	2483	11.10	2601	7.66	2161	34.90	1157	0.00	0
IRWIN GA	0.00	0	6.47	880	12.75	2282	14.05	2328	4.73	2730	29.49	1457	0.00	0
JACKSON GA	0.27	521	7.60	796	9.44	2616	13.88	2341	9.30	1987	28.63	1502	0.00	0

TOTAL FALLOUT ACTIVITY (µCi/sq meter) by SHOT SERIES WITH COUNTY RANK

COUNTY AND STATE	RANGER SERIES	RANGER RANK	B-J SERIES	B-J RANK	T-S SERIES	T-S RANK	U-K SERIES	UK RANK	TEAPOT SERIES	TP RANK	PLUMBBOB SERIES	PB RANK	SHOT SEDAN	SEDAN RANK
	1951		1951		1952		1953		1955		1957		1962	
JASPER GA	0.00	0	7.57	799	6.19	2931	10.13	2735	6.09	2438	34.59	1171	0.00	0
JEFF DAVIS GA	0.00	0	4.84	1038	14.08	2149	18.77	1925	4.99	2662	32.88	1271	0.00	0
JEFFERSON GA	0.27	522	8.88	721	14.81	2085	11.80	2531	6.02	2451	33.68	1223	0.00	0
JENKINS GA	0.27	523	15.38	408	14.09	2148	11.56	2548	6.93	2284	33.09	1261	0.00	0
JOHNSON GA	0.00	0	8.86	723	14.37	2125	10.72	2659	6.52	2357	34.27	1194	0.00	0
JONES GA	0.47	478	7.45	814	5.84	2959	9.98	2750	6.42	2376	34.29	1191	0.00	0
LAMAR GA	0.27	524	7.40	816	6.99	2854	9.93	2755	5.73	2503	34.85	1159	0.00	0
LANIER GA	0.00	0	6.94	843	11.20	2455	14.70	2263	5.94	2469	34.98	1153	0.00	0
LAURENS GA	0.00	0	6.73	859	11.19	2456	11.14	2599	5.06	2645	24.14	1828	0.00	0
LEE GA	0.00	0	11.46	567	18.14	1857	10.73	2658	5.71	2507	31.46	1337	0.00	0
LIBERTY GA	0.00	0	12.63	500	22.53	1629	14.34	2300	4.39	2797	48.05	859	0.00	0
LINCOLN GA	0.27	525	10.57	621	3.72	3046	12.91	2432	6.97	2277	27.44	1565	0.00	0
LONG GA	0.00	0	11.95	538	22.51	1630	14.78	2250	4.15	2850	46.85	879	0.00	0
LOWNDES GA	0.00	0	2.23	1546	10.66	2507	14.62	2271	4.90	2687	34.06	1206	0.00	0
LUMPKIN GA	0.71	436	5.57	958	9.22	2642	13.37	2389	12.23	1691	23.90	1856	0.00	0
MACON GA	0.00	0	8.47	746	3.54	3052	9.51	2796	5.65	2521	30.95	1369	0.00	0
MADISON GA	0.27	526	10.22	643	8.16	2749	13.51	2382	8.54	2050	27.79	1543	0.00	0
MARION GA	0.00	0	8.72	730	6.63	2890	9.70	2779	5.44	2561	27.80	1541	0.00	0
MCDUFFIE GA	0.00	0	11.92	541	7.12	2843	12.94	2427	6.53	2353	32.14	1299	0.00	0
MCINTOSH GA	0.00	0	13.87	459	17.56	1903	13.35	2394	4.37	2801	27.50	1560	0.00	0
MERIWETHER GA	0.00	0	7.71	791	7.80	2778	10.12	2736	5.79	2496	34.86	1158	0.00	0
MILLER GA	0.00	0	3.80	1249	12.35	2326	10.24	2714	6.46	2365	20.00	2252	0.00	0
MITCHELL GA	0.29	501	8.53	741	13.66	2193	9.80	2766	6.53	2354	35.82	1128	0.00	0
MONROE GA	0.00	0	8.04	774	10.74	2496	9.85	2764	7.50	2183	38.49	1055	0.00	0
MONTGOMERY GA	0.00	0	12.61	502	26.67	1491	15.55	2177	3.76	2918	30.19	1413	0.00	0
MORGAN GA	0.00	0	12.31	517	5.80	2965	12.03	2508	8.11	2108	33.70	1222	0.00	0

TOTAL FALLOUT ACTIVITY (µCi/sq meter) by SHOT SERIES WITH COUNTY RANK

COUNTY AND STATE	RANGER SERIES	RANGER RANK	B-J SERIES	B-J RANK	T-S SERIES	T-S RANK	U-K SERIES	UK RANK	TEAPOT SERIES	TP RANK	PLUMBBOB SERIES	PB RANK	SHOT SEDAN	SEDAN RANK
	1951		1951		1952		1953		1955		1957		1962	
MURRAY GA	0.00	0	11.44	570	11.78	2384	14.93	2231	12.19	1696	19.23	2326	0.00	0
NEWTON GA	0.00	0	4.87	1033	4.98	3011	10.61	2673	6.87	2291	34.96	1155	0.00	0
OCONEE GA	0.00	0	8.35	755	6.83	2865	11.81	2529	8.98	2015	29.55	1453	0.00	0
OGLETHORPE GA	0.00	0	10.03	656	4.76	3017	13.07	2414	6.49	2361	32.06	1304	0.00	0
PAULDING GA	0.00	0	6.20	902	7.60	2792	10.35	2703	10.36	1884	40.21	1024	0.00	0
PEACH GA	0.00	0	8.96	716	8.64	2698	10.25	2713	5.12	2631	45.71	912	0.00	0
PICKENS GA	0.00	0	5.89	931	9.14	2647	12.70	2451	11.29	1791	24.45	1810	0.00	0
PIERCE GA	0.00	0	7.51	806	14.40	2118	13.56	2374	4.99	2661	22.03	2048	0.00	0
PIKE GA	0.00	0	7.56	803	7.86	2771	10.23	2719	5.61	2527	36.82	1099	0.00	0
POLK GA	0.00	0	5.47	966	8.43	2725	11.46	2562	9.68	1957	22.30	2022	0.00	0
PULASKI GA	0.00	0	14.08	450	5.36	2989	11.03	2615	4.41	2793	46.41	892	0.00	0
PUTNAM GA	0.00	0	7.58	797	6.26	2922	10.70	2660	6.34	2393	33.43	1239	0.00	0
QUITMAN GA	0.27	527	15.77	388	11.12	2468	8.96	2859	7.58	2174	40.09	1026	0.00	0
RABUN GA	0.00	0	9.30	694	9.85	2586	15.94	2139	20.98	920	50.06	810	0.00	0
RANDOLPH GA	0.00	0	5.49	965	22.17	1651	13.53	2380	8.35	2079	34.08	1204	0.00	0
RICHMOND GA	0.29	502	11.24	580	6.08	2936	12.06	2503	7.78	2143	26.16	1661	0.00	0
ROCKDALE GA	0.27	528	4.90	1024	4.63	3020	11.33	2575	6.24	2411	32.25	1289	0.00	0
SCHLEY GA	0.00	0	14.07	451	6.68	2881	10.39	2697	5.65	2522	44.43	938	0.00	0
SCREVEN GA	0.27	529	15.79	387	21.71	1677	11.95	2519	5.54	2543	30.04	1422	0.00	0
SEMINOLE GA	0.00	0	1.23	1975	14.41	2117	9.63	2785	5.27	2595	22.64	1981	0.00	0
SPALDING GA	0.29	503	7.49	809	6.83	2866	10.23	2717	6.23	2416	37.06	1093	0.00	0
STEPHENS GA	0.00	0	7.94	782	7.34	2821	16.94	2060	7.81	2137	25.89	1689	0.00	0
STEWART GA	0.00	0	14.76	431	11.33	2434	8.99	2854	6.55	2348	32.37	1285	0.00	0
SUMTER GA	0.27	530	11.52	562	13.56	2199	9.45	2805	5.25	2608	34.58	1173	0.00	0
TALBOT GA	0.00	0	8.16	763	7.11	2844	9.47	2800	5.47	2555	30.79	1381	0.00	0
TALIAFERRO GA	0.00	0	10.37	630	6.70	2879	12.20	2489	6.56	2346	34.38	1185	0.00	0

TOTAL FALLOUT ACTIVITY (µCi/sq meter) by SHOT SERIES WITH COUNTY RANK

COUNTY AND STATE	RANGER SERIES	RANGER RANK	B-J SERIES	B-J RANK	T-S SERIES	T-S RANK	U-K SERIES	UK RANK	TEAPOT SERIES	TP RANK	PLUMBBOB SERIES	PB RANK	SHOT SEDAN	SEDAN RANK
	1951		1951		1952		1953		1955		1957		1962	
TATTNALL GA	0.00	0	10.34	634	17.15	1932	15.42	2190	5.58	2536	31.70	1319	0.00	0
TAYLOR GA	0.00	0	8.73	729	3.66	3049	9.60	2789	5.98	2457	35.53	1137	0.00	0
TELFAIR GA	0.29	504	10.11	648	11.46	2424	14.97	2227	4.48	2778	29.67	1446	0.00	0
TERRELL GA	0.00	0	12.61	504	18.22	1850	10.61	2674	6.22	2420	39.21	1043	0.00	0
THOMAS GA	0.00	0	6.31	893	14.80	2088	14.82	2243	6.34	2394	26.74	1615	0.00	0
TIFT GA	0.27	531	9.61	680	11.62	2405	15.52	2178	4.25	2835	29.97	1424	0.00	0
TOOMBS GA	0.00	0	12.32	515	17.29	1924	15.83	2151	3.80	2912	28.35	1516	0.00	0
TOWNS GA	0.00	0	6.69	863	10.48	2529	20.98	1771	17.78	1182	37.76	1074	0.00	0
TREUTLEN GA	0.00	0	6.18	903	11.01	2476	12.02	2509	5.10	2636	22.76	1963	0.00	0
TROUP GA	0.27	532	8.37	752	13.00	2262	10.10	2738	7.97	2126	33.60	1230	0.00	0
TURNER GA	0.00	0	11.09	589	16.40	1976	10.61	2672	3.86	2901	28.44	1512	0.00	0
TWIGGS GA	0.00	0	11.56	560	11.52	2416	11.58	2544	7.61	2172	34.52	1175	0.00	0
UNION GA	0.27	533	6.12	911	15.22	2052	19.76	1854	14.65	1467	22.73	1966	0.00	0
UPSON GA	0.00	0	6.08	913	8.11	2755	9.88	2759	6.39	2388	30.96	1367	0.00	0
WALKER GA	0.00	0	12.15	527	9.02	2658	13.26	2398	13.42	1577	22.29	2024	0.00	0
WALTON GA	0.00	0	6.49	879	3.86	3040	12.18	2493	6.92	2285	33.38	1241	0.00	0
WARE GA	0.00	0	6.69	864	11.31	2438	13.08	2412	5.83	2487	19.71	2283	0.00	0
WARREN GA	0.00	0	11.05	591	6.89	2861	12.71	2448	6.58	2335	33.52	1233	0.00	0
WASHINGTON GA	0.00	0	9.10	704	5.50	2980	11.37	2571	6.26	2408	35.76	1131	0.00	0
WAYNE GA	0.27	534	9.49	685	21.07	1711	13.98	2335	4.93	2680	41.61	1001	0.00	0
WEBSTER GA	0.00	0	12.78	496	13.19	2238	11.17	2597	5.70	2509	39.36	1039	0.00	0
WHEELER GA	0.00	0	10.08	651	28.54	1448	15.02	2223	4.01	2872	31.01	1362	0.00	0
WHITE GA	0.00	0	6.84	854	14.87	2081	17.12	2052	12.29	1684	28.88	1487	0.00	0
WHITFIELD GA	0.00	0	12.08	532	13.81	2175	13.02	2421	9.81	1941	18.39	2409	0.00	0
WILCOX GA	0.00	0	11.34	575	14.03	2161	10.79	2651	5.05	2646	32.65	1276	0.00	0
WILKES GA	0.00	0	8.47	747	5.80	2966	13.09	2410	6.39	2387	30.75	1384	0.00	0

TOTAL FALLOUT ACTIVITY (µCi/sq meter) by SHOT SERIES WITH COUNTY RANK

COUNTY AND STATE	RANGER SERIES	RANGER RANK	B-J SERIES	B-J RANK	T-S SERIES	T-S RANK	U-K SERIES	UK RANK	TEAPOT SERIES	TP RANK	PLUMBBOB SERIES	PB RANK	SHOT SEDAN	SEDAN RANK
	1951		1951		1952		1953		1955		1957		1962	
WILKINSON GA	0.00	0	10.31	638	15.18	2057	11.76	2534	7.17	2248	25.46	1725	0.00	0
WORTH GA	0.00	0	5.98	924	13.18	2239	11.81	2530	5.23	2611	32.93	1270	0.00	0
ADAIR IA	0.00	0	1.93	1643	56.62	830	42.91	803	17.34	1224	135.29	114	345.56	77
ADAMS IA	0.00	0	1.93	1644	71.44	605	40.24	889	18.73	1077	113.43	199	345.56	82
ALLAMAKEE IA	0.00	0	0.60	2375	45.09	1018	74.51	342	17.73	1186	47.39	870	139.44	302
APPANOOSE IA	0.00	0	0.60	2371	199.36	253	38.89	941	23.80	760	54.28	730	345.56	99
AUDUBON IA	0.00	0	2.00	1620	27.80	1466	70.94	380	17.54	1204	124.04	165	345.56	79
BENTON IA	0.00	0	0.35	2754	51.49	899	73.63	353	16.62	1290	101.94	237	527.44	21
BLACK HAWK IA	0.00	0	0.35	2755	42.68	1064	63.86	461	16.48	1302	101.39	240	252.68	125
BOONE IA	0.00	0	0.43	2598	39.32	1137	65.08	451	18.68	1082	111.76	203	527.44	20
BREMER IA	0.00	0	1.28	1955	25.22	1525	74.65	341	18.77	1073	98.83	253	220.22	141
BUCHANAN IA	0.00	0	0.18	2917	20.04	1740	65.65	444	14.66	1466	86.22	341	220.22	144
BUENA VISTA IA	0.00	0	1.59	1802	68.98	637	49.91	647	15.59	1381	32.97	1267	527.44	35
BUTLER IA	0.00	0	0.68	2297	40.73	1104	77.64	314	15.17	1425	40.57	1019	208.70	196
CALHOUN IA	0.00	0	1.89	1670	53.42	878	57.00	540	18.40	1107	29.41	1461	527.44	40
CARROLL IA	0.00	0	1.64	1772	73.44	581	51.76	625	16.33	1315	112.12	201	527.44	19
CASS IA	0.00	0	2.04	1601	77.00	537	48.77	671	17.98	1162	131.93	123	345.56	78
CEDAR IA	0.00	0	0.66	2320	32.97	1316	58.82	515	17.72	1187	75.72	422	146.81	287
CERRO GORDO IA	0.00	0	0.43	2599	32.57	1330	74.12	349	13.31	1586	26.46	1638	208.70	198
CHEROKEE IA	0.00	0	1.64	1773	61.11	753	45.20	743	14.06	1526	30.71	1387	527.44	38
CHICKASAW IA	0.00	0	1.28	1956	38.82	1151	77.01	318	12.86	1630	55.54	703	0.00	0
CLARKE IA	0.00	0	1.93	1645	61.93	742	40.03	894	18.83	1064	92.99	291	345.56	88
CLAY IA	0.00	0	2.53	1478	51.96	890	44.56	758	14.00	1535	30.90	1374	527.44	37
CLAYTON IA	0.00	0	0.50	2525	37.26	1195	68.85	407	16.78	1275	75.24	428	139.13	311
CLINTON IA	0.00	0	0.50	2533	49.00	941	47.39	699	11.41	1777	54.73	720	172.89	252
CRAWFORD IA	0.00	0	1.84	1688	33.41	1300	50.62	638	17.25	1232	38.73	1051	527.44	30

TOTAL FALLOUT ACTIVITY (µCi/sq meter) by SHOT SERIES WITH COUNTY RANK

COUNTY AND STATE	RANGER SERIES	RANGER RANK	B-J SERIES	B-J RANK	T-S SERIES	T-S RANK	U-K SERIES	UK RANK	TEAPOT SERIES	TP RANK	PLUMBBOB SERIES	PB RANK	SHOT SEDAN	SEDAN RANK
	1951		1951		1952		1953		1955		1957		1962	
DALLAS IA	0.00	0	1.91	1666	43.19	1051	59.76	506	21.19	909	108.21	211	139.13	303
DAVIS IA	0.00	0	0.52	2484	89.14	426	40.53	878	23.73	766	136.16	109	345.56	76
DECATUR IA	0.00	0	1.93	1646	50.80	909	31.69	1216	22.16	851	114.98	191	345.56	81
DELAWARE IA	0.00	0	0.52	2507	28.69	1445	79.88	296	15.29	1409	88.34	331	345.56	90
DES MOINES IA	0.00	0	0.41	2686	25.35	1522	35.95	1035	15.89	1349	76.23	418	139.13	310
DICKINSON IA	0.00	0	0.35	2756	44.90	1022	48.65	675	14.36	1501	26.19	1657	527.44	45
DUBUQUE IA	0.00	0	0.67	2314	21.00	1712	68.29	412	14.35	1502	74.23	439	172.89	248
EMMET IA	0.00	0	0.64	2342	42.64	1065	52.19	616	14.42	1496	21.29	2106	527.44	50
FAYETTE IA	0.00	0	0.56	2422	48.89	943	62.65	470	16.33	1314	89.11	325	208.70	189
FLOYD IA	0.00	0	1.38	1893	45.15	1017	53.30	598	16.81	1273	33.16	1254	527.44	34
FRANKLIN IA	0.00	0	0.68	2298	37.86	1176	69.96	388	14.19	1517	108.23	210	208.70	187
FREMONT IA	0.00	0	2.04	1602	127.92	319	40.72	865	17.37	1222	37.64	1076	76.52	509
GREENE IA	0.00	0	1.56	1815	32.61	1327	46.97	704	20.23	965	82.44	374	527.44	23
GRUNDY IA	0.00	0	0.18	2918	39.88	1130	66.07	436	18.70	1080	62.04	615	0.00	0
GUTHRIE IA	0.00	0	1.93	1647	21.51	1686	54.03	584	18.17	1135	82.50	373	345.56	91
HAMILTON IA	0.00	0	0.18	2919	34.52	1273	66.69	431	27.25	618	77.96	406	527.44	25
HANCOCK IA	0.00	0	0.68	2299	51.62	895	67.43	424	14.10	1523	22.48	2002	527.44	48
HARDIN IA	0.00	0	0.68	2300	43.01	1057	68.80	409	17.42	1212	76.14	419	208.70	190
HARRISON IA	0.00	0	1.77	1729	69.41	627	47.42	698	16.52	1301	39.31	1042	139.13	326
HENRY IA	0.00	0	0.41	2687	16.49	1967	26.01	1474	22.03	862	93.22	288	345.56	87
HOWARD IA	0.00	0	1.28	1957	31.84	1349	80.53	289	10.59	1861	46.24	900	950.00	1
HUMBOLDT IA	0.00	0	0.18	2920	51.02	906	75.93	328	15.42	1393	29.58	1450	527.44	39
IDA IA	0.00	0	1.64	1774	65.87	682	49.34	654	16.31	1317	31.09	1354	548.72	17
IOWA IA	0.00	0	0.56	2423	22.86	1617	56.58	543	16.55	1297	71.66	464	527.44	27
JACKSON IA	0.00	0	0.74	2257	22.43	1633	55.58	558	11.31	1786	86.47	339	172.89	247
JASPER IA	0.00	0	0.27	2839	41.45	1089	51.48	627	17.78	1183	126.82	159	527.44	18

TOTAL FALLOUT ACTIVITY (µCi/sq meter) by SHOT SERIES WITH COUNTY RANK

COUNTY AND STATE	RANGER SERIES	RANGER RANK	B-J SERIES	B-J RANK	T-S SERIES	T-S RANK	U-K SERIES	UK RANK	TEAPOT SERIES	TP RANK	PLUMBBOB SERIES	PB RANK	SHOT SEDAN	SEDAN RANK
	1951		1951		1952		1953		1955		1957		1962	
JEFFERSON IA	0.00	0	0.41	2688	15.78	2008	40.34	884	23.37	783	54.72	721	621.15	9
JOHNSON IA	0.00	0	0.58	2404	19.77	1753	52.01	621	17.85	1175	39.89	1030	345.56	102
JONES IA	0.00	0	0.88	2162	27.67	1469	56.56	544	15.86	1355	52.94	755	139.13	319
KEOKUK IA	0.00	0	0.41	2689	24.99	1533	52.53	612	22.61	818	97.63	257	345.56	86
KOSSUTH IA	0.00	0	0.60	2376	45.31	1014	62.88	468	13.68	1565	22.48	2000	527.44	47
LEE IA	0.00	0	0.56	2420	59.26	782	38.88	942	27.38	611	61.62	620	621.15	8
LINN IA	0.00	0	0.67	2315	41.30	1092	59.67	507	15.48	1387	88.73	329	207.58	201
LOUISA IA	0.00	0	0.50	2534	28.22	1456	33.27	1148	22.18	848	61.44	626	345.56	98
LUCAS IA	0.00	0	1.93	1648	18.35	1844	49.04	660	22.52	824	107.58	215	0.00	0
LYON IA	0.00	0	2.28	1533	68.52	644	34.90	1081	7.22	2235	35.13	1150	527.44	31
MADISON IA	0.00	0	1.93	1649	44.47	1030	48.34	681	18.83	1063	139.51	103	345.56	75
MAHASKA IA	0.00	0	0.56	2424	63.83	712	53.32	595	27.09	622	80.99	383	345.56	92
MARION IA	0.00	0	0.52	2508	72.89	588	48.72	672	23.76	763	100.59	246	345.56	85
MARSHALL IA	0.00	0	0.43	2600	60.18	774	75.86	329	20.21	966	82.94	369	527.44	22
MILLS IA	0.00	0	2.04	1603	43.85	1040	47.32	701	17.84	1176	58.44	660	172.89	251
MITCHELL IA	0.00	0	1.38	1894	58.20	796	89.80	253	8.46	2059	23.41	1893	950.00	2
MONONA IA	0.00	0	1.64	1775	73.88	573	43.44	786	16.35	1312	33.34	1245	527.44	33
MONROE IA	0.00	0	0.98	2103	172.42	266	51.03	633	26.61	635	62.72	605	345.56	97
MONTGOMERY IA	0.00	0	2.04	1604	54.41	865	43.36	790	19.17	1042	121.10	174	172.89	246
MUSCATINE IA	0.00	0	0.50	2535	28.75	1441	36.14	1029	23.29	788	55.27	707	139.13	315
O BRIEN IA	0.00	0	2.53	1484	78.91	515	46.38	714	14.97	1438	28.84	1488	527.44	42
OSCEOLA IA	0.00	0	2.56	1473	48.66	947	44.56	759	13.89	1541	28.59	1504	551.97	16
PAGE IA	0.00	0	2.04	1605	28.06	1460	32.45	1184	18.22	1128	70.09	488	200.06	208
PALO ALTO IA	0.00	0	0.60	2377	50.51	915	58.79	516	13.89	1540	22.10	2040	527.44	49
PLYMOUTH IA	0.00	0	2.03	1612	56.73	827	42.52	811	6.07	2442	30.48	1398	553.32	15
POCAHONTAS IA	0.00	0	2.31	1529	59.46	781	59.84	505	15.36	1398	27.28	1573	527.44	44

TOTAL FALLOUT ACTIVITY (µCi/sq meter) by SHOT SERIES WITH COUNTY RANK

COUNTY AND STATE	RANGER SERIES	RANGER RANK	B-J SERIES	B-J RANK	T-S SERIES	T-S RANK	U-K SERIES	UK RANK	TEAPOT SERIES	TP RANK	PLUMBBOB SERIES	PB RANK	SHOT SEDAN	SEDAN RANK
	1951		1951		1952		1953		1955		1957		1962	
POLK IA	0.00	0	0.52	2509	40.08	1123	35.01	1076	15.80	1362	102.78	234	139.13	305
POTTAWATTAMI IA	0.00	0	1.83	1694	38.61	1159	51.22	630	16.14	1329	118.42	179	146.81	286
POWESHIEK IA	0.00	0	0.18	2921	51.49	898	57.52	531	19.80	995	106.86	220	208.70	188
RINGGOLD IA	0.00	0	2.04	1606	97.34	371	32.49	1180	24.26	741	108.17	213	345.56	83
SAC IA	0.00	0	1.81	1713	57.37	814	52.30	615	15.89	1350	29.39	1462	527.44	41
SCOTT IA	0.00	0	0.65	2333	26.49	1499	32.93	1161	9.77	1947	52.63	763	139.13	320
SHELBY IA	0.00	0	1.75	1740	82.08	486	53.41	594	18.03	1150	119.58	178	345.56	80
SIOUX IA	0.00	0	2.56	1474	70.60	614	46.01	722	5.82	2492	33.90	1214	527.44	32
STORY IA	0.00	0	0.43	2601	57.50	812	68.11	417	20.96	921	82.09	377	527.44	24
TAMA IA	0.00	0	0.56	2425	95.20	384	70.83	382	18.74	1076	113.55	196	208.70	186
TAYLOR IA	0.00	0	2.15	1575	32.83	1319	38.76	947	23.00	803	118.41	180	200.06	203
UNION IA	0.00	0	1.93	1650	66.50	673	46.77	706	19.84	992	107.80	214	345.56	84
VAN BUREN IA	0.00	0	0.60	2372	34.39	1276	39.60	910	23.51	773	97.40	261	621.15	6
WAPELLO IA	0.00	0	0.60	2373	43.04	1054	53.61	591	24.68	716	143.59	92	345.56	74
WARREN IA	0.00	0	2.18	1563	27.46	1474	53.69	589	22.43	831	105.10	225	139.13	304
WASHINGTON IA	0.00	0	0.75	2253	24.43	1562	42.26	819	21.36	898	48.74	841	621.15	10
WAYNE IA	0.00	0	1.93	1651	98.41	366	38.90	938	29.23	575	66.31	551	345.56	96
WEBSTER IA	0.00	0	1.48	1852	40.65	1108	51.42	629	17.39	1218	31.03	1361	208.70	197
WINNEBAGO IA	0.00	0	0.68	2301	39.92	1126	71.28	379	8.09	2112	20.10	2241	527.44	51
WINNESHIEK IA	0.00	0	1.10	2036	57.95	800	67.69	420	10.76	1842	44.05	944	527.44	29
WOODBURY IA	0.00	0	1.67	1766	78.86	516	43.77	779	14.76	1456	34.63	1170	665.68	5
WORTH IA	0.00	0	1.38	1895	30.91	1378	80.42	290	17.90	1171	20.93	2151	950.00	3
WRIGHT IA	0.00	0	0.43	2602	30.76	1384	62.67	469	14.91	1444	22.97	1942	208.70	200
ADA ID	0.00	0	60.55	25	17419.10	26	12.54	2463	32.55	507	55.82	697	0.00	0
ADAMS ID	0.00	0	50.73	35	4211.37	74	9.35	2820	27.28	616	49.07	833	0.00	0
BANNOCK ID	0.00	0	2.56	1472	2904.05	80	56.39	547	14.05	1527	74.11	442	0.00	0

TOTAL FALLOUT ACTIVITY (µCi/sq meter) by SHOT SERIES WITH COUNTY RANK

COUNTY AND STATE	RANGER SERIES	RANGER RANK	B-J SERIES	B-J RANK	T-S SERIES	T-S RANK	U-K SERIES	UK RANK	TEAPOT SERIES	TP RANK	PLUMBBOB SERIES	PB RANK	SHOT SEDAN	SEDAN RANK
	1951		1951		1952		1953		1955		1957		1962	
BEAR LAKE ID	0.00	0	1.81	1704	1275.63	112	469.72	59	24.16	747	107.57	217	0.00	0
BENEWAH ID	0.00	0	1.40	1883	4329.45	72	11.68	2539	5.71	2508	48.83	839	0.00	0
BINGHAM ID	0.00	0	1.81	1705	2452.17	85	16.69	2088	12.95	1621	57.23	677	0.00	0
BLAINE ID	0.00	0	0.56	2428	60936.57	3	17.89	1988	5.28	2593	54.66	723	0.00	0
BOISE ID	0.00	0	45.22	48	22648.86	12	8.20	2901	25.10	690	46.21	901	0.00	0
BONNER ID	0.00	0	1.00	2077	1095.91	118	24.28	1563	6.59	2334	52.04	777	0.00	0
BONNEVILLE ID	0.00	0	1.81	1706	2452.06	86	55.19	567	9.06	2006	54.91	716	0.00	0
BOUNDARY ID	0.00	0	0.87	2168	623.97	155	16.12	2126	5.68	2514	46.31	895	0.00	0
BUTTE ID	0.00	0	0.56	2429	16541.29	29	16.13	2125	5.11	2633	50.35	806	0.00	0
CAMAS ID	0.00	0	29.80	99	26334.32	8	8.33	2894	16.82	1272	48.49	844	0.00	0
CANYON ID	0.00	0	46.15	46	6143.08	66	8.96	2857	27.15	620	48.17	856	0.00	0
CARIBOU ID	0.00	0	1.81	1707	2465.60	84	196.00	114	12.05	1712	69.63	497	0.00	0
CASSIA ID	0.00	0	10.32	637	12321.25	40	22.32	1678	14.17	1519	84.04	356	0.00	0
CLARK ID	0.00	0	0.96	2119	3101.20	78	16.33	2108	14.63	1470	127.04	158	0.00	0
CLEARWATER ID	0.00	0	53.35	32	13414.48	37	8.28	2897	4.78	2719	48.83	838	0.00	0
CUSTER ID	0.00	0	33.63	77	64272.71	2	6.26	2976	14.62	1473	56.20	691	0.00	0
ELMORE ID	0.00	0	46.26	45	18339.35	24	14.83	2238	27.92	600	58.88	656	0.00	0
FRANKLIN ID	0.00	0	1.81	1708	1892.14	92	287.43	90	18.17	1136	106.94	219	0.00	0
FREMONT ID	0.00	0	0.82	2198	4149.23	75	16.68	2090	21.16	912	95.78	273	0.00	0
GEM ID	0.00	0	48.40	41	67575.76	1	7.01	2951	21.91	867	45.46	915	0.00	0
GOODING ID	0.00	0	19.07	290	24452.81	10	14.83	2237	19.89	989	48.63	842	0.00	0
IDAHO ID	0.00	0	48.44	40	26815.43	7	8.12	2905	23.96	753	55.70	700	0.00	0
JEFFERSON ID	0.00	0	0.56	2430	14355.40	34	14.96	2228	8.42	2068	122.13	168	0.00	0
JEROME ID	0.00	0	9.93	662	12277.84	42	14.69	2265	19.01	1049	58.14	664	0.00	0
KOOTENAI ID	0.00	0	1.00	2078	1477.84	101	17.42	2031	5.60	2532	45.13	922	0.00	0
LATAH ID	0.00	0	1.00	2079	5765.74	68	17.37	2035	7.42	2197	62.52	608	0.00	0

TOTAL FALLOUT ACTIVITY (µCi/sq meter) by SHOT SERIES WITH COUNTY RANK

COUNTY AND STATE	RANGER SERIES	RANGER RANK	B-J SERIES	B-J RANK	T-S SERIES	T-S RANK	U-K SERIES	UK RANK	TEAPOT SERIES	TP RANK	PLUMBBOB SERIES	PB RANK	SHOT SEDAN	SEDAN RANK
	1951		1951		1952		1953		1955		1957		1962	
LEMHI ID	0.00	0	0.56	2431	43722.44	4	9.33	2823	4.68	2743	67.11	537	0.00	0
LEWIS ID	0.00	0	1.00	2080	19297.33	20	9.21	2834	9.05	2009	46.25	899	0.00	0
LINCOLN ID	0.00	0	10.05	653	18350.26	22	13.84	2345	18.91	1058	55.03	713	0.00	0
MADISON ID	0.00	0	0.82	2199	7550.52	59	15.77	2154	14.31	1506	121.45	173	0.00	0
MINIDOKA ID	0.00	0	10.05	654	12263.69	43	16.37	2107	11.98	1718	61.46	625	0.00	0
NEZ PERCE ID	0.00	0	1.00	2081	8606.45	53	7.54	2927	6.31	2399	49.63	822	0.00	0
ONEIDA ID	0.00	0	2.71	1448	2528.36	82	105.62	185	17.99	1159	93.93	284	0.00	0
OWYHEE ID	0.00	0	92.86	6	18343.64	23	14.72	2261	26.61	636	62.11	614	0.00	0
PAYETTE ID	0.00	0	36.87	64	13602.07	36	5.72	2998	17.99	1160	49.82	815	0.00	0
POWER ID	0.00	0	2.71	1449	4303.57	73	23.13	1631	12.19	1697	75.75	421	0.00	0
SHOSHONE ID	0.00	0	1.58	1807	11603.55	46	8.79	2867	8.25	2093	49.03	835	0.00	0
TETON ID	0.00	0	0.82	2200	1636.43	100	18.12	1970	9.64	1959	146.16	86	0.00	0
TWIN FALLS ID	0.00	0	19.98	265	12320.12	41	18.05	1976	19.79	998	56.33	690	0.00	0
VALLEY ID	0.00	0	49.56	39	29504.28	6	8.43	2889	27.88	602	63.37	591	0.00	0
WASHINGTON ID	0.00	0	34.69	68	19876.01	18	6.60	2967	20.52	953	49.19	829	0.00	0
ADAMS IL	2.82	304	0.49	2548	139.47	306	25.59	1494	24.97	696	52.75	761	200.06	215
ALEXANDER IL	0.00	0	3.44	1318	29.45	1427	35.59	1048	18.57	1091	22.49	1998	0.00	0
BOND IL	0.00	0	0.52	2485	43.26	1050	30.80	1248	14.83	1450	20.09	2243	90.94	411
BOONE IL	0.00	0	0.73	2258	27.63	1471	31.26	1230	11.66	1747	49.09	832	146.81	291
BROWN IL	0.00	0	0.41	2690	36.44	1215	19.37	1887	21.46	893	67.94	520	200.06	209
BUREAU IL	0.00	0	0.65	2330	23.98	1573	33.13	1151	9.82	1940	37.47	1082	0.00	0
CALHOUN IL	0.00	0	0.52	2486	47.13	980	19.11	1902	16.04	1334	48.41	846	90.94	407
CARROLL IL	0.00	0	0.91	2140	17.36	1917	48.33	682	8.36	2076	44.90	925	0.00	0
CASS IL	0.00	0	0.52	2478	49.46	936	18.67	1937	41.32	399	48.24	849	200.06	217
CHAMPAIGN IL	0.00	0	0.50	2536	82.62	479	24.41	1555	19.73	1001	16.00	2621	99.74	387
CHRISTIAN IL	0.00	0	0.40	2708	146.72	292	24.13	1575	18.04	1148	46.13	906	76.52	506

TOTAL FALLOUT ACTIVITY (µCi/sq meter) by SHOT SERIES WITH COUNTY RANK

COUNTY AND STATE	RANGER SERIES	RANGER RANK	B-J SERIES	B-J RANK	T-S SERIES	T-S RANK	U-K SERIES	UK RANK	TEAPOT SERIES	TP RANK	PLUMBBOB SERIES	PB RANK	SHOT SEDAN	SEDAN RANK
	1951		1951		1952		1953		1955		1957		1962	
CLARK IL	0.00	0	0.52	2487	50.18	924	23.00	1643	15.47	1389	23.89	1858	99.74	382
CLAY IL	0.00	0	0.52	2488	76.97	538	26.46	1454	24.67	718	25.21	1747	92.52	400
CLINTON IL	0.00	0	0.44	2591	40.71	1106	23.26	1624	15.24	1413	25.81	1698	0.00	0
COLES IL	0.00	0	0.60	2374	77.87	525	30.89	1245	16.58	1295	23.78	1869	200.06	221
COOK IL	1.27	366	0.71	2277	29.98	1410	18.67	1938	10.76	1841	15.70	2641	0.00	0
CRAWFORD IL	0.00	0	0.84	2188	64.87	692	27.77	1396	34.91	473	27.05	1594	0.00	0
CUMBERLAND IL	0.00	0	0.69	2296	22.16	1652	24.41	1556	25.49	676	20.66	2179	80.75	486
DE KALB IL	0.00	0	0.73	2259	32.07	1340	24.97	1521	11.78	1735	64.13	580	0.00	0
DE WITT IL	0.00	0	0.52	2479	75.77	550	16.59	2095	17.06	1252	14.93	2700	139.13	327
DOUGLAS IL	0.00	0	0.41	2691	140.05	303	20.92	1777	21.63	879	22.92	1948	80.75	485
DU PAGE IL	0.00	0	0.63	2351	56.61	831	19.95	1842	12.27	1687	13.23	2808	0.00	0
EDGAR IL	0.00	0	1.21	1989	43.48	1046	20.67	1800	14.41	1497	23.97	1845	80.75	484
EDWARDS IL	0.00	0	0.32	2778	48.37	958	28.06	1380	19.70	1006	20.44	2208	0.00	0
EFFINGHAM IL	0.00	0	0.52	2489	130.55	315	21.28	1744	16.00	1336	30.12	1417	80.75	483
FAYETTE IL	0.00	0	0.52	2490	32.83	1320	27.75	1399	18.06	1145	20.44	2207	90.94	409
FORD IL	0.00	0	0.50	2537	156.47	282	22.15	1690	14.95	1440	15.94	2627	0.00	0
FRANKLIN IL	0.00	0	0.32	2779	18.47	1835	17.78	2000	16.37	1308	28.84	1489	0.00	0
FULTON IL	0.00	0	0.41	2685	38.38	1163	31.46	1221	23.43	778	48.23	851	345.56	101
GALLATIN IL	0.00	0	0.32	2780	22.33	1642	18.90	1915	14.57	1478	23.88	1860	0.00	0
GREENE IL	0.00	0	0.41	2692	47.90	969	20.16	1833	18.93	1057	58.12	666	200.06	214
GRUNDY IL	0.71	423	0.64	2340	82.88	475	33.63	1127	14.64	1468	15.99	2622	0.00	0
HAMILTON IL	0.00	0	0.32	2781	20.96	1713	23.53	1608	15.91	1347	20.72	2175	0.00	0
HANCOCK IL	0.00	0	0.50	2538	37.00	1201	39.44	919	20.07	975	74.48	435	621.15	7
HARDIN IL	0.00	0	0.32	2782	17.43	1912	19.18	1898	16.13	1330	20.96	2143	0.00	0
HENDERSON IL	1.11	402	0.72	2271	56.40	832	34.92	1079	17.47	1208	53.85	739	139.13	317
HENRY IL	0.00	0	0.44	2597	20.00	1743	34.33	1105	10.87	1832	53.22	749	139.13	318

TOTAL FALLOUT ACTIVITY (µCi/sq meter) by SHOT SERIES WITH COUNTY RANK

COUNTY AND STATE	RANGER SERIES	RANGER RANK	B-J SERIES	B-J RANK	T-S SERIES	T-S RANK	U-K SERIES	UK RANK	TEAPOT SERIES	TP RANK	PLUMBBOB SERIES	PB RANK	SHOT SEDAN	SEDAN RANK
	1951		1951		1952		1953		1955		1957		1962	
IROQUOIS IL	0.00	0	0.88	2160	96.21	379	23.18	1628	15.20	1418	17.06	2538	0.00	0
JACKSON IL	0.00	0	3.64	1279	19.30	1789	25.70	1487	17.85	1174	30.64	1388	0.00	0
JASPER IL	0.00	0	0.52	2491	129.89	317	23.46	1612	26.20	649	27.35	1569	99.74	381
JEFFERSON IL	0.00	0	0.44	2592	41.67	1084	28.46	1362	15.19	1419	20.33	2218	37.08	634
JERSEY IL	15.59	116	0.52	2482	55.35	852	18.66	1939	16.73	1280	46.51	890	90.94	408
JO DAVIESS IL	0.00	0	0.57	2415	21.91	1666	53.63	590	9.95	1924	21.44	2096	0.00	0
JOHNSON IL	0.00	0	0.54	2449	18.07	1865	26.44	1455	16.30	1320	72.73	452	0.00	0
KANE IL	0.00	0	0.63	2352	45.42	1013	22.91	1646	11.34	1782	46.66	886	0.00	0
KANKAKEE IL	0.00	0	0.58	2405	96.76	376	21.11	1756	18.06	1144	16.32	2597	0.00	0
KENDALL IL	0.00	0	0.42	2681	46.46	998	20.24	1828	12.35	1677	13.02	2822	0.00	0
KNOX IL	0.00	0	0.47	2560	32.66	1325	31.21	1234	18.45	1101	41.89	993	139.13	324
LA SALLE IL	0.78	416	0.64	2339	75.26	553	29.87	1287	13.68	1563	47.01	875	0.00	0
LAKE IL	0.00	0	0.63	2353	61.53	747	27.77	1397	10.31	1887	17.02	2542	0.00	0
LAWRENCE IL	0.00	0	0.59	2401	101.78	354	19.74	1860	16.37	1309	18.59	2385	0.00	0
LEE IL	0.00	0	0.50	2524	60.30	769	31.25	1231	11.73	1736	37.93	1067	0.00	0
LIVINGSTON IL	0.00	0	0.68	2307	74.55	564	29.70	1299	19.09	1045	15.64	2647	0.00	0
LOGAN IL	0.00	0	0.47	2561	54.68	861	28.59	1355	26.69	633	65.31	566	76.52	503
MACON IL	0.00	0	0.55	2442	142.04	299	27.97	1387	19.71	1004	24.96	1769	76.52	511
MACOUPIN IL	0.00	0	0.41	2693	44.11	1036	21.86	1709	17.42	1213	46.09	909	200.06	218
MADISON IL	0.00	0	0.52	2492	36.58	1213	17.34	2036	15.34	1402	42.79	978	0.00	0
MARION IL	0.00	0	0.44	2593	35.81	1236	25.88	1481	17.17	1238	20.93	2152	0.00	0
MARSHALL IL	0.74	418	0.52	2471	22.94	1614	33.01	1157	12.92	1623	45.54	913	172.89	256
MASON IL	0.00	0	0.47	2567	45.50	1012	31.18	1238	26.81	627	47.84	863	139.13	323
MASSAC IL	0.00	0	0.76	2239	18.97	1809	22.21	1686	14.47	1490	26.30	1648	0.00	0
MCDONOUGH IL	0.00	0	0.57	2417	33.11	1312	30.43	1267	19.60	1011	70.19	485	345.56	94
MCHENRY IL	0.00	0	0.42	2682	25.02	1530	28.95	1340	10.26	1893	46.27	898	0.00	0

TOTAL FALLOUT ACTIVITY (µCi/sq meter) by SHOT SERIES WITH COUNTY RANK

COUNTY AND STATE	RANGER SERIES	RANGER RANK	B-J SERIES	B-J RANK	T-S SERIES	T-S RANK	U-K SERIES	UK RANK	TEAPOT SERIES	TP RANK	PLUMBBOB SERIES	PB RANK	SHOT SEDAN	SEDAN RANK
	1951		1951		1952		1953		1955		1957		1962	
MCLEAN IL	0.00	0	0.58	2410	57.40	813	25.60	1493	16.35	1311	14.98	2695	146.81	295
MENARD IL	0.00	0	0.37	2745	46.82	989	25.19	1511	20.44	956	43.76	950	76.52	507
MERCER IL	0.00	0	0.40	2709	23.86	1576	32.50	1179	14.34	1503	54.97	714	139.13	316
MONROE IL	0.00	0	0.52	2493	24.80	1544	17.23	2047	30.23	552	36.97	1095	0.00	0
MONTGOMERY IL	0.00	0	0.41	2694	37.93	1174	21.93	1703	17.39	1219	43.38	962	200.06	220
MORGAN IL	0.00	0	0.40	2710	34.61	1270	23.58	1605	19.72	1002	67.28	532	76.52	500
MOULTRIE IL	0.00	0	0.41	2695	138.43	308	19.48	1876	16.39	1307	19.97	2258	200.06	223
OGLE IL	0.00	0	0.49	2549	38.36	1165	32.31	1193	10.65	1856	63.96	584	0.00	0
PEORIA IL	0.00	0	0.25	2852	34.93	1260	33.46	1136	22.09	855	49.64	821	146.81	290
PERRY IL	0.00	0	0.44	2594	20.62	1720	26.21	1467	16.45	1305	20.66	2178	0.00	0
PIATT IL	0.00	0	0.57	2418	149.65	290	19.76	1855	15.76	1367	15.25	2680	80.75	488
PIKE IL	0.00	0	0.41	2696	81.88	487	25.67	1490	25.80	664	70.78	477	200.06	207
POPE IL	0.00	0	0.32	2783	23.26	1603	30.43	1266	15.31	1407	28.64	1501	0.00	0
PULASKI IL	0.00	0	3.55	1292	21.35	1697	37.40	997	18.96	1053	26.59	1629	0.00	0
PUTNAM IL	0.00	0	0.58	2403	59.58	779	24.80	1529	11.96	1720	37.47	1081	0.00	0
RANDOLPH IL	10.09	166	0.63	2358	20.39	1730	25.74	1485	24.35	735	43.67	955	0.00	0
RICHLAND IL	0.00	0	0.52	2494	110.42	339	20.54	1806	15.13	1426	19.65	2290	35.48	643
ROCK ISLAND IL	1.49	353	0.56	2419	24.89	1540	42.07	826	11.58	1760	58.14	665	139.13	314
SALINE IL	0.00	0	0.43	2611	15.37	2044	25.98	1476	16.89	1267	25.15	1755	0.00	0
SANGAMON IL	0.00	0	0.45	2576	35.65	1238	21.03	1764	24.71	710	49.80	817	200.06	216
SCHUYLER IL	0.00	0	0.40	2711	33.31	1303	20.01	1838	23.09	797	43.49	959	200.06	219
SCOTT IL	0.00	0	0.41	2697	63.37	721	21.79	1713	27.69	607	66.64	542	76.52	501
SHELBY IL	0.00	0	0.52	2495	133.60	311	19.62	1867	16.34	1313	21.12	2127	200.06	222
ST CLAIR IL	0.00	0	0.52	2496	37.78	1179	28.03	1384	14.80	1452	41.17	1008	0.00	0
STARK IL	0.37	485	0.52	2472	24.11	1568	38.52	958	11.93	1723	51.13	789	139.13	321
STEPHENSON IL	0.00	0	1.05	2060	27.11	1483	40.03	893	12.62	1654	51.25	788	0.00	0

TOTAL FALLOUT ACTIVITY (µCi/sq meter) by SHOT SERIES WITH COUNTY RANK

COUNTY AND STATE	RANGER SERIES	RANGER RANK	B-J SERIES	B-J RANK	T-S SERIES	T-S RANK	U-K SERIES	UK RANK	TEAPOT SERIES	TP RANK	PLUMBBOB SERIES	PB RANK	SHOT SEDAN	SEDAN RANK
	1951		1951		1952		1953		1955		1957		1962	
TAZEWELL IL	0.00	0	0.80	2218	31.38	1360	31.73	1214	32.16	512	41.11	1011	139.13	325
UNION IL	0.00	0	3.44	1319	29.74	1419	32.43	1187	18.61	1086	23.35	1897	0.00	0
VERMILION IL	0.00	0	0.50	2539	55.73	845	27.91	1391	19.77	999	21.71	2079	99.74	385
WABASH IL	0.00	0	0.40	2712	46.90	986	39.63	909	24.76	707	20.94	2149	0.00	0
WARREN IL	0.00	0	0.54	2448	19.19	1791	38.89	940	15.94	1342	68.66	512	345.56	95
WASHINGTON IL	0.00	0	0.44	2595	31.87	1348	27.93	1388	17.59	1197	22.03	2047	0.00	0
WAYNE IL	0.00	0	0.53	2470	39.27	1138	25.69	1488	16.29	1321	26.52	1630	0.00	0
WHITE IL	0.00	0	0.32	2784	28.21	1457	26.64	1443	15.88	1352	18.21	2423	0.00	0
WHITESIDE IL	0.00	0	0.55	2444	45.73	1008	38.62	953	10.77	1840	54.88	717	0.00	0
WILL IL	0.49	472	0.88	2164	87.87	435	26.02	1473	15.12	1428	16.69	2567	0.00	0
WILLIAMSON IL	0.00	0	0.66	2327	21.50	1687	26.95	1433	17.57	1200	20.38	2215	0.00	0
WINNEBAGO IL	0.00	0	0.63	2356	27.31	1479	36.38	1019	11.82	1733	50.06	809	146.81	289
WOODFORD IL	0.00	0	0.51	2513	42.12	1074	24.10	1578	13.80	1546	46.68	885	172.89	255
ADAMS IN	0.00	0	0.94	2127	10.17	2563	27.21	1419	13.08	1604	19.75	2277	121.04	350
ALLEN IN	0.99	409	0.65	2329	14.94	2075	26.21	1465	17.94	1168	20.16	2232	308.53	118
BARTHOLOMEW IN	0.00	0	1.03	2067	23.78	1580	21.96	1701	22.05	859	23.88	1859	85.56	446
BENTON IN	0.00	0	0.49	2550	64.82	693	21.52	1730	13.59	1567	17.91	2452	0.00	0
BLACKFORD IN	0.00	0	0.46	2572	13.21	2237	22.28	1680	15.03	1435	17.97	2446	0.00	0
BOONE IN	0.00	0	0.44	2596	54.24	867	27.66	1405	12.14	1705	25.45	1727	0.00	0
BROWN IN	0.00	0	1.21	1988	41.19	1095	34.62	1098	14.24	1513	19.76	2276	0.00	0
CARROLL IN	0.00	0	0.64	2343	74.35	566	23.28	1622	15.13	1427	16.18	2606	80.75	487
CASS IN	0.00	0	0.55	2439	70.05	622	34.83	1087	15.92	1345	19.93	2266	103.59	369
CLARK IN	2.23	320	1.00	2095	14.01	2166	18.16	1966	19.67	1008	16.95	2546	108.54	356
CLAY IN	0.71	424	0.29	2815	40.77	1100	28.21	1377	26.28	648	22.18	2033	0.00	0
CLINTON IN	0.00	0	0.46	2574	38.86	1150	23.90	1587	15.24	1414	19.55	2302	99.74	386
CRAWFORD IN	1.75	340	0.79	2222	16.44	1971	20.84	1782	28.80	586	14.14	2756	157.12	268

TOTAL FALLOUT ACTIVITY (µCi/sq meter) by SHOT SERIES WITH COUNTY RANK

COUNTY AND STATE	RANGER SERIES	RANGER RANK	B-J SERIES	B-J RANK	T-S SERIES	T-S RANK	U-K SERIES	UK RANK	TEAPOT SERIES	TP RANK	PLUMBBOB SERIES	PB RANK	SHOT SEDAN	SEDAN RANK
	1951		1951		1952		1953		1955		1957		1962	
DAVIESS IN	0.00	0	0.37	2741	51.08	904	25.19	1512	40.88	401	25.48	1722	0.00	0
DE KALB IN	0.00	0	0.94	2131	9.37	2625	20.82	1783	9.39	1979	18.48	2403	0.00	0
DEARBORN IN	0.00	0	0.74	2254	15.64	2020	23.35	1616	11.28	1792	54.62	724	0.00	0
DECATUR IN	0.00	0	0.53	2461	14.02	2163	24.40	1557	12.03	1716	43.47	961	35.48	642
DELAWARE IN	9.31	177	1.13	2020	12.05	2353	35.13	1072	9.93	1926	20.87	2161	162.09	262
DUBOIS IN	0.00	0	0.62	2361	66.59	672	20.77	1791	52.57	296	16.32	2596	37.44	626
ELKHART IN	1.11	403	0.59	2398	36.33	1220	20.44	1815	10.08	1910	18.69	2376	0.00	0
FAYETTE IN	0.00	0	0.70	2291	15.73	2014	23.58	1604	9.28	1990	24.48	1800	0.00	0
FLOYD IN	1.84	327	1.42	1876	8.83	2677	18.95	1912	26.04	653	13.79	2778	108.54	357
FOUNTAIN IN	0.00	0	0.35	2757	56.16	837	35.81	1040	13.83	1544	18.34	2413	0.00	0
FRANKLIN IN	19.06	79	1.35	1914	12.39	2322	27.67	1403	16.08	1332	42.92	975	90.29	417
FULTON IN	0.00	0	0.47	2562	20.12	1736	34.35	1104	19.46	1025	17.48	2496	103.59	374
GIBSON IN	0.00	0	0.40	2713	37.55	1185	23.75	1593	18.23	1127	23.71	1876	0.00	0
GRANT IN	9.31	178	0.87	2176	12.99	2263	37.45	994	9.90	1930	19.04	2348	208.78	168
GREENE IN	14.48	143	0.37	2736	66.28	675	21.24	1748	47.66	334	19.70	2285	76.52	514
HAMILTON IN	0.00	0	0.42	2680	20.62	1719	34.80	1089	9.48	1968	26.33	1647	208.78	154
HANCOCK IN	0.00	0	0.55	2437	26.46	1500	29.25	1321	8.19	2100	26.63	1623	0.00	0
HARRISON IN	1.84	328	1.42	1877	9.33	2628	19.66	1864	26.45	643	14.07	2758	95.67	389
HENDRICKS IN	0.71	425	0.83	2193	18.87	1815	34.90	1080	30.79	544	25.14	1760	141.72	298
HENRY IN	0.39	483	1.10	2040	18.44	1837	38.99	933	21.69	876	26.11	1669	125.77	338
HOWARD IN	0.00	0	0.69	2294	17.36	1915	31.93	1207	16.17	1328	20.13	2236	208.78	161
HUNTINGTON IN	0.00	0	1.21	1986	13.49	2212	29.62	1302	12.67	1649	18.90	2360	76.52	516
JACKSON IN	0.71	426	1.51	1834	17.46	1909	19.97	1840	32.82	503	23.38	1894	166.81	258
JASPER IN	0.00	0	0.83	2191	68.66	642	35.90	1037	23.75	765	15.68	2645	0.00	0
JAY IN	11.06	161	1.26	1962	14.34	2132	31.72	1215	12.06	1711	20.98	2141	0.00	0
JEFFERSON IN	1.75	341	3.71	1268	11.21	2454	18.77	1924	19.55	1016	17.09	2534	157.12	265

TOTAL FALLOUT ACTIVITY (µCi/sq meter) by SHOT SERIES WITH COUNTY RANK

COUNTY AND STATE	RANGER SERIES	RANGER RANK	B-J SERIES	B-J RANK	T-S SERIES	T-S RANK	U-K SERIES	UK RANK	TEAPOT SERIES	TP RANK	PLUMBBOB SERIES	PB RANK	SHOT SEDAN	SEDAN RANK
	1951		1951		1952		1953		1955		1957		1962	
JENNINGS IN	0.00	0	0.47	2559	18.62	1827	18.14	1969	14.82	1451	21.20	2115	0.00	0
JOHNSON IN	15.59	117	0.53	2463	30.43	1398	27.03	1428	11.32	1784	21.81	2070	0.00	0
KNOX IN	10.09	167	0.29	2814	107.66	343	26.26	1461	23.73	767	22.66	1979	0.00	0
KOSCIUSKO IN	0.00	0	0.50	2532	33.96	1286	29.59	1304	14.39	1500	18.26	2417	0.00	0
LA PORTE IN	1.82	337	1.51	1839	51.34	902	24.67	1535	9.74	1949	22.57	1989	0.00	0
LAGRANGE IN	1.42	355	0.83	2192	30.86	1380	16.13	2124	9.29	1988	21.48	2093	0.00	0
LAKE IN	0.00	0	0.50	2526	63.89	709	33.05	1154	12.98	1616	19.09	2344	0.00	0
LAWRENCE IN	15.19	140	0.78	2227	46.66	994	20.28	1826	29.00	581	21.17	2124	203.34	202
MADISON IN	13.36	145	0.58	2402	14.64	2103	34.48	1101	17.26	1229	27.19	1583	90.29	419
MARION IN	0.71	427	0.51	2520	32.90	1317	45.33	737	30.82	543	19.90	2269	141.72	299
MARSHALL IN	1.23	367	0.49	2546	35.82	1235	22.53	1667	9.40	1978	19.38	2320	0.00	0
MARTIN IN	11.68	154	0.62	2360	46.94	984	22.56	1665	38.39	429	23.92	1851	76.52	512
MIAMI IN	0.00	0	0.69	2295	13.27	2226	33.60	1128	10.65	1855	16.55	2584	76.52	518
MONROE IN	0.00	0	0.37	2742	41.29	1093	23.05	1636	31.27	526	23.28	1909	99.74	384
MONTGOMERY IN	0.49	473	0.18	2913	52.18	888	40.55	877	12.33	1679	23.04	1935	0.00	0
MORGAN IN	0.00	0	0.62	2362	33.15	1311	27.06	1427	29.63	567	28.66	1498	0.00	0
NEWTON IN	0.00	0	0.65	2334	67.49	661	32.43	1186	13.40	1578	19.36	2321	0.00	0
NOBLE IN	0.78	417	1.08	2048	19.99	1744	22.37	1674	9.32	1984	18.22	2422	0.00	0
OHIO IN	0.00	0	0.59	2387	12.23	2336	17.41	2033	10.83	1835	36.49	1110	0.00	0
ORANGE IN	1.84	329	1.21	1987	36.07	1227	20.66	1801	37.08	443	16.12	2613	95.67	388
OWEN IN	0.71	428	0.37	2740	48.18	960	26.07	1471	35.42	464	22.26	2027	103.59	363
PARKE IN	0.00	0	0.35	2758	48.49	952	28.70	1350	12.01	1717	26.35	1645	0.00	0
PERRY IN	1.75	342	0.37	2738	13.52	2205	18.64	1940	59.87	257	14.91	2703	39.12	616
PIKE IN	1.84	330	0.37	2737	32.68	1324	22.63	1661	20.47	955	21.07	2136	0.00	0
PORTER IN	1.11	404	0.66	2325	63.95	707	30.02	1282	10.35	1885	21.22	2114	0.00	0
POSEY IN	1.84	331	0.63	2355	26.01	1510	23.69	1594	13.28	1588	27.33	1571	39.12	613

TOTAL FALLOUT ACTIVITY (µCi/sq meter) by SHOT SERIES WITH COUNTY RANK

COUNTY AND STATE	RANGER SERIES	RANGER RANK	B-J SERIES	B-J RANK	T-S SERIES	T-S RANK	U-K SERIES	UK RANK	TEAPOT SERIES	TP RANK	PLUMBBOB SERIES	PB RANK	SHOT SEDAN	SEDAN RANK
	1951		1951		1952		1953		1955		1957		1962	
PULASKI IN	0.00	0	0.92	2135	69.87	626	35.44	1056	21.40	895	19.59	2297	0.00	0
PUTNAM IN	0.71	429	0.75	2246	19.00	1807	34.39	1103	33.67	492	27.20	1582	0.00	0
RANDOLPH IN	0.99	410	1.19	1997	11.53	2415	34.14	1112	18.41	1106	17.87	2457	85.56	466
RIPLEY IN	0.00	0	0.54	2445	17.53	1906	25.10	1517	11.94	1722	55.40	706	0.00	0
RUSH IN	0.00	0	0.81	2209	19.55	1771	38.15	971	24.54	724	27.49	1561	0.00	0
SCOTT IN	1.75	343	3.35	1337	17.13	1933	19.82	1850	16.53	1300	18.80	2365	85.56	463
SHELBY IN	15.59	118	0.48	2555	19.36	1783	26.26	1462	12.98	1615	22.48	2001	90.29	426
SPENCER IN	1.75	344	0.37	2739	15.72	2016	18.88	1917	56.88	272	15.25	2679	39.12	615
ST JOSEPH IN	1.23	368	0.42	2683	39.62	1131	15.43	2188	8.96	2016	23.30	1907	99.74	383
STARKE IN	0.00	0	0.40	2706	35.33	1244	22.76	1652	9.99	1917	21.73	2078	0.00	0
STEUBEN IN	0.00	0	0.79	2220	31.14	1370	21.58	1725	10.15	1898	18.92	2359	0.00	0
SULLIVAN IN	0.00	0	0.67	2313	38.56	1160	33.27	1147	41.21	400	33.04	1265	0.00	0
SWITZERLAND IN	15.23	122	3.72	1266	10.71	2501	18.71	1932	16.31	1319	20.04	2249	157.12	263
TIPPECANOE IN	13.36	146	0.58	2407	64.41	702	33.01	1158	29.81	561	19.55	2300	0.00	0
TIPTON IN	0.71	430	0.31	2796	20.60	1721	27.65	1406	11.62	1752	21.69	2080	208.78	158
UNION IN	0.71	431	0.83	2190	16.99	1940	23.13	1632	8.93	2020	24.73	1785	123.00	343
VANDERBURGH IN	2.23	321	0.32	2767	22.38	1639	35.54	1052	19.45	1027	22.78	1960	45.80	586
VERMILLION IN	0.00	0	0.58	2406	49.29	939	26.54	1450	19.84	991	19.63	2293	76.52	515
VIGO IN	0.78	415	0.52	2483	33.20	1308	33.45	1137	41.55	396	33.14	1257	0.00	0
WABASH IN	11.06	162	0.65	2328	13.08	2251	30.79	1249	11.86	1729	16.72	2563	76.52	517
WARREN IN	0.00	0	0.56	2421	91.80	400	35.21	1067	14.14	1522	19.04	2347	0.00	0
WARRICK IN	2.23	322	0.65	2332	26.57	1497	26.40	1456	42.45	386	17.79	2468	39.12	614
WASHINGTON IN	0.00	0	0.36	2747	25.51	1516	19.61	1868	24.18	746	19.86	2271	76.52	513
WAYNE IN	0.71	432	1.05	2058	16.97	1943	30.17	1275	14.02	1532	22.45	2007	299.72	122
WELLS IN	9.31	179	1.66	1767	10.05	2572	29.21	1323	12.46	1665	17.76	2470	208.78	173
WHITE IN	0.00	0	0.34	2764	70.11	621	35.55	1050	18.54	1095	14.31	2745	0.00	0

TOTAL FALLOUT ACTIVITY (µCi/sq meter) by SHOT SERIES WITH COUNTY RANK

COUNTY AND STATE	RANGER SERIES	RANGER RANK	B-J SERIES	B-J RANK	T-S SERIES	T-S RANK	U-K SERIES	UK RANK	TEAPOT SERIES	TP RANK	PLUMBBOB SERIES	PB RANK	SHOT SEDAN	SEDAN RANK
	1951		1951		1952		1953		1955		1957		1962	
WHITLEY IN	0.00	0	1.11	2033	14.30	2136	19.89	1845	12.73	1646	18.89	2361	0.00	0
ALLEN KS	0.00	0	7.48	811	65.51	689	29.48	1308	17.21	1236	48.82	840	0.00	0
ANDERSON KS	0.00	0	4.37	1127	49.34	938	32.79	1167	18.12	1140	52.96	754	0.00	0
ATCHISON KS	0.00	0	2.40	1504	48.18	961	31.30	1228	36.27	455	61.65	619	0.00	0
BARBER KS	0.00	0	4.34	1136	73.10	585	43.28	792	24.96	699	56.37	688	0.00	0
BARTON KS	0.00	0	3.75	1262	75.09	556	44.96	750	49.65	316	88.82	328	0.00	0
BOURBON KS	0.00	0	7.80	787	46.84	988	28.60	1354	19.45	1026	43.30	966	0.00	0
BROWN KS	0.00	0	2.47	1494	35.28	1247	30.68	1255	30.52	548	63.29	594	0.00	0
BUTLER KS	0.00	0	7.71	790	68.14	648	35.06	1074	16.96	1264	74.91	432	0.00	0
CHASE KS	0.00	0	3.59	1289	101.06	358	32.34	1190	15.95	1340	66.57	547	0.00	0
CHAUTAUQUA KS	0.00	0	11.86	546	66.13	679	33.73	1123	40.08	412	50.84	799	0.00	0
CHEROKEE KS	0.00	0	11.48	564	81.50	490	37.78	984	27.92	601	46.29	896	0.00	0
CHEYENNE KS	0.00	0	1.45	1864	87.18	441	34.82	1088	121.34	117	50.80	800	0.00	0
CLARK KS	0.00	0	4.42	1116	86.43	453	53.30	597	25.45	679	58.88	657	0.00	0
CLAY KS	0.00	0	2.47	1495	32.16	1336	40.08	891	27.78	604	67.91	521	0.00	0
CLOUD KS	0.00	0	2.25	1540	29.41	1428	39.86	900	21.22	908	60.27	641	0.00	0
COFFEY KS	0.00	0	4.39	1123	56.31	833	27.92	1390	18.62	1085	62.31	610	0.00	0
COMANCHE KS	0.00	0	2.74	1436	60.24	772	43.39	789	24.87	702	42.63	982	0.00	0
COWLEY KS	0.00	0	10.24	640	75.50	551	33.86	1120	19.98	986	71.16	470	0.00	0
CRAWFORD KS	0.00	0	10.93	599	85.95	457	31.36	1227	22.08	857	56.38	687	0.00	0
DECATUR KS	0.00	0	1.90	1668	78.83	518	37.92	979	120.73	119	49.62	823	0.00	0
DICKINSON KS	0.00	0	3.94	1225	63.36	722	38.77	946	34.62	476	80.98	385	0.00	0
DONIPHAN KS	0.00	0	2.92	1405	25.71	1514	35.25	1063	25.25	684	45.97	910	0.00	0
DOUGLAS KS	0.00	0	4.12	1191	63.70	714	29.77	1296	33.94	487	55.60	701	0.00	0
EDWARDS KS	0.00	0	3.95	1222	68.13	649	43.16	796	30.93	538	73.71	447	0.00	0
ELK KS	0.00	0	8.14	765	70.99	611	33.37	1142	18.18	1134	59.29	651	0.00	0

TOTAL FALLOUT ACTIVITY (µCi/sq meter) by SHOT SERIES WITH COUNTY RANK

COUNTY AND STATE	RANGER SERIES	RANGER RANK	B-J SERIES	B-J RANK	T-S SERIES	T-S RANK	U-K SERIES	UK RANK	TEAPOT SERIES	TP RANK	PLUMBBOB SERIES	PB RANK	SHOT SEDAN	SEDAN RANK
	1951		1951		1952		1953		1955		1957		1962	
ELLIS KS	0.00	0	2.09	1583	69.18	631	42.85	804	47.33	338	91.74	308	0.00	0
ELLSWORTH KS	0.00	0	3.64	1277	48.44	954	47.57	695	42.23	387	71.40	466	0.00	0
FINNEY KS	0.00	0	1.63	1781	60.73	756	52.95	605	151.50	83	45.45	916	0.00	0
FORD KS	7.80	231	3.28	1343	68.57	643	46.70	708	128.96	96	76.34	416	0.00	0
FRANKLIN KS	17.85	95	4.12	1190	30.53	1393	26.32	1458	25.40	680	56.96	681	0.00	0
GEARY KS	0.00	0	2.74	1438	70.84	612	36.63	1014	30.86	542	67.31	531	0.00	0
GOVE KS	0.00	0	1.18	2000	81.51	489	48.78	669	124.00	108	54.45	726	0.00	0
GRAHAM KS	2.23	323	2.07	1593	70.56	615	44.32	767	126.31	100	67.80	526	0.00	0
GRANT KS	0.00	0	1.47	1856	76.29	543	63.39	464	129.99	92	55.16	710	0.00	0
GRAY KS	0.00	0	2.50	1486	65.68	687	46.04	720	133.99	89	49.75	820	0.00	0
GREELEY KS	0.00	0	1.00	2096	57.18	822	54.15	582	113.96	147	53.15	751	0.00	0
GREENWOOD KS	0.00	0	4.90	1025	90.61	411	30.48	1265	19.48	1020	68.62	515	0.00	0
HAMILTON KS	0.00	0	1.12	2030	53.97	869	65.86	442	114.89	144	55.88	695	0.00	0
HARPER KS	0.00	0	7.95	780	51.01	907	42.28	817	22.77	813	65.52	565	0.00	0
HARVEY KS	0.00	0	4.58	1085	65.85	683	36.59	1016	20.01	980	75.62	424	0.00	0
HASKELL KS	0.00	0	4.00	1211	66.04	680	51.99	622	122.51	112	51.08	790	0.00	0
HODGEMAN KS	0.00	0	2.42	1501	77.18	534	45.18	745	129.23	94	62.15	613	0.00	0
JACKSON KS	0.00	0	2.55	1476	50.43	917	31.74	1213	29.08	578	63.87	585	0.00	0
JEFFERSON KS	0.00	0	2.57	1471	41.39	1090	29.07	1332	34.59	478	66.20	555	0.00	0
JEWELL KS	0.00	0	1.96	1629	27.71	1467	39.47	918	30.98	536	60.59	636	0.00	0
JOHNSON KS	0.00	0	4.17	1176	33.23	1306	25.66	1491	34.17	483	83.79	359	0.00	0
KEARNY KS	0.00	0	1.35	1913	72.86	589	62.10	478	124.31	107	60.65	634	0.00	0
KINGMAN KS	0.00	0	6.09	912	55.98	840	43.12	797	30.45	549	70.85	473	0.00	0
KIOWA KS	0.00	0	4.21	1169	153.32	286	60.42	497	27.95	599	52.85	759	0.00	0
LABETTE KS	0.00	0	8.81	726	55.37	851	27.00	1430	29.75	562	66.11	558	0.00	0
LANE KS	0.00	0	2.95	1397	76.41	541	48.05	691	123.12	110	62.24	611	0.00	0

TOTAL FALLOUT ACTIVITY (µCi/sq meter) by SHOT SERIES WITH COUNTY RANK

COUNTY AND STATE	RANGER SERIES	RANGER RANK	B-J SERIES	B-J RANK	T-S SERIES	T-S RANK	U-K SERIES	UK RANK	TEAPOT SERIES	TP RANK	PLUMBBOB SERIES	PB RANK	SHOT SEDAN	SEDAN RANK
	1951		1951		1952		1953		1955		1957		1962	
LEAVENWORTH KS	0.00	0	4.03	1206	48.43	955	32.93	1162	55.48	279	69.57	499	0.00	0
LINCOLN KS	0.00	0	2.17	1567	56.19	835	41.92	832	40.58	406	92.60	298	0.00	0
LINN KS	0.00	0	4.64	1077	40.70	1107	36.87	1008	25.23	685	50.32	807	0.00	0
LOGAN KS	3.34	298	1.08	2049	67.29	665	45.49	731	120.73	118	56.71	684	0.00	0
LYON KS	0.00	0	4.30	1149	71.60	603	31.01	1242	26.52	641	68.60	516	0.00	0
MARION KS	0.00	0	4.15	1180	70.37	617	35.57	1049	31.11	532	76.67	415	0.00	0
MARSHALL KS	2.33	317	2.40	1505	33.76	1294	38.57	955	27.05	624	62.91	601	0.00	0
MCPHERSON KS	0.00	0	4.15	1181	57.29	817	37.77	985	23.21	791	91.20	310	0.00	0
MEADE KS	0.00	0	4.13	1186	62.29	734	55.87	556	84.47	191	61.27	627	0.00	0
MIAMI KS	0.00	0	4.19	1172	39.37	1136	31.07	1240	48.42	327	47.78	866	0.00	0
MITCHELL KS	0.00	0	2.17	1568	27.62	1472	56.24	550	35.60	459	85.00	346	0.00	0
MONTGOMERY KS	0.00	0	8.94	718	51.53	897	32.75	1169	24.25	743	55.52	705	0.00	0
MORRIS KS	0.00	0	4.12	1193	95.42	383	33.96	1117	17.10	1246	72.91	451	0.00	0
MORTON KS	0.00	0	0.92	2137	59.24	783	82.71	279	89.02	180	49.76	819	0.00	0
NEMAHA KS	0.00	0	2.40	1506	37.51	1188	32.54	1178	31.88	517	56.17	692	0.00	0
NEOSHO KS	0.00	0	8.04	775	65.77	686	39.44	920	17.83	1178	53.30	747	0.00	0
NESS KS	0.00	0	3.00	1390	78.55	521	46.75	707	129.88	93	78.47	403	0.00	0
NORTON KS	1.84	332	2.07	1594	78.01	524	36.35	1022	128.67	97	73.55	448	0.00	0
OSAGE KS	0.00	0	3.36	1330	49.16	940	27.58	1408	21.64	878	61.49	624	0.00	0
OSBORNE KS	0.00	0	2.30	1530	87.02	443	38.41	961	45.12	354	82.65	372	0.00	0
OTTAWA KS	0.00	0	2.36	1517	51.28	903	41.82	836	35.01	470	79.56	394	0.00	0
PAWNEE KS	0.00	0	2.64	1458	62.87	727	45.45	734	31.15	531	79.97	391	0.00	0
PHILLIPS KS	2.23	324	1.62	1786	70.30	619	36.26	1027	36.18	456	74.15	441	0.00	0
POTTAWATOMIE KS	0.00	0	2.58	1469	48.57	949	38.55	956	29.44	571	70.03	489	0.00	0
PRATT KS	0.00	0	4.82	1039	81.07	494	44.16	769	28.34	592	67.15	534	0.00	0
RAWLINS KS	0.00	0	1.08	2050	74.87	558	36.78	1011	122.03	114	49.27	828	0.00	0

TOTAL FALLOUT ACTIVITY (µCi/sq meter) by SHOT SERIES WITH COUNTY RANK

COUNTY AND STATE	RANGER SERIES	RANGER RANK	B-J SERIES	B-J RANK	T-S SERIES	T-S RANK	U-K SERIES	UK RANK	TEAPOT SERIES	TP RANK	PLUMBBOB SERIES	PB RANK	SHOT SEDAN	SEDAN RANK
	1951		1951		1952		1953		1955		1957		1962	
RENO KS	0.00	0	2.82	1423	28.48	1450	77.49	315	25.86	662	73.86	445	0.00	0
REPUBLIC KS	2.23	325	2.37	1513	49.51	933	36.30	1023	26.18	650	78.69	400	0.00	0
RICE KS	0.00	0	3.96	1221	24.01	1571	71.51	376	27.84	603	71.23	467	0.00	0
RILEY KS	0.00	0	2.49	1489	56.81	825	37.47	993	30.64	546	70.23	484	0.00	0
ROOKS KS	2.23	326	2.22	1555	71.22	609	48.85	666	46.46	345	77.95	407	0.00	0
RUSH KS	0.00	0	4.65	1075	72.39	596	42.84	805	48.05	330	78.97	398	0.00	0
RUSSELL KS	0.00	0	3.53	1297	99.65	363	61.10	489	46.76	342	89.02	326	0.00	0
SALINE KS	0.00	0	2.47	1496	77.29	532	45.57	727	35.95	457	84.26	351	0.00	0
SCOTT KS	0.00	0	1.08	2051	83.53	470	48.08	689	119.19	127	58.18	662	0.00	0
SEDGWICK KS	0.00	0	6.41	884	40.33	1119	32.78	1168	18.45	1103	72.34	457	0.00	0
SEWARD KS	0.00	0	5.32	978	62.06	738	61.35	487	88.49	181	41.52	1003	0.00	0
SHAWNEE KS	0.00	0	2.09	1585	73.38	582	36.10	1032	35.72	458	73.05	450	0.00	0
SHERIDAN KS	0.00	0	1.96	1628	77.57	528	42.21	821	123.10	111	43.38	963	0.00	0
SHERMAN KS	2.46	316	1.28	1948	50.46	916	31.26	1229	59.22	258	59.31	650	0.00	0
SMITH KS	0.00	0	1.85	1680	72.26	598	40.71	866	41.71	394	63.35	592	0.00	0
STAFFORD KS	0.00	0	2.93	1403	64.35	704	48.30	685	30.71	545	75.17	429	0.00	0
STANTON KS	0.00	0	0.81	2214	61.79	744	72.10	371	125.54	104	70.26	483	0.00	0
STEVENS KS	0.00	0	1.31	1940	59.70	778	73.04	361	259.75	47	47.53	869	0.00	0
SUMNER KS	0.00	0	9.73	672	50.58	912	35.55	1051	22.40	833	84.26	350	0.00	0
THOMAS KS	0.00	0	1.36	1908	70.14	620	40.57	875	122.42	113	46.56	889	0.00	0
TREGO KS	3.10	301	2.22	1554	71.91	601	44.82	751	131.39	91	72.10	458	0.00	0
WABAUNSEE KS	0.00	0	4.21	1167	82.58	480	35.27	1062	33.48	497	79.85	393	0.00	0
WALLACE KS	0.00	0	1.28	1953	83.97	469	44.45	763	115.35	141	52.26	776	0.00	0
WASHINGTON KS	0.00	0	2.48	1491	34.74	1265	40.26	887	26.80	628	69.62	498	0.00	0
WICHITA KS	0.00	0	1.08	2052	59.81	777	51.07	632	119.68	124	54.26	731	0.00	0
WILSON KS	0.00	0	10.85	601	64.69	697	29.69	1300	17.03	1256	60.86	631	0.00	0

TOTAL FALLOUT ACTIVITY (µCi/sq meter) by SHOT SERIES WITH COUNTY RANK

COUNTY AND STATE	RANGER SERIES	RANGER RANK	B-J SERIES	B-J RANK	T-S SERIES	T-S RANK	U-K SERIES	UK RANK	TEAPOT SERIES	TP RANK	PLUMBBOB SERIES	PB RANK	SHOT SEDAN	SEDAN RANK
	1951		1951		1952		1953		1955		1957		1962	
WOODSON KS	0.00	0	7.98	777	59.04	784	31.51	1219	16.18	1327	56.88	683	0.00	0
WYANDOTTE KS	0.00	0	4.16	1179	39.94	1125	39.73	902	26.53	640	69.14	501	0.00	0
ADAIR KY	0.00	0	2.80	1429	8.93	2670	13.85	2343	13.80	1547	28.00	1530	0.00	0
ALLEN KY	0.00	0	1.32	1933	10.39	2540	15.00	2226	19.01	1050	18.72	2371	0.00	0
ANDERSON KY	0.00	0	12.10	528	10.15	2566	25.21	1509	10.64	1857	14.27	2748	157.12	267
BALLARD KY	0.00	0	3.55	1293	21.23	1704	37.69	989	17.54	1205	22.55	1990	0.00	0
BARREN KY	0.00	0	1.82	1695	9.00	2659	12.71	2449	14.63	1471	20.92	2153	17.11	675
BATH KY	0.00	0	0.81	2210	14.06	2151	14.04	2329	9.70	1954	17.61	2489	244.26	134
BELL KY	0.00	0	4.24	1158	8.54	2710	20.48	1813	18.37	1110	30.03	1423	43.75	590
BOONE KY	15.23	123	2.89	1408	8.50	2715	17.42	2032	18.19	1133	27.61	1553	90.29	418
BOURBON KY	5.83	248	12.99	488	11.56	2411	15.25	2201	10.78	1838	14.97	2697	90.94	414
BOYD KY	0.00	0	5.20	988	11.05	2474	13.09	2409	14.92	1443	38.26	1058	208.78	147
BOYLE KY	0.00	0	10.84	602	9.33	2629	15.91	2141	12.56	1660	29.27	1468	39.12	612
BRACKEN KY	15.23	124	4.94	1015	9.55	2610	15.87	2147	9.78	1946	18.75	2369	123.00	344
BREATHITT KY	0.00	0	11.71	553	8.50	2717	30.13	1278	12.70	1647	54.62	725	164.78	259
BRECKINRIDGE KY	1.75	345	1.20	1995	12.30	2333	17.94	1982	35.54	462	12.26	2849	0.00	0
BULLITT KY	1.84	333	2.23	1553	9.69	2593	30.60	1259	12.96	1619	12.12	2853	39.12	617
BUTLER KY	0.00	0	0.66	2323	9.69	2594	15.01	2224	37.96	431	15.22	2682	17.11	680
CALDWELL KY	0.00	0	0.54	2450	14.46	2115	21.09	1759	12.80	1637	34.06	1207	0.00	0
CALLOWAY KY	0.00	0	0.90	2149	14.17	2143	30.57	1261	14.48	1488	21.30	2105	0.00	0
CAMPBELL KY	7.80	232	3.20	1365	11.59	2409	18.20	1964	15.36	1397	29.25	1469	299.72	121
CARLISLE KY	0.00	0	3.72	1264	9.40	2621	44.15	770	17.63	1192	15.87	2631	0.00	0
CARROLL KY	4.06	262	4.05	1205	10.44	2535	15.68	2162	16.36	1310	15.89	2629	299.72	123
CARTER KY	0.00	0	5.99	923	7.42	2813	10.88	2638	12.57	1659	26.46	1636	208.78	153
CASEY KY	0.00	0	3.84	1241	8.97	2662	20.30	1824	13.94	1538	25.14	1759	18.06	664
CHRISTIAN KY	0.00	0	0.45	2588	12.88	2274	19.65	1865	25.16	688	36.74	1105	0.00	0

TOTAL FALLOUT ACTIVITY (µCi/sq meter) by SHOT SERIES WITH COUNTY RANK

COUNTY AND STATE	RANGER SERIES	RANGER RANK	B-J SERIES	B-J RANK	T-S SERIES	T-S RANK	U-K SERIES	UK RANK	TEAPOT SERIES	TP RANK	PLUMBBOB SERIES	PB RANK	SHOT SEDAN	SEDAN RANK
	1951		1951		1952		1953		1955		1957		1962	
CLARK KY	15.23	125	0.64	2338	14.03	2160	14.56	2276	10.48	1872	17.23	2523	240.07	137
CLAY KY	0.00	0	9.02	713	7.85	2773	15.64	2165	17.02	1258	25.17	1751	43.75	599
CLINTON KY	0.00	0	1.12	2027	7.88	2770	14.74	2258	18.97	1052	26.77	1612	0.00	0
CRITTENDEN KY	0.00	0	0.32	2785	18.42	1840	18.48	1950	13.07	1606	22.85	1956	0.00	0
CUMBERLAND KY	0.00	0	0.50	2531	11.31	2439	14.87	2236	12.91	1625	43.63	956	43.75	588
DAVIESS KY	0.00	0	0.54	2446	19.48	1776	23.05	1638	38.83	424	17.83	2464	31.92	650
EDMONSON KY	0.00	0	0.66	2319	12.76	2279	18.02	1977	15.94	1341	17.71	2474	17.11	678
ELLIOTT KY	0.00	0	6.29	897	9.75	2590	14.02	2331	16.20	1326	19.66	2287	208.78	164
ESTILL KY	0.00	0	12.19	524	6.25	2927	14.16	2316	14.56	1481	35.43	1143	94.45	390
FAYETTE KY	0.00	0	0.62	2363	13.51	2207	19.90	1844	13.10	1602	17.87	2458	90.94	412
FLEMING KY	5.83	249	6.64	868	15.00	2070	15.46	2186	9.47	1970	15.38	2670	208.78	178
FLOYD KY	0.00	0	9.79	670	12.93	2269	32.97	1159	15.64	1373	39.36	1040	149.13	272
FRANKLIN KY	0.00	0	10.73	610	11.97	2365	16.92	2065	13.04	1612	14.50	2737	240.07	139
FULTON KY	0.00	0	3.17	1368	21.23	1705	68.15	416	26.53	639	17.83	2466	0.00	0
GALLATIN KY	6.15	247	4.14	1185	10.57	2514	16.51	2102	14.77	1454	19.35	2322	208.78	166
GARRARD KY	0.00	0	5.84	933	10.67	2506	16.44	2104	10.66	1854	28.28	1519	37.08	632
GRANT KY	15.23	126	8.13	766	11.13	2466	16.88	2069	13.09	1603	19.91	2268	176.50	244
GRAVES KY	0.00	0	0.60	2383	19.51	1775	32.69	1171	18.15	1139	15.53	2658	43.75	608
GRAYSON KY	0.00	0	0.68	2308	10.58	2512	17.72	2007	15.18	1421	14.74	2716	0.00	0
GREEN KY	0.00	0	2.68	1453	9.29	2635	14.48	2287	18.20	1132	20.00	2256	0.00	0
GREENUP KY	15.23	127	10.48	623	10.68	2504	13.98	2334	17.25	1231	20.91	2156	359.52	60
HANCOCK KY	4.06	263	7.96	779	8.84	2676	20.68	1797	55.06	280	14.67	2722	31.92	651
HARDIN KY	1.84	334	1.22	1984	9.88	2584	16.44	2103	23.40	780	14.74	2717	0.00	0
HARLAN KY	0.00	0	8.56	737	8.73	2685	23.65	1600	18.80	1071	25.58	1713	77.90	492
HARRISON KY	15.23	128	10.61	618	12.06	2351	15.34	2196	14.17	1518	22.94	1944	246.22	127
HART KY	0.00	0	4.09	1198	9.32	2630	15.57	2172	15.73	1368	15.42	2665	0.00	0

TOTAL FALLOUT ACTIVITY (µCi/sq meter) by SHOT SERIES WITH COUNTY RANK

COUNTY AND STATE	RANGER SERIES	RANGER RANK	B-J SERIES	B-J RANK	T-S SERIES	T-S RANK	U-K SERIES	UK RANK	TEAPOT SERIES	TP RANK	PLUMBBOB SERIES	PB RANK	SHOT SEDAN	SEDAN RANK
	1951		1951		1952		1953		1955		1957		1962	
HENDERSON KY	0.00	0	0.43	2612	33.93	1290	21.81	1712	15.42	1392	26.47	1634	0.00	0
HENRY KY	0.00	0	4.72	1060	8.95	2667	17.69	2011	13.74	1552	11.87	2867	244.26	135
HICKMAN KY	0.00	0	3.17	1369	22.05	1661	70.26	385	27.05	625	17.59	2492	0.00	0
HOPKINS KY	0.00	0	0.32	2786	9.89	2582	17.77	2002	16.95	1265	25.42	1731	0.00	0
JACKSON KY	0.00	0	12.06	534	8.69	2690	17.25	2046	17.04	1253	23.33	1900	164.78	261
JEFFERSON KY	1.84	335	0.56	2433	9.65	2600	17.28	2043	14.66	1463	14.75	2715	37.44	627
JESSAMINE KY	0.00	0	14.81	430	10.19	2558	36.37	1020	14.07	1525	14.61	2727	90.94	415
JOHNSON KY	0.00	0	10.89	600	10.68	2503	20.16	1835	17.80	1181	19.62	2294	184.61	232
KENTON KY	5.83	250	1.34	1923	13.69	2189	19.75	1856	13.56	1568	43.33	965	244.26	132
KNOTT KY	0.00	0	9.85	667	15.65	2019	20.90	1780	16.75	1278	29.22	1470	94.45	393
KNOX KY	0.00	0	4.15	1182	9.25	2640	16.26	2115	20.23	964	22.45	2006	43.75	603
LARUE KY	0.00	0	1.35	1911	12.30	2332	15.35	2194	28.76	588	13.54	2794	0.00	0
LAUREL KY	0.00	0	9.07	705	8.97	2664	19.16	1899	21.09	915	28.37	1515	0.00	0
LAWRENCE KY	0.00	0	9.87	665	8.51	2713	14.10	2324	16.33	1316	23.15	1924	359.52	59
LEE KY	0.00	0	6.32	892	6.59	2895	19.30	1893	14.24	1512	28.18	1523	0.00	0
LESLIE KY	0.00	0	9.63	677	7.64	2788	16.31	2109	16.61	1291	37.84	1071	125.14	340
LETCHER KY	0.00	0	4.87	1035	8.40	2728	22.15	1689	16.20	1325	29.16	1472	94.45	394
LEWIS KY	6.22	246	5.86	932	9.39	2622	15.56	2174	8.12	2107	18.66	2377	359.52	61
LINCOLN KY	0.00	0	0.53	2469	9.20	2644	14.75	2254	8.91	2021	27.92	1534	37.08	633
LIVINGSTON KY	0.00	0	0.32	2787	23.13	1608	30.28	1272	14.73	1461	26.18	1658	0.00	0
LOGAN KY	0.00	0	0.92	2136	9.63	2602	17.26	2045	21.90	869	22.80	1959	0.00	0
LYON KY	0.00	0	0.76	2240	13.46	2216	20.52	1810	11.09	1813	28.20	1521	0.00	0
MADISON KY	0.00	0	6.36	888	10.45	2533	16.92	2064	10.58	1862	29.70	1444	94.45	392
MAGOFFIN KY	0.00	0	11.55	561	11.01	2477	18.78	1923	18.38	1109	31.91	1312	184.61	231
MARION KY	0.00	0	1.10	2039	10.15	2564	14.77	2251	20.91	922	16.17	2607	0.00	0
MARSHALL KY	0.00	0	0.64	2336	16.60	1960	22.38	1672	14.55	1482	16.58	2580	43.75	606

U.S. FALLOUT ATLAS : TOTAL FALLOUT

TOTAL FALLOUT ACTIVITY (µCi/sq meter) by SHOT SERIES WITH COUNTY RANK

COUNTY AND STATE	RANGER SERIES	RANGER RANK	B-J SERIES	B-J RANK	T-S SERIES	T-S RANK	U-K SERIES	UK RANK	TEAPOT SERIES	TP RANK	PLUMBBOB SERIES	PB RANK	SHOT SEDAN	SEDAN RANK
	1951		1951		1952		1953		1955		1957		1962	
MARTIN KY	0.00	0	5.64	952	13.13	2248	11.99	2511	17.55	1203	38.51	1054	57.04	559
MASON KY	5.83	251	5.22	984	9.36	2627	15.32	2199	9.43	1975	18.52	2394	208.78	169
MCCRACKEN KY	0.00	0	0.43	2613	21.45	1695	34.23	1107	17.26	1230	27.59	1556	0.00	0
MCCREARY KY	0.00	0	6.02	917	8.58	2707	16.16	2122	18.82	1067	18.98	2353	17.11	677
MCLEAN KY	0.00	0	0.46	2568	10.33	2544	19.54	1870	26.79	629	23.43	1892	0.00	0
MEADE KY	1.84	336	0.53	2466	24.73	1546	19.19	1897	13.76	1549	18.17	2426	0.00	0
MENIFEE KY	0.00	0	12.83	494	11.68	2394	25.37	1502	17.06	1250	52.32	772	184.61	229
MERCER KY	0.00	0	6.57	875	9.41	2619	18.84	1920	13.81	1545	28.60	1503	37.08	631
METCALFE KY	0.00	0	2.24	1545	19.55	1768	15.15	2209	13.28	1589	29.28	1467	0.00	0
MONROE KY	0.00	0	0.30	2797	19.98	1746	12.96	2425	8.30	2083	33.17	1252	17.11	668
MONTGOMERY KY	0.00	0	0.79	2221	14.58	2107	17.16	2050	10.47	1873	17.73	2473	157.12	264
MORGAN KY	0.00	0	6.63	871	10.20	2556	15.41	2191	15.57	1384	18.77	2366	184.61	233
MUHLENBERG KY	0.00	0	1.35	1915	12.20	2339	17.07	2053	26.73	632	20.60	2192	0.00	0
NELSON KY	0.00	0	4.36	1129	11.71	2393	17.45	2026	8.69	2037	13.08	2815	0.00	0
NICHOLAS KY	15.23	129	12.54	506	11.65	2401	17.43	2030	10.52	1869	15.58	2653	246.22	131
OHIO KY	0.00	0	0.97	2113	11.32	2435	22.70	1657	36.84	448	18.06	2439	18.06	666
OLDHAM KY	4.06	264	0.53	2468	10.54	2517	16.81	2075	15.90	1348	13.08	2814	157.12	269
OWEN KY	15.23	130	4.47	1105	11.66	2398	21.18	1751	10.95	1824	15.82	2635	299.72	124
OWSLEY KY	0.00	0	5.71	943	6.54	2902	21.01	1769	14.46	1493	29.08	1476	164.78	260
PENDLETON KY	15.23	131	7.61	794	12.07	2347	16.89	2068	14.59	1476	25.56	1715	123.00	342
PERRY KY	0.00	0	10.35	632	10.04	2573	31.44	1222	17.59	1195	26.13	1664	94.45	395
PIKE KY	0.00	0	5.59	956	15.45	2038	16.91	2066	14.88	1446	41.15	1009	149.13	270
POWELL KY	0.00	0	11.98	536	6.24	2928	14.29	2307	14.58	1477	36.06	1125	354.59	68
PULASKI KY	0.00	0	8.55	740	6.47	2906	16.10	2129	18.02	1152	23.21	1917	43.75	602
ROBERTSON KY	15.23	132	5.69	946	9.59	2606	17.44	2028	10.01	1914	18.23	2419	208.78	171
ROCKCASTLE KY	0.00	0	10.01	658	8.35	2733	17.34	2037	17.64	1190	19.67	2286	94.45	397

TOTAL FALLOUT ACTIVITY (µCi/sq meter) by SHOT SERIES WITH COUNTY RANK

COUNTY AND STATE	RANGER SERIES	RANGER RANK	B-J SERIES	B-J RANK	T-S SERIES	T-S RANK	U-K SERIES	UK RANK	TEAPOT SERIES	TP RANK	PLUMBBOB SERIES	PB RANK	SHOT SEDAN	SEDAN RANK
	1951		1951		1952		1953		1955		1957		1962	
ROWAN KY	0.00	0	6.25	898	11.92	2367	15.05	2219	18.24	1126	29.45	1459	299.72	120
RUSSELL KY	0.00	0	4.00	1212	9.16	2646	16.29	2111	12.44	1670	28.70	1496	43.75	593
SCOTT KY	15.23	133	6.64	867	11.77	2387	16.64	2092	11.50	1766	15.98	2624	240.07	138
SHELBY KY	0.00	0	4.93	1016	12.71	2285	19.10	1903	17.04	1254	14.23	2753	184.61	234
SIMPSON KY	0.00	0	0.97	2111	5.82	2962	16.30	2110	15.85	1357	32.14	1297	0.00	0
SPENCER KY	0.00	0	4.44	1112	8.95	2666	18.00	1978	13.54	1570	14.06	2759	128.02	336
TAYLOR KY	0.00	0	0.84	2187	10.28	2548	17.68	2012	19.98	985	16.61	2575	0.00	0
TODD KY	0.00	0	0.45	2589	12.79	2278	19.41	1882	25.45	678	34.78	1162	21.81	654
TRIGG KY	0.00	0	0.76	2241	18.09	1863	25.34	1505	13.69	1562	29.00	1482	0.00	0
TRIMBLE KY	4.06	265	2.03	1614	11.82	2377	17.40	2034	19.64	1010	14.38	2743	359.52	63
UNION KY	0.00	0	0.49	2543	19.08	1802	19.96	1841	25.54	673	29.29	1466	21.81	655
WARREN KY	0.00	0	0.95	2122	23.80	1578	20.82	1784	25.07	692	22.68	1975	0.00	0
WASHINGTON KY	0.00	0	4.72	1061	8.50	2714	14.57	2275	14.73	1459	16.88	2552	37.08	635
WAYNE KY	0.00	0	2.48	1490	9.93	2578	14.81	2245	17.33	1225	28.73	1495	17.11	670
WEBSTER KY	0.00	0	0.32	2788	19.57	1766	20.72	1795	19.60	1013	21.02	2139	0.00	0
WHITLEY KY	0.00	0	6.75	858	7.07	2847	20.52	1809	20.84	925	17.31	2515	17.11	679
WOLFE KY	0.00	0	11.47	566	8.11	2756	29.14	1328	13.23	1591	37.61	1077	184.61	230
WOODFORD KY	0.00	0	13.42	471	10.53	2518	18.57	1942	12.19	1695	19.84	2273	240.07	136
ACADIA LA	0.00	0	0.51	2521	8.64	2697	96.25	214	17.95	1166	132.43	122	0.00	0
ALLEN LA	0.00	0	0.29	2816	4.60	3022	104.30	189	27.11	621	83.33	365	0.00	0
ASCENSION LA	0.00	0	0.95	2121	3.69	3048	71.40	378	7.80	2139	90.22	316	0.00	0
ASSUMPTION LA	0.00	0	1.55	1821	3.29	3061	54.87	571	8.99	2014	127.31	157	0.00	0
AVOYELLES LA	0.00	0	0.61	2367	6.60	2894	23.65	1601	15.18	1423	136.38	107	0.00	0
BEAUREGARD LA	0.00	0	2.82	1416	14.05	2155	110.73	182	21.46	894	130.85	131	0.00	0
BIENVILLE LA	0.00	0	4.66	1072	6.74	2873	30.76	1252	28.86	582	29.11	1475	0.00	0
BOSSIER LA	0.00	0	4.66	1073	38.29	1167	27.74	1400	19.20	1039	20.95	2146	0.00	0

TOTAL FALLOUT ACTIVITY (µCi/sq meter) by SHOT SERIES WITH COUNTY RANK

COUNTY AND STATE	RANGER SERIES	RANGER RANK	B-J SERIES	B-J RANK	T-S SERIES	T-S RANK	U-K SERIES	UK RANK	TEAPOT SERIES	TP RANK	PLUMBBOB SERIES	PB RANK	SHOT SEDAN	SEDAN RANK
	1951		1951		1952		1953		1955		1957		1962	
CADDO LA	0.00	0	3.09	1379	24.29	1566	39.96	896	24.76	706	30.93	1373	0.00	0
CALCASIEU LA	0.00	0	2.82	1417	9.09	2655	98.10	209	18.94	1055	125.48	162	0.00	0
CALDWELL LA	0.00	0	3.25	1349	5.23	2994	45.31	739	16.98	1261	17.29	2519	0.00	0
CAMERON LA	0.00	0	0.29	2817	7.47	2806	93.86	226	19.04	1048	131.12	130	0.00	0
CATAHOULA LA	0.00	0	0.72	2268	6.72	2878	47.05	703	18.68	1081	88.45	330	0.00	0
CLAIBORNE LA	0.00	0	3.13	1373	13.81	2177	28.78	1346	20.49	954	22.93	1945	0.00	0
CONCORDIA LA	0.00	0	0.83	2197	6.58	2898	50.18	645	18.21	1131	89.17	324	0.00	0
DE SOTO LA	0.00	0	4.77	1051	13.51	2209	25.67	1489	26.01	655	27.97	1532	0.00	0
EAST BATON R LA	0.00	0	0.36	2748	3.98	3036	56.36	548	8.09	2110	136.11	110	0.00	0
EAST CARROLL LA	0.00	0	2.88	1411	5.86	2956	47.37	700	18.60	1088	17.35	2508	0.00	0
EAST FELICIA LA	0.00	0	1.23	1969	4.16	3033	75.30	334	9.15	2001	95.43	278	0.00	0
EVANGELINE LA	0.00	0	0.40	2714	4.86	3016	97.54	212	22.50	827	129.43	144	0.00	0
FRANKLIN LA	0.00	0	1.04	2061	4.87	3015	45.12	746	21.39	897	18.35	2412	0.00	0
GRANT LA	0.00	0	3.04	1386	11.39	2430	129.59	163	23.75	764	128.66	150	0.00	0
IBERIA LA	0.00	0	0.59	2400	8.72	2687	69.45	399	10.04	1912	130.13	135	0.00	0
IBERVILLE LA	0.00	0	1.02	2071	3.92	3038	57.53	530	7.41	2201	135.41	113	0.00	0
JACKSON LA	0.00	0	3.46	1313	8.71	2688	43.47	785	18.74	1075	24.32	1820	0.00	0
JEFFERSON LA	0.00	0	1.23	1973	19.10	1798	57.49	532	9.77	1948	36.91	1098	0.00	0
JEFFERSON DA LA	0.00	0	0.29	2818	4.62	3021	100.19	201	27.35	613	83.10	367	0.00	0
LA SALLE LA	0.00	0	4.77	1052	15.16	2059	52.19	617	22.72	814	131.36	129	0.00	0
LAFAYETTE LA	0.00	0	0.40	2715	19.18	1792	114.36	173	17.64	1189	128.71	148	0.00	0
LAFOURCHE LA	0.00	0	1.29	1946	3.39	3057	52.81	607	7.90	2132	112.03	202	0.00	0
LINCOLN LA	0.00	0	3.52	1300	8.75	2683	39.71	903	19.85	990	24.40	1814	0.00	0
LIVINGSTON LA	0.00	0	1.12	2026	3.72	3045	74.87	339	8.00	2124	92.82	294	0.00	0
MADISON LA	0.00	0	1.68	1758	14.40	2120	58.78	517	22.28	843	24.46	1808	0.00	0
MOREHOUSE LA	0.00	0	1.82	1696	6.06	2938	41.47	844	20.59	945	26.60	1627	0.00	0

TOTAL FALLOUT ACTIVITY (µCi/sq meter) by SHOT SERIES WITH COUNTY RANK

COUNTY AND STATE	RANGER SERIES	RANGER RANK	B-J SERIES	B-J RANK	T-S SERIES	T-S RANK	U-K SERIES	UK RANK	TEAPOT SERIES	TP RANK	PLUMBBOB SERIES	PB RANK	SHOT SEDAN	SEDAN RANK
	1951		1951		1952		1953		1955		1957		1962	
NATCHITOCHES LA	0.00	0	1.04	2063	13.01	2260	41.94	830	29.37	572	15.57	2655	0.00	0
ORLEANS LA	0.00	0	1.48	1845	6.56	2899	34.84	1085	7.52	2180	29.66	1447	0.00	0
OUACHITA LA	0.00	0	3.48	1310	5.72	2971	42.98	802	22.34	839	32.42	1284	0.00	0
PLAQUEMINES LA	0.00	0	1.96	1630	3.11	3068	54.76	575	8.29	2084	87.67	335	0.00	0
POINTE COUPE LA	0.00	0	0.61	2368	6.03	2941	56.40	546	9.74	1951	136.98	104	0.00	0
RAPIDES LA	0.00	0	2.93	1401	5.35	2990	124.11	166	28.38	591	128.34	154	0.00	0
RED RIVER LA	0.00	0	4.56	1089	14.34	2131	31.93	1205	22.31	840	20.48	2201	0.00	0
RICHLAND LA	0.00	0	3.57	1291	17.22	1929	53.11	602	30.99	534	17.69	2478	0.00	0
SABINE LA	0.00	0	2.94	1400	11.50	2419	105.28	186	27.61	608	30.26	1409	0.00	0
ST BERNARD LA	0.00	0	1.82	1697	5.52	2978	52.64	610	8.70	2036	78.59	401	0.00	0
ST CHARLES LA	0.00	0	1.23	1974	3.56	3050	53.55	592	8.41	2069	111.02	204	0.00	0
ST HELENA LA	0.00	0	1.11	2031	3.38	3059	82.53	280	8.47	2058	136.34	108	0.00	0
ST JAMES LA	0.00	0	0.94	2125	3.73	3044	64.21	460	8.78	2028	65.99	559	0.00	0
ST JOHN THE LA	0.00	0	1.46	1857	3.78	3042	66.03	438	9.14	2002	67.55	530	0.00	0
ST LANDRY LA	0.00	0	0.61	2369	3.71	3047	55.94	554	9.21	1994	87.82	334	0.00	0
ST MARTIN LA	0.00	0	1.55	1822	3.18	3065	54.02	585	8.47	2055	128.61	152	0.00	0
ST MARY LA	0.00	0	1.31	1936	3.12	3066	54.99	569	8.36	2075	128.61	151	0.00	0
ST TAMMANY LA	0.00	0	1.82	1698	4.29	3030	55.53	559	11.08	1814	83.59	361	0.00	0
TANGIPAHOA LA	0.00	0	1.10	2038	3.25	3062	79.42	300	8.25	2092	133.88	119	0.00	0
TENSAS LA	0.00	0	1.15	2013	13.51	2208	57.40	534	22.17	850	132.73	121	0.00	0
TERREBONNE LA	0.00	0	1.55	1823	3.56	3051	59.28	512	7.21	2239	130.34	133	0.00	0
UNION LA	0.00	0	3.79	1253	9.11	2654	49.42	652	19.33	1034	20.45	2206	0.00	0
VERMILION LA	0.00	0	0.72	2272	7.55	2798	97.63	210	20.76	933	92.31	301	0.00	0
VERNON LA	0.00	0	2.82	1418	8.95	2669	112.45	177	23.30	787	24.17	1825	0.00	0
WASHINGTON LA	0.00	0	1.33	1928	3.40	3056	53.90	586	10.46	1874	84.48	349	0.00	0
WEBSTER LA	0.00	0	2.95	1396	36.34	1219	29.66	1301	22.27	844	24.38	1817	0.00	0

TOTAL FALLOUT ACTIVITY (µCi/sq meter) by SHOT SERIES WITH COUNTY RANK

COUNTY AND STATE	RANGER SERIES	RANGER RANK	B-J SERIES	B-J RANK	T-S SERIES	T-S RANK	U-K SERIES	UK RANK	TEAPOT SERIES	TP RANK	PLUMBBOB SERIES	PB RANK	SHOT SEDAN	SEDAN RANK
	1951		1951		1952		1953		1955		1957		1962	
WEST BATON R LA	0.00	0	1.12	2021	3.85	3041	58.25	521	7.43	2195	136.54	106	0.00	0
WEST CARROLL LA	0.00	0	3.89	1235	6.05	2940	43.18	793	19.52	1018	17.20	2526	0.00	0
WEST FELICIA LA	0.00	0	0.72	2269	6.55	2901	55.29	563	10.28	1890	136.81	105	0.00	0
WINN LA	0.00	0	3.25	1350	6.31	2917	45.49	730	22.05	858	23.68	1879	0.00	0
BARNS MA	6.78	240	15.69	390	22.64	1626	103.22	191	6.56	2345	10.24	2923	0.00	0
BERKSHIRE MA	11.65	156	21.88	215	37.03	1200	60.24	498	4.80	2714	24.47	1803	0.00	0
BRISTOL MA	6.41	244	21.25	233	26.63	1492	165.41	126	5.64	2524	12.19	2852	0.00	0
DUKES MA	6.98	239	17.33	345	18.33	1846	104.90	187	7.08	2260	10.69	2911	0.00	0
ESSEX MA	37.16	11	19.77	271	15.41	2039	103.04	192	6.58	2336	7.25	2982	0.00	0
FRANKLIN MA	12.84	148	25.12	177	31.18	1368	34.16	1111	5.76	2497	14.18	2754	0.00	0
HAMPDEN MA	29.27	23	21.77	216	42.78	1063	37.92	978	5.36	2578	17.89	2454	0.00	0
HAMPSHIRE MA	12.73	149	22.96	203	29.20	1434	33.90	1119	5.55	2541	35.66	1132	0.00	0
MIDDLESEX MA	37.00	13	24.57	182	17.27	1925	145.40	142	6.41	2378	7.03	2984	0.00	0
NANTUCKET MA	6.78	241	10.32	636	23.56	1587	58.11	524	6.20	2423	10.72	2909	0.00	0
NORFOLK MA	36.05	16	21.66	220	15.83	2005	200.82	112	6.50	2359	11.89	2866	0.00	0
PLYMOUTH MA	6.63	242	21.10	235	17.56	1904	221.10	103	7.48	2187	11.69	2875	0.00	0
SUFFOLK MA	36.26	15	20.57	248	15.84	2002	132.64	157	5.40	2569	6.53	2989	0.00	0
WORCESTER MA	18.43	85	25.62	167	26.59	1494	64.93	453	6.02	2449	10.65	2912	0.00	0
ALLEGANY MD	2.54	306	7.63	793	7.72	2783	15.97	2137	12.78	1638	22.16	2036	57.04	567
ANNE ARUNDEL MD	0.00	0	19.36	282	13.54	2202	16.84	2073	5.35	2582	24.15	1827	0.00	0
BALTIMORE MD	1.35	364	14.98	423	12.33	2329	14.02	2332	4.58	2764	21.99	2054	73.06	520
CALVERT MD	0.00	0	20.01	262	14.91	2078	16.70	2087	8.07	2114	23.31	1905	0.00	0
CAROLINE MD	0.00	0	16.48	369	15.09	2064	17.90	1987	7.14	2251	16.23	2603	0.00	0
CARROLL MD	1.17	384	10.59	620	11.55	2413	14.82	2242	7.67	2158	17.16	2527	0.00	0
CECIL MD	2.72	305	19.98	264	13.36	2223	21.77	1714	4.58	2761	19.91	2267	0.00	0
CHARLES MD	0.00	0	25.32	173	13.43	2219	16.52	2101	8.76	2031	14.96	2698	0.00	0

U.S. FALLOUT ATLAS : TOTAL FALLOUT

TOTAL FALLOUT ACTIVITY (µCi/sq meter) by SHOT SERIES WITH COUNTY RANK

COUNTY AND STATE	RANGER SERIES	RANGER RANK	B-J SERIES	B-J RANK	T-S SERIES	T-S RANK	U-K SERIES	UK RANK	TEAPOT SERIES	TP RANK	PLUMBBOB SERIES	PB RANK	SHOT SEDAN	SEDAN RANK
	1951		1951		1952		1953		1955		1957		1962	
DORCHESTER MD	0.00	0	19.76	272	12.28	2334	18.32	1959	6.22	2418	20.00	2253	0.00	0
FREDERICK MD	1.23	369	13.33	473	11.22	2453	17.62	2016	8.36	2074	22.75	1964	0.00	0
GARRETT MD	2.54	307	6.18	905	5.79	2967	17.14	2051	12.05	1713	23.26	1911	103.59	360
HARFORD MD	1.17	385	19.31	283	14.69	2099	23.01	1641	7.07	2263	20.64	2184	60.19	547
HOWARD MD	1.17	386	18.66	303	12.16	2340	15.70	2161	6.27	2407	24.17	1826	0.00	0
KENT MD	1.17	387	17.40	342	13.27	2229	21.22	1750	5.75	2501	20.34	2217	0.00	0
MONTGOMERY MD	1.17	388	20.16	257	14.54	2112	18.38	1955	7.78	2145	22.11	2038	0.00	0
PRINCE GEORG MD	0.90	414	27.99	124	14.03	2159	20.82	1785	6.79	2305	15.45	2663	0.00	0
QUEEN ANNES MD	1.17	389	19.38	280	13.08	2252	19.37	1888	5.60	2531	13.61	2788	0.00	0
SOMERSET MD	0.00	0	33.01	81	14.02	2162	20.69	1796	7.26	2225	21.42	2098	0.00	0
ST MARYS MD	0.00	0	21.51	224	14.36	2129	19.75	1858	6.71	2314	23.50	1888	0.00	0
TALBOT MD	0.19	544	18.10	323	13.67	2190	21.73	1716	6.15	2431	19.29	2323	0.00	0
WASHINGTON MD	1.79	339	8.73	728	15.07	2066	17.78	2001	9.10	2004	21.79	2073	0.00	0
WICOMICO MD	0.34	490	19.51	279	12.34	2327	23.41	1614	7.41	2200	21.09	2133	0.00	0
WORCESTER MD	0.00	0	19.67	274	12.10	2346	25.43	1498	7.14	2252	27.00	1598	0.00	0
ANDROSCOGGIN ME	7.96	228	24.95	179	19.10	1800	23.53	1609	5.25	2606	15.63	2648	0.00	0
AROOSTOOK ME	0.57	444	86.45	11	19.71	1756	44.70	754	5.37	2576	23.62	1883	0.00	0
CUMBERLAND ME	6.62	243	33.76	76	21.86	1668	23.61	1602	4.31	2813	12.12	2854	0.00	0
FRANKLIN ME	0.30	491	33.23	80	17.55	1905	22.02	1696	4.99	2663	16.06	2616	0.00	0
HANCOCK ME	7.80	230	37.54	61	17.80	1889	28.41	1366	6.90	2286	18.81	2364	0.00	0
KENNEBEC ME	8.01	225	26.77	146	17.07	1938	22.69	1658	6.26	2410	17.71	2475	0.00	0
KNOX ME	18.36	86	25.84	164	19.93	1749	15.70	2160	7.39	2204	19.56	2299	0.00	0
LINCOLN ME	18.36	87	22.61	206	17.90	1878	21.07	1762	7.37	2207	14.84	2707	0.00	0
OXFORD ME	1.70	349	33.47	79	22.38	1638	24.27	1564	5.16	2625	15.97	2625	0.00	0
PENOBSCOT ME	8.49	222	53.60	31	17.99	1872	33.80	1122	6.19	2424	22.39	2012	0.00	0
PISCATAQUIS ME	0.53	458	50.86	33	24.16	1567	33.04	1155	7.22	2236	17.47	2497	0.00	0

TOTAL FALLOUT ACTIVITY (µCi/sq meter) by SHOT SERIES WITH COUNTY RANK

COUNTY AND STATE	RANGER SERIES	RANGER RANK	B-J SERIES	B-J RANK	T-S SERIES	T-S RANK	U-K SERIES	UK RANK	TEAPOT SERIES	TP RANK	PLUMBBOB SERIES	PB RANK	SHOT SEDAN	SEDAN RANK
	1951		1951		1952		1953		1955		1957		1962	
SAGADAHOC ME	6.41	245	34.38	71	18.11	1859	20.46	1814	4.89	2694	13.27	2805	0.00	0
SOMERSET ME	1.36	362	21.27	232	16.87	1948	26.36	1457	5.73	2504	16.59	2579	0.00	0
WALDO ME	8.04	224	30.02	97	19.13	1797	18.76	1927	7.45	2194	20.69	2177	0.00	0
WASHINGTON ME	7.83	229	47.73	42	15.40	2043	34.65	1096	7.17	2249	18.98	2352	0.00	0
YORK ME	14.56	142	26.31	156	23.36	1598	23.05	1637	5.42	2564	13.08	2813	0.00	0
ALCONA MI	0.00	0	0.70	2288	17.87	1880	23.49	1610	12.55	1662	14.36	2744	0.00	0
ALGER MI	0.00	0	0.45	2575	16.13	1994	52.06	618	10.67	1853	9.62	2939	0.00	0
ALLEGAN MI	0.00	0	0.20	2885	57.27	818	12.10	2499	6.66	2323	15.40	2666	0.00	0
ALPENA MI	0.00	0	0.56	2432	7.56	2797	13.54	2379	9.70	1955	13.02	2821	0.00	0
ANTRIM MI	0.00	0	0.50	2527	40.48	1114	23.94	1585	9.96	1922	16.40	2594	0.00	0
ARENAC MI	0.00	0	0.58	2411	31.58	1355	20.92	1778	9.31	1986	12.00	2861	0.00	0
BARAGA MI	0.00	0	0.53	2462	19.60	1763	52.73	609	11.97	1719	12.67	2834	0.00	0
BARRY MI	0.00	0	0.18	2937	77.18	535	15.48	2183	7.94	2128	19.12	2340	0.00	0
BAY MI	0.00	0	0.29	2819	34.20	1280	19.36	1889	11.78	1734	9.74	2936	0.00	0
BENZIE MI	0.00	0	0.09	2981	28.42	1451	22.02	1697	3.86	2904	12.93	2828	0.00	0
BERRIEN MI	0.00	0	0.46	2571	39.16	1142	23.89	1588	8.47	2056	19.75	2280	0.00	0
BRANCH MI	0.00	0	0.53	2467	57.21	820	19.07	1905	12.88	1629	17.69	2480	0.00	0
CALHOUN MI	0.00	0	0.43	2677	97.41	369	14.58	2274	8.75	2032	15.38	2669	0.00	0
CASS MI	0.00	0	0.40	2705	29.81	1416	21.12	1754	8.21	2095	23.72	1875	0.00	0
CHARLEVOIX MI	0.00	0	0.38	2733	40.57	1110	27.34	1416	10.00	1916	18.71	2373	0.00	0
CHEBOYGAN MI	0.00	0	0.28	2831	36.46	1214	23.56	1607	14.70	1462	14.62	2724	0.00	0
CHIPPEWA MI	0.00	0	0.06	2994	55.57	849	28.76	1347	11.47	1768	13.41	2799	0.00	0
CLARE MI	0.00	0	0.11	2966	21.83	1672	22.11	1691	5.27	2598	13.80	2776	0.00	0
CLINTON MI	0.00	0	0.28	2830	82.71	477	14.74	2257	23.94	754	13.38	2801	0.00	0
CRAWFORD MI	0.00	0	0.11	2967	22.79	1620	20.80	1787	10.29	1889	14.98	2694	0.00	0
DELTA MI	0.00	0	0.66	2322	10.51	2525	45.48	732	8.44	2063	12.79	2831	0.00	0

TOTAL FALLOUT ACTIVITY (µCi/sq meter) by SHOT SERIES WITH COUNTY RANK

COUNTY AND STATE	RANGER SERIES	RANGER RANK	B-J SERIES	B-J RANK	T-S SERIES	T-S RANK	U-K SERIES	UK RANK	TEAPOT SERIES	TP RANK	PLUMBBOB SERIES	PB RANK	SHOT SEDAN	SEDAN RANK
	1951		1951		1952		1953		1955		1957		1962	
DICKINSON MI	0.00	0	0.39	2726	6.74	2872	44.45	764	10.70	1847	16.75	2560	0.00	0
EATON MI	0.00	0	0.22	2856	81.14	493	14.38	2297	43.66	376	16.13	2611	68.45	535
EMMET MI	0.00	0	0.43	2606	12.53	2305	20.53	1808	8.06	2117	17.54	2495	0.00	0
GENESEE MI	0.00	0	0.40	2716	83.30	473	22.03	1695	37.02	446	12.26	2848	0.00	0
GLADWIN MI	0.00	0	0.37	2743	22.21	1650	17.62	2015	7.33	2214	9.61	2941	0.00	0
GOGEBIC MI	0.00	0	1.17	2005	41.62	1086	55.93	555	12.04	1715	27.73	1545	0.00	0
GRAND TRAVER MI	0.00	0	0.17	2944	29.14	1437	20.32	1822	4.58	2763	12.59	2838	0.00	0
GRATIOT MI	0.00	0	0.17	2945	60.59	761	17.56	2019	9.42	1976	14.40	2742	0.00	0
HILLSDALE MI	0.00	0	0.65	2331	63.67	716	14.79	2248	13.05	1610	16.50	2588	35.48	646
HOUGHTON MI	0.00	0	2.28	1535	32.48	1332	57.38	536	12.34	1678	24.72	1787	0.00	0
HURON MI	0.00	0	7.34	821	26.40	1503	18.50	1948	51.07	309	9.51	2943	0.00	0
INGHAM MI	0.00	0	0.28	2832	98.89	365	23.95	1583	11.14	1807	14.58	2730	68.45	536
IONIA MI	0.00	0	0.14	2960	31.32	1363	11.49	2557	6.58	2338	13.83	2775	0.00	0
IOSCO MI	0.00	0	0.48	2552	25.47	1517	22.37	1673	15.92	1344	14.16	2755	0.00	0
IRON MI	0.00	0	0.87	2167	15.82	2006	48.53	678	13.95	1536	18.60	2382	0.00	0
ISABELLA MI	0.00	0	0.17	2946	37.70	1181	17.84	1994	6.93	2282	13.32	2803	0.00	0
JACKSON MI	0.00	0	0.46	2573	55.20	853	18.54	1945	12.31	1681	10.14	2924	0.00	0
KALAMAZOO MI	0.00	0	0.39	2724	91.78	401	13.68	2366	13.71	1555	17.13	2530	0.00	0
KALKASKA MI	0.00	0	0.38	2734	7.43	2811	17.01	2054	8.20	2097	12.57	2839	0.00	0
KENT MI	0.00	0	0.17	2943	47.26	979	16.08	2132	4.67	2745	17.87	2459	57.79	558
KEWEENAW MI	0.00	0	2.09	1584	19.98	1745	63.63	463	12.14	1704	12.65	2835	0.00	0
LAKE MI	0.00	0	0.00	0	31.22	1366	20.38	1820	3.93	2887	11.58	2881	0.00	0
LAPEER MI	0.00	0	8.35	754	67.64	656	29.41	1315	22.60	821	13.77	2780	0.00	0
LEELANAU MI	0.00	0	0.18	2922	27.18	1482	24.30	1562	4.33	2811	8.12	2969	0.00	0
LENAWEE MI	0.00	0	0.81	2215	17.94	1875	18.10	1974	13.35	1582	12.34	2846	90.94	416
LIVINGSTON MI	0.00	0	0.60	2381	97.75	368	21.84	1711	18.01	1155	12.83	2830	0.00	0

TOTAL FALLOUT ACTIVITY (µCi/sq meter) by SHOT SERIES WITH COUNTY RANK

COUNTY AND STATE	RANGER SERIES	RANGER RANK	B-J SERIES	B-J RANK	T-S SERIES	T-S RANK	U-K SERIES	UK RANK	TEAPOT SERIES	TP RANK	PLUMBBOB SERIES	PB RANK	SHOT SEDAN	SEDAN RANK
	1951		1951		1952		1953		1955		1957		1962	
LUCE MI	0.00	0	0.45	2590	35.27	1248	48.37	679	7.36	2209	10.37	2920	0.00	0
MACKINAC MI	0.00	0	0.07	2988	55.19	854	28.67	1352	11.37	1779	11.86	2868	0.00	0
MACOMB MI	0.00	0	11.74	550	16.84	1949	18.69	1933	66.83	236	9.47	2945	0.00	0
MANISTEE MI	0.00	0	0.06	2991	14.07	2150	24.32	1561	3.62	2934	9.47	2944	0.00	0
MARQUETTE MI	0.00	0	0.73	2265	6.79	2869	44.47	762	7.24	2229	13.68	2786	0.00	0
MASON MI	0.00	0	0.08	2985	49.48	935	15.60	2168	4.07	2863	14.53	2734	0.00	0
MECOSTA MI	0.00	0	0.06	2992	29.55	1422	19.33	1890	4.56	2766	9.82	2932	0.00	0
MENOMINEE MI	0.00	0	1.12	2022	8.05	2761	39.27	924	13.33	1584	31.40	1338	0.00	0
MIDLAND MI	0.00	0	0.48	2554	50.11	926	18.18	1965	9.01	2010	10.78	2906	0.00	0
MISSAUKEE MI	0.00	0	0.18	2923	26.91	1487	18.56	1943	5.39	2571	14.88	2705	0.00	0
MONROE MI	0.00	0	0.81	2208	12.07	2349	18.80	1921	16.96	1263	11.14	2895	85.56	473
MONTCALM MI	0.00	0	0.09	2982	50.81	908	19.22	1896	5.04	2648	7.43	2980	0.00	0
MONTMORENCY MI	0.00	0	0.36	2749	16.69	1955	14.88	2235	11.27	1794	12.52	2841	0.00	0
MUSKEGON MI	0.00	0	0.20	2886	48.96	942	18.69	1934	3.51	2947	15.34	2674	0.00	0
NEWAYGO MI	0.00	0	0.07	2989	34.96	1259	19.92	1843	3.68	2928	15.52	2659	0.00	0
OAKLAND MI	0.00	0	5.40	972	43.69	1043	15.66	2164	23.43	777	10.04	2926	68.45	537
OCEANA MI	0.00	0	0.08	2986	44.60	1029	21.30	1742	3.68	2929	15.58	2652	0.00	0
OGEMAW MI	0.00	0	0.29	2820	14.37	2126	19.53	1872	10.28	1891	13.05	2818	0.00	0
ONTONAGON MI	0.00	0	1.30	1941	22.66	1624	59.58	510	8.62	2043	21.44	2095	0.00	0
OSCEOLA MI	0.00	0	0.00	0	29.79	1417	19.29	1894	3.98	2881	8.13	2968	0.00	0
OSCODA MI	0.00	0	0.30	2799	16.45	1970	17.31	2041	9.88	1931	9.92	2930	0.00	0
OTSEGO MI	0.00	0	0.37	2746	20.01	1742	19.02	1907	9.83	1938	15.38	2668	0.00	0
OTTAWA MI	0.00	0	0.16	2951	83.53	471	17.85	1990	4.70	2739	17.32	2513	0.00	0
PRESQUE ISLE MI	0.00	0	0.78	2234	14.35	2130	16.87	2070	10.56	1865	13.90	2771	0.00	0
ROSCOMMON MI	0.00	0	0.11	2968	14.78	2092	23.17	1629	7.25	2226	15.38	2667	0.00	0
SAGINAW MI	0.00	0	0.29	2821	66.77	671	19.16	1900	12.28	1686	12.25	2850	0.00	0

TOTAL FALLOUT ACTIVITY (µCi/sq meter) by SHOT SERIES WITH COUNTY RANK

COUNTY AND STATE	RANGER SERIES	RANGER RANK	B-J SERIES	B-J RANK	T-S SERIES	T-S RANK	U-K SERIES	UK RANK	TEAPOT SERIES	TP RANK	PLUMBBOB SERIES	PB RANK	SHOT SEDAN	SEDAN RANK
	1951		1951		1952		1953		1955		1957		1962	
SANILAC MI	0.00	0	6.64	866	30.01	1409	27.18	1421	42.45	385	9.59	2942	0.00	0
SCHOOLCRAFT MI	0.00	0	0.15	2955	48.59	948	55.44	560	13.77	1548	11.65	2879	0.00	0
SHIAWASSEE MI	0.00	0	0.28	2833	78.59	520	18.47	1951	11.68	1741	10.87	2902	21.68	658
ST CLAIR MI	0.00	0	9.12	703	17.31	1922	24.80	1530	26.43	645	14.89	2704	0.00	0
ST JOSEPH MI	0.00	0	0.64	2341	32.13	1338	20.11	1836	5.65	2520	20.43	2211	0.00	0
TUSCOLA MI	0.00	0	0.51	2522	29.53	1424	29.87	1288	18.12	1141	13.71	2785	0.00	0
VAN BUREN MI	0.00	0	0.47	2565	90.76	410	21.51	1731	7.19	2243	18.28	2416	0.00	0
WASHTENAW MI	0.00	0	0.75	2247	25.42	1518	19.54	1871	19.80	996	11.57	2882	0.00	0
WAYNE MI	0.00	0	12.22	521	13.16	2243	16.23	2118	91.83	174	9.47	2946	0.00	0
WEXFORD MI	0.00	0	0.11	2969	16.98	1942	22.57	1664	4.26	2829	9.77	2933	0.00	0
AITKIN MN	0.00	0	1.78	1722	32.24	1334	32.22	1197	8.75	2034	31.75	1317	0.00	0
ANOKA MN	0.00	0	1.00	2097	29.94	1412	48.09	688	6.67	2322	31.56	1329	0.00	0
BECKER MN	0.00	0	1.78	1723	26.54	1498	11.92	2520	7.12	2255	25.06	1766	0.00	0
BELTRAMI MN	0.00	0	1.48	1846	24.68	1551	16.42	2105	9.66	1958	25.82	1697	0.00	0
BENTON MN	0.00	0	1.22	1976	35.94	1230	32.42	1188	7.46	2191	29.85	1434	0.00	0
BIG STONE MN	0.00	0	1.18	2002	16.61	1959	21.62	1723	5.98	2456	22.66	1980	0.00	0
BLUE EARTH MN	0.00	0	1.06	2055	23.48	1592	52.44	614	6.81	2298	25.67	1708	325.19	114
BROWN MN	0.00	0	0.58	2408	24.00	1572	48.32	683	8.77	2030	28.00	1529	325.19	111
CARLTON MN	0.00	0	1.86	1678	31.80	1350	49.09	658	18.21	1130	34.83	1160	0.00	0
CARVER MN	0.00	0	0.68	2302	29.47	1426	45.21	742	6.79	2303	34.05	1208	0.00	0
CASS MN	0.00	0	2.01	1619	30.39	1399	23.66	1597	8.59	2046	28.07	1528	0.00	0
CHIPPEWA MN	0.00	0	0.35	2759	26.13	1508	29.36	1319	5.45	2558	27.17	1584	0.00	0
CHISAGO MN	0.00	0	1.00	2098	31.76	1351	53.26	600	6.27	2406	22.71	1970	0.00	0
CLAY MN	0.00	0	0.54	2447	7.31	2826	14.63	2268	7.73	2152	20.78	2165	0.00	0
CLEARWATER MN	0.00	0	2.05	1596	17.69	1896	17.27	2044	8.26	2090	20.91	2158	0.00	0
COOK MN	0.00	0	1.04	2062	18.24	1849	43.82	778	7.74	2148	23.08	1931	0.00	0

TOTAL FALLOUT ACTIVITY (µCi/sq meter) by SHOT SERIES WITH COUNTY RANK

COUNTY AND STATE	RANGER SERIES	RANGER RANK	B-J SERIES	B-J RANK	T-S SERIES	T-S RANK	U-K SERIES	UK RANK	TEAPOT SERIES	TP RANK	PLUMBBOB SERIES	PB RANK	SHOT SEDAN	SEDAN RANK
	1951		1951		1952		1953		1955		1957		1962	
COTTONWOOD MN	0.00	0	0.18	2924	37.48	1189	44.01	773	8.47	2054	23.25	1912	132.13	331
CROW WING MN	0.00	0	1.78	1724	32.21	1335	27.54	1410	8.53	2051	27.89	1536	0.00	0
DAKOTA MN	0.00	0	1.81	1715	34.41	1275	57.93	525	6.28	2403	33.39	1240	0.00	0
DODGE MN	0.00	0	0.89	2157	58.71	788	72.98	363	9.90	1929	18.97	2354	181.88	242
DOUGLAS MN	0.00	0	1.38	1896	8.68	2691	19.68	1863	4.39	2795	30.93	1372	0.00	0
FARIBAULT MN	0.00	0	1.45	1866	44.71	1027	64.99	452	7.39	2205	25.19	1748	584.61	12
FILLMORE MN	0.00	0	0.20	2884	34.43	1274	84.45	270	10.91	1831	21.91	2060	327.38	108
FREEBORN MN	0.00	0	1.38	1897	29.81	1415	51.59	626	6.59	2333	28.39	1514	0.00	0
GOODHUE MN	0.00	0	1.22	1977	39.57	1132	69.66	397	10.50	1871	30.05	1421	181.88	238
GRANT MN	0.00	0	0.41	2698	8.55	2709	20.25	1827	3.55	2942	26.68	1619	0.00	0
HENNEPIN MN	0.00	0	1.58	1809	30.09	1407	42.06	828	4.22	2838	15.19	2686	0.00	0
HOUSTON MN	0.00	0	1.27	1960	47.48	976	69.90	390	10.76	1843	58.03	669	139.44	300
HUBBARD MN	0.00	0	1.33	1926	27.99	1462	15.57	2171	4.89	2691	24.45	1812	0.00	0
ISANTI MN	0.00	0	1.00	2099	31.02	1373	45.60	726	6.41	2379	25.37	1734	0.00	0
ITASCA MN	0.00	0	1.15	2014	23.68	1584	24.21	1568	12.07	1710	30.11	1418	0.00	0
JACKSON MN	0.00	0	0.35	2760	37.44	1190	43.95	775	9.20	1995	21.38	2100	173.92	245
KANABEC MN	0.00	0	1.68	1760	30.73	1385	38.17	970	6.76	2308	25.15	1757	0.00	0
KANDIYOHI MN	0.00	0	0.60	2378	24.65	1555	32.29	1195	8.46	2060	30.58	1389	0.00	0
KITTSON MN	0.00	0	1.25	1963	20.65	1717	11.07	2607	15.25	1412	18.57	2388	0.00	0
KOOCHICHING MN	0.00	0	1.70	1756	25.41	1519	15.94	2140	14.49	1486	31.50	1331	0.00	0
LAC QUI PARL MN	0.00	0	0.58	2409	23.40	1596	27.84	1393	6.44	2370	28.10	1526	0.00	0
LAKE MN	0.00	0	1.15	2016	25.06	1528	41.87	834	7.78	2144	29.50	1456	0.00	0
LAKE OF THE MN	0.00	0	1.21	1990	17.98	1873	9.06	2846	8.96	2017	22.02	2050	0.00	0
LE SUEUR MN	0.00	0	0.91	2139	34.84	1262	57.42	533	7.66	2160	34.27	1192	325.19	109
LINCOLN MN	0.00	0	0.27	2840	37.71	1180	32.94	1160	8.45	2062	23.97	1846	181.88	240
LYON MN	0.00	0	0.18	2925	35.58	1241	37.56	992	6.90	2287	28.54	1506	181.88	239

TOTAL FALLOUT ACTIVITY (µCi/sq meter) by SHOT SERIES WITH COUNTY RANK

COUNTY AND STATE	RANGER SERIES	RANGER RANK	B-J SERIES	B-J RANK	T-S SERIES	T-S RANK	U-K SERIES	UK RANK	TEAPOT SERIES	TP RANK	PLUMBBOB SERIES	PB RANK	SHOT SEDAN	SEDAN RANK
	1951		1951		1952		1953		1955		1957		1962	
MAHNOMEN MN	0.00	0	1.78	1725	8.48	2720	15.23	2203	7.50	2184	26.59	1628	0.00	0
MARSHALL MN	0.00	0	0.64	2335	11.87	2373	10.14	2734	10.26	1892	20.47	2204	0.00	0
MARTIN MN	0.00	0	0.35	2761	33.59	1297	50.58	640	8.37	2071	28.74	1494	327.38	105
MCLEOD MN	0.00	0	1.38	1898	36.43	1216	43.86	777	7.67	2156	19.64	2292	0.00	0
MEEKER MN	0.00	0	1.38	1899	26.26	1504	40.07	892	7.65	2164	30.35	1404	0.00	0
MILLE LACS MN	0.00	0	1.68	1761	32.58	1328	32.46	1182	7.06	2264	24.80	1780	0.00	0
MORRISON MN	0.00	0	0.49	2541	26.23	1506	27.70	1401	7.35	2212	29.67	1445	0.00	0
MOWER MN	0.00	0	0.89	2158	55.86	842	83.81	272	10.36	1883	20.82	2164	584.61	13
MURRAY MN	0.00	0	0.64	2344	47.70	972	38.89	939	5.91	2477	21.93	2058	327.38	107
NICOLLET MN	0.00	0	2.37	1512	31.50	1357	50.93	636	6.94	2280	27.23	1578	325.19	113
NOBLES MN	0.00	0	1.94	1638	57.60	809	41.78	837	6.03	2447	23.68	1878	208.70	199
NORMAN MN	0.00	0	1.78	1726	5.36	2988	15.02	2222	7.62	2170	21.89	2063	0.00	0
OLMSTED MN	0.00	0	1.48	1853	19.41	1778	74.37	344	6.23	2415	26.36	1643	0.00	0
OTTER TAIL MN	0.00	0	0.77	2236	19.64	1761	20.94	1776	6.21	2421	25.17	1753	0.00	0
PENNINGTON MN	0.00	0	0.81	2211	15.71	2017	15.02	2220	11.00	1822	22.03	2049	0.00	0
PINE MN	0.00	0	1.78	1727	32.41	1333	48.14	687	7.23	2231	31.77	1316	0.00	0
PIPESTONE MN	0.00	0	1.56	1816	54.79	859	33.11	1152	7.40	2203	28.16	1524	327.38	106
POLK MN	0.00	0	2.05	1597	15.63	2021	13.35	2395	10.56	1864	24.83	1779	0.00	0
POPE MN	0.00	0	2.47	1493	14.69	2101	26.20	1468	4.68	2742	25.45	1728	0.00	0
RAMSEY MN	0.00	0	1.68	1762	31.22	1365	44.74	753	16.60	1294	29.80	1437	0.00	0
RED LAKE MN	0.00	0	0.81	2212	10.40	2539	12.89	2435	11.31	1785	23.20	1918	0.00	0
REDWOOD MN	0.00	0	0.52	2510	27.43	1475	35.22	1065	9.47	1969	26.06	1675	73.41	519
RENVILLE MN	0.00	0	0.60	2379	28.91	1439	37.99	976	8.52	2053	27.16	1585	0.00	0
RICE MN	0.00	0	0.77	2237	29.74	1420	68.16	414	6.72	2313	30.38	1402	325.19	110
ROCK MN	0.00	0	1.56	1817	64.40	703	35.24	1064	7.61	2171	28.53	1508	527.44	43
ROSEAU MN	0.00	0	1.25	1964	17.46	1908	9.94	2754	14.56	1480	18.94	2357	0.00	0

TOTAL FALLOUT ACTIVITY (µCi/sq meter) by SHOT SERIES WITH COUNTY RANK

COUNTY AND STATE	RANGER SERIES	RANGER RANK	B-J SERIES	B-J RANK	T-S SERIES	T-S RANK	U-K SERIES	UK RANK	TEAPOT SERIES	TP RANK	PLUMBBOB SERIES	PB RANK	SHOT SEDAN	SEDAN RANK
	1951		1951		1952		1953		1955		1957		1962	
SCOTT MN	0.00	0	0.77	2238	30.60	1391	47.73	694	6.49	2360	34.43	1184	0.00	0
SHERBURNE MN	0.00	0	1.22	1978	40.86	1098	36.70	1013	7.09	2259	30.52	1393	0.00	0
SIBLEY MN	0.00	0	0.68	2303	29.21	1432	42.17	823	7.34	2213	30.48	1397	181.88	237
ST LOUIS MN	0.00	0	1.15	2015	29.62	1421	38.07	974	12.09	1709	32.55	1281	0.00	0
STEARNS MN	0.00	0	1.48	1854	23.29	1602	30.33	1269	8.14	2105	28.92	1486	0.00	0
STEELE MN	0.00	0	0.68	2304	29.96	1411	72.22	370	7.19	2244	32.15	1296	181.88	236
STEVENS MN	0.00	0	0.18	2926	14.94	2074	22.08	1693	6.37	2391	23.96	1847	0.00	0
SWIFT MN	0.00	0	0.60	2380	28.58	1447	24.20	1569	5.73	2505	20.92	2154	0.00	0
TODD MN	0.00	0	1.58	1810	23.70	1582	25.36	1504	8.31	2082	34.27	1195	0.00	0
TRAVERSE MN	0.00	0	1.28	1958	17.23	1928	19.24	1895	4.88	2696	21.50	2092	0.00	0
WABASHA MN	0.00	0	0.72	2266	38.49	1162	76.95	320	10.67	1851	24.94	1772	325.19	115
WADENA MN	0.00	0	1.68	1759	30.44	1396	21.28	1745	4.18	2844	34.81	1161	0.00	0
WASECA MN	0.00	0	0.68	2305	30.45	1395	62.11	477	7.62	2167	27.91	1535	325.19	112
WASHINGTON MN	0.00	0	1.22	1979	31.91	1346	47.53	697	16.75	1277	31.09	1356	0.00	0
WATONWAN MN	0.00	0	0.35	2762	34.91	1261	50.46	641	7.70	2154	30.83	1379	584.61	11
WILKIN MN	0.00	0	0.43	2603	11.66	2400	12.07	2501	4.28	2823	26.09	1674	0.00	0
WINONA MN	0.00	0	1.10	2037	44.97	1021	78.91	306	17.22	1234	20.74	2172	181.88	241
WRIGHT MN	0.00	0	1.92	1658	28.42	1452	34.45	1102	6.97	2275	34.18	1200	0.00	0
YELLOW MEDIC MN	0.00	0	0.52	2511	26.02	1509	27.44	1413	5.92	2475	21.33	2104	0.00	0
ADAIR MO	0.49	474	0.76	2242	101.15	357	34.84	1086	27.39	610	74.16	440	76.52	496
ANDREW MO	0.00	0	2.42	1502	28.18	1458	32.87	1166	21.09	914	48.17	857	0.00	0
ATCHISON MO	0.00	0	2.23	1548	24.81	1543	40.46	879	20.41	959	44.81	927	99.74	380
AUDRAIN MO	37.86	9	0.63	2357	165.84	270	23.67	1596	17.99	1161	73.89	444	35.48	639
BARRY MO	0.00	0	11.35	574	65.79	685	37.40	996	32.43	508	59.62	649	0.00	0
BARTON MO	0.00	0	10.72	611	74.69	561	26.66	1442	19.46	1022	49.51	826	0.00	0
BATES MO	0.00	0	4.32	1144	33.90	1292	41.64	842	19.60	1014	62.23	612	0.00	0

TOTAL FALLOUT ACTIVITY (µCi/sq meter) by SHOT SERIES WITH COUNTY RANK

COUNTY AND STATE	RANGER SERIES	RANGER RANK	B-J SERIES	B-J RANK	T-S SERIES	T-S RANK	U-K SERIES	UK RANK	TEAPOT SERIES	TP RANK	PLUMBBOB SERIES	PB RANK	SHOT SEDAN	SEDAN RANK
	1951		1951		1952		1953		1955		1957		1962	
BENTON MO	28.72	28	3.91	1231	24.64	1556	33.35	1143	54.02	286	80.26	389	0.00	0
BOLLINGER MO	0.00	0	3.92	1228	15.84	2003	23.09	1633	21.59	881	31.50	1332	0.00	0
BOONE MO	0.00	0	0.82	2205	36.85	1202	28.21	1376	19.44	1029	81.51	379	0.00	0
BUCHANAN MO	0.00	0	2.63	1459	23.92	1575	35.37	1059	18.43	1105	46.20	902	0.00	0
BUTLER MO	0.00	0	4.21	1166	19.10	1799	41.22	850	34.54	479	46.27	897	0.00	0
CALDWELL MO	0.00	0	2.34	1520	33.09	1313	31.46	1220	23.03	801	94.40	280	90.94	402
CALLAWAY MO	26.39	29	0.71	2278	35.45	1242	25.06	1519	51.19	305	92.03	305	0.00	0
CAMDEN MO	0.00	0	3.72	1263	43.03	1055	33.81	1121	78.46	211	90.68	312	0.00	0
CAPE GIRARDE MO	0.00	0	3.81	1245	30.97	1374	33.67	1125	16.45	1304	24.74	1784	0.00	0
CARROLL MO	0.00	0	2.34	1521	28.73	1443	24.08	1581	22.78	812	70.43	482	35.48	640
CARTER MO	0.00	0	5.92	929	19.38	1781	39.64	908	32.70	505	52.50	764	0.00	0
CASS MO	0.00	0	4.21	1168	36.11	1224	29.28	1320	61.12	253	69.99	490	0.00	0
CEDAR MO	0.00	0	4.23	1159	52.04	889	42.19	822	19.23	1038	66.58	545	0.00	0
CHARITON MO	17.85	96	2.15	1574	42.33	1070	28.39	1368	25.55	672	73.79	446	90.94	404
CHRISTIAN MO	3.34	299	8.04	773	56.29	834	38.14	972	19.60	1012	68.90	507	0.00	0
CLARK MO	0.00	0	0.52	2497	64.26	705	32.66	1175	34.15	484	96.56	268	358.08	65
CLAY MO	0.00	0	2.86	1412	50.75	910	32.68	1173	23.65	768	57.03	680	0.00	0
CLINTON MO	0.00	0	2.53	1482	35.93	1231	27.97	1386	23.60	771	51.69	782	35.48	641
COLE MO	0.00	0	3.51	1303	23.58	1586	27.12	1423	53.83	287	96.51	269	0.00	0
COOPER MO	0.00	0	2.25	1541	40.75	1103	26.22	1464	50.72	311	80.49	387	0.00	0
CRAWFORD MO	12.25	150	4.18	1173	28.69	1444	25.93	1478	34.15	485	65.63	563	0.00	0
DADE MO	0.00	0	10.96	597	97.18	372	26.21	1466	18.79	1072	35.47	1140	0.00	0
DALLAS MO	0.00	0	4.74	1057	21.70	1678	31.18	1237	57.21	270	67.12	536	0.00	0
DAVIESS MO	0.00	0	2.23	1549	28.79	1440	30.94	1244	20.59	946	91.85	307	200.06	205
DE KALB MO	0.00	0	2.34	1522	37.85	1177	27.08	1426	41.98	391	95.58	276	35.48	638
DENT MO	23.97	33	4.66	1071	30.43	1397	24.11	1577	47.13	340	67.90	522	0.00	0

TOTAL FALLOUT ACTIVITY (µCi/sq meter) by SHOT SERIES WITH COUNTY RANK

COUNTY AND STATE	RANGER SERIES	RANGER RANK	B-J SERIES	B-J RANK	T-S SERIES	T-S RANK	U-K SERIES	UK RANK	TEAPOT SERIES	TP RANK	PLUMBBOB SERIES	PB RANK	SHOT SEDAN	SEDAN RANK
	1951		1951		1952		1953		1955		1957		1962	
DOUGLAS MO	0.00	0	6.22	901	14.39	2122	48.59	677	71.93	225	41.05	1012	0.00	0
DUNKLIN MO	0.00	0	4.51	1098	21.53	1683	69.75	395	32.17	511	24.28	1821	0.00	0
FRANKLIN MO	10.09	168	0.85	2181	83.21	474	22.74	1654	29.64	566	71.94	461	0.00	0
GASCONADE MO	31.35	19	0.85	2180	22.24	1648	24.62	1541	31.65	520	71.19	468	0.00	0
GENTRY MO	0.00	0	2.23	1550	22.65	1625	28.51	1360	19.79	997	92.81	295	0.00	0
GREENE MO	0.00	0	7.28	825	33.32	1302	49.05	659	31.16	529	70.18	486	0.00	0
GRUNDY MO	0.00	0	2.12	1580	30.68	1386	39.71	904	25.31	682	99.92	250	76.52	494
HARRISON MO	0.00	0	2.35	1518	35.06	1255	28.05	1382	22.61	817	94.28	282	80.75	477
HENRY MO	0.00	0	2.74	1437	35.30	1246	31.41	1224	65.78	238	94.51	279	0.00	0
HICKORY MO	0.00	0	6.94	845	22.93	1615	29.80	1293	20.41	960	72.60	455	0.00	0
HOLT MO	0.00	0	2.34	1523	36.68	1208	28.23	1375	18.60	1089	49.86	814	0.00	0
HOWARD MO	0.00	0	2.15	1576	49.61	931	29.84	1290	20.14	972	80.99	384	0.00	0
HOWELL MO	0.00	0	4.33	1142	32.16	1337	52.74	608	49.19	320	51.77	780	0.00	0
IRON MO	0.00	0	3.80	1248	21.09	1708	27.10	1424	29.28	573	50.92	796	0.00	0
JACKSON MO	0.00	0	11.29	578	19.70	1757	30.73	1253	24.12	748	57.18	678	0.00	0
JASPER MO	0.00	0	10.63	616	64.47	701	29.57	1305	46.18	346	51.51	784	0.00	0
JEFFERSON MO	0.00	0	0.63	2359	67.11	667	22.23	1683	26.40	646	49.87	813	0.00	0
JOHNSON MO	0.00	0	4.00	1213	35.08	1253	29.93	1285	25.23	686	92.78	296	0.00	0
KNOX MO	0.00	0	0.52	2498	260.78	223	27.55	1409	36.31	454	68.93	506	358.08	66
LACLEDE MO	18.93	83	4.49	1100	21.49	1689	29.84	1291	74.88	216	61.51	623	0.00	0
LAFAYETTE MO	0.00	0	3.89	1236	53.79	873	27.19	1420	40.48	408	96.80	266	0.00	0
LAWRENCE MO	2.33	318	10.96	596	68.49	645	37.85	981	49.85	315	53.73	742	0.00	0
LEWIS MO	0.00	0	0.60	2385	242.38	236	32.08	1200	30.07	558	66.29	552	200.06	210
LINCOLN MO	0.00	0	0.52	2499	89.50	422	20.55	1805	15.59	1382	50.85	798	0.00	0
LINN MO	0.00	0	2.04	1607	88.99	429	37.80	983	28.16	596	80.82	386	99.74	378
LIVINGSTON MO	0.00	0	2.23	1551	34.71	1266	27.69	1402	33.91	488	95.59	275	35.48	637

TOTAL FALLOUT ACTIVITY (µCi/sq meter) by SHOT SERIES WITH COUNTY RANK

COUNTY AND STATE	RANGER SERIES	RANGER RANK	B-J SERIES	B-J RANK	T-S SERIES	T-S RANK	U-K SERIES	UK RANK	TEAPOT SERIES	TP RANK	PLUMBBOB SERIES	PB RANK	SHOT SEDAN	SEDAN RANK
	1951		1951		1952		1953		1955		1957		1962	
MACON MO	0.00	0	0.71	2279	161.64	273	27.81	1394	42.03	390	72.60	456	46.20	585
MADISON MO	0.00	0	3.92	1229	22.11	1659	22.23	1684	16.74	1279	39.73	1034	0.00	0
MARIES MO	0.00	0	0.82	2206	41.26	1094	28.05	1381	23.03	802	97.49	260	0.00	0
MARION MO	0.00	0	0.52	2500	163.77	272	25.41	1500	26.75	630	71.00	471	76.52	497
MCDONALD MO	0.00	0	10.42	627	93.48	391	34.72	1091	48.60	326	42.00	992	0.00	0
MERCER MO	0.00	0	2.04	1608	30.34	1401	38.92	937	24.44	731	89.47	320	0.00	0
MILLER MO	0.00	0	3.61	1283	27.06	1486	34.06	1115	56.19	277	96.33	271	0.00	0
MISSISSIPPI MO	0.00	0	4.11	1197	24.59	1557	59.63	508	30.11	557	20.70	2176	0.00	0
MONITEAU MO	0.00	0	3.61	1284	23.11	1610	25.89	1480	22.35	838	97.05	264	0.00	0
MONROE MO	0.00	0	0.52	2501	31.02	1372	29.80	1294	28.64	589	78.10	405	90.94	403
MONTGOMERY MO	0.00	0	0.52	2502	30.88	1379	24.25	1566	18.16	1137	67.83	524	0.00	0
MORGAN MO	0.00	0	2.94	1398	40.49	1113	32.33	1192	74.65	218	92.64	297	0.00	0
NEW MADRID MO	0.00	0	4.11	1194	24.66	1553	64.54	458	39.05	422	27.48	1563	0.00	0
NEWTON MO	0.00	0	11.22	582	76.22	545	39.66	907	58.09	267	46.60	888	0.00	0
NODAWAY MO	0.00	0	2.23	1552	25.93	1511	33.59	1129	18.32	1115	100.83	243	0.00	0
OREGON MO	0.00	0	5.43	969	23.93	1574	58.24	522	58.13	266	50.87	797	0.00	0
OSAGE MO	0.00	0	0.82	2207	24.06	1570	25.28	1506	70.11	232	113.86	194	0.00	0
OZARK MO	0.00	0	4.34	1138	18.39	1842	65.32	447	68.43	233	46.97	876	0.00	0
PEMISCOT MO	0.00	0	3.93	1226	19.00	1806	67.20	429	34.79	474	20.64	2183	0.00	0
PERRY MO	0.00	0	3.70	1269	23.54	1589	19.47	1878	22.29	841	27.82	1540	0.00	0
PETTIS MO	0.00	0	3.81	1247	47.00	982	27.37	1415	78.92	208	98.58	255	0.00	0
PHELPS MO	23.97	34	3.30	1342	29.77	1418	24.58	1544	52.08	300	65.64	562	0.00	0
PIKE MO	37.86	10	0.52	2480	170.18	268	22.24	1682	16.24	1324	60.05	643	90.94	406
PLATTE MO	0.00	0	2.63	1460	24.37	1565	31.87	1209	26.54	638	69.64	496	0.00	0
POLK MO	0.00	0	4.56	1088	50.13	925	30.20	1273	29.66	565	97.61	258	0.00	0
PULASKI MO	0.00	0	4.13	1188	24.70	1549	28.96	1338	58.40	264	95.49	277	0.00	0

TOTAL FALLOUT ACTIVITY (µCi/sq meter) by SHOT SERIES WITH COUNTY RANK

COUNTY AND STATE	RANGER SERIES	RANGER RANK	B-J SERIES	B-J RANK	T-S SERIES	T-S RANK	U-K SERIES	UK RANK	TEAPOT SERIES	TP RANK	PLUMBBOB SERIES	PB RANK	SHOT SEDAN	SEDAN RANK
	1951		1951		1952		1953		1955		1957		1962	
PUTNAM MO	1.11	405	1.93	1642	119.16	326	46.35	716	51.91	301	65.30	567	200.06	211
RALLS MO	25.61	30	0.52	2481	172.03	267	23.26	1625	33.06	502	59.85	647	200.06	213
RANDOLPH MO	0.00	0	0.71	2280	48.83	945	27.09	1425	19.09	1046	70.18	487	90.94	405
RAY MO	0.00	0	2.53	1483	67.59	657	29.53	1306	24.66	720	93.60	285	0.00	0
REYNOLDS MO	0.00	0	4.90	1026	19.97	1747	25.47	1497	44.25	366	52.28	775	0.00	0
RIPLEY MO	0.00	0	4.05	1203	17.37	1914	67.49	423	31.55	523	44.17	941	0.00	0
SALINE MO	0.00	0	2.34	1524	50.34	920	29.72	1297	24.68	717	92.20	303	0.00	0
SCHUYLER MO	0.71	433	0.60	2370	114.78	330	44.14	771	36.34	453	64.46	573	200.06	212
SCOTLAND MO	0.00	0	0.52	2503	203.68	250	39.22	925	22.78	811	53.93	737	358.08	67
SCOTT MO	0.00	0	4.22	1165	20.51	1725	23.57	1606	31.32	525	24.04	1837	0.00	0
SHANNON MO	0.00	0	4.91	1023	24.48	1559	24.55	1546	47.80	333	54.79	719	0.00	0
SHELBY MO	0.00	0	0.52	2504	95.75	381	30.52	1264	27.73	605	77.75	410	200.06	206
ST CHARLES MO	0.00	0	0.75	2245	31.96	1343	18.69	1935	24.29	738	57.62	674	0.00	0
ST CLAIR MO	0.00	0	0.52	2505	40.63	1109	31.04	1241	19.55	1017	49.04	834	0.00	0
ST FRANCOIS MO	0.00	0	3.96	1219	29.38	1429	28.41	1364	24.90	701	51.05	793	0.00	0
ST LOUIS MO	0.00	0	0.60	2386	41.97	1076	21.65	1721	14.02	1533	46.34	894	0.00	0
STE GENEVIEV MO	0.00	0	2.64	1455	22.71	1622	27.14	1422	17.61	1193	36.14	1121	0.00	0
STODDARD MO	0.00	0	4.11	1195	19.90	1751	48.32	684	33.41	498	28.08	1527	0.00	0
STONE MO	0.00	0	11.22	583	58.28	795	35.72	1043	24.76	708	41.67	1000	0.00	0
SULLIVAN MO	0.00	0	2.04	1609	131.44	314	43.03	799	22.57	823	70.63	479	76.52	498
TANEY MO	0.00	0	4.74	1059	20.63	1718	38.48	959	35.13	468	34.16	1201	0.00	0
TEXAS MO	16.71	102	4.55	1094	22.34	1641	27.77	1398	53.05	291	74.44	437	0.00	0
VERNON MO	0.00	0	7.58	798	57.56	810	30.28	1271	20.79	932	70.84	474	0.00	0
WARREN MO	0.00	0	0.52	2506	25.07	1527	21.17	1752	18.22	1129	70.80	476	0.00	0
WASHINGTON MO	0.00	0	3.60	1286	21.48	1691	24.54	1548	29.87	560	51.92	778	0.00	0
WAYNE MO	0.00	0	4.30	1150	23.44	1594	24.95	1523	39.48	418	53.13	752	0.00	0

U.S. FALLOUT ATLAS : TOTAL FALLOUT

TOTAL FALLOUT ACTIVITY (µCi/sq meter) by SHOT SERIES WITH COUNTY RANK

COUNTY AND STATE	RANGER SERIES	RANGER RANK	B-J SERIES	B-J RANK	T-S SERIES	T-S RANK	U-K SERIES	UK RANK	TEAPOT SERIES	TP RANK	PLUMBBOB SERIES	PB RANK	SHOT SEDAN	SEDAN RANK
	1951		1951		1952		1953		1955		1957		1962	
WEBSTER MO	0.00	0	3.38	1328	25.01	1531	31.37	1226	72.73	220	63.69	589	0.00	0
WORTH MO	0.00	0	2.12	1581	46.37	1000	35.84	1039	18.82	1068	114.20	193	200.06	204
WRIGHT MO	16.71	103	4.80	1041	22.88	1616	31.39	1225	71.75	226	63.05	597	0.00	0
ADAMS MS	0.00	0	0.93	2132	5.62	2973	61.63	485	16.60	1292	129.00	146	0.00	0
ALCORN MS	0.00	0	4.34	1137	11.67	2395	48.84	667	15.29	1408	22.72	1969	0.00	0
AMITE MS	0.00	0	4.48	1102	7.29	2828	40.36	883	17.23	1233	50.53	803	0.00	0
ATTALA MS	0.00	0	3.91	1230	24.47	1560	18.00	1979	21.47	891	30.49	1396	0.00	0
BENTON MS	0.00	0	3.71	1267	14.31	2134	63.21	466	13.17	1595	26.67	1620	0.00	0
BOLIVAR MS	0.00	0	3.79	1251	21.47	1692	57.35	537	17.80	1180	26.20	1655	0.00	0
CALHOUN MS	0.00	0	3.47	1312	13.59	2198	38.28	967	13.06	1608	34.30	1190	0.00	0
CARROLL MS	0.00	0	4.17	1175	15.90	2000	42.05	829	25.90	660	37.43	1084	0.00	0
CHICKASAW MS	0.00	0	3.48	1307	16.48	1968	54.27	580	15.82	1360	45.79	911	0.00	0
CHOCTAW MS	0.00	0	5.41	971	7.54	2800	16.63	2093	17.93	1170	34.25	1197	0.00	0
CLAIBORNE MS	0.00	0	1.45	1860	6.20	2930	62.49	473	22.04	861	60.95	630	0.00	0
CLARKE MS	0.00	0	3.44	1317	3.77	3043	12.74	2446	21.32	901	64.76	571	0.00	0
CLAY MS	0.00	0	3.21	1364	11.73	2389	18.88	1916	20.56	949	38.48	1056	0.00	0
COAHOMA MS	0.00	0	4.70	1065	13.44	2217	61.71	484	18.10	1143	30.95	1370	0.00	0
COPIAH MS	0.00	0	1.88	1673	14.62	2105	26.31	1459	15.33	1403	64.85	569	0.00	0
COVINGTON MS	0.00	0	1.45	1861	6.44	2908	13.62	2369	16.99	1259	48.48	845	0.00	0
DE SOTO MS	0.00	0	3.35	1336	15.01	2069	75.74	331	17.81	1179	27.01	1596	0.00	0
FORREST MS	0.00	0	3.55	1294	3.42	3055	50.61	639	13.40	1579	69.90	492	0.00	0
FRANKLIN MS	0.00	0	1.23	1970	5.29	2991	37.68	991	18.05	1146	131.53	127	0.00	0
GEORGE MS	0.00	0	4.65	1076	2.67	3075	48.71	673	11.57	1762	83.26	366	0.00	0
GREENE MS	0.00	0	3.26	1347	2.95	3070	48.77	670	11.68	1740	59.86	646	0.00	0
GRENADA MS	0.00	0	4.41	1118	11.29	2443	46.59	712	13.02	1613	43.70	953	0.00	0
HANCOCK MS	0.00	0	2.71	1450	19.53	1774	50.74	637	10.68	1849	86.24	340	0.00	0

TOTAL FALLOUT ACTIVITY (µCi/sq meter) by SHOT SERIES WITH COUNTY RANK

COUNTY AND STATE	RANGER SERIES	RANGER RANK	B-J SERIES	B-J RANK	T-S SERIES	T-S RANK	U-K SERIES	UK RANK	TEAPOT SERIES	TP RANK	PLUMBBOB SERIES	PB RANK	SHOT SEDAN	SEDAN RANK
	1951		1951		1952		1953		1955		1957		1962	
HARRISON MS	0.00	0	3.10	1377	2.95	3071	48.65	676	11.61	1754	61.52	622	0.00	0
HINDS MS	0.00	0	1.66	1768	22.55	1628	45.92	723	20.71	937	89.98	319	0.00	0
HOLMES MS	0.00	0	3.52	1299	7.90	2766	18.36	1956	21.18	910	24.45	1811	0.00	0
HUMPHREYS MS	0.00	0	4.01	1210	7.15	2840	39.48	917	25.67	666	30.19	1414	0.00	0
ISSAQUENA MS	0.00	0	3.66	1276	22.43	1632	59.46	511	18.91	1059	19.46	2314	0.00	0
ITAWAMBA MS	0.00	0	4.13	1187	6.02	2943	30.16	1276	14.03	1531	28.67	1497	0.00	0
JACKSON MS	0.00	0	3.70	1270	2.33	3078	49.09	657	14.05	1528	61.24	628	0.00	0
JASPER MS	0.00	0	1.50	1843	7.61	2791	13.61	2370	17.64	1191	50.93	795	0.00	0
JEFFERSON MS	0.00	0	1.23	1971	6.05	2939	60.21	499	21.58	884	60.98	629	0.00	0
JEFFERSON DA MS	0.00	0	2.85	1413	8.34	2734	15.13	2210	17.84	1177	23.91	1853	0.00	0
JONES MS	0.00	0	1.85	1681	6.65	2888	12.38	2476	21.36	899	33.61	1229	0.00	0
KEMPER MS	0.00	0	2.18	1564	19.32	1786	27.79	1395	18.15	1138	16.72	2564	0.00	0
LAFAYETTE MS	0.00	0	4.54	1095	14.01	2165	52.59	611	12.05	1714	25.30	1736	0.00	0
LAMAR MS	0.00	0	3.22	1360	8.90	2672	14.15	2318	17.13	1243	75.63	423	0.00	0
LAUDERDALE MS	0.00	0	2.15	1573	24.73	1547	29.43	1313	16.78	1276	84.07	354	0.00	0
LAWRENCE MS	0.00	0	1.55	1824	34.70	1267	18.36	1957	16.65	1286	30.86	1377	0.00	0
LEAKE MS	0.00	0	3.77	1256	24.51	1558	16.58	2096	25.95	658	25.55	1716	0.00	0
LEE MS	0.00	0	4.38	1124	11.87	2372	51.42	628	14.88	1447	31.18	1350	0.00	0
LEFLORE MS	0.00	0	4.05	1204	6.37	2913	38.62	954	21.31	904	33.26	1251	0.00	0
LINCOLN MS	0.00	0	1.45	1862	5.17	2998	32.88	1164	17.41	1214	33.47	1235	0.00	0
LOWNDES MS	0.00	0	3.79	1252	5.58	2975	16.52	2100	17.20	1237	22.87	1954	0.00	0
MADISON MS	0.00	0	3.33	1339	6.67	2882	50.40	642	19.46	1024	18.51	2397	0.00	0
MARION MS	0.00	0	2.18	1561	8.88	2675	19.74	1859	15.83	1359	99.04	252	0.00	0
MARSHALL MS	0.00	0	3.48	1309	14.85	2084	72.24	369	14.98	1437	29.79	1439	0.00	0
MONROE MS	0.00	0	4.60	1083	7.33	2823	29.09	1331	14.04	1529	29.93	1428	0.00	0
MONTGOMERY MS	0.00	0	3.94	1224	7.55	2799	38.24	969	33.66	493	38.28	1057	0.00	0

TOTAL FALLOUT ACTIVITY (µCi/sq meter) by SHOT SERIES WITH COUNTY RANK

COUNTY AND STATE	RANGER SERIES	RANGER RANK	B-J SERIES	B-J RANK	T-S SERIES	T-S RANK	U-K SERIES	UK RANK	TEAPOT SERIES	TP RANK	PLUMBBOB SERIES	PB RANK	SHOT SEDAN	SEDAN RANK
	1951		1951		1952		1953		1955		1957		1962	
NESHOBA MS	0.00	0	3.44	1320	6.61	2892	16.10	2130	23.81	759	18.53	2392	0.00	0
NEWTON MS	0.00	0	2.02	1615	16.79	1951	26.82	1437	17.94	1169	47.90	860	0.00	0
NOXUBEE MS	0.00	0	2.18	1565	8.29	2740	16.56	2099	17.16	1239	25.00	1767	0.00	0
OKTIBBEHA MS	0.00	0	3.75	1261	16.14	1993	37.80	982	18.50	1096	31.04	1360	0.00	0
PANOLA MS	0.00	0	4.41	1119	15.82	2007	70.66	383	17.40	1215	22.26	2028	0.00	0
PEARL RIVER MS	0.00	0	2.61	1464	8.07	2758	58.36	520	9.80	1945	83.99	357	0.00	0
PERRY MS	0.00	0	3.43	1322	2.77	3074	49.11	656	13.45	1575	40.59	1018	0.00	0
PIKE MS	0.00	0	1.33	1929	22.28	1645	45.24	741	16.31	1318	85.21	344	0.00	0
PONTOTOC MS	0.00	0	4.69	1066	16.20	1990	53.82	588	11.71	1737	31.66	1326	0.00	0
PRENTISS MS	0.00	0	3.59	1287	9.73	2591	55.03	568	11.67	1743	26.77	1613	0.00	0
QUITMAN MS	0.00	0	3.72	1265	14.57	2108	74.90	338	18.82	1065	25.41	1732	0.00	0
RANKIN MS	0.00	0	1.34	1920	7.79	2779	16.87	2071	20.83	926	29.93	1426	0.00	0
SCOTT MS	0.00	0	1.40	1882	7.37	2818	15.10	2213	20.73	935	44.58	934	0.00	0
SHARKEY MS	0.00	0	3.84	1240	7.57	2796	48.26	686	18.35	1113	30.07	1419	0.00	0
SIMPSON MS	0.00	0	1.45	1863	6.64	2889	16.93	2061	17.52	1206	27.70	1548	0.00	0
SMITH MS	0.00	0	1.84	1686	5.02	3004	15.63	2166	16.09	1331	43.02	970	0.00	0
STONE MS	0.00	0	3.43	1323	18.43	1839	49.48	651	11.66	1744	64.19	578	0.00	0
SUNFLOWER MS	0.00	0	3.33	1340	6.86	2863	38.71	949	20.71	938	30.88	1375	0.00	0
TALLAHATCHIE MS	0.00	0	3.99	1216	15.31	2047	61.00	491	17.42	1211	30.75	1383	0.00	0
TATE MS	0.00	0	4.47	1103	19.69	1759	68.23	413	16.03	1335	25.18	1750	0.00	0
TIPPAH MS	0.00	0	4.09	1200	17.10	1935	69.80	393	13.30	1587	23.32	1903	0.00	0
TISHOMINGO MS	0.00	0	4.78	1048	14.36	2128	50.09	646	12.86	1631	26.00	1681	0.00	0
TUNICA MS	0.00	0	4.24	1157	18.39	1841	54.16	581	21.48	890	24.56	1794	0.00	0
UNION MS	0.00	0	4.11	1196	17.73	1895	53.30	596	10.56	1863	20.51	2200	0.00	0
WALTHALL MS	0.00	0	1.43	1872	3.90	3039	73.88	351	8.55	2048	37.17	1091	0.00	0
WARREN MS	0.00	0	1.88	1674	22.27	1646	61.83	483	18.32	1117	129.82	139	0.00	0

TOTAL FALLOUT ACTIVITY (µCi/sq meter) by SHOT SERIES WITH COUNTY RANK

COUNTY AND STATE	RANGER SERIES	RANGER RANK	B-J SERIES	B-J RANK	T-S SERIES	T-S RANK	U-K SERIES	UK RANK	TEAPOT SERIES	TP RANK	PLUMBBOB SERIES	PB RANK	SHOT SEDAN	SEDAN RANK
	1951		1951		1952		1953		1955		1957		1962	
WASHINGTON MS	0.00	0	4.16	1178	7.22	2833	45.44	735	19.43	1031	24.83	1778	0.00	0
WAYNE MS	0.00	0	3.39	1326	3.22	3063	11.99	2512	21.47	892	62.65	607	0.00	0
WEBSTER MS	0.00	0	3.02	1388	12.47	2313	19.43	1880	20.62	942	42.52	983	0.00	0
WILKINSON MS	0.00	0	3.77	1258	10.65	2508	43.99	774	18.00	1158	129.97	137	0.00	0
WINSTON MS	0.00	0	7.46	813	24.43	1563	17.84	1993	18.84	1062	22.67	1977	0.00	0
YALOBUSHA MS	0.00	0	3.83	1242	14.55	2110	67.41	425	12.36	1676	28.15	1525	0.00	0
YAZOO MS	0.00	0	3.51	1302	13.25	2232	44.05	772	24.74	709	31.68	1322	0.00	0
BEAVERHEAD MT	0.00	0	0.93	2133	24096.33	11	9.60	2787	3.99	2876	78.27	404	0.00	0
BIG HORN MT	0.00	0	1.92	1654	364.92	194	66.69	433	13.70	1561	67.98	519	0.00	0
BLAINE MT	0.00	0	0.62	2365	5907.66	67	54.37	579	9.72	1953	83.02	368	0.00	0
BROADWATER MT	0.00	0	0.43	2614	19948.70	17	2.84	3070	3.31	2971	67.70	528	0.00	0
CARBON MT	0.00	0	0.51	2514	2895.99	81	40.87	862	15.58	1383	88.11	332	0.00	0
CARTER MT	0.00	0	1.01	2072	109.10	340	79.00	305	4.45	2781	96.07	272	0.00	0
CASCADE MT	0.00	0	0.54	2451	7989.33	56	6.42	2970	4.66	2747	78.58	402	0.00	0
CHOUTEAU MT	0.00	0	0.43	2615	25380.09	9	9.10	2842	5.75	2499	75.61	425	0.00	0
CUSTER MT	0.00	0	1.12	2023	195.07	259	242.01	97	4.44	2783	62.75	604	0.00	0
DANIELS MT	0.00	0	1.03	2064	346.01	196	99.68	203	4.18	2843	75.02	431	0.00	0
DAWSON MT	0.00	0	1.48	1847	298.47	213	202.77	110	4.49	2775	100.44	248	0.00	0
DEER LODGE MT	0.00	0	0.43	2616	38686.36	5	5.53	3007	3.94	2884	68.86	509	0.00	0
FALLON MT	0.00	0	0.79	2224	48.81	946	141.12	148	4.45	2782	102.79	233	0.00	0
FERGUS MT	0.00	0	1.12	2024	7130.36	60	57.57	529	8.65	2041	100.59	247	0.00	0
FLATHEAD MT	0.00	0	0.43	2617	9890.98	50	4.54	3030	2.65	3038	51.78	779	0.00	0
GALLATIN MT	0.00	0	0.54	2452	12688.89	39	5.87	2994	4.73	2733	93.44	286	0.00	0
GARFIELD MT	0.00	0	2.03	1611	834.66	133	87.35	263	17.02	1257	60.60	635	0.00	0
GLACIER MT	0.00	0	0.43	2618	3564.49	77	6.92	2958	2.98	3010	57.66	673	0.00	0
GOLDEN VALLE MT	0.00	0	0.51	2515	701.37	150	35.60	1047	8.25	2091	97.24	263	0.00	0

TOTAL FALLOUT ACTIVITY (µCi/sq meter) by SHOT SERIES WITH COUNTY RANK

COUNTY AND STATE	RANGER SERIES	RANGER RANK	B-J SERIES	B-J RANK	T-S SERIES	T-S RANK	U-K SERIES	UK RANK	TEAPOT SERIES	TP RANK	PLUMBBOB SERIES	PB RANK	SHOT SEDAN	SEDAN RANK
	1951		1951		1952		1953		1955		1957		1962	
GRANITE MT	0.00	0	0.43	2619	7092.94	61	6.19	2981	3.15	2990	81.16	382	0.00	0
HILL MT	0.00	0	0.43	2620	10597.03	48	9.25	2832	5.76	2498	58.26	661	0.00	0
JEFFERSON MT	0.00	0	0.43	2621	21764.19	16	2.81	3072	2.92	3013	67.84	523	0.00	0
JUDITH BASIN MT	0.00	0	0.43	2622	22409.18	14	6.92	2959	3.63	2933	103.22	229	0.00	0
LAKE MT	0.00	0	0.54	2453	6800.47	62	4.01	3041	2.89	3017	59.21	653	0.00	0
LEWIS AND CL MT	0.00	0	0.43	2623	12092.64	45	3.50	3052	3.21	2981	74.54	434	0.00	0
LIBERTY MT	0.00	0	0.43	2624	7907.62	57	7.15	2948	4.36	2802	56.37	689	0.00	0
LINCOLN MT	0.00	0	0.43	2625	4700.40	71	6.93	2956	2.81	3025	44.63	931	0.00	0
MADISON MT	0.00	0	0.43	2626	16152.83	31	7.21	2945	4.98	2665	86.73	338	0.00	0
MCCONE MT	0.00	0	1.12	2025	191.88	261	144.50	144	4.25	2833	92.08	304	0.00	0
MEAGHER MT	0.00	0	0.54	2454	17385.22	27	5.42	3009	4.84	2707	89.46	321	0.00	0
MINERAL MT	0.00	0	28.68	114	12933.57	38	4.79	3026	3.71	2925	51.01	794	0.00	0
MISSOULA MT	0.00	0	0.43	2627	17678.51	25	3.62	3048	3.17	2986	58.16	663	0.00	0
MUSSELSHELL MT	0.00	0	0.51	2516	428.15	192	59.62	509	11.82	1732	64.24	577	0.00	0
PARK MT	0.00	0	0.54	2455	8603.47	54	5.62	3001	4.90	2688	98.49	256	0.00	0
PETROLEUM MT	0.00	0	0.51	2517	1854.85	95	65.18	449	12.11	1706	68.31	518	0.00	0
PHILLIPS MT	0.00	0	0.79	2225	1666.35	98	57.39	535	11.47	1767	62.76	603	0.00	0
PONDERA MT	0.00	0	0.43	2628	5529.72	69	6.93	2957	3.72	2923	54.69	722	0.00	0
POWDER RIVER MT	0.00	0	1.92	1655	85.31	460	226.44	100	27.31	615	53.55	745	0.00	0
POWELL MT	0.00	0	0.43	2629	22359.87	15	10.64	2666	3.44	2956	79.92	392	0.00	0
PRAIRIE MT	0.00	0	0.62	2366	199.17	254	142.79	146	5.07	2643	71.17	469	0.00	0
RAVALLI MT	0.00	0	0.43	2630	8586.33	55	6.65	2965	3.85	2905	55.11	712	0.00	0
RICHLAND MT	0.00	0	1.14	2017	80.40	502	174.51	124	4.14	2851	101.63	238	0.00	0
ROOSEVELT MT	0.00	0	1.22	1980	796.00	142	138.99	150	3.89	2894	83.39	362	0.00	0
ROSEBUD MT	0.00	0	1.01	2073	145.27	294	69.70	396	16.97	1262	63.02	598	0.00	0
SANDERS MT	0.00	0	0.43	2631	8799.00	52	5.33	3011	3.22	2980	48.21	853	0.00	0

TOTAL FALLOUT ACTIVITY (µCi/sq meter) by SHOT SERIES WITH COUNTY RANK

COUNTY AND STATE	RANGER SERIES	RANGER RANK	B-J SERIES	B-J RANK	T-S SERIES	T-S RANK	U-K SERIESI	UK RANK	TEAPOT SERIES	TP RANK	PLUMBBOB SERIES	PB RANK	SHOT SEDAN	SEDAN RANK
	1951		1951		1952		1953		1955		1957		1962	
SHERIDAN MT	0.00	0	0.94	2123	713.18	149	133.98	156	3.79	2914	72.69	453	0.00	0
SILVER BOW MT	0.00	0	0.43	2632	22637.79	13	4.41	3033	3.52	2944	70.53	481	0.00	0
STILLWATER MT	0.00	0	0.70	2283	928.40	125	38.72	948	7.23	2232	92.90	293	0.00	0
SWEET GRASS MT	0.00	0	0.43	2633	869.29	129	26.55	1449	6.90	2288	91.14	311	0.00	0
TETON MT	0.00	0	0.43	2634	10689.88	47	6.03	2985	3.87	2898	60.80	632	0.00	0
TOOLE MT	0.00	0	0.43	2635	7854.55	58	6.94	2955	3.38	2959	68.97	505	0.00	0
TREASURE MT	0.00	0	1.01	2074	173.45	265	51.11	631	16.72	1281	63.82	586	0.00	0
VALLEY MT	0.00	0	0.79	2226	872.70	128	101.20	198	4.80	2715	61.96	617	0.00	0
WHEATLAND MT	0.00	0	0.43	2636	14642.77	33	37.74	986	5.45	2559	92.22	302	0.00	0
WIBAUX MT	0.00	0	1.92	1656	51.55	896	139.00	149	4.23	2836	126.34	160	0.00	0
YELLOWSTONE MT	0.00	0	0.68	2309	518.43	165	24.88	1527	7.92	2129	44.16	942	0.00	0
ALAMANCE NC	0.00	0	30.46	93	6.73	2875	12.25	2485	4.32	2812	27.21	1580	0.00	0
ALEXANDER NC	0.00	0	9.87	664	14.56	2109	12.10	2500	7.05	2266	42.33	986	0.00	0
ALLEGHANY NC	0.00	0	15.88	384	13.24	2234	10.81	2649	5.85	2486	24.89	1776	17.11	673
ANSON NC	0.00	0	20.73	241	6.88	2862	10.38	2699	4.63	2755	19.65	2291	0.00	0
ASHE NC	0.00	0	11.70	554	7.58	2794	18.22	1962	14.95	1439	43.67	954	18.06	662
AVERY NC	0.00	0	9.74	671	12.07	2348	16.99	2058	16.65	1284	34.46	1179	0.00	0
BEAUFORT NC	0.19	545	35.50	65	24.71	1548	10.34	2704	3.19	2984	29.15	1473	38.35	619
BERTIE NC	0.00	0	32.79	82	17.73	1893	12.47	2469	3.18	2985	18.18	2424	0.00	0
BLADEN NC	0.00	0	25.27	175	8.57	2708	9.96	2751	4.38	2800	21.43	2097	0.00	0
BRUNSWICK NC	0.41	482	19.92	267	15.56	2027	8.72	2872	4.60	2760	21.76	2076	0.00	0
BUNCOMBE NC	0.23	538	12.07	533	10.70	2502	18.76	1926	15.92	1346	33.66	1224	0.00	0
BURKE NC	0.00	0	9.36	690	21.77	1674	13.56	2375	14.47	1491	37.47	1083	0.00	0
CABARRUS NC	0.19	546	15.87	385	7.82	2775	14.27	2310	3.94	2886	16.80	2556	0.00	0
CALDWELL NC	0.00	0	16.51	365	9.36	2626	14.29	2305	13.05	1609	34.71	1164	0.00	0
CAMDEN NC	0.19	547	28.01	122	18.14	1858	9.59	2790	3.05	2999	25.46	1726	191.00	227

TOTAL FALLOUT ACTIVITY (µCi/sq meter) by SHOT SERIES WITH COUNTY RANK

COUNTY AND STATE	RANGER SERIES	RANGER RANK	B-J SERIES	B-J RANK	T-S SERIES	T-S RANK	U-K SERIES	UK RANK	TEAPOT SERIES	TP RANK	PLUMBBOB SERIES	PB RANK	SHOT SEDAN	SEDAN RANK
	1951		1951		1952		1953		1955		1957		1962	
CARTERET NC	0.19	548	33.77	75	17.09	1936	9.67	2782	3.57	2940	23.14	1926	0.00	0
CASWELL NC	0.56	445	21.28	230	3.38	3058	8.01	2910	7.31	2218	20.77	2168	0.00	0
CATAWBA NC	0.19	549	16.29	374	10.21	2555	12.58	2461	6.27	2405	33.29	1249	0.00	0
CHATHAM NC	0.00	0	26.05	163	7.58	2795	11.54	2551	4.94	2677	24.95	1771	43.75	600
CHEROKEE NC	0.19	550	12.82	495	14.30	2135	21.09	1760	19.48	1021	34.46	1180	0.00	0
CHOWAN NC	0.00	0	30.06	96	15.61	2022	9.45	2806	4.54	2772	26.95	1599	84.15	475
CLAY NC	0.00	0	6.95	842	16.55	1962	24.48	1551	19.45	1028	40.51	1021	0.00	0
CLEVELAND NC	0.19	551	16.09	377	8.50	2716	13.24	2400	12.10	1708	26.16	1660	0.00	0
COLUMBUS NC	0.00	0	26.80	143	10.22	2553	9.48	2799	4.36	2804	21.81	2072	0.00	0
CRAVEN NC	0.19	552	33.86	73	17.82	1885	10.85	2642	3.34	2966	20.98	2140	0.00	0
CUMBERLAND NC	0.00	0	24.30	185	12.48	2309	10.00	2748	2.91	3016	27.58	1557	0.00	0
CURRITUCK NC	0.56	446	29.12	109	18.47	1836	9.46	2803	3.04	3003	25.29	1738	191.00	228
DARE NC	0.00	0	33.53	78	16.26	1983	11.11	2600	3.04	3001	26.09	1673	0.00	0
DAVIDSON NC	0.19	553	23.99	190	25.90	1512	8.68	2874	4.98	2666	16.86	2553	0.00	0
DAVIE NC	0.00	0	20.04	261	7.03	2852	9.74	2775	3.34	2967	22.73	1967	0.00	0
DUPLIN NC	0.00	0	28.03	121	21.79	1673	11.95	2517	3.36	2961	15.44	2664	0.00	0
DURHAM NC	0.00	0	19.07	291	10.20	2557	9.41	2813	5.29	2591	23.78	1871	21.81	657
EDGECOMBE NC	0.00	0	20.97	237	14.14	2145	11.05	2612	4.01	2873	17.33	2512	0.00	0
FORSYTH NC	0.00	0	23.61	196	10.91	2485	9.77	2769	6.70	2318	24.47	1806	0.00	0
FRANKLIN NC	0.00	0	18.08	325	5.72	2970	11.82	2528	3.33	2968	18.84	2363	0.00	0
GASTON NC	0.00	0	14.99	422	8.25	2742	14.02	2330	9.17	1997	20.06	2248	0.00	0
GATES NC	0.00	0	29.44	104	19.53	1773	11.44	2564	2.27	3057	18.12	2431	0.00	0
GRAHAM NC	0.00	0	12.98	489	15.03	2068	24.19	1571	17.89	1172	24.06	1833	0.00	0
GRANVILLE NC	0.00	0	18.87	297	12.66	2292	7.02	2950	7.23	2230	25.77	1701	50.92	583
GREENE NC	0.00	0	17.94	327	13.92	2171	8.93	2860	4.91	2686	24.61	1792	0.00	0
GUILFORD NC	0.00	0	19.68	273	10.23	2551	10.65	2665	5.36	2577	20.32	2219	18.06	665

TOTAL FALLOUT ACTIVITY (µCi/sq meter) by SHOT SERIES WITH COUNTY RANK

COUNTY AND STATE	RANGER SERIES	RANGER RANK	B-J SERIES	B-J RANK	T-S SERIES	T-S RANK	U-K SERIES	UK RANK	TEAPOT SERIES	TP RANK	PLUMBBOB SERIES	PB RANK	SHOT SEDAN	SEDAN RANK
	1951		1951		1952		1953		1955		1957		1962	
HALIFAX NC	0.00	0	18.61	307	16.52	1964	11.57	2545	3.04	3000	18.99	2351	0.00	0
HARNETT NC	0.00	0	27.06	139	7.47	2807	8.68	2875	4.86	2703	22.82	1958	0.00	0
HAYWOOD NC	0.00	0	8.80	727	14.65	2102	21.43	1734	19.32	1035	39.89	1031	0.00	0
HENDERSON NC	0.00	0	12.74	499	14.89	2079	19.09	1904	18.29	1120	35.95	1126	0.00	0
HERTFORD NC	0.00	0	31.53	88	18.37	1843	11.35	2572	2.37	3054	18.15	2428	0.00	0
HOKE NC	0.00	0	21.57	223	8.20	2747	9.02	2851	5.35	2584	21.65	2084	0.00	0
HYDE NC	0.00	0	29.79	100	27.32	1478	10.52	2684	3.01	3004	33.59	1231	0.00	0
IREDELL NC	0.00	0	15.97	381	18.33	1845	10.98	2626	7.11	2256	29.73	1442	0.00	0
JACKSON NC	0.00	0	6.96	841	11.66	2399	32.59	1176	18.35	1112	38.00	1065	0.00	0
JOHNSTON NC	0.00	0	19.97	266	6.90	2860	11.09	2605	2.86	3020	21.23	2113	0.00	0
JONES NC	0.00	0	23.55	197	30.64	1390	8.35	2893	2.45	3051	20.77	2170	27.88	652
LEE NC	0.00	0	26.09	160	7.46	2809	8.23	2900	5.04	2650	22.89	1950	0.00	0
LENOIR NC	0.00	0	23.96	192	23.79	1579	9.04	2850	2.80	3026	21.09	2134	0.00	0
LINCOLN NC	0.00	0	14.26	447	7.45	2810	12.83	2441	13.71	1557	23.65	1882	0.00	0
MACON NC	0.00	0	7.73	789	18.22	1852	20.75	1792	21.54	887	42.98	973	0.00	0
MADISON NC	0.00	0	10.10	650	8.98	2660	24.47	1552	19.11	1044	29.06	1477	0.00	0
MARTIN NC	0.00	0	33.83	74	17.85	1883	11.96	2515	2.75	3031	15.53	2657	38.35	620
MCDOWELL NC	0.00	0	10.24	641	6.00	2945	16.75	2081	15.28	1411	37.21	1089	0.00	0
MECKLENBURG NC	0.00	0	12.42	511	5.83	2961	11.22	2591	7.43	2196	24.47	1802	0.00	0
MITCHELL NC	0.00	0	12.36	513	13.74	2182	16.03	2133	17.35	1223	36.30	1114	0.00	0
MONTGOMERY NC	0.00	0	16.94	355	14.10	2147	9.15	2836	5.26	2599	23.33	1899	0.00	0
MOORE NC	0.00	0	24.41	184	24.66	1554	9.31	2826	4.60	2758	25.24	1744	0.00	0
NASH NC	0.00	0	18.08	326	17.24	1926	10.14	2732	4.04	2869	19.23	2327	0.00	0
NEW HANOVER NC	0.00	0	25.13	176	17.17	1930	9.31	2825	5.02	2653	32.88	1272	0.00	0
NORTHAMPTON NC	0.00	0	29.08	110	16.41	1974	8.62	2879	4.58	2762	27.96	1533	57.04	561
ONSLOW NC	0.00	0	19.86	269	25.53	1515	10.93	2633	2.24	3059	31.10	1353	0.00	0

TOTAL FALLOUT ACTIVITY (µCi/sq meter) by SHOT SERIES WITH COUNTY RANK

COUNTY AND STATE	RANGER SERIES	RANGER RANK	B-J SERIES	B-J RANK	T-S SERIES	T-S RANK	U-K SERIES	UK RANK	TEAPOT SERIES	TP RANK	PLUMBBOB SERIES	PB RANK	SHOT SEDAN	SEDAN RANK
	1951		1951		1952		1953		1955		1957		1962	
ORANGE NC	0.00	0	19.12	287	7.50	2803	7.75	2919	4.66	2746	18.54	2391	43.75	605
PAMLICO NC	0.00	0	28.46	117	23.24	1605	8.31	2896	4.70	2740	26.88	1606	0.00	0
PASQUOTANK NC	0.00	0	28.35	119	18.04	1868	9.55	2793	3.10	2996	25.49	1721	191.00	226
PENDER NC	0.00	0	26.15	159	22.39	1637	15.02	2221	2.71	3035	18.05	2440	0.00	0
PERQUIMANS NC	0.00	0	29.64	101	18.06	1866	9.61	2786	3.04	3002	25.71	1705	191.00	225
PERSON NC	0.00	0	22.12	212	8.23	2744	8.40	2891	4.26	2828	18.48	2404	73.06	524
PITT NC	0.00	0	31.80	87	17.73	1894	11.10	2603	3.25	2974	20.77	2169	0.00	0
POLK NC	0.00	0	12.08	531	10.72	2500	13.97	2336	17.16	1240	31.64	1327	0.00	0
RANDOLPH NC	0.00	0	20.20	256	9.98	2575	22.41	1670	4.91	2683	23.48	1890	0.00	0
RICHMOND NC	0.00	0	17.43	340	8.30	2739	8.82	2865	5.94	2470	24.01	1841	0.00	0
ROBESON NC	0.00	0	21.41	227	14.54	2111	9.41	2814	5.80	2494	22.22	2031	0.00	0
ROCKINGHAM NC	0.00	0	17.81	331	7.22	2832	7.91	2915	5.95	2468	19.05	2345	0.00	0
ROWAN NC	0.00	0	20.63	246	10.23	2550	10.01	2746	7.67	2157	20.90	2159	0.00	0
RUTHERFORD NC	0.00	0	10.79	606	9.13	2648	13.73	2360	17.32	1226	25.49	1720	0.00	0
SAMPSON NC	0.00	0	19.36	281	14.87	2082	12.65	2454	3.34	2964	18.73	2370	0.00	0
SCOTLAND NC	0.00	0	20.42	252	8.96	2665	9.05	2849	5.37	2575	22.07	2043	0.00	0
STANLY NC	0.00	0	12.31	516	12.94	2267	9.32	2824	5.38	2573	26.10	1670	0.00	0
STOKES NC	0.00	0	20.32	255	6.48	2905	8.50	2884	4.85	2705	26.23	1651	0.00	0
SURRY NC	0.00	0	20.40	253	8.47	2721	9.29	2828	6.80	2302	26.23	1652	15.20	683
SWAIN NC	0.00	0	13.89	458	13.64	2194	25.86	1482	18.63	1084	34.65	1167	0.00	0
TRANSYLVANIA NC	0.00	0	8.94	719	14.01	2167	19.03	1906	18.82	1066	43.95	947	0.00	0
TYRRELL NC	0.00	0	21.37	229	10.51	2524	13.25	2399	2.86	3021	15.04	2693	0.00	0
UNION NC	0.00	0	15.64	391	9.30	2633	9.99	2749	6.61	2330	25.96	1683	0.00	0
VANCE NC	0.00	0	21.43	225	7.39	2814	7.51	2930	3.94	2885	16.44	2592	73.06	525
WAKE NC	0.00	0	16.01	380	12.45	2315	9.90	2758	3.89	2893	25.53	1718	21.81	656
WARREN NC	0.00	0	20.67	244	19.91	1750	8.48	2887	5.40	2568	22.69	1972	0.00	0

TOTAL FALLOUT ACTIVITY (µCi/sq meter) by SHOT SERIES WITH COUNTY RANK

COUNTY AND STATE	RANGER SERIES	RANGER RANK	B-J SERIES	B-J RANK	T-S SERIES	T-S RANK	U-K SERIES	UK RANK	TEAPOT SERIES	TP RANK	PLUMBBOB SERIES	PB RANK	SHOT SEDAN	SEDAN RANK
	1951		1951		1952		1953		1955		1957		1962	
WASHINGTON NC	0.00	0	31.35	89	19.74	1754	9.16	2835	3.52	2943	16.52	2587	94.45	398
WATAUGA NC	0.00	0	9.83	668	8.14	2752	16.25	2116	15.35	1400	28.77	1493	18.06	663
WAYNE NC	0.00	0	23.01	202	23.65	1585	10.05	2743	2.92	3012	19.04	2346	0.00	0
WILKES NC	0.00	0	10.50	622	8.44	2724	11.02	2617	6.56	2344	36.79	1102	43.75	589
WILSON NC	0.00	0	18.29	318	18.43	1838	8.82	2866	5.48	2553	24.95	1770	0.00	0
YADKIN NC	0.00	0	16.42	371	16.17	1991	20.21	1831	7.66	2159	27.72	1547	0.00	0
YANCEY NC	0.00	0	10.94	598	8.95	2668	20.40	1817	20.21	968	34.36	1186	0.00	0
ADAMS ND	0.00	0	0.83	2194	84.05	468	75.22	335	4.03	2870	125.89	161	0.00	0
BARNES ND	0.19	554	1.11	2035	7.97	2762	11.91	2522	4.92	2682	152.69	74	0.00	0
BENSON ND	0.00	0	0.64	2348	34.61	1269	13.01	2422	9.43	1972	130.33	134	0.00	0
BILLINGS ND	0.00	0	1.31	1938	41.37	1091	137.56	152	4.35	2808	219.51	42	0.00	0
BOTTINEAU ND	0.19	555	0.75	2248	35.19	1250	175.32	122	5.12	2632	80.08	390	0.00	0
BOWMAN ND	0.00	0	0.68	2310	47.06	981	72.96	364	4.94	2678	114.73	192	0.00	0
BURKE ND	0.00	0	0.85	2182	74.21	569	163.62	129	4.78	2721	97.56	259	0.00	0
BURLEIGH ND	0.00	0	0.51	2512	33.93	1289	87.97	261	4.72	2736	121.46	172	0.00	0
CASS ND	0.00	0	0.43	2604	5.00	3007	12.82	2443	5.70	2510	173.97	56	0.00	0
CAVALIER ND	0.00	0	1.32	1930	48.38	956	12.05	2506	11.51	1765	100.85	242	0.00	0
DICKEY ND	0.19	556	0.77	2235	40.24	1121	17.84	1995	6.03	2446	167.27	61	0.00	0
DIVIDE ND	0.00	0	0.85	2183	221.31	245	130.95	162	4.13	2854	129.92	138	0.00	0
DUNN ND	0.00	0	0.93	2134	63.66	718	147.71	138	4.94	2676	142.46	94	0.00	0
EDDY ND	0.00	0	0.72	2273	36.68	1207	11.27	2583	5.92	2476	134.98	116	0.00	0
EMMONS ND	0.00	0	0.53	2464	37.92	1175	21.76	1715	4.09	2859	182.99	50	0.00	0
FOSTER ND	0.00	0	0.82	2202	39.55	1133	10.95	2631	7.59	2173	162.32	63	0.00	0
GOLDEN VALLE ND	0.00	0	1.92	1657	62.34	733	135.86	155	4.65	2750	135.00	115	0.00	0
GRAND FORKS ND	0.00	0	1.22	1981	10.40	2538	12.85	2437	11.24	1798	135.72	111	0.00	0
GRANT ND	0.00	0	0.73	2262	43.17	1052	99.70	202	4.23	2837	103.65	226	0.00	0

TOTAL FALLOUT ACTIVITY (µCi/sq meter) by SHOT SERIES WITH COUNTY RANK

COUNTY AND STATE	RANGER SERIES	RANGER RANK	B-J SERIES	B-J RANK	T-S SERIES	T-S RANK	U-K SERIES	UK RANK	TEAPOT SERIES	TP RANK	PLUMBBOB SERIES	PB RANK	SHOT SEDAN	SEDAN RANK
	1951		1951		1952		1953		1955		1957		1962	
GRIGGS ND	0.00	0	0.72	2274	7.53	2801	12.92	2430	8.37	2073	140.69	98	0.00	0
HETTINGER ND	0.00	0	0.99	2101	41.15	1096	102.00	196	4.14	2852	117.56	183	0.00	0
KIDDER ND	0.00	0	0.64	2349	34.97	1257	13.10	2408	5.26	2601	120.16	177	0.00	0
LA MOURE ND	0.00	0	0.70	2281	36.75	1205	13.73	2361	6.61	2329	150.52	80	0.00	0
LOGAN ND	0.00	0	0.45	2577	34.99	1256	16.93	2062	5.42	2565	143.91	91	0.00	0
MCHENRY ND	0.00	0	0.83	2195	62.49	730	155.33	136	7.62	2168	97.00	265	0.00	0
MCINTOSH ND	0.00	0	0.45	2578	43.96	1038	19.73	1861	8.29	2086	135.48	112	0.00	0
MCKENZIE ND	0.00	0	1.31	1939	68.81	639	165.34	127	4.54	2769	108.92	208	0.00	0
MCLEAN ND	0.00	0	1.01	2076	315.01	205	99.06	206	5.63	2525	128.94	147	0.00	0
MERCER ND	0.00	0	0.84	2189	38.04	1172	131.56	159	4.72	2735	129.69	140	0.00	0
MORTON ND	0.00	0	0.82	2203	37.96	1173	98.69	208	5.17	2624	120.31	176	0.00	0
MOUNTRAIL ND	0.00	0	1.03	2065	114.50	332	130.97	161	4.48	2777	109.27	206	0.00	0
NELSON ND	0.00	0	1.22	1982	10.45	2532	12.36	2478	11.92	1724	131.61	126	0.00	0
OLIVER ND	0.19	557	0.64	2347	156.48	281	131.31	160	4.82	2709	129.09	145	0.00	0
PEMBINA ND	0.00	0	0.99	2100	13.13	2246	14.14	2320	16.70	1283	115.51	189	0.00	0
PIERCE ND	0.00	0	0.75	2249	33.77	1293	11.41	2568	8.21	2096	87.15	336	0.00	0
RAMSEY ND	0.00	0	0.88	2161	50.32	921	12.30	2483	11.84	1731	115.84	187	0.00	0
RANSOM ND	0.00	0	0.52	2473	10.58	2513	13.36	2391	6.15	2430	188.18	47	0.00	0
RENVILLE ND	0.00	0	0.85	2184	32.62	1326	132.25	158	5.09	2637	103.38	228	0.00	0
RICHLAND ND	0.00	0	0.43	2605	8.62	2701	14.76	2252	3.61	2935	231.47	37	0.00	0
ROLETTE ND	0.00	0	0.75	2250	34.68	1268	14.08	2327	4.54	2771	90.50	314	0.00	0
SARGENT ND	0.00	0	0.60	2384	14.05	2154	16.17	2121	4.44	2784	174.94	53	0.00	0
SHERIDAN ND	0.00	0	0.64	2350	36.14	1223	84.07	271	5.61	2529	109.72	205	0.00	0
SIOUX ND	0.19	558	0.73	2261	146.69	293	68.16	415	4.16	2849	84.09	353	0.00	0
SLOPE ND	0.00	0	1.01	2075	99.04	364	117.50	169	4.27	2825	121.69	170	0.00	0
STARK ND	0.00	0	1.12	2029	94.24	386	137.40	153	4.68	2741	168.18	60	0.00	0

TOTAL FALLOUT ACTIVITY (µCi/sq meter) by SHOT SERIES WITH COUNTY RANK

COUNTY AND STATE	RANGER SERIES	RANGER RANK	B-J SERIES	B-J RANK	T-S SERIES	T-S RANK	U-K SERIES	UK RANK	TEAPOT SERIES	TP RANK	PLUMBBOB SERIES	PB RANK	SHOT SEDAN	SEDAN RANK
	1951		1951		1952		1953		1955		1957		1962	
STEELE ND	0.00	0	1.22	1983	9.90	2581	13.27	2397	6.22	2419	129.68	141	0.00	0
STUTSMAN ND	0.00	0	0.96	2116	35.09	1252	14.54	2278	6.09	2439	141.75	95	0.00	0
TOWNER ND	0.00	0	0.75	2251	33.19	1310	12.07	2502	6.33	2395	96.71	267	0.00	0
TRAILL ND	0.00	0	0.87	2166	6.31	2916	13.81	2347	8.09	2113	149.29	83	0.00	0
WALSH ND	0.00	0	2.13	1578	12.64	2293	11.03	2614	12.59	1658	141.01	97	0.00	0
WARD ND	0.00	0	0.75	2252	34.38	1277	164.84	128	4.54	2770	105.47	222	0.00	0
WELLS ND	0.00	0	0.82	2204	37.60	1184	13.05	2415	5.26	2604	102.94	231	0.00	0
WILLIAMS ND	0.00	0	1.03	2066	139.53	305	62.23	475	3.59	2938	100.31	249	0.00	0
ADAMS NE	0.00	0	2.20	1558	78.55	522	38.63	951	28.15	597	65.90	560	76.52	502
ANTELOPE NE	0.00	0	1.57	1814	54.20	868	46.68	709	5.27	2594	92.98	292	220.22	143
ARTHUR NE	0.00	0	0.11	2970	78.96	514	33.53	1132	38.11	430	74.77	433	1.20	729
BANNER NE	0.00	0	0.19	2889	90.98	407	20.39	1819	31.68	519	66.27	553	2.00	715
BLAINE NE	0.00	0	2.62	1461	58.85	786	32.36	1189	7.57	2175	116.25	186	0.27	746
BOONE NE	0.00	0	1.62	1784	38.92	1148	41.71	839	21.58	883	90.55	313	0.00	0
BOX BUTTE NE	0.21	540	0.19	2888	56.70	829	22.22	1685	37.61	438	59.99	644	2.00	716
BOYD NE	0.00	0	3.58	1290	88.24	433	41.42	845	5.28	2592	78.78	399	0.27	749
BROWN NE	0.00	0	2.23	1547	77.81	526	36.11	1031	6.34	2392	113.46	198	0.67	732
BUFFALO NE	0.19	559	1.51	1830	89.53	421	42.56	809	27.25	617	83.87	358	76.52	495
BURT NE	0.00	0	1.87	1675	55.68	847	43.17	795	24.70	713	42.98	972	0.00	0
BUTLER NE	0.00	0	1.94	1635	42.20	1072	40.37	882	21.86	871	72.09	459	139.13	312
CASS NE	0.00	0	1.88	1671	31.66	1352	43.02	800	22.63	816	38.21	1060	146.81	293
CEDAR NE	0.00	0	1.83	1690	68.01	651	42.40	814	4.38	2798	86.19	342	170.40	257
CHASE NE	0.19	560	0.86	2177	88.82	431	29.01	1336	117.67	134	68.44	517	0.00	0
CHERRY NE	0.45	479	0.19	2887	58.90	785	78.04	309	12.84	1633	174.46	55	1.20	726
CHEYENNE NE	0.00	0	0.19	2890	85.64	459	31.81	1211	37.03	445	91.63	309	0.67	737
CLAY NE	0.00	0	2.08	1589	42.15	1073	39.52	914	25.56	670	63.02	599	99.74	379

TOTAL FALLOUT ACTIVITY (µCi/sq meter) by SHOT SERIES WITH COUNTY RANK

COUNTY AND STATE	RANGER SERIES	RANGER RANK	B-J SERIES	B-J RANK	T-S SERIES	T-S RANK	U-K SERIES	UK RANK	TEAPOT SERIES	TP RANK	PLUMBBOB SERIES	PB RANK	SHOT SEDAN	SEDAN RANK
	1951		1951		1952		1953		1955		1957		1962	
COLFAX NE	0.00	0	1.61	1795	40.17	1122	48.35	680	21.06	916	52.34	771	172.89	253
CUMING NE	0.00	0	1.61	1796	82.21	484	41.28	847	20.74	934	51.52	783	208.70	193
CUSTER NE	0.45	480	1.76	1731	68.80	640	38.35	965	25.11	689	81.46	381	126.34	337
DAKOTA NE	0.00	0	1.64	1776	77.69	527	41.92	831	5.16	2627	25.12	1762	527.44	46
DAWES NE	0.00	0	0.19	2891	129.99	316	20.75	1793	32.14	513	46.18	903	2.00	717
DAWSON NE	0.00	0	1.57	1813	63.67	715	36.35	1021	112.25	151	82.82	370	0.00	0
DEUEL NE	0.19	561	0.11	2962	89.25	424	34.62	1097	44.25	365	66.65	541	0.27	750
DIXON NE	0.00	0	2.28	1537	81.26	491	43.17	794	4.01	2871	55.26	708	557.97	14
DODGE NE	0.19	562	1.62	1783	40.75	1102	49.86	648	24.26	742	51.73	781	0.00	0
DOUGLAS NE	0.19	563	1.75	1738	44.90	1024	43.95	776	22.44	829	34.34	1187	146.81	294
DUNDY NE	0.00	0	1.08	2053	67.95	652	30.11	1279	119.10	129	60.37	639	0.00	0
FILLMORE NE	0.00	0	2.08	1590	44.08	1037	38.63	952	24.04	750	49.00	836	76.52	505
FRANKLIN NE	0.00	0	1.92	1659	75.91	547	36.89	1007	32.01	516	66.43	550	0.00	0
FRONTIER NE	0.00	0	2.07	1595	72.50	595	32.05	1201	116.35	139	74.09	443	0.00	0
FURNAS NE	0.00	0	1.51	1832	100.40	361	39.71	905	114.09	146	53.95	735	0.00	0
GAGE NE	0.00	0	2.37	1515	34.82	1263	38.99	932	24.60	722	64.36	575	80.75	479
GARDEN NE	0.00	0	0.11	2971	73.53	579	31.18	1236	35.08	469	76.29	417	1.47	724
GARFIELD NE	0.00	0	2.53	1479	67.70	654	41.23	849	6.02	2448	107.22	218	220.22	140
GOSPER NE	0.00	0	1.62	1787	89.55	420	35.21	1066	116.38	138	58.72	659	0.00	0
GRANT NE	0.00	0	0.11	2972	54.70	860	26.89	1435	62.10	248	75.93	420	1.20	728
GREELEY NE	0.00	0	2.09	1582	80.22	506	37.98	977	22.26	846	66.75	540	172.89	250
HALL NE	0.00	0	2.21	1557	84.57	465	44.58	757	24.97	697	83.61	360	0.00	0
HAMILTON NE	0.00	0	2.20	1559	36.62	1209	37.12	1003	24.28	739	58.08	667	80.75	480
HARLAN NE	0.00	0	1.51	1833	79.50	510	39.42	921	31.56	522	77.40	412	0.00	0
HAYES NE	0.00	0	1.56	1819	87.33	439	29.11	1330	117.51	135	53.95	736	0.00	0
HITCHCOCK NE	0.00	0	1.45	1865	64.67	698	32.56	1177	119.21	125	63.75	588	0.00	0

TOTAL FALLOUT ACTIVITY (µCi/sq meter) by SHOT SERIES WITH COUNTY RANK

COUNTY AND STATE	RANGER SERIES	RANGER RANK	B-J SERIES	B-J RANK	T-S SERIES	T-S RANK	U-K SERIES	UK RANK	TEAPOT SERIES	TP RANK	PLUMBBOB SERIES	PB RANK	SHOT SEDAN	SEDAN RANK
	1951		1951	1952			1953		1955		1957		1962	
HOLT NE	0.00	0	2.53	1480	92.62	395	41.82	835	6.13	2434	71.68	463	398.55	55
HOOKER NE	0.00	0	0.11	2973	57.72	804	31.14	1239	30.99	535	105.40	223	0.67	734
HOWARD NE	0.00	0	2.04	1599	91.69	402	31.99	1202	24.28	740	89.99	318	345.56	89
JEFFERSON NE	0.45	481	2.37	1514	38.20	1168	41.27	848	24.47	726	43.79	949	37.44	624
JOHNSON NE	0.00	0	2.60	1465	40.33	1118	36.23	1028	23.91	755	56.44	685	80.75	481
KEARNEY NE	0.00	0	1.69	1757	77.56	529	40.73	864	28.78	587	81.51	380	0.00	0
KEITH NE	0.00	0	0.11	2974	75.43	552	27.50	1411	119.21	126	77.34	413	0.00	0
KEYA PAHA NE	0.19	564	1.51	1835	89.92	418	32.34	1191	4.16	2848	74.32	438	0.67	739
KIMBALL NE	0.00	0	0.88	2165	49.52	932	34.20	1109	29.52	570	72.68	454	0.67	740
KNOX NE	0.00	0	2.72	1439	48.84	944	37.68	990	5.20	2621	83.36	363	0.27	748
LANCASTER NE	0.00	0	2.44	1499	35.86	1234	42.74	807	23.04	800	42.99	971	76.52	508
LINCOLN NE	0.00	0	1.77	1730	76.17	546	33.14	1150	118.40	132	102.83	232	0.00	0
LOGAN NE	0.00	0	1.51	1836	79.52	509	30.84	1247	109.16	159	93.21	289	0.67	735
LOUP NE	0.00	0	1.51	1837	71.85	602	35.19	1068	6.63	2327	69.12	502	436.77	53
MADISON NE	0.00	0	0.95	2120	45.88	1004	47.18	702	21.31	905	129.54	143	208.70	185
MCPHERSON NE	0.00	0	0.11	2975	40.33	1120	25.25	1508	109.83	156	59.87	645	0.67	742
MERRICK NE	0.00	0	2.76	1434	49.39	937	41.48	843	22.51	826	85.85	343	139.13	306
MORRILL NE	0.00	0	0.19	2892	97.91	367	19.64	1866	55.63	278	77.78	408	2.00	712
NANCE NE	0.00	0	1.62	1785	61.40	749	42.77	806	22.13	853	77.76	409	139.13	308
NEMAHA NE	0.00	0	2.25	1542	30.95	1376	36.28	1026	24.00	751	46.13	907	80.75	482
NUCKOLLS NE	0.19	565	2.08	1588	38.17	1169	40.25	888	26.09	652	82.33	375	37.44	621
OTOE NE	0.00	0	2.48	1492	37.31	1193	36.28	1025	26.57	637	36.80	1101	76.52	510
PAWNEE NE	0.19	566	2.80	1430	37.20	1197	33.46	1134	24.78	704	58.01	670	0.00	0
PERKINS NE	0.00	0	0.11	2976	86.94	445	29.50	1307	119.12	128	56.38	686	0.00	0
PHELPS NE	0.19	567	1.51	1831	103.59	348	38.54	957	30.17	553	105.17	224	38.93	618
PIERCE NE	0.19	568	2.55	1477	37.18	1198	47.80	693	5.43	2562	84.95	347	0.00	0

TOTAL FALLOUT ACTIVITY (µCi/sq meter) by SHOT SERIES WITH COUNTY RANK

COUNTY AND STATE	RANGER SERIES	RANGER RANK	B-J SERIES	B-J RANK	T-S SERIES	T-S RANK	U-K SERIES	UK RANK	TEAPOT SERIES	TP RANK	PLUMBBOB SERIES	PB RANK	SHOT SEDAN	SEDAN RANK
	1951		1951		1952		1953		1955		1957		1962	
PLATTE NE	0.00	0	1.61	1797	48.37	957	45.50	728	21.23	907	79.34	396	139.13	307
POLK NE	0.19	569	1.94	1634	49.93	929	39.54	913	22.26	845	69.24	500	139.13	313
RED WILLOW NE	0.00	0	1.79	1721	84.97	462	33.40	1140	119.08	130	58.06	668	0.00	0
RICHARDSON NE	0.21	541	2.31	1527	42.36	1069	41.00	859	26.44	644	47.88	862	0.00	0
ROCK NE	0.00	0	3.67	1273	61.32	751	38.00	975	5.70	2513	64.06	581	0.67	741
SALINE NE	0.00	0	2.55	1475	42.92	1061	39.06	928	24.34	736	81.73	378	80.75	478
SARPY NE	0.00	0	2.00	1621	50.21	923	40.63	871	23.99	752	33.00	1266	0.00	0
SAUNDERS NE	0.00	0	1.94	1636	37.29	1194	39.02	931	21.79	873	50.60	802	146.81	288
SCOTTS BLUFF NE	0.00	0	0.19	2893	62.10	737	15.52	2179	18.81	1070	47.61	867	2.66	707
SEWARD NE	0.00	0	2.17	1569	39.06	1146	42.06	827	23.06	798	69.10	503	76.52	499
SHERIDAN NE	0.00	0	0.19	2894	92.15	398	28.38	1369	32.38	509	69.67	495	2.00	714
SHERMAN NE	0.00	0	1.94	1639	66.93	669	35.15	1071	25.20	687	75.38	426	345.56	93
SIOUX NE	0.00	0	0.19	2895	208.21	248	20.68	1798	37.82	435	64.01	583	2.66	706
STANTON NE	0.00	0	1.82	1701	46.70	993	48.96	663	20.18	970	53.85	738	527.44	28
THAYER NE	0.00	0	2.19	1560	34.59	1271	38.36	964	25.54	674	63.76	587	37.44	622
THOMAS NE	0.00	0	2.62	1462	51.87	893	30.87	1246	109.30	158	92.49	300	0.67	736
THURSTON NE	0.00	0	1.85	1682	83.37	472	44.34	765	4.97	2669	31.69	1320	527.44	36
VALLEY NE	0.00	0	2.04	1600	74.94	557	36.30	1024	23.39	781	76.91	414	139.13	309
WASHINGTON NE	0.19	570	1.75	1739	38.74	1153	33.30	1146	21.80	872	34.21	1198	345.56	103
WAYNE NE	0.00	0	1.78	1728	62.23	736	45.01	748	4.93	2679	55.14	711	208.70	191
WEBSTER NE	0.00	0	2.08	1591	76.28	544	37.01	1005	30.12	556	60.71	633	37.44	623
WHEELER NE	0.00	0	1.59	1803	88.46	432	40.63	873	6.40	2385	68.71	510	0.00	0
YORK NE	0.00	0	1.97	1626	38.74	1154	38.86	943	23.52	772	64.28	576	76.52	504
BELKNAP NH	15.55	119	28.00	123	16.57	1961	29.39	1316	4.42	2791	16.81	2555	0.00	0
CARROLL NH	36.35	14	39.14	58	17.86	1881	22.83	1649	4.06	2868	12.62	2837	0.00	0
CHESHIRE NH	18.06	88	23.11	201	18.88	1814	40.63	870	5.35	2585	20.11	2239	0.00	0

TOTAL FALLOUT ACTIVITY (µCi/sq meter) by SHOT SERIES WITH COUNTY RANK

COUNTY AND STATE	RANGER SERIES	RANGER RANK	B-J SERIES	B-J RANK	T-S SERIES	T-S RANK	U-K SERIES	UK RANK	TEAPOT SERIES	TP RANK	PLUMBBOB SERIES	PB RANK	SHOT SEDAN	SEDAN RANK
	1951		1951		1952		1953		1955		1957		1962	
COOS NH	1.73	346	23.66	195	27.22	1480	24.55	1547	6.17	2425	20.20	2229	0.00	0
GRAFTON NH	9.28	180	27.62	130	27.06	1485	24.52	1550	5.43	2563	19.09	2343	0.00	0
HILLSBOROUGH NH	15.06	141	22.72	205	16.40	1975	43.40	788	5.82	2490	17.04	2540	0.00	0
MERRIMACK NH	15.55	120	21.63	221	22.67	1623	34.86	1084	4.28	2822	10.74	2907	0.00	0
ROCKINGHAM NH	36.05	17	14.64	436	15.76	2010	62.56	472	6.41	2383	12.51	2842	0.00	0
STRAFFORD NH	36.05	18	23.96	191	24.42	1564	28.58	1356	6.15	2429	9.86	2931	0.00	0
SULLIVAN NH	15.98	113	22.24	210	22.16	1653	32.45	1183	5.60	2530	15.31	2675	0.00	0
ATLANTIC NJ	0.00	0	15.17	417	18.83	1819	71.83	373	9.25	1992	13.54	2792	45.80	587
BERGEN NJ	3.50	274	15.89	383	36.16	1222	46.67	710	23.14	796	12.97	2826	0.00	0
BURLINGTON NJ	3.50	275	17.88	329	18.87	1816	66.88	430	10.72	1844	14.25	2749	79.17	489
CAMDEN NJ	0.00	0	18.86	298	15.59	2024	55.27	564	7.87	2135	11.46	2886	0.00	0
CAPE MAY NJ	0.00	0	25.10	178	17.66	1899	57.22	538	8.81	2025	21.25	2111	0.00	0
CUMBERLAND NJ	0.00	0	18.32	316	18.01	1871	48.06	690	8.62	2044	13.84	2774	0.00	0
ESSEX NJ	3.50	276	26.07	162	35.25	1249	49.39	653	18.57	1092	13.54	2793	0.00	0
GLOUCESTER NJ	0.00	0	17.12	349	19.15	1796	47.55	696	9.59	1963	15.66	2646	0.00	0
HUDSON NJ	9.14	189	14.68	434	40.46	1115	52.06	619	22.95	806	11.16	2894	0.00	0
HUNTERDON NJ	3.50	277	16.83	358	17.76	1892	37.45	995	10.45	1875	13.72	2784	0.00	0
MERCER NJ	3.50	278	17.05	351	26.58	1495	55.82	557	12.16	1701	13.97	2769	0.00	0
MIDDLESEX NJ	3.50	279	17.48	338	31.93	1344	54.43	578	12.11	1707	15.27	2677	0.00	0
MONMOUTH NJ	3.50	280	16.53	364	30.96	1375	69.23	402	14.74	1458	15.99	2623	20.31	661
MORRIS NJ	3.50	281	29.84	98	42.86	1062	31.94	1204	12.17	1699	11.74	2873	0.00	0
OCEAN NJ	3.50	282	16.77	360	33.94	1288	69.25	401	14.48	1489	16.68	2569	37.08	636
PASSAIC NJ	3.50	283	26.92	142	41.92	1079	34.56	1099	12.28	1685	13.54	2795	0.00	0
SALEM NJ	1.23	370	18.94	295	13.93	2170	40.93	860	9.61	1962	15.35	2672	0.00	0
SOMERSET NJ	3.50	284	15.36	410	28.93	1438	47.81	692	9.97	1921	13.09	2812	0.00	0
SUSSEX NJ	3.50	285	25.51	169	37.12	1199	23.09	1634	10.11	1902	14.71	2719	0.00	0

TOTAL FALLOUT ACTIVITY (µCi/sq meter) by SHOT SERIES WITH COUNTY RANK

COUNTY AND STATE	RANGER SERIES	RANGER RANK	B-J SERIES	B-J RANK	T-S SERIES	T-S RANK	U-K SERIES	UK RANK	TEAPOT SERIES	TP RANK	PLUMBBOB SERIES	PB RANK	SHOT SEDAN	SEDAN RANK
	1951		1951		1952		1953		1955		1957		1962	
UNION NJ	3.50	286	19.08	289	35.90	1232	50.33	643	18.05	1147	13.02	2823	0.00	0
WARREN NJ	3.50	287	27.52	132	38.71	1155	28.34	1371	11.90	1727	10.41	2918	0.00	0
BERNALILLO NM	0.00	0	0.00	0	74.09	572	619.73	37	33.87	489	115.57	188	0.00	0
CATRON NM	0.00	0	0.00	0	76.66	539	177.97	121	32.24	510	58.90	655	0.00	0
CHAVES NM	0.00	0	0.00	0	53.01	882	448.80	62	45.10	355	48.24	850	0.00	0
COLFAX NM	0.00	0	0.00	0	43.00	1059	185.02	118	217.97	56	102.44	235	0.00	0
CURRY NM	0.00	0	0.00	0	36.61	1211	318.69	84	72.06	224	65.09	568	0.00	0
DE BACA NM	0.00	0	0.00	0	64.64	699	337.29	80	51.73	303	67.12	535	0.00	0
DONA ANA NM	0.00	0	0.00	0	69.04	633	9.35	2819	5.67	2517	44.10	943	0.00	0
EDDY NM	0.00	0	0.00	0	53.93	870	114.69	170	22.17	849	46.81	881	0.00	0
GRANT NM	0.00	0	0.00	0	73.31	583	16.60	2094	7.76	2146	52.49	766	0.00	0
GUADALUPE NM	0.00	0	0.00	0	64.81	694	405.85	70	58.94	260	94.33	281	0.00	0
HARDING NM	0.00	0	0.00	0	43.77	1041	216.08	104	161.16	77	103.45	227	0.00	0
HIDALGO NM	0.00	0	0.00	0	69.04	634	6.22	2980	4.76	2725	52.49	768	0.00	0
LEA NM	0.00	0	0.00	0	36.19	1221	114.58	171	29.10	576	46.81	880	0.00	0
LINCOLN NM	0.00	0	0.00	0	64.64	700	426.28	69	43.30	378	56.13	693	0.00	0
LOS ALAMOS NM	0.00	0	0.00	0	85.07	461	473.85	57	37.57	439	85.05	345	0.00	0
LUNA NM	0.00	0	0.00	0	69.04	635	10.35	2701	7.17	2247	52.49	767	0.00	0
MCKINLEY NM	0.00	0	0.00	0	101.69	355	961.51	16	51.17	306	77.44	411	0.00	0
MORA NM	0.00	0	0.00	0	62.39	732	238.27	99	151.36	84	102.94	230	0.00	0
OTERO NM	0.00	0	0.00	0	60.54	763	23.29	1620	18.01	1156	46.14	905	0.00	0
QUAY NM	0.00	0	0.00	0	43.77	1042	295.16	88	112.70	150	93.41	287	0.00	0
RIO ARRIBA NM	0.00	0	0.00	0	76.34	542	429.05	68	37.57	440	37.94	1066	0.00	0
ROOSEVELT NM	0.00	0	0.00	0	39.16	1141	310.36	86	58.43	263	64.81	570	0.00	0
SAN JUAN NM	0.00	0	0.00	0	90.28	413	514.75	51	64.13	241	52.29	774	0.00	0
SAN MIGUEL NM	0.00	0	0.00	0	61.51	748	322.55	83	92.40	172	113.31	200	0.00	0

TOTAL FALLOUT ACTIVITY (µCi/sq meter) by SHOT SERIES WITH COUNTY RANK

COUNTY AND STATE	RANGER SERIES	RANGER RANK	B-J SERIES	B-J RANK	T-S SERIES	T-S RANK	U-K SERIES	UK RANK	TEAPOT SERIES	TP RANK	PLUMBBOB SERIES	PB RANK	SHOT SEDAN	SEDAN RANK
	1951		1951		1952		1953		1955		1957		1962	
SANDOVAL NM	0.00	0	0.00	0	86.63	450	803.90	19	42.78	384	99.16	251	0.00	0
SANTA FE NM	0.00	0	0.00	0	74.10	571	621.53	36	38.60	425	116.91	185	0.00	0
SIERRA NM	0.00	0	0.00	0	86.75	448	54.88	570	20.02	979	53.96	734	0.00	0
SOCORRO NM	0.00	0	0.00	0	74.69	560	247.27	96	41.51	397	58.84	658	0.00	0
TAOS NM	0.00	0	0.00	0	72.97	587	324.59	82	38.95	423	84.07	355	0.00	0
TORRANCE NM	0.00	0	0.00	0	64.81	695	627.36	35	34.19	481	113.62	195	0.00	0
UNION NM	0.00	0	0.00	0	33.21	1307	171.03	125	220.49	53	102.16	236	0.00	0
VALENCIA NM	0.00	0	0.00	0	90.18	414	680.50	28	49.36	319	101.19	241	0.00	0
CARSON CITY NV	0.00	0	0.00	0	1278.58	109	28.96	1339	10.09	1904	370.71	23	0.00	0
CHURCHILL NV	0.19	571	2.82	1422	1276.46	111	29.86	1289	11.57	1761	511.33	14	0.00	0
CLARK1 NV	0.00	0	0.00	0	0.00	0	2744.79	4	545.07	7	123.66	166	0.00	0
CLARK2 NV	0.00	0	0.00	0	0.18	3088	274.69	93	169.83	71	1.39	3093	0.00	0
CLARK3 NV	0.00	0	0.00	0	0.00	0	57.72	526	104.23	163	0.00	0	0.00	0
DOUGLAS NV	0.00	0	0.00	0	649.80	154	17.91	1985	10.09	1907	382.36	21	0.00	0
ELKO NV	0.00	0	113.79	3	16442.04	30	17.90	1986	24.92	700	149.38	82	0.00	0
ESMERALDA1 NV	0.00	0	0.00	0	0.00	0	0.00	0	0.00	0	366.58	24	0.00	0
ESMERALDA2 NV	0.00	0	0.00	0	0.00	0	0.00	0	41.77	393	624.12	6	0.00	0
EUREKA NV	0.00	0	111.48	4	6178.47	65	21.71	1718	67.84	234	606.19	8	3.54	699
HUMBOLDT NV	0.00	0	27.53	131	12139.76	44	20.53	1807	14.94	1441	84.78	348	0.00	0
LANDER1 NV	0.00	0	56.41	30	18528.88	21	15.01	2225	11.16	1804	121.50	171	0.00	0
LANDER2 NV	0.00	0	7.56	805	15638.24	32	26.67	1441	16.79	1274	424.13	17	0.00	0
LINCOLN1 NV	0.00	0	0.42	2684	302.37	211	1966.00	8	615.88	4	491.39	15	0.50	743
LINCOLN2 NV	0.00	0	89.03	9	617.36	156	501.33	52	641.20	2	257.55	32	1.61	718
LYON NV	0.00	0	0.00	0	653.81	153	23.33	1618	10.09	1906	391.65	19	0.00	0
MINERAL NV	0.00	0	0.00	0	336.78	200	9.13	2839	6.21	2422	615.87	7	0.00	0
NYE1 NV	0.00	0	0.00	0	1665.53	99	0.00	0	0.00	0	578.03	11	0.00	0

TOTAL FALLOUT ACTIVITY (µCi/sq meter) by SHOT SERIES WITH COUNTY RANK

COUNTY AND STATE	RANGER SERIES	RANGER RANK	B-J SERIES	B-J RANK	T-S SERIES	T-S RANK	U-K SERIES	UK RANK	TEAPOT SERIES	TP RANK	PLUMBBOB SERIES	PB RANK	SHOT SEDAN	SEDAN RANK
	1951		1951		1952		1953		1955		1957		1962	
NYE2 NV	0.00	0	232.73	1	16555.44	28	269.69	94	638.70	3	1714.91	1	120.28	353
NYE3 NV	0.00	0	0.00	0	0.00	0	2.44	3083	22.37	837	39.56	1036	0.00	0
PERSHING NV	0.00	0	18.62	306	1867.20	94	32.67	1174	11.53	1763	254.61	33	0.00	0
STOREY NV	0.00	0	0.00	0	1278.81	108	28.17	1379	10.09	1905	363.28	25	0.00	0
WASHOE NV	0.00	0	0.90	2148	1195.74	116	23.65	1598	9.33	1983	379.75	22	0.00	0
WHITE PINE1 NV	0.00	0	86.99	10	1306.73	104	111.02	180	370.98	21	480.52	16	77.06	493
WHITE PINE2 NV	0.00	0	129.04	2	1965.49	89	155.62	134	678.09	1	806.44	4	143.84	297
WHITE PINE3 NV	0.00	0	65.94	21	73.65	578	552.55	47	392.81	17	589.28	9	117.11	355
ALBANY NY	7.57	233	17.86	330	23.70	1583	196.75	113	3.99	2877	17.84	2463	0.00	0
ALLEGANY NY	8.98	202	62.80	22	21.97	1664	6.53	2968	20.00	982	27.59	1554	0.00	0
BRONX NY	3.50	288	23.33	199	34.14	1283	49.24	655	11.51	1764	10.98	2900	0.00	0
BROOME NY	8.56	219	28.36	118	8.36	2732	7.94	2913	3.57	2939	16.28	2601	0.00	0
CATTARAUGUS NY	9.17	185	50.55	37	22.11	1657	10.19	2723	24.69	715	24.24	1822	0.00	0
CAYUGA NY	8.40	223	43.63	50	14.38	2124	4.69	3027	7.71	2153	18.13	2429	0.00	0
CHAUTAUQUA NY	9.41	173	59.56	27	19.55	1769	7.26	2942	52.83	294	19.45	2315	0.00	0
CHEMUNG NY	8.99	192	49.98	38	8.13	2753	6.00	2987	5.03	2651	20.91	2155	0.00	0
CHENANGO NY	8.99	193	39.81	56	9.30	2634	10.42	2695	4.36	2805	23.24	1914	0.00	0
CLINTON NY	0.99	411	18.99	294	55.83	843	48.95	664	3.16	2989	20.03	2250	0.00	0
COLUMBIA NY	12.02	152	24.04	188	39.08	1145	191.86	115	4.25	2832	24.57	1793	0.00	0
CORTLAND NY	8.99	194	39.05	59	11.52	2417	5.60	3002	6.80	2299	29.89	1430	0.00	0
DELAWARE NY	20.28	70	47.30	43	25.30	1524	19.42	1881	5.10	2634	22.53	1993	0.00	0
DUTCHESS NY	11.65	157	20.68	243	35.99	1229	33.46	1135	4.77	2724	20.24	2227	0.00	0
ERIE NY	1.17	379	96.29	5	12.56	2302	9.73	2777	35.17	467	27.43	1566	0.00	0
ESSEX NY	1.42	356	32.16	85	39.88	1129	88.50	257	3.89	2896	28.53	1507	0.00	0
FRANKLIN NY	1.42	357	37.05	63	41.63	1085	32.73	1170	4.42	2789	18.75	2368	0.00	0
FULTON NY	8.56	220	32.60	83	26.14	1507	303.87	87	5.79	2495	24.69	1790	0.00	0

TOTAL FALLOUT ACTIVITY (µCi/sq meter) by SHOT SERIES WITH COUNTY RANK

COUNTY AND STATE	RANGER SERIES	RANGER RANK	B-J SERIES	B-J RANK	T-S SERIES	T-S RANK	U-K SERIES	UK RANK	TEAPOT SERIES	TP RANK	PLUMBBOB SERIES	PB RANK	SHOT SEDAN	SEDAN RANK
	1951		1951		1952		1953		1955		1957		1962	
GENESEE NY	1.17	380	91.63	7	14.73	2095	8.03	2908	26.98	626	31.05	1358	0.00	0
GREENE NY	28.94	25	29.41	105	28.60	1446	79.38	301	5.24	2609	21.62	2088	0.00	0
HAMILTON NY	8.98	203	43.40	51	28.12	1459	69.06	405	5.30	2587	22.84	1957	0.00	0
HERKIMER NY	8.99	195	35.46	66	20.51	1724	29.84	1292	7.54	2178	31.55	1330	0.00	0
JEFFERSON NY	1.17	381	79.32	13	26.79	1489	5.58	3004	6.57	2340	16.59	2577	0.00	0
KINGS NY	3.69	273	14.51	441	40.77	1101	65.51	446	15.64	1376	11.50	2885	0.00	0
LEWIS NY	1.42	358	43.86	49	20.92	1714	11.54	2553	6.57	2341	19.66	2288	0.00	0
LIVINGSTON NY	8.75	205	71.87	19	30.65	1389	4.66	3028	15.86	1356	26.76	1614	0.00	0
MADISON NY	9.17	186	40.17	54	19.70	1758	8.48	2888	5.81	2493	18.51	2396	0.00	0
MONROE NY	1.67	350	57.91	29	9.41	2620	5.87	2993	11.20	1799	9.97	2928	0.00	0
MONTGOMERY NY	7.57	234	28.07	120	17.81	1886	103.96	190	4.86	2701	21.35	2103	0.00	0
NASSAU NY	3.50	289	24.09	187	47.99	966	63.16	467	15.64	1375	12.06	2857	0.00	0
NEW YORK NY	3.50	290	18.70	300	36.74	1206	54.85	572	15.43	1390	11.12	2896	0.00	0
NIAGARA NY	1.17	382	90.17	8	14.78	2090	11.42	2566	25.81	663	13.79	2777	0.00	0
ONEIDA NY	9.41	174	38.07	60	15.78	2009	12.38	2474	7.30	2220	20.74	2173	0.00	0
ONONDAGA NY	8.75	206	40.98	53	11.66	2396	3.96	3043	5.09	2638	14.82	2709	0.00	0
ONTARIO NY	8.75	207	73.00	17	20.43	1728	6.26	2975	14.51	1484	27.23	1576	0.00	0
ORANGE NY	11.65	158	21.74	218	37.67	1183	26.49	1453	14.25	1511	14.61	2728	0.00	0
ORLEANS NY	0.00	0	81.19	12	9.66	2598	8.54	2882	21.59	882	13.40	2800	0.00	0
OSWEGO NY	8.75	208	74.08	16	18.55	1829	6.17	2982	7.22	2237	17.33	2510	0.00	0
OTSEGO NY	20.71	68	50.60	36	21.56	1682	21.69	1719	5.57	2538	23.91	1855	0.00	0
PUTNAM NY	9.14	190	34.63	70	46.76	991	30.56	1262	10.80	1837	11.51	2884	0.00	0
QUEENS NY	3.50	291	13.32	474	42.23	1071	62.07	480	24.50	725	9.13	2951	0.00	0
RENSSELAER NY	29.93	21	17.77	333	24.83	1542	499.21	54	4.30	2815	22.77	1962	0.00	0
RICHMOND NY	3.50	292	14.81	428	29.82	1414	54.11	583	19.95	987	11.86	2869	0.00	0
ROCKLAND NY	3.50	293	17.21	348	35.60	1240	35.67	1045	13.11	1600	10.44	2917	0.00	0

U.S. FALLOUT ATLAS : TOTAL FALLOUT

TOTAL FALLOUT ACTIVITY (µCi/sq meter) by SHOT SERIES WITH COUNTY RANK

COUNTY AND STATE	RANGER SERIES	RANGER RANK	B-J SERIES	B-J RANK	T-S SERIES	T-S RANK	U-K SERIES	UK RANK	TEAPOT SERIES	TP RANK	PLUMBBOB SERIES	PB RANK	SHOT SEDAN	SEDAN RANK
	1951		1951		1952		1953		1955		1957		1962	
SARATOGA NY	8.56	221	29.21	107	31.89	1347	312.89	85	4.22	2840	19.40	2319	0.00	0
SCHENECTADY NY	7.99	226	15.25	415	28.40	1453	239.80	98	2.87	3019	16.04	2618	0.00	0
SCHOHARIE NY	7.57	235	27.22	137	24.91	1539	81.94	283	5.66	2518	25.09	1764	0.00	0
SCHUYLER NY	0.00	0	34.95	67	12.91	2272	4.88	3023	7.19	2242	15.57	2654	0.00	0
SENECA NY	8.99	196	61.49	24	17.65	1900	4.28	3037	8.33	2081	25.25	1743	0.00	0
ST LAWRENCE NY	1.42	359	32.33	84	80.66	498	10.94	2632	6.07	2441	23.14	1927	0.00	0
STEUBEN NY	8.99	197	60.43	26	19.31	1788	7.40	2934	14.54	1483	29.89	1431	0.00	0
SUFFOLK NY	3.50	294	17.01	353	42.00	1075	67.29	428	12.45	1667	11.95	2864	0.00	0
SULLIVAN NY	28.94	26	39.85	55	31.33	1362	16.78	2078	7.22	2233	19.95	2262	0.00	0
TIOGA NY	8.99	198	30.78	91	8.41	2726	7.22	2944	4.95	2672	19.17	2334	0.00	0
TOMPKINS NY	8.99	199	61.71	23	14.04	2156	5.92	2989	4.95	2673	25.80	1699	0.00	0
ULSTER NY	28.94	27	43.23	52	30.83	1381	32.31	1194	5.35	2583	22.92	1947	0.00	0
WARREN NY	7.99	227	26.44	153	41.53	1087	516.10	50	3.65	2932	14.92	2701	0.00	0
WASHINGTON NY	7.57	236	20.69	242	33.36	1301	542.51	48	3.27	2973	24.45	1813	0.00	0
WAYNE NY	9.17	187	77.35	14	25.05	1529	6.82	2962	15.54	1385	15.63	2650	0.00	0
WESTCHESTER NY	9.14	191	26.74	148	44.64	1028	43.29	791	15.11	1430	13.08	2816	0.00	0
WYOMING NY	7.57	237	76.86	15	30.19	1404	8.02	2909	20.42	958	27.14	1586	0.00	0
YATES NY	0.00	0	59.52	28	16.73	1953	5.23	3015	11.45	1772	9.32	2948	0.00	0
ADAMS OH	5.83	252	5.11	999	9.42	2618	14.30	2304	9.74	1950	27.77	1544	208.78	151
ALLEN OH	21.22	66	1.63	1779	14.40	2121	24.46	1553	13.70	1560	17.77	2469	208.78	172
ASHLAND OH	20.05	75	13.92	455	9.22	2641	24.15	1574	22.39	836	10.25	2922	103.59	377
ASHTABULA OH	9.22	181	17.65	336	14.92	2077	22.35	1675	24.45	729	16.73	2562	0.00	0
ATHENS OH	37.06	12	13.49	466	8.51	2711	14.78	2249	12.37	1674	24.64	1791	129.45	332
AUGLAIZE OH	0.00	0	1.70	1755	21.73	1676	33.39	1141	13.70	1559	17.60	2491	85.56	468
BELMONT OH	46.59	5	4.66	1074	4.64	3019	12.22	2488	18.56	1093	18.13	2430	90.29	430
BROWN OH	15.23	134	0.85	2185	10.61	2511	16.28	2112	14.09	1524	31.15	1351	134.17	329

TOTAL FALLOUT ACTIVITY (µCi/sq meter) by SHOT SERIES WITH COUNTY RANK

COUNTY AND STATE	RANGER SERIES	RANGER RANK	B-J SERIES	B-J RANK	T-S SERIES	T-S RANK	U-K SERIES	UK RANK	TEAPOT SERIES	TP RANK	PLUMBBOB SERIES	PB RANK	SHOT SEDAN	SEDAN RANK
	1951		1951		1952		1953		1955		1957		1962	
BUTLER OH	16.31	105	1.58	1811	13.37	2221	30.04	1281	12.61	1655	30.99	1364	299.72	119
CARROLL OH	18.06	89	12.36	514	4.92	3014	19.50	1875	15.34	1401	16.67	2570	0.00	0
CHAMPAIGN OH	17.21	99	1.38	1900	18.22	1851	25.36	1503	24.39	734	17.34	2509	208.78	175
CLARK OH	20.05	71	2.05	1598	16.24	1985	28.56	1358	11.64	1750	16.68	2568	246.22	130
CLERMONT OH	15.23	135	3.06	1384	12.91	2271	19.75	1857	11.27	1793	41.12	1010	121.04	348
CLINTON OH	16.70	104	3.44	1321	13.00	2261	24.67	1536	13.36	1580	17.98	2444	246.22	129
COLUMBIANA OH	22.27	41	14.01	454	4.55	3025	19.71	1862	23.14	795	21.19	2119	85.56	456
COSHOCTON OH	19.06	80	13.89	457	5.39	2985	21.10	1758	17.95	1167	25.86	1690	121.04	349
CRAWFORD OH	0.00	0	13.44	468	11.80	2379	23.58	1603	19.18	1041	10.54	2914	90.29	441
CUYAHOGA OH	20.05	76	6.29	896	14.76	2094	22.08	1694	96.30	166	13.74	2782	0.00	0
DARKE OH	0.71	434	1.73	1743	16.67	1956	29.97	1283	13.08	1605	18.09	2434	129.45	335
DEFIANCE OH	1.11	406	0.75	2244	13.02	2258	23.76	1590	12.64	1651	15.63	2649	0.00	0
DELAWARE OH	20.05	72	14.31	445	10.19	2559	21.63	1722	15.06	1431	11.23	2891	90.29	439
ERIE OH	0.00	0	19.00	292	31.38	1361	20.86	1781	63.71	242	9.37	2947	35.48	648
FAIRFIELD OH	16.31	106	3.63	1281	10.36	2542	14.33	2301	13.50	1572	17.45	2500	90.29	433
FAYETTE OH	16.31	107	4.14	1183	10.68	2505	18.90	1914	13.10	1601	16.19	2605	90.29	436
FRANKLIN OH	0.00	0	2.62	1463	15.59	2023	22.41	1669	12.17	1700	17.30	2516	35.48	644
FULTON OH	0.00	0	0.87	2175	19.05	1804	21.68	1720	13.16	1596	15.88	2630	208.78	176
GALLIA OH	15.23	136	8.28	759	16.38	1977	15.08	2215	10.59	1860	44.61	932	123.00	341
GEAUGA OH	22.27	42	16.35	373	8.98	2661	22.16	1688	62.91	246	14.80	2712	0.00	0
GREENE OH	0.00	0	3.18	1367	15.24	2051	20.16	1834	12.43	1672	20.12	2237	90.94	410
GUERNSEY OH	46.20	6	12.76	498	4.50	3026	14.30	2303	15.18	1422	17.69	2479	103.59	373
HAMILTON OH	17.21	100	1.53	1827	15.08	2065	22.90	1647	11.61	1756	19.62	2296	107.68	358
HANCOCK OH	0.00	0	1.44	1871	65.89	681	21.11	1755	25.50	675	14.83	2708	157.12	266
HARDIN OH	0.00	0	1.58	1806	14.80	2087	25.11	1514	12.50	1664	16.53	2586	37.44	625
HARRISON OH	18.06	90	12.23	520	6.08	2935	20.34	1821	18.93	1056	19.44	2316	85.56	462

U.S. FALLOUT ATLAS : TOTAL FALLOUT

TOTAL FALLOUT ACTIVITY (µCi/sq meter) by SHOT SERIES WITH COUNTY RANK

COUNTY AND STATE	RANGER SERIES	RANGER RANK	B-J SERIES	B-J RANK	T-S SERIES	T-S RANK	U-K SERIES	UK RANK	TEAPOT SERIES	TP RANK	PLUMBBOB SERIES	PB RANK	SHOT SEDAN	SEDAN RANK
	1951		1951		1952		1953		1955		1957		1962	
HENRY OH	13.36	147	1.02	2070	19.55	1770	22.99	1644	12.78	1641	20.43	2209	0.00	0
HIGHLAND OH	15.23	137	4.47	1104	10.12	2568	20.99	1770	12.96	1617	35.84	1127	208.78	149
HOCKING OH	17.26	97	3.91	1232	10.83	2492	12.72	2447	12.96	1620	32.14	1298	85.56	442
HOLMES OH	18.06	91	13.47	467	5.22	2995	22.33	1677	23.16	794	18.22	2421	0.00	0
HURON OH	0.00	0	19.10	288	10.61	2510	18.62	1941	51.12	308	8.82	2960	35.48	649
JACKSON OH	15.62	114	8.10	769	7.86	2772	17.71	2009	10.50	1870	24.87	1777	208.78	155
JEFFERSON OH	0.39	484	18.34	315	7.06	2849	17.66	2013	20.00	981	16.15	2609	0.00	0
KNOX OH	19.55	77	13.27	478	7.17	2837	22.69	1659	18.02	1151	11.25	2890	208.78	182
LAKE OH	0.00	0	25.28	174	14.72	2096	21.07	1761	47.07	341	13.73	2783	72.20	532
LAWRENCE OH	0.00	0	5.47	967	16.25	1984	13.70	2363	14.91	1445	44.53	936	176.50	243
LICKING OH	19.06	81	17.33	344	10.81	2493	26.98	1431	15.87	1353	10.83	2904	208.78	184
LOGAN OH	15.39	121	2.08	1592	16.24	1986	28.24	1373	13.55	1569	18.65	2379	129.45	334
LORAIN OH	0.00	0	17.74	334	9.39	2623	22.09	1692	52.74	295	8.36	2965	37.44	629
LUCAS OH	0.00	0	1.59	1801	8.78	2680	16.99	2057	17.11	1245	12.09	2856	123.00	346
MADISON OH	16.31	108	2.53	1481	16.97	1944	26.79	1439	12.62	1653	16.45	2591	147.48	285
MAHONING OH	0.00	0	10.71	613	5.01	3005	22.58	1663	28.85	583	16.43	2593	103.59	376
MARION OH	20.05	73	14.83	427	7.21	2834	28.34	1370	15.64	1374	10.45	2916	85.56	474
MEDINA OH	0.00	0	14.96	424	10.57	2515	23.28	1621	107.71	161	18.33	2414	208.78	170
MEIGS OH	15.23	138	8.15	764	9.98	2576	21.36	1739	16.90	1266	20.62	2187	90.29	428
MERCER OH	0.00	0	1.46	1858	12.57	2298	28.57	1357	10.22	1895	15.85	2633	35.48	647
MIAMI OH	20.05	74	2.83	1415	13.77	2180	29.97	1284	14.25	1510	19.55	2301	208.78	165
MONROE OH	21.83	48	5.01	1009	4.56	3024	11.22	2590	15.80	1363	18.40	2408	103.59	371
MONTGOMERY OH	20.76	67	2.01	1618	47.63	975	24.93	1524	12.81	1636	19.95	2261	244.26	133
MORGAN OH	21.83	49	12.96	490	8.41	2727	11.32	2576	10.93	1827	17.42	2503	208.78	174
MORROW OH	22.27	43	14.67	435	5.01	3006	21.01	1767	16.98	1260	12.04	2859	85.56	472
MUSKINGUM OH	18.06	92	16.19	375	9.46	2613	14.16	2317	13.75	1550	17.88	2456	125.77	339

TOTAL FALLOUT ACTIVITY (µCi/sq meter) by SHOT SERIES WITH COUNTY RANK

COUNTY AND STATE	RANGER SERIES	RANGER RANK	B-J SERIES	B-J RANK	T-S SERIES	T-S RANK	U-K SERIES	UK RANK	TEAPOT SERIES	TP RANK	PLUMBBOB SERIES	PB RANK	SHOT SEDAN	SEDAN RANK
	1951		1951		1952		1953		1955		1957		1962	
NOBLE OH	0.00	0	13.10	482	9.25	2638	9.25	2833	15.31	1405	24.04	1836	85.56	445
OTTAWA OH	0.00	0	6.91	848	14.41	2116	15.74	2155	23.81	758	13.00	2824	37.44	628
PAULDING OH	0.00	0	0.55	2438	30.52	1394	21.37	1738	13.73	1554	19.23	2329	129.45	333
PERRY OH	18.06	93	19.21	284	8.31	2736	12.64	2455	12.46	1666	19.11	2341	208.78	167
PICKAWAY OH	16.31	109	3.96	1220	10.07	2571	15.89	2142	13.68	1564	15.06	2692	0.00	0
PIKE OH	15.23	139	4.56	1087	10.12	2567	17.55	2022	11.30	1789	21.52	2091	208.78	159
PORTAGE OH	0.00	0	15.25	416	8.62	2700	19.11	1901	54.35	283	15.80	2638	121.04	351
PREBLE OH	0.00	0	1.53	1828	10.32	2545	29.41	1314	12.24	1689	20.19	2230	208.78	160
PUTNAM OH	0.00	0	1.25	1966	79.80	508	22.00	1698	15.47	1388	15.36	2671	0.00	0
RICHLAND OH	22.27	44	15.55	400	12.56	2301	26.57	1447	19.99	984	10.50	2915	194.53	224
ROSS OH	17.26	98	7.17	831	8.72	2686	19.41	1883	11.60	1757	19.17	2335	90.29	429
SANDUSKY OH	0.00	0	7.61	795	12.21	2338	16.08	2131	52.20	298	10.62	2913	90.29	440
SCIOTO OH	15.60	115	5.39	973	16.34	1978	13.78	2351	9.32	1985	16.25	2602	359.52	62
SENECA OH	22.27	45	9.72	674	10.08	2569	24.83	1528	81.63	199	10.98	2899	208.78	183
SHELBY OH	1.42	360	2.84	1414	12.94	2268	26.59	1445	14.66	1465	16.66	2572	0.00	0
STARK OH	18.06	94	9.04	708	6.07	2937	22.63	1660	24.97	698	17.41	2505	90.29	434
SUMMIT OH	0.00	0	19.20	285	6.41	2910	20.39	1818	22.82	810	17.27	2521	35.48	645
TRUMBULL OH	11.14	159	13.78	463	7.13	2841	18.21	1963	43.80	374	15.46	2662	90.29	437
TUSCARAWAS OH	18.55	84	13.43	470	4.97	3012	21.32	1741	17.27	1228	18.12	2432	103.59	372
UNION OH	17.21	101	2.38	1509	19.53	1772	25.41	1501	13.07	1607	17.74	2472	123.00	345
VAN WERT OH	0.00	0	1.33	1927	12.27	2335	26.57	1446	11.15	1805	15.79	2639	90.94	413
VINTON OH	7.38	238	2.03	1610	7.33	2822	15.59	2169	11.04	1817	27.59	1555	85.56	443
WARREN OH	16.31	110	1.83	1689	11.73	2390	27.93	1389	13.71	1556	21.28	2107	246.22	128
WASHINGTON OH	21.83	50	5.80	936	13.10	2250	13.45	2386	18.26	1121	26.92	1603	90.29	420
WAYNE OH	19.06	82	10.14	647	5.28	2992	21.97	1700	22.09	856	25.72	1702	0.00	0
WILLIAMS OH	0.00	0	0.63	2354	19.16	1794	16.71	2084	9.93	1927	17.01	2544	0.00	0

TOTAL FALLOUT ACTIVITY (µCi/sq meter) by SHOT SERIES WITH COUNTY RANK

COUNTY AND STATE	RANGER SERIES	RANGER RANK	B-J SERIES	B-J RANK	T-S SERIES	T-S RANK	U-K SERIES	UK RANK	TEAPOT SERIES	TP RANK	PLUMBBOB SERIES	PB RANK	SHOT SEDAN	SEDAN RANK
	1951		1951		1952		1953		1955		1957		1962	
WOOD OH	0.00	0	1.23	1972	11.81	2378	22.39	1671	22.29	842	11.77	2870	123.00	347
WYANDOT OH	0.00	0	17.48	339	9.59	2607	16.68	2091	15.38	1395	11.64	2880	208.78	181
ADAIR OK	0.00	0	4.24	1156	79.45	511	34.90	1082	34.75	475	45.22	920	0.00	0
ALFALFA OK	0.00	0	8.07	772	179.87	263	69.21	403	18.44	1104	55.17	709	0.00	0
ATOKA OK	0.00	0	4.79	1042	51.63	894	62.12	476	72.15	223	35.64	1133	0.00	0
BEAVER OK	0.00	0	3.08	1381	157.09	280	55.33	562	84.82	189	33.36	1243	0.00	0
BECKHAM OK	0.00	0	2.78	1431	69.24	629	66.44	434	80.77	203	22.66	1978	0.00	0
BLAINE OK	0.00	0	6.73	860	104.19	346	41.16	856	25.27	683	42.69	980	0.00	0
BRYAN OK	0.00	0	5.98	925	53.43	877	69.12	404	43.70	375	31.08	1357	0.00	0
CADDO OK	0.00	0	4.94	1014	301.08	212	34.68	1094	42.80	383	33.33	1246	0.00	0
CANADIAN OK	0.00	0	7.10	834	58.59	789	36.94	1006	42.05	389	40.89	1015	0.00	0
CARTER OK	0.00	0	4.79	1043	44.90	1023	46.62	711	50.26	314	25.47	1724	0.00	0
CHEROKEE OK	0.00	0	9.73	673	87.18	440	39.80	901	49.39	318	55.77	699	0.00	0
CHOCTAW OK	0.00	0	2.93	1404	61.04	754	43.08	798	65.41	239	34.43	1183	0.00	0
CIMARRON OK	0.00	0	1.80	1718	63.33	723	142.76	147	114.15	145	25.72	1704	0.00	0
CLEVELAND OK	0.00	0	5.14	995	54.92	856	34.98	1077	38.52	427	26.81	1609	0.00	0
COAL OK	0.00	0	2.35	1519	52.66	884	32.17	1198	39.21	419	30.71	1386	0.00	0
COMANCHE OK	0.00	0	3.50	1305	540.51	161	38.24	968	54.85	282	22.88	1953	0.00	0
COTTON OK	0.00	0	2.82	1424	198.89	255	29.16	1326	44.66	359	24.79	1782	0.00	0
CRAIG OK	3.34	300	10.76	607	77.30	531	66.02	439	51.20	304	55.54	704	0.00	0
CREEK OK	0.00	0	7.37	819	84.83	463	84.67	268	28.28	593	53.17	750	0.00	0
CUSTER OK	0.00	0	4.67	1070	160.79	275	38.85	944	29.55	569	29.81	1436	0.00	0
DELAWARE OK	0.00	0	11.26	579	92.75	394	36.82	1010	44.43	362	49.81	816	0.00	0
DEWEY OK	0.00	0	6.50	877	82.15	485	56.89	541	25.08	691	34.97	1154	0.00	0
ELLIS OK	0.00	0	4.78	1049	71.08	610	70.17	386	92.68	170	31.95	1311	0.00	0
GARFIELD OK	0.00	0	9.32	693	30.60	1392	40.32	885	43.03	381	52.45	770	0.00	0

TOTAL FALLOUT ACTIVITY (µCi/sq meter) by SHOT SERIES WITH COUNTY RANK

COUNTY AND STATE	RANGER SERIES	RANGER RANK	B-J SERIES	B-J RANK	T-S SERIES	T-S RANK	U-K SERIES	UK RANK	TEAPOT SERIES	TP RANK	PLUMBBOB SERIES	PB RANK	SHOT SEDAN	SEDAN RANK
	1951		1951		1952		1953		1955		1957		1962	
GARVIN OK	0.00	0	3.81	1246	77.10	536	34.49	1100	40.72	403	24.01	1840	0.00	0
GRADY OK	0.00	0	7.22	829	45.22	1016	31.93	1206	39.55	415	24.11	1832	0.00	0
GRANT OK	0.00	0	6.40	885	57.93	801	38.39	963	17.44	1209	64.45	574	0.00	0
GREER OK	0.00	0	1.95	1633	253.91	229	61.62	486	84.70	190	17.68	2481	0.00	0
HARMON OK	0.00	0	0.43	2637	71.39	606	57.57	528	80.83	202	16.30	2599	0.00	0
HARPER OK	0.00	0	3.46	1316	57.10	823	48.81	668	22.42	832	47.19	872	0.00	0
HASKELL OK	0.00	0	3.99	1215	103.64	347	39.68	906	35.23	466	29.17	1471	0.00	0
HUGHES OK	0.00	0	7.44	815	53.45	876	67.34	426	44.02	370	34.49	1176	0.00	0
JACKSON OK	0.00	0	3.89	1234	93.15	392	48.98	661	34.17	482	18.70	2375	0.00	0
JEFFERSON OK	0.00	0	3.67	1274	47.74	971	32.88	1163	50.41	312	25.47	1723	0.00	0
JOHNSTON OK	0.00	0	5.03	1005	41.70	1083	34.31	1106	52.89	293	32.23	1290	0.00	0
KAY OK	0.00	0	9.37	689	62.01	741	37.27	1000	37.85	434	62.66	606	0.00	0
KINGFISHER OK	0.00	0	5.75	939	33.49	1299	43.61	782	39.54	416	46.40	893	0.00	0
KIOWA OK	0.00	0	4.32	1145	189.56	262	39.50	916	44.02	371	18.92	2358	0.00	0
LATIMER OK	0.00	0	3.36	1332	55.65	848	39.08	926	59.02	259	29.53	1454	0.00	0
LE FLORE OK	0.00	0	4.45	1107	40.71	1105	42.39	815	43.05	380	36.12	1122	0.00	0
LINCOLN OK	0.00	0	8.85	724	55.78	844	44.22	768	25.97	657	47.81	864	0.00	0
LOGAN OK	0.00	0	8.42	749	34.18	1281	37.01	1004	27.37	612	43.73	952	0.00	0
LOVE OK	0.00	0	4.71	1063	71.38	607	42.38	816	51.12	307	31.32	1344	0.00	0
MAJOR OK	0.00	0	10.33	635	67.68	655	51.00	634	16.64	1287	43.59	957	0.00	0
MARSHALL OK	0.00	0	4.82	1040	37.53	1187	36.54	1018	45.91	347	27.98	1531	0.00	0
MAYES OK	0.00	0	9.98	660	86.32	454	38.34	966	48.64	325	53.77	741	0.00	0
MCCLAIN OK	0.00	0	6.00	920	86.54	452	34.12	1113	37.85	433	26.02	1679	0.00	0
MCCURTAIN OK	0.00	0	3.54	1296	54.92	857	33.25	1149	58.64	262	42.94	974	0.00	0
MCINTOSH OK	0.00	0	7.27	826	66.19	677	46.13	719	47.84	332	44.56	935	0.00	0
MURRAY OK	0.00	0	4.90	1027	31.92	1345	35.51	1055	39.08	420	29.05	1480	0.00	0

TOTAL FALLOUT ACTIVITY (µCi/sq meter) by SHOT SERIES WITH COUNTY RANK

COUNTY AND STATE	RANGER SERIES	RANGER RANK	B-J SERIES	B-J RANK	T-S SERIES	T-S RANK	U-K SERIES	UK RANK	TEAPOT SERIES	TP RANK	PLUMBBOB SERIES	PB RANK	SHOT SEDAN	SEDAN RANK
	1951		1951		1952		1953		1955		1957		1962	
MUSKOGEE OK	0.00	0	9.04	709	92.62	396	46.29	717	32.59	506	50.79	801	0.00	0
NOBLE OK	0.00	0	8.85	725	50.06	927	40.44	880	38.50	428	52.31	773	0.00	0
NOWATA OK	0.00	0	12.94	491	84.29	467	43.60	783	18.47	1100	53.65	744	0.00	0
OKFUSKEE OK	0.00	0	6.94	844	51.95	891	43.67	781	26.74	631	39.74	1033	0.00	0
OKLAHOMA OK	0.00	0	3.89	1237	56.71	828	36.60	1015	26.39	647	34.76	1163	0.00	0
OKMULGEE OK	0.00	0	8.10	770	86.64	449	41.01	857	27.43	609	40.62	1017	0.00	0
OSAGE OK	0.00	0	9.51	684	60.13	775	42.46	813	40.61	405	53.97	733	0.00	0
OTTAWA OK	0.00	0	7.82	786	90.41	412	53.86	587	40.55	407	51.37	786	0.00	0
PAWNEE OK	0.00	0	8.71	731	54.61	862	41.17	855	31.20	527	54.17	732	0.00	0
PAYNE OK	0.00	0	9.03	712	32.57	1329	37.17	1002	24.70	711	50.41	805	0.00	0
PITTSBURG OK	0.00	0	5.62	953	67.14	666	66.69	432	56.20	276	43.05	969	0.00	0
PONTOTOC OK	0.00	0	4.55	1092	32.52	1331	35.15	1070	84.19	192	25.91	1685	0.00	0
POTTAWATOMIE OK	0.00	0	6.35	889	80.45	500	42.26	818	44.36	363	31.99	1306	0.00	0
PUSHMATAHA OK	0.00	0	3.91	1233	43.14	1053	38.98	935	63.28	245	38.10	1062	0.00	0
ROGER MILLS OK	0.00	0	6.04	915	69.99	624	99.18	205	82.59	197	21.99	2053	0.00	0
ROGERS OK	0.00	0	10.36	631	80.33	504	41.91	833	21.39	896	59.08	654	0.00	0
SEMINOLE OK	0.00	0	6.53	876	44.77	1025	59.97	502	53.17	290	34.49	1177	0.00	0
SEQUOYAH OK	0.00	0	3.46	1315	95.71	382	36.76	1012	28.82	585	46.80	883	0.00	0
STEPHENS OK	0.00	0	3.25	1352	80.22	507	32.88	1165	52.24	297	24.39	1816	0.00	0
TEXAS OK	0.00	0	1.63	1778	62.43	731	82.18	282	229.67	51	30.39	1401	0.00	0
TILLMAN OK	0.00	0	2.89	1409	147.56	291	36.82	1009	33.16	501	17.10	2532	0.00	0
TULSA OK	0.00	0	7.83	785	119.21	325	73.01	362	34.54	480	57.98	671	0.00	0
WAGONER OK	0.00	0	9.25	698	113.68	334	39.96	897	29.67	564	54.36	728	0.00	0
WASHINGTON OK	2.33	319	9.29	696	63.87	711	38.10	973	41.57	395	55.81	698	0.00	0
WASHITA OK	0.00	0	3.21	1361	254.03	227	56.69	542	77.06	213	26.84	1607	0.00	0
WOODS OK	0.00	0	6.34	890	151.59	289	54.61	576	23.06	799	47.88	861	0.00	0

TOTAL FALLOUT ACTIVITY (µCi/sq meter) by SHOT SERIES WITH COUNTY RANK

COUNTY AND STATE	RANGER SERIES	RANGER RANK	B-J SERIES	B-J RANK	T-S SERIES	T-S RANK	U-K SERIES	UK RANK	TEAPOT SERIES	TP RANK	PLUMBBOB SERIES	PB RANK	SHOT SEDAN	SEDAN RANK
	1951		1951		1952		1953		1955		1957		1962	
WOODWARD OK	0.00	0	7.35	820	67.72	653	40.62	874	22.60	822	39.48	1038	0.00	0
BAKER OR	0.00	0	1.58	1808	13750.25	35	8.39	2892	21.56	885	62.35	609	0.00	0
BENTON OR	0.00	0	0.43	2638	3.00	3069	6.15	2983	5.12	2630	5.02	3007	0.00	0
CLACKAMAS OR	0.00	0	0.43	2639	17.98	1874	11.24	2588	8.02	2122	5.63	2997	0.00	0
CLATSOP OR	0.00	0	0.43	2640	13.62	2195	7.73	2921	8.09	2111	5.46	2999	0.00	0
COLUMBIA OR	0.00	0	0.43	2641	5.61	2974	9.36	2818	6.77	2306	4.52	3022	0.00	0
COOS OR	0.00	0	0.43	2642	5.06	3001	7.39	2935	5.24	2610	7.81	2975	0.00	0
CROOK OR	0.00	0	1.00	2082	933.09	123	5.25	3014	3.74	2920	6.29	2991	0.00	0
CURRY OR	0.00	0	0.43	2643	7.81	2776	8.14	2903	5.25	2607	8.60	2963	0.00	0
DESCHUTES OR	0.00	0	0.43	2644	201.08	251	3.01	3066	2.83	3023	4.57	3019	0.00	0
DOUGLAS OR	0.00	0	0.43	2645	8.67	2692	8.41	2890	4.79	2718	8.05	2971	0.00	0
GILLIAM OR	0.00	0	1.00	2083	466.65	190	5.52	3008	4.06	2866	5.22	3004	0.00	0
GRANT OR	0.00	0	1.49	1844	4877.41	70	7.76	2918	24.70	712	62.95	600	0.00	0
HARNEY OR	0.00	0	18.28	319	614.37	158	10.30	2706	23.16	793	39.11	1045	0.00	0
HOOD RIVER OR	0.00	0	0.43	2646	15.56	2025	9.95	2753	6.44	2369	3.86	3043	0.00	0
JACKSON OR	0.00	0	0.43	2647	18.73	1823	5.06	3020	2.39	3053	2.97	3060	0.00	0
JEFFERSON OR	0.00	0	1.09	2041	340.87	198	5.97	2988	3.15	2992	5.26	3003	0.00	0
JOSEPHINE OR	0.00	0	0.43	2648	8.45	2723	6.31	2973	3.23	2978	7.87	2973	0.00	0
KLAMATH OR	0.00	0	0.43	2649	124.70	321	5.56	3006	4.63	2757	5.45	3000	0.00	0
LAKE OR	0.00	0	0.43	2650	845.38	132	3.57	3050	5.22	2613	5.43	3001	0.00	0
LANE OR	0.00	0	0.43	2651	7.75	2782	8.77	2869	5.70	2512	7.61	2976	0.00	0
LINCOLN OR	0.00	0	0.43	2652	6.84	2864	10.56	2681	7.14	2253	7.53	2979	0.00	0
LINN OR	0.00	0	0.43	2653	3.20	3064	8.06	2907	7.21	2240	6.74	2987	0.00	0
MALHEUR OR	0.00	0	19.18	286	1236.96	115	11.09	2604	23.16	792	49.91	811	0.00	0
MARION OR	0.00	0	0.43	2654	12.34	2328	10.61	2675	7.46	2190	5.87	2996	0.00	0
MORROW OR	0.00	0	1.40	1884	930.63	124	5.11	3019	4.71	2737	7.00	2985	0.00	0

TOTAL FALLOUT ACTIVITY (µCi/sq meter) by SHOT SERIES WITH COUNTY RANK

COUNTY AND STATE	RANGER SERIES	RANGER RANK	B-J SERIES	B-J RANK	T-S SERIES	T-S RANK	U-K SERIES	UK RANK	TEAPOT SERIES	TP RANK	PLUMBBOB SERIES	PB RANK	SHOT SEDAN	SEDAN RANK
	1951		1951		1952		1953		1955		1957		1962	
MULTNOMAH OR	0.00	0	0.43	2655	4.98	3009	10.21	2720	3.92	2889	4.42	3028	0.00	0
POLK OR	0.00	0	0.54	2456	2.19	3082	5.57	3005	6.61	2331	5.12	3005	0.00	0
SHERMAN OR	0.00	0	1.00	2084	307.10	209	5.71	2999	3.01	3005	5.10	3006	0.00	0
TILLAMOOK OR	0.00	0	0.43	2656	41.86	1080	7.64	2923	6.29	2402	6.53	2988	0.00	0
UMATILLA OR	0.00	0	0.67	2317	2255.56	87	7.23	2943	6.54	2351	55.86	696	0.00	0
UNION OR	0.00	0	1.00	2085	6311.57	64	8.26	2899	7.64	2165	63.20	595	0.00	0
WALLOWA OR	0.00	0	1.00	2086	19535.70	19	7.50	2932	7.47	2189	49.46	827	0.00	0
WASCO OR	0.00	0	1.18	1999	195.67	258	5.92	2990	3.95	2883	4.61	3014	0.00	0
WASHINGTON OR	0.00	0	0.43	2657	4.98	3010	7.17	2947	5.96	2465	3.89	3042	0.00	0
WHEELER OR	0.00	0	1.00	2087	721.99	148	4.13	3040	5.67	2516	6.31	2990	0.00	0
YAMHILL OR	0.00	0	0.43	2658	2.05	3084	6.07	2984	5.47	2554	4.13	3034	0.00	0
ADAMS PA	1.17	390	11.45	569	12.71	2286	15.12	2211	6.79	2304	14.52	2736	73.06	529
ALLEGHENY PA	22.32	38	9.13	702	6.28	2921	18.31	1960	11.00	1821	14.55	2732	72.20	531
ARMSTRONG PA	21.83	51	8.55	739	5.39	2986	23.00	1642	20.68	939	22.96	1943	90.29	424
BEAVER PA	21.83	52	12.49	508	4.70	3018	19.39	1886	17.98	1163	20.37	2216	103.59	368
BEDFORD PA	1.17	391	9.81	669	11.37	2432	20.30	1825	11.90	1726	20.94	2148	57.04	569
BERKS PA	2.54	308	16.07	378	13.06	2254	28.00	1385	15.87	1354	14.65	2723	0.00	0
BLAIR PA	22.22	46	10.45	625	12.12	2345	19.32	1891	14.04	1530	23.31	1904	0.00	0
BRADFORD PA	9.22	182	26.52	150	9.06	2656	9.81	2765	5.68	2515	22.49	1995	0.00	0
BUCKS PA	3.94	266	16.71	361	14.98	2072	34.88	1083	10.40	1881	13.31	2804	0.00	0
BUTLER PA	0.00	0	14.30	446	5.51	2979	21.92	1705	21.73	875	20.74	2174	85.56	458
CAMBRIA PA	22.22	47	10.71	612	9.20	2643	18.87	1918	14.49	1487	21.97	2056	90.29	427
CAMERON PA	22.54	35	18.16	322	6.66	2885	9.68	2780	14.63	1469	22.49	1996	46.95	584
CARBON PA	1.17	392	26.61	149	14.22	2140	21.97	1699	13.84	1543	13.76	2781	0.00	0
CENTRE PA	21.83	53	13.91	456	10.34	2543	10.77	2653	11.20	1801	18.37	2411	57.04	571
CHESTER PA	1.17	393	18.69	301	16.72	1954	31.86	1210	13.12	1599	15.87	2632	73.06	526

TOTAL FALLOUT ACTIVITY (µCi/sq meter) by SHOT SERIES WITH COUNTY RANK

COUNTY AND STATE	RANGER SERIES	RANGER RANK	B-J SERIES	B-J RANK	T-S SERIES	T-S RANK	U-K SERIES	UK RANK	TEAPOT SERIES	TP RANK	PLUMBBOB SERIES	PB RANK	SHOT SEDAN	SEDAN RANK
	1951		1951		1952		1953		1955		1957		1962	
CLARION PA	9.45	172	18.31	317	5.27	2993	26.16	1469	28.26	594	22.46	2004	85.56	453
CLEARFIELD PA	21.83	54	15.43	405	6.38	2912	19.98	1839	18.02	1154	21.07	2135	103.59	366
CLINTON PA	8.74	210	16.51	366	14.13	2146	9.52	2795	11.04	1816	19.96	2260	73.06	523
COLUMBIA PA	8.74	211	15.50	402	12.47	2312	10.37	2700	8.47	2057	17.26	2522	0.00	0
CRAWFORD PA	10.21	163	13.37	472	8.37	2730	16.71	2085	45.16	353	18.23	2418	85.56	465
CUMBERLAND PA	1.23	371	4.95	1013	13.70	2188	14.54	2277	8.13	2106	20.09	2244	0.00	0
DAUPHIN PA	1.17	394	10.39	629	7.37	2819	14.75	2253	7.80	2142	18.51	2398	0.00	0
DELAWARE PA	2.54	309	16.82	359	15.33	2046	39.02	930	10.93	1828	14.70	2720	57.04	578
ELK PA	22.54	36	39.17	57	6.82	2867	14.26	2311	17.55	1202	23.35	1896	132.81	330
ERIE PA	12.13	151	21.60	222	15.75	2011	17.30	2042	44.96	357	19.28	2324	21.07	660
FAYETTE PA	21.83	55	5.42	970	5.73	2969	19.85	1848	12.53	1663	26.95	1600	118.49	354
FOREST PA	8.74	212	33.99	72	7.29	2829	14.63	2267	49.17	321	17.85	2460	85.56	467
FRANKLIN PA	1.11	407	12.62	501	13.29	2225	16.58	2097	9.43	1973	20.26	2226	0.00	0
FULTON PA	0.00	0	11.72	552	13.83	2174	16.03	2134	9.81	1942	20.65	2181	149.13	281
GREENE PA	21.83	56	5.60	955	4.97	3013	12.71	2450	14.31	1507	23.70	1877	85.56	448
HUNTINGDON PA	21.83	57	10.98	595	12.01	2357	20.79	1788	6.44	2368	17.46	2498	57.04	573
INDIANA PA	22.32	39	11.40	571	8.60	2704	20.43	1816	15.89	1351	24.53	1796	90.29	422
JEFFERSON PA	9.93	170	13.43	469	6.68	2880	19.51	1874	26.68	634	17.84	2462	90.29	432
JUNIATA PA	8.74	213	9.35	691	12.64	2294	13.73	2359	7.36	2210	13.25	2807	0.00	0
LACKAWANNA PA	16.20	111	31.03	90	18.90	1813	5.86	2995	9.12	2003	16.16	2608	40.85	609
LANCASTER PA	1.17	395	14.95	426	15.91	1999	17.49	2025	6.98	2272	18.06	2438	60.19	550
LAWRENCE PA	0.00	0	9.70	676	5.86	2957	21.01	1768	25.05	694	16.65	2573	103.59	375
LEBANON PA	1.17	396	11.37	572	13.05	2256	17.64	2014	6.54	2350	10.31	2921	0.00	0
LEHIGH PA	2.54	310	20.50	249	17.83	1884	24.20	1570	15.00	1436	14.25	2750	0.00	0
LUZERNE PA	9.22	183	27.83	127	19.65	1760	9.49	2798	9.16	1998	23.65	1880	16.72	681
LYCOMING PA	8.74	214	14.71	432	9.66	2597	9.87	2761	7.10	2258	17.03	2541	0.00	0

TOTAL FALLOUT ACTIVITY (µCi/sq meter) by SHOT SERIES WITH COUNTY RANK

COUNTY AND STATE	RANGER SERIES	RANGER RANK	B-J SERIES	B-J RANK	T-S SERIES	T-S RANK	U-K SERIES	UK RANK	TEAPOT SERIES	TP RANK	PLUMBBOB SERIES	PB RANK	SHOT SEDAN	SEDAN RANK
	1951		1951		1952		1953		1955		1957		1962	
MCKEAN PA	10.21	164	47.27	44	8.59	2706	14.29	2308	19.43	1030	20.43	2210	0.00	0
MERCER PA	8.94	204	13.10	483	6.25	2926	19.52	1873	50.30	313	16.61	2576	85.56	470
MIFFLIN PA	9.22	184	10.27	639	12.31	2331	13.75	2356	6.41	2381	16.31	2598	0.00	0
MONROE PA	11.68	155	31.86	86	23.24	1604	19.40	1885	9.82	1939	13.86	2773	0.00	0
MONTGOMERY PA	1.17	397	16.51	367	14.99	2071	37.90	980	10.14	1899	14.58	2731	0.00	0
MONTOUR PA	21.83	58	10.21	645	11.31	2441	10.54	2683	4.17	2846	23.06	1934	0.00	0
NORTHAMPTON PA	4.66	256	29.17	108	17.12	1934	22.28	1681	9.86	1933	13.57	2790	0.00	0
NORTHUMBERLA PA	0.00	0	15.41	406	12.74	2284	11.84	2525	5.93	2471	20.61	2188	73.06	522
PERRY PA	1.23	372	8.67	733	13.81	2176	14.13	2322	8.63	2042	13.13	2811	0.00	0
PHILADELPHIA PA	0.23	539	13.04	486	15.74	2012	36.11	1030	6.57	2339	8.04	2972	0.00	0
PIKE PA	14.48	144	25.42	172	40.78	1099	15.56	2175	9.16	1999	17.29	2520	0.00	0
POTTER PA	9.91	171	50.82	34	7.15	2839	12.15	2495	18.49	1098	19.97	2259	7.38	691
SCHUYLKILL PA	1.17	398	15.05	420	15.21	2053	21.26	1746	12.26	1688	17.92	2450	0.00	0
SNYDER PA	21.83	59	10.10	649	12.02	2356	12.61	2456	8.38	2070	19.70	2284	149.13	282
SOMERSET PA	2.54	311	7.15	832	6.94	2859	21.72	1717	13.51	1571	23.87	1862	85.56	447
SULLIVAN PA	21.83	60	22.93	204	7.89	2769	8.49	2885	5.01	2658	18.06	2437	0.00	0
SUSQUEHANNA PA	21.83	61	26.43	154	6.18	2932	11.29	2581	4.63	2754	22.06	2045	0.00	0
TIOGA PA	21.83	62	27.44	135	5.97	2950	10.16	2729	10.21	1896	21.04	2138	0.00	0
UNION PA	21.83	63	9.62	678	11.06	2473	10.35	2702	7.21	2238	17.30	2517	0.00	0
VENANGO PA	11.14	160	12.91	493	6.58	2897	20.32	1823	46.55	343	20.12	2238	208.78	162
WARREN PA	8.74	215	45.65	47	9.13	2649	17.78	1999	28.25	595	22.09	2042	21.07	659
WASHINGTON PA	22.54	37	6.87	851	5.21	2996	25.20	1510	17.49	1207	21.52	2090	85.56	455
WAYNE PA	30.97	20	37.22	62	19.31	1787	15.99	2136	5.13	2629	17.32	2514	8.99	689
WESTMORELAND PA	22.32	40	8.59	736	6.39	2911	19.82	1851	17.08	1247	22.74	1965	90.29	425
WYOMING PA	16.20	112	34.69	69	7.50	2804	9.09	2844	4.01	2874	26.52	1631	16.19	682
YORK PA	1.17	399	10.21	644	15.56	2026	15.57	2170	8.37	2072	20.62	2186	73.06	521

TOTAL FALLOUT ACTIVITY (µCi/sq meter) by SHOT SERIES WITH COUNTY RANK

COUNTY AND STATE	RANGER SERIES	RANGER RANK	B-J SERIES	B-J RANK	T-S SERIES	T-S RANK	U-K SERIES	UK RANK	TEAPOT SERIES	TP RANK	PLUMBBOB SERIES	PB RANK	SHOT SEDAN	SEDAN RANK
	1951		1951		1952		1953		1955		1957		1962	
BRISTOL RI	0.21	542	18.54	309	33.75	1295	110.87	181	5.41	2566	9.10	2954	0.00	0
KENT RI	3.87	269	26.75	147	29.21	1433	94.38	224	5.88	2483	12.11	2855	0.00	0
NEWPORT RI	3.87	270	18.69	302	23.77	1581	76.53	324	4.89	2693	8.23	2967	0.00	0
PROVIDENCE RI	3.87	271	21.40	228	23.43	1595	81.16	287	5.02	2654	9.12	2952	0.00	0
WASHINGTON RI	3.71	272	26.77	145	29.14	1435	111.54	179	6.43	2374	12.31	2847	0.00	0
ABBEVILLE SC	0.00	0	12.43	510	6.67	2883	14.13	2321	7.88	2133	22.63	1983	0.00	0
AIKEN SC	0.27	535	12.16	525	4.59	3023	11.39	2569	6.68	2321	22.71	1971	0.00	0
ALLENDALE SC	0.00	0	14.81	429	9.60	2604	11.77	2533	6.16	2427	23.56	1887	0.00	0
ANDERSON SC	0.27	536	12.10	529	8.59	2705	15.51	2181	9.62	1961	25.40	1733	0.00	0
BAMBERG SC	0.00	0	15.85	386	16.94	1945	10.83	2645	5.09	2639	25.58	1714	0.00	0
BARNWELL SC	0.00	0	15.07	419	9.79	2588	11.97	2514	6.16	2426	23.95	1850	0.00	0
BEAUFORT SC	0.00	0	16.15	376	17.66	1898	9.92	2757	4.07	2861	26.64	1621	0.00	0
BERKELEY SC	0.00	0	18.44	310	21.31	1699	9.43	2809	4.28	2821	27.13	1587	0.00	0
CALHOUN SC	0.00	0	16.50	368	28.53	1449	12.53	2464	4.87	2700	25.22	1745	0.00	0
CHARLESTON SC	0.00	0	13.31	477	28.40	1454	7.58	2926	2.56	3045	30.27	1408	0.00	0
CHEROKEE SC	0.00	0	13.10	484	6.00	2946	13.68	2365	14.47	1492	24.71	1789	0.00	0
CHESTER SC	0.00	0	14.54	440	7.38	2815	11.83	2527	8.89	2022	23.77	1872	0.00	0
CHESTERFIELD SC	0.00	0	22.20	211	12.13	2343	10.99	2624	3.87	2897	21.85	2067	0.00	0
CLARENDON SC	0.00	0	20.39	254	34.36	1278	10.45	2692	4.25	2830	22.58	1987	0.00	0
COLLETON SC	0.00	0	21.05	236	17.78	1890	9.78	2767	3.16	2988	29.61	1448	0.00	0
DARLINGTON SC	0.00	0	24.01	189	15.83	2004	10.88	2640	5.58	2533	21.63	2085	0.00	0
DILLON SC	0.00	0	19.00	293	23.46	1593	10.21	2721	4.07	2862	19.56	2298	0.00	0
DORCHESTER SC	0.00	0	20.44	250	23.07	1612	9.74	2776	4.11	2857	24.47	1805	0.00	0
EDGEFIELD SC	0.00	0	13.85	460	7.13	2842	11.60	2542	6.57	2342	22.48	1999	0.00	0
FAIRFIELD SC	0.00	0	14.23	448	8.18	2748	12.29	2484	6.64	2326	21.82	2069	0.00	0
FLORENCE SC	0.00	0	24.80	180	27.71	1468	10.55	2682	5.21	2615	41.54	1002	0.00	0

TOTAL FALLOUT ACTIVITY (µCi/sq meter) by SHOT SERIES WITH COUNTY RANK

COUNTY AND STATE	RANGER SERIES	RANGER RANK	B-J SERIES	B-J RANK	T-S SERIES	T-S RANK	U-K SERIES	UK RANK	TEAPOT SERIES	TP RANK	PLUMBBOB SERIES	PB RANK	SHOT SEDAN	SEDAN RANK
	1951		1951		1952		1953		1955		1957		1962	
GEORGETOWN SC	0.00	0	17.72	335	13.49	2211	10.23	2716	3.47	2952	23.82	1865	0.00	0
GREENVILLE SC	0.00	0	11.88	544	11.39	2431	15.32	2197	16.07	1333	31.47	1335	0.00	0
GREENWOOD SC	0.00	0	13.05	485	8.15	2750	13.36	2390	7.37	2206	24.71	1788	0.00	0
HAMPTON SC	0.00	0	18.62	305	9.30	2632	10.11	2737	4.16	2847	27.47	1564	0.00	0
HORRY SC	0.00	0	25.60	168	20.66	1716	11.69	2538	4.39	2796	23.17	1921	0.00	0
JASPER SC	0.00	0	17.36	343	18.92	1811	9.30	2827	3.45	2954	36.80	1100	0.00	0
KERSHAW SC	0.00	0	15.59	398	11.84	2375	11.66	2540	4.48	2776	20.59	2193	7.67	690
LANCASTER SC	0.00	0	20.09	259	7.89	2768	11.28	2582	5.36	2579	20.61	2190	0.00	0
LAURENS SC	0.27	537	9.61	679	6.31	2918	13.83	2346	12.44	1669	27.34	1570	0.00	0
LEE SC	0.00	0	21.42	226	32.72	1321	11.80	2532	5.27	2597	20.48	2202	0.00	0
LEXINGTON SC	0.29	505	14.64	437	5.05	3002	12.44	2471	6.11	2435	21.44	2094	0.00	0
MARION SC	0.00	0	26.79	144	23.48	1591	10.65	2664	5.01	2657	22.14	2037	0.00	0
MARLBORO SC	0.00	0	18.21	321	23.15	1607	9.46	2802	5.17	2623	23.20	1919	0.00	0
MCCORMICK SC	0.00	0	15.37	409	6.97	2856	13.35	2393	5.52	2548	26.63	1622	0.00	0
NEWBERRY SC	0.00	0	12.61	505	7.77	2781	12.65	2453	7.62	2169	24.52	1798	0.00	0
OCONEE SC	0.00	0	10.61	619	12.05	2354	18.71	1931	14.94	1442	48.18	855	0.00	0
ORANGEBURG SC	0.00	0	15.48	404	27.34	1477	11.29	2580	4.74	2729	25.89	1688	0.00	0
PICKENS SC	0.00	0	15.35	411	12.38	2323	18.72	1930	15.62	1377	30.25	1411	0.00	0
RICHLAND SC	0.00	0	9.88	663	7.64	2789	12.03	2507	5.00	2660	19.94	2264	0.00	0
SALUDA SC	0.00	0	14.03	453	4.26	3032	12.85	2438	7.33	2217	23.32	1902	0.00	0
SPARTANBURG SC	0.00	0	13.31	476	10.02	2574	14.29	2306	15.35	1399	26.10	1671	0.00	0
SUMTER SC	0.00	0	15.32	413	26.62	1493	11.25	2585	5.05	2647	26.79	1611	0.00	0
UNION SC	0.00	0	15.29	414	7.67	2786	13.78	2352	8.05	2119	21.19	2121	0.00	0
WILLIAMSBURG SC	0.00	0	20.63	245	19.97	1748	10.19	2725	4.64	2752	21.76	2075	0.00	0
YORK SC	0.00	0	16.43	370	11.32	2436	11.88	2524	8.84	2023	21.41	2099	0.00	0
AURORA SD	0.00	0	1.81	1714	74.51	565	28.86	1341	10.23	1894	156.71	70	0.27	744

TOTAL FALLOUT ACTIVITY (µCi/sq meter) by SHOT SERIES WITH COUNTY RANK

COUNTY AND STATE	RANGER SERIES	RANGER RANK	B-J SERIES	B-J RANK	T-S SERIES	T-S RANK	U-K SERIES	U-K RANK	TEAPOT SERIES	TP RANK	PLUMBBOB SERIES	PB RANK	SHOT SEDAN	SEDAN RANK
	1951		1951		1952		1953		1955		1957		1962	
BEADLE SD	0.00	0	1.63	1780	57.33	815	18.32	1958	6.66	2324	139.61	102	0.00	0
BENNETT SD	0.00	0	0.11	2977	57.65	807	69.47	398	6.70	2317	120.82	175	1.60	720
BON HOMME SD	0.00	0	3.25	1348	40.07	1124	34.69	1092	4.99	2664	66.53	548	0.27	751
BROOKINGS SD	0.00	0	1.48	1848	46.57	996	28.99	1337	6.94	2279	133.24	120	181.88	235
BROWN SD	0.00	0	0.52	2474	36.82	1204	22.34	1676	6.14	2433	124.38	163	0.00	0
BRULE SD	0.00	0	1.59	1800	87.56	436	29.13	1329	7.80	2141	149.59	81	0.67	731
BUFFALO SD	0.00	0	1.65	1769	73.03	586	28.19	1378	5.99	2455	122.08	169	0.27	745
BUTTE SD	0.00	0	0.38	2731	58.01	798	69.86	391	5.55	2540	95.62	274	0.00	0
CAMPBELL SD	0.00	0	0.45	2579	38.63	1157	26.05	1472	4.36	2803	151.88	77	0.00	0
CHARLES MIX SD	0.00	0	2.59	1467	89.35	423	35.90	1038	7.52	2181	92.53	299	0.27	747
CLARK SD	0.00	0	1.81	1703	24.95	1538	23.23	1627	6.10	2437	128.28	155	0.00	0
CLAY SD	0.00	0	1.64	1777	43.40	1048	44.60	756	4.77	2722	52.50	765	208.70	192
CODINGTON SD	0.00	0	1.48	1849	18.04	1867	23.75	1592	5.51	2549	144.25	90	0.00	0
CORSON SD	0.00	0	0.72	2267	45.29	1015	74.25	346	4.42	2787	131.65	124	0.00	0
CUSTER SD	0.00	0	0.69	2292	226.14	242	64.74	456	8.16	2103	60.29	640	434.17	54
DAVISON SD	0.00	0	1.59	1804	47.98	967	27.64	1407	14.15	1521	98.76	254	208.97	145
DAY SD	0.00	0	1.65	1770	15.35	2045	20.74	1794	7.98	2125	145.52	87	0.00	0
DEUEL SD	0.00	0	0.18	2927	22.96	1613	29.01	1335	6.15	2432	18.52	2393	0.00	0
DEWEY SD	0.00	0	0.79	2223	52.31	887	75.94	327	4.48	2779	148.28	84	0.00	0
DOUGLAS SD	0.00	0	1.84	1684	101.79	353	33.52	1133	7.06	2265	79.16	397	0.00	0
EDMUNDS SD	0.00	0	0.45	2580	39.05	1147	27.67	1404	5.27	2596	129.63	142	0.00	0
FALL RIVER SD	0.00	0	0.19	2896	110.64	338	63.75	462	9.29	1989	53.70	743	345.60	73
FAULK SD	0.00	0	0.45	2581	58.79	787	24.60	1543	4.66	2748	171.26	58	0.00	0
GRANT SD	0.00	0	0.18	2928	21.94	1665	21.90	1707	3.80	2913	20.26	2224	0.00	0
GREGORY SD	0.00	0	1.90	1669	92.24	397	40.65	868	4.17	2845	88.03	333	0.67	738
HAAKON SD	0.00	0	0.19	2897	91.53	403	79.56	299	4.87	2698	140.54	100	1.20	727

TOTAL FALLOUT ACTIVITY (µCi/sq meter) by SHOT SERIES WITH COUNTY RANK

COUNTY AND STATE	RANGER SERIES	RANGER RANK	B-J SERIES	B-J RANK	T-S SERIES	T-S RANK	U-K SERIES	UK RANK	TEAPOT SERIES	TP RANK	PLUMBBOB SERIES	PB RANK	SHOT SEDAN	SEDAN RANK
	1951		1951		1952		1953		1955		1957		1962	
HAMLIN SD	0.00	0	1.48	1850	30.32	1402	26.50	1452	6.77	2307	122.49	167	451.85	52
HAND SD	0.00	0	0.52	2475	67.44	662	24.10	1579	7.73	2151	113.54	197	0.00	0
HANSON SD	0.00	0	1.75	1735	51.35	901	26.91	1434	11.01	1819	93.12	290	220.22	142
HARDING SD	0.00	0	0.46	2569	87.01	444	73.31	357	4.35	2807	141.63	96	0.00	0
HUGHES SD	0.00	0	1.34	1918	78.73	519	40.64	869	4.40	2794	60.51	637	0.27	752
HUTCHINSON SD	0.00	0	2.28	1536	80.85	496	33.33	1145	8.26	2088	92.02	306	346.94	70
HYDE SD	0.00	0	0.45	2582	75.18	555	28.66	1353	4.97	2668	131.62	125	0.00	0
JACKSON SD	0.00	0	0.11	2978	72.57	593	77.48	316	5.50	2551	130.05	136	1.60	719
JERAULD SD	0.00	0	1.96	1631	71.56	604	29.05	1334	11.36	1780	109.25	207	145.87	296
JONES SD	0.00	0	0.27	2835	85.69	458	31.90	1208	4.60	2759	117.05	184	1.47	723
KINGSBURY SD	0.00	0	1.82	1699	34.55	1272	26.61	1444	8.04	2121	148.25	85	361.42	56
LAKE SD	0.00	0	1.75	1736	56.87	824	30.71	1254	7.91	2131	118.37	182	327.38	104
LAWRENCE SD	0.00	0	1.71	1751	61.36	750	69.85	392	7.74	2149	75.14	430	2.00	713
LINCOLN SD	0.00	0	2.80	1426	70.38	616	39.32	923	6.86	2293	50.52	804	208.70	195
LYMAN SD	0.00	0	1.40	1880	100.71	360	31.63	1217	6.48	2362	150.84	79	0.67	730
MARSHALL SD	0.00	0	0.52	2476	11.78	2383	15.71	2159	3.44	2955	151.35	78	0.00	0
MCCOOK SD	0.00	0	1.98	1622	55.46	850	33.55	1131	8.29	2085	100.65	245	346.94	69
MCPHERSON SD	0.00	0	0.45	2583	37.22	1196	23.48	1611	5.26	2602	124.31	164	0.00	0
MEADE SD	0.00	0	0.36	2751	81.18	492	71.47	377	5.00	2659	94.06	283	0.00	0
MELLETTE SD	0.00	0	0.27	2836	58.07	797	78.36	308	5.44	2560	140.06	101	1.47	721
MINER SD	0.00	0	2.08	1587	39.24	1139	27.42	1414	8.20	2098	128.43	153	309.57	117
MINNEHAHA SD	0.00	0	1.65	1771	56.09	839	34.68	1093	6.55	2349	108.18	212	138.24	328
MOODY SD	0.00	0	1.56	1818	48.04	964	30.76	1251	7.18	2246	21.25	2110	361.42	57
PENNINGTON SD	0.00	0	0.69	2293	82.67	478	38.83	945	4.65	2749	82.67	371	2.00	710
PERKINS SD	0.00	0	0.38	2732	36.39	1217	77.89	312	4.12	2856	107.57	216	0.00	0
POTTER SD	0.00	0	0.45	2584	56.14	838	30.19	1274	4.88	2695	156.29	71	0.00	0

TOTAL FALLOUT ACTIVITY (µCi/sq meter) by SHOT SERIES WITH COUNTY RANK

COUNTY AND STATE	RANGER SERIES	RANGER RANK	B-J SERIES	B-J RANK	T-S SERIES	T-S RANK	U-K SERIES	UK RANK	TEAPOT SERIES	TP RANK	PLUMBBOB SERIES	PB RANK	SHOT SEDAN	SEDAN RANK
	1951		1951		1952		1953		1955		1957		1962	
ROBERTS SD	0.00	0	0.35	2763	11.24	2449	17.99	1980	4.73	2731	183.87	49	0.00	0
SANBORN SD	0.00	0	1.59	1805	46.92	985	24.43	1554	9.06	2007	131.53	128	0.00	0
SHANNON SD	0.00	0	0.19	2898	87.46	438	67.31	427	5.57	2537	57.88	672	345.60	72
SPINK SD	0.00	0	1.48	1851	58.49	791	27.25	1418	6.32	2398	145.16	89	0.00	0
STANLEY SD	0.00	0	0.45	2585	74.81	559	76.45	325	4.31	2814	145.26	88	345.60	71
SULLY SD	0.00	0	0.45	2586	75.25	554	42.50	812	4.28	2819	134.30	117	0.00	0
TODD SD	0.00	0	0.11	2979	53.04	881	75.22	336	4.81	2710	118.41	181	1.47	722
TRIPP SD	0.00	0	2.58	1470	81.61	488	37.27	1001	4.27	2824	108.46	209	0.67	733
TURNER SD	0.00	0	2.70	1451	82.37	481	33.58	1130	5.96	2463	73.05	449	527.44	26
UNION SD	0.00	0	2.80	1427	73.76	576	40.55	876	6.71	2316	51.28	787	208.70	194
WALWORTH SD	0.00	0	0.45	2587	43.35	1049	29.44	1312	4.79	2717	159.46	67	0.00	0
WASHABAUGH SD	0.00	0	0.11	2980	68.88	638	74.16	347	5.93	2473	90.43	315	860.52	4
YANKTON SD	0.00	0	1.44	1869	36.10	1225	39.55	912	4.98	2667	75.32	427	252.68	126
ZIEBACH SD	0.00	0	0.36	2752	40.51	1111	77.65	313	4.07	2864	142.94	93	0.00	0
ANDERSON TN	0.00	0	16.54	363	13.02	2259	28.03	1383	22.11	854	26.93	1602	0.00	0
BEDFORD TN	0.00	0	1.34	1922	11.22	2452	18.41	1954	18.02	1153	32.10	1300	0.00	0
BENTON TN	0.00	0	0.73	2263	16.65	1957	25.10	1515	12.18	1698	15.91	2628	0.00	0
BLEDSOE TN	0.00	0	10.41	628	12.90	2273	21.03	1766	21.31	903	30.55	1390	0.00	0
BLOUNT TN	0.00	0	11.93	540	9.50	2612	24.71	1533	20.57	947	32.09	1302	0.00	0
BRADLEY TN	0.00	0	11.59	558	13.16	2242	14.92	2234	19.92	988	32.03	1305	0.00	0
CAMPBELL TN	0.00	0	3.06	1383	10.44	2534	24.91	1525	22.93	807	21.11	2131	43.75	604
CANNON TN	0.00	0	1.23	1968	7.91	2764	20.51	1811	15.29	1410	21.96	2057	0.00	0
CARROLL TN	0.00	0	1.42	1879	20.32	1731	26.80	1438	21.91	868	21.63	2086	0.00	0
CARTER TN	0.00	0	12.42	512	6.00	2947	17.45	2027	17.77	1184	23.88	1861	43.75	601
CHEATHAM TN	0.00	0	0.67	2318	11.79	2380	23.37	1615	19.29	1036	32.74	1275	0.00	0
CHESTER TN	0.00	0	2.98	1394	16.47	1969	54.81	573	10.86	1834	16.53	2585	0.00	0

TOTAL FALLOUT ACTIVITY (µCi/sq meter) by SHOT SERIES WITH COUNTY RANK

COUNTY AND STATE	RANGER SERIES	RANGER RANK	B-J SERIES	B-J RANK	T-S SERIES	T-S RANK	U-K SERIES	UK RANK	TEAPOT SERIES	TP RANK	PLUMBBOB SERIES	PB RANK	SHOT SEDAN	SEDAN RANK
	1951		1951		1952		1953		1955		1957		1962	
CLAIBORNE TN	0.00	0	3.85	1238	10.08	2570	25.50	1496	18.33	1114	26.01	1680	43.75	597
CLAY TN	0.00	0	1.71	1750	5.43	2981	13.20	2401	19.70	1007	29.92	1429	43.75	591
COCKE TN	0.00	0	13.15	480	9.12	2652	28.67	1351	19.06	1047	30.28	1406	0.00	0
COFFEE TN	0.00	0	9.70	675	10.21	2554	16.26	2113	18.12	1142	41.18	1006	0.00	0
CROCKETT TN	0.00	0	3.28	1344	20.06	1739	76.86	322	15.78	1366	17.40	2506	0.00	0
CUMBERLAND TN	0.00	0	10.62	617	10.72	2499	18.56	1944	19.64	1009	33.81	1215	0.00	0
DAVIDSON TN	0.00	0	0.48	2553	11.36	2433	25.02	1520	13.33	1585	34.03	1209	0.00	0
DE KALB TN	0.00	0	1.20	1991	8.12	2754	21.12	1753	19.26	1037	27.12	1588	0.00	0
DECATUR TN	0.00	0	0.97	2114	18.79	1820	25.85	1483	16.56	1296	22.00	2052	0.00	0
DICKSON TN	0.00	0	0.52	2477	14.47	2114	28.26	1372	25.75	665	26.12	1667	0.00	0
DYER TN	0.00	0	3.20	1366	15.28	2049	87.18	264	14.62	1472	22.98	1941	0.00	0
FAYETTE TN	0.00	0	1.42	1878	16.99	1941	73.19	359	13.14	1598	24.46	1807	0.00	0
FENTRESS TN	0.00	0	8.44	748	15.20	2055	16.83	2074	20.05	978	30.87	1376	0.00	0
FRANKLIN TN	0.00	0	7.68	792	10.42	2536	14.94	2230	14.49	1485	18.50	2399	0.00	0
GIBSON TN	0.00	0	4.03	1207	18.20	1854	80.34	291	30.92	539	25.29	1739	0.00	0
GILES TN	0.00	0	1.05	2057	8.30	2738	19.88	1846	15.61	1380	23.29	1908	0.00	0
GRAINGER TN	0.00	0	3.66	1275	7.32	2825	23.65	1599	18.98	1051	20.57	2195	0.00	0
GREENE TN	0.00	0	4.61	1080	6.65	2887	30.78	1250	18.81	1069	30.06	1420	0.00	0
GRUNDY TN	0.00	0	11.36	573	11.41	2429	13.78	2350	20.64	940	37.91	1068	0.00	0
HAMBLEN TN	0.00	0	3.77	1259	7.47	2808	24.64	1539	18.76	1074	27.11	1590	0.00	0
HAMILTON TN	0.00	0	2.59	1466	11.18	2458	15.68	2163	19.12	1043	36.25	1116	0.00	0
HANCOCK TN	0.00	0	4.08	1201	8.77	2681	21.58	1726	18.00	1157	29.06	1478	43.75	592
HARDEMAN TN	0.00	0	3.13	1371	12.70	2288	57.68	527	14.40	1498	31.29	1346	0.00	0
HARDIN TN	0.00	0	1.17	2007	10.93	2482	29.45	1309	15.04	1434	23.82	1864	0.00	0
HAWKINS TN	0.00	0	4.17	1177	9.25	2639	24.02	1582	17.59	1196	22.01	2051	17.11	674
HAYWOOD TN	0.00	0	2.77	1433	20.11	1737	79.10	303	16.63	1289	22.47	2003	0.00	0

TOTAL FALLOUT ACTIVITY (µCi/sq meter) by SHOT SERIES WITH COUNTY RANK

COUNTY AND STATE	RANGER SERIES	RANGER RANK	B-J SERIES	B-J RANK	T-S SERIES	T-S RANK	U-K SERIES	UK RANK	TEAPOT SERIES	TP RANK	PLUMBBOB SERIES	PB RANK	SHOT SEDAN	SEDAN RANK
	1951		1951	1952		1953		1955		1957		1962		
HENDERSON TN	0.00	0	1.20	1992	15.10	2061	26.23	1463	18.58	1090	21.82	2068	0.00	0
HENRY TN	0.00	0	1.08	2046	28.28	1455	24.53	1549	20.72	936	19.21	2332	0.00	0
HICKMAN TN	0.00	0	1.26	1961	18.83	1818	24.10	1580	15.65	1372	21.20	2116	0.00	0
HOUSTON TN	0.00	0	0.68	2311	18.58	1828	23.80	1589	11.61	1755	19.10	2342	0.00	0
HUMPHREYS TN	0.00	0	0.94	2124	16.76	1952	25.63	1492	20.82	928	23.35	1898	0.00	0
JACKSON TN	0.00	0	1.37	1904	5.62	2972	13.53	2381	20.12	973	31.00	1363	0.00	0
JEFFERSON TN	0.00	0	10.45	624	11.61	2406	21.94	1702	17.87	1173	23.91	1854	0.00	0
JOHNSON TN	0.00	0	7.33	822	6.28	2920	16.86	2072	17.37	1221	20.24	2228	17.11	676
KNOX TN	0.00	0	13.17	479	8.90	2673	28.79	1343	15.23	1415	15.56	2656	0.00	0
LAKE TN	0.00	0	4.02	1209	23.34	1600	65.86	443	34.62	477	25.19	1749	0.00	0
LAUDERDALE TN	0.00	0	3.27	1345	22.84	1619	73.26	358	25.55	671	21.15	2125	0.00	0
LAWRENCE TN	0.00	0	1.21	1985	14.17	2144	21.92	1706	17.40	1216	19.83	2274	0.00	0
LEWIS TN	0.00	0	0.96	2118	5.98	2949	22.96	1645	12.94	1622	17.97	2445	0.00	0
LINCOLN TN	0.00	0	1.28	1954	15.52	2032	17.79	1998	15.19	1420	18.49	2401	0.00	0
LOUDON TN	0.00	0	12.78	497	8.65	2696	21.93	1704	18.39	1108	26.03	1678	0.00	0
MACON TN	0.00	0	0.94	2130	9.26	2637	15.50	2182	11.31	1788	20.01	2251	0.00	0
MADISON TN	0.00	0	3.48	1311	15.71	2018	83.23	276	19.20	1040	22.18	2034	0.00	0
MARION TN	0.00	0	10.81	604	6.95	2858	12.54	2462	12.96	1618	22.59	1985	0.00	0
MARSHALL TN	0.00	0	1.19	1998	10.74	2497	19.76	1853	19.56	1015	31.61	1328	0.00	0
MAURY TN	0.00	0	0.97	2112	13.53	2203	24.16	1573	12.88	1628	28.40	1513	0.00	0
MCMINN TN	0.00	0	11.84	547	8.08	2757	21.36	1740	21.73	874	32.59	1277	0.00	0
MCNAIRY TN	0.00	0	3.98	1217	14.19	2142	55.22	565	12.32	1680	24.46	1809	0.00	0
MEIGS TN	0.00	0	11.30	576	9.71	2592	16.39	2106	21.53	889	22.29	2025	0.00	0
MONROE TN	0.00	0	10.02	657	15.07	2067	23.02	1640	21.30	906	26.46	1635	0.00	0
MONTGOMERY TN	0.00	0	0.86	2178	16.28	1981	21.89	1708	22.39	835	35.18	1148	0.00	0
MOORE TN	0.00	0	1.20	1996	13.43	2218	16.79	2077	11.59	1758	22.09	2041	0.00	0

TOTAL FALLOUT ACTIVITY (µCi/sq meter) by SHOT SERIES WITH COUNTY RANK

COUNTY AND STATE	RANGER SERIES	RANGER RANK	B-J SERIES	B-J RANK	T-S SERIES	T-S RANK	U-K SERIES	UK RANK	TEAPOT SERIES	TP RANK	PLUMBBOB SERIES	PB RANK	SHOT SEDAN	SEDAN RANK
	1951		1951		1952		1953		1955		1957		1962	
MORGAN TN	0.00	0	10.80	605	10.48	2530	23.24	1626	20.09	974	25.15	1758	0.00	0
OBION TN	0.00	0	3.82	1244	21.08	1709	65.99	441	33.75	490	23.36	1895	0.00	0
OVERTON TN	0.00	0	11.14	587	6.26	2924	15.07	2217	20.55	950	25.64	1711	0.00	0
PERRY TN	0.00	0	0.97	2115	12.43	2317	25.92	1479	16.88	1269	21.35	2102	0.00	0
PICKETT TN	0.00	0	12.27	518	11.48	2421	14.79	2247	23.27	789	29.33	1464	17.11	669
POLK TN	0.00	0	12.09	530	11.53	2414	18.12	1971	17.58	1199	30.81	1380	0.00	0
PUTNAM TN	0.00	0	10.82	603	6.26	2925	14.44	2293	19.75	1000	34.48	1178	0.00	0
RHEA TN	0.00	0	10.68	614	10.15	2565	17.57	2017	18.26	1122	34.06	1205	0.00	0
ROANE TN	0.00	0	11.74	551	13.17	2241	24.66	1538	20.44	957	30.25	1410	0.00	0
ROBERTSON TN	0.00	0	0.73	2260	8.37	2731	19.82	1852	14.15	1520	34.53	1174	0.00	0
RUTHERFORD TN	0.00	0	1.02	2069	8.60	2703	18.48	1949	9.88	1932	21.11	2130	0.00	0
SCOTT TN	0.00	0	4.69	1068	7.82	2774	18.99	1909	18.73	1078	16.35	2595	43.75	607
SEQUATCHIE TN	0.00	0	14.50	442	6.61	2893	14.10	2323	19.72	1003	36.07	1124	0.00	0
SEVIER TN	0.00	0	12.44	509	15.41	2040	29.37	1318	19.99	983	35.53	1138	0.00	0
SHELBY TN	0.00	0	4.59	1084	7.03	2851	70.46	384	12.90	1627	19.16	2337	0.00	0
SMITH TN	0.00	0	1.17	2004	8.97	2663	14.62	2270	12.42	1673	32.94	1269	0.00	0
STEWART TN	0.00	0	0.68	2312	18.71	1826	23.33	1617	12.30	1682	20.95	2147	0.00	0
SULLIVAN TN	0.00	0	5.25	981	7.42	2812	41.66	841	18.56	1094	25.82	1696	17.11	672
SUMNER TN	0.00	0	0.57	2413	13.72	2185	17.55	2020	9.81	1943	34.68	1166	0.00	0
TIPTON TN	0.00	0	3.21	1362	22.73	1621	75.85	330	24.21	745	22.35	2019	0.00	0
TROUSDALE TN	0.00	0	1.39	1889	9.04	2657	16.10	2128	9.98	1919	19.23	2328	0.00	0
UNICOI TN	0.00	0	10.76	608	6.74	2874	21.26	1747	22.52	825	25.78	1700	0.00	0
UNION TN	0.00	0	3.54	1295	12.67	2291	21.23	1749	18.25	1125	20.94	2150	0.00	0
VAN BUREN TN	0.00	0	10.43	626	7.96	2763	13.93	2339	20.85	924	32.54	1283	0.00	0
WARREN TN	0.00	0	8.34	756	7.38	2816	15.30	2200	17.07	1248	36.98	1094	0.00	0
WASHINGTON TN	0.00	0	5.13	998	6.79	2868	24.40	1558	17.96	1165	26.60	1626	77.90	491

TOTAL FALLOUT ACTIVITY (µCi/sq meter) by SHOT SERIES WITH COUNTY RANK

COUNTY AND STATE	RANGER SERIES	RANGER RANK	B-J SERIES	B-J RANK	T-S SERIES	T-S RANK	U-K SERIES	UK RANK	TEAPOT SERIES	TP RANK	PLUMBBOB SERIES	PB RANK	SHOT SEDAN	SEDAN RANK
	1951		1951		1952		1953		1955		1957		1962	
WAYNE TN	0.00	0	1.20	1993	8.62	2699	22.90	1648	13.04	1611	24.92	1773	0.00	0
WEAKLEY TN	0.00	0	4.43	1113	25.37	1521	72.29	368	24.46	728	24.02	1839	0.00	0
WHITE TN	0.00	0	7.56	802	7.33	2824	14.40	2296	18.32	1116	31.80	1315	0.00	0
WILLIAMSON TN	0.00	0	0.94	2126	5.85	2958	23.94	1584	15.17	1424	27.49	1562	0.00	0
WILSON TN	0.00	0	1.05	2059	9.51	2611	17.73	2006	10.12	1901	31.47	1336	0.00	0
ANDERSON TX	0.00	0	3.76	1260	11.47	2423	26.28	1460	26.51	642	20.51	2199	0.00	0
ANDREWS TX	0.00	0	1.35	1916	89.07	428	92.71	235	30.86	540	31.33	1342	0.00	0
ANGELINA TX	0.00	0	1.33	1925	7.23	2831	113.80	174	15.49	1386	19.98	2257	0.00	0
ARANSAS TX	0.00	0	4.88	1032	26.72	1490	11.21	2594	3.98	2880	11.57	2883	0.00	0
ARCHER TX	0.00	0	1.03	2068	536.38	163	31.54	1218	20.27	963	20.00	2254	0.00	0
ARMSTRONG TX	0.00	0	0.84	2186	126.65	320	81.26	286	92.98	169	22.10	2039	0.00	0
ATASCOSA TX	0.00	0	2.33	1526	497.11	171	10.28	2708	2.59	3042	7.82	2974	0.00	0
AUSTIN TX	0.00	0	2.25	1544	16.07	1995	90.44	248	25.63	669	8.97	2957	0.00	0
BAILEY TX	0.00	0	0.72	2270	112.51	335	144.89	143	85.34	188	44.20	940	0.00	0
BANDERA TX	0.00	0	1.71	1754	1305.02	105	5.89	2991	11.67	1742	13.26	2806	0.00	0
BASTROP TX	0.00	0	3.77	1257	20.04	1741	13.44	2387	3.83	2908	13.16	2810	0.00	0
BAYLOR TX	0.00	0	0.53	2465	536.80	162	32.12	1199	17.60	1194	14.72	2718	0.00	0
BEE TX	0.00	0	4.48	1101	26.58	1496	10.51	2685	2.22	3060	15.08	2690	0.00	0
BELL TX	0.00	0	0.39	2722	19.74	1755	17.85	1991	21.93	865	19.44	2317	0.00	0
BEXAR TX	0.00	0	2.12	1579	793.98	145	10.59	2678	3.86	2902	12.64	2836	0.00	0
BLANCO TX	0.00	0	0.48	2557	801.55	136	14.42	2294	10.12	1900	16.06	2615	0.00	0
BORDEN TX	0.00	0	0.40	2707	61.02	755	52.94	606	14.83	1449	17.21	2524	0.00	0
BOSQUE TX	0.00	0	0.39	2723	20.48	1726	28.43	1363	22.44	830	20.30	2221	0.00	0
BOWIE TX	0.00	0	4.78	1047	38.34	1166	23.14	1630	50.98	310	29.93	1427	0.00	0
BRAZORIA TX	0.00	0	4.45	1106	15.10	2062	90.19	250	21.69	877	18.42	2405	0.00	0
BRAZOS TX	0.00	0	1.93	1640	4.10	3034	20.67	1799	21.17	911	17.82	2467	0.00	0

TOTAL FALLOUT ACTIVITY (µCi/sq meter) by SHOT SERIES WITH COUNTY RANK

COUNTY AND STATE	RANGER SERIES	RANGER RANK	B-J SERIES	B-J RANK	T-S SERIES	T-S RANK	U-K SERIES	UK RANK	TEAPOT SERIES	TP RANK	PLUMBBOB SERIES	PB RANK	SHOT SEDAN	SEDAN RANK
	1951		1951		1952		1953		1955		1957		1962	
BREWSTER TX	0.00	0	1.93	1653	65.79	684	11.05	2610	11.32	1783	13.23	2809	0.00	0
BRISCOE TX	0.00	0	0.43	2678	496.08	172	98.76	207	110.69	155	14.11	2757	0.00	0
BROOKS TX	0.00	0	2.41	1503	27.93	1463	10.81	2648	1.63	3084	15.21	2683	0.00	0
BROWN TX	0.00	0	0.27	2837	488.75	182	20.78	1789	15.04	1433	24.13	1830	0.00	0
BURLESON TX	0.00	0	4.33	1140	7.18	2836	15.84	2150	20.81	930	15.81	2636	0.00	0
BURNET TX	0.00	0	0.37	2744	12.61	2296	20.78	1790	10.78	1839	16.55	2583	0.00	0
CALDWELL TX	0.00	0	2.02	1616	20.29	1732	12.85	2439	3.60	2937	15.96	2626	0.00	0
CALHOUN TX	0.00	0	4.43	1114	23.07	1611	11.32	2577	4.43	2785	14.24	2752	0.00	0
CALLAHAN TX	0.00	0	0.18	2938	790.72	147	35.52	1053	12.19	1693	20.26	2225	0.00	0
CAMERON TX	0.00	0	2.75	1435	40.50	1112	10.86	2641	1.22	3088	15.69	2643	0.00	0
CAMP TX	0.00	0	4.92	1019	68.31	646	27.49	1412	34.99	471	24.40	1815	0.00	0
CARSON TX	0.00	0	0.49	2547	143.93	297	79.78	298	93.62	168	22.32	2021	0.00	0
CASS TX	0.00	0	3.04	1385	60.35	767	21.03	1765	39.61	414	25.09	1763	0.00	0
CASTRO TX	0.00	0	0.40	2703	117.68	328	128.17	164	90.64	176	37.16	1092	0.00	0
CHAMBERS TX	0.00	0	5.82	935	7.51	2802	92.30	237	20.98	919	22.89	1949	0.00	0
CHEROKEE TX	0.00	0	4.55	1091	13.72	2184	33.08	1153	29.05	579	27.70	1549	0.00	0
CHILDRESS TX	0.00	0	0.32	2793	66.94	668	74.14	348	78.64	210	14.85	2706	0.00	0
CLAY TX	0.00	0	2.81	1425	63.73	713	34.75	1090	39.07	421	26.46	1637	0.00	0
COCHRAN TX	0.00	0	0.40	2704	63.47	720	157.92	133	40.21	411	37.91	1069	0.00	0
COKE TX	0.00	0	0.26	2844	248.72	234	18.96	1911	11.24	1796	15.81	2637	0.00	0
COLEMAN TX	0.00	0	0.26	2845	486.78	186	19.48	1877	13.19	1594	18.08	2435	0.00	0
COLLIN TX	0.00	0	4.60	1082	57.30	816	75.50	333	36.92	447	30.54	1391	0.00	0
COLLINGSWORT TX	0.00	0	0.32	2794	67.39	663	79.05	304	78.27	212	19.94	2265	0.00	0
COLORADO TX	0.00	0	4.33	1141	29.34	1431	12.98	2424	5.64	2523	20.48	2203	0.00	0
COMAL TX	0.00	0	2.02	1617	794.52	144	10.66	2663	3.81	2911	16.05	2617	0.00	0
COMANCHE TX	0.00	0	0.29	2827	63.04	724	24.69	1534	16.53	1299	18.95	2355	0.00	0

U.S. FALLOUT ATLAS : TOTAL FALLOUT

TOTAL FALLOUT ACTIVITY (µCi/sq meter) by SHOT SERIES WITH COUNTY RANK

COUNTY AND STATE	RANGER SERIES	RANGER RANK	B-J SERIES	B-J RANK	T-S SERIES	T-S RANK	U-K SERIES	UK RANK	TEAPOT SERIES	TP RANK	PLUMBBOB SERIES	PB RANK	SHOT SEDAN	SEDAN RANK
	1951		1951		1952		1953		1955		1957		1962	
CONCHO TX	0.00	0	0.26	2846	490.81	178	17.34	2038	16.43	1306	21.77	2074	0.00	0
COOKE TX	0.00	0	4.67	1069	36.35	1218	66.04	437	58.17	265	30.97	1365	0.00	0
CORYELL TX	0.00	0	0.29	2828	13.47	2215	18.09	1975	22.04	860	17.65	2487	0.00	0
COTTLE TX	0.00	0	0.43	2679	101.33	356	45.74	724	25.87	661	11.44	2887	0.00	0
CRANE TX	0.00	0	1.61	1793	73.83	574	21.42	1735	14.39	1499	9.05	2956	0.00	0
CROCKETT TX	0.00	0	1.61	1794	63.89	710	14.49	2286	12.81	1635	9.69	2938	0.00	0
CROSBY TX	0.00	0	0.36	2750	84.78	464	68.84	408	72.60	221	16.79	2557	0.00	0
CULBERSON TX	0.00	0	1.82	1700	97.08	374	87.40	262	30.59	547	24.77	1783	0.00	0
DALLAM TX	46.74	3	0.94	2129	60.22	773	146.59	140	173.99	68	43.76	951	0.00	0
DALLAS TX	0.00	0	2.28	1534	23.35	1599	84.95	267	37.11	442	25.28	1740	0.00	0
DAWSON TX	0.00	0	0.30	2800	82.37	482	68.90	406	12.78	1640	16.46	2590	0.00	0
DE WITT TX	0.00	0	4.31	1147	35.37	1243	10.98	2627	3.82	2910	14.62	2726	0.00	0
DEAF SMITH TX	0.00	0	0.30	2801	133.99	310	158.47	132	85.98	186	37.32	1087	0.00	0
DELTA TX	0.00	0	2.38	1510	43.91	1039	30.05	1280	33.28	500	30.52	1394	0.00	0
DENTON TX	0.00	0	3.99	1214	41.73	1082	72.73	367	40.86	402	29.50	1455	0.00	0
DICKENS TX	0.00	0	0.21	2875	266.33	221	53.46	593	20.53	952	18.56	2389	0.00	0
DIMMIT TX	0.00	0	1.31	1937	493.36	174	5.77	2996	10.94	1825	9.24	2950	0.00	0
DONLEY TX	0.00	0	0.32	2789	69.41	628	89.47	255	87.59	182	24.47	1801	0.00	0
DUVAL TX	0.00	0	1.83	1691	14.78	2091	11.60	2543	1.74	3080	11.21	2892	0.00	0
EASTLAND TX	0.00	0	0.17	2948	489.82	181	22.80	1651	16.86	1270	22.35	2018	0.00	0
ECTOR TX	0.00	0	0.30	2802	60.63	758	65.15	450	29.56	568	11.65	2878	0.00	0
EDWARDS TX	0.00	0	1.43	1873	57.64	808	5.20	3016	11.41	1775	10.80	2905	0.00	0
EL PASO TX	0.00	0	1.40	1881	169.13	269	83.33	274	30.41	550	22.63	1982	0.00	0
ELLIS TX	0.00	0	0.29	2822	22.34	1640	55.96	553	35.56	461	24.14	1829	0.00	0
ERATH TX	0.00	0	0.18	2929	10.18	2560	24.62	1540	24.47	727	20.07	2247	0.00	0
FALLS TX	0.00	0	0.18	2930	9.93	2579	21.04	1763	23.38	782	14.30	2746	0.00	0

TOTAL FALLOUT ACTIVITY (µCi/sq meter) by SHOT SERIES WITH COUNTY RANK

COUNTY AND STATE	RANGER SERIES	RANGER RANK	B-J SERIES	B-J RANK	T-S SERIES	T-S RANK	U-K SERIES	UK RANK	TEAPOT SERIES	TP RANK	PLUMBBOB SERIES	PB RANK	SHOT SEDAN	SEDAN RANK
	1951		1951		1952		1953		1955		1957		1962	
FANNIN TX	0.00	0	2.38	1511	57.72	805	88.53	256	43.93	373	45.48	914	0.00	0
FAYETTE TX	0.00	0	4.31	1148	21.22	1706	13.04	2417	6.43	2372	20.65	2180	0.00	0
FISHER TX	0.00	0	0.16	2954	505.18	168	28.80	1342	19.80	994	24.90	1775	0.00	0
FLOYD TX	0.00	0	0.21	2876	82.87	476	79.21	302	81.62	200	16.30	2600	0.00	0
FOARD TX	0.00	0	1.71	1752	96.56	377	39.07	927	31.16	530	15.15	2687	0.00	0
FORT BEND TX	0.00	0	4.22	1161	15.48	2035	89.96	251	15.97	1339	16.61	2574	0.00	0
FRANKLIN TX	0.00	0	1.40	1885	33.27	1305	28.78	1345	32.13	514	28.66	1499	0.00	0
FREESTONE TX	0.00	0	0.70	2284	10.53	2520	24.73	1532	22.49	828	22.69	1973	0.00	0
FRIO TX	0.00	0	1.29	1942	1304.74	106	5.38	3010	11.69	1739	11.03	2897	0.00	0
GAINES TX	0.00	0	0.30	2803	80.54	499	120.08	168	33.49	495	33.16	1255	0.00	0
GALVESTON TX	0.00	0	2.31	1528	8.06	2759	90.35	249	14.57	1479	16.13	2612	0.00	0
GARZA TX	0.00	0	0.19	2899	100.77	359	60.09	501	14.44	1495	19.47	2312	0.00	0
GILLESPIE TX	0.00	0	1.36	1906	793.75	146	7.31	2941	12.90	1626	17.75	2471	0.00	0
GLASSCOCK TX	0.00	0	0.30	2804	59.51	780	25.76	1484	7.84	2136	12.55	2840	0.00	0
GOLIAD TX	0.00	0	2.27	1538	24.87	1541	11.00	2623	3.35	2963	9.26	2949	0.00	0
GONZALES TX	0.00	0	6.22	900	21.52	1684	12.19	2491	3.79	2915	12.21	2851	0.00	0
GRAY TX	0.00	0	0.51	2518	74.21	568	59.00	513	90.76	175	17.44	2501	0.00	0
GRAYSON TX	0.00	0	5.76	938	38.37	1164	92.58	236	40.71	404	41.79	997	0.00	0
GREGG TX	0.00	0	4.56	1090	13.30	2224	24.33	1560	29.27	574	33.06	1263	0.00	0
GRIMES TX	0.00	0	6.14	907	6.45	2907	97.27	213	17.11	1244	16.93	2547	0.00	0
GUADALUPE TX	0.00	0	1.91	1660	11.31	2442	10.95	2629	2.84	3022	9.93	2929	0.00	0
HALE TX	0.00	0	0.46	2570	73.75	577	95.32	221	90.50	177	42.13	989	0.00	0
HALL TX	0.00	0	0.32	2790	853.85	130	69.77	394	80.55	205	16.22	2604	0.00	0
HAMILTON TX	0.00	0	0.18	2931	18.52	1832	21.45	1733	23.35	784	19.73	2281	0.00	0
HANSFORD TX	0.00	0	2.89	1410	75.89	549	69.38	400	170.27	70	29.77	1441	0.00	0
HARDEMAN TX	0.00	0	0.32	2791	89.16	425	49.65	650	37.06	444	17.70	2476	0.00	0

TOTAL FALLOUT ACTIVITY (µCi/sq meter) by SHOT SERIES WITH COUNTY RANK

COUNTY AND STATE	RANGER SERIES	RANGER RANK	B-J SERIES	B-J RANK	T-S SERIES	T-S RANK	U-K SERIES	UK RANK	TEAPOT SERIES	TP RANK	PLUMBBOB SERIES	PB RANK	SHOT SEDAN	SEDAN RANK
	1951		1951		1952		1953		1955		1957		1962	
HARDIN TX	0.00	0	2.71	1440	8.67	2694	93.50	230	18.50	1097	17.07	2537	0.00	0
HARRIS TX	0.00	0	4.22	1162	14.53	2113	92.88	233	18.45	1102	19.46	2313	0.00	0
HARRISON TX	0.00	0	2.92	1406	23.32	1601	22.74	1655	28.82	584	30.84	1378	0.00	0
HARTLEY TX	0.00	0	1.09	2045	63.56	719	153.86	137	129.21	95	37.57	1079	0.00	0
HASKELL TX	0.00	0	0.21	2877	516.68	166	28.40	1367	20.06	977	17.60	2490	0.00	0
HAYS TX	0.00	0	1.91	1661	18.07	1864	11.31	2578	4.12	2855	16.56	2582	0.00	0
HEMPHILL TX	0.00	0	1.84	1685	80.42	501	55.22	566	83.87	194	22.37	2015	0.00	0
HENDERSON TX	0.00	0	0.81	2213	9.42	2617	35.06	1073	31.68	518	22.42	2010	0.00	0
HIDALGO TX	0.00	0	3.43	1324	21.57	1681	11.70	2537	1.14	3092	9.77	2934	0.00	0
HILL TX	0.00	0	0.29	2823	15.87	2001	30.53	1263	23.60	770	23.98	1844	0.00	0
HOCKLEY TX	0.00	0	0.30	2805	69.01	636	114.42	172	82.79	196	36.44	1111	0.00	0
HOOD TX	0.00	0	0.29	2824	21.85	1669	35.05	1075	24.54	723	20.61	2189	0.00	0
HOPKINS TX	0.00	0	2.16	1570	58.44	792	31.44	1223	35.36	465	28.48	1510	0.00	0
HOUSTON TX	0.00	0	2.71	1441	7.38	2817	22.20	1687	25.03	695	22.05	2046	0.00	0
HOWARD TX	0.00	0	0.19	2900	63.00	725	37.71	988	11.45	1771	16.90	2549	0.00	0
HUDSPETH TX	0.00	0	1.72	1747	92.01	399	88.38	258	30.14	555	23.99	1842	0.00	0
HUNT TX	0.00	0	3.04	1387	39.89	1128	38.39	962	37.66	437	30.94	1371	0.00	0
HUTCHINSON TX	0.00	0	0.32	2792	72.84	590	71.70	374	174.45	67	28.93	1485	0.00	0
IRION TX	0.00	0	0.30	2806	492.26	177	18.41	1953	7.87	2134	13.98	2767	0.00	0
JACK TX	0.00	0	0.40	2717	31.98	1342	31.21	1235	27.09	623	27.80	1542	0.00	0
JACKSON TX	0.00	0	4.20	1170	33.93	1291	11.21	2593	5.15	2628	14.96	2699	0.00	0
JASPER TX	0.00	0	2.71	1442	9.64	2601	106.85	184	21.02	918	23.13	1929	0.00	0
JEFF DAVIS TX	0.00	0	1.72	1748	94.32	385	45.00	749	30.16	554	40.42	1022	0.00	0
JEFFERSON TX	0.00	0	2.71	1443	5.00	3008	45.33	736	10.59	1859	22.37	2014	0.00	0
JIM HOGG TX	0.00	0	1.83	1692	8.49	2719	11.24	2587	1.15	3091	15.15	2688	0.00	0
JIM WELLS TX	0.00	0	2.29	1531	21.45	1694	10.40	2696	2.41	3052	17.58	2493	0.00	0

TOTAL FALLOUT ACTIVITY (μCi/sq meter) by SHOT SERIES WITH COUNTY RANK

COUNTY AND STATE	RANGER SERIES	RANGER RANK	B-J SERIES	B-J RANK	T-S SERIES	T-S RANK	U-K SERIES	UK RANK	TEAPOT SERIES	TP RANK	PLUMBBOB SERIES	PB RANK	SHOT SEDAN	SEDAN RANK
	1951		1951		1952		1953		1955		1957		1962	
JOHNSON TX	0.00	0	0.18	2932	13.47	2214	46.39	713	27.18	619	25.91	1686	0.00	0
JONES TX	0.00	0	0.06	2993	494.83	173	21.29	1743	13.43	1576	18.17	2427	0.00	0
KARNES TX	0.00	0	3.35	1333	21.16	1707	11.34	2573	2.54	3047	12.05	2858	0.00	0
KAUFMAN TX	0.00	0	2.16	1571	52.83	883	43.60	784	18.85	1060	18.66	2378	0.00	0
KENDALL TX	0.00	0	1.36	1907	490.01	180	6.37	2972	11.64	1751	18.04	2442	0.00	0
KENEDY TX	0.00	0	3.34	1338	32.12	1339	10.99	2625	1.34	3087	10.92	2901	0.00	0
KENT TX	0.00	0	0.21	2878	72.35	597	45.50	729	18.60	1087	18.41	2406	0.00	0
KERR TX	0.00	0	1.29	1943	796.64	141	6.48	2969	12.60	1657	13.05	2819	0.00	0
KIMBLE TX	0.00	0	1.29	1944	487.90	184	6.98	2953	12.29	1683	13.41	2798	0.00	0
KING TX	0.00	0	0.26	2849	87.08	442	39.94	899	16.88	1268	11.94	2865	0.00	0
KINNEY TX	0.00	0	1.45	1859	57.99	799	4.82	3025	10.80	1836	9.10	2955	0.00	0
KLEBERG TX	0.00	0	6.08	914	21.24	1703	10.15	2730	1.92	3077	11.77	2871	0.00	0
KNOX TX	0.00	0	0.21	2879	515.03	167	33.96	1118	25.06	693	16.01	2620	0.00	0
LA SALLE TX	0.00	0	1.84	1687	20.10	1738	10.14	2733	2.53	3048	8.45	2964	0.00	0
LAMAR TX	0.00	0	3.25	1351	67.35	664	53.26	599	47.45	336	38.00	1064	0.00	0
LAMB TX	0.00	0	0.47	2566	293.86	215	121.30	167	87.59	183	36.93	1097	0.00	0
LAMPASAS TX	0.00	0	0.18	2933	500.89	169	23.67	1595	10.71	1846	13.08	2817	0.00	0
LAVACA TX	0.00	0	4.20	1171	20.55	1723	11.89	2523	5.16	2626	13.02	2820	0.00	0
LEE TX	0.00	0	3.35	1334	15.47	2036	15.24	2202	20.56	948	14.04	2762	0.00	0
LEON TX	0.00	0	2.71	1444	8.81	2678	23.93	1586	21.92	866	20.27	2223	0.00	0
LIBERTY TX	0.00	0	2.71	1445	11.66	2397	94.90	223	20.81	929	18.54	2390	0.00	0
LIMESTONE TX	0.00	0	0.70	2285	19.56	1767	30.60	1260	24.86	703	22.53	1994	0.00	0
LIPSCOMB TX	0.00	0	3.95	1223	74.16	570	82.94	278	119.74	123	39.77	1032	0.00	0
LIVE OAK TX	0.00	0	3.36	1331	24.66	1552	10.26	2711	3.08	2998	10.02	2927	0.00	0
LLANO TX	0.00	0	0.27	2841	801.19	137	16.21	2120	10.67	1852	13.32	2802	0.00	0
LOVING TX	0.00	0	1.50	1840	140.26	302	95.18	222	30.94	537	27.12	1589	0.00	0

TOTAL FALLOUT ACTIVITY (µCi/sq meter) by SHOT SERIES WITH COUNTY RANK

COUNTY AND STATE	RANGER SERIES	RANGER RANK	B-J SERIES	B-J RANK	T-S SERIES	T-S RANK	U-K SERIES	UK RANK	TEAPOT SERIES	TP RANK	PLUMBBOB SERIES	PB RANK	SHOT SEDAN	SEDAN RANK
	1951		1951		1952		1953		1955		1957		1962	
LUBBOCK TX	0.00	0	0.30	2807	72.21	599	89.95	252	75.64	214	39.14	1044	0.00	0
LYNN TX	0.00	0	0.30	2808	82.28	483	88.35	259	71.52	227	24.53	1797	0.00	0
MADISON TX	0.00	0	4.34	1131	6.23	2929	58.37	519	22.60	820	16.83	2554	0.00	0
MARION TX	0.00	0	2.94	1399	33.71	1296	21.55	1728	34.93	472	21.18	2123	0.00	0
MARTIN TX	0.00	0	0.30	2809	76.55	540	50.20	644	9.95	1923	20.16	2233	0.00	0
MASON TX	0.00	0	0.19	2901	798.66	139	14.80	2246	10.52	1867	18.50	2400	0.00	0
MATAGORDA TX	0.00	0	3.35	1335	12.67	2290	12.50	2467	4.80	2716	19.01	2350	0.00	0
MAVERICK TX	0.00	0	1.36	1905	59.87	776	5.29	3013	10.42	1879	11.70	2874	0.00	0
MCCULLOCH TX	0.00	0	0.19	2902	799.36	138	20.91	1779	11.42	1773	20.15	2235	0.00	0
MCLENNAN TX	0.00	0	0.18	2934	25.84	1513	24.61	1542	22.97	804	21.11	2129	0.00	0
MCMULLEN TX	0.00	0	1.91	1662	1312.75	102	9.74	2774	2.36	3055	7.07	2983	0.00	0
MEDINA TX	0.00	0	1.29	1945	1308.75	103	5.58	3003	11.66	1745	14.05	2761	0.00	0
MENARD TX	0.00	0	0.18	2940	256.78	225	14.25	2312	9.63	1960	16.59	2578	0.00	0
MIDLAND TX	0.00	0	0.30	2810	66.13	678	35.16	1069	8.68	2038	12.83	2829	0.00	0
MILAM TX	0.00	0	1.83	1693	7.48	2805	15.73	2158	22.96	805	17.68	2483	0.00	0
MILLS TX	0.00	0	0.18	2935	253.96	228	20.55	1804	9.93	1928	15.21	2684	0.00	0
MITCHELL TX	0.00	0	0.19	2903	60.72	757	28.79	1344	11.24	1797	14.80	2711	0.00	0
MONTAGUE TX	0.00	0	3.12	1374	37.38	1192	45.47	733	38.54	426	31.67	1324	0.00	0
MONTGOMERY TX	0.00	0	4.22	1163	10.17	2561	95.94	217	15.85	1358	17.11	2531	0.00	0
MOORE TX	0.00	0	0.38	2727	81.01	495	99.61	204	94.78	167	34.58	1172	0.00	0
MORRIS TX	0.00	0	4.55	1093	35.13	1251	24.67	1537	31.19	528	24.05	1835	0.00	0
MOTLEY TX	0.00	0	0.21	2880	266.52	220	65.22	448	89.60	178	15.19	2685	0.00	0
NACOGDOCHES TX	0.00	0	4.34	1132	9.11	2653	21.54	1729	19.70	1005	29.88	1432	0.00	0
NAVARRO TX	0.00	0	0.70	2286	27.39	1476	33.41	1139	23.78	761	18.63	2381	0.00	0
NEWTON TX	0.00	0	2.82	1419	9.65	2599	104.75	188	27.32	614	19.47	2311	0.00	0
NOLAN TX	0.00	0	0.15	2957	795.54	143	21.42	1736	11.59	1759	17.46	2499	0.00	0

TOTAL FALLOUT ACTIVITY (µCi/sq meter) by SHOT SERIES WITH COUNTY RANK

COUNTY AND STATE	RANGER SERIES	RANGER RANK	B-J SERIES	B-J RANK	T-S SERIES	T-S RANK	U-K SERIES	UK RANK	TEAPOT SERIES	TP RANK	PLUMBBOB SERIES	PB RANK	SHOT SEDAN	SEDAN RANK
	1951		1951		1952		1953		1955		1957		1962	
NUECES TX	0.00	0	1.11	2032	18.78	1821	6.76	2963	2.60	3041	14.97	2696	0.00	0
OCHILTREE TX	0.00	0	2.50	1488	158.66	279	60.44	496	165.48	75	32.59	1279	0.00	0
OLDHAM TX	0.00	0	0.70	2289	78.38	523	155.49	135	219.19	55	36.21	1119	0.00	0
ORANGE TX	0.00	0	2.71	1446	7.11	2845	93.26	231	18.31	1118	128.69	149	0.00	0
PALO PINTO TX	0.00	0	0.29	2825	160.81	274	31.98	1203	14.44	1494	31.32	1343	0.00	0
PANOLA TX	0.00	0	4.61	1079	19.60	1764	32.28	1196	22.60	819	25.84	1695	0.00	0
PARKER TX	0.00	0	0.40	2718	25.01	1532	30.98	1243	29.08	577	29.06	1479	0.00	0
PARMER TX	0.00	0	0.19	2904	70.34	618	159.58	130	92.33	173	36.10	1123	0.00	0
PECOS TX	0.00	0	1.61	1790	73.23	584	12.61	2457	13.15	1597	14.53	2735	0.00	0
POLK TX	0.00	0	4.34	1133	7.18	2835	96.20	215	20.86	923	23.22	1916	0.00	0
POTTER TX	0.00	0	0.21	2881	74.63	562	96.00	216	220.48	54	35.50	1139	0.00	0
PRESIDIO TX	0.00	0	1.74	1741	60.63	759	11.07	2608	11.65	1749	24.80	1781	0.00	0
RAINS TX	0.00	0	2.27	1539	72.62	592	34.67	1095	26.00	656	24.73	1786	0.00	0
RANDALL TX	0.00	0	0.21	2882	133.08	312	91.57	241	55.04	281	38.75	1050	0.00	0
REAGAN TX	0.00	0	1.39	1886	492.97	175	14.45	2292	12.21	1692	13.55	2791	0.00	0
REAL TX	0.00	0	1.61	1788	158.68	278	5.18	3017	11.12	1809	10.14	2925	0.00	0
RED RIVER TX	0.00	0	3.24	1355	36.62	1210	28.73	1348	42.91	382	31.38	1339	0.00	0
REEVES TX	0.00	0	1.61	1791	67.54	660	62.61	471	31.09	533	29.37	1463	0.00	0
REFUGIO TX	1067.00	1	4.17	1174	46.76	992	11.22	2592	3.69	2926	12.41	2844	0.00	0
ROBERTS TX	0.00	0	4.40	1121	84.37	466	61.06	490	86.84	185	24.19	1824	0.00	0
ROBERTSON TX	0.00	0	4.22	1164	14.03	2158	17.75	2004	21.32	902	19.48	2310	0.00	0
ROCKWALL TX	0.00	0	1.24	1967	41.95	1077	48.66	674	33.73	491	21.81	2071	0.00	0
RUNNELS TX	0.00	0	0.15	2958	490.08	179	17.52	2024	13.22	1592	18.37	2410	0.00	0
RUSK TX	0.00	0	2.82	1420	16.43	1973	40.09	890	25.66	667	31.97	1307	0.00	0
SABINE TX	0.00	0	4.45	1109	11.17	2459	102.33	194	21.35	900	19.82	2275	0.00	0
SAN AUGUSTIN TX	0.00	0	4.45	1110	10.36	2541	100.76	200	25.94	659	17.65	2486	0.00	0

TOTAL FALLOUT ACTIVITY (µCi/sq meter) by SHOT SERIES WITH COUNTY RANK

COUNTY AND STATE	RANGER SERIES	RANGER RANK	B-J SERIES	B-J RANK	T-S SERIES	T-S RANK	U-K SERIES	UK RANK	TEAPOT SERIES	TP RANK	PLUMBBOB SERIES	PB RANK	SHOT SEDAN	SEDAN RANK
	1951		1951		1952		1953		1955		1957		1962	
SAN JACINTO TX	0.00	0	4.34	1134	7.24	2830	93.74	229	14.73	1460	15.08	2691	0.00	0
SAN PATRICIO TX	0.00	0	3.21	1363	18.20	1853	10.19	2724	2.70	3036	11.95	2863	0.00	0
SAN SABA TX	0.00	0	0.27	2842	492.86	176	18.51	1947	11.17	1803	17.85	2461	0.00	0
SCHLEICHER TX	0.00	0	1.38	1892	60.25	771	8.66	2876	13.33	1583	16.78	2559	0.00	0
SCURRY TX	0.00	0	0.19	2905	73.81	575	39.56	911	14.75	1457	19.22	2330	0.00	0
SHACKELFORD TX	0.00	0	0.07	2987	487.23	185	23.07	1635	13.20	1593	22.78	1961	0.00	0
SHELBY TX	0.00	0	2.99	1391	11.79	2382	22.56	1666	22.25	847	18.59	2383	0.00	0
SHERMAN TX	0.00	0	1.20	1994	60.59	762	100.98	199	85.57	187	21.05	2137	0.00	0
SMITH TX	0.00	0	2.82	1421	19.40	1780	44.54	760	29.01	580	36.62	1107	0.00	0
SOMERVELL TX	0.00	0	0.18	2936	14.79	2089	25.56	1495	25.66	668	19.21	2331	0.00	0
STARR TX	0.00	0	1.94	1637	12.51	2306	11.95	2518	1.65	3083	11.65	2877	0.00	0
STEPHENS TX	0.00	0	0.17	2949	153.44	285	24.25	1565	16.63	1288	27.05	1595	0.00	0
STERLING TX	0.00	0	0.19	2906	60.28	770	23.30	1619	8.23	2094	24.54	1795	0.00	0
STONEWALL TX	0.00	0	0.21	2883	498.99	170	33.45	1138	24.31	737	17.09	2533	0.00	0
SUTTON TX	0.00	0	1.34	1919	155.43	283	6.91	2960	13.01	1614	9.62	2940	0.00	0
SWISHER TX	0.00	0	0.38	2728	86.81	447	89.50	254	89.28	179	44.74	928	0.00	0
TARRANT TX	0.00	0	0.40	2719	27.06	1484	60.86	492	39.48	417	32.15	1295	0.00	0
TAYLOR TX	0.00	0	0.05	2995	152.93	287	10.63	2668	8.27	2087	19.73	2282	0.00	0
TERRELL TX	0.00	0	1.61	1792	60.52	766	7.50	2931	11.85	1730	6.80	2986	0.00	0
TERRY TX	0.00	0	0.19	2907	87.48	437	112.76	176	78.66	209	23.79	1867	0.00	0
THROCKMORTON TX	0.00	0	0.18	2939	808.77	135	26.74	1440	12.23	1690	14.91	2702	0.00	0
TITUS TX	0.00	0	3.23	1357	57.90	802	25.41	1499	37.88	432	23.07	1933	0.00	0
TOM GREEN TX	0.00	0	0.18	2941	88.83	430	17.33	2039	9.93	1925	20.65	2182	0.00	0
TRAVIS TX	0.00	0	1.91	1663	13.27	2227	15.39	2192	4.84	2706	16.91	2548	0.00	0
TRINITY TX	0.00	0	2.71	1447	7.04	2850	101.27	197	14.62	1474	17.66	2485	0.00	0
TYLER TX	0.00	0	0.70	2287	8.77	2682	102.75	193	14.32	1505	17.68	2482	0.00	0

TOTAL FALLOUT ACTIVITY (µCi/sq meter) by SHOT SERIES WITH COUNTY RANK

COUNTY AND STATE	RANGER SERIES	RANGER RANK	B-J SERIES	B-J RANK	T-S SERIES	T-S RANK	U-K SERIES	UK RANK	TEAPOT SERIES	TP RANK	PLUMBBOB SERIES	PB RANK	SHOT SEDAN	SEDAN RANK
	1951		1951		1952		1953		1955		1957		1962	
UPSHUR TX	0.00	0	1.80	1719	29.14	1436	41.20	851	31.54	524	24.98	1768	0.00	0
UPTON TX	0.00	0	1.39	1887	68.24	647	14.82	2241	13.47	1574	8.62	2962	0.00	0
UVALDE TX	0.00	0	1.61	1789	159.58	277	5.17	3018	10.87	1833	14.42	2741	0.00	0
VAL VERDE TX	0.00	0	0.98	2104	57.27	819	4.27	3038	5.85	2485	8.68	2961	0.00	0
VAN ZANDT TX	0.00	0	2.16	1572	47.91	968	56.18	551	28.57	590	27.37	1568	0.00	0
VICTORIA TX	0.00	0	3.24	1354	15.28	2048	11.45	2563	3.31	2970	7.57	2978	0.00	0
WALKER TX	0.00	0	4.34	1135	10.47	2531	95.54	219	21.55	886	23.79	1868	0.00	0
WALLER TX	0.00	0	6.14	908	9.91	2580	91.11	243	16.65	1285	17.20	2525	0.00	0
WARD TX	0.00	0	1.50	1841	60.59	760	59.88	504	29.92	559	22.36	2016	0.00	0
WASHINGTON TX	0.00	0	6.14	909	18.96	1810	90.86	245	15.93	1343	11.00	2898	0.00	0
WEBB TX	0.00	0	1.86	1676	7.81	2777	10.39	2698	1.99	3075	7.26	2981	0.00	0
WHARTON TX	0.00	0	8.94	717	28.04	1461	13.04	2416	4.89	2692	17.88	2455	0.00	0
WHEELER TX	0.00	0	5.21	985	80.81	497	72.87	365	108.71	160	21.87	2065	0.00	0
WICHITA TX	0.00	0	0.51	2523	525.12	164	34.22	1108	21.88	870	22.44	2008	0.00	0
WILBARGER TX	0.00	0	0.51	2519	102.64	351	40.02	895	17.32	1227	17.02	2543	0.00	0
WILLACY TX	0.00	0	2.50	1487	42.59	1066	11.03	2613	1.20	3089	11.20	2893	0.00	0
WILLIAMSON TX	0.00	0	0.27	2843	29.87	1413	13.67	2367	21.97	864	19.85	2272	0.00	0
WILSON TX	0.00	0	1.91	1664	5.83	2960	10.59	2677	3.24	2977	10.40	2919	0.00	0
WINKLER TX	0.00	0	1.50	1842	61.83	743	80.20	293	30.32	551	22.33	2020	0.00	0
WISE TX	0.00	0	0.40	2720	35.76	1237	43.02	801	43.44	377	28.99	1483	0.00	0
WOOD TX	0.00	0	2.93	1402	54.43	864	35.74	1042	30.86	541	35.80	1130	0.00	0
YOAKUM TX	0.00	0	0.41	2701	91.43	405	143.10	145	35.60	460	25.26	1742	0.00	0
YOUNG TX	0.00	0	0.29	2826	250.39	232	29.21	1324	12.85	1632	22.88	1952	0.00	0
ZAPATA TX	0.00	0	1.86	1677	9.44	2615	11.23	2589	1.18	3090	7.59	2977	0.00	0
ZAVALA TX	0.00	0	1.39	1888	488.33	183	5.63	3000	10.72	1845	11.41	2888	0.00	0
BEAVER UT	0.00	0	0.00	0	266.01	222	330.02	81	574.68	5	83.34	364	0.00	0

TOTAL FALLOUT ACTIVITY (µCi/sq meter) by SHOT SERIES WITH COUNTY RANK

COUNTY AND STATE	RANGER SERIES	RANGER RANK	B-J SERIES	B-J RANK	T-S SERIES	T-S RANK	U-K SERIES	UK RANK	TEAPOT SERIES	TP RANK	PLUMBBOB SERIES	PB RANK	SHOT SEDAN	SEDAN RANK
	1951		1951		1952		1953		1955		1957		1962	
BOX ELDER1 UT	0.00	0	11.86	545	9343.23	51	102.18	195	64.67	240	130.35	132	0.00	0
BOX ELDER2 UT	0.00	0	6.44	881	1288.41	107	471.49	58	97.69	165	181.63	51	11.17	688
CACHE UT	0.00	0	1.81	1709	1880.49	93	563.29	46	79.20	206	156.89	69	2.18	709
CARBON UT	0.00	0	0.00	0	268.68	218	441.01	64	115.00	143	152.11	75	0.00	0
DAGGETT UT	0.00	0	0.91	2141	286.56	216	71.88	372	355.79	26	389.20	20	0.00	0
DAVIS UT	0.00	0	1.76	1732	952.56	119	755.11	21	264.44	46	221.10	39	11.17	685
DUCHESNE UT	0.00	0	0.91	2142	313.47	206	95.42	220	346.26	28	219.58	41	0.00	0
EMERY UT	0.00	0	0.00	0	235.10	238	379.52	75	420.15	14	156.22	72	0.00	0
GARFIELD UT	0.00	0	0.00	0	143.11	298	693.37	26	208.30	57	1309.97	2	0.00	0
GRAND UT	0.00	0	0.00	0	228.52	241	401.02	71	439.60	13	97.39	262	0.00	0
IRON1 UT	0.00	0	0.00	0	607.30	159	453.26	61	235.11	49	180.60	52	0.00	0
IRON2 UT	0.00	0	0.00	0	224.91	243	366.50	77	573.18	6	101.45	239	0.00	0
IRON3 UT	0.00	0	0.00	0	131.69	313	611.73	38	359.01	25	244.64	35	0.00	0
JUAB UT	0.00	0	22.36	207	277.96	217	347.06	78	362.09	23	227.84	38	0.00	0
KANE1 UT	0.00	0	0.00	0	140.68	300	2516.34	6	175.68	65	48.86	837	0.00	0
KANE2 UT	0.00	0	0.00	0	137.93	309	2717.35	5	189.19	60	127.74	156	0.00	0
MILLARD UT	0.00	0	1.71	1749	246.56	235	475.24	56	354.15	27	170.75	59	0.00	0
MORGAN UT	0.00	0	1.76	1733	827.94	134	757.75	20	287.92	44	210.86	44	11.17	686
PIUTE UT	0.00	0	0.00	0	222.22	244	285.29	91	512.98	9	115.41	190	0.00	0
RICH UT	0.00	0	1.81	1710	940.95	122	652.79	32	79.11	207	153.40	73	3.27	704
SALT LAKE UT	0.00	0	1.71	1753	900.31	127	645.58	33	310.70	39	235.36	36	11.17	684
SAN JUAN UT	0.00	0	0.00	0	129.43	318	712.16	23	103.66	164	70.53	480	0.00	0
SANPETE UT	0.00	0	0.00	0	209.91	247	445.89	63	141.21	87	159.39	68	0.00	0
SEVIER UT	0.00	0	0.00	0	233.15	239	284.32	92	379.36	20	105.81	221	0.00	0
SUMMIT UT	0.00	0	1.81	1711	796.71	140	501.30	53	287.97	43	212.56	43	0.00	0
TOOELE1 UT	0.00	0	14.61	438	2492.21	83	136.77	154	126.65	99	280.81	29	2.18	708

TOTAL FALLOUT ACTIVITY (µCi/sq meter) by SHOT SERIES WITH COUNTY RANK

COUNTY AND STATE	RANGER SERIES	RANGER RANK	B-J SERIES	B-J RANK	T-S SERIES	T-S RANK	U-K SERIES	UK RANK	TEAPOT SERIES	TP RANK	PLUMBBOB SERIES	PB RANK	SHOT SEDAN	SEDAN RANK
	1951		1951		1952		1953		1955		1957		1962	
TOOELE2 UT	0.00	0	5.54	960	1272.26	113	670.98	30	222.38	52	248.53	34	22.33	653
UINTAH UT	0.00	0	0.91	2143	322.81	203	92.23	238	367.76	22	401.50	18	0.00	0
UTAH UT	0.00	0	0.86	2179	341.73	197	433.72	67	478.66	12	351.70	26	0.00	0
WASATCH UT	0.00	0	0.91	2144	309.85	208	371.38	76	417.32	16	273.02	30	0.00	0
WASHINGTON1 UT	0.00	0	0.00	0	18.49	1834	2822.95	3	167.07	73	527.50	13	0.00	0
WASHINGTON2 UT	0.00	0	0.00	0	0.44	3087	5767.46	1	361.37	24	671.34	5	0.00	0
WASHINGTON3 UT	0.00	0	0.00	0	0.00	0	4249.18	2	139.95	88	568.04	12	0.00	0
WAYNE UT	0.00	0	0.00	0	207.71	249	382.59	74	167.50	72	163.17	62	0.00	0
WEBER UT	0.00	0	1.76	1734	947.32	121	709.00	24	235.48	48	198.11	46	11.17	687
ACCOMACK VA	0.00	0	28.91	113	9.56	2609	15.18	2207	5.21	2620	16.12	2614	103.95	359
ALBEMARLE VA	0.00	0	13.74	464	6.54	2903	9.43	2810	5.58	2535	16.01	2619	0.00	0
ALLEGHANY VA	0.00	0	7.04	836	5.40	2984	8.86	2864	5.89	2480	17.05	2539	85.56	469
AMELIA VA	0.00	0	18.09	324	18.11	1860	7.38	2936	5.89	2481	25.90	1687	149.13	277
AMHERST VA	0.00	0	12.93	492	6.56	2900	8.12	2904	5.22	2612	14.55	2733	73.06	528
APPOMATTOX VA	0.00	0	17.40	341	6.72	2877	15.34	2195	4.81	2711	13.94	2770	60.19	556
ARLINGTON VA	0.00	0	29.52	103	14.23	2138	16.80	2076	7.25	2228	17.37	2507	0.00	0
AUGUSTA VA	0.00	0	11.48	565	8.46	2722	11.30	2579	4.78	2720	22.63	1984	60.19	545
BATH VA	0.00	0	6.03	916	9.45	2614	6.64	2966	6.47	2363	21.75	2077	208.78	157
BEDFORD VA	0.00	0	19.54	278	11.15	2460	7.34	2939	6.39	2386	23.80	1866	60.19	543
BLAND VA	0.00	0	6.37	887	10.54	2516	9.27	2830	4.57	2765	24.90	1774	0.00	0
BOTETOURT VA	0.00	0	8.99	715	5.37	2987	8.89	2861	4.55	2768	20.08	2245	208.78	163
BRUNSWICK VA	0.00	0	20.58	247	22.12	1655	7.67	2922	4.25	2834	23.13	1928	0.00	0
BUCHANAN VA	0.00	0	3.37	1329	13.84	2173	23.03	1639	15.61	1379	23.85	1863	149.13	278
BUCKINGHAM VA	0.00	0	17.58	337	8.66	2695	8.49	2886	5.52	2546	18.76	2367	0.00	0
CAMPBELL VA	0.00	0	17.78	332	8.27	2741	5.75	2997	5.07	2642	21.58	2089	57.04	568
CAROLINE VA	0.00	0	17.26	347	17.32	1920	14.46	2289	6.07	2440	14.24	2751	57.04	579

TOTAL FALLOUT ACTIVITY (µCi/sq meter) by SHOT SERIES WITH COUNTY RANK

COUNTY AND STATE	RANGER SERIES	RANGER RANK	B-J SERIES	B-J RANK	T-S SERIES	T-S RANK	U-K SERIES	UK RANK	TEAPOT SERIES	TP RANK	PLUMBBOB SERIES	PB RANK	SHOT SEDAN	SEDAN RANK
	1951		1951		1952		1953		1955		1957		1962	
CARROLL VA	0.00	0	15.34	412	6.66	2886	9.06	2847	5.06	2644	34.15	1202	94.45	391
CHARLES CITY VA	0.00	0	27.47	134	22.29	1644	11.02	2616	5.04	2649	14.50	2738	60.19	555
CHARLOTTE VA	0.00	0	24.29	186	5.42	2983	7.59	2925	3.69	2927	18.18	2425	0.00	0
CHESTERFIELD VA	0.00	0	26.41	155	13.37	2222	8.65	2877	3.76	2917	15.26	2678	149.13	284
CLARKE VA	0.00	0	9.56	681	11.78	2386	12.53	2465	9.35	1981	19.75	2279	0.00	0
CRAIG VA	0.00	0	8.51	744	6.11	2934	7.84	2916	4.90	2690	17.95	2447	90.29	431
CULPEPER VA	0.00	0	20.10	258	32.69	1323	10.62	2669	7.74	2150	22.73	1968	0.00	0
CUMBERLAND VA	0.00	0	18.72	299	18.73	1824	9.35	2822	5.25	2605	16.75	2561	0.00	0
DICKENSON VA	0.00	0	9.05	707	14.40	2119	18.28	1961	15.12	1429	23.32	1901	94.45	396
DINWIDDIE VA	0.00	0	23.35	198	22.02	1662	8.27	2898	4.20	2841	23.78	1870	0.00	0
ESSEX VA	0.56	447	15.11	418	18.90	1812	12.80	2444	6.31	2400	17.67	2484	0.00	0
FAIRFAX VA	0.00	0	25.48	171	14.01	2164	15.56	2173	7.42	2198	17.08	2535	60.19	551
FAUQUIER VA	0.00	0	14.59	439	20.19	1735	15.73	2156	9.38	1980	15.69	2644	57.04	575
FLOYD VA	0.00	0	18.57	308	11.25	2448	9.70	2778	6.56	2347	28.65	1500	60.19	539
FLUVANNA VA	0.00	0	11.65	556	16.49	1966	10.49	2687	4.25	2831	20.52	2198	0.00	0
FRANKLIN VA	0.00	0	19.65	275	6.14	2933	6.83	2961	5.83	2489	19.17	2336	149.13	283
FREDERICK VA	1.23	373	8.13	767	9.37	2624	15.09	2214	8.01	2123	15.09	2689	57.04	577
GILES VA	0.00	0	6.99	838	11.97	2366	8.99	2855	4.92	2681	20.07	2246	57.04	570
GLOUCESTER VA	0.00	0	24.61	181	14.60	2106	14.41	2295	3.87	2899	15.47	2661	0.00	0
GOOCHLAND VA	0.00	0	17.04	352	21.42	1696	11.25	2586	3.93	2888	20.96	2144	149.13	280
GRAYSON VA	0.19	572	7.93	783	18.03	1869	10.08	2741	5.72	2506	26.35	1644	43.75	595
GREENE VA	0.00	0	11.45	568	7.09	2846	9.41	2812	5.29	2590	15.70	2640	60.19	554
GREENSVILLE VA	0.00	0	30.09	95	21.28	1701	11.64	2541	3.67	2931	15.69	2642	0.00	0
HALIFAX VA	0.00	0	18.65	304	19.17	1793	7.35	2937	3.83	2909	17.33	2511	0.00	0
HAMPTON VA	0.00	0	23.86	193	12.04	2355	17.53	2023	4.55	2767	17.57	2494	0.00	0
HANOVER VA	0.00	0	15.53	401	22.00	1663	9.64	2784	6.73	2311	22.67	1976	149.13	279

TOTAL FALLOUT ACTIVITY (µCi/sq meter) by SHOT SERIES WITH COUNTY RANK

COUNTY AND STATE	RANGER SERIES	RANGER RANK	B-J SERIES	B-J RANK	T-S SERIES	T-S RANK	U-K SERIES	UK RANK	TEAPOT SERIES	TP RANK	PLUMBBOB SERIES	PB RANK	SHOT SEDAN	SEDAN RANK
	1951		1951		1952		1953		1955		1957		1962	
HENRICO VA	0.00	0	26.94	141	22.84	1618	9.41	2811	7.65	2163	23.95	1848	0.00	0
HENRY VA	0.56	448	21.73	219	11.22	2451	9.58	2791	5.95	2467	23.77	1873	0.00	0
HIGHLAND VA	0.00	0	4.76	1055	6.26	2923	9.05	2848	7.07	2261	22.55	1991	85.56	452
ISLE OF WIGH VA	0.00	0	26.44	152	13.38	2220	10.48	2688	5.36	2580	23.57	1886	0.00	0
JAMES CITY VA	0.00	0	27.93	126	15.49	2034	10.30	2707	4.96	2670	14.62	2725	0.00	0
KING AND QUE VA	0.00	0	27.64	129	21.51	1685	11.09	2606	6.80	2300	24.49	1799	57.04	564
KING GEORGE VA	0.00	0	20.85	239	21.49	1690	15.47	2185	7.14	2250	14.00	2766	0.00	0
KING WILLIAM VA	0.00	0	29.02	112	22.13	1654	10.84	2643	8.35	2078	27.23	1577	57.04	562
LANCASTER VA	0.56	449	29.64	102	17.09	1937	13.15	2405	4.68	2744	18.71	2372	0.00	0
LEE VA	0.00	0	4.27	1154	7.67	2785	21.42	1737	18.25	1124	25.28	1741	43.75	598
LOUDOUN VA	0.00	0	16.07	379	12.46	2314	16.26	2114	8.42	2066	23.91	1852	60.19	542
LOUISA VA	0.00	0	14.05	452	18.26	1848	8.76	2870	4.82	2708	21.24	2112	0.00	0
LUNENBURG VA	0.00	0	20.90	238	16.27	1982	9.44	2808	4.14	2853	16.89	2551	57.04	574
MADISON VA	0.00	0	9.16	701	16.22	1989	11.51	2554	5.36	2581	22.53	1992	60.19	546
MATHEWS VA	0.00	0	26.20	158	16.89	1947	13.18	2402	3.91	2890	17.42	2502	0.00	0
MECKLENBURG VA	0.00	0	27.05	140	7.79	2780	9.00	2853	5.61	2528	22.99	1939	57.04	565
MIDDLESEX VA	0.00	0	29.36	106	17.30	1923	12.96	2426	4.76	2726	23.15	1925	0.00	0
MONTGOMERY VA	0.00	0	9.53	683	10.74	2498	9.45	2804	5.30	2588	20.54	2197	60.19	548
NELSON VA	0.00	0	11.78	549	6.77	2870	11.73	2535	5.21	2618	19.66	2289	60.19	549
NEW KENT VA	0.00	0	27.13	138	21.66	1679	10.58	2679	5.08	2640	16.72	2565	60.19	552
NEWPORT NEWS VA	0.00	0	23.85	194	11.91	2370	17.93	1983	3.10	2994	25.51	1719	0.00	0
NORFOLK/CHES VA	0.00	0	25.79	165	11.29	2444	11.00	2621	2.82	3024	28.82	1490	0.00	0
NORTHAMPTON VA	0.00	0	25.72	166	17.45	1911	14.75	2256	3.86	2903	14.47	2739	0.00	0
NORTHUMBERLA VA	0.00	0	30.22	94	17.31	1921	14.81	2244	4.77	2723	18.52	2395	0.00	0
NOTTOWAY VA	0.00	0	18.26	320	17.80	1888	7.34	2938	6.06	2445	27.63	1551	0.00	0
ORANGE VA	0.00	0	14.50	443	19.34	1785	12.31	2482	9.72	1952	16.48	2589	60.19	553

TOTAL FALLOUT ACTIVITY (µCi/sq meter) by SHOT SERIES WITH COUNTY RANK

COUNTY AND STATE	RANGER SERIES	RANGER RANK	B-J SERIES	B-J RANK	T-S SERIES	T-S RANK	U-K SERIES	UK RANK	TEAPOT SERIES	TP RANK	PLUMBBOB SERIES	PB RANK	SHOT SEDAN	SEDAN RANK
	1951		1951		1952		1953		1955		1957		1962	
PAGE VA	0.00	0	10.04	655	6.52	2904	11.38	2570	5.55	2542	25.85	1693	0.00	0
PATRICK VA	0.00	0	21.27	231	8.30	2737	8.32	2895	6.22	2417	32.26	1288	37.08	630
PITTSYLVANIA VA	0.00	0	16.89	356	11.78	2385	7.78	2917	6.15	2428	25.53	1717	60.19	540
POWHATAN VA	0.00	0	16.70	362	18.09	1861	8.64	2878	6.32	2396	30.45	1400	149.13	276
PRINCE EDWAR VA	0.00	0	19.92	268	7.66	2787	7.32	2940	6.89	2289	23.44	1891	60.19	544
PRINCE GEORG VA	0.00	0	28.48	115	22.42	1634	10.93	2634	5.75	2500	25.15	1756	0.00	0
PRINCE WILLI VA	0.00	0	22.08	213	10.50	2526	12.34	2479	10.93	1829	21.87	2064	0.00	0
PULASKI VA	0.00	0	10.34	633	12.00	2359	9.65	2783	5.32	2586	28.50	1509	57.04	560
RAPPAHANNOCK VA	0.00	0	8.88	720	17.24	1927	14.95	2229	9.00	2012	18.00	2443	57.04	572
RICHMOND VA	0.19	573	14.99	421	19.07	1803	13.36	2392	6.30	2401	17.15	2528	0.00	0
ROANOKE VA	0.00	0	18.36	314	10.49	2528	7.94	2914	6.53	2355	25.30	1737	60.19	541
ROCKBRIDGE VA	0.00	0	9.25	699	8.51	2712	7.47	2933	6.43	2371	22.49	1997	57.04	566
ROCKINGHAM VA	0.00	0	7.99	776	8.79	2679	9.57	2792	5.90	2478	19.94	2263	82.07	476
RUSSELL VA	0.00	0	5.03	1006	19.02	1805	14.15	2319	15.81	1361	26.16	1659	43.75	596
SCOTT VA	0.00	0	4.49	1099	6.98	2855	18.68	1936	17.39	1217	26.45	1640	43.75	594
SHENANDOAH VA	0.00	0	9.54	682	7.59	2793	13.86	2342	4.94	2675	14.05	2760	57.04	580
SMYTH VA	0.00	0	6.88	850	11.31	2440	11.71	2536	5.26	2603	31.49	1333	77.90	490
SOUTHAMPTON VA	0.00	0	29.07	111	16.44	1972	10.24	2715	3.12	2993	14.76	2714	0.00	0
SPOTSYLVANIA VA	0.00	0	16.37	372	17.16	1931	14.46	2290	8.44	2065	15.24	2681	57.04	576
STAFFORD VA	0.00	0	20.06	260	18.49	1833	11.49	2558	10.44	1876	20.57	2194	0.00	0
SUFFOLK/NANS VA	0.00	0	26.31	157	16.32	1979	11.34	2574	3.19	2983	22.93	1946	40.47	611
SURRY VA	0.00	0	27.71	128	15.40	2041	9.88	2760	5.02	2652	19.88	2270	0.00	0
SUSSEX VA	0.00	0	30.71	92	16.94	1946	10.10	2739	3.51	2945	11.75	2872	0.00	0
TAZEWELL VA	0.00	0	9.27	697	13.59	2197	11.06	2609	5.96	2462	50.21	808	60.19	538
VIRGINIA BEA VA	0.00	0	15.38	407	11.32	2437	10.23	2718	2.52	3049	24.47	1804	40.47	610
WARREN VA	0.00	0	10.99	594	18.55	1830	13.42	2388	8.67	2040	15.35	2673	73.06	527

TOTAL FALLOUT ACTIVITY (µCi/sq meter) by SHOT SERIES WITH COUNTY RANK

COUNTY AND STATE	RANGER SERIES	RANGER RANK	B-J SERIES	B-J RANK	T-S SERIES	T-S RANK	U-K SERIES	UK RANK	TEAPOT SERIES	TP RANK	PLUMBBOB SERIES	PB RANK	SHOT SEDAN	SEDAN RANK
	1951		1951		1952		1953		1955		1957		1962	
WASHINGTON VA	0.00	0	5.36	975	14.64	2104	24.89	1526	17.04	1255	26.49	1633	17.11	671
WESTMORELAND VA	0.00	0	17.11	350	19.40	1779	14.52	2281	7.31	2219	17.83	2465	0.00	0
WISE VA	0.00	0	4.69	1067	8.22	2746	19.59	1869	18.48	1099	33.79	1216	17.11	667
WYTHE VA	0.00	0	7.50	808	14.21	2141	10.74	2656	4.43	2786	25.60	1712	0.00	0
YORK VA	0.00	0	22.35	208	14.77	2093	12.05	2505	5.21	2616	18.22	2420	0.00	0
ADDISON VT	9.97	169	20.43	251	45.99	1002	184.54	119	3.60	2936	27.22	1579	0.00	0
BENNINGTON VT	20.71	69	26.48	151	34.14	1284	267.45	95	3.75	2919	29.33	1465	0.00	0
CALEDONIA VT	1.80	338	21.77	217	30.65	1388	25.10	1516	4.94	2674	20.60	2191	0.00	0
CHITTENDEN VT	1.42	361	14.36	444	38.79	1152	110.27	183	3.90	2891	41.82	995	0.00	0
ESSEX VT	1.29	365	21.95	214	27.92	1464	21.85	1710	5.40	2570	33.63	1225	0.00	0
FRANKLIN VT	1.72	347	23.17	200	66.43	674	32.69	1172	4.41	2792	14.58	2729	0.00	0
GRAND ISLE VT	1.72	348	24.47	183	69.20	630	41.67	840	3.89	2892	20.30	2220	0.00	0
LAMOILLE VT	0.29	492	18.90	296	48.00	965	28.49	1361	4.64	2753	26.22	1653	0.00	0
ORANGE VT	8.99	200	17.89	328	41.06	1097	26.13	1470	4.86	2702	22.06	2044	0.00	0
ORLEANS VT	0.29	493	25.50	170	47.36	977	17.76	2003	4.38	2799	20.55	2196	0.00	0
RUTLAND VT	8.99	201	21.22	234	37.55	1186	92.84	234	3.73	2922	21.26	2109	0.00	0
WASHINGTON VT	1.54	352	19.55	277	46.62	995	31.80	1212	5.29	2589	23.98	1843	0.00	0
WINDHAM VT	46.68	4	27.26	136	29.52	1425	45.33	738	4.27	2826	33.26	1250	0.00	0
WINDSOR VT	8.75	209	28.47	116	37.68	1182	37.74	987	5.21	2617	13.64	2787	0.00	0
ADAMS WA	0.00	0	0.87	2169	484.76	187	9.39	2816	3.22	2979	4.18	3033	0.00	0
ASOTIN WA	0.00	0	1.00	2088	6563.29	63	6.95	2954	7.07	2262	52.86	758	0.00	0
BENTON WA	0.00	0	0.87	2170	337.15	199	7.99	2911	3.67	2930	5.50	2998	0.00	0
CHELAN WA	0.00	0	0.43	2659	15.47	2037	15.88	2146	3.71	2924	3.60	3053	0.00	0
CLALLAM WA	0.00	0	0.54	2457	3.12	3067	7.00	2952	3.24	2975	4.94	3008	0.00	0
CLARK WA	0.00	0	0.64	2337	2.36	3077	8.70	2873	6.07	2443	4.64	3013	0.00	0
COLUMBIA WA	0.00	0	1.00	2089	2971.51	79	6.39	2971	6.52	2356	54.30	729	0.00	0

TOTAL FALLOUT ACTIVITY (µCi/sq meter) by SHOT SERIES WITH COUNTY RANK

COUNTY AND STATE	RANGER SERIES	RANGER RANK	B-J SERIES	B-J RANK	T-S SERIES	T-S RANK	U-K SERIES	UK RANK	TEAPOT SERIES	TP RANK	PLUMBBOB SERIES	PB RANK	SHOT SEDAN	SEDAN RANK
	1951		1951		1952		1953		1955		1957		1962	
COWLITZ WA	0.00	0	0.43	2660	15.73	2013	10.61	2676	7.20	2241	4.66	3010	0.00	0
DOUGLAS WA	0.00	0	1.00	2090	56.78	826	19.47	1879	4.07	2865	3.58	3055	0.00	0
FERRY WA	0.00	0	0.87	2171	152.36	288	41.17	854	4.22	2839	4.64	3012	0.00	0
FRANKLIN WA	0.00	0	0.87	2172	297.66	214	6.29	2974	3.43	2957	4.59	3016	0.00	0
GARFIELD WA	0.00	0	0.87	2173	10147.86	49	14.52	2280	5.93	2474	45.25	919	0.00	0
GRANT WA	0.00	0	1.09	2042	122.91	322	11.00	2619	3.34	2965	4.00	3039	0.00	0
GRAYS HARBOR WA	0.00	0	0.54	2458	2.81	3073	10.88	2637	3.16	2987	4.50	3026	0.00	0
ISLAND WA	0.00	0	0.43	2661	2.02	3085	4.60	3029	2.18	3066	2.52	3067	0.00	0
JEFFERSON WA	0.00	0	0.43	2662	3.35	3060	7.64	2924	2.95	3011	4.85	3009	0.00	0
KING WA	0.00	0	0.43	2663	30.01	1408	12.36	2477	2.73	3032	2.57	3066	0.00	0
KITSAP WA	0.00	0	0.43	2664	2.25	3079	8.16	2902	1.76	3079	4.57	3018	0.00	0
KITTITAS WA	0.00	0	0.43	2665	29.36	1430	12.77	2445	4.19	2842	4.10	3037	0.00	0
KLICKITAT WA	0.00	0	1.00	2091	121.95	323	8.56	2880	3.89	2895	3.83	3044	0.00	0
LEWIS WA	0.00	0	0.43	2666	15.51	2033	11.10	2602	2.98	3009	4.51	3024	0.00	0
LINCOLN WA	0.00	0	1.00	2092	230.67	240	13.07	2413	4.10	2858	4.64	3011	0.00	0
MASON WA	0.00	0	0.54	2459	2.11	3083	7.73	2920	2.98	3008	4.61	3015	0.00	0
OKANOGAN WA	0.00	0	0.43	2667	45.50	1011	30.67	1256	4.28	2820	4.11	3036	0.00	0
PACIFIC WA	0.00	0	0.43	2668	2.44	3076	7.51	2929	7.48	2188	3.79	3045	0.00	0
PEND OREILLE WA	0.00	0	1.00	2093	1105.59	117	34.07	1114	5.93	2472	51.38	785	0.00	0
PIERCE WA	0.00	0	0.43	2669	42.99	1060	13.73	2358	3.24	2976	5.37	3002	0.00	0
SAN JUAN WA	0.00	0	0.43	2670	13.24	2233	4.87	3024	2.14	3068	2.92	3061	0.00	0
SKAGIT WA	0.00	0	0.43	2671	2.21	3081	10.82	2647	2.92	3014	4.13	3035	0.00	0
SKAMANIA WA	0.00	0	0.43	2672	3.98	3035	13.77	2353	7.51	2182	4.51	3025	0.00	0
SNOHOMISH WA	0.00	0	0.43	2673	30.30	1403	14.45	2291	3.10	2995	4.33	3032	0.00	0
SPOKANE WA	0.00	0	0.87	2174	160.20	276	12.00	2510	3.51	2946	52.91	755	0.00	0
STEVENS WA	0.00	0	1.00	2094	391.95	193	29.21	1322	4.91	2685	6.21	2992	0.00	0

TOTAL FALLOUT ACTIVITY (µCi/sq meter) by SHOT SERIES WITH COUNTY RANK

COUNTY AND STATE	RANGER SERIES	RANGER RANK	B-J SERIES	B-J RANK	T-S SERIES	T-S RANK	U-K SERIES	UK RANK	TEAPOT SERIES	TP RANK	PLUMBBOB SERIES	PB RANK	SHOT SEDAN	SEDAN RANK
	1951		1951		1952		1953		1955		1957		1962	
THURSTON WA	0.00	0	0.54	2460	15.53	2031	7.18	2946	2.57	3044	3.55	3056	0.00	0
WAHKIAKUM WA	0.00	0	0.43	2674	1.22	3086	10.95	2630	8.55	2049	3.72	3047	0.00	0
WALLA WALLA WA	0.00	0	1.09	2043	927.85	126	8.51	2883	6.24	2413	56.88	682	0.00	0
WHATCOM WA	0.00	0	0.43	2675	2.82	3072	9.50	2797	2.99	3007	4.48	3027	0.00	0
WHITMAN WA	0.00	0	1.09	2044	1277.02	110	12.38	2475	5.56	2539	47.20	871	0.00	0
YAKIMA WA	0.00	0	0.43	2676	69.06	632	10.77	2654	2.62	3039	4.55	3020	0.00	0
ADAMS WI	0.00	0	0.71	2275	18.20	1855	46.24	718	33.50	494	38.71	1052	0.00	0
ASHLAND WI	0.00	0	2.13	1577	22.26	1647	64.85	454	17.06	1251	32.96	1268	0.00	0
BARRON WI	0.00	0	1.68	1763	27.62	1473	73.55	354	6.42	2375	29.43	1460	0.00	0
BAYFIELD WI	0.00	0	0.66	2321	25.39	1520	62.09	479	10.64	1858	31.68	1323	0.00	0
BROWN WI	0.00	0	0.47	2558	11.14	2462	15.78	2153	6.41	2380	33.16	1253	0.00	0
BUFFALO WI	0.00	0	1.93	1641	48.53	950	76.97	319	8.17	2102	21.93	2059	325.19	116
BURNETT WI	0.00	0	1.68	1764	31.06	1371	58.48	518	12.63	1652	32.09	1301	0.00	0
CALUMET WI	0.00	0	0.76	2243	62.25	735	20.48	1812	11.71	1738	33.10	1260	0.00	0
CHIPPEWA WI	0.00	0	0.55	2434	24.99	1535	76.93	321	11.07	1815	21.14	2126	0.00	0
CLARK WI	0.00	0	0.64	2345	45.61	1010	60.17	500	24.69	714	20.00	2255	0.00	0
COLUMBIA WI	0.00	0	0.73	2264	46.40	999	41.00	858	20.53	951	48.37	847	0.00	0
CRAWFORD WI	0.00	0	1.35	1912	44.28	1033	68.57	411	25.35	681	53.31	746	56.35	581
DANE WI	0.00	0	0.50	2528	20.24	1733	35.40	1057	10.31	1888	66.84	539	70.48	533
DODGE WI	0.00	0	0.17	2947	15.55	2028	29.39	1317	11.30	1790	35.60	1134	0.00	0
DOOR WI	0.00	0	0.57	2416	36.58	1212	19.85	1849	33.48	496	11.95	2862	0.00	0
DOUGLAS WI	0.00	0	1.90	1667	34.97	1258	62.06	481	15.32	1404	32.33	1286	0.00	0
DUNN WI	0.00	0	0.64	2346	55.69	846	81.48	285	10.02	1913	26.13	1666	0.00	0
EAU CLAIRE WI	0.00	0	1.60	1798	50.56	914	70.83	381	13.75	1551	20.97	2142	0.00	0
FLORENCE WI	0.00	0	0.56	2426	11.42	2428	40.32	886	4.50	2774	21.68	2081	0.00	0
FOND DU LAC WI	0.00	0	0.32	2795	14.34	2133	25.95	1477	16.71	1282	35.46	1142	0.00	0

TOTAL FALLOUT ACTIVITY (µCi/sq meter) by SHOT SERIES WITH COUNTY RANK

COUNTY AND STATE	RANGER SERIES	RANGER RANK	B-J SERIES	B-J RANK	T-S SERIES	T-S RANK	U-K SERIES	UK RANK	TEAPOT SERIES	TP RANK	PLUMBBOB SERIES	PB RANK	SHOT SEDAN	SEDAN RANK
	1951		1951		1952		1953		1955		1957		1962	
FOREST WI	0.00	0	0.41	2699	16.23	1987	35.52	1054	23.76	762	12.36	2845	0.00	0
GRANT WI	0.00	0	0.67	2316	43.00	1058	59.97	503	20.06	976	70.75	478	172.89	249
GREEN WI	0.00	0	0.50	2529	23.19	1606	42.22	820	9.41	1977	40.21	1025	120.50	352
GREEN LAKE WI	0.00	0	0.24	2854	62.52	729	30.14	1277	7.66	2162	37.41	1085	0.00	0
IOWA WI	0.00	0	0.78	2231	24.69	1550	50.98	635	20.82	927	49.12	831	345.56	100
IRON WI	0.00	0	1.17	2006	33.02	1314	60.79	494	23.43	779	26.61	1625	0.00	0
JACKSON WI	0.00	0	1.14	2019	39.52	1134	64.66	457	24.45	730	17.92	2451	0.00	0
JEFFERSON WI	0.00	0	0.55	2441	28.75	1442	29.90	1286	12.36	1675	41.33	1004	70.48	534
JUNEAU WI	0.00	0	0.98	2108	30.67	1387	53.08	603	33.39	499	41.73	998	0.00	0
KENOSHA WI	0.00	0	0.30	2798	63.67	717	24.55	1545	11.25	1795	13.42	2797	0.00	0
KEWAUNEE WI	0.00	0	0.37	2735	11.10	2470	13.77	2355	21.13	913	34.14	1203	0.00	0
LA CROSSE WI	0.00	0	1.73	1742	40.42	1116	72.77	366	6.00	2453	49.87	812	0.00	0
LAFAYETTE WI	0.00	0	0.78	2232	42.58	1067	48.96	662	11.41	1776	45.31	918	146.81	292
LANGLADE WI	0.00	0	0.55	2443	23.81	1577	40.80	863	17.21	1235	17.08	2536	0.00	0
LINCOLN WI	0.00	0	0.78	2228	58.55	790	46.37	715	26.15	651	21.27	2108	0.00	0
MANITOWOC WI	0.00	0	1.15	2012	23.53	1590	15.52	2180	11.89	1728	33.13	1258	0.00	0
MARATHON WI	0.00	0	0.98	2107	39.91	1127	45.20	744	16.55	1298	26.70	1617	0.00	0
MARINETTE WI	0.00	0	1.31	1934	18.52	1831	35.33	1061	15.61	1378	39.36	1041	0.00	0
MARQUETTE WI	0.00	0	0.33	2765	18.09	1862	41.18	853	31.65	521	40.99	1013	0.00	0
MENOMINEE WI	0.00	0	0.74	2255	21.08	1710	30.39	1268	16.24	1323	36.22	1117	0.00	0
MILWAUKEE WI	0.00	0	0.08	2983	23.36	1597	26.98	1432	6.26	2409	10.71	2910	0.00	0
MONROE WI	0.00	0	2.17	1566	38.11	1171	56.55	545	7.75	2147	43.17	968	0.00	0
OCONTO WI	0.00	0	1.31	1935	11.26	2446	29.61	1303	22.39	834	34.99	1152	0.00	0
ONEIDA WI	0.00	0	0.78	2229	43.60	1044	44.65	755	24.77	705	23.74	1874	0.00	0
OUTAGAMIE WI	0.00	0	0.50	2530	13.04	2257	20.62	1802	6.80	2301	32.75	1274	0.00	0
OZAUKEE WI	0.00	0	0.24	2855	15.53	2030	22.74	1653	7.00	2270	35.09	1151	0.00	0

TOTAL FALLOUT ACTIVITY (µCi/sq meter) by SHOT SERIES WITH COUNTY RANK

COUNTY AND STATE	RANGER SERIES	RANGER RANK	B-J SERIES	B-J RANK	T-S SERIES	T-S RANK	U-K SERIESI	UK RANK	TEAPOT SERIES	TP RANK	PLUMBBOB SERIES	PB RANK	SHOT SEDAN	SEDAN RANK
	1951		1951		1952		1953		1955		1957		1962	
PEPIN WI	0.00	0	1.45	1867	58.35	793	74.92	337	24.67	719	31.05	1359	0.00	0
PIERCE WI	0.00	0	1.81	1716	27.89	1465	67.56	421	6.32	2397	31.66	1325	0.00	0
POLK WI	0.00	0	1.68	1765	33.94	1287	61.98	482	6.93	2281	32.23	1291	0.00	0
PORTAGE WI	0.00	0	0.71	2276	17.34	1918	36.09	1033	23.87	757	38.04	1063	0.00	0
PRICE WI	0.00	0	0.55	2435	21.61	1680	70.11	387	11.95	1721	23.90	1857	0.00	0
RACINE WI	0.00	0	0.16	2952	33.58	1298	25.18	1513	12.66	1650	15.60	2651	0.00	0
RICHLAND WI	0.00	0	0.98	2109	31.61	1353	60.85	493	24.61	721	52.82	760	56.35	582
ROCK WI	0.00	0	0.40	2721	40.40	1117	32.48	1181	12.19	1694	48.50	843	139.13	322
RUSK WI	0.00	0	0.33	2766	24.74	1545	80.04	295	7.49	2185	30.30	1405	0.00	0
SAUK WI	0.00	0	0.88	2163	47.64	974	53.00	604	8.10	2109	66.57	546	59.46	557
SAWYER WI	0.00	0	1.45	1868	54.99	855	83.62	273	17.75	1185	26.83	1608	0.00	0
SHAWANO WI	0.00	0	0.74	2256	21.75	1675	29.78	1295	16.47	1303	36.78	1103	0.00	0
SHEBOYGAN WI	0.00	0	0.36	2753	57.21	821	17.72	2008	17.07	1249	33.46	1237	0.00	0
ST CROIX WI	0.00	0	1.81	1717	26.25	1505	69.96	389	6.58	2337	31.86	1314	0.00	0
TAYLOR WI	0.00	0	1.60	1799	46.80	990	64.81	455	17.14	1242	20.83	2162	0.00	0
TREMPEALEAU WI	0.00	0	1.08	2047	57.74	803	84.52	269	24.05	749	19.62	2295	0.00	0
VERNON WI	0.00	0	1.43	1874	53.08	880	67.49	422	17.96	1164	47.55	868	139.44	301
VILAS WI	0.00	0	0.78	2230	46.86	987	51.97	623	15.31	1406	24.02	1838	0.00	0
WALWORTH WI	0.00	0	0.78	2233	45.94	1003	29.05	1333	11.46	1770	46.90	878	172.89	254
WASHBURN WI	0.00	0	3.07	1382	27.64	1470	83.24	275	7.18	2245	30.72	1385	0.00	0
WASHINGTON WI	0.00	0	0.16	2953	15.96	1997	23.42	1613	17.58	1198	35.23	1146	0.00	0
WAUKESHA WI	0.00	0	0.24	2853	22.41	1635	29.44	1310	10.94	1826	37.36	1086	0.00	0
WAUPACA WI	0.00	0	0.60	2382	14.39	2123	29.71	1298	24.43	733	45.17	921	0.00	0
WAUSHARA WI	0.00	0	0.55	2436	16.23	1988	37.29	999	34.11	486	37.31	1088	0.00	0
WINNEBAGO WI	0.00	0	0.57	2412	19.16	1795	24.19	1572	12.73	1645	33.91	1213	0.00	0
WOOD WI	0.00	0	0.49	2542	25.34	1523	49.82	649	23.89	756	46.65	887	0.00	0

TOTAL FALLOUT ACTIVITY (µCi/sq meter) by SHOT SERIES WITH COUNTY RANK

COUNTY AND STATE	RANGER SERIES 1951	RANGER RANK	B-J SERIES 1951	B-J RANK	T-S SERIES 1952	T-S RANK	U-K SERIES 1953	UK RANK	TEAPOT SERIES 1955	TP RANK	PLUMBBOB SERIES 1957	PB RANK	SHOT SEDAN 1962	SEDAN RANK
BARBOUR WV	0.00	0	4.74	1058	9.13	2651	20.20	1832	12.56	1661	18.11	2433	0.00	0
BERKELEY WV	1.23	374	9.30	695	14.06	2153	11.15	2598	9.00	2013	17.42	2504	0.00	0
BOONE WV	0.00	0	11.21	585	10.89	2489	14.36	2299	8.06	2115	12.44	2843	359.52	64
BRAXTON WV	0.00	0	6.31	894	10.52	2523	22.62	1662	12.61	1656	19.52	2306	85.56	461
BROOKE WV	8.74	216	6.00	919	6.02	2942	18.54	1946	13.94	1537	15.28	2676	85.56	471
CABELL WV	0.00	0	8.12	768	13.12	2249	12.14	2496	16.83	1271	45.37	917	0.00	0
CALHOUN WV	0.00	0	7.29	824	10.25	2549	13.03	2420	12.15	1703	22.58	1986	103.59	361
CLAY WV	0.00	0	6.78	856	11.12	2467	18.75	1928	6.45	2366	14.00	2765	90.29	438
DODDRIDGE WV	1.17	400	5.20	987	10.31	2546	15.06	2218	14.33	1504	18.29	2415	0.00	0
FAYETTE WV	0.00	0	7.57	800	12.76	2280	15.07	2216	7.33	2215	19.01	2349	92.18	401
GILMER WV	0.00	0	5.53	962	9.77	2589	16.69	2089	10.31	1886	22.57	1988	103.59	362
GRANT WV	0.00	0	5.50	963	5.43	2982	17.83	1996	12.78	1639	20.64	2185	85.56	459
GREENBRIER WV	0.00	0	6.33	891	17.86	1882	13.35	2396	6.72	2312	21.66	2083	103.59	364
HAMPSHIRE WV	0.00	0	7.19	830	14.06	2152	11.55	2549	8.15	2104	20.10	2242	0.00	0
HANCOCK WV	8.74	217	14.13	449	5.81	2964	19.88	1847	22.88	808	18.05	2441	0.00	0
HARDY WV	0.00	0	7.37	818	5.53	2977	13.18	2404	6.54	2352	25.43	1730	57.04	563
HARRISON WV	1.17	401	4.86	1036	10.90	2487	16.21	2119	15.43	1391	18.70	2374	103.59	370
JACKSON WV	0.00	0	6.98	839	11.00	2478	20.80	1786	20.29	962	19.54	2303	0.00	0
JEFFERSON WV	1.23	375	9.87	666	13.21	2235	13.79	2348	8.36	2077	20.78	2166	0.00	0
KANAWHA WV	0.00	0	8.40	751	11.92	2368	16.78	2080	7.92	2130	22.38	2013	208.78	156
LEWIS WV	0.00	0	4.93	1017	10.77	2495	20.23	1829	17.15	1241	23.17	1923	85.56	450
LINCOLN WV	0.00	0	11.05	590	10.30	2547	18.11	1972	14.83	1448	36.19	1120	208.78	148
LOGAN WV	0.00	0	12.00	535	14.71	2098	12.59	2459	15.21	1416	33.76	1219	359.52	58
MARION WV	2.54	312	4.97	1011	6.59	2896	14.75	2255	20.63	941	28.33	1518	72.20	530
MARSHALL WV	21.83	64	5.15	993	3.94	3037	13.03	2418	19.48	1019	20.41	2213	103.59	367
MASON WV	0.00	0	9.46	687	12.70	2289	16.78	2079	19.38	1032	42.66	981	208.78	146

U.S. FALLOUT ATLAS : TOTAL FALLOUT

TOTAL FALLOUT ACTIVITY (µCi/sq meter) by SHOT SERIES WITH COUNTY RANK

COUNTY AND STATE	RANGER SERIES 1951	RANGER RANK	B-J SERIES 1951	B-J RANK	T-S SERIES 1952	T-S RANK	U-K SERIES 1953	UK RANK	TEAPOT SERIES 1955	TP RANK	PLUMBBOB SERIES 1957	PB RANK	SHOT SEDAN 1962	SEDAN RANK
MCDOWELL WV	0.00	0	8.24	760	14.88	2080	42.54	810	6.40	2384	39.91	1029	149.13	271
MERCER WV	0.00	0	12.21	522	16.31	1980	13.55	2377	6.71	2315	33.61	1228	149.13	274
MINERAL WV	2.54	313	6.58	874	4.39	3027	13.85	2344	11.42	1774	19.25	2325	0.00	0
MINGO WV	0.00	0	5.45	968	12.42	2318	20.97	1772	15.80	1364	35.56	1136	149.13	273
MONONGALIA WV	2.91	303	5.30	979	5.55	2976	13.60	2371	13.48	1573	22.42	2011	85.56	454
MONROE WV	0.00	0	8.28	758	18.74	1822	8.78	2868	6.11	2436	14.28	2747	208.78	179
MORGAN WV	1.23	376	7.51	807	13.71	2186	19.32	1892	7.63	2166	20.43	2212	0.00	0
NICHOLAS WV	0.00	0	8.31	757	12.82	2276	15.21	2205	7.02	2268	21.19	2120	85.56	457
OHIO WV	46.20	7	8.22	761	5.99	2948	18.97	1910	18.64	1083	16.66	2571	0.00	0
PENDLETON WV	0.00	0	5.14	994	5.89	2955	16.91	2067	5.74	2502	15.83	2634	208.78	177
PLEASANTS WV	39.78	8	6.30	895	8.31	2735	14.36	2298	15.98	1338	20.28	2222	0.00	0
POCAHONTAS WV	0.00	0	4.92	1018	7.35	2820	10.51	2686	6.82	2296	24.36	1818	90.29	423
PRESTON WV	2.54	314	5.03	1007	6.43	2909	17.85	1989	13.61	1566	25.65	1710	0.00	0
PUTNAM WV	0.00	0	11.95	539	9.31	2631	13.10	2407	15.40	1394	18.64	2380	85.56	464
RALEIGH WV	0.00	0	8.07	771	11.51	2418	12.94	2428	6.62	2328	14.02	2763	208.78	180
RANDOLPH WV	0.00	0	2.99	1392	7.30	2827	23.26	1623	8.26	2089	23.08	1932	85.56	451
RITCHIE WV	0.00	0	5.64	950	9.96	2577	13.96	2337	17.65	1188	18.40	2407	0.00	0
ROANE WV	0.00	0	6.70	862	8.92	2671	12.88	2436	11.10	1811	20.19	2231	0.00	0
SUMMERS WV	0.00	0	9.46	688	18.72	1825	11.54	2552	7.37	2208	27.20	1581	208.78	152
TAYLOR WV	2.54	315	5.38	974	5.03	3003	17.93	1984	11.66	1746	20.15	2234	0.00	0
TUCKER WV	0.00	0	5.07	1003	5.07	3000	22.83	1650	15.66	1371	23.23	1915	85.56	449
TYLER WV	21.83	65	7.47	812	10.52	2521	15.11	2212	14.76	1455	25.17	1752	90.29	421
UPSHUR WV	0.00	0	4.62	1078	13.48	2213	18.99	1908	18.72	1079	21.11	2132	103.59	365
WAYNE WV	0.00	0	9.01	714	13.13	2247	16.56	2098	18.03	1149	29.60	1449	208.78	150
WEBSTER WV	0.00	0	4.92	1021	9.26	2636	14.92	2233	6.41	2382	16.78	2558	90.29	435
WETZEL WV	8.74	218	5.24	982	6.97	2857	12.23	2486	20.16	971	24.19	1823	85.56	444

TOTAL FALLOUT ACTIVITY (μCi/sq meter) by SHOT SERIES WITH COUNTY RANK

COUNTY AND STATE	RANGER SERIES	RANGER RANK	B-J SERIES	B-J RANK	T-S SERIES	T-S RANK	U-K SERIES	UK RANK	TEAPOT SERIES	TP RANK	PLUMBBOB SERIES	PB RANK	SHOT SEDAN	SEDAN RANK
	1951		1951		1952		1953		1955		1957		1962	
WIRT WV	0.00	0	6.40	886	9.86	2585	21.50	1732	17.37	1220	17.29	2518	93.66	399
WOOD WV	19.29	78	9.21	700	9.89	2583	15.55	2176	15.78	1365	20.40	2214	85.56	460
WYOMING WV	0.00	0	8.59	735	13.96	2168	12.60	2458	5.18	2622	31.96	1309	149.13	275
ALBANY WY	0.00	0	0.98	2102	103.36	349	58.17	523	149.11	85	49.80	818	5.91	692
BIG HORN WY	0.00	0	0.49	2544	461.55	191	40.69	867	9.20	1996	53.10	753	0.00	0
CAMPBELL WY	0.00	0	0.57	2414	80.37	503	222.75	101	37.29	441	65.53	564	2.66	705
CARBON WY	0.00	0	1.82	1702	481.50	188	93.86	227	275.42	45	324.10	28	0.00	0
CONVERSE WY	0.00	0	0.30	2811	176.73	264	215.42	107	24.44	732	71.74	462	4.66	698
CROOK WY	0.00	0	0.91	2147	70.01	623	73.41	356	7.54	2177	82.14	376	2.00	711
FREMONT WY	0.00	0	0.91	2145	1746.73	97	631.80	34	70.57	230	345.59	27	0.00	0
GOSHEN WY	0.00	0	0.19	2908	91.41	406	30.61	1258	32.10	515	66.62	543	3.33	701
HOT SPRINGS WY	0.00	0	0.70	2282	260.14	224	42.08	825	6.93	2283	52.73	762	4.73	696
JOHNSON WY	0.00	0	0.30	2812	89.13	427	222.59	102	22.15	852	60.38	638	4.73	695
LARAMIE WY	0.00	0	1.37	1903	63.95	708	22.29	1679	84.17	193	31.37	1340	1.33	725
LINCOLN WY	0.00	0	1.81	1712	698.21	151	601.23	39	45.36	350	173.26	57	5.45	693
NATRONA WY	0.00	0	0.30	2813	114.59	331	113.77	175	14.02	1534	84.19	352	4.73	694
NIOBRARA WY	0.00	0	0.19	2909	333.42	201	24.79	1531	36.59	452	79.37	395	3.33	700
PARK WY	0.00	0	0.62	2364	617.31	157	35.69	1044	10.69	1848	100.70	244	0.00	0
PLATTE WY	0.00	0	0.19	2910	85.96	456	26.51	1451	164.45	76	48.15	858	3.33	703
SHERIDAN WY	0.00	0	0.49	2545	89.68	419	66.18	435	15.04	1432	66.59	544	0.00	0
SUBLETTE WY	0.00	0	0.91	2146	1774.48	96	587.68	41	39.72	413	185.34	48	0.00	0
SWEETWATER WY	0.00	0	2.46	1497	569.57	160	80.15	294	230.93	50	582.33	10	0.00	0
TETON WY	0.00	0	0.82	2201	950.17	120	20.22	1830	11.15	1806	90.21	317	0.00	0
UINTA WY	0.00	0	3.51	1301	689.06	152	342.66	79	159.24	80	220.63	40	0.00	0
WASHAKIE WY	0.00	0	0.41	2702	97.38	370	214.70	108	9.15	2000	48.23	852	4.73	697
WESTON WY	0.00	0	0.80	2217	65.25	690	68.66	410	8.05	2118	66.19	556	3.33	702

SECTION 10

COUNTIES RANKED BY RAINOUT POTENTIAL

1951-1962

COUNTIES RANKED BY RAINOUT POTENTIAL

(RAINFALL RANK x RADIOACTIVITY RANK, NORMALIZED).
LOW RANK VALUES INDICATE
HIGHER OVERALL RAINFALL-FALLOUT COMBINATION
1951-1962

COUNTY	RANK	COUNTY	RANK	COUNTY	RANK
NYE2 NV	1	UINTA WY	33	TRANSYLVANIA NC	65
WHITE PINE2 NV	2	MILLARD UT	34	RAPIDES LA	66
LINCOLN2 NV	3	MAYES OK	35	CRAIG OK	67
WHITE PINE1 NV	4	CHOCTAW OK	36	DUCHESNE UT	68
WHITE PINE3 NV	5	OUACHITA AR	37	DENT MO	69
DAVIS UT	6	BOX ELDER1 UT	38	LINCOLN WY	70
JUAB UT	7	CAMP TX	39	POTTAWATOMIE OK	71
TOOELE2 UT	8	CARROLL GA	40	EUREKA NV	72
WEBER UT	9	TULSA OK	41	GIBSON TN	73
TOOELE1 UT	10	PAWNEE OK	42	LINCOLN OK	74
SALT LAKE UT	11	OKMULGEE OK	43	SHELBY IA	75
WAGONER OK	12	MUSKOGEE OK	44	WEAKLEY TN	76
CHEROKEE OK	13	HENDERSON NC	45	BENNINGTON VT	77
BEAUREGARD LA	14	PUSHMATAHA OK	46	TIPPAH MS	78
DALLAS AR	15	LITTLE RIVER AR	47	DELAWARE OK	79
INYO3 CA	16	GRANT AR	48	MONTGOMERY AR	80
WASATCH UT	17	EMERY UT	49	SWEETWATER WY	81
BOX ELDER2 UT	18	PRAIRIE AR	50	PITTSBURG OK	82
MORGAN UT	19	MCINTOSH OK	51	CUSTER CO	83
UTAH UT	20	IRON1 UT	52	LA SALLE LA	84
SUMMIT UT	21	SEVIER AR	53	QUITMAN MS	85
LINCOLN1 NV	22	HOT SPRING AR	54	IBERIA LA	86
MCCURTAIN OK	23	TIPTON TN	55	PIKE AR	87
CACHE UT	24	GREENE AR	56	LAWRENCE AR	88
GRANT LA	25	IRON3 UT	57	TITUS TX	89
ATOKA OK	26	CARBON UT	58	RANDOLPH AR	90
HUGHES OK	27	OKFUSKEE OK	59	CARBON WY	91
HOWARD AR	28	MONROE IA	60	OSAGE OK	92
CALCASIEU LA	29	RICH UT	61	ROGERS OK	93
CREEK OK	30	SANPETE UT	62	WOOD TX	94
CLARK AR	31	NOBLE OK	63	COAHOMA MS	95
LAMAR TX	32	MACON NC	64	HAYWOOD TN	96

COUNTIES RANKED BY RAINOUT POTENTIAL

(RAINFALL RANK x RADIOACTIVITY RANK, NORMALIZED).
LOW RANK VALUES INDICATE
HIGHER OVERALL RAINFALL-FALLOUT COMBINATION
1951-1962

COUNTY	RANK	COUNTY	RANK	COUNTY	RANK
SULLIVAN MO	97	LAFAYETTE AR	129	HINSDALE CO	161
WILKINSON MS	98	SEQUOYAH OK	130	WOODS OK	162
FREMONT WY	99	BENTON IA	131	FAULKNER AR	163
UINTAH UT	100	SEVIER UT	132	QUEENS NY	164
CROSS AR	101	GARFIELD UT	133	RED RIVER TX	165
TUNICA MS	102	BOWIE TX	134	TISHOMINGO MS	166
TAMA IA	103	GARFIELD CO	135	CADDO OK	167
TALLAHATCHIE MS	104	RIO BLANCO CO	136	FANNIN TX	168
PICKENS SC	105	MISSISSIPPI MO	137	OTTAWA OK	169
ARKANSAS AR	106	OBION TN	138	LAFAYETTE LA	170
CRAIGHEAD AR	107	TOWNS GA	139	CROCKETT TN	171
VERNON MO	108	EFFINGHAM GA	140	FULTON AR	172
IRON2 UT	109	SHELBY MO	141	COWLEY KS	173
CHICKASAW MS	110	MILLER AR	142	POPE AR	174
NEW HAVEN CT	111	JACKSON AR	143	ESSEX NJ	175
WASHINGTON OK	112	VERNON LA	144	KNOX MO	176
SEMINOLE OK	113	MAHASKA IA	145	LAUDERDALE TN	177
KAY OK	114	MARSHALL MS	146	DAGGETT UT	178
SUBLETTE WY	115	VAN BUREN AR	147	SEDGWICK CO	179
ALFALFA OK	116	IOWA WI	148	YANCEY NC	180
MOHAVE2 AZ	117	OREGON MO	149	SHARP AR	181
NEWTON AR	118	NEW MADRID MO	150	SABINE LA	182
CHAUTAUQUA KS	119	DAVIS IA	151	PHELPS MO	183
CASS TX	120	BEAVER UT	152	HAMPSHIRE MA	184
CARROLL MS	121	BLAINE OK	153	SALINE AR	185
RABUN GA	122	DUNKLIN MO	154	FRANKLIN AR	186
MISSISSIPPI AR	123	WAYNE IA	155	BERGEN NJ	187
NOWATA OK	124	CLAY AR	156	ELKO NV	188
POLK AR	125	BRYAN OK	157	BALDWIN GA	189
PUTNAM MO	126	HARDEMAN TN	158	BUTLER AL	190
RICHLAND WI	127	SHELBY TN	159	LATIMER OK	191
CRAWFORD KS	128	HANCOCK IL	160	BENTON MS	192

COUNTIES RANKED BY RAINOUT POTENTIAL

**(RAINFALL RANK x RADIOACTIVITY RANK, NORMALIZED).
LOW RANK VALUES INDICATE
HIGHER OVERALL RAINFALL-FALLOUT COMBINATION
1951-1962**

COUNTY	RANK	COUNTY	RANK	COUNTY	RANK
GREENVILLE SC	193	LINN MO	225	BRADLEY AR	257
SEBASTIAN AR	194	WASHINGTON AR	226	CHITTENDEN VT	258
MORRIS TX	195	ORANGE TX	227	MADISON AR	259
PIUTE UT	196	YALOBUSHA MS	228	WESTCHESTER NY	260
UNION IL	197	ALEXANDER IL	229	ANDERSON TN	261
CARROLL IA	198	PHELPS NE	230	WATAUGA NC	262
BLECKLEY GA	199	KANE2 UT	231	WILKINSON GA	263
YELL AR	200	WINDHAM VT	232	PULASKI MO	264
ST FRANCIS AR	201	TATE MS	233	GRAND CO	265
FAYETTE TN	202	JOHNSTON OK	234	CARBON PA	266
COAL OK	203	JEFFERSON AR	235	BOLIVAR MS	267
WHITE AR	204	BENTON AR	236	LAFAYETTE MS	268
WAPELLO IA	205	AUDUBON IA	237	MCNAIRY TN	269
LEE MS	206	VAN ZANDT TX	238	HOWARD NE	270
HARTFORD CT	207	NEWTON MO	239	ADAIR OK	271
LEWIS MO	208	PERRY AR	240	FAIRFIELD CT	272
RICHMOND NY	209	DESHA AR	241	CASS IA	273
HUERFANO CO	210	OCEAN NJ	242	SCHUYLER MO	274
MCDOWELL WV	211	GARFIELD OK	243	CHARITON MO	275
GARLAND AR	212	WHITE GA	244	LEHIGH PA	276
INDEPENDENCE AR	213	EL PASO CO	245	NASSAU NY	277
GRUNDY TN	214	MONMOUTH NJ	246	FRANKLIN ID	278
HASKELL OK	215	UNION NJ	247	GRAYSON TX	279
LE FLORE OK	216	MOFFAT CO	248	VERNON WI	280
CADDO LA	217	CONEJOS CO	249	WABAUNSEE KS	281
NEVADA AR	218	CASS MO	250	LAFAYETTE WI	282
WAYNE UT	219	AVERY NC	251	HOUSTON MN	283
HUDSON NJ	220	PENDLETON KY	252	CLAY NC	284
RAINS TX	221	PULASKI AR	253	YAZOO MS	285
PONTOTOC MS	222	MARION IA	254	LESLIE KY	286
BEAR LAKE ID	223	RIPLEY MO	255	UNION MS	287
COLLIN TX	224	RUSSELL KS	256	BREMER IA	288

COUNTIES RANKED BY RAINOUT POTENTIAL

(RAINFALL RANK x RADIOACTIVITY RANK, NORMALIZED).
LOW RANK VALUES INDICATE
HIGHER OVERALL RAINFALL-FALLOUT COMBINATION
1951-1962

COUNTY	RANK	COUNTY	RANK	COUNTY	RANK
FANNIN GA	289	GRANT WI	321	LINN IA	353
LIPSCOMB TX	290	RINGGOLD IA	322	PITKIN CO	354
HANCOCK MS	291	HUNT TX	323	TOLLAND CT	355
ADAIR MO	292	MADISON TN	324	JOHNSON KY	356
DE SOTO LA	293	KIOWA KS	325	POLK NC	357
MINERAL CO	294	FULTON NY	326	HARPER KS	358
RENSSELAER NY	295	FORD KS	327	RUSK TX	359
CLEBURNE AR	296	CHEROKEE KS	328	CARTER MO	360
MCDONALD MO	297	GREGG TX	329	BARTON KS	361
GRENADA MS	298	BUFFALO NE	330	MCDONOUGH IL	362
TWIGGS GA	299	WARREN NJ	331	GORDON GA	363
CHESTER TN	300	MITCHELL NC	332	CANADIAN OK	364
CALHOUN MS	301	COLUMBIA NY	333	HARDIN TX	365
PANOLA MS	302	BUTLER MO	334	LAKE CO	366
MIDDLESEX CT	303	MORA NM	335	RUSH KS	367
ARCHULETA CO	304	JOHNSON AR	336	CALHOUN IA	368
SEVIER TN	305	CHEROKEE IA	337	GREENE MO	369
MARSHALL IA	306	GRAND UT	338	LOGAN OK	370
UNION IA	307	SAGUACHE CO	339	SALINE KS	371
ALEXANDER NC	308	CAMDEN MO	340	SMITH TX	372
BUENA VISTA IA	309	CLARK MO	341	LEE AR	373
BARBOUR AL	310	ORANGE NY	342	ERIE PA	374
HOWELL MO	311	LITCHFIELD CT	343	DEWEY OK	375
KINGS NY	312	MENIFEE KY	344	LEFLORE MS	376
NESS KS	313	MIAMI KS	345	BURLINGTON NJ	377
ST FRANCOIS MO	314	OKLAHOMA OK	346	SARATOGA NY	378
PASSAIC NJ	315	GREENE NY	347	SCOTLAND MO	379
BERKSHIRE MA	316	JASPER TX	348	BARRY MO	380
CRAWFORD WI	317	RANDOLPH AL	349	MORRIS NJ	381
MACON MO	318	PEPIN WI	350	MONTGOMERY KS	382
HOPKINS TX	319	TREGO KS	351	CAPE MAY NJ	383
RALLS MO	320	LANE KS	352	LAKE TN	384

COUNTIES RANKED BY RAINOUT POTENTIAL

(RAINFALL RANK x RADIOACTIVITY RANK, NORMALIZED).
LOW RANK VALUES INDICATE
HIGHER OVERALL RAINFALL-FALLOUT COMBINATION
1951-1962

COUNTY	RANK	COUNTY	RANK	COUNTY	RANK
CHEROKEE TX	385	HENDERSON IL	417	SUSSEX NJ	449
WALLACE KS	386	FERGUS MT	418	DE SOTO MS	450
VALLEY ID	387	DOUGLAS KS	419	YADKIN NC	451
MARTIN IN	388	KEOKUK IA	420	SCHUYLKILL PA	452
COOKE TX	389	PULASKI IL	421	KNOTT KY	453
NEW YORK NY	390	HUNTERDON NJ	422	HOUSTON TX	454
BUTLER KS	391	KINGMAN KS	423	PUEBLO CO	455
BIENVILLE LA	392	DECATUR TN	424	GRUNDY MO	456
LOGAN AR	393	WARREN PA	425	LOVE OK	457
NEWTON TX	394	WAYNE MO	426	WYANDOTTE KS	458
ULSTER NY	395	ADDISON VT	427	LINCOLN AR	459
WRIGHT MO	396	ATTALA MS	428	ALLEGHANY NC	460
CAMPBELL TN	397	LOGAN WV	429	GRANT KS	461
POINSETT AR	398	PUTNAM NY	430	OKTIBBEHA MS	462
TEXAS MO	399	SUFFOLK NY	431	AMITE MS	463
STONE MS	400	SCREVEN GA	432	SHAWNEE KS	464
BERKS PA	401	PEACH GA	433	MCDOWELL NC	465
SUNFLOWER MS	402	MARION MO	434	CHESTER PA	466
SCOTT AR	403	MIDDLESEX NJ	435	HICKMAN KY	467
WARREN NY	404	TELLER CO	436	FREMONT CO	468
SUMNER KS	405	MAJOR OK	437	MAGOFFIN KY	469
UNION GA	406	COMANCHE OK	438	LA PLATA CO	470
O BRIEN IA	407	SAUK WI	439	WOODRUFF AR	471
HOUSTON GA	408	MCCLAIN OK	440	MONROE PA	472
HAMPDEN MA	409	MIDDLESEX MA	441	BRONX NY	473
CRAWFORD GA	410	GEORGE MS	442	LIBERTY FL	474
HOLMES FL	411	LONOKE AR	443	HUTCHINSON SD	475
LEAVENWORTH KS	412	MONTGOMERY MD	444	ONEIDA ID	476
APPANOOSE IA	413	ADAMS ID	445	LOUDOUN VA	477
PRENTISS MS	414	MONTGOMERY PA	446	ARMSTRONG TX	478
WINDHAM CT	415	MONTGOMERY MS	447	CALDWELL MO	479
STORY IA	416	CRAWFORD AR	448	CLEBURNE AL	480

COUNTIES RANKED BY RAINOUT POTENTIAL

(RAINFALL RANK x RADIOACTIVITY RANK, NORMALIZED).
LOW RANK VALUES INDICATE
HIGHER OVERALL RAINFALL-FALLOUT COMBINATION
1951-1962

COUNTY	RANK	COUNTY	RANK	COUNTY	RANK
WARREN MS	481	AUDRAIN MO	513	OTTAWA KS	545
LEE IA	482	ACADIA LA	514	GREENWOOD KS	546
CRITTENDEN AR	483	RAY MO	515	MONTROSE CO	547
MARLBORO SC	484	LAKE SD	516	SUMMIT CO	548
RANDOLPH GA	485	BULLOCH GA	517	SCOTT KS	549
CEDAR IA	486	POLK MO	518	WARREN GA	550
PAYNE OK	487	DAKOTA MN	519	GARFIELD MT	551
REYNOLDS MO	488	CLINTON IA	520	MERCER NJ	552
PANOLA TX	489	DYER TN	521	HOWARD IA	553
VENANGO PA	490	FRANKLIN IA	522	KEARNY KS	554
MESA CO	491	WILCOX AL	523	FRANKLIN KS	555
PHILLIPS AR	492	IZARD AR	524	WEBSTER LA	556
WASHINGTON NY	493	STONE AR	525	JOHNSON IL	557
NEOSHO KS	494	MERRICK NE	526	CHESHIRE NH	558
GUNNISON CO	495	COLFAX NM	527	JACKSON NC	559
GRANT OK	496	LEE VA	528	HODGEMAN KS	560
CARIBOU ID	497	ELK PA	529	JASPER IL	561
TERREBONNE LA	498	AUSTIN TX	530	IRON MO	562
HOWARD MD	499	SCHOHARIE NY	531	FRANKLIN MA	563
COLUMBIA WI	500	STODDARD MO	532	SOMERSET NJ	564
PEMISCOT MO	501	EDWARDS KS	533	BRYAN GA	565
BOONE IA	502	CHATHAM GA	534	CHAUTAUQUA NY	566
DECATUR IA	503	SUFFOLK MA	535	SEARCY AR	567
POWESHIEK IA	504	POINTE COUPE LA	536	NEW LONDON CT	568
MURRAY GA	505	MCDUFFIE GA	537	ALCORN MS	569
TROUP GA	506	LAWRENCE MO	538	WINDSOR VT	570
WILKES NC	507	WASHITA OK	539	MARENGO AL	571
LAMOILLE VT	508	HAYWOOD NC	540	VIGO IN	572
ELK KS	509	MARSHALL OK	541	PETTIS MO	573
ROGER MILLS OK	510	HAMILTON NY	542	MORGAN MO	574
COLUMBIA AR	511	BLEDSOE TN	543	WINNESHIEK IA	575
STANTON KS	512	CALHOUN FL	544	IDAHO ID	576

COUNTIES RANKED BY RAINOUT POTENTIAL

(RAINFALL RANK x RADIOACTIVITY RANK, NORMALIZED).
LOW RANK VALUES INDICATE
HIGHER OVERALL RAINFALL-FALLOUT COMBINATION
1951-1962

COUNTY	RANK	COUNTY	RANK	COUNTY	RANK
EFFINGHAM IL	577	WHITE TN	609	MERCER WV	641
GOODHUE MN	578	UPSHUR TX	610	PERSHING NV	642
ELMORE ID	579	ADA ID	611	PRATT KS	643
LOGAN CO	580	KINGFISHER OK	612	MEADE KS	644
NEW CASTLE DE	581	BAXTER AR	613	SAC IA	645
MORRIS KS	582	CRAWFORD MO	614	CHAMBERS AL	646
SUSSEX DE	583	BAILEY TX	615	PICKETT TN	647
HARLAN KY	584	CONCORDIA LA	616	STRAFFORD NH	648
HASKELL KS	585	OLDHAM TX	617	AROOSTOOK ME	649
GREELEY NE	586	BENTON MO	618	STEELE MN	650
OSBORNE KS	587	BUCKS PA	619	MOBILE AL	651
WHEELER TX	588	CLAYTON IA	620	CALVERT MD	652
MINGO WV	589	LAMAR MS	621	LOGAN NE	653
ADAMS IA	590	MONTGOMERY IA	622	NORTON KS	654
BERNALILLO NM	591	PIKE MS	623	PHILLIPS CO	655
DALLAM TX	592	MONROE MO	624	ROCKLAND NY	656
LINN KS	593	SAN AUGUSTIN TX	625	PIKE PA	657
ETOWAH AL	594	BALTIMORE MD	626	NICOLLET MN	658
CLAY IL	595	MARION KS	627	MONROE AL	659
ERIE NY	596	EDWARDS IL	628	RUTLAND VT	660
FOREST PA	597	STEPHENS OK	629	EAGLE CO	661
HARRISON IA	598	ROANE TN	630	GREENE GA	662
GLASCOCK GA	599	JOHNSON MO	631	BURT NE	663
GOSPER NE	600	WORCESTER MA	632	DELTA CO	664
DOUGLAS CO	601	MERRIMACK NH	633	ALLAMAKEE IA	665
BOISE ID	602	CLAY MO	634	SHANNON MO	666
WHEELER GA	603	CLEVELAND AR	635	DADE MO	667
OWYHEE ID	604	ALLEN LA	636	DELAWARE PA	668
MCCOOK SD	605	MONONA IA	637	MADISON IA	669
HAMILTON KS	606	MARION TX	638	LINCOLN NE	670
CLARK KS	607	HARRISON TX	639	HARRISON KY	671
SCOTT MO	608	JACKSON IL	640	CHEROKEE NC	672

COUNTIES RANKED BY RAINOUT POTENTIAL

(RAINFALL RANK x RADIOACTIVITY RANK, NORMALIZED).
LOW RANK VALUES INDICATE
HIGHER OVERALL RAINFALL-FALLOUT COMBINATION
1951-1962

COUNTY	RANK	COUNTY	RANK	COUNTY	RANK
LANDER2 NV	673	MORGAN TN	705	SUMTER FL	737
GARVIN OK	674	MCMINN TN	706	CLEVELAND OK	738
JACKSON MS	675	FRANKLIN MS	707	DALLAS IA	739
COTTON OK	676	WILCOX GA	708	CLEAR CREEK CO	740
WISE VA	677	JEFFERSON KS	709	ELLSWORTH KS	741
DOUGLAS MO	678	SULLIVAN NH	710	POTTAWATTAMI IA	742
ADAMS IL	679	ORLEANS LA	711	BREATHITT KY	743
LAUREL KY	680	JEFFERSON LA	712	KANE1 UT	744
GRAFTON NH	681	CLINTON MO	713	WYOMING WV	745
MONROE MS	682	COSTILLA CO	714	SALINE MO	746
DELAWARE IA	683	MOORE NC	715	SANDOVAL NM	747
HENRY MO	684	ADAIR IA	716	CLARION PA	748
DENTON TX	685	MCKEAN PA	717	HARRISON MO	749
WOLFE KY	686	AVOYELLES LA	718	MCKINLEY NM	750
PARK CO	687	FLOYD KY	719	CHATTOOGA GA	751
OCHILTREE TX	688	GOVE KS	720	MITCHELL IA	752
ST MARY LA	689	PISCATAQUIS ME	721	MONTGOMERY NY	753
METCALFE KY	690	CUSTER OK	722	MORGAN GA	754
ALBANY NY	691	MONROE AR	723	MCLEAN ND	755
OSCEOLA IA	692	ATLANTIC NJ	724	CLAIBORNE TN	756
JEFFERSON DA MS	693	CAPE GIRARDE MO	725	FRANKLIN TX	757
NORTHAMPTON PA	694	HOWARD MO	726	CHEYENNE KS	758
ELLIS KS	695	APACHE AZ	727	CUMBERLAND IL	759
BOURBON KS	696	TIPPECANOE IN	728	ASHE NC	760
JASPER MO	697	NACOGDOCHES TX	729	DICKINSON KS	761
ESSEX NY	698	CONWAY AR	730	SAN MIGUEL NM	762
JEFFERSON CO	699	JOHNSON KS	731	PONTOTOC OK	763
WAYNE MS	700	ROCK ISLAND IL	732	MARION MS	764
POWDER RIVER MT	701	MONTEZUMA CO	733	WAYNE IL	765
ST BERNARD LA	702	CATTARAUGUS NY	734	CHATTAHOOCHE GA	766
DECATUR KS	703	WILSON KS	735	WASHINGTON1 UT	767
EVANGELINE LA	704	PERRY KY	736	JENKINS GA	768

COUNTIES RANKED BY RAINOUT POTENTIAL

(RAINFALL RANK x RADIOACTIVITY RANK, NORMALIZED).
LOW RANK VALUES INDICATE
HIGHER OVERALL RAINFALL-FALLOUT COMBINATION
1951-1962

COUNTY	RANK	COUNTY	RANK	COUNTY	RANK
VALENCIA NM	769	ONSLOW NC	801	CRAWFORD IA	833
SWISHER TX	770	FAYETTE IA	802	CALHOUN AL	834
MASON WV	771	WORTH IA	803	LIVINGSTON LA	835
OLIVER ND	772	CARLISLE KY	804	MARION TN	836
CHOCTAW MS	773	SAN JUAN CO	805	DAWSON MT	837
COOPER MO	774	LINCOLN CO	806	COLES IL	838
LAS ANIMAS CO	775	MOWER MN	807	FLOYD GA	839
LACLEDE MO	776	WASHINGTON MS	808	BOULDER CO	840
CRAWFORD PA	777	STEPHENSON IL	809	COCONINO1 AZ	841
WASHINGTON IA	778	ST TAMMANY LA	810	FRANKLIN MO	842
IDA IA	779	CLARKE IA	811	JEFFERSON PA	843
HITCHCOCK NE	780	WORTH MO	812	PEARL RIVER MS	844
JASPER IA	781	WASHINGTON2 UT	813	KIT CARSON CO	845
BARBER KS	782	HARDING NM	814	CUMBERLAND TN	846
VAN BUREN IA	783	ROOKS KS	815	WAYNE WV	847
CEDAR NE	784	SWAIN NC	816	MOHAVE1 AZ	848
SAN JUAN UT	785	ADAMS NE	817	SABINE TX	849
GLYNN GA	786	WINNEBAGO IL	818	EAST FELICIA LA	850
COOS NH	787	SULLIVAN NY	819	GRAHAM KS	851
CLAYTON GA	788	LAFAYETTE MO	820	BATES MO	852
FILLMORE MN	789	TALIAFERRO GA	821	LANDER1 NV	853
KNOX IN	790	CLARKE AL	822	UNION AR	854
HARFORD MD	791	FULTON KY	823	CONECUH AL	855
THOMAS NE	792	TAZEWELL IL	824	ROCKWALL TX	856
BEAVER OK	793	ITAWAMBA MS	825	DE KALB MO	857
HARPER OK	794	HINDS MS	826	LAMB TX	858
CAMDEN NJ	795	LINCOLN MS	827	GREENE MS	859
JUNEAU WI	796	BRADLEY TN	828	CAMBRIA PA	860
OZARK MO	797	LARIMER CO	829	GRAY KS	861
GEARY KS	798	SANTA FE NM	830	SAN MIGUEL CO	862
MORGAN IN	799	TURNER SD	831	STEVENS KS	863
GEAUGA OH	800	HYDE NC	832	ALBANY WY	864

COUNTIES RANKED BY RAINOUT POTENTIAL

(RAINFALL RANK x RADIOACTIVITY RANK, NORMALIZED).
LOW RANK VALUES INDICATE
HIGHER OVERALL RAINFALL-FALLOUT COMBINATION
1951-1962

COUNTY	RANK	COUNTY	RANK	COUNTY	RANK
ST MARTIN LA	865	CHESTERFIELD SC	897	KNOX KY	929
FLOYD IA	866	STAFFORD KS	898	CASTRO TX	930
SULLIVAN IN	867	LIVINGSTON MO	899	FRANKLIN VT	931
HARVEY KS	868	GREENE IA	900	POLK NE	932
ELLIS OK	869	IRON WI	901	DOUGLAS IL	933
CARROLL TN	870	SALINE NE	902	OURAY CO	934
DAVIESS IN	871	DE KALB AL	903	TILLMAN OK	935
VALLEY NE	872	MERCER MO	904	GUTHRIE IA	936
ASSUMPTION LA	873	OXFORD ME	905	WICOMICO MD	937
CIMARRON OK	874	LUCAS IA	906	WASHINGTON MO	938
JASPER SC	875	DUBOIS IN	907	UNION NM	939
RIO GRANDE CO	876	OUACHITA LA	908	GLOUCESTER NJ	940
BURKE NC	877	ORANGE IN	909	CUSTER NE	941
CARROLL AR	878	IREDELL NC	910	MADISON FL	942
BACA CO	879	KNOX TN	911	TERRELL GA	943
CECIL MD	880	MARION IL	912	KIOWA OK	944
STANTON NE	881	FURNAS NE	913	SEWARD KS	945
CHASE NE	882	HERKIMER NY	914	MADISON MS	946
MADISON MO	883	WYOMING NY	915	HUMPHREYS TN	947
CHRISTIAN MO	884	HUMPHREYS MS	916	HALL NE	948
PEORIA IL	885	ALAMOSA CO	917	ANDROSCOGGIN ME	949
RANDOLPH MO	886	GEM ID	918	WALKER TX	950
HARTLEY TX	887	BOSSIER LA	919	KING WILLIAM VA	951
TENSAS LA	888	HICKORY MO	920	WOODSON KS	952
JEFFERSON IA	889	LUZERNE PA	921	WASHINGTON DC	953
RILEY KS	890	NORFOLK MA	922	POTTAWATOMIE KS	954
BLACK HAWK IA	891	VERMILION LA	923	MORTON KS	955
ROCKINGHAM NH	892	BOONE NE	924	ORANGE VT	956
COSHOCTON OH	893	TAZEWELL VA	925	SHARKEY MS	957
RHEA TN	894	CARTER TN	926	BIG HORN MT	958
JERAULD SD	895	DODGE GA	927	SCHENECTADY NY	959
BLAIR PA	896	SHERMAN NE	928	ANDERSON KS	960

COUNTIES RANKED BY RAINOUT POTENTIAL

(RAINFALL RANK x RADIOACTIVITY RANK, NORMALIZED).
LOW RANK VALUES INDICATE
HIGHER OVERALL RAINFALL-FALLOUT COMBINATION
1951-1962

COUNTY	RANK	COUNTY	RANK	COUNTY	RANK
MCPHERSON KS	961	DIVIDE ND	993	PIKE IL	1025
CARTER OK	962	TAYLOR IA	994	RED WILLOW NE	1026
INDIANA PA	963	STAFFORD VA	995	ARLINGTON VA	1027
CALDWELL NC	964	ST CROIX WI	996	WASHINGTON3 UT	1028
DAVISON SD	965	WEBSTER GA	997	STARK ND	1029
LEE IL	966	BOLLINGER MO	998	GENEVA AL	1030
DREW AR	967	PERRY MO	999	CHICKASAW IA	1031
PUTNAM TN	968	CHURCHILL NV	1000	TAYLOR GA	1032
CROWLEY CO	969	WEST FELICIA LA	1001	WASHINGTON VT	1033
OCONEE SC	970	IBERVILLE LA	1002	WASHINGTON ID	1034
ANDERSON TX	971	LYON KS	1003	DIXON NE	1035
DUTCHESS NY	972	PAWNEE KS	1004	GILMER GA	1036
ST JOHN THE LA	973	NANCE NE	1005	FENTRESS TN	1037
POCAHONTAS IA	974	HEMPSTEAD AR	1006	WILLIAMSON TN	1038
ATCHISON KS	975	RED RIVER LA	1007	MONROE TN	1039
BARTON MO	976	QUAY NM	1008	DEAF SMITH TX	1040
FULTON IL	977	GREEN WI	1009	LIBERTY GA	1041
BUNCOMBE NC	978	PRINCE WILLI VA	1010	BURNETT WI	1042
LINCOLN ME	979	CUMING NE	1011	POLK TX	1043
WEST BATON R LA	980	DAWSON NE	1012	CEDAR MO	1044
HOLT NE	981	SCHLEY GA	1013	NAVAJO AZ	1045
DELTA TX	982	PRINCE GEORG MD	1014	HILLSBOROUGH NH	1046
ROUTT CO	983	COLUMBUS GA	1015	BEAUFORT NC	1047
RUTHERFORD NC	984	JACKSON KS	1016	FAYETTE IL	1048
ESSEX MA	985	CHEYENNE CO	1017	FRONTIER NE	1049
JASPER IN	986	MONROE WI	1018	CONVERSE WY	1050
PULASKI GA	987	TAOS NM	1019	DUNN ND	1051
JEFFERSON WV	988	CAMDEN GA	1020	CLAIBORNE MS	1052
DUVAL FL	989	WHITFIELD GA	1021	MILLER MO	1053
ESCAMBIA AL	990	ISSAQUENA MS	1022	TALLAPOOSA AL	1054
HUMBOLDT NV	991	DAVIESS MO	1023	CLAY IA	1055
POLK IA	992	WHITESIDE IL	1024	JEFFERSON GA	1056

U.S. FALLOUT ATLAS : TOTAL FALLOUT

COUNTIES RANKED BY RAINOUT POTENTIAL

(RAINFALL RANK x RADIOACTIVITY RANK, NORMALIZED).
LOW RANK VALUES INDICATE
HIGHER OVERALL RAINFALL-FALLOUT COMBINATION
1951-1962

COUNTY	RANK	COUNTY	RANK	COUNTY	RANK
CUMBERLAND ME	1057	YUMA CO	1089	CAMPBELL WY	1121
PLATTE NE	1058	STONE MO	1090	LOGAN KS	1122
WOODWARD OK	1059	WABASH IL	1091	PATRICK VA	1123
CARROLL NH	1060	MONTGOMERY GA	1092	MCCONE MT	1124
KENT DE	1061	CURRITUCK NC	1093	FREDERICK MD	1125
GRAHAM NC	1062	DORCHESTER MD	1094	GIBSON IN	1126
GREER OK	1063	ROOSEVELT MT	1095	CATAWBA NC	1127
HARDIN IA	1064	MURRAY OK	1096	MARSHALL AL	1128
MEIGS TN	1065	CAMDEN NC	1097	LEWIS NY	1129
BALLARD KY	1066	ADAMS MS	1098	CLEARFIELD PA	1130
WEBSTER MO	1067	LABETTE KS	1099	UNION SD	1131
LEE GA	1068	CLEARWATER ID	1100	CHEATHAM TN	1132
THOMAS KS	1069	WALKER GA	1101	MCINTOSH GA	1133
PIKE MO	1070	JEFFERSON DA LA	1102	SHELBY TX	1134
BEDFORD VA	1071	CALLAWAY MO	1103	THURSTON NE	1135
HANCOCK GA	1072	MUSCATINE IA	1104	JACKSON OK	1136
SCOTT IA	1073	TUCKER WV	1105	DANE WI	1137
ASHTABULA OH	1074	WINSTON MS	1106	SEQUATCHIE TN	1138
KEARNEY NE	1075	FAYETTE WV	1107	LOGAN IL	1139
KIOWA CO	1076	PRINCE GEORG VA	1108	PLYMOUTH MA	1140
MARTIN KY	1077	HEMPHILL TX	1109	LAKE FL	1141
COCONINO2 AZ	1078	CAROLINE MD	1110	FORREST MS	1142
CARSON TX	1079	STARK IL	1111	PASCO FL	1143
BENT CO	1080	SHERIDAN KS	1112	BELKNAP NH	1144
PLAQUEMINES LA	1081	JEFFERSON MS	1113	DARLINGTON SC	1145
MADISON NE	1082	FAIRFAX VA	1114	CANYON ID	1146
GRANT KY	1083	MCCREARY KY	1115	ANNE ARUNDEL MD	1147
DALLAS MO	1084	GUADALUPE NM	1116	WOODFORD IL	1148
CRAWFORD IL	1085	WEBSTER MS	1117	BLAINE MT	1149
JACKSON GA	1086	AURORA SD	1118	CHAFFEE CO	1150
KENT MD	1087	SURRY NC	1119	LINCOLN SD	1151
WOODBURY IA	1088	NEWTON IN	1120	MADISON LA	1152

COUNTIES RANKED BY RAINOUT POTENTIAL

(RAINFALL RANK x RADIOACTIVITY RANK, NORMALIZED).
LOW RANK VALUES INDICATE
HIGHER OVERALL RAINFALL-FALLOUT COMBINATION
1951-1962

COUNTY	RANK	COUNTY	RANK	COUNTY	RANK
THOMAS GA	1153	SIOUX IA	1185	EAST CARROLL LA	1217
WAYNE NE	1154	SEDGWICK KS	1186	CLARK1 NV	1218
RAWLINS KS	1155	PROWERS CO	1187	RANDOLPH WV	1219
MATHEWS VA	1156	ANDREWS TX	1188	STEUBEN NY	1220
WASHBURN WI	1157	NEMAHA KS	1189	UNICOI TN	1221
PERRY MS	1158	PHILADELPHIA PA	1190	LOS ALAMOS NM	1222
SOMERSET PA	1159	LETCHER KY	1191	DICKSON TN	1223
HARRISON MS	1160	MERCER PA	1192	RICE KS	1224
LAUDERDALE MS	1161	CUMBERLAND NJ	1193	TORRANCE NM	1225
HANSON SD	1162	BACON GA	1194	LINCOLN KS	1226
ST MARYS MD	1163	LOUISA IA	1195	COLLETON SC	1227
RUSH IN	1164	SAWYER WI	1196	ROSEBUD MT	1228
HANSFORD TX	1165	BOND IL	1197	CHEYENNE NE	1229
ROBERTS TX	1166	BIBB GA	1198	PIERCE WI	1230
MARION FL	1167	GRAND ISLE VT	1199	BANNOCK ID	1231
OSAGE KS	1168	ASHLAND WI	1200	RENO KS	1232
MARION AR	1169	LAWRENCE MS	1201	HENRY VA	1233
WINONA MN	1170	FRANKLIN IN	1202	SAGADAHOC ME	1234
JEFFERSON IN	1171	LE SUEUR MN	1203	CHRISTIAN IL	1235
POLK TN	1172	LAWRENCE IL	1204	CLAIBORNE LA	1236
BRAZORIA TX	1173	TALBOT MD	1205	NUCKOLLS NE	1237
WASHOE NV	1174	BUTLER NE	1206	ANTELOPE NE	1238
HAMILTON IA	1175	SEWARD NE	1207	OTERO CO	1239
CLAY MS	1176	MCKENZIE ND	1208	ST LANDRY LA	1240
DADE GA	1177	HAYES NE	1209	CLARK ID	1241
SCOTT IL	1178	WARREN VA	1210	CLARK WI	1242
LAWRENCE SD	1179	BUTLER IA	1211	MOORE TX	1243
WICHITA KS	1180	MADISON VA	1212	PROVIDENCE RI	1244
LONG GA	1181	BROWN IL	1213	TEXAS OK	1245
BUREAU IL	1182	CAMERON LA	1214	PERRY TN	1246
MONTGOMERY KY	1183	JOHNSON IA	1215	CLAY KS	1247
MONROE GA	1184	ASCENSION LA	1216	DE KALB IL	1248

COUNTIES RANKED BY RAINOUT POTENTIAL

(RAINFALL RANK x RADIOACTIVITY RANK, NORMALIZED).
LOW RANK VALUES INDICATE
HIGHER OVERALL RAINFALL-FALLOUT COMBINATION
1951-1962

COUNTY	RANK	COUNTY	RANK	COUNTY	RANK
MARION IN	1249	CHEROKEE GA	1281	MACON IL	1313
JONES NC	1250	TREASURE MT	1282	SUMMERS WV	1314
SLOPE ND	1251	GREENE IN	1283	MINER SD	1315
WARREN IA	1252	BUCHANAN IA	1284	JACKSON CO	1316
BOONE AR	1253	WINN LA	1285	JACKSON IN	1317
CARROLL IL	1254	TREUTLEN GA	1286	KNOX IL	1318
SNYDER PA	1255	POLK WI	1287	KENT RI	1319
LAWRENCE IN	1256	MOUNTRAIL ND	1288	GILCHRIST FL	1320
MONTGOMERY TN	1257	TAYLOR FL	1289	JACKSON OH	1321
DOLORES CO	1258	SPARTANBURG SC	1290	IROQUOIS IL	1322
WASECA MN	1259	WASHINGTON LA	1291	ARMSTRONG PA	1323
HAMPTON SC	1260	DODGE NE	1292	HENDERSON TX	1324
CAMAS ID	1261	PINE MN	1293	PERQUIMANS NC	1325
LINCOLN LA	1262	WABASHA MN	1294	BEAUFORT SC	1326
PIKE KY	1263	DOUGHERTY GA	1295	GALLIA OH	1327
KING AND QUE VA	1264	WAKULLA FL	1296	GILPIN CO	1328
NORTHUMBERLA PA	1265	KEITH NE	1297	HUMBOLDT IA	1329
PULASKI IN	1266	YORK ME	1298	LOUDON TN	1330
EAU CLAIRE WI	1267	CASSIA ID	1299	NORTHAMPTON VA	1331
GOLDEN VALLE ND	1268	PETROLEUM MT	1300	PENOBSCOT ME	1332
DEUEL NE	1269	CLINTON IN	1301	ST CHARLES LA	1333
ALLEN KS	1270	HICKMAN TN	1302	MIDDLESEX VA	1334
JEFFERSON OK	1271	WASHINGTON RI	1303	CLAY TX	1335
OTSEGO NY	1272	YORK PA	1304	BLOUNT AL	1336
COVINGTON AL	1273	FAUQUIER VA	1305	JEFF DAVIS GA	1337
TREMPEALEAU WI	1274	DELAWARE NY	1306	FRANKLIN PA	1338
MARION SC	1275	JACKSON IA	1307	GENESEE NY	1339
HANCOCK ME	1276	CHASE KS	1308	LA PORTE IN	1340
CATAHOULA LA	1277	FREMONT ID	1309	PHILLIPS MT	1341
IOWA IA	1278	MADISON NC	1310	JACKSON AL	1342
DORCHESTER SC	1279	COFFEY KS	1311	ASHLEY AR	1343
GARFIELD NE	1280	BOYD NE	1312	SULLIVAN TN	1344

COUNTIES RANKED BY RAINOUT POTENTIAL

(RAINFALL RANK x RADIOACTIVITY RANK, NORMALIZED).
LOW RANK VALUES INDICATE
HIGHER OVERALL RAINFALL-FALLOUT COMBINATION
1951-1962

COUNTY	RANK	COUNTY	RANK	COUNTY	RANK
LOVING TX	1345	CULPEPER VA	1377	LA SALLE IL	1409
BRULE SD	1346	CAMERON PA	1378	TWIN FALLS ID	1410
JACKSON MO	1347	CHARLES MIX SD	1379	SUMTER SC	1411
HALE TX	1348	COLE MO	1380	COLFAX NE	1412
WASHINGTON NE	1349	ESSEX VT	1381	GRADY OK	1413
CALEDONIA VT	1350	JEFFERSON MO	1382	WEBSTER NE	1414
SHERIDAN MT	1351	CLINTON OH	1383	CHARLESTON SC	1415
FORSYTH NC	1352	SENECA OH	1384	HARLAN NE	1416
KINGSBURY SD	1353	HERNANDO FL	1385	HANCOCK KY	1417
HEARD GA	1354	LEAKE MS	1386	MARIES MO	1418
GREELEY KS	1355	ISANTI MN	1387	OSAGE MO	1419
PUTNAM IN	1356	OWEN IN	1388	LIVINGSTON NY	1420
BOONE IL	1357	JONES MS	1389	SCOTT MN	1421
CARLTON MN	1358	DOUGLAS SD	1390	HANOVER VA	1422
CARROLL MD	1359	WEBSTER IA	1391	BAMBERG SC	1423
ELMORE AL	1360	HARRIS TX	1392	PLATTE MO	1424
ELBERT CO	1361	PAULDING GA	1393	ARTHUR NE	1425
MONO CA	1362	MADISON KY	1394	ROCKCASTLE KY	1426
PASQUOTANK NC	1363	ORANGEBURG SC	1395	CLAY AL	1427
DUBUQUE IA	1364	HARDIN TN	1396	POTTER PA	1428
BRISCOE TX	1365	PAYETTE ID	1397	WASHINGTON GA	1429
MINNEHAHA SD	1366	TALLADEGA AL	1398	COCONINO3 AZ	1430
MACOUPIN IL	1367	DUNDY NE	1399	ADAMS CO	1431
GOODING ID	1368	CUSTER MT	1400	TIOGA PA	1432
DICKENSON VA	1369	FINNEY KS	1401	KNOX ME	1433
HENRICO VA	1370	CITRUS FL	1402	WORTH GA	1434
UNION TN	1371	CLARENDON SC	1403	VAN BUREN TN	1435
GOOCHLAND VA	1372	BECKHAM OK	1404	ROCK MN	1436
CALHOUN SC	1373	RIPLEY IN	1405	CHOUTEAU MT	1437
RICHARDSON NE	1374	RICHLAND MT	1406	BILLINGS ND	1438
SMITH KS	1375	GREENE IL	1407	CARBON MT	1439
PULASKI KY	1376	DIXIE FL	1408	POWHATAN VA	1440

COUNTIES RANKED BY RAINOUT POTENTIAL

(RAINFALL RANK x RADIOACTIVITY RANK, NORMALIZED).
LOW RANK VALUES INDICATE
HIGHER OVERALL RAINFALL-FALLOUT COMBINATION
1951-1962

COUNTY	RANK	COUNTY	RANK	COUNTY	RANK
WALDO ME	1441	WILKES GA	1473	SIERRA CA	1505
WARREN IL	1442	COLUMBIANA OH	1474	GWINNETT GA	1506
DICKINSON IA	1443	CHICOT AR	1475	DE BACA NM	1507
MAHONING OH	1444	MALHEUR OR	1476	STE GENEVIEV MO	1508
LANCASTER PA	1445	WASHINGTON CO	1477	MASON IL	1509
HENDERSON TN	1446	HENDERSON KY	1478	ALLEGANY NY	1510
WESTMORELAND VA	1447	MORGAN CO	1479	SCOTLAND NC	1511
HARRIS GA	1448	HILLSBOROUGH FL	1480	CHARLES MD	1512
CUSTER ID	1449	PIKE AL	1481	DE KALB TN	1513
SCOTT VA	1450	MILLS IA	1482	EDGAR IL	1514
AUTAUGA AL	1451	JACKSON WI	1483	TRIPP SD	1515
LUNENBURG VA	1452	PERRY AL	1484	FAYETTE GA	1516
JUNIATA PA	1453	SHENANDOAH VA	1485	CASS IN	1517
WYOMING PA	1454	CARROLL MO	1486	HARALSON GA	1518
MAURY TN	1455	VERMILLION IN	1487	FLORENCE SC	1519
LIBERTY TX	1456	JONES IA	1488	WIBAUX MT	1520
JEFFERSON FL	1457	WAYNE PA	1489	WASHINGTON VA	1521
CHERRY NE	1458	MITCHELL GA	1490	ROWAN NC	1522
KIMBALL NE	1459	PIPESTONE MN	1491	DOUGLAS WI	1523
LINCOLN WV	1460	MARSHALL KS	1492	DAKOTA NE	1524
GRAYSON VA	1461	TRIMBLE KY	1493	LYON IA	1525
LINCOLN WI	1462	LEBANON PA	1494	CLEVELAND NC	1526
ONEIDA NY	1463	SUMTER GA	1495	CLARK IL	1527
SHERBURNE MN	1464	WASHINGTON MN	1496	BURKE GA	1528
SAN JUAN NM	1465	PIKE IN	1497	MOREHOUSE LA	1529
JOHNSON GA	1466	PHILLIPS KS	1498	WARRICK IN	1530
CASCADE MT	1467	FLOYD VA	1499	FRANKLIN FL	1531
CUSTER SD	1468	DONIPHAN KS	1500	CATRON NM	1532
POTTER TX	1469	KAUFMAN TX	1501	ORANGE VA	1533
BROWN KS	1470	HALL GA	1502	SUWANNEE FL	1534
JOHNSON TN	1471	MADISON AL	1503	MACON GA	1535
BRISTOL RI	1472	SHELBY IL	1504	BROOKINGS SD	1536

COUNTIES RANKED BY RAINOUT POTENTIAL

(RAINFALL RANK x RADIOACTIVITY RANK, NORMALIZED).
LOW RANK VALUES INDICATE
HIGHER OVERALL RAINFALL-FALLOUT COMBINATION
1951-1962

COUNTY	RANK	COUNTY	RANK	COUNTY	RANK
SALEM NJ	1537	MONTAGUE TX	1569	LANCASTER VA	1601
BUFFALO WI	1538	CARROLL VA	1570	BALDWIN AL	1602
ORLEANS NY	1539	BANKS GA	1571	COCKE TN	1603
VOLUSIA FL	1540	NOBLES MN	1572	STEPHENS GA	1604
PERKINS NE	1541	DUKES MA	1573	DAVIDSON NC	1605
PALO ALTO IA	1542	AMELIA VA	1574	FARIBAULT MN	1606
ST HELENA LA	1543	FAYETTE TX	1575	BELL KY	1607
MONITEAU MO	1544	JEFFERSON TX	1576	HENRY IL	1608
BRAZOS TX	1545	MORGAN AL	1577	STARK OH	1609
CARTER MT	1546	LAURENS GA	1578	CLARKE VA	1610
JACKSON KY	1547	WARREN TN	1579	MEEKER MN	1611
JOHNSON IN	1548	LENOIR NC	1580	CASS IL	1612
MITCHELL KS	1549	GAGE NE	1581	BOONE IN	1613
RAMSEY MN	1550	JOHNSON NE	1582	CLAY KY	1614
BARNS MA	1551	WILLIAMS ND	1583	MADISON TX	1615
VALLEY MT	1552	WARREN MO	1584	WALTHALL MS	1616
WASHINGTON ME	1553	DANIELS MT	1585	STOREY NV	1617
QUITMAN GA	1554	WALWORTH WI	1586	WHITLEY KY	1618
BURKE ND	1555	HENRY IN	1587	HALIFAX VA	1619
HOOKER NE	1556	ARAPAHOE CO	1588	SOCORRO NM	1620
HENDRICKS IN	1557	CARSON CITY NV	1589	DOUGLAS NE	1621
HUTCHINSON TX	1558	WELD CO	1590	ONEIDA WI	1622
PORTAGE OH	1559	ROANOKE VA	1591	PARKE IN	1623
OVERTON TN	1560	WHITE IN	1592	BELL TX	1624
WHEELER NE	1561	STANLEY SD	1593	YORK NE	1625
ADAMS ND	1562	COOSA AL	1594	VANDERBURGH IN	1626
ERIE OH	1563	RANDALL TX	1595	NEWTON MS	1627
LACKAWANNA PA	1564	SOMERSET ME	1596	PIERCE NE	1628
RUSSELL VA	1565	GREGORY SD	1597	JACKSON LA	1629
PLYMOUTH IA	1566	GARDEN NE	1598	WASHINGTON TX	1630
ELLIS TX	1567	BROOKS GA	1599	CLAY FL	1631
EMMET IA	1568	SMITH MS	1600	RICE MN	1632

COUNTIES RANKED BY RAINOUT POTENTIAL

(RAINFALL RANK x RADIOACTIVITY RANK, NORMALIZED).
LOW RANK VALUES INDICATE
HIGHER OVERALL RAINFALL-FALLOUT COMBINATION
1951-1962

COUNTY	RANK	COUNTY	RANK	COUNTY	RANK
GRUNDY IA	1633	BUTLER OH	1665	SCOTT IN	1697
NESHOBA MS	1634	CALLOWAY KY	1666	NIAGARA NY	1698
JEFF DAVIS TX	1635	RICHLAND LA	1667	SAUNDERS NE	1699
NATRONA WY	1636	MONTGOMERY OH	1668	CARTERET NC	1700
ABBEVILLE SC	1637	KENNEBEC ME	1669	LAWRENCE PA	1701
BLOUNT TN	1638	SHERMAN TX	1670	MORGAN IL	1702
NOTTOWAY VA	1639	NIOBRARA WY	1671	HOWARD IN	1703
ALLEN IN	1640	GENTRY MO	1672	DES MOINES IA	1704
JESSAMINE KY	1641	ONTARIO NY	1673	RENVILLE MN	1705
GADSDEN FL	1642	WASHABAUGH SD	1674	COWETA GA	1706
LEE AL	1643	COLQUITT GA	1675	FULTON PA	1707
RANDOLPH NC	1644	CURRY NM	1676	LYMAN SD	1708
HOKE NC	1645	CORTLAND NY	1677	ROCK WI	1709
CABELL WV	1646	LARAMIE WY	1678	BIBB AL	1710
BONNEVILLE ID	1647	CLERMONT OH	1679	HANCOCK IA	1711
WALTON FL	1648	MEADE KY	1680	CULBERSON TX	1712
UNION LA	1649	TELFAIR GA	1681	HAMILTON NE	1713
RIO ARRIBA NM	1650	PARMER TX	1682	DILLON SC	1714
WAYNE NC	1651	PRAIRIE MT	1683	MEDINA OH	1715
SHELBY KY	1652	WORCESTER MD	1684	LIVINGSTON IL	1716
ARCHER TX	1653	SANGAMON IL	1685	MADISON GA	1717
MUSSELSHELL MT	1654	MADISON ID	1686	HARDING SD	1718
COLBERT AL	1655	LAWRENCE OH	1687	FRANKLIN NE	1719
KOSSUTH IA	1656	NORTHUMBERLA VA	1688	BAYFIELD WI	1720
LEE SC	1657	WATONWAN MN	1689	JEROME ID	1721
QUEEN ANNES MD	1658	CLAY NE	1690	HOLMES OH	1722
HALL TX	1659	FRANKLIN ME	1691	BENTON TN	1723
NODAWAY MO	1660	MARQUETTE WI	1692	LA CROSSE WI	1724
PUTNAM IL	1661	RICHLAND IL	1693	LINCOLN MO	1725
OLMSTED MN	1662	DECATUR IN	1694	HART GA	1726
POWELL KY	1663	TETON WY	1695	MONTGOMERY AL	1727
PERRY PA	1664	KANAWHA WV	1696	TANGIPAHOA LA	1728

COUNTIES RANKED BY RAINOUT POTENTIAL

(RAINFALL RANK x RADIOACTIVITY RANK, NORMALIZED).
LOW RANK VALUES INDICATE
HIGHER OVERALL RAINFALL-FALLOUT COMBINATION
1951-1962

COUNTY	RANK	COUNTY	RANK	COUNTY	RANK
DUNN WI	1729	TARRANT TX	1761	LOUP NE	1793
TRINITY TX	1730	GLOUCESTER VA	1762	GRANT NE	1794
POSEY IN	1731	SIBLEY MN	1763	JONES GA	1795
WAYNE TN	1732	HABERSHAM GA	1764	ESSEX VA	1796
COBB GA	1733	TRUMBULL OH	1765	BROWN NE	1797
HURON OH	1734	MOTLEY TX	1766	JERSEY IL	1798
SCOTT MS	1735	ROWAN KY	1767	ANGELINA TX	1799
GRANT OR	1736	EARLY GA	1768	BRUNSWICK NC	1800
TETON ID	1737	SHERIDAN NE	1769	WILLIAMSON TX	1801
DOOLY GA	1738	FREEBORN MN	1770	LICKING OH	1802
DODGE MN	1739	PAWNEE NE	1771	FLEMING KY	1803
DURHAM NC	1740	MIAMI OH	1772	MECKLENBURG VA	1804
SAN JACINTO TX	1741	LAKE OH	1773	BEN HILL GA	1805
HUDSPETH TX	1742	CLARKE MS	1774	NASSAU FL	1806
LANCASTER NE	1743	GREENE AL	1775	HAMILTON TN	1807
HAAKON SD	1744	BENTON MN	1776	OGLE IL	1808
WESTMORELAND PA	1745	SULLY SD	1777	HERTFORD NC	1809
HAWKINS TN	1746	WOOD WI	1778	NATCHITOCHES LA	1810
MOHAVE3 AZ	1747	DALLAS TX	1779	GASCONADE MO	1811
FREMONT IA	1748	MCCORMICK SC	1780	RICHMOND VA	1812
TURNER GA	1749	MOULTRIE IL	1781	LEWIS WV	1813
WARREN NC	1750	TAYLOR WI	1782	BAKER OR	1814
DOUGLAS GA	1751	CUMBERLAND PA	1783	CALHOUN GA	1815
SWITZERLAND IN	1752	DENVER CO	1784	COLUMBIA GA	1816
BROWN IN	1753	KANABEC MN	1785	PARK WY	1817
NEMAHA NE	1754	LINCOLN ID	1786	ATCHISON MO	1818
MOHAVE4 AZ	1755	HENRY IA	1787	JEFFERSON WI	1819
WRIGHT MN	1756	SOMERSET MD	1788	ORLEANS VT	1820
GRANVILLE NC	1757	CARROLL IN	1789	LINCOLN NM	1821
CHESTERFIELD VA	1758	RICHLAND OH	1790	LAFOURCHE LA	1822
BERTIE NC	1759	BOONE MO	1791	STILLWATER MT	1823
DOUGLAS NV	1760	DARE NC	1792	BREVARD FL	1824

COUNTIES RANKED BY RAINOUT POTENTIAL

(RAINFALL RANK x RADIOACTIVITY RANK, NORMALIZED).
LOW RANK VALUES INDICATE
HIGHER OVERALL RAINFALL-FALLOUT COMBINATION
1951-1962

COUNTY	RANK	COUNTY	RANK	COUNTY	RANK
ST JOHNS FL	1825	HOCKLEY TX	1857	BAKER GA	1889
SUMMIT OH	1826	HAMILTON IL	1858	STEWART TN	1890
ALACHUA FL	1827	CUMBERLAND KY	1859	BOYD KY	1891
JOHNSON WY	1828	EDGEFIELD SC	1860	GRAY TX	1892
KANE IL	1829	GREENUP KY	1861	CHATHAM NC	1893
WASHINGTON PA	1830	SANBORN SD	1862	COLORADO TX	1894
RUSSELL AL	1831	CHAMBERS TX	1863	ATKINSON GA	1895
EMANUEL GA	1832	OKALOOSA FL	1864	CRAWFORD IN	1896
HANCOCK TN	1833	HOUSTON AL	1865	NORFOLK/CHES VA	1897
MCLEAN IL	1834	BUCHANAN VA	1866	MELLETTE SD	1898
CROOK WY	1835	WILBARGER TX	1867	SPOTSYLVANIA VA	1899
LIMESTONE TX	1836	HART KY	1868	REEVES TX	1900
WILLIAMSON IL	1837	DONLEY TX	1869	LINCOLN GA	1901
COPIAH MS	1838	CLINTON NY	1870	WRIGHT IA	1902
COMANCHE KS	1839	MONTOUR PA	1871	GULF FL	1903
POLK FL	1840	CLINTON IL	1872	SHERMAN KS	1904
RUSSELL KY	1841	HIGHLAND VA	1873	HIGHLAND OH	1905
COLUMBIA PA	1842	ANDERSON KY	1874	FOARD TX	1906
SANTA ROSA FL	1843	WAYNE GA	1875	PICKENS GA	1907
JACKSON FL	1844	MCHENRY ND	1876	ST CLAIR MO	1908
LAUDERDALE AL	1845	JEFFERSON NE	1877	FORD IL	1909
ROBERTSON TX	1846	JEFFERSON IL	1878	INYO1 CA	1910
CATOOSA GA	1847	MARATHON WI	1879	BON HOMME SD	1911
WAYNE OH	1848	MINIDOKA ID	1880	WASHINGTON KS	1912
LOWNDES AL	1849	BRISTOL MA	1881	MARION GA	1913
CUYAHOGA OH	1850	SIOUX ND	1882	JASPER MS	1914
THAYER NE	1851	LEWIS KY	1883	WAYNE KY	1915
RANKIN MS	1852	COVINGTON MS	1884	FRANKLIN KY	1916
LAFAYETTE FL	1853	TALBOT GA	1885	INYO2 CA	1917
NICHOLAS WV	1854	CLINTON PA	1886	COFFEE TN	1918
PUTNAM GA	1855	WILL IL	1887	SHERIDAN WY	1919
CHEROKEE AL	1856	MARTIN MN	1888	DAWSON GA	1920

COUNTIES RANKED BY RAINOUT POTENTIAL

(RAINFALL RANK x RADIOACTIVITY RANK, NORMALIZED).
LOW RANK VALUES INDICATE
HIGHER OVERALL RAINFALL-FALLOUT COMBINATION
1951-1962

COUNTY	RANK	COUNTY	RANK	COUNTY	RANK
CLAY SD	1921	MORROW OH	1953	PLEASANTS WV	1985
BARTOW GA	1922	LYNN TX	1954	DAUPHIN PA	1986
VILAS WI	1923	RAPPAHANNOCK VA	1955	DEWEY SD	1987
TETON MT	1924	MERCER ND	1956	JO DAVIESS IL	1988
ALLEN OH	1925	KNOX NE	1957	TANEY MO	1989
TIFT GA	1926	IRWIN GA	1958	CARVER MN	1990
CALHOUN AR	1927	NORTHAMPTON NC	1959	OCONEE GA	1991
KING GEORGE VA	1928	WINNEBAGO IA	1960	OSWEGO NY	1992
BROWN MN	1929	MENARD IL	1961	WICHITA TX	1993
MINERAL MT	1930	HAMLIN SD	1962	GRUNDY IL	1994
LAVACA TX	1931	MEIGS OH	1963	ACCOMACK VA	1995
VERMILION IL	1932	FOSTER ND	1964	AMHERST VA	1996
CLAY GA	1933	FAYETTE PA	1965	STEWART GA	1997
BARTHOLOMEW IN	1934	CHARLOTTE VA	1966	GRADY GA	1998
JACKSON SD	1935	GREENWOOD SC	1967	CUMBERLAND NC	1999
BENEWAH ID	1936	TATTNALL GA	1968	BERKELEY SC	2000
GATES NC	1937	HUNTINGDON PA	1969	GALLATIN IL	2001
WASHINGTON IN	1938	LINCOLN KY	1970	CLAY WV	2002
EAST BATON R LA	1939	WARREN IN	1971	MONTGOMERY TX	2003
SARPY NE	1940	BUCHANAN MO	1972	NEWPORT RI	2004
CULLMAN AL	1941	GREENE TN	1973	TYLER WV	2005
PORTER IN	1942	BOONE WV	1974	SHOSHONE ID	2006
ST LAWRENCE NY	1943	MARSHALL IL	1975	ALBEMARLE VA	2007
CLARKE GA	1944	FRANKLIN NY	1976	ORANGE FL	2008
LAKE IN	1945	WISE TX	1977	WAYNE NY	2009
EVANS GA	1946	BUTLER PA	1978	TUSCARAWAS OH	2010
SALUDA SC	1947	MORRILL NE	1979	BROWN TX	2011
LYON NV	1948	BRUNSWICK VA	1980	TIPTON IN	2012
FULTON GA	1949	WHEATLAND MT	1981	GOSHEN WY	2013
WIRT WV	1950	BONNER ID	1982	BAY FL	2014
FLAGLER FL	1951	KENTON KY	1983	CERRO GORDO IA	2015
ROBESON NC	1952	FORSYTH GA	1984	ALAMANCE NC	2016

COUNTIES RANKED BY RAINOUT POTENTIAL

(RAINFALL RANK x RADIOACTIVITY RANK, NORMALIZED).
LOW RANK VALUES INDICATE
HIGHER OVERALL RAINFALL-FALLOUT COMBINATION
1951-1962

COUNTY	RANK	COUNTY	RANK	COUNTY	RANK
BURLESON TX	2017	DELAWARE IN	2049	MADISON NY	2081
WHITE IL	2018	MONROE IN	2050	FILLMORE NE	2082
POWER ID	2019	STONEWALL TX	2051	LORAIN OH	2083
TRIGG KY	2020	EL PASO TX	2052	HARNEY OR	2084
SANILAC MI	2021	WHARTON TX	2053	TERRY TX	2085
AITKIN MN	2022	OWEN KY	2054	FALLON MT	2086
SIMPSON MS	2023	FREDERICK VA	2055	GRAINGER TN	2087
HORRY SC	2024	CAMPBELL KY	2056	ADAMS PA	2088
WEST CARROLL LA	2025	WASHINGTON MD	2057	BRAXTON WV	2089
HANCOCK OH	2026	CLARK IN	2058	VIRGINIA BEA VA	2090
BARRON WI	2027	LEON FL	2059	OCONTO WI	2091
VAN BUREN MI	2028	JEFFERSON NY	2060	KNOX OH	2092
ROOSEVELT NM	2029	ROCK NE	2061	GRAVES KY	2093
WASHINGTON FL	2030	HARNETT NC	2062	NANTUCKET MA	2094
MCCRACKEN KY	2031	WILSON TN	2063	PRESTON WV	2095
CLAY IN	2032	LEE NC	2064	MEADE SD	2096
HUGHES SD	2033	KANDIYOHI MN	2065	BERRIEN GA	2097
WASHINGTON IL	2034	ESTILL KY	2066	SAN BERNADIN CA	2098
STANLY NC	2035	MONTGOMERY IL	2067	ANDERSON SC	2099
SIOUX NE	2036	CHILTON AL	2068	DADE FL	2100
CHIPPEWA WI	2037	LAWRENCE KY	2069	PALO PINTO TX	2101
CALDWELL LA	2038	JACKSON WV	2070	FRANKLIN TN	2102
MCLEOD MN	2039	GOLDEN VALLE MT	2071	CASEY KY	2103
SPENCER KY	2040	CLINCH GA	2072	PAMLICO NC	2104
LEWIS TN	2041	REPUBLIC KS	2073	UPSHUR WV	2105
MARINETTE WI	2042	CRAVEN NC	2074	KANKAKEE IL	2106
HOLT MO	2043	FULTON IN	2075	YORK SC	2107
FALL RIVER SD	2044	SMYTH VA	2076	ST CLAIR AL	2108
SCIOTO OH	2045	SCHUYLER IL	2077	GUERNSEY OH	2109
SANDUSKY OH	2046	CROSBY TX	2078	BOX BUTTE NE	2110
SPINK SD	2047	HETTINGER ND	2079	SMITH TN	2111
YANKTON SD	2048	ASHLAND OH	2080	GAINES TX	2112

COUNTIES RANKED BY RAINOUT POTENTIAL

(RAINFALL RANK x RADIOACTIVITY RANK, NORMALIZED).
LOW RANK VALUES INDICATE
HIGHER OVERALL RAINFALL-FALLOUT COMBINATION
1951-1962

COUNTY	RANK	COUNTY	RANK	COUNTY	RANK
BEDFORD PA	2113	HOLMES MS	2145	LEON TX	2177
LUBBOCK TX	2114	COOK GA	2146	CALUMET WI	2178
PAGE IA	2115	HENNEPIN MN	2147	WAUSHARA WI	2179
PEND OREILLE WA	2116	CHEROKEE SC	2148	CLARK KY	2180
DE WITT IL	2117	PIERCE GA	2149	BLUE EARTH MN	2181
POLK GA	2118	ST LOUIS MO	2150	SHEBOYGAN WI	2182
GOGEBIC MI	2119	WARREN OH	2151	NICHOLAS KY	2183
GRIMES TX	2120	CORSON SD	2152	KEYA PAHA NE	2184
WASHINGTON NC	2121	CAROLINE VA	2153	PAGE VA	2185
TOOLE MT	2122	WARREN KY	2154	LUMPKIN GA	2186
TODD MN	2123	CHARLES CITY VA	2155	NEW KENT VA	2187
ANDREW MO	2124	CRENSHAW AL	2156	GREENSVILLE VA	2188
HAMILTON FL	2125	JEWELL KS	2157	ONONDAGA NY	2189
TYLER TX	2126	HARRISON OH	2158	WALSH ND	2190
BLAINE NE	2127	GLADES FL	2159	PRINCE EDWAR VA	2191
MUSKINGUM OH	2128	BROWN OH	2160	ADAMS WI	2192
ST JAMES LA	2129	MONTGOMERY IN	2161	GALLATIN MT	2193
ATHENS OH	2130	CASS MI	2162	OLDHAM KY	2194
BEAVER PA	2131	WETZEL WV	2163	TOOMBS GA	2195
JENNINGS IN	2132	LINCOLN TN	2164	MACON TN	2196
DINWIDDIE VA	2133	GONZALES TX	2165	RANDOLPH IL	2197
ANOKA MN	2134	SULLIVAN PA	2166	WESTON WY	2198
NELSON VA	2135	HANCOCK IN	2167	REDWOOD MN	2199
PLATTE WY	2136	YORK VA	2168	BRADFORD PA	2200
BANNER NE	2137	COCHRAN TX	2169	MERCER IL	2201
BEADLE SD	2138	GEORGETOWN SC	2170	CHOWAN NC	2202
HENRY TN	2139	SEMINOLE GA	2171	CHAMPAIGN OH	2203
MORTON ND	2140	LANIER GA	2172	GREEN LAKE WI	2204
FRANKLIN AL	2141	WARE GA	2173	COFFEE GA	2205
MORGAN WV	2142	CHAVES NM	2174	HARMON OK	2206
JEFFERSON TN	2143	MOODY SD	2175	MIFFLIN PA	2207
YOAKUM TX	2144	DEARBORN IN	2176	HANCOCK WV	2208

COUNTIES RANKED BY RAINOUT POTENTIAL

(RAINFALL RANK x RADIOACTIVITY RANK, NORMALIZED).
LOW RANK VALUES INDICATE
HIGHER OVERALL RAINFALL-FALLOUT COMBINATION
1951-1962

COUNTY	RANK	COUNTY	RANK	COUNTY	RANK
ST LOUIS MN	2209	MERCER KY	2241	MURRAY MN	2273
PONDERA MT	2210	GREENBRIER WV	2242	MILLER GA	2274
MUHLENBERG KY	2211	FLUVANNA VA	2243	COLLINGSWORT TX	2275
WASHINGTON TN	2212	JEFFERSON ID	2244	BARREN KY	2276
SHANNON SD	2213	WASHINGTON OH	2245	UNION PA	2277
GRANT ND	2214	TODD KY	2246	LAPEER MI	2278
EASTLAND TX	2215	BUFFALO SD	2247	JACKSON TX	2279
LIBERTY MT	2216	RICHMOND GA	2248	MADISON IL	2280
OHIO KY	2217	GILES TN	2249	COFFEE AL	2281
MILLE LACS MN	2218	MARION WV	2250	BROOME NY	2282
SEMINOLE FL	2219	LEE TX	2251	NEVADA CA	2283
KOOCHICHING MN	2220	SUMNER TN	2252	ST CHARLES MO	2284
BULLOCK AL	2221	KOSCIUSKO IN	2253	MILAM TX	2285
WILLIAMSBURG SC	2222	SUSSEX VA	2254	CASS NE	2286
GUILFORD NC	2223	SCOTT KY	2255	WARD ND	2287
MARION OH	2224	LAWRENCE TN	2256	LANCASTER SC	2288
WASHINGTON AL	2225	COTTONWOOD MN	2257	MARTIN NC	2289
CLOUD KS	2226	DAVIDSON TN	2258	WALLER TX	2290
ECHOLS GA	2227	MCCULLOCH TX	2259	LOUISA VA	2291
NELSON KY	2228	ELBERT GA	2260	WADENA MN	2292
TUSCALOOSA AL	2229	BOWMAN ND	2261	DALE AL	2293
HILL MT	2230	SCOTT TN	2262	EDGECOMBE NC	2294
PITT NC	2231	PENNINGTON SD	2263	TAYLOR WV	2295
MONTGOMERY NC	2232	CHISAGO MN	2264	CENTRE PA	2296
GREENE OH	2233	PUTNAM FL	2265	WYANDOT OH	2297
UNION KY	2234	LAMPASAS TX	2266	FISHER TX	2298
PIKE GA	2235	WINKLER TX	2267	WAYNE MI	2299
POCAHONTAS WV	2236	BRANTLEY GA	2268	PIKE OH	2300
FORT BEND TX	2237	FAYETTE AL	2269	BURLEIGH ND	2301
SWEET GRASS MT	2238	PERRY OH	2270	CHESTER SC	2302
HURON MI	2239	PRICE WI	2271	HARDEMAN TX	2303
PITTSYLVANIA VA	2240	HENDRY FL	2272	YELLOWSTONE MT	2304

COUNTIES RANKED BY RAINOUT POTENTIAL

(RAINFALL RANK x RADIOACTIVITY RANK, NORMALIZED).
LOW RANK VALUES INDICATE
HIGHER OVERALL RAINFALL-FALLOUT COMBINATION
1951-1962

COUNTY	RANK	COUNTY	RANK	COUNTY	RANK
CHOCTAW AL	2305	BURNET TX	2337	RALEIGH WV	2369
RENVILLE ND	2306	ST CLAIR MI	2338	LOWNDES GA	2370
BARROW GA	2307	LATAH ID	2339	MCINTOSH ND	2371
BARNWELL SC	2308	SHACKELFORD TX	2340	BOTTINEAU ND	2372
JUDITH BASIN MT	2309	AIKEN SC	2341	FRANKLIN VA	2373
HARDIN KY	2310	FLOYD TX	2342	KENDALL IL	2374
PIATT IL	2311	BEAVERHEAD MT	2343	CHENANGO NY	2375
ST CLAIR IL	2312	MADISON IN	2344	MADISON OH	2376
LINCOLN MN	2313	OWSLEY KY	2345	CANDLER GA	2377
OTTAWA OH	2314	CALHOUN IL	2346	UNION OH	2378
APPLING GA	2315	YATES NY	2347	DAVIE NC	2379
PEMBINA ND	2316	BULLITT KY	2348	ALLEGHENY PA	2380
KERSHAW SC	2317	HAMBLEN TN	2349	WAYNE IN	2381
DAWES NE	2318	STARKE IN	2350	LOWNDES MS	2382
SENECA NY	2319	GRANT WV	2351	MERIWETHER GA	2383
WOODFORD KY	2320	TAYLOR KY	2352	BOSQUE TX	2384
DICKEY ND	2321	FRANKLIN IL	2353	BENTON IN	2385
LEWIS ID	2322	WOOD WV	2354	HAMILTON OH	2386
CANNON TN	2323	ROBERTSON KY	2355	PUTNAM WV	2387
RICHMOND NC	2324	OGLETHORPE GA	2356	BOONE KY	2388
ALLENDALE SC	2325	WYTHE VA	2357	LLANO TX	2389
PARKER TX	2326	HARDIN IL	2358	ROSS OH	2390
BUTTE SD	2327	MARSHALL KY	2359	EATON MI	2391
PORTAGE WI	2328	LYON MN	2360	JAY IN	2392
NEWBERRY SC	2329	HOUGHTON MI	2361	CUMBERLAND VA	2393
SHERIDAN ND	2330	WAUPACA WI	2362	LEWIS AND CL MT	2394
PENDLETON WV	2331	NYE1 NV	2363	ORANGE NC	2395
LAURENS SC	2332	DARKE OH	2364	PUTNAM OH	2396
MONTGOMERY MO	2333	DAVIESS KY	2365	BECKER MN	2397
MINERAL NV	2334	SHELBY IN	2366	CALDWELL TX	2398
RUSK WI	2335	POPE IL	2367	HYDE SD	2399
SHAWANO WI	2336	HUNTINGTON IN	2368	JEFFERSON AL	2400

COUNTIES RANKED BY RAINOUT POTENTIAL

(RAINFALL RANK x RADIOACTIVITY RANK, NORMALIZED).
LOW RANK VALUES INDICATE
HIGHER OVERALL RAINFALL-FALLOUT COMBINATION
1951-1962

COUNTY	RANK	COUNTY	RANK	COUNTY	RANK
BELMONT OH	2401	SAN SABA TX	2433	MORGAN OH	2465
GILMER WV	2402	POTTER SD	2434	CHILDRESS TX	2466
LANGLADE WI	2403	POPE MN	2435	BROOKE WV	2467
HAMPSHIRE WV	2404	CLARK OH	2436	WAUKESHA WI	2468
SALINE IL	2405	MASSAC IL	2437	CONCHO TX	2469
BATH KY	2406	PERKINS SD	2438	JEFFERSON OH	2470
SUFFOLK/NANS VA	2407	JASPER GA	2439	HILL TX	2471
BATH VA	2408	LYON KY	2440	HARRISON IN	2472
FLOYD IN	2409	HOCKING OH	2441	KNOX TX	2473
ZIEBACH SD	2410	JEFFERSON KY	2442	BAKER FL	2474
CRISP GA	2411	MOORE TN	2443	LINCOLN NC	2475
OTOE NE	2412	WALWORTH SD	2444	BOYLE KY	2476
UNION SC	2413	LOGAN OH	2445	SOUTHAMPTON VA	2477
YOUNG TX	2414	DICKENS TX	2446	LAKE MN	2478
SHELBY AL	2415	RAMSEY ND	2447	GREENE VA	2479
MARSHALL WV	2416	MEAGHER MT	2448	HALIFAX NC	2480
HAMILTON IN	2417	WINSTON AL	2449	CHAMPAIGN IL	2481
JONES SD	2418	GLACIER MT	2450	CHARLTON GA	2482
ALLEGANY MD	2419	JOHNSON TX	2451	ELLIOTT KY	2483
HARDEE FL	2420	HOT SPRINGS WY	2452	AUGUSTA VA	2484
BINGHAM ID	2421	COMANCHE TX	2453	SIERRA NM	2485
ST LUCIE FL	2422	FOUNTAIN IN	2454	SPALDING GA	2486
MCHENRY IL	2423	BENSON ND	2455	SARGENT ND	2487
NOXUBEE MS	2424	TUOLUMNE CA	2456	KEMPER MS	2488
ESCAMBIA FL	2425	UNION OR	2457	ALPINE CA	2489
MASON TX	2426	LIVINGSTON MI	2458	STUTSMAN ND	2490
WARD TX	2427	MACOMB MI	2459	CAVALIER ND	2491
WASHAKIE WY	2428	OSCEOLA FL	2460	LAWRENCE AL	2492
CASS MN	2429	WELLS IN	2461	OHIO WV	2493
PERSON NC	2430	JAMES CITY VA	2462	JACK TX	2494
BENNETT SD	2431	FRANKLIN LA	2463	HAND SD	2495
CALLAHAN TX	2432	WASHINGTON WI	2464	NEW HANOVER NC	2496

COUNTIES RANKED BY RAINOUT POTENTIAL

(RAINFALL RANK x RADIOACTIVITY RANK, NORMALIZED).
LOW RANK VALUES INDICATE
HIGHER OVERALL RAINFALL-FALLOUT COMBINATION
1951-1962

COUNTY	RANK	COUNTY	RANK	COUNTY	RANK
ROCKINGHAM VA	2497	PALM BEACH FL	2529	GARRARD KY	2561
MCLENNAN TX	2498	DELAWARE OH	2530	TOMPKINS NY	2562
BASTROP TX	2499	DOOR WI	2531	GENESEE MI	2563
CRAIG VA	2500	MECKLENBURG NC	2532	MONROE WV	2564
COLLIER FL	2501	OKEECHOBEE FL	2533	WALKER AL	2565
KALAMAZOO MI	2502	SAMPSON NC	2534	WALLOWA OR	2566
MANATEE FL	2503	LA MOURE ND	2535	KERR TX	2567
ADAMS IN	2504	JACKSON MN	2536	HAMILTON TX	2568
COTTLE TX	2505	UNION IN	2537	RUNNELS TX	2569
GILLESPIE TX	2506	NAVARRO TX	2538	PULASKI VA	2570
MONROE NY	2507	MARSHALL TN	2539	ADAMS OH	2571
FAULK SD	2508	STOKES NC	2540	CAMPBELL VA	2572
FRIO TX	2509	JACKSON TN	2541	LAKE IL	2573
OZAUKEE WI	2510	DECATUR GA	2542	MEDINA TX	2574
CALHOUN WV	2511	MARION AL	2543	CRAWFORD OH	2575
ELKHART IN	2512	WASHINGTON KY	2544	KEWEENAW MI	2576
CARROLL OH	2513	CHRISTIAN KY	2545	BRADFORD FL	2577
TROUSDALE TN	2514	HENRY GA	2546	BAYLOR TX	2578
WILSON NC	2515	CASWELL NC	2547	SUMTER AL	2579
COOK IL	2516	GILES VA	2548	GALLATIN KY	2580
GARRETT MD	2517	SURRY VA	2549	BIG HORN WY	2581
HALE AL	2518	HASKELL TX	2550	PAULDING OH	2582
DAY SD	2519	RANDOLPH IN	2551	FAYETTE KY	2583
PINELLAS FL	2520	EMMONS ND	2552	CALDWELL KY	2584
MONROE KY	2521	STEARNS MN	2553	MARSHALL IN	2585
ONTONAGON MI	2522	WHITLEY IN	2554	HENRY KY	2586
CLARK SD	2523	SARASOTA FL	2555	RITCHIE WV	2587
MONTGOMERY VA	2524	CAYUGA NY	2556	WALLA WALLA WA	2588
PENDER NC	2525	POWELL MT	2557	VANCE NC	2589
CARROLL KY	2526	KOOTENAI ID	2558	STEPHENS TX	2590
SUSQUEHANNA PA	2527	WEBSTER WV	2559	ADAIR KY	2591
TODD SD	2528	PARK MT	2560	COLUMBIA WA	2592

COUNTIES RANKED BY RAINOUT POTENTIAL

(RAINFALL RANK x RADIOACTIVITY RANK, NORMALIZED).
LOW RANK VALUES INDICATE
HIGHER OVERALL RAINFALL-FALLOUT COMBINATION
1951-1962

COUNTY	RANK	COUNTY	RANK	COUNTY	RANK
ANSON NC	2593	CARTER KY	2625	BOURBON KY	2657
BROWN SD	2594	LOGAN KY	2626	GRIGGS ND	2658
GREENE PA	2595	DUPLIN NC	2627	MONROE IL	2659
ST JOSEPH IN	2596	OTTAWA MI	2628	RICHLAND ND	2660
BERRIEN MI	2597	AUGLAIZE OH	2629	BOUNDARY ID	2661
LEA NM	2598	BEDFORD TN	2630	LOGAN ND	2662
ALLEN KY	2599	LAKE MT	2631	CROW WING MN	2663
MACON AL	2600	KIDDER ND	2632	NOBLE IN	2664
CAMPBELL SD	2601	HIGHLANDS FL	2633	STEUBEN IN	2665
HARRISON WV	2602	MONROE FL	2634	UNION NC	2666
BARRY MI	2603	GARZA TX	2635	ERATH TX	2667
LYCOMING PA	2604	MISSOULA MT	2636	BARAGA MI	2668
BANDERA TX	2605	LIMESTONE AL	2637	CLINTON KY	2669
HOOD TX	2606	EDMUNDS SD	2638	BLAND VA	2670
REFUGIO TX	2607	ROCKINGHAM NC	2639	PREBLE OH	2671
VINTON OH	2608	BRACKEN KY	2640	OAKLAND MI	2672
MONONGALIA WV	2609	WAKE NC	2641	PLACER CA	2673
PERRY IL	2610	MENOMINEE WI	2642	CRITTENDEN KY	2674
LEMHI ID	2611	BLAINE ID	2643	ASOTIN WA	2675
INGHAM MI	2612	HARDY WV	2644	REAGAN TX	2676
MADISON MT	2613	HENRY AL	2645	NEWPORT NEWS VA	2677
KENOSHA WI	2614	LUCE MI	2646	ROCKBRIDGE VA	2678
CLAY TN	2615	FLATHEAD MT	2647	DODDRIDGE WV	2679
WABASH IN	2616	ITASCA MN	2648	EDDY ND	2680
EDDY NM	2617	MCPHERSON NE	2649	WALTON GA	2681
KENDALL TX	2618	MILLS TX	2650	ROANE WV	2682
BERKELEY WV	2619	BEXAR TX	2651	VAN WERT OH	2683
DU PAGE IL	2620	UMATILLA OR	2652	COLUMBUS NC	2684
MCPHERSON SD	2621	FRANKLIN GA	2653	LAGRANGE IN	2685
WELLS ND	2622	DALLAS AL	2654	LIVINGSTON KY	2686
WEBSTER KY	2623	CLINTON MI	2655	BLANCO TX	2687
GASTON NC	2624	FAYETTE OH	2656	SILVER BOW MT	2688

COUNTIES RANKED BY RAINOUT POTENTIAL

(RAINFALL RANK x RADIOACTIVITY RANK, NORMALIZED).
LOW RANK VALUES INDICATE
HIGHER OVERALL RAINFALL-FALLOUT COMBINATION
1951-1962

COUNTY	RANK	COUNTY	RANK	COUNTY	RANK
UPSON GA	2689	COLUMBIA FL	2721	HOPKINS KY	2753
SHELBY OH	2690	BARBOUR WV	2722	KIMBLE TX	2754
GALVESTON TX	2691	RANSOM ND	2723	DIMMIT TX	2755
SIMPSON KY	2692	BLADEN NC	2724	OHIO IN	2756
BARNES ND	2693	UVALDE TX	2725	NEZ PERCE ID	2757
CHEMUNG NY	2694	DE WITT TX	2726	SCHOOLCRAFT MI	2758
LEE KY	2695	REAL TX	2727	KENT TX	2759
CABARRUS NC	2696	MORGAN KY	2728	JOHNSTON NC	2760
APPOMATTOX VA	2697	GRANT IN	2729	MARION KY	2761
BROADWATER MT	2698	HILLSDALE MI	2730	EDMONSON KY	2762
SPENCER IN	2699	MENOMINEE MI	2731	GRAYSON KY	2763
FAIRFIELD OH	2700	COMAL TX	2732	ESMERALDA2 NV	2764
WOOD OH	2701	PERRY IN	2733	MATAGORDA TX	2765
BUCKINGHAM VA	2702	THROCKMORTON TX	2734	MARSHALL SD	2766
PICKENS AL	2703	GARFIELD WA	2735	FALLS TX	2767
RICHLAND SC	2704	WHITMAN WA	2736	SOMERVELL TX	2768
GREENE NC	2705	BELTRAMI MN	2737	TOWNER ND	2769
LARUE KY	2706	MASON KY	2738	FAIRFIELD SC	2770
KEWAUNEE WI	2707	LEXINGTON SC	2739	BRECKINRIDGE KY	2771
NELSON ND	2708	IRON MI	2740	NOBLE OH	2772
HOUSTON TN	2709	ROBERTS SD	2741	ROLETTE ND	2773
COLEMAN TX	2710	CLARK2 NV	2742	LAC QUI PARL MN	2774
JEFFERSON MT	2711	CHIPPEWA MN	2743	LAMAR GA	2775
MANITOWOC WI	2712	STEELE ND	2744	STEVENS WA	2776
MINERAL WV	2713	MCMULLEN TX	2745	ISLE OF WIGH VA	2777
GRANITE MT	2714	DEER LODGE MT	2746	NASH NC	2778
TIOGA NY	2715	SHIAWASSEE MI	2747	BUTTE ID	2779
FREESTONE TX	2716	MILWAUKEE WI	2748	BRANCH MI	2780
CODINGTON SD	2717	MIAMI IN	2749	RUTHERFORD TN	2781
BUTLER KY	2718	HARDIN OH	2750	MONROE OH	2782
BUTTS GA	2719	FOND DU LAC WI	2751	KING TX	2783
MAHNOMEN MN	2720	CHEBOYGAN MI	2752	WINNEBAGO WI	2784

COUNTIES RANKED BY RAINOUT POTENTIAL

**(RAINFALL RANK x RADIOACTIVITY RANK, NORMALIZED).
LOW RANK VALUES INDICATE
HIGHER OVERALL RAINFALL-FALLOUT COMBINATION
1951-1962**

COUNTY	RANK	COUNTY	RANK	COUNTY	RANK
UNION FL	2785	FERRY WA	2817	SAGINAW MI	2849
GRAND FORKS ND	2786	COOK MN	2818	MIDLAND MI	2850
ANTRIM MI	2787	MERCER OH	2819	FULTON OH	2851
WILLACY TX	2788	RAVALLI MT	2820	JACKSON MI	2852
PIMA AZ	2789	BOTETOURT VA	2821	HAMPTON VA	2853
GREEN KY	2790	HUBBARD MN	2822	JONES TX	2854
SCOTTS BLUFF NE	2791	CHARLEVOIX MI	2823	LA SALLE TX	2855
SCHUYLER NY	2792	ZAVALA TX	2824	ARENAC MI	2856
DODGE WI	2793	POLK MN	2825	GREENLEE AZ	2857
ST JOSEPH MI	2794	GILA AZ	2826	INDIAN RIVER FL	2858
ROBERTSON TN	2795	BIG STONE MN	2827	LINCOLN MT	2859
MARTIN FL	2796	ALCONA MI	2828	FLORENCE WI	2860
CORYELL TX	2797	DAWSON TX	2829	NOLAN TX	2861
MENARD TX	2798	ALGER MI	2830	HENRY OH	2862
PICKAWAY OH	2799	SANDERS MT	2831	GRAHAM AZ	2863
FRANKLIN OH	2800	BROWARD FL	2832	ALLEGHANY VA	2864
MCLEAN KY	2801	YAVAPAI AZ	2833	STEVENS MN	2865
MORRISON MN	2802	LAMAR AL	2834	TUSCOLA MI	2866
SANTA CRUZ AZ	2803	VAL VERDE TX	2835	CROCKETT TX	2867
CLEARWATER MN	2804	TYRRELL NC	2836	OTTER TAIL MN	2868
FOREST WI	2805	RACINE WI	2837	CRANE TX	2869
DE KALB GA	2806	TRAILL ND	2838	CASS ND	2870
FRANKLIN NC	2807	CHIPPEWA MI	2839	GUADALUPE TX	2871
SPOKANE WA	2808	KENT MI	2840	VICTORIA TX	2872
DEFIANCE OH	2809	COKE TX	2841	MARICOPA AZ	2873
SKAMANIA WA	2810	FAYETTE IN	2842	SCURRY TX	2874
LEVY FL	2811	PINAL AZ	2843	SWIFT MN	2875
SUTTON TX	2812	LEE FL	2844	KARNES TX	2876
BLACKFORD IN	2813	ROCKDALE GA	2845	DOUGLAS MN	2877
CALHOUN MI	2814	ECTOR TX	2846	MARQUETTE MI	2878
ATASCOSA TX	2815	PIERCE ND	2847	NEWTON GA	2879
PLUMAS CA	2816	DELTA MI	2848	GOLIAD TX	2880

SECTION 11

RADIONUCLIDES

1951-1962

RADIONUCLIDES

Each nuclear test produces a number of radioactive elements called radionuclides. Chemically, they are similar to their non-radioactive counterparts, but they vary in the amount and type of radiation they emit. They are also known as radioactive isotopes or radioisotopes.

ISOTOPES

An atom is comprised of neutrons, protons and electrons. The protons supply a positive charge, the electrons supply a negative charge, and the neutrons supply no charge at all, but contribute to the mass—how heavy the element is. In classical atomic theory, there are an equal number of electrons for each proton, and the number of protons (and corresponding electrons) determines the element's chemical characteristics. If there are six protons, then the element will be known forever as carbon and will always behave chemically as carbon. Add a proton and a corresponding electron and the element becomes nitrogen—and will behave chemically as nitrogen.

Neutrons contribute essentially nothing to the chemical characteristics of the element, but they contribute significantly to the element's mass. In fact, how "heavy" an element is depends on how many neutrons and protons it has. For example, two atoms of iodine might be chemically similar, but because one has more neutrons, it would have a higher mass number. Thus, while the two atoms would both be the same element and equivalent behavior chemically, they have different masses, and are thus *different isotopes of the same element.*

Iodine 132 is slightly heavier than iodine-131 and they are both heavier than iodine-127. But the three atoms are all chemically the same. All three isotopes are chemically indistinguishable from one another. Carbon-14 is slightly heavier than carbon-12, but they both behave *chemically* just like any carbon anywhere.

While all isotopes of an element act the same chemically, they are unique from the perspective of nuclear physics.

Radioactive Iodine-131, for example, has a half-life of 8.04 days—which means, half of the original mass decays in 8.04 days, then half of that decays in another 8.04 days, and so on. In the process of decay, this radionuclide (another name for radioactive isotope) emits some powerful beta particles and gamma rays. A close, somewhat lighter cousin to iodine-131, iodine-*130* is somewhat hotter radiologically. It has a half-life of only 12½ hours and emits a gamma ray as well as a beta particle that, measured in electron volts, is more powerful than the one emitted by iodine-131. Iodine-128, by contrast emits a gamma ray and a beta particle that is just *one-tenth* as hot as the one emitted by iodine-131. But iodine-128 has a half-life of 17 million years.

The radioisotopes discussed in this book are generally referred by their chemical symbols. For example, iodine-131 is written as I-131; the radioactive isotope manganese-24 is Mn-24, and Beryllium-7 is written as Be-7.

ESTIMATING THE RADIONUCLIDES

Nuclear fallout is composed of more than one element. A nuclear blast is usually hot enough to vaporize just about anything, reducing it to the size of small pellets no more than the size of fog particles. Some elements, however, resist vaporization and pretty much stay in solid form throughout the experience. Scientists term these materials *refractory*. Others, like the 700-ton steel towers supporting the nuclear canister, will vaporize easily, and then solidify into fog-sized spherical particles that can be carried east with the wind. Since the atomic fireball usually touches the ground shortly after detonation, material from the desert often joins the atomic debris cloud.

Elements caught up in the debris cloud are also made radioactive through a process called neutron activation. As a result, the debris cloud is a mixture of upwards of 145 different isotopes of elements, from beryllium to americium. And, of course, since a nuclear explosion is not particularly efficient, there might even be some plutonium trailing along amid the debris.

The problem of guessing what is in the cloud is complicated by several factors. For one thing, a particular radioactive element will change from one element to another over time. For example, the iodine-131 that the NCI evaluated began its life in the nuclear explosion as an entirely different element, *indium*-131. In a complicated series of transformations, the indium-131 atoms change to tin-131, then to antimony-131—all while the nuclear fireball is still churning above ground zero. About an hour after detonation, while the debris cloud is still visible, the atoms that were once indium-131 are now tellurium-131. After another hour, most of these atoms will change from tellurium-131 to iodine-131. Since I-131 has a half-life of eight days, this means that the fallout that reaches the ground will contain a substantial amount of this element.

All the elements found in fallout have similarly complicated histories. This means that the nuclear fallout coming to earth on the first day after the detonation will have a different mixture of isotopes than the fallout coming down on day two. Each day post-shot has it's own unique mixture of radioactive elements. In addition, the mixture of radioactive elements varies drastically with the particular detonation. The debris cloud from the underground shot SEDAN, for example, contained elements from the desert soil, while the debris cloud from HARRY probably contained a significant amount of the steel tower supporting the device. Some airborne detonations produced relatively little fallout. One shot, Teapot HA (for high-altitude) resulted in such a small amount of fallout that it was not included in the NCI's study.

All values found in this book are calculated estimates based upon ratios between Hicks Tables values for I-131 and the particular radionuclide. Since the original I-131 values were based upon the lognormal distribution, the range of values was often quite wide. It should be understood that this is also the case with the derived values appearing in this book for both total fallout and the various radionuclides

For this Volume the following radionuclides were evaluated for each of the major aboveground tests from 1951-1962:

Beryllium-7, Sodium-24, Manganese-54, Iron-59, Cobalt-60, Tungsten-185, Gold-198, Gold-199, Lead-203, Uranium-237, Uranium-240, Neptunium-237, Americium-241, Curium-242, Bromine-82, Bromine-83, Strontium-89, Strontium-90, Yttrium-90, Yttrium-91, Zirconium-95, Niobium-95m, Niobium-95, Molybdenum-99, Technetium-99m, Ruthenium-106, Rhodium-106, Silver-109m, Silver-111, Palladium-112, Cadmium-115, Indium-115m, Indium-117, Tellurium-127m, Iodine-130, Iodine-131, Iodine-132, Iodine-133, Iodine-135, Cesium-137, Barium-137m, Lanthanum-141, Cerium-144, Praesodymium-144, Europium-155, Terbium-161.

Generally, deposition calculations are made for the day of deposition. However, for those radionuclides marked with an asterisk (*) the deposition value is estimated for H+ 365 days, or one year post-shot. These radionuclides include: **Be7, Mn54, Fe59, Co60, Sr89, Sr90, Y90, Y91, Zr95, Nb95m, Nb95, Rh106, Ru106, Te127m, Cs137, Ba137m, Eu155, W185, Am241, and Cu242**.

In the book the following radionuclides are given as their chemical symbol:

SYMBOL	RADIONUCLIDE
Ag109m	Silver-109m (meta)
Ag111	Silver-111
Am241*	Americium-241
Au198, 199	Gold 198, 199
Ba137m*	Barium-137m
Be7*	Beryllium-7
Br82, Br83	Bromine-82, 83
Cd115	Cadmium-115

Ce144	Cerium-144
Cm242*	Curium-242
Co60*	Cobalt-60
Cs137*	Cesium-137
Cu24	Copper-24
Eu155*	Europium-155
Fe59*	Iron-59
I-130, I-131, I-132, I-133, I135	Iodine-130, 131, 132, 133, 135
In115m	Indium-115m
In117	Indium-117
Kr85, Kr87, Kr88	Krypton-85, 87, 88
La141	Lanthanum-141
Mn54*	Manganese 54
Mo99	Molybdenum-99
Na24	Sodium-24
Nb95*	Niobium-95
Nb95m*	Niobium-95m (meta)
Np237	Neptunium-237
Pb203	Lead-203
Pd111, Pd112	Palladium, 111, 112
Pr144	Praesodymium-144

Rh106*	Rhodium-106
Ru106*	Ruthenium-106
Sm156	Samarium-156
Sr89*, Sr90*	Strontium 89, 90
Tb161	Terbium-161
Tc99m	Technetium-99m
Te127m*	Tellurium-127m
U237	Uranium-237
W181, W185*, W188	Tungsten 181,185, 188
Y90*, Y91*	Yttrium-90, 91
Zr95*	Zirconium-95

As with all radionuclides one can calculate the activity at any time post-detonation given the half-life.

The formula: $$A_t = A_o e^{-\left(\frac{0.693t}{T}\right)}$$

Where: A_t = Activity at time t
A_o = Activity at original time (time of detonation or fallout deposition)
T = Published Half-Life of the radionuclide
t = time of interest. Note that T and t must be in the same units of measure.

MAPS AND TABLES OF ESTIMATED RADIONUCLIDE DEPOSITION

The following maps and tables are estimates of deposition values based upon Hicks Tables ratios applied to the National Cancer Institute's deposition values. Unless otherwise indicated, the fallout values are expressed in nanoCuries per square meter. One Curie/sq meter is equivalent to 3.7×10^{10} disintegrations or radioactive events per second. Each of these events can produce a particle or ray. Unless otherwise indicated fallout radioactivity values in this book and in this section are expressed in nanoCuries/square meter, a level equal to one billionth of a Curie.

Maps displaying radionuclide values as columns in the z-axis only show relative values for that particular radionuclide. Estimates of for different radionuclides may vary by many orders of magnitude, thus it is inappropriate to compare the relative fallout values based upon these illustrations.

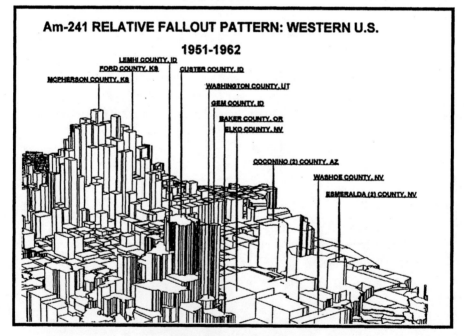

Americium-241 (Am-241)

Americium-241 is considered by the International Atomic Energy Commission to be a Class I very highly toxic radionuclide[1][2]. It is produced primarily by the disintegration of plutonium-241. Radiation includes alpha particles, gamma rays and X-rays[3]. It is a long-lived radionuclide with a half-life of 432.2 years.[4] For that reason, americium-241 will have retained 92% of its activity 50 years after deposition. Americium-241 was produced primarily by shots Tumbler-Snapper HOW (TS8), Plumbbob WHITNEY (PB16) and shot SEDAN. Counties affected included the group located northwest of the Nevada Test Site as well as an area in the Midwest surrounding McPherson County, Kansas.

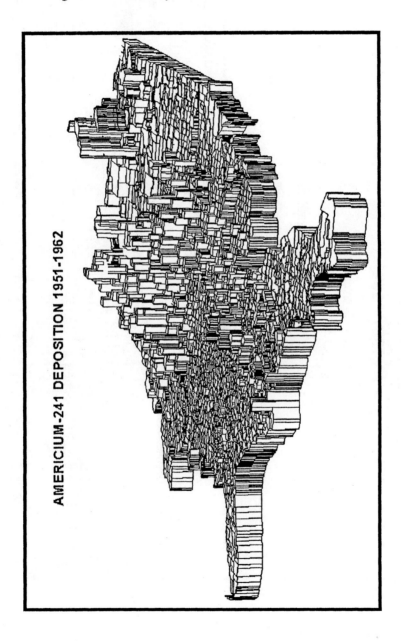

AMERICIUM-241 DEPOSITION 1951-1962

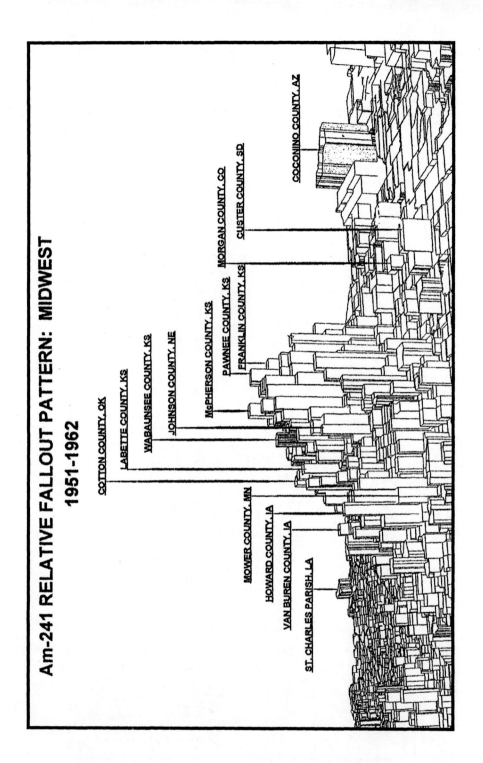

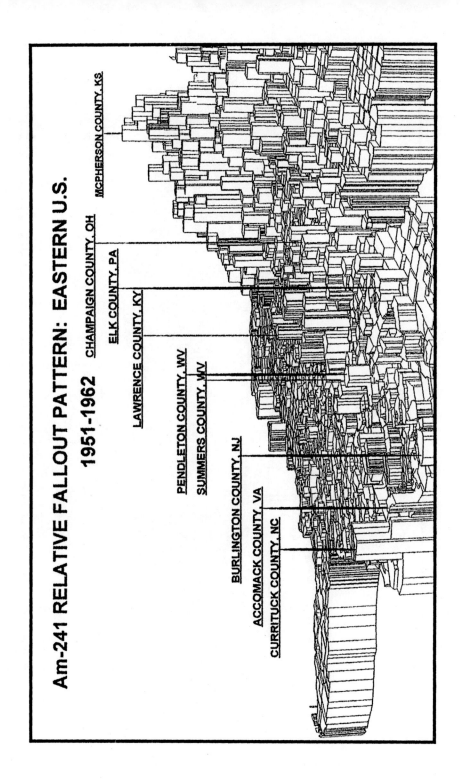

U.S. FALLOUT ATLAS : TOTAL FALLOUT

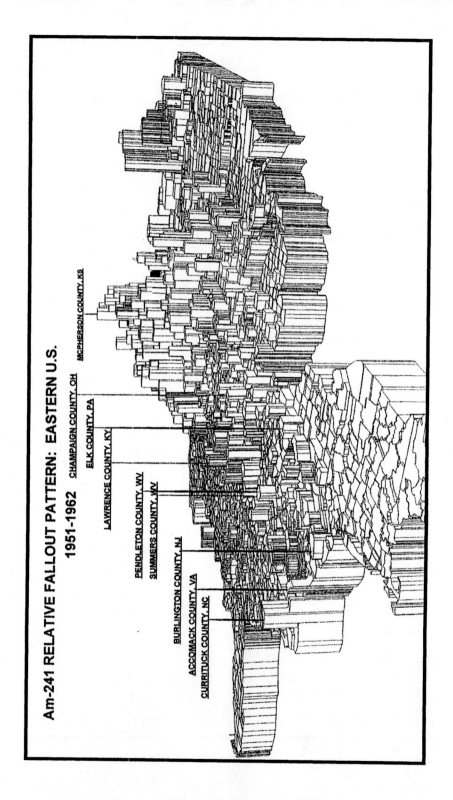

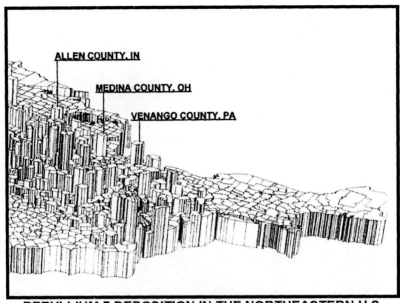

BERYLLIUM-7 DEPOSITION IN THE NORTHEASTERN U.S.

Beryllium-7 (Be-7)

Beryllium-7 is considered by the International Atomic Energy Commission to be a Class IV (slightly toxic) radioisotope[5,6]. It is produced from the transformation of lithium-6. It has a relatively short half-life, approximately 53.1 days[7] Thus, after 1 year the activity would be 0.008 of that at the time of deposition. However, the Hicks Table for shot SEDAN notes that Be7 activity at 1 year is 5.58E-04 vs 6.38E-2 at one day post shot.[8] This suggests that the material is being replenished through the transformation of other radionuclides in fallout. Beryllium-7 produces low-energy electrons (through electron capture) and gamma radiation.

The primary source of beryllium-7 from U.S. nuclear tests was shot SEDAN, followed by shot Plumbbob HOOD, Upshot-Knothole SIMON and Tumbler-Snapper HOW. Interestingly, shot SEDAN produced more Be-7 than all the other shots combined. For this reason, the fallout pattern for Be-7 is associated with that for SEDAN. Counties with the highest Be-7 fallout included the Iowa counties of Howard, Worth and Mitchell.

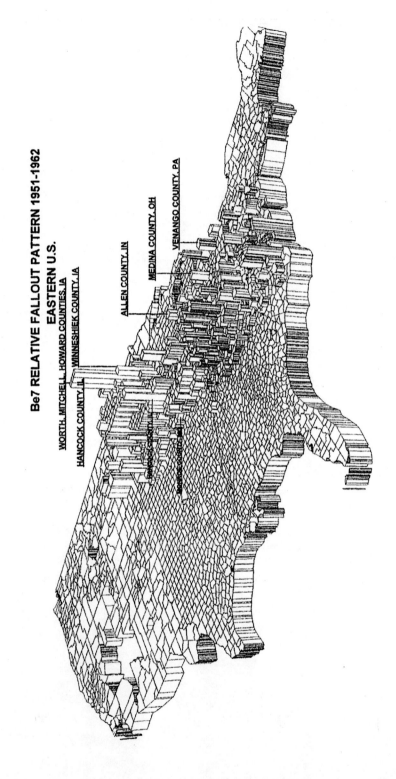

U.S. FALLOUT ATLAS : TOTAL FALLOUT

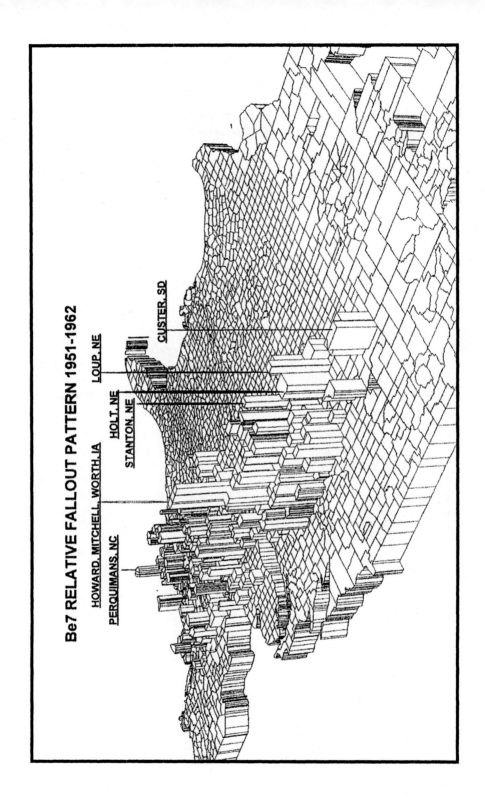

U.S. FALLOUT ATLAS : TOTAL FALLOUT

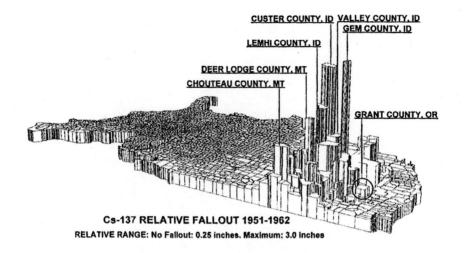

Cs-137 RELATIVE FALLOUT 1951-1962
RELATIVE RANGE: No Fallout: 0.25 inches. Maximum: 3.0 inches

Cesium-137 (Cs-137)

Cesium-137 is considered to be a Class III or moderately toxic radionuclide[9]. [10] It is a fission product with a half-life of approximately 30 years. Twenty years after deposition, cesium-137 retains 63% of its original activity, while after 50 years the activity reduces to 31.5% of the deposition level. Cesium-137 produces beta particles, Auger electrons and X-rays.[11]

Most of the 3D maps in this volume were calculated using the zero rate equal to 0.25 virtual inches in height (z-axis) against a maximum of 1.5 inches virtual height. This is equal to a range ratio of 1:6. Even by expanding the range from 0.25-1.5 to 0.25 to 5.0—an expansion from 1:6 to 1:20, it is clear that most of the Cs-137 remained in the Northwest U.S. The same kind of pattern is seen with strontium-89, strontium-90, yttrium 90 and barium 137m. Nuclear tests high in production of Cs137 include Tumbler-Snapper HOW, Plumbbob BOLTZMANN and Tumbler-Snapper EASY.

Co-60 RELATIVE FALLOUT PATTERN 1951-1962

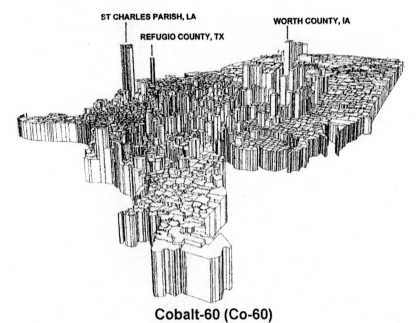

Cobalt-60 (Co-60)

Cobalt-60 is considered to be a Class III (moderately toxic) radionuclide[12][13]. It has a relatively long half-life of 5.27 years. During the decay process Co-60 produces beta and gamma radiation   Ten years after deposition the activity of Co60 has decreased to 0.072 of the original activity. The nuclear detonation highest in Co-60 was shot Galileo (PB-12), followed by shot SEDAN. For that reason, the primary deposition areas for this radionuclide are found in the Southwest and Midwestern United States.

EASTERN U.S.
Co-60 RELATIVE FALLOUT PATTERN 1951-1962

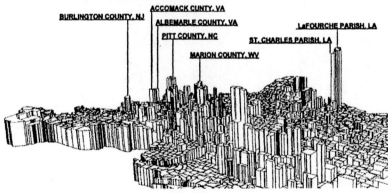

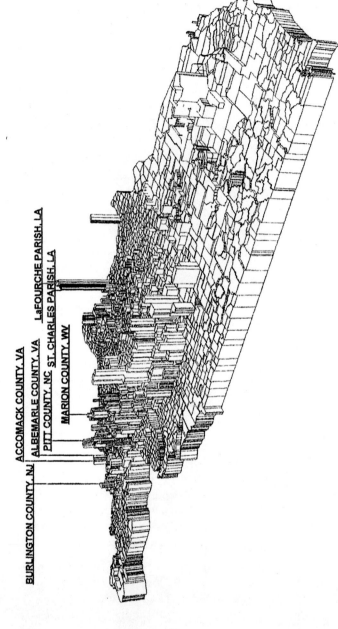

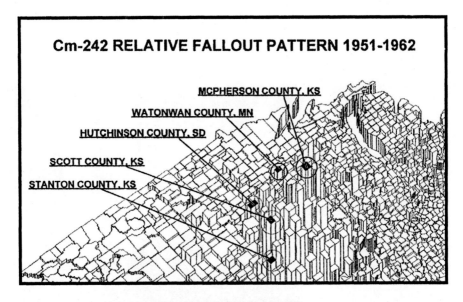

Curium-242 (Cm-242)

Curium-242 is considered to be a Class I (very highly toxic) radionuclide by the International Atomic Energy Commission[14, 15]. It can be produced either through transformation of Americium-241 or by neutron capture from uranium-238 and plutonium-239. It produces alpha particles, gamma rays, X-rays and electrons. It has a half-life of approximately 163 days. After 1 year the activity level of this curium-242 has dropped to approximately 1/5 of the original. Shots producing relatively high levels of curium-242 include SEDAN, Plumbbob CHARLESTON and Plumbbob WHITNEY.

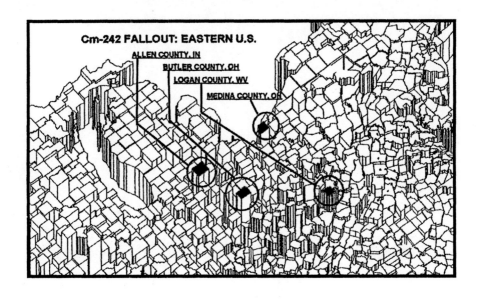

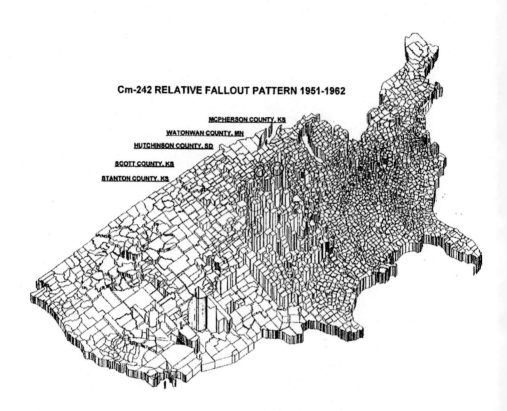

U.S. FALLOUT ATLAS : TOTAL FALLOUT

Europium-155 (Eu-155)

Europium-155 is produced from Sm-154. It has a half-life of 1.811 years. Five years after deposition only 0.147 of the original Eu-155 remains. Eu-155 produces beta particles, electrons and X-rays[16]. Most of the Eu-155 deposited on counties in the U.S. was from shot Plumbbob OWENS, followed by Plumbbob DOPPLER and Upshot-Knothole SIMON. Plumbbob OWENS produced approximately twice as much Eu-155 as did Plumbbob KEPLER.

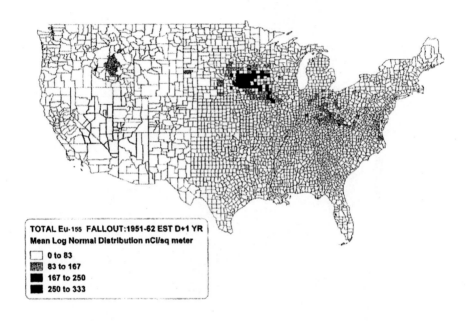

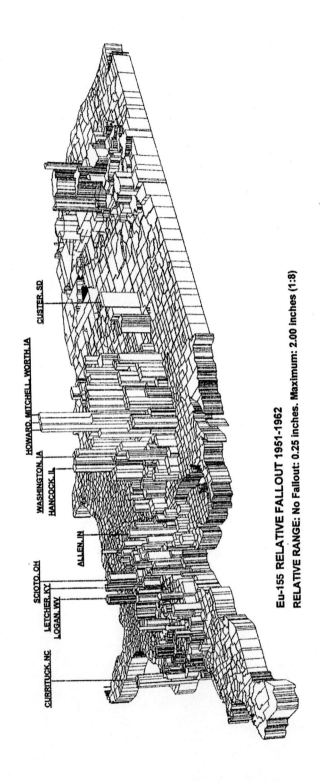

11-22 U.S. FALLOUT ATLAS : TOTAL FALLOUT

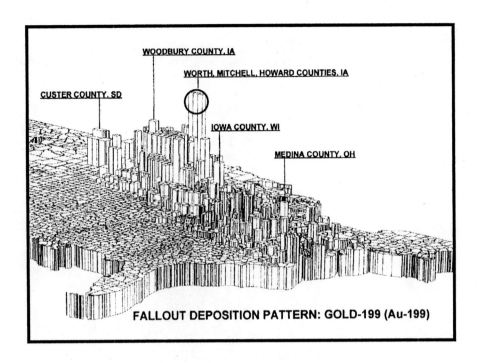

Gold-199 (Au-199)

Gold 199 is classified as a Class III (moderately toxic) radionuclide by the International Atomic Energy Commission[17,18]. It is produced in a nuclear detonation as a daughter product of gold-197 and platinum-199. It produces beta particles, electrons and gamma rays, and has a half life of approximately 2.68 days. Thus, ten days after detonation, the radionuclide will reduced to 0.075 of the original level. Interestingly, only two nuclear tests produced either Au-198 or Au-199: Plumbbob HOOD and shot SEDAN. And while HOOD deposited fallout over more counties than SEDAN, the SEDAN shot deposited far more radioactive gold—11.2 times as much as the HOOD shot.

Iron-59 (Fe-59)

Iron-59 is a Class II (highly toxic) radionuclide.[19,20] It is produced from iron-58 in a nuclear reaction and has a half-life of approximately 45 days. Thus, 90 days after deposition, the amount of Fe-59 has dropped to 25% of the original level. Plumbbob GALILEO distributed the most Fe-59 across the country, followed by Upshot-Knothole SIMON and Storax SEDAN. Some nuclear tests, such as Tumbler-Snapper HOW, Jangle SUGAR, Jangle UNCLE, Tumbler-Snapper EASY and Tumbler-Snapper FOX produced negligible Fe-59.

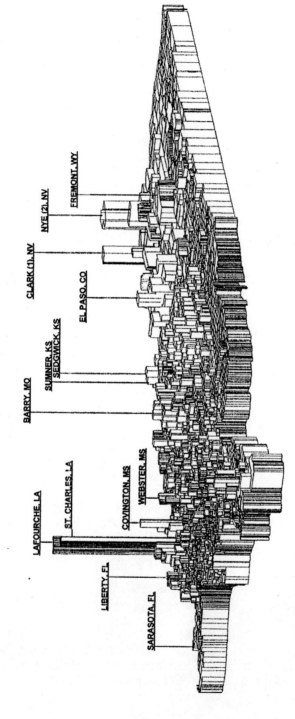

11-24 U.S. FALLOUT ATLAS : TOTAL FALLOUT

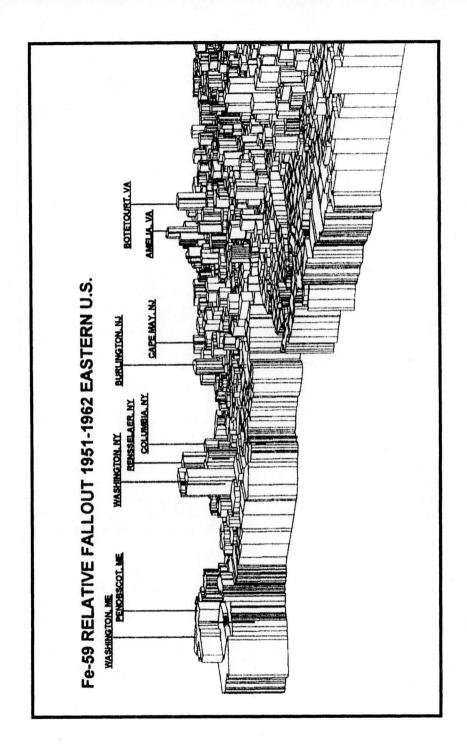

U.S. FALLOUT ATLAS : TOTAL FALLOUT

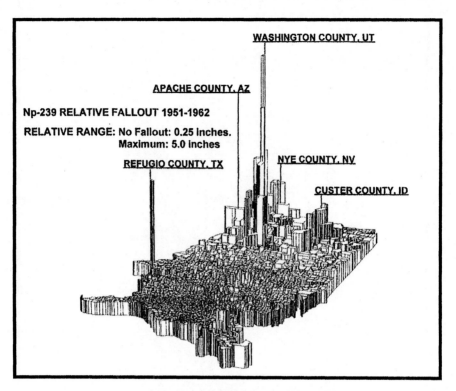

Neptunium-239 (Np-239)

Neptunium is a daughter product of uranium-238 and uranium-239. It produces beta particles, electrons and gamma radiation (as X-rays)[21]. It has a half-life of approximately 2.34 days. The nuclear test with the greatest ratio of Np239 in fallout was Teapot ESS, followed by Tumbler-Snapper ABLE, Tumbler-Snapper BAKER and Ranger BAKER and BAKER-2. Plumbbob shots OWEN and COULOMB B produced *no* Np239.

Interestingly, even though Upshot-Knothole HARRY ranked 20th among the nuclear tests for percent of Np239 in fallout, because it produced so much fallout, HARRY ranked number one in deposition of Np239. Other shots depositing high levels of this radionuclide included Tumbler-Snapper GEORGE (TS7) and Upshot-Knothole SIMON (UK7). As the expanded range (1:20) 3D map above shows, the vast majority of Np239 was deposited on the counties surrounding the Nevada Test Site. Still, thanks primarily to the 1953 nuclear test Upshot-Knothole SIMON, significant amounts of this short-lived radionuclide was deposited in the Eastern United States.

Pd-111m RELATIVE FALLOUT PATTERN 1951-1962

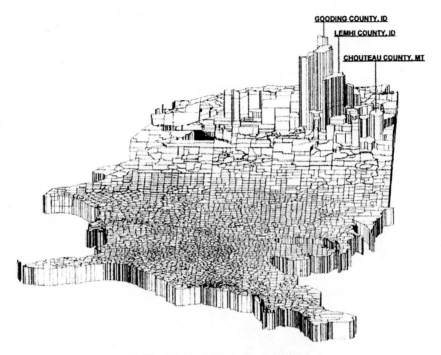

Palladium-111m (Pd-111m)

Palladium-111m is a fission product of palladium-110. It produces beta particles, electrons and gamma rays and has a relatively short half-life of only 5.5 hours. Thus, within 24 hours of detonation, less than 0.05 of the original amount of Pd-111m remains in the fallout. For that reason, exposure to Pd111m was generally limited to counties close to the Nevada Test Site. Shots Plumbbob KEPLER, Teapot POST and Plumbbob OWENS had the greatest ratio of Pd111m in fallout, while the Upshot-Knothole shots HARRY, NANCY and SIMON were responsible for the greatest amounts of Pd111m deposited.

Strontium-90

Strontium-90 is considered a Class I (very highly toxic) radionuclide by the International Atomic Energy Commission[22]. [23] It is a fission product with a half-life of approximately 28 years.[24] Thus, 40 years after deposition, over a third of the Sr-90 deposited is still active. Strontium-90 produces no gamma radiation, only beta particles. Most of the Sr-90 deposited across the country came from Upshot-Knothole SIMON, followed by Tumbler-Snapper GEORGE and Upshot-Knothole HARRY. Still, as the expanded range (1:20) 3D map on the next page suggests, the vast majority of Sr-90 was deposited in the counties north of the Nevada Test Site. Sr-90 deposition levels are similar to those for Sr-89, Y-90, and Y-91.

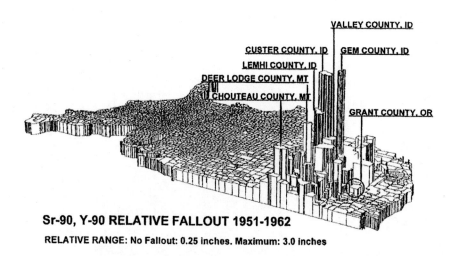

Sr-90, Y-90 RELATIVE FALLOUT 1951-1962

RELATIVE RANGE: No Fallout: 0.25 inches. Maximum: 3.0 inches

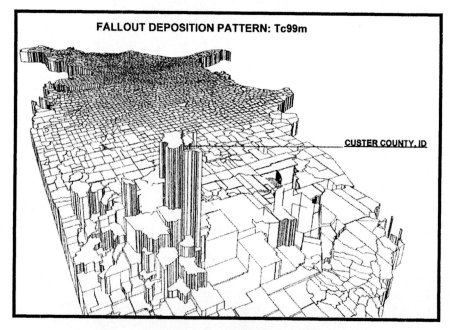

Technetium-99m (Tc99m)

Technectium-99m has an extremely short half-life of only 6 hours, but because it is constantly being produced from the longer-lived Mo-99, (half-life: 66 hours)[25] it remains in the environment until the parent radionuclides are depleted. Technectium-99m produces X-rays and electrons. Shots Buster

BAKER and Plumbbob OWENS produced the greatest ratio of Tc99m in fallout, while shot Tumbler-Snapper HOW and Upshot-Knothole HARRY produced the greatest deposition of this radionuclide.

Uranium-237 (U-237)

Uranium 237 is a fission product of U-238. It produces beta particles, electrons and gamma radiation. It has a half-life of only 6.75 days, yet according to the Hicks Tables the overall activity level for U-237 from many detonations such as shot TESLA remained relatively constant for up to fifty days post shot[26]. For that reason, despite it's short half-life, U-237 was deposited in areas far from the Nevada Test Site. Upshot-Knothole BADGER had the greatest percentage of U-237 in fallout, followed by Jangle UNCLE and Plumbbob HOOD. Shot SEDAN, however, deposited the greatest amount of U-237, followed by Upshot-Knothole SIMON and Tumbler-Snapper HOW.

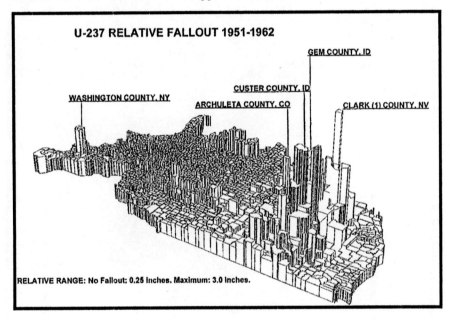

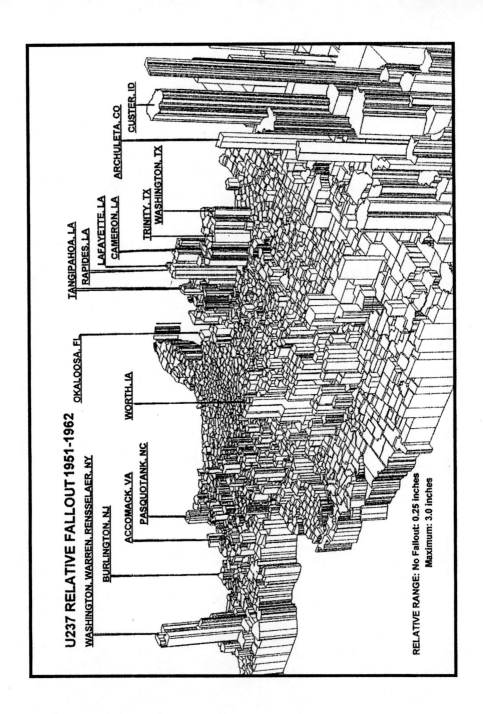

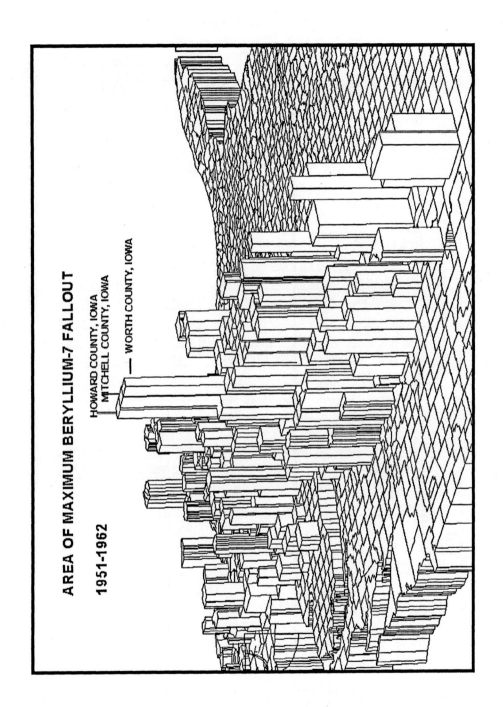

U.S. FALLOUT ATLAS : TOTAL FALLOUT 11-31

COUNTIES WITH HIGHEST
RADIONUCLIDE DEPOSITION AMOUNTS
U.S. RANK WITH ASSOCIATED DEPOSITION ACTIVITY: nCi/Sq meter

RANK	Be7*		Na24		Mn54*	
1	HOWARD IA	17.6	NYE2 NV	1849.1	MITCHELL IA	17.4
2	WORTH IA	17.6	GARFIELD UT	1197.5	HOWARD IA	17.4
3	MITCHELL IA	17.6	INYO3 CA	853.7	WORTH IA	17.3
4	LEE IA	11.5	NYE1 NV	849.1	VAN BUREN IA	11.8
5	HANCOCK IL	11.5	WHITE PINE2 NV	796.1	HANCOCK IL	11.6
6	WASHINGTON IA	11.5	MINERAL NV	765.9	JEFFERSON IA	11.6
7	JEFFERSON IA	11.5	WASHABAUGH SD	739.8	LEE IA	11.6
8	VAN BUREN IA	11.5	WASHINGTON1 UT	673.5	WASHINGTON IA	11.5
9	WOODBURY IA	11.5	WASHINGTON2 UT	672.5	WOODBURY IA	11.5
10	WATONWAN MN	10.8	CHURCHILL NV	636.9	WATONWAN MN	10.8
11	FARIBAULT MN	10.8	WASHINGTON3 UT	588.8	FARIBAULT MN	10.8
12	MOWER MN	10.8	LINCOLN1 NV	584.0	MOWER MN	10.8
13	DIXON NE	10.5	SWEETWATER WY	526.7	PLYMOUTH IA	10.5
14	WASHABAUGH SD	10.4	LANDER2 NV	519.8	DIXON NE	10.5
15	PLYMOUTH IA	10.4	LYON NV	516.9	WASHABAUGH SD	10.4

RANK	Fe59*		Co60*		Cu64	
1	LAFOURCHE LA	0.16	MITCHELL IA	7.4	WASHINGTON2 UT	52784
2	ST CHARLES LA	0.15	HOWARD IA	7.3	WASHINGTON3 UT	42969
3	CLARK1 NV	0.07	WORTH IA	7.3	WASHINGTON1 UT	32574
4	NYE2 NV	0.07	LAFOURCHE LA	6.1	KANE1 UT	31455
5	COVINGTON MS	0.06	ST CHARLES LA	5.8	KANE2 UT	29399
6	SEDGWICK KS	0.05	VAN BUREN IA	5.3	MOHAVE2 AZ	27960
7	SUMNER KS	0.05	WOODBURY IA	5.1	CLARK1 NV	27907
8	BARRY MO	0.05	HANCOCK IL	5.1	LINCOLN1 NV	26943
9	EL PASO CO	0.05	LEE IA	5.0	COCONINO1 AZ	23722
10	BALDWIN AL	0.05	JEFFERSON IA	4.9	MOHAVE1 AZ	18376
11	WEBSTER MS	0.05	WASHINGTON IA	4.9	GUNNISON CO	17836
12	VERNON MO	0.05	PLYMOUTH IA	4.7	CONEJOS CO	16612
13	SHAWNEE KS	0.05	CRAWFORD IA	4.7	GARFIELD UT	15251
14	WILSON KS	0.05	IDA IA	4.7	COCONINO2 AZ	11594
15	ROGERS OK	0.05	MOWER MN	4.6	ARCHULETA CO	10973

RANK	Br82		Br83		Sr89*	
1	GEM ID	360	GEM ID	11334	GEM ID	1071
2	CUSTER ID	343	CUSTER ID	10776	CUSTER ID	1019
3	BLAINE ID	325	BLAINE ID	10218	BLAINE ID	965
4	LEMHI ID	233	LEMHI ID	7329	LEMHI ID	694
5	DEER LODGE MT	206	DEER LODGE MT	6485	DEER LODGE MT	615
6	VALLEY ID	158	VALLEY ID	4945	VALLEY ID	473
7	IDAHO ID	143	IDAHO ID	4494	IDAHO ID	431
8	CAMAS ID	141	CAMAS ID	4414	CAMAS ID	421
9	CHOUTEAU MT	136	CHOUTEAU MT	4254	CHOUTEAU MT	405
10	GOODING ID	131	GOODING ID	4114	GOODING ID	387
11	BEAVERHEAD MT	129	BEAVERHEAD MT	4036	BEAVERHEAD MT	385
12	BOISE ID	121	BOISE ID	3797	BOISE ID	364
13	SILVER BOW MT	121	SILVER BOW MT	3793	SILVER BOW MT	361
14	JUDITH BASIN MT	120	JUDITH BASIN MT	3754	JUDITH BASIN MT	359
15	POWELL MT	119	POWELL MT	3746	POWELL MT	358

RANK	Sr90*-Y90*		Y91*		Nb95*	
1	GEM ID	919	GEM ID	2020	GEM ID	4161
2	CUSTER ID	874	CUSTER ID	1922	CUSTER ID	3959
3	BLAINE ID	828	BLAINE ID	1820	BLAINE ID	3753
4	LEMHI ID	596	LEMHI ID	1309	LEMHI ID	2700
5	DEER LODGE MT	527	DEER LODGE MT	1159	DEER LODGE MT	2394
6	VALLEY ID	406	VALLEY ID	893	VALLEY ID	1841
7	IDAHO ID	369	IDAHO ID	812	IDAHO ID	1675
8	CAMAS ID	361	CAMAS ID	794	CAMAS ID	1638
9	CHOUTEAU MT	347	CHOUTEAU MT	764	CHOUTEAU MT	1580
10	GOODING ID	332	GOODING ID	729	BEAVERHEAD MT	1502
11	BEAVERHEAD MT	330	BEAVERHEAD MT	727	GOODING ID	1502
12	BOISE ID	312	BOISE ID	686	BOISE ID	1415
13	SILVER BOW MT	310	SILVER BOW MT	682	SILVER BOW MT	1410
14	JUDITH BASIN MT	308	JUDITH BASIN MT	677	JUDITH BASIN MT	1399
15	POWELL MT	307	POWELL MT	675	POWELL MT	1397

RANK	Nb95m*		Zr95*		Mo99	
1	GEM ID	48	GEM ID	1940	GEM ID	1.84E+06
2	CUSTER ID	45	CUSTER ID	1846	CUSTER ID	1.75E+06
3	BLAINE ID	37	BLAINE ID	1750	BLAINE ID	1.66E+06
4	LEMHI ID	27	LEMHI ID	1259	LEMHI ID	1.19E+06
5	VALLEY ID	26	DEER LODGE MT	1116	DEER LODGE MT	1.06E+06
6	IDAHO ID	24	VALLEY ID	858	VALLEY ID	8.10E+05
7	DEER LODGE MT	24	IDAHO ID	781	IDAHO ID	7.36E+05
8	CAMAS ID	22	CAMAS ID	764	CAMAS ID	7.22E+05
9	BOISE ID	21	CHOUTEAU MT	737	CHOUTEAU MT	6.96E+05
10	WASHINGTON ID	19	BEAVERHEAD MT	700	GOODING ID	6.67E+05
11	NYE2 NV	18	GOODING ID	700	BEAVERHEAD MT	6.62E+05
12	CLEARWATER ID	17	BOISE ID	660	BOISE ID	6.22E+05
13	GOODING ID	16	SILVER BOW MT	658	SILVER BOW MT	6.21E+05
14	CHOUTEAU MT	16	JUDITH BASIN MT	652	JUDITH BASIN MT	6.16E+05
15	OWYHEE ID	15	POWELL MT	652	POWELL MT	6.14E+05

RANK	Tc99m		Ru106* Rh106*		Ag109m	
1	GEM ID	1.62E+06	GEM ID	5926	GEM ID	2.41E+05
2	CUSTER ID	1.55E+06	CUSTER ID	5643	CUSTER ID	2.29E+05
3	BLAINE ID	1.46E+06	BLAINE ID	5347	BLAINE ID	2.17E+05
4	LEMHI ID	1.05E+06	LEMHI ID	3855	LEMHI ID	1.56E+05
5	DEER LODGE MT	9.32E+05	DEER LODGE MT	3413	DEER LODGE MT	1.38E+05
6	VALLEY ID	7.14E+05	VALLEY ID	2622	VALLEY ID	1.05E+05
7	IDAHO ID	6.49E+05	IDAHO ID	2384	IDAHO ID	9.58E+04
8	CAMAS ID	6.36E+05	CAMAS ID	2333	CAMAS ID	9.40E+04
9	CHOUTEAU MT	6.14E+05	CHOUTEAU MT	2254	CHOUTEAU MT	9.07E+04
10	GOODING ID	5.88E+05	BEAVERHEAD MT	2150	GOODING ID	8.75E+04
11	BEAVERHEAD MT	5.84E+05	GOODING ID	2140	BEAVERHEAD MT	8.61E+04
12	BOISE ID	5.49E+05	BOISE ID	2013	BOISE ID	8.09E+04
13	SILVER BOW MT	5.48E+05	SILVER BOW MT	2011	SILVER BOW MT	8.09E+04
14	JUDITH BASIN MT	5.43E+05	POWELL MT	1995	JUDITH BASIN MT	8.01E+04
15	POWELL MT	5.42E+05	JUDITH BASIN MT	1995	POWELL MT	7.99E+04

RANK	Pd111m		Pd112		Ag112	
1	GEM ID	5.69E+04	GEM ID	1.01E+05	GEM ID	1.16E+05
2	CUSTER ID	5.41E+04	CUSTER ID	9.57E+04	CUSTER ID	1.11E+05
3	BLAINE ID	5.13E+04	BLAINE ID	9.07E+04	BLAINE ID	1.05E+05
4	LEMHI ID	3.68E+04	LEMHI ID	6.52E+04	LEMHI ID	7.54E+04
5	DEER LODGE MT	3.26E+04	DEER LODGE MT	5.77E+04	DEER LODGE MT	6.67E+04
6	VALLEY ID	2.49E+04	VALLEY ID	4.41E+04	VALLEY ID	5.10E+04
7	IDAHO ID	2.26E+04	IDAHO ID	4.01E+04	IDAHO ID	4.63E+04
8	CAMAS ID	2.22E+04	CAMAS ID	3.93E+04	CAMAS ID	4.55E+04
9	CHOUTEAU MT	2.14E+04	CHOUTEAU MT	3.79E+04	CHOUTEAU MT	4.39E+04
10	GOODING ID	2.07E+04	GOODING ID	3.65E+04	GOODING ID	4.22E+04
11	BEAVERHEAD MT	2.03E+04	BEAVERHEAD MT	3.61E+04	BEAVERHEAD MT	4.17E+04
12	BOISE ID	1.91E+04	SILVER BOW MT	3.38E+04	SILVER BOW MT	3.92E+04
13	SILVER BOW MT	1.91E+04	BOISE ID	3.38E+04	BOISE ID	3.92E+04
14	JUDITH BASIN MT	1.89E+04	JUDITH BASIN MT	3.36E+04	JUDITH BASIN MT	3.89E+04
15	POWELL MT	1.88E+04	POWELL MT	3.35E+04	POWELL MT	3.87E+04

RANK	Cd115		In115m		In117	
1	GEM ID	32950	GEM ID	34826	GEM ID	2857
2	CUSTER ID	31384	CUSTER ID	33175	CUSTER ID	2716
3	BLAINE ID	29723	BLAINE ID	31418	BLAINE ID	2576
4	LEMHI ID	21408	LEMHI ID	22639	LEMHI ID	1847
5	DEER LODGE MT	18962	DEER LODGE MT	20049	DEER LODGE MT	1634
6	VALLEY ID	14545	VALLEY ID	15387	VALLEY ID	1247
7	IDAHO ID	13216	IDAHO ID	13976	IDAHO ID	1133
8	CAMAS ID	12944	CAMAS ID	13688	CAMAS ID	1112
9	CHOUTEAU MT	12539	CHOUTEAU MT	13259	CHOUTEAU MT	1072
10	GOODING ID	11948	BEAVERHEAD MT	12636	GOODING ID	1038
11	BEAVERHEAD MT	11941	GOODING ID	12626	BEAVERHEAD MT	1017
12	NYE2 NV	11414	NYE2 NV	12057	NYE2 NV	991
13	SILVER BOW MT	11180	SILVER BOW MT	11822	BOISE ID	957
14	BOISE ID	11166	BOISE ID	11810	SILVER BOW MT	956
15	JUDITH BASIN MT	11137	JUDITH BASIN MT	11780	JUDITH BASIN MT	946

RANK	Sb126		Te127m*		I130	
1	GEM ID	2726	GEM ID	178	GEM ID	7950
2	CUSTER ID	2597	CUSTER ID	170	CUSTER ID	7559
3	BLAINE ID	2460	BLAINE ID	161	BLAINE ID	7166
4	LEMHI ID	1773	LEMHI ID	116	LEMHI ID	5142
5	DEER LODGE MT	1571	DEER LODGE MT	102	DEER LODGE MT	4550
6	VALLEY ID	1207	VALLEY ID	79	VALLEY ID	3472
7	IDAHO ID	1098	IDAHO ID	72	IDAHO ID	3155
8	CAMAS ID	1074	CAMAS ID	70	CAMAS ID	3097
9	CHOUTEAU MT	1040	CHOUTEAU MT	68	CHOUTEAU MT	2986
10	BEAVERHEAD MT	991	BEAVERHEAD MT	64	GOODING ID	2884
11	GOODING ID	985	GOODING ID	64	BEAVERHEAD MT	2833
12	BOISE ID	927	BOISE ID	61	BOISE ID	2666
13	SILVER BOW MT	927	SILVER BOW MT	60	SILVER BOW MT	2662
14	JUDITH BASIN MT	922	JUDITH BASIN MT	60	JUDITH BASIN MT	2638
15	POWELL MT	920	POWELL MT	60	POWELL MT	2630

RANK	I131		I132		I133	
1	GEM ID	8.78E+05	GEM ID	2.60E+06	GEM ID	7.45E+06
2	CUSTER ID	8.36E+05	CUSTER ID	2.47E+06	CUSTER ID	7.08E+06
3	BLAINE ID	7.91E+05	BLAINE ID	2.34E+06	BLAINE ID	6.71E+06
4	LEMHI ID	5.69E+05	LEMHI ID	1.68E+06	LEMHI ID	4.82E+06
5	DEER LODGE MT	5.04E+05	DEER LODGE MT	1.49E+06	DEER LODGE MT	4.26E+06
6	VALLEY ID	3.87E+05	VALLEY ID	1.14E+06	VALLEY ID	3.26E+06
7	IDAHO ID	3.52E+05	IDAHO ID	1.04E+06	IDAHO ID	2.96E+06
8	CAMAS ID	3.45E+05	CAMAS ID	1.02E+06	CAMAS ID	2.90E+06
9	CHOUTEAU MT	3.32E+05	CHOUTEAU MT	9.79E+05	CHOUTEAU MT	2.80E+06
10	GOODING ID	3.17E+05	GOODING ID	9.39E+05	GOODING ID	2.70E+06
11	BEAVERHEAD MT	3.16E+05	BEAVERHEAD MT	9.31E+05	BEAVERHEAD MT	2.66E+06
12	BOISE ID	2.97E+05	BOISE ID	8.76E+05	BOISE ID	2.50E+06
13	SILVER BOW MT	2.96E+05	SILVER BOW MT	8.73E+05	SILVER BOW MT	2.49E+06
14	JUDITH BASIN MT	2.94E+05	JUDITH BASIN MT	8.67E+05	JUDITH BASIN MT	2.47E+06
15	POWELL MT	2.93E+05	POWELL MT	8.64E+05	POWELL MT	2.46E+06

RANK	I135		Cs137*		Ba137m*	
1	GEM ID	3.66E+06	GEM ID	1156	GEM ID	1085
2	CUSTER ID	3.48E+06	CUSTER ID	1100	CUSTER ID	1032
3	BLAINE ID	3.30E+06	BLAINE ID	1042	BLAINE ID	978
4	LEMHI ID	2.37E+06	LEMHI ID	750	LEMHI ID	704
5	DEER LODGE MT	2.09E+06	DEER LODGE MT	664	DEER LODGE MT	623
6	VALLEY ID	1.60E+06	VALLEY ID	511	VALLEY ID	479
7	IDAHO ID	1.45E+06	IDAHO ID	465	IDAHO ID	436
8	CAMAS ID	1.43E+06	CAMAS ID	455	CAMAS ID	427
9	CHOUTEAU MT	1.37E+06	CHOUTEAU MT	438	CHOUTEAU MT	411
10	GOODING ID	1.33E+06	GOODING ID	417	GOODING ID	392
11	BEAVERHEAD MT	1.30E+06	BEAVERHEAD MT	417	BEAVERHEAD MT	391
12	BOISE ID	1.23E+06	BOISE ID	393	BOISE ID	368
13	SILVER BOW MT	1.23E+06	SILVER BOW MT	391	SILVER BOW MT	367
14	JUDITH BASIN MT	1.21E+06	JUDITH BASIN MT	388	JUDITH BASIN MT	364
15	POWELL MT	1.21E+06	POWELL MT	387	POWELL MT	363

RANK	La141		Ce144, Pr144		Sm153	
1	GEM ID	7.19E+05	GEM ID	6806	GEM ID	99049
2	CUSTER ID	6.83E+05	CUSTER ID	6476	CUSTER ID	94269
3	BLAINE ID	6.48E+05	BLAINE ID	6139	BLAINE ID	89329
4	LEMHI ID	4.65E+05	LEMHI ID	4417	LEMHI ID	64220
5	DEER LODGE MT	4.11E+05	DEER LODGE MT	3915	DEER LODGE MT	56851
6	VALLEY ID	3.14E+05	VALLEY ID	3012	VALLEY ID	43530
7	IDAHO ID	2.85E+05	IDAHO ID	2741	IDAHO ID	39540
8	CAMAS ID	2.80E+05	CAMAS ID	2680	CAMAS ID	38780
9	CHOUTEAU MT	2.70E+05	CHOUTEAU MT	2585	CHOUTEAU MT	37432
10	GOODING ID	2.61E+05	GOODING ID	2457	GOODING ID	35909
11	BEAVERHEAD MT	2.56E+05	BEAVERHEAD MT	2457	BEAVERHEAD MT	35614
12	BOISE ID	2.41E+05	BOISE ID	2315	BOISE ID	33413
13	SILVER BOW MT	2.41E+05	SILVER BOW MT	2307	SILVER BOW MT	33403
14	JUDITH BASIN MT	2.38E+05	JUDITH BASIN MT	2288	JUDITH BASIN MT	33099
15	POWELL MT	2.38E+05	POWELL MT	2285	POWELL MT	33061

RANK	Eu155*		Tb161		W181	
1	HOWARD IA	333	GEM ID	386	MITCHELL IA	8.35E+04
2	MITCHELL IA	333	CUSTER ID	368	WORTH IA	8.35E+04
3	WORTH IA	333	BLAINE ID	349	HOWARD IA	8.35E+04
4	VAN BUREN IA	220	LEMHI ID	252	WASHINGTON IA	5.46E+04
5	HANCOCK IL	219	DEER LODGE MT	223	VAN BUREN IA	5.46E+04
6	LEE IA	219	VALLEY ID	172	JEFFERSON IA	5.46E+04
7	JEFFERSON IA	219	IDAHO ID	156	LEE IA	5.46E+04
8	WASHINGTON IA	219	CAMAS ID	152	HANCOCK IL	5.46E+04
9	WOODBURY IA	217	CHOUTEAU MT	149	WOODBURY IA	5.43E+04
10	MOWER MN	205	BEAVERHEAD MT	142	WATONWAN MN	5.14E+04
11	FARIBAULT MN	205	GOODING ID	140	FARIBAULT MN	5.14E+04
12	WATONWAN MN	205	SILVER BOW MT	132	MOWER MN	5.14E+04
13	DIXON NE	200	POWELL MT	132	DIXON NE	4.99E+04
14	WASHABAUGH SD	198	BOISE ID	132	PLYMOUTH IA	4.93E+04
15	PLYMOUTH IA	197	JUDITH BASIN MT	131	WASHABAUGH SD	4.93E+04

RANK	W185*		Au198		Au199	
1	WORTH IA	6716	HOWARD IA	3371	HOWARD IA	16.26
2	MITCHELL IA	6716	WORTH IA	3371	WORTH IA	16.26
3	HOWARD IA	6716	MITCHELL IA	3371	MITCHELL IA	16.25
4	LEE IA	4391	WASHABAUGH SD	2570	WASHABAUGH SD	12.23
5	WASHINGTON IA	4391	WOODBURY IA	2285	WOODBURY IA	10.99
6	VAN BUREN IA	4391	LEE IA	2204	LEE IA	10.64
7	HANCOCK IL	4391	WASHINGTON IA	2204	WASHINGTON IA	10.63
8	JEFFERSON IA	4391	HANCOCK IL	2204	HANCOCK IL	10.63
9	WOODBURY IA	4366	JEFFERSON IA	2204	JEFFERSON IA	10.63
10	WATONWAN MN	4133	VAN BUREN IA	2204	VAN BUREN IA	10.63
11	FARIBAULT MN	4133	FARIBAULT MN	2075	FARIBAULT MN	10.01
12	MOWER MN	4133	MOWER MN	2075	MOWER MN	10.01
13	DIXON NE	4013	WATONWAN MN	2075	WATONWAN MN	10.01
14	PLYMOUTH IA	3970	DIXON NE	1982	DIXON NE	9.59
15	WASHABAUGH SD	3966	PLYMOUTH IA	1965	PLYMOUTH IA	9.50

RANK	U237		Np239	
1	CLARK1 NV	15517	WASHINGTON2 UT	1.07E+06
2	GEM ID	14962	WASHINGTON3 UT	7.81E+05
3	CUSTER ID	14284	WASHINGTON1 UT	5.34E+05
4	BLAINE ID	13362	KANE2 UT	5.09E+05
5	LEMHI ID	9626	REFUGIO TX	4.90E+05
6	NYE2 NV	9008	MOHAVE2 AZ	4.86E+05
7	WASHINGTON2 UT	8942	LINCOLN1 NV	4.79E+05
8	COCONINO2 AZ	8915	KANE1 UT	4.75E+05
9	DEER LODGE MT	8540	CLARK1 NV	4.49E+05
10	APACHE AZ	7313	ARCHULETA CO	3.68E+05
11	ARCHULETA CO	7098	COCONINO1 AZ	3.45E+05
12	VALLEY ID	6983	GUNNISON CO	3.44E+05
13	WASHINGTON3 UT	6388	CONEJOS CO	3.39E+05
14	IDAHO ID	6351	NYE2 NV	3.27E+05
15	NAVAJO AZ	6158	MOHAVE1 AZ	2.71E+05

RANK	U240		Am241*	
1	WASHINGTON2 UT	20116	MCPHERSON KS	0.028
2	WASHINGTON3 UT	15728	PAWNEE KS	0.028
3	CLARK1 NV	12401	OTTAWA KS	0.026
4	WASHINGTON1 UT	11863	STAFFORD KS	0.026
5	KANE2 UT	10707	BARTON KS	0.026
6	KANE1 UT	10191	MITCHELL KS	0.026
7	MOHAVE2 AZ	9545	RUSSELL KS	0.026
8	LINCOLN1 NV	8884	SALINE KS	0.025
9	COCONINO2 AZ	7773	LINCOLN KS	0.025
10	COCONINO1 AZ	7284	RICE KS	0.025
11	APACHE AZ	7206	CLOUD KS	0.024
12	GUNNISON CO	6376	FORD KS	0.024
13	ARCHULETA CO	6252	DICKINSON KS	0.024
14	CONEJOS CO	6065	ELLSWORTH KS	0.024
15	MCKINLEY NM	5490	WABAUNSEE KS	0.024

COUNTY RANK: Cm242* in TOTAL FALLOUT
nCi/sq meter

#	County	Value	#	County	Value	#	County	Value
1	MCPHERSON KS	0.821	29	HARVEY KS	0.603	57	HASKELL KS	0.519
2	PAWNEE KS	0.811	30	POTTAWATOMIE KS	0.602	58	LINCOLN OK	0.518
3	OTTAWA KS	0.775	31	SCOTT KS	0.600	59	CHAUTAUQUA KS	0.516
4	BARTON KS	0.768	32	MEADE KS	0.589	60	MAYES OK	0.515
5	MITCHELL KS	0.763	33	LANE KS	0.584	61	CHEROKEE OK	0.515
6	STAFFORD KS	0.758	34	HARPER OK	0.582	62	JACKSON KS	0.514
7	RUSSELL KS	0.758	35	COWLEY KS	0.577	63	LABETTE KS	0.512
8	SALINE KS	0.746	36	MARION KS	0.567	64	ELK KS	0.511
9	LINCOLN KS	0.744	37	CREEK OK	0.562	65	HALL NE	0.510
10	RICE KS	0.731	38	WASHINGTON OK	0.558	66	SEDGWICK KS	0.500
11	CLOUD KS	0.725	39	NOBLE OK	0.552	67	JEFFERSON KS	0.495
12	FORD KS	0.705	40	ALFALFA OK	0.551	68	JOHNSON NE	0.494
13	ELLSWORTH KS	0.703	41	WASHINGTON KS	0.550	69	ROGERS OK	0.486
14	DICKINSON KS	0.699	42	OSAGE OK	0.548	70	GAGE NE	0.484
15	WABAUNSEE KS	0.697	43	SHAWNEE KS	0.546	71	PRATT KS	0.484
16	GRANT OK	0.695	44	PAWNEE OK	0.542	72	KINGFISHER OK	0.481
17	CLAY KS	0.680	45	BUFFALO NE	0.542	73	LOGAN OK	0.478
18	EDWARDS KS	0.671	46	MARSHALL KS	0.542	74	KEARNEY NE	0.477
19	CLARK KS	0.660	47	PAYNE OK	0.539	75	WAGONER OK	0.474
20	RUSH KS	0.655	48	TULSA OK	0.539	76	DELAWARE OK	0.472
21	GEARY KS	0.655	49	MAJOR OK	0.535	77	HARPER KS	0.471
22	NESS KS	0.652	50	GRANT KS	0.530	78	OKMULGEE OK	0.469
23	RILEY KS	0.641	51	BUTLER KS	0.528	79	PAWNEE NE	0.465
24	ELLIS KS	0.616	52	GREENWOOD KS	0.524	80	MORRIS KS	0.463
25	TREGO KS	0.616	53	GARFIELD OK	0.523			
26	RENO KS	0.610	54	KAY OK	0.520			
27	SUMNER KS	0.606	55	CHASE KS	0.520			
28	OSBORNE KS	0.603	56	WOODS OK	0.520			

COUNTY RANK:
U237 in TOTAL FALLOUT
nCi/sq meter

#	County	Value	#	County	Value	#	County	Value
1	CLARK1 NV	15517	29	RENSSELAER NY	5096	57	MISSOULA MT	3986
2	GEM ID	14962	30	JEFFERSON MT	5065	58	LANDER2 NV	3984
3	CUSTER ID	14284	31	WORTH IA	5062	59	CARROLL IA	3978
4	BLAINE ID	13362	32	LINCOLN1 NV	4985	60	WOODBURY IA	3966
5	LEMHI ID	9626	33	WASHINGTON1 UT	4960	61	DIXON NE	3907
6	NYE2 NV	9008	34	GARFIELD UT	4901	62	BEAUREGARD LA	3904
7	WASHINGTON2 UT	8942	35	VALENCIA NM	4737	63	EUREKA NV	3888
8	COCONINO2 AZ	8915	36	WASHINGTON ID	4737	64	INYO3 CA	3860
9	DEER LODGE MT	8540	37	BROADWATER MT	4689	65	JASPER IA	3839
10	APACHE AZ	7313	38	MEAGHER MT	4686	66	BUTTE ID	3839
11	ARCHULETA CO	7098	39	SANDOVAL NM	4497	67	LEE IA	3832
12	VALLEY ID	6983	40	ELKO NV	4458	68	VAN BUREN IA	3817
13	WASHINGTON3 UT	6388	41	MOHAVE2 AZ	4447	69	TURNER SD	3769
14	IDAHO ID	6351	42	KANE2 UT	4388	70	JEFFERSON ID	3766
15	NAVAJO AZ	6158	43	WALLOWA OR	4355	71	WHEATLAND MT	3757
16	CAMAS ID	6107	44	LANDER1 NV	4348	72	KANE1 UT	3755
17	CHOUTEAU MT	5760	45	OWYHEE ID	4311	73	HANCOCK IL	3751
18	WASHINGTON NY	5564	46	LEWIS ID	4295	74	CONEJOS CO	3716
19	MCKINLEY NM	5463	47	GRANT LA	4234	75	BOONE IA	3709
20	BEAVERHEAD MT	5447	48	WASHABAUGH SD	4214	76	CALCASIEU LA	3685
21	GOODING ID	5421	49	ELMORE ID	4182	77	ACADIA LA	3681
22	BOISE ID	5396	50	RAPIDES LA	4161	78	EVANGELINE LA	3658
23	JUDITH BASIN MT	5374	51	WHITE PINE2 NV	4152	79	KNOX MO	3656
24	WARREN NY	5291	52	LINCOLN ID	4108	80	MARSHALL IA	3624
25	HOWARD IA	5175	53	WHITE PINE3 NV	4068			
26	SILVER BOW MT	5167	54	MADISON MT	4035			
27	MITCHELL IA	5145	55	ADA ID	4017			
28	POWELL MT	5131	56	LAFAYETTE LA	3999			

NOTES TO CHAPTER 11

[1] International Atomic Energy Commission, Vienna, *Safe Handling of Radioactive Nuclides*, Safety Series No. 1.1973.
[2] Wilson, William Feathergail. NORM: A Guide to Naturally Occurring Radioactive Material. Pennwell Books. Pennwell Pub. Tulsa, OK. p. 147.
[3] Radiological Health Handbook. US Department of Health, Education and Welfare, Public Health Service. Jan. 1970. p.374.
[4] Emsley, John, The Elements, 3rd Edition. Clarendon Press, Oxford. 1998. p.21.
[5] International Atomic Energy Commission. Op cit.
[6] Wilson. Op cit. p. 147.
[7] Radiological Health Handbook. Ibid. p. 232.
[8] Hicks, Harry G. *Results of Calculations of External Gamma Radiation Exposure Rates from Fallout and the Related Radionuclide Compositions. Operation Nougat Through Bowline, 1962-1968.* Lawrence Livermore National Lab, CA Jul 1981. DE82007515. p. B-8, 10
[9] International Atomic Energy Commission Op cit.
[10] Wilson. Op cit. p. 147.
[11] Radiological Health Handbook. Op cit. p. 303.
[12] International Atomic Energy Commission. Op cit.
[13] Wilson, Op cit. p. 147.
[14] International Atomic Energy Commission. Op cit.
[15] Wilson. Op cit. p. 147.
[16] Radiological Health Handbook. Op cit. p. 316.
[17] International Atomic Energy Commisssion. Op cit.
[18] Wilson. Op cit. p. 147.
[19] International Atomic Energy Commission. Op cit.
[20] Wilson. Op cit. p. 147.
[21] Radiological Health Handbook. Op cit. p. 371.
[22] International Atomic Energy Commission. Op cit.
[23] Wilson. Op cit. p. 147.
[24] Radiological Health Handbook. Op cit. p. 266.
[25] Radiological Health Handbook. Op cit. pp 273-275.
[26] Hicks, Op cit.DE82005601. p.C-8

SECTION 12

NUCLEAR FALLOUT AND CANCER

NUCLEAR FALLOUT AND CANCER

Since the time of the first nuclear test series in 1951, there was public concern that fallout caused cancer. Certainly the *components* of fallout can cause cancer (see Table 4 in Appendix G.) The question in the minds of many scientists seems to be not whether the fallout was carcinogenic but whether there was a sufficient amount to cause damage.

CORRELATION COEFFICIENTS

A basic question involves how the cancer rates are correlated with fallout deposition and exposure rates. The first statistical tool of choice for this procedure is the *correlation coefficient*. The correlation coefficient compares two sets of numbers. If the numbers fluctuate in a similar manner, they are said to be positively correlated. If one set of numbers decreases as the other set increases (for example) then they are said to be negatively correlated.

Consider three sets of numbers, A B and C.

A	B	C
1	8	1
2	7	2
3	6	3
4	5	4
5	4	5
6	3	6
7	2	7
8	1	8

Group A consists of a set of numbers between 1 and 8 inclusive. Group B consists of the same set of numbers, however, their order is reversed from those in Group A. Group C consists of a set of numbers that is ordered similar to those in Group A.

Group A and Group C are considered to be *positively correlated*. In fact, the two groups are as correlated as two groups can get: They are exactly similar And a correlation coefficient—usually termed R or "rho" describing their relationship would be **R=+1.**

Group A and B are considered to be negatively correlated. The correlation coefficient describing these two groups would be **R=-1.**

If we added a fourth group and jumbled the numbers at random, then we might get a correlation coefficient of **R=0**.

There are primarily two types of correlation coefficient, the *Pearson Product Moment Correlation Coefficient,* and the *Spearman (Ranked Order)Correlation Coefficient.*

The Pearson Product Moment Correlation Coefficient is fairly common in business and is included in the popular Microsoft Excel spreadsheet package. Unfortunately, the Pearson requires that the data be normally distributed. That is, the numbers should fall into what is commonly known as the bell-shaped curve. Fallout deposition is *not* normally distributed, and thus the Pearson Product Moment Correlation Coefficient is not the tool to use for fallout.

The Spearman Correlation Coefficient, on the other hand does not rely on the normal distribution, rather on how the data is ranked. While the formula and the calculation procedures are vastly different, the end result is a numerical description of how well the two sets of numbers match up.

The Correlation coefficients can be applied to just about any set of numbers, and there is a technique available called a T-test. that can be used to test whether the numbers arrived at are the result of random processes or are the result of something else. While it's beyond the scope of this book to discuss the mathematics of statistical probabilities, it should be sufficient to know that the higher the positive T-score, the greater the chance that the correlation is not due to chance. Put another way, a T-score is a value used to describe the strength of an association.

In general, for sets of numbers involving 500 or more cases (in this case, counties) a T-score of 1.96 suggests that the probability of the two sets of values being similar by reasons other than chance is 0.95% or 1 in 20. A T-score of 2.58 is associated with a 99%ile, while a T-score of 3.29 indicates the probability of the two numbers being similar by reasons other than chance is 99.9995%.

A T-SCORE CAVEAT

While T-Scores are an important means of evaluating data arrays (pairwise means comparisons), it should be understood that the T-Score assumes that only two pairs are being evaluated. When a large number of value pairs are being evaluated, a process called *multiple comparison*[1], the T-scores and probability values associated with the correlations become less valid. The reason is because the greater the number of groups compared, the greater the chance of finding two that result in a significantly high T-score. For example, for 2 pairs of values tested, at the 95% confidence interval, the chances of finding significance is:

$$1-(0.95)^2 = 0.0975$$

But if we increase the number of pairs of values tested at the 95% confidence interval to 8, the chances of finding significance is

$$1-(0.95)^8 = 0.337$$

In other words, the greater the number of tests, the greater the chances of finding significance *somewhere*.

There are, however, statistical techniques that can be used with multiple comparison procedures that help determine which members of a set of comparisons are significantly different. Some of these techniques include Scheffe's test, Tukey's test, Fisher's least significant difference test, and the **Bonferroni Pairwise procedure.**

THE BONFERRONI PROCEDURE

The Bonferroni procedure is very simple. When analyzing probabilities within a confidence interval for multiple comparisons, one simply divides α (the probability for committing a Type I error) by the number of comparisons made, then using the adjusted test statistic as the confidence interval associated with the original α.

For example, suppose we performed 10 comparisons on a large (N=500+) population and wanted the values associated with a 95% confidence interval. This corresponds to an $.\alpha = 0.05$: (1-0.05 = 95% ile). We look this value up in the table of Z scores and find it is associated with a two-tailed test statistic of **1.96**. However, that Z score is appropriate only if one comparison is made. The Bonferroni procedure involves dividing the $\alpha/10$, then using that quotient to calculate the Z score associated with the 95% confidence interval. [2]

The resultant quotient—the "new" α equals not 0.05 but 0.05/10 = 0.005. This in turn, corresponds to a test statistic of approximately 2.79. In other words, when making multiple comparisons, the confidence probabilities must be more stringent[3].

Some statisticians suggest the Bonferroni procedure is too conservative. Robert P. Hirsch and Richard K. Riegelman, in their book Statistical First Aid have this to say about the Bonferroni:

> "The cost of using the Bonferroni method is quite high. It does not take many pairwise comparisons to require a testwise α so small that it becomes virtually impossible to reject null hypothesis in any of the pairwise comparisons. that is because reducing α increases β (the probability of making a Type II error). Because of its low statistical power, we do not recommend the Bonferroni method."[4]

Rebecca G. Knapp and M. Clinton Miller III, in their book, Clinical Epidemiology and Biostatistics had this to say about the Bonferroni Procedure:

> "When the number of comparisons is large, the Bonferroni procedure is very conservative (i.e. it has a very low false positive rate but a high false negative rate). As the number of comparisons increases, so does the tabulated critical value. As a result, it is increasingly difficult to detect significant differences between the pairs of means; often, it is more difficult than really necessary. Under these circumstances, a less stringent multiple comparison procedure may be preferred."[5]

THE BONFERRONI CRITICAL VALUES
FOR THE MULTIPLE COMPARISONS IN THIS SECTION

In this section are a number of tables comparing fallout components and cancer rates. Included in most tables are test statistics and associated probabilities. The probabilities and test statistics given were calculated using statistical software and assumed a single comparison was made.

DATA SET	NUMBER OF COMPARISONS (k)	95% C.I.	α/k	RESULTANT TEST STATISTIC (APPROXIMATE)
NCI 2000 CANCER ATLAS	68	0.05	0.0007	3.4
CDC WONDER DATA (FOR MIDWEST GROUP)	108	0.05	0.0005	3.5
FALLOUT COMPONENTS	66	0.05	0.0008	3.4

The reader may evaluate each set of multiple comparisons against the above critical values, however, as a practical matter, one may assume for all the data in this section that a test statistic greater than 6.0 is associated with a significance level of 95% or greater ($p<0.05$).

That said, there can be no absolute assurance that the high correlations and test statistics are *not* associated with random processes, or undetected variables.. Spatial epidemiological studies can involve innumerable confounders, even when the test statistics reflect a zero probability that the association is due to random processes.

For that reason, these statistical procedures and results should be viewed with caution and primarily as a first step in isolating these variable pairs for further study.

Thunderstorms can scavenge suspended particles from as high as 5 kilometers and bring them to earth with the rain.[6][7][8][9][10][11] Fallout has been associated with rainfall, so it is no surprise that some cancers are highly correlated with rainfall during periods of nuclear testing. (See Tables in Appendix G)

The table below suggests very strong correlations between the initial nuclear test series, Ranger in 1951 with cancer rates subsequent to that time. Others, such as the Teapot series in 1955 show an overall negative correlation with cancer rates for fallout. The Ranger series, however, deposited fallout on relatively few counties and those counties may have had other factors associated with cancer rates. The important thing to remember is that simple correlations between variables *cannot* show causation. However, the statistical technique is useful in the overall analysis of fallout and cancer rates.

TABLE 12-1
SUM OF T-SCORES AMONG SELECTED CANCER RATES FOR TOTAL FALLOUT AND RAINFALL BY TEST SERIES

RANGER FALLOUT	1951	424.2602
RANGER RAIN	1951	399.8767
BJ FALLOUT	1951	109.5705
BJ RAIN	1951	73.1222
TS FALLOUT	1952	-36.8557
TS RAIN	1952	263.167
UK FALLOUT	1953	-31.4113
UK RAIN	1953	289.7156
TP FALLOUT	1955	-233.58
TP RAIN	1955	61.6673
PB FALLOUT	1957	-360.887
PB RAIN	1957	-63.3679
SEDAN FALLOUT	1962	117.1471
SEDAN RAIN	1962	145.9984
TOTAL FALLOUT	1951-62	-78.077
TOTAL RAIN	1951-62	137.8018

In Table 12-2, cancer rates from the 1999 NCI Cancer Atlas were correlated (Spearman) against fallout components for 3094 U.S. counties. T-scores (n-2) and probabilities were calculated. Positive correlations would produce positive T-scores while negative correlations would result in negative T-scores. The T-scores were summed for each cancer group. The results suggest that, indeed, cancers of the lower digest tract were, overall, positively correlated with fallout components.

TABLE 12-2 CANCER VARIABLES WITH HIGH T-SCORE SUMS

CANCER	T SUM
COLON	272.7894
BREAST	176.6074
OVARIAN	155.4477
KIDNEY	136.4534
STOMACH	126.0654
LEUKEMIA	120.7585
NON HODGKINS LYMPHOMA	113.5372
RECTUM	92.90464

To determine if there were specific components of fallout that were *consistently* correlated with the above cancers, the T-scores were summed for each radionuclide component and then ranked from 1 to 126 with 1 corresponding to the T score with the highest positive value and 126 corresponding to the T score with the highest negative value. Thus, 49 fallout components were associated with overall positive correlations with the above cancer rates. The list is included as Table G-3 in Appendix G.

HIGHLY CORRELATED VARIABLES

Among the fallout components associated with the cancer rates were 392 correlated fallout-cancer rate pairs with T-Scores corresponding to probability levels of zero (N=3094). There were 81 pairs associated with Spearman R-values of 0.2 or greater. They are listed in Table G-3 in Appendix G. There were also 81 pairs of variables with T-scores greater than 10. Had these pairs been the only ones evaluated the probability level would have been less than <0.0001. These highly correlated variables are listed in Table G-4 in Appendix G.

The full list for highly correlated variables for White Female Breast Cancer, 1970-94 is found in Table G-5 Appendix G. The high Total Fallout-Colon Cancer correlations are found in Table G-6 in Appendix G.

CONFOUNDERS

While the data from the Spearman correlations suggest that cancer rates vary with fallout deposition, it does not show that the fallout *caused* the cancer. For example, in the four-state area of Texas, Louisiana, Arkansas and Oklahoma, precipitation during the nuclear test period is highly correlated with rainfall (R:

0.4). Confounders, however, could include the fact that more people tend to live in areas that happen to receive high rainfall amounts or be exposed to chemical toxins.[12,13,14,15,16,17]

Using another example from an earlier NCI Cancer Atlas, Shot SEDAN happens to be highly correlated with Female Colon Cancer during the 1950s. This could be due to chance and it also could be due to the fact that the SEDAN trajectory happens to resemble trajectories from earlier nuclear tests.

Clearly, other statistical techniques should be used in addition to correlation coefficients. One technique involves the use of the pooled standard error.

POOLED STANDARD ERROR

While correlation techniques are used to determine if two or more variables are similar, other statistical techniques can be used to determine if two groups are significantly different. To determine if groups are different at a given significance level:

1. Divide the variance of each sample by the sample size (N).

2. Add the results together and take the square root of the result to get the Pooled Standard Error.

3. A two-tailed test for the 0.9995-ile of the normal distribution is 3.29.

4. If the difference between the means is greater than 3.29 times the Pooled Standard Error, then the results are significant at a probability level of 0.0005.

Using this technique, in combination with the Bonferroni procedure, one can determine if the average rate for a particular cancer is significantly higher in a group of counties with higher levels of fallout vs those counties with lower levels of fallout. One can compare counties with fallout deposition totals in the lower quartile vs counties in the upper quartile and, similarly, by decile. Some shots, such as SEDAN deposited a large amount of fallout on relatively few counties. For this reason, the SEDAN data was considered for this technique.

Comparing the two groups: Non-SEDAN fallout vs SEDAN fallout, and a criteria of $p<0.0005$, only *one* cancer rate is seen as significantly greater in the SEDAN group: all Cancers for White Females, 1950-1969.(PSE x 3.29)-Difference between Means = 2.132.

The problem with this technique is that the counties in the No Fallout category associated with shot SEDAN may have received significant fallout from *other* nuclear tests. When we limit the analysis to a small region that equally divides the counties receiving fallout from SEDAN from those that did not, then the results are different.

THE MIDWEST GROUP

Nuclear fallout from the Nevada Test Site reached virtually every county in the United States. The states of Nevada, Utah, Idaho, Montana and Colorado received a significant amount of fallout from the Upshot-Knothole and Tumbler-Snapper series; the Northeastern states received fallout from Buster-Jangle and Upshot-Knothole group and the Midwest received fallout primarily from Tumbler-Snapper GEORGE and the 1962 shot SEDAN. The 513-county group found in the states of Iowa, Illinois, Kansas, Missouri and Nebraska were almost equally divided between those that received fallout from SEDAN—246 counties—and those that received no fallout—267 counties. Using the Pooled Standard Error and the same cancer rates (NCI Cancer Atlas 1999) and criteria—$p<0.0005$— a Bonferroni equivalent to $p<0.05$ for 513 Midwestern counties, the results suggest higher cancer rates in counties exposed to fallout from shot SEDAN.

TABLE 12-3

**POOLED STANDARD ERROR: MIDWEST REGION
SEDAN vs NCI CANCER ATLAS (1999-2000)
$p<0.0005$**

CANCER TYPE	[Difference in Means (SEDAN Fallout-No SEDAN Fallout)]-[(3.29)(Pooled Standard Error)]
All Cancer WF 1950-69	1.339347
All Cancer WM 1950-69	1.154118
Prostate Cancer WM 1950-69	0.750005
Colon Cancer WF 1950-69	0.603175
Colon Cancer WM 1950-69	0.503419
Colon Cancer WF 1970-94	0.499046
Breast Cancer WF 1950-69	0.41645
Bladder Cancer WM 1950-69	0.287201
Stomach Cancer WM 1950-69	0.28321
Cancer of Rectum WM 1950-69	0.263099
Prostate Cancer WM 1970-94	0.187885

When we use the same procedures and criteria ($p>0.0005$) using a different source of cancer information, the Centers for Disease Control WONDER site, we get remarkably similar results (Table 12-8). However, a word of caution: cancer data for low population counties, such as is found in the Midwest, is generally unstable. That is, it can change drastically with a single extra case. For example, consider a county with only 1000 people. Suppose one of them—one in a thousand—is diagnosed with cancer. Since the rates for cancer are given in terms of cancers per 100,000 individuals, the cancer rate for this county is listed—theoretically—as 100, or 100 in 100,000. Suppose, that someone else in the county contracts the same kind of cancer. The rate would then double to 200. Statisticians generally consider low-population county cancer rates subject to large fluctuations.

TABLE 12-4
POOLED STANDARD ERROR: MIDWEST REGION
SEDAN: CDC WONDER CANCER RATES
$p<0.0005$

CANCER TYPE	[Difference in Means (SEDAN Fallout-No SEDAN Fallout)]-[(3.29)(Pooled Standard Error)]
All Cancers WM 1950-59	3.624345
Colon Cancer WF 1979-96	1.525694
Brain Cancer WF 1979-96	0.973327
All Cancer WF 1950-59	0.714689
Colon Cancer WM 1950-59	0.663209
Breast Cancer WF 1950-59	0.436061
Colon Cancer WF 1950-59	0.2596
All Cancers WF 1960-69	0.144875
Lymphosarcoma WF 1950-59	0.045668
Colon Cancer WF 1960-69	0.019108
Lymphosarcoma WF 1979-96	0.000198

The problem with attributing causation to the SEDAN fallout for many of these cancer rates should be obvious: Most of the cancer rates here are associated with the decade of the 1950s—and the SEDAN detonation occurred in 1962.

One possible explanation might be that the cancer rates are associated with other nuclear tests that had a trajectory similar to SEDAN. As it turns out, SEDAN's trajectory was quite similar to many other high-fallout nuclear tests (TABLE 12-8) For example, SEDAN's total fallout pattern was strongly correlated with the pattern formed on the second day after detonation of shot DIABLO, day 3 of shot TESLA, day 8 of shot HOW, and day 11 of shot HOOD.

Interestingly, when the 246 counties in the Midwest that received SEDAN fallout are further grouped by upper (N= 56) vs lower (N= 59) quartiles of fallout and then evaluated using the Pooled Standard Error technique, only one cancer rate shows up with a significantly higher rate in the upper quartile counties: Colon cancer White Females, 1979-1996. (+0.549, $p<0.01$).

TABLE 12-5
NUCLEAR TESTS FROM 1951-1957 WITH FALLOUT PATTERNS SIMILAR TO SEDAN (JULY 6, 1962)

RANKED MOST SIMILAR TO LEAST SIMILAR

	NUCLEAR TEST AND ASSOCIATIED DAY AFTER DEPOSITION		CORRELATION COEFFICIENT
RANK	SHOT	DAY	SPEARMAN R
1	PB- DIABLO	2	0.581
2	TP TESLA	3	0.559
3	TS HOW	8	0.537
4	PB HOOD	11	0.531
5	UK SIMON	7	0.516
6	PB SHASTA	2	0.512
7	UK HARRY	5	0.488
8	UK SIMON	5	0.459
9	TS DOG	3	0.449
10	PB MORGAN	10	0.44
11	TS HOW	10	0.426
12	PB DOPPLER	6	0.414
13	PB-KEPLER-OWENS	7	0.401
14	PB-WCL	10	0.401
15	UK HARRY	7	0.4

M.A.R.S.

Multiple regression analysis can be used to determine predictors of a target (dependent) variable such as a cancer rate in a population. Given the multiplicity of independent variables and the potential for interaction between the variables, statistical techniques have been developed to narrow the list of potential predictors for further analysis. One software package that is useful in the evaluation of fallout components against cancer rates is Salford Systems' Multivariate Adaptive Regression Splines or MARS.[18] After choosing a target (dependent) variable such as U.S. Female Colon Cancer Rates 1970-1995, and then choosing the entire list of potential predictor variables, the software uses genetic algorithms to choose the independent variables that best predict the curve for the target variable. In the example above, the target variable, White Female Colon Cancer in the United States 1970-1995 (3094 counties) was found to have the following predictor variables:

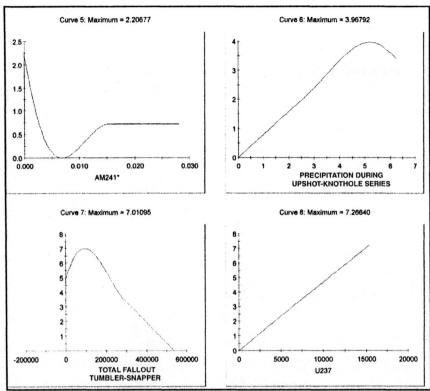

MARS PREDICTOR VARIABLES: WHITE FEMALE COLON CANCER 1970-95 (NCI CANCER ATLAS)

Many positive linear predictor variables among fallout components have been identified using Salford's M.A.R.S. software. Note, however that precipitation during the Upshot-Knothole series is a strong predictor for white female colon cancer. In the table below you will find this same variable as a predictor for a number of different cancers. Interestingly, during the Upshot-Knothole test series it rained practically everywhere in the United States. In fact, only 0.65 percent of the counties in the U.S. at that time received little or no rain. Perhaps a future study can determine whether widespread precipitation is a confounder or a clue to the high correlation rate between the Upshot-Knothole series and cancer rates.

TABLE 12-6
MARS ANALYSIS SHOWING POSITIVE LINEAR PREDICTORS

CANCER VARIABLE	POSITIVE LINEAR PREDICTOR VARIABLE
BREAST WF 1950-69	TOTAL FALLOUT: BUSTER JANGLE SERIES RAINFALL: UPSHOT-KNOTHOLE SERIES RADIONUCLIDE: Pb203
BREAST WF 1970-95	RAINFALL: UPSHOT-KNOTHOLE SERIES TOTAL FALLOUT: 1951-1962
KIDNEY WF 1970-95	RADIONUCLIDE: Nb95m
KIDNEY WM 1970-95	TOTAL FALLOUT: PLUMBBOB SERIES
LEUKEMIA WF 1950-69	RAINFALL: UPSHOT-KNOTHOLE SERIES TOTAL FALLOUT: 1951-1962
LEUKEMIA WF 1970-95	RADIONUCLIDES: Br82, Am-241
LEUKEMIA WM 1950-69	TOTAL FALLOUT: BUSTER-JANGLE SERIES
LEUKEMIA WM 1970-95	RAINFALL: UPSHOT-KNOTHOLE SERIES RAINFALL: TEAPOT SERIES TOTAL FALLOUT: 1951-1962
NON-HODGKINS LYMPHOMA WF 1950-1969	TOTAL FALLOUT: SEDAN
NON-HODGKINS LYMPHOMA WF 1970-95	RAINFALL: UPSHOT-KNOTHOLE SERIES RADIONUCLIDE: Pd111m
NON-HODGKINS LYMPHOMA WM 1950-69	RADIONUCLIDES: Cu64, Nb95 TOTAL FALLOUT: TUMBLER-SNAPPER SERIES
NON-HODGKINS LYMPHOMA WM 1970-95	RAINFALL: UPSHOT-KNOTHOLE SERIES RADIONUCLIDES: Cu64, In115m, Fe59
STOMACH WM 1950-1969	RAINFALL: UPSHOT-KNOTHOLE SERIES RADIONUCLIDE: Na24
STOMACH WM 1970-95	RADIONUCLIDES: Fe59, Tb161, U240
RECTUM WF 1970-95	RAINFALL: UPSHOT-KNOTHOLE SERIES RADIONUCLIDE Tb161 TOTAL FALLOUT: UPSHOT-KNOTHOLE SERIES
RECTUM WM 1970-95	RAINFALL: UPSHOT-KNOTHOLE SERIES
PANCREAS WM 1950-69	TOTAL FALLOUT: SEDAN TOTAL FALLOUT: TUMBLER-SNAPPER SERIES TOTAL FALLUT: TEAPOT SERIES
PROSTATE WM 1950-69	RAINFALL: UPSHOT-KNOTHOLE SERIES TOTAL FALLOUT: 1951-1962
PROSTATE WM 1970-95	TOTAL RAINFALL DURING NUCLEAR TESTING 1951-1962

POISSON REGRESSION

The Poisson distribution is used to describe the occurrence of rare, isolated events in a continuum of time.[19] The formula is given as:

$$p(x=k) = \frac{e^{-\lambda}\lambda^k}{k!}$$

Where e = the base of natural logarithm, 2.7183. . . .and the mean and variance of the distribution are both equal to λ.

The Poisson distribution is extraordinarily useful when evaluating exposure[20]. When used with background rates, the Poisson probability function can be used to determine the likelihood an event will occur based on random processes. Suppose for example, a cluster of leukemia cases occurs in a community. If the background rates are known, then Poisson probability can be used to determine how likely it is that the cluster occurred by chance.

While cancer is relatively common, specific types of cancer can be relatively rare within a given population. As mentioned earlier, one or two cases in a small population can expand the rate to the point where rates become misleading and evaluation by standard means becomes difficult. To deal with this problem statisticians came up with a technique that uses actual incidence numbers rather than rates in the calculation. Termed Poisson regression, the procedure applies the somewhat obscure Poisson Probability function to multivariate analysis[21]. Poisson regression assumes that the magnitude of the rate is an exponential function of linear covariates and parameters.[22] That is,

$$\text{Rate} = e^{(b0+b1x1+b2x2+\ldots+bkxk)}$$

The disadvantage to Poisson regression is the requirement that the independent variable data be categorized. Theoretically, the values can remain continuous, but since Poisson regression essentially calculates rate ratios between the categories, continuous data generally yields very small rate ratios. In application, Poisson regression is quite similar to logistic regression.

As indicated earlier, one interesting characteristic of the Poisson distribution is that the mean and the variance are generally the same number. If the variance is much larger than the mean—the technical term is *overdispersion*—then Poisson regression may not work very well. Unfortunately, much of the data associated with nuclear testing and cancer rates *is* characterized by overdispersion. To solve the problem, some workarounds were necessary.

The software used for Poisson regression analysis for this volume was *Egret* by Cytel Software, Inc. It evaluates the data using the Newton Raphson algorithm to develop a likelihood ratio test. To correct for overdispersion in the data, a correction factor was used:

$$C = [(\text{deviance value})/(\text{degrees of freedom})]^{1/2}$$

The standard error for the independent variable (for example, the radionuclide) is then multiplied by this correction factor to obtain a *revised* estimate of the standard error. This value of the test statistic, z, is calculated by dividing the coefficient of the independent variable (in this case, the radionuclide) by the revised estimate of the standard error. From the new test statistic one obtains the appropriate probability values.

This procedure, though somewhat cumbersome, works particularly well if the independent variable, i.e. the fallout component, can be categorized into say, five or ten ranges of values. Unfortunately, many of the deposition patterns don't lend themselves to this procedure. For example, if one groups the deposition values of the radionuclide Np-239 into four equal intervals, one would discover only 15 counties in the highest *three* intervals. It is almost the same situation with strontium-90 (and strontium-89 for that matter). Over 99 per cent of the U.S. counties received less than 230.18 nCi/sq meter of strontium-90 during the entire test period (1951-1962). Only 29 counties received between 230.18 and 919.13 nCi/sq meter. Clearly, 29 counties is not a large enough group to evaluate properly. Unfortunately, even expanding the number of categories doesn't help. Even with 10 categories, the vast majority of the United States—98 per cent—is still in the 1st category, having received less than 92 nCi/sq meter of Sr-90 during the period of nuclear testing. Even with Poisson regression techniques, it might be difficult to make much sense of rate ratios based on so few counties.

The alternative, to leave the data as a continuous set, can result in a rate ratios (between increments) that is essentially zero. In other words, the rate ratios one obtains from a Poisson regression will depend on how the data was initially categorized. To make matters worse, software programs that may divide the data into equal sections may assume an even distribution when none exists. As a result, there may be some classes with no data (imagine 99 counties each receiving exactly 1 microCurie of fallout with the hundredth country receiving 10 microCuries. One *could* divide the data into ten equal intervals, but eight intervals would have no counties in them. And of course, if there are no counties, there is no rate ratio.

Nonetheless, Poisson regression can be used to effectively evaluate relative risk between independent variable classes. The next tables show how Poisson regression can be used to evaluate cancer incidence against various fallout components among 513 Midwestern counties. Dependent variable data were obtained from the U.S. Environmental Protection Agency-National Cancer Institute U.S. Cancer Mortality Rates and Trends 1950-1979[23] and the Centers For Disease Control WONDER site. Both the CDC WONDER and NCI 1950-1979 data was age-adjusted and the specific population for 1970 was used as the standard for the total study period. The table on the next page lists Poisson regression rate ratios for various cancers measured against total fallout in a 513-county area of the Midwest.

The first column ranks the cancer parameters from highest to lowest Z scores, and thus higher to lower chance that the rate ratios were due to non-random processes. In evaluating the Z scores one may want to apply the Bonferroni procedure and evaluate only the highest values.

The second column lists the cancer parameter. The symbol key for the cancer parameters is found in Appendix G. The third through sixth columns list the Poisson regression factors of Coefficient Estimate, uncorrected Estimate of Standard Error, Degrees of Freedom and the Deviance Value.

The seventh column lists the Rate Ratio between categories corrected for overdispersion of the data. The eighth and ninth columns list the corrected lower confidence limit and the upper confidence limit at the 95th percentile (the calculations indicate there is a 95% certainty that the Rate Ratio falls somewhere between the lower confidence limit and the upper confidence limit.) The tenth column lists the corrected Z score and the final column lists the probability that the Rate Ratio is due to chance alone. The entire list can be found in Table G-7 in Appendix G.

TABLE 12-7
TOTAL FALLOUT AND CANCER
POISSON REGRESSION RATE RATIOS FOR 25 VARIABLES
CLASSES: 7 EQUAL INTERVALS
TOTAL FALLOUT FOR 513 COUNTIES IN IA, IL, KS, MO, NE 1951-1962

RANK	CANCER PARAMETER	COEFF	STD. ERROR	df	DEV VALUE	RATE RATIO	LCI	UCI	Z SCORE	PROB DUE TO CHANCE
1	FLS95	0.28080	0.01296	511	4.86E+03	1.32	1.29	1.36	7.03	0.00000
2	FCLN96	0.07805	0.00290	511	1.39E+04	1.08	1.08	1.09	5.16	0.00000
3	FBRAIN96	0.14680	0.00627	511	1.07E+04	1.16	1.14	1.17	5.12	0.00000
4	FBRST60	0.05684	0.00244	511	2.09E+04	1.06	1.05	1.06	3.65	0.00026
5	MEYE96	0.30350	0.08150	511	5.49E+02	1.36	1.16	1.59	3.59	0.00033
6	BMHST96	0.29070	0.06910	511	7.52E+02	1.34	1.17	1.53	3.47	0.00054
7	BNOD96	0.17630	0.03188	511	1.66E+03	1.19	1.12	1.27	3.07	0.00214
8	FCLN70	0.04183	0.00298	511	1.17E+04	1.04	1.04	1.05	2.94	0.00328
9	MMM60	0.09186	0.00793	511	9.54E+03	1.10	1.08	1.11	2.68	0.00736
10	MCONSAR96	0.21930	0.05654	511	1.15E+03	1.25	1.12	1.39	2.59	0.00960
11	MCLN60	0.03743	0.00304	511	1.18E+04	1.04	1.03	1.04	2.56	0.01047
12	MBONE96	0.11370	0.01497	511	4.84E+03	1.12	1.09	1.15	2.47	0.01351
13	MLIV96	0.07934	0.00975	511	5.56E+03	1.08	1.06	1.10	2.47	0.01351
14	FBRST96	0.02020	0.00227	511	6.78E+03	1.02	1.02	1.03	2.44	0.01469
15	FLS70	0.05841	0.00584	511	9.56E+03	1.06	1.05	1.07	2.31	0.02089

df = degrees of freedom; LCL, UCL = Lower and Upper Confidence Intervals

POISSON REGRESSION AND RADIONUCLIDES

Just as Poisson regression can be used to determine rate ratios between Total Fallout categories, the technique can also be applied to individual radionuclides. The following two tables beginning on the next page include analyses of various cancer parameters and the radionuclides beryllium-7 and americium-241. Estimated deposition values are for one year post-shot.

As with the Total Fallout table, dependent variable data was taken from the U.S. Environmental Protection Agency-National Cancer Institute U.S. Cancer Mortality Rates and Trends 1950-1979[24] and the Centers For Disease Control WONDER site. Both the CDC WONDER and NCI 1950-1979 data was age-adjusted and the specific population for 1970 was used as the standard for the total study period.

While data from the 2000 NCI Cancer Atlas was used for correlation coefficient and pooled standard error evaluations, the 1983 NCI data and current CDC WONDER data was used for Poisson. Here is why: the NCI 2000 data was divided into two time segments: 1950-1969 and 1970-1995. The first group encompassed 19 years while the second group encompassed 25 years. The 1983 NCI Cancer Atlas divided the rates into ten-year segments: 1950-59, 1960-69 and 1970-79. The CDC WONDER data carried the data forward from 1979 through 1995 and for some cases through 1996. The difference between the outcome can be illustrated by the following table.

NCI 2000 CANCER ATLAS	POISSON REGRESSION ANALYSIS	
	REL RATE	Z-SCORE
LEUKEMIA WF 1950-69	0.7821	-4.37
LEUKEMIA WF 1970-95	0.7569	-4.93
NCI 1983 CANCER ATLAS AND CDC WONDER DATA	POISSON REGRESSION ANALYSIS	
	REL RATE	Z-SCORE
LEUKEMIA WF 1950-59	1.00	0.39
LEUKEMIA WF 1960-69	0.98	-0.79
LEUKE MIA WF 1970-79	0.99	0.07
LEUKEMIA WF 1979-95	1.04	2.53

Note that the WF Leukemia rates in 1979-1995 show a rate ratio increase significant at the Z-score level of 2.53 (p<0.011 not Bonferroni corrected).

TABLE 12-8
BERYLLIUM-7 AND CANCER
POISSON REGRESSION RATE RATIOS FOR 25 VARIABLES
CLASSES: 7 EQUAL INTERVALS
TOTAL Be7 FALLOUT, FOR 513 COUNTIES IN IA, IL, KS, MO, NE
1951-1962

RANK	CANCER PARAMETER	COEFF	STD. ERROR	df	DEV VALUE	RATE RATIO	LCI	UCI	Z SCORE	PROB DUE TO CHANCE
1	FLS95	0.19270	0.00848	511	4830	1.21	1.19	1.23	7.39	0.00
2	FCLN96	0.06643	0.00194	511	13520	1.07	1.07	1.07	6.65	0.00
3	FBRAIN96	0.09359	0.00425	511	10730	1.10	1.09	1.11	4.81	0.00
4	BNOD96	0.15500	0.02044	511	1639	1.17	1.12	1.22	4.23	0.00
5	MEYE96	0.20390	0.05339	511	548.9	1.23	1.10	1.36	3.68	0.000233
6	MBONE96	0.09340	0.00992	511	4813	1.10	1.08	1.12	3.07	0.002141
7	MMM60	0.06796	0.00535	511	9521	1.07	1.06	1.08	2.94	0.003282
8	FBRST50	0.02619	0.00174	511	14630	1.03	1.02	1.03	2.81	0.004954
9	FLS70	0.04786	0.00396	511	9520	1.05	1.04	1.06	2.80	0.00511
10	MBRAIN96	0.03662	0.00356	511	7072	1.04	1.03	1.05	2.76	0.00578
11	BLCL96	0.13960	0.04100	511	801.2	1.15	1.06	1.25	2.72	0.006528
12	MCLN96	0.01838	0.00190	511	6657	1.02	1.02	1.02	2.68	0.007362
13	MLS60	0.04224	0.00353	511	10770	1.04	1.04	1.05	2.61	0.009054
14	UT96	0.04792	0.00517	511	6551	1.05	1.04	1.06	2.59	0.009598
15	MCONSAR96	0.14570	0.03759	511	1147	1.16	1.08	1.25	2.59	0.009598
16	MMM50	0.06943	0.00700	511	8116	1.07	1.06	1.09	2.49	0.012774
17	MMM70	0.04000	0.00491	511	9490	1.04	1.03	1.05	1.89	0.058758
18	FKDNY96	0.02675	0.00480	511	5004	1.03	1.02	1.04	1.78	0.075076
19	FSTOM96	0.02698	0.00495	511	4852	1.03	1.02	1.04	1.77	0.076727
20	MLS70	0.02603	0.00341	511	10240	1.03	1.02	1.03	1.70	0.089131
21	BOTHLR96	0.07478	0.03479	511	1047	1.01	1.15	0.00	1.50	0.131811
22	MMM96	0.02173	0.00449	511	6312	1.02	1.01	1.03	1.38	0.167587
23	MLEUK50	0.01866	0.00290	511	11520	1.02	1.01	1.03	1.36	0.17383
24	FMM60	0.00962	0.00732	511	7352	1.00	1.02	0.00	0.35	0.726339

df = degrees of freedom; LCL, UCL = Lower and Upper Confidence Intervals

TABLE 12-9
AMERICIUM-241 AND CANCER
POISSON REGRESSION RATE RATIOS FOR 25 VARIABLES
CLASSES: 7 EQUAL INTERVALS
TOTAL Am241 FALLOUT, FOR 513 COUNTIES IN IA, IL, KS, MO, NE
1951-1962

KEY TO CANCER RATE ABBREVIATIONS IN TABLE G-9

RANK	CANCER PARAMETER	COEFF	STD. ERROR	df	DEV VALUE	RATE RATIO	LCI	UCI	Z SCORE	PROB DUE TO CHANCE
1	FCLN96	0.04807	0.00201	511	14070.00	1.05	1.05	1.05	4.55	0.00
2	BOTHLR96	0.18300	0.03253	511	1023.00	1.20	1.13	1.28	3.98	0.00
3	MMM96	0.05791	0.00435	511	6164.00	1.06	1.05	1.07	3.84	0.000123
4	MCONSAR96	0.21110	0.03753	511	1132.00	1.24	1.15	1.33	3.78	0.000157
5	MMM60	0.08183	0.00539	511	9454.00	1.09	1.07	1.10	3.53	0.000416
6	BLCL96	0.18030	0.04158	511	794.50	1.20	1.10	1.30	3.48	0.000501
7	MLEUK50	0.04255	0.00283	511	11340.00	1.04	1.04	1.05	3.19	0.001423
8	FBRAIN70	0.06340	0.00458	511	10040.00	1.07	1.06	1.08	3.13	0.001748
9	FLS95	0.09504	0.00976	511	5190.00	1.10	1.08	1.12	3.06	0.002213
10	MBRAIN96	0.03715	0.00358	511	7069.00	1.04	1.03	1.05	2.79	0.005271
11	FMM96	0.04104	0.00471	511	5135.00	1.04	1.03	1.05	2.75	0.00596
12	MLS70	0.04130	0.00337	511	10150.00	1.04	1.04	1.05	2.75	0.00596
13	MEYE96	0.15730	0.05895	511	555.00	1.17	1.04	1.31	2.56	0.010467
14	FLEUK50	0.04032	0.00349	511	10670.00	1.04	1.03	1.05	2.53	0.011406
15	FLS50	0.05465	0.00504	511	9760.00	1.06	1.05	1.07	2.48	0.013138
16	FKDNY70	0.05245	0.00552	511	7574.00	1.05	1.04	1.07	2.47	0.013511
17	BNOD96	0.09702	0.02257	511	1673.00	1.10	1.05	1.15	2.38	0.017313
18	MTHYM95	0.13390	0.03804	511	1151.00	1.14	1.06	1.23	2.35	0.018773
19	MBRAIN50	0.04070	0.00430	511	10310.00	1.04	1.03	1.05	2.11	0.034858
20	MPAN96	0.01771	0.00277	511	6308.00	1.02	1.01	1.02	1.82	0.068759
21	MLEUK96	0.01759	0.00282	511	6300.00	1.02	1.02	1.02	1.78	0.075076
22	FLS70	0.03124	0.00407	511	9602.00	1.03	1.03	1.04	1.77	0.076727
23	MLS60	0.02740	0.00361	511	10850.00	1.03	1.03	1.04	1.65	0.098943
24	FSTOM96	0.02357	0.00498	511	4858.00	1.02	1.02	1.03	1.54	0.12356

df = degrees of freedom; LCL, UCL = Lower and Upper Confidence Intervals

PRACTICAL CONSIDERATIONS

The information in this book, while based upon actual samples taken in the 1950s and 1960s, is also largely theoretical. The National Cancer Institute fallout data for I-131 involved estimates with extraordinarily wide ranges. Yet while the numbers may vary, the important thing, in my view, is the elucidation of relative fallout deposition patterns for wide areas of the country. It is probably less important to know the exact milliCuries of radiation for a particular county than it is to know the composition of the fallout that fell there. Americium-241 behaves in the body differently than, say cobalt-60 or strontium-90. If these radionuclides are associated with disease, it is reasonable to assume that the diseases would appear primarily in areas associated with deposition of the particular radionuclide In other words, from an environmental and epidemiological perspective, any disease risk from fallout would depend in a large part on the fallout pattern of the disease-linked radionuclide.

CLUSTERS

Disease clusters, and particularly clusters associated with cancer occur with depressing regularity across the countryside. Yet, official investigations rarely point to a cause[25]. In fact, many epidemiologists and state health officials consider cluster investigation a waste of resources[26]. [27] [28] Even if the evidence supported causation by some independent variable such as fallout, many scientists approach the subject by first demanding extraordinarily high standards of proof. [29] [30] [31] or evaluate the cluster based on known risks[32].. As one researcher put it, reasons for closure can include: (a) *the type of cancer under study does not appear to be associated with known risk factors* and (b) *the facts discovered do not present any cohesive picture of common exposure and disease.*[33]

In most cluster investigations, however, the researchers have failed to include the one factor associated with common exposure to carcinogenic agents: Nuclear rainouts.

Rain was a primary factor in fallout deposition. In fact, in the 1997 I-131 study, the NCI helpfully included precipitation indices along with the fallout data. What is not generally known is that there is an extraordinarily high correlation between cancer rates and rainfall, particularly during the Upshot-Knothole series. While confounders are not in short supply—for example, it rained over most of the U.S. in the spring of 1953—the strong correlation deserves further research, and should be taken into account during every investigation of cancer clusters.

Of particular interest is the fact that rainstorms can scavenge *most* of the radioactivity in a nuclear cloud, and that thunderstorms typically drop rain on relatively small areas—and small populations. If fallout causes cancer, it will likely result in clusters of particular cancers: leukemia, lymphoma, or any of the cancers found by Poisson regression to show high rate ratios.

Clusters of these cancers have been reported in high fallout areas such as Missouri, Iowa and northern New York. Yet it is rare that fallout is considered as a potential factor.

The unfortunate fact is that data collection and research into population exposures uses up time and resources. Perhaps for this reason, detailed onsite investigations are rare; the researchers prefer to concentrate on a review of morbidity statistics for the area. But assumptions going into the investigation affect the outcome.

In 1998 and 1999, shortly after the NCI released its I-131 study detailing fallout and exposure rates for every county in the United States, I phoned several state health departments at random. I asked if they included I-131 deposition information when investigating cancer clusters. Not one said they did. Some told me that cluster investigations were a waste of time. Interestingly, none were familiar with the NCI I-131 data[34]. An evaluation of the scientific literature (since 1985) addressing community disease clusters rarely if ever finds nuclear fallout discussed as a potential factor. [35 36 37 38 39 40 41]

It turns out, however, that there are practical means of including nuclear fallout as part of health investigations: radioisotope analysis of local watersheds.

Many of the radionuclides that were deposited in the 1950s and early 1960s are still in the environment. And if the radionuclide happened to land in a field or forest, it would become part of the local watershed. Since each nuclear test was associated with its own unique compliment of fallout, simple core analysis of selected ponds and lakes—a relatively inexpensive procedure— may give us a better idea of the actual fallout level for a given area. Research published by Harry Hicks, Philip Krey and others lend theoretical support for such an analysis [42 43 44]

CONCLUSION

Even when using the conservative Bonferroni procedure, the statistical evidence suggests strong associations between fallout components and certain cancers, particularly female colon cancer and the cancer designated by the NCI circa 1983 as lymphosarcoma. The evidence, I would hope, will make a strong case for further research, particularly on a regional basis.

In fact, by statistically identifying the fallout matrix most associated with certain cancers, it may be possible to predict which cancers in a given region will be associated with which fallout events.

The tools for this research are available. Salford Systems MARS™ and CART™, Spotfire™, Crystal Ball Predictor™ and Neurosolutions™ can all be used to evaluate the data and make predictions for cancer rates based upon fallout levels, composition and time of deposition. At the local level, researchers can include radioisotope characterization of watersheds as an integral part of their investigation.

As this book goes to press, the National Cancer Institute is planning to release data on a number of radionuclides that will possibly include detailed Poisson regression analysis against various cancers.

Thus, is presented a unique opportunity to offer the following predictions—all in the form of testable hypotheses:

- The pooled standard error analysis of 500-contiguous-county regions in different parts of the country will all show significantly higher rates of female colon cancer 1979-1995 in areas of higher total fallout at a probability level of $p<0.05$. Using the Bonferroni:procedure: $p<0.005$.

- Similar analysis of 500 randomly-chosen counties will show results at the $p<0.10$ probability level.

- Poisson regression analysis of 500 randomly chosen counties will show rate ratios of female lymphosarcoma (1979-1995) at 1.1 or greater with a lower rate of 1.09 or greater (95%ile). Bonferroni Z score: 4.0.

- The radionuclide-cancer associations listed in Table G-8 of this book will be the same ones found in areas of high radionuclide deposition.

If the above hypotheses are found to be false, then we can view the nuclear decades as an interesting time in our history with little practical health effect on our lives today.

If any of the hypotheses are found to be *true*, then we should make concerted efforts to more precisely characterize the exposure to the population in terms of intensity and specific kinds of radionuclides. Hopefully, there will be interest among epidemiologists and state health officials in the outcome.

REFERENCES SECTION 12

[1] Knapp, Rebecca G. and M. Clinton Miller, *Clinical Epidemiology and Biostatistics,*

[2] For those interested in the background and procedures involving confidence intervals and the null hypothesis, an excellent discussion is found in the chapters on Inference and P-Values in: Hirsch, Robert P. and Richard K. Riegelman: *Statistical First Aid: Interpretation of Health Research Data.* Blackwell Science. Cambridge, MA 1992. pp 51-59.

[3] Snedecor, George W. and William G. Cochran. *Statistical Methods*, Eithth Edition.Iowa State University Press, Ames, IA. 1989. p. 167

[4] Hirsch, Robert P and Richard K. Riegelman, *Statistical First Aid: Interpretation of health Research Data*, Blackwell Science, Cambridge, MA 1992. p235.

[5] Knapp, Rebecca G. and M. Clinton Miller, III *Clinical Epidemiology and Biostatistics* Williams and Wilkins, Philadeplhia: 1992. p. 304.

[6] Brown, R.A. *Initiation and propagation of thunderstorm mesocyclones,* PhD. dissertation, University of Oklahoma. (University microfilms, Ann Arbor MI. order #89-19983, 321 pp. Referenced in Doviak op cit.

[7] Knupp, K.R. and Cotton, W.R. "An intense, quasi-steady thunderstorm over mountainous terrain, II, Doppler radar observations of the storm morphological structure." J. Atmos. Sci. 39. 1983. pp 343-358.

[8] A good discussion of storm movement and propagation can be found in: Cotton, William R. and Richard A. Anthes, *Storm and Cloud Dynamics.* Academic Press. New York: 1989. pp 497-520.

[9] A discussion of thunderstorm climatology along with maps of thunderstorm distribution can be found in Kessler, Edwin, *Thunderstorm Morphology and Dynamics*, University of Oklahoma Press, Norman, OK 1988, 1992. pp 9-39.

[10] Houze, Robert A. Jr. *Cloud Dynamics.* Academic Press, Inc. New York, 1993. pp. 268-295.

[11] Doviak, Richard J. and Dušan Zrnić, Doppler Radar and Weather Observations. 2nd Ed. Academic Press., Harcourt, Brace and Jovanovich, San Diego, 1993. pp. 275-288.

[12] *Cancer Epidemiology and Prevention* 2nd Ed. Editors: David Schttenfeld and Joseph F. Fraumeni, Jr. Oxford University Press. Oxford, UK. 1994.

[13] *General and Applied Toxicology.* Editors: Bryan Ballantyne, Timothy Marrs and Paul Turner. Stockton Press, New York NY. 1995.

[14] *Occupational, Industrial, and Environmental Toxicology*, Editor-In-Chief: Michael I Greenberg. Mosby Publishing Co. New York NY: 1997.

[15] Keller, Edward A. *Environmental Geology* 2nd Ed.Charles E. Merrill Publishing Co. Columbus, OH. 1979. pp 321-359.

[16] Francis, B. Magnus *Toxic Substances in the Environment.* John Wiley and Sons, Inc. New York. pp. 261-269.

[17] *Hormonally Active Agents in the Environment.* National Research Council, National Academy of Sciences. National Academy Press. 1999.pp. 243-273.

[18] Salford Systems, San Diego, CA.

[19] Last, James M. *Dictionary of Epidemiology* Third Edition. Oxford University Press. 1995. p. 125

[20] Ash, Carol. *The Probability Tutoring Book.* IEEE Press. The Institute of Electrical and Electronics Engineers, Inc. New York. 1993. pp 63-67

[21] *EGRET for Windows.* Cytel Software Corp.p161

[22] Nieto, F. Javier and Moyses Szklo. *Epidemiology Beyond The Basics.* Aspen. Gaithersburg, MD.. 2000. p. 310.

[23] *.U.S. Cancer Mortality Rates and Trends, 1950-1979* Wilson B. Riggan PhD and Thomas J. Mason PhD Ed., National Cancer Institute, Environmental Protection Agency. EPA-600/1-83-015a. September, 1983.

[24] *.U.S. Cancer Mortality Rates and Trends*, 1950-1979 Wilson B. Riggan PhD and Thomas J. Mason PhD Ed., National Cancer Institute, Environmental Protection Agency. EPA-600/1-83-015a. September, 1983.

[25] An excellent discussion of disease clustering can be found in Walter, SD et al "Section III: Disease Mapping and Clustering" in *Spatial Epidemiology Methods and Applications.* Edited by P. Elliot, J.C. Wakefield, N.G. Best and D.J. Briggs. Oxford University Press, Oxford OX26DP, 2000. pp 221-317.

[26] Rothman, Kenneth J. "A Sobering Start For the Cluster Busters' Conference." Keynote Presentation presented at the National Conference on Clustering of Health Events, Atlanta GA, February, 16, 1989. in: *American Journal of Epidemiology.* Vol 132. Suppl. No. 1. 1990.

[27] Warner, Stephanie C. and Timothy E. Aldrich. "The Status of Cancer Cluster Investigations Undertaken by State Health Departments" *Public Health Briefs* AJPH March 1988 vol 28. No. 3.pp 306-307

[28] Devier, Janice R. et al. "A Public Health Response to Cancer Clusters in Missouri," *American Journal of Epidemiology* Vol 132. Suppl. 1. 1990. pp S23-31..

[29] Rothman, Kenneth J. and Sander Greenland. *Modern Epidemiology* Second Edition. Lippincott Williams and Wilkins. Philadelphia PA. 1998. p. 22.

[30] Anderson, Henry A. "Evolution of Environmental Epidemiologic Risk Assessment" *Environmental Health Perspectives* Vol 62, 1985. p. 389-392.

[31] Neutra, Raymond R. "Epidemiology for and with a Distrustful Community" *Environmental Health Perspectives* Vol 62. 1985. pp 393-397.

[32] Beral, Valerie "Childhood Leukemia Near Nuclear Plants in the United Kingdom: The Evolution of a Systematic Approach to Studying Rare Disease in Small Geographic Areas," *American Journal of Epidemiology.* Vol 132, Suppl. 1 1990.

[33] Devier, Op cit.

[34] Phone interviews with state health department epidemiologists and officials were conducted in 1998 and 1999.

[35] Thomas, Duncan C. "The Problem of Multiple Inference in Identifying Point-Source Environmental Hazards," *Environmental Health Perspectives.* Vol. 62, 1985. pp 407-414

[36] Cantor, Kenneth P. and Aaron Blair. "Farming and Mortality from Multiple Myeloma: A Case-Control Study with the Use of Death Certificates," *Journal of the National Cancer Institute,* Vol 72, No. 2, February, 1984. pp. 251-255.

[37] Cantor, Kenneth P. "Farming and Mortality from Non-Hodgkin's Lymphoma: A Case-Control Study." *Int. J. Cancer.* 29.1982..pp 239-247,

[38] Mallin, Katherine. "Investigation of a Bladder Cancer Cluster in Northwestern Illinois," *American Journal of Epidemiology.* Vol 132.Suppl I, 1990.

[39] Wartenburg, Daniel and Michael Greenberg. "Detecting Disease Clusters: The Importance of Statistical Power" *American Journal of Epidemiology.* Vol 132 Suppl. No 1. S I. 1990.

[40] Blair, Aaron and Terry L. Thomas "Leukemia Among Nebraska Farmers: A Death Certificate Study." *American Journal of Epidemiology.* Vol 110. No 3. 1979. pp 264-273.

[41] Burmeister, Leon F. et al. "Selected Cancer Mortality and Farm Practices in Iowa," *American Journal of Epidemiology,"* Vol 118 No 1. 1983. pp 72-77.

[42] Hicks, Harry G. "Additional Calculations of Radionuclide Production Following Nuclear Explosions and Pu Isotopic Ratios for Nevada Test Site Events." *Health Physics* Vol. 59 No. 5 (November) pp 515-523.

[43] Krey, Philip W, Merrill Heit and Kevin M. Miller: "Radioactive Fallout Reconstruction From Contemporary Measurements of Reservoir Sediments" *Health Physics.* Vol 59, No. 5 (November) 1990. pp 541-544

[44] Beck, Harold L., Irene K. Helfer, Andre Bouville, Mona Dreicer. "Estimates of Fallout in the Continental U.S.From Nevada Weapons Testing Based on Gummed-Film Monitoring Data" *Health Physics.* Vol. 59. No 5. (November 1990. pp. 565-576.

APPENDICES

A. CALCULATION PROCEDURES
B. I-131 ESTIMATE AND TOTAL FALLOUT CURVES
C. ESTIMATED VALUES FOR I-131
D. :MULTIPLIERS : I-131 TO TOTAL FALLOUT
E. UNCERTAINTY AND ERROR PROPAGATION
F. FORMATTING PROCEDURES
G. TABLES: NUCLEAR FALLOUT AND CANCER

APPENDIX A.
CALCULATION PROCEDURES

CONVERTING I-131 TO TOTAL FALLOUT

Sometime in the late 1970s, Harry Hicks of the Lawrence Livermore National Laboratory began an ambitious project. He wanted to develop a method to calculate the ground concentration and activity of *each* radionuclide from any nuclear test.

In principle, it was relatively simple: Each radionuclide has its own particular decay scheme. For example, as the nucleus of an atom of Uranium-235 releases energy, the atom transforms to less energetic and thus more stable elements. The first transformation changes the atom from Uranium-235 to Thorium-231. Then the Thorium-231 nucleus releases energy and the atom becomes Protoactinium-231. From there, it becomes Actinium-237, Thorium-237 and so on until it finally stops at Lead-207—which is not radioactive.

Complicating the calculations, however, was the fact that each nuclear test involved unique combinations of materials. Some, like Tumbler-Snapper How, were constructed with a significant amount of beryllium. Others, like Upshot-Knothole Dixie, involved lithium deuteride, a fusion booster. Shot Badger was a scaled-down version of the original thermonuclear device, "Mike." In essence, it was a mini-H-bomb like the later shot Hood.

To make matters even more complicated, scientists sometimes added elements such as tungsten and zirconium to the mix in order to track the nuclear cloud's progress across the country. Weapons designer Ted Taylor once claimed he could—in theory—design a nuclear bomb that would spew green paint.

And there was the problem of the shot towers. A shot tower weighs about 500 tons. With rare exceptions, a nuclear detonation vaporizes every last foot of the tower, sending a cloud of tiny micron-sized radioactive iron balls east with the wind. Any calculation of fallout components from tower shots would likely have to include the significant amount of radioactive iron and cobalt from the tower. Surface and subsurface shots such as Jangle Sugar and Uncle and Teapot Ess were worse still. Even a small subsurface nuclear device can result in an exceptionally high quantity of fallout. Given all the possible combinations, and the fact that the tests had taken place more than 20 years before, the task looked particularly difficult if not impossible.

Fortunately, however, the scientists involved in the detonations had kept excellent notes, not only about the materials inside each nuclear device, but also data about the composition of the fallout. And Hicks had access to the notes.

So, after considerable time with Livermore's mainframe computer and endless calculations, Hicks had his data: the relative fallout composition for every major aboveground shot in the fifties and early sixties. Hicks had calculated the relative abundance in fallout for 128 radionuclides, from beryllium-7 through terbium-161. Not only that, he calculated the relative abundance of many of these fallout components from the instant of detonation up to 50 years post-shot. That wasn't all. On the final row, he indicated what the total fallout level would be for that particular shot for days 1, 2, 5 and 10—and then from day 10 through day 300.

Hicks published his data, which came to be known as the Hicks Tables, in July, 1981, under the title: *Results of Calculations of External Gamma Radiation Exposure Rates from Fallout and The Related Radionuclide Compositions*. While it was a work of extraordinary importance, it offered little explanatory material. To the casual reader, it appears to be merely row upon row of numbers in scientific notation format. In fact, the Hicks Tables represents a singularly important tool for characterizing nuclear fallout when one radionuclide is known.

RATIOS

It has been long known that I-131 comprised approximately 2 percent of the total fallout load from any given test. But prior to 1997 no one had any idea how much I-131 had actually been deposited. After the publication of the I-131 levels by the NCI on October 4, 1997, it was possible to use the ratios to determine not only total fallout but the amounts of any of the 127 other radionuclides found in the Hicks Tables.

But there was a problem: The NCI provided data for shot days 1 through 10 inclusive. The Hicks Tables included only days 1, 2, 5 and 10. Some way would have to be found to fill in the blanks for the missing days.

FILLING IN THE BLANKS

The problem was solved using a basic spreadsheet technique found in Microsoft Excel 97 and Excel 2000. The procedure:

1. Create a 10-row, 2-column table.

2. Insert the Hicks Table I-131 data for day 1 in the first row, the I-131 data for the second day in the second row, the I-131 data for the fifth day in the fifth row, and the I-131 data for the tenth day in the tenth row. Leave the cells in rows 3, 4, 6, 7, 8 and 9 blank.

3. Create a graph using the table data, including the blank spaces. The y axis represents rows (days) 1 through 10, while the X axis represents the I-131 values.

4. First, right-click on one of the dots on the graph. A menu appears. Choose *Add Trend line*.

5. Choose Exponential from the menu of trend lines (Linear, Logarithmic, Polynomial, Power, Exponential and Moving Average).

6. Click on the Options tab. Click the following two boxes on the next menu:

    •Display equation on chart

    •Display R-squared value on chart.

7. A line connecting the dots will appear on the chart. An equation next to the line describes the relationship between the dots on the x and y coordinates. An R-squared value appears next to the equation. The closer this value is to 1, the better the curve. If the R-squared (R^2) value is less than 0.91, then consider using another type of trend line or rechecking the data to make sure there were no typos.

The equation should describe how the I-131 activity changes with respect to the day post-shot. By plugging in missing days 3, 4, 6, 7, 8 and 9, one can obtain values that reasonably describe the expected values for I-131.

AN EXAMPLE: SEDAN

Suppose one wanted to determine the Total Fallout from the Sedan test based upon the NCI's I-131 data for the Sedan shot. The first step would be to obtain a copy of the Hicks Tables for Sedan. This is found in the Lawrence Livermore National Laboratory Document No. DE82007575, *Results of Calculations of External Gamma-Radiation Exposure Rates From Fallout and the Related Radionuclide Compositions. Operations Nougat Through Bowline, 1962-1968.* This document covers the Sedan shot of July 6, 1962. On page B-11, one can find the Total Fallout for days 1, 2, 5 and 10 subsequent to the Sedan detonation:

DEBRIS DECAY FROM 1–300 DAYS					
	ZERO TIME	1.00+00	2.00+00	5.00E+00	1.00E+01
TOTAL	2.32E+02	8.73E+01	4.65E+01	1.61E+01	1.03E+01

We then construct a table in Excel displaying the Hicks Table values against shot days.

SHOT DAY	HICKS VALUE
1	8.73E+01 (87.3)
2	4.65E+01 (46.5)
5	1.61E+01 (16.1)
10	1.03E+01 (10.3)

By analyzing this set of values using the techniques described earlier, we find the values correspond to a curve with this formula: $y=86.457x^{0.9595}$ where x = day post-shot. Solving for y in the equation, we arrive at this set of values for Sedan Total Fallout:

DAY POST-SHOT	TOTAL FALLOUT µCI/SQ METER		DAY POST-SHOT	TOTAL FALLOUT µCI/SQ METER
1	86.457		1	8.73E+01
2	44.459	Next, substitute the derived values for the Hicks Tables values →	2	4.65E+01
3	30.130		3	30.130
4	22.862		4	22.862
5	18.456		5	1.61E+01
6	15.494		6	15.494
7	13.364		7	13.364
8	11.757		8	11.757
9	10.500		9	10.500
10	9.490		10	1.03E+01

Now we go to the Hicks Table data for Sedan's I-131 values. We find the following information:

DEBRIS DECAY FROM 1–300 DAYS					
	ZERO TIME	1.00+00	2.00+00	5.00E+00	1.00E+01
I-131	6.84E-0.3	3.69E-01	3.49E-01	2.80E-01	1.85E-01

As with the Total Fallout data, a curve is developed from the available I-131 values, missing values are filled in from the associated equation, and the equation values are replaced by Hicks Tables values where Hicks values are already known. As it turns out, the equation for the I-131 curve, like all radionuclide curves, is *exponential*, whereas the total fallout curve is a *power* curve.

As a final step, ratios are determined between the Total Fallout values and the I-131 values:

DAY POST SHOT	TOTAL FALLOUT µCI/SQ METER	I-131 µCI/SQ METER	RATIO
1	86.457	0.369	236.58
2	44.459	0.349	133.24
3	30.130	0.321	93.87
4	22.862	0.297	76.97
5	18.456	0.280	57.5
6	15.494	0.254	60.91
7	13.364	0.235	56.76
8	11.757	0.217	53.96
9	10.500	0.201	52.08
10	9.490	0.185	55.68

The relationship between the I-131 and Total Fallout values is in the form of a power curve,

$y = 207.23x^{-0.6577}$. While the fit to the values is not perfect ($R^2=0.9433$), it still allows a reasonable guess of the total fallout based upon the I-131 values.

This book includes tables that list the Hicks Table values, as well as the calculated values for I-131 for each shot, from days 1 through 10. Also included is a table showing the specific equation used to obtain the I-131 values for each shot.

TABLE OF MULTIPLIERS

The third table in the example above is actually a table of multipliers. When presented with I-131 activity data for a given county on a specific day post-shot, one can now determine the Total Fallout by multiplying the I-131 activity level times the appropriate multiplier. The Table of Multipliers can be found in Appendix D.

APPENDIX B:

I-131 ESTIMATES AND TOTAL FALLOUT CURVES

X = DAY POSTSHOT

SHOT	y = I-131	R^2	y = TOTAL FALLOUT	R^2
R-BAKER	$y=0.6848e^{-0.0777x}$	0.9974	$Y=98.44X^{-1.1364}$	0.9954
R-BAKER-2	$y=0.6908e^{-0.0779x}$	0.9973	$y=95.607x^{-1.1374}$	0.9893
BJ BAKER	$y=0.8299e^{-0.777x}$	0.9986	$y=65.298x^{-1.1392}$	0.9999
BJ CHARLIE	$y=0.5994e^{-0.0777x}$	0.9976	$y=81.408e^{-1.1413x}$	0.9463
BJ DOG	$y=0.6156e^{-0.0778x}$	0.9971	$y=87.655x^{-1.1388}$	0.9933
BJ EASY	$y=0.6391e^{-0.0777x}$	0.9974	$y=71.048x^{-1.1441}$	0.9996
BJ SUGAR	$y=0.7697e^{-0.0779x}$	0.9975	$y=70.783x^{-1.1061}$	0.9977
BJ UNCLE	$y=0.769e^{-0.0778x}$	0.9973	$y=71.081x^{-1.099}$	0.9977
TS ABLE	$y=0.5235e^{-0.0779x}$	0.9975	$y=111.29x^{-1.1406}$	0.9801
TS BAKER	$y=0.5259e^{-0.0775x}$	0.9974	$y=108.23x^{-1.1437}$	0.9823
TS CHARLIE	$y=0.6057e^{-0.0777x}$	0.9975	$y=90.978x^{-1.1578}$	0.9927
TS DOG	$y=0.6541e^{-0.0777x}$	0.9976	$y=76.7x^{-1.151}$	0.998
TS EASY	$y=0.8217e^{-0.0777x}$	0.9972	$y=60.038x^{-1.1192}$	0.9996
TS FOX	$y=0.7949e^{-0.0777x}$	0.9987	$y=71.604x^{-1.116}$	0.996
TS GEORGE	$y=0.7959e^{-0.0777x}$	0.9986	$y=71.015x^{-1.1163}$	0.9965
TS HOW	$y=0.844e^{-0.0827x}$	0.9987	$y=56.498x^{-1.1188}$	0.9992
UK ANNIE	$y=0.7994e^{-0.0777x}$	0.9974	$y=69.792x^{-1.117}$	0.9987

APPENDIX B:

I-131 ESTIMATES AND TOTAL FALLOUT CURVES

X = DAY POSTSHOT

SHOT	y = I-131	R^2	y = TOTAL FALLOUT	R^2
UK NANCY	$y=0.8139e^{-0.0779x}$	0.9973	$y=69.341x^{-1.1174}$	0.9988
UK RUTH	$y=0.7038e^{-0.0778x}$	0.9973	$y=85.858x^{-1.1334}$	0.9913
UK DIXIE	$y=0.649e^{-0.0776x}$	0.9974	$y=76.869x^{-1.1432}$	0.9978
UK RAY	$y=0.7074e^{-0.0778x}$	0.9974	$y=83.299x^{-1.1205}$	0.9925
UK BADGER	$y=0.797e^{-0.0777x}$	0.9975	$y=82.081x^{-1.1076}$	0.9947
UK SIMON	$y=0.7735e^{-0.0778x}$	0.9973	$y=66.794x^{-1.1165}$	0.9994
UK ENCORE	$y=0.5843e^{-0.0778x}$	0.9974	$y=82.524x^{-1.1624}$	0.9967
UK HARRY	$y=0.7759e^{-0.0778x}$	0.9972	$y=71.154x^{-1.1253}$	0.9985
UK GRABLE	$y=0.5623e^{-0.07742}$	0.9974	$y=63.45x^{-1.2001}$	0.9994
UK CLIMAX	$y=0.6093e^{-0.0779x}$	0.9975	$y=66.626x^{-1.1495}$	1.0
TP WASP	$y=0.7516e^{-0.0778x}$	0.9973	$y=66.076x^{-1.151}$	0.999
TP MOTH	$y=0.9165e^{-0.0776x}$	0.9975	$y=61.271x^{-1.1181}$	0.9998
TP TESLA	$y=0.9643e^{-0.0777x}$	0.9974	$y=71.206x^{-1.1121}$	0.9983
TP TURK	$y=0.8124e^{-0.0777x}$	0.9974	$y=59.554x^{-1.1208}$	0.9996
TEAPOT HORNET	$y=0.8139e^{-0.0776x}$	1.0	$y=65.607x^{-1.1745}$	1.0
TEAPOT BEE	$y=0.8566e^{-0.0777x}$	0.9973	$y=57.911x^{-1.1188}$	0.9991
TEAPOT ESS	$y=0.6032e^{-0.0779x}$	0.9973	$y=148.75x^{-1.1143}$	0.9916
TP APPLE I	$y=0.7865e^{-0.0777x}$	0.9973	$y=60.998x^{-1.1232}$	0.9997

APPENDIX B:

I-131 ESTIMATES AND TOTAL FALLOUT CURVES

X = DAY POSTSHOT

SHOT	y = I-131	R^2	y = TOTAL FALLOUT	R^2
TP POST	$y=0.9863e^{-0.0777x}$	0.9974	$y=65.862x^{-1.1117}$	0.9997
TP MET	$y=0.7471e^{-0.0777x}$	0.9973	$y=60.524x^{-1.1246}$	0.9997
TP APPLE II	$y=0.7796e^{-0.0777x}$	0.9975	$y=60.84x^{-1.1212}$	0.9998
TP ZUCCHINI	$y=0.782e^{-0.0777x}$	0.9973	$y=58.685x^{-1.1229}$	0.9993
PB BOLTZMANN	$y=0.8771e^{-0.0776x}$	0.9975	$y=58.205x^{-1.1209}$	0.9993
PB WILSON	$y=0.683e^{-0.0778x}$	0.9975	$y=61.32x^{-1.15}$	1.0
PB PRISCILLA	$y=0.6126e^{-0.0777x}$	0.9974	$y=63.593x^{-1.1492}$	0.9998
PB HOOD	$y=0.6379e^{-0.0775x}$	0.9973	$y=62.516x^{-1.1435}$	0.9995
PB DIABLO	$y=0.7994e^{-0.0777x}$	0.9974	$y=66.774x^{-1.1175}$	0.9994
PB KEPLER	$y=0.9797e^{-0.0777x}$	0.9973	$y=62.787x^{-1.1216}$	0.9999
PB OWENS	$y=0.8337e^{-0.0776x}$	0.9973	$y=64.59x^{-1.1469}$	0.9999
PB STOKES	$y=0.6301e^{-0.0777x}$	0.9974	$y=59.97x^{-1.1573}$	0.9989
PB SHASTA	$y=0.7934e^{-0.0778x}$	0.9972	$y=67.004x^{-1.1191}$	0.9994
PB DOPPLER	$y=0.6686e^{-0.0779x}$	0.9975	$y=61.935x^{-1.1522}$	0.9996
PB FP	$y=0.6306e^{-0.0778x}$	0.9974	$y=60.062x^{-1.1588}$	0.9990
PB SMOKY	$y=0.7766e^{-0.0779x}$	0.9973	$y=60.062x^{-1.1206}$	0.9998
PB GALILEO	$y=0.9189e^{-0.0776x}$	0.9974	$y=82.044x^{-1.1181}$	0.9934
PB COULOMB-B	$y=0.8025e^{-0.0778x}$	0.9974	$y=53.738x^{-1.1116}$	0.9975

APPENDIX B:

I-131 ESTIMATES AND TOTAL FALLOUT CURVES

X = DAY POSTSHOT

SHOT	y = I-131	R^2	y = TOTAL FALLOUT	R^2
PB WHEELER	$y=0.6012e^{-0.0776x}$	0.9975	$y=60.896x^{-1.1546}$	0.9989
PB LAPLACE	$y=0.606e^{-0.0778x}$	0.9975	$y=64.781x^{-1.1532}$	0.9999
PB FIZEAU	$y=0.8707e^{-0.0778x}$	0.9974	$y=57.767x^{-1.1222}$	0.9991
PB NEWTON	$y=0.8115e^{-0.0779x}$	0.9974	$y=73.909x^{-1.1382}$	0.9987
PB WHITNEY	$y=0.8358e^{-0.0776x}$	0.9974	$y=71.455x^{-1.1203}$	0.9982
PB CHARLESTON	$y=0.6665e^{-0.0779x}$	0.9974	$y=63.037x^{-1.1519}$	0.9996
PB MORGAN	$y=0.6703e^{-0.0778x}$	0.9972	$y=61.003x^{-1.1547}$	0.9992
SEDAN	$y=0.405e^{-0.0775x}$	0.9974	$y=86.457x^{-0.9595}$	0.9904

APPENDIX C
ESTIMATED I-131 LEVELS FROM HICKS TABLES
µCi/sq meter

TEST, SHOT DAY AND EST. LEVEL		TEST, SHOT DAY AND EST. LEVEL		TEST, SHOT DAY AND EST. LEVEL	
R-BAKER		9	0.412405	7	0.370985
1	6.24E-01	10	3.78E-01	8	0.343251
2	5.89E-01	BUSTER CHARLIE		9	0.31759
3	0.474066	1	5.46E-01	10	2.91E-01
4	0.436046	2	5.16E-01	JANGLE SUGAR	
5	4.74E-01	3	0.474769	1	7.01E-01
6	0.368908	4	0.439277	2	6.62E-01
7	0.339321	5	4.14E-01	3	0.609294
8	0.312107	6	0.376053	4	0.563632
9	0.287076	7	0.34794	5	5.32E-01
10	3.12E-01	8	0.321928	6	0.482317
R-BAKER-2		9	0.297862	7	0.44617
1	6.29E-01	10	2.73E-01	8	0.412733
2	5.94E-01	BUSTER DOG		9	0.381801
3	0.546837	1	5.60E-01	10	3.50E-01
4	0.505855	2	5.30E-01	JANGLE UNCLE	
5	4.78E-01	3	0.487455	1	7.00E-01
6	0.432876	4	0.450968	2	6.62E-01
7	0.400435	5	4.26E-01	3	0.608922
8	0.370425	6	0.385985	4	0.563344
9	0.342664	7	0.357093	5	5.32E-01
10	3.14E-01	8	0.330365	6	0.482167
BUSTER BAKER		9	0.305637	7	0.446077
1	7.56E-01	10	2.80E-01	8	0.412688
2	7.14E-01	BUSTER EASY		9	0.381798
3	0.657343	1	5.82E-01	10	3.50E-01
4	0.608201	2	5.50E-01		
5	5.74E-01	3	0.506215	TS ABLE	
6	0.520664	4	0.468371	1	4.77E-01
7	0.48174	5	4.42E-01	2	4.50E-01
8	0.445726	6	0.40096	3	0.414

APPENDIX C

ESTIMATED I-131 LEVELS FROM HICKS TABLES
µCi/sq meter

TEST, SHOT DAY AND EST. LEVEL		TEST, SHOT DAY AND EST. LEVEL		TEST, SHOT DAY AND EST. LEVEL	
4	0.383	2	5.63E-01	TS GEORGE	
5	3.62E-01	3	0.518096	1	7.25E-01
6	0.328	4	0.479364	2	6.85E-01
7	0.303	5	4.52E-01	3	0.598246
8	0.281	6	0.41037	4	0.514812
9	0.26	7	0.379692	5	5.51E-01
10	2.38E-01	8	0.351307	6	0.38123
TS BAKER		9	0.325044	7	0.328062
1	4.79E-01	10	2.98E-01	8	0.282309
2	4.53E-01	TS EASY		9	0.242937
3	0.416802	1	7.48E-01	10	3.63E-01
4	0.38572	2	7.07E-01	TS HOW	
5	3.64E-01	3	0.650848	1	7.55E-01
6	0.330336	4	0.602191	2	7.13E-01
7	0.305702	5	5.69E-01	3	0.646776
8	0.282905	6	0.51552	4	0.595439
9	0.261808	7	0.47698	5	5.73E-01
10	2.40E-01	8	0.441322	6	0.504667
TS CHARLIE		9	0.40833	7	0.464611
1	5.52E-01	10	3.74E-01	8	0.427733
2	5.21E-01	TS FOX		9	0.393783
3	0.479759	1	7.24E-01	10	3.77E-01
4	0.443894	2	6.84E-01	UK ANNIE	
5	4.19E-01	3	0.62962	1	7.28E-01
6	0.380005	4	0.582551	2	6.88E-01
7	0.351597	5	5.50E-01	3	0.633184
8	0.325312	6	0.498706	4	0.585849
9	0.300992	7	0.461424	5	5.53E-01
10	2.76E-01	8	0.426928	6	0.501529
TS DOG		9	0.395012	7	0.464036
1	5.96E-01	10	3.62E-01	8	0.429345

APPENDIX C

ESTIMATED I-131 LEVELS FROM HICKS TABLES
µCi/sq meter

TEST, SHOT DAY AND EST. LEVEL		TEST, SHOT DAY AND EST. LEVEL		TEST, SHOT DAY AND EST. LEVEL	
9	0.397248	6	0.407415	5	5.35E-01
10	3.64E-01	7	0.376995	6	0.484989
		8	0.348847	7	0.448687
UK NANCY		9	0.3228	8	0.415103
1	7.41E-01	10	2.96E-01	9	0.384032
2	7.00E-01	UK RAY		10	3.52E-01
3	0.644283	1	6.44E-01	UK ENCORE	
4	0.595998	2	6.09E-01	1	5.32E-01
5	5.63E-01	3	0.560145	2	5.03E-01
6	0.510014	4	0.518218	3	0.46267
7	0.471792	5	4.89E-01	4	0.428039
8	0.436434	6	0.443544	5	4.04E-01
9	0.403726	7	0.410344	6	0.366359
10	3.70E-01	8	0.37963	7	0.338937
UK RUTH		9	0.351214	8	0.313567
1	6.41E-01	10	3.22E-01	9	0.290097
2	6.05E-01	UK BADGER		10	2.66E-01
3	0.557295	1	7.26E-01	UK HARRY	
4	0.515581	2	6.86E-01	1	7.06E-01
5	4.87E-01	3	0.631283	2	6.68E-01
6	0.441286	4	0.58409	3	0.614386
7	0.408256	5	5.51E-01	4	0.568399
8	0.377698	6	0.500023	5	5.37E-01
9	0.349427	7	0.462643	6	0.486494
10	3.20E-01	8	0.428056	7	0.450079
UK DIXIE		9	0.396056	8	0.416391
1	5.91E-01	10	3.63E-01	9	0.385223
2	5.59E-01	UK SIMON		10	3.53E-01
3	0.514211	1	7.04E-01	UK GRABLE	
4	0.475817	2	6.66E-01	1	5.12E-01
5	4.49E-01	3	0.612486	2	4.84E-01
		4	0.566641	3	0.445383

APPENDIX C

ESTIMATED I-131 LEVELS FROM HICKS TABLES
µCi/sq meter

TEST, SHOT DAY AND EST. LEVEL		TEST, SHOT DAY AND EST. LEVEL		TEST, SHOT DAY AND EST. LEVEL	
4	0.412087	3	0.726154	2	7.01E-01
5	3.89E-01	4	0.671935	3	0.644863
6	0.352777	5	6.34E-01	4	0.596714
7	0.326404	6	0.575341	5	5.63E-01
8	0.302003	7	0.532382	6	0.510933
9	0.279425	8	0.492632	7	0.472783
10	2.56E-01	9	0.455849	8	0.437483
UK CLIMAX		10	4.18E-01	9	0.404818
1	5.55E-01	TEAPOT TESLA		10	3.71E-01
2	5.24E-01	1	8.78E-01	TEAPOT BEE	
3	0.482321	2	8.30E-01	1	7.80E-01
4	0.446175	3	0.763797	2	7.37E-01
5	4.21E-01	4	0.706697	3	0.678491
6	0.381805	5	6.67E-01	4	0.627768
7	0.353192	6	0.604984	5	5.93E-01
8	0.326722	7	0.559757	6	0.537415
9	0.302237	8	0.517911	7	0.497239
10	2.77E-01	9	0.479193	8	0.460067
TEAPOT WASP		10	4.39E-01	9	0.425673
1	6.84E-01	TEAPOT TURK		10	3.90E-01
2	6.47E-01	1	7.40E-01	TEAPOT ESS	
3	0.595145	2	6.99E-01	1	5.49E-01
4	0.550598	3	0.643481	2	5.19E-01
5	5.20E-01	4	0.595376	3	0.477493
6	0.471257	5	5.62E-01	4	0.441708
7	0.435983	6	0.509685	5	4.17E-01
8	0.40335	7	0.471582	6	0.377983
9	0.373159	8	0.436327	7	0.349656
10	3.42E-01	9	0.403708	8	0.323451
TEAPOT MOTH		10	3.70E-01	9	0.299211
1	8.35E-01	TEAPOT HORNET		10	2.74E-01
2	7.89E-01	1	7.41E-01	TEAPOT APPLE-1	

APPENDIX C

ESTIMATED I-131 LEVELS FROM HICKS TABLES
µCi/sq meter

TEST, SHOT DAY AND EST. LEVEL		TEST, SHOT DAY AND EST. LEVEL		TEST, SHOT DAY AND EST. LEVEL	
	1 7.16E-01	TEAPOT APPLE-2			10 4.00E-01
	2 6.77E-01		1 7.10E-01	PB WILSON	
	3 0.622966		2 6.71E-01		1 6.22E-01
	4 0.576395		3 0.617501		2 5.88E-01
	5 5.44E-01		4 0.571338		3 0.540825
	6 0.493436		5 5.39E-01		4 0.500343
	7 0.456547		6 0.489107		5 4.72E-01
	8 0.422417		7 0.452542		6 0.428245
	9 0.390838		8 0.418711		7 0.39619
	10 3.58E-01		9 0.387409		8 0.366535
TEAPOT POST			10 3.55E-01		9 0.3391
	1 8.98E-01	TEAPOT ZUCCHINI			10 3.11E-01
	2 8.49E-01		1 7.12E-01	PB PRISCILLA	
	3 0.781223		2 6.73E-01		1 5.58E-01
	4 0.72282		3 0.619402		2 5.27E-01
	5 6.82E-01		4 0.573097		3 0.485225
	6 0.618787		5 5.41E-01		4 0.44895
	7 0.572527		6 0.490613		5 4.24E-01
	8 0.529726		7 0.453935		6 0.384334
	9 0.490125		8 0.42		7 0.355602
	10 4.49E-01		9 0.388602		8 0.329018
TEAPOT MET			10 3.56E-01		9 0.304421
	1 6.80E-01	PB BOLTZMANN			10 2.79E-01
	2 6.43E-01		1 7.99E-01	PB HOOD	
	3 0.591759		2 7.55E-01		1 5.81E-01
	4 0.54752		3 0.694937		2 5.49E-01
	5 5.17E-01		4 0.643049		3 0.505567
	6 0.468717		5 6.07E-01		4 0.467866
	7 0.433677		6 0.550607		5 4.42E-01
	8 0.401256		7 0.509495		6 0.400687
	9 0.371259		8 0.471454		7 0.370807
	10 3.40E-01		9 0.436252		8 0.343155

APPENDIX C

ESTIMATED I-131 LEVELS FROM HICKS TABLES
µCi/sq meter

TEST, SHOT DAY AND EST. LEVEL		TEST, SHOT DAY AND EST. LEVEL		TEST, SHOT DAY AND EST. LEVEL	
9	0.317565	8	0.448126	7	0.387566
10	2.91E-01	9	0.414666	8	0.358521
PB DIABLO		10	3.80E-01	9	0.331652
1	7.28E-01	PB STOKES		10	3.04E-01
2	6.88E-01	1	5.74E-01	PB SMOKY	
3	0.633184	2	5.42E-01	1	7.07E-01
4	0.585849	3	0.499086	2	6.68E-01
5	5.53E-01	4	0.461775	3	0.614756
6	0.501529	5	4.36E-01	4	0.568684
7	0.464036	6	0.395313	5	5.37E-01
8	0.429345	7	0.36576	6	0.48664
9	0.397248	8	0.338417	7	0.45017
10	3.64E-01	9	0.313118	8	0.416433
PB KEPLER		10	2.87E-01	9	0.385224
1	8.92E-01	PB SHASTA		10	3.53E-01
2	8.43E-01	1	7.22E-01	GALILEO	
3	0.718174	2	6.83E-01	1	8.37E-01
4	0.664485	3	0.628243	2	7.91E-01
5	6.78E-01	4	0.581219	3	0.728056
6	0.568847	5	5.49E-01	4	0.673695
7	0.526321	6	0.497466	5	6.36E-01
8	0.486974	7	0.460231	6	0.576847
9	0.450569	8	0.425782	7	0.533777
10	4.46E-01	9	0.393912	8	0.493922
PB OWENS		10	3.61E-01	9	0.457043
1	7.59E-01	PB DOPPLER		10	4.19E-01
2	7.18E-01	1	6.09E-01	WHEELER	
3	0.660551	2	5.75E-01	1	5.48E-01
4	0.61123	3	0.529263	2	5.17E-01
5	5.77E-01	4	0.489599	3	0.476338
6	0.523362	5	4.62E-01	4	0.440772
7	0.484285	6	0.418964	5	4.16E-01

U.S. FALLOUT ATLAS : TOTAL FALLOUT

APPENDIX C

ESTIMATED I-131 LEVELS FROM HICKS TABLES
µCi/sq meter

TEST, SHOT DAY AND EST. LEVEL		TEST, SHOT DAY AND EST. LEVEL		TEST, SHOT DAY AND EST. LEVEL	
	6	0.377408	4	0.637846	3	0.527601
	7	0.349229	5	6.02E-01	4	0.488061
	8	0.323154	6	0.545934	5	4.61E-01
	9	0.299025	7	0.50507	6	0.417648
	10	2.74E-01	8	0.467265	7	0.386349
COULOMB B			9	0.43229	8	0.357394
	1	7.31E-01	10	3.96E-01	9	0.33061
	2	6.90E-01	NEWTON		10	3.03E-01
	3	0.635449	1	7.39E-01	MORGAN	
	4	0.587885	2	6.98E-01	1	6.10E-01
	5	5.55E-01	3	0.642383	2	5.77E-01
	6	0.503172	4	0.594241	3	0.530768
	7	0.465509	5	5.61E-01	4	0.49104
	8	0.430666	6	0.50851	5	4.64E-01
	9	0.39843	7	0.4704	6	0.420282
	10	3.65E-01	8	0.435147	7	0.388823
LAPLACE			9	0.402536	8	0.35972
	1	5.52E-01	10	3.69E-01	9	0.332795
	2	5.22E-01	WHITNEY		10	3.05E-01
	3	0.480091	1	7.61E-01	SEDAN	
	4	0.444156	2	7.20E-01	1	3.69E-01
	5	4.19E-01	3	0.662214	2	3.49E-01
	6	0.380153	4	0.61277	3	0.320983
	7	0.351699	5	5.78E-01	4	0.297046
	8	0.325374	6	0.52468	5	2.80E-01
	9	0.301019	7	0.485505	6	0.254395
	10	2.76E-01	8	0.449254	7	0.235424
FIZEAU			9	0.415711	8	0.217867
	1	7.93E-01	10	3.81E-01	9	0.20162
	2	7.49E-01	CHARLESTON		10	1.85E-01
	3	0.689452	1	6.07E-01		
			2	5.73E-01		

APPENDIX D:

MULTIPLIERS: I-131 TO TOTAL FALLOUT
(nCi/sq meter)

SHOT	SHOT NAME	DAY 1	DAY 2	DAY 3	DAY 4	DAY 5	DAY 6	DAY 7	DAY 8	DAY 9	DAY 10
R-1	BAKER	143.43	78.61	50.08	38.91	34.18	28.54	25.85	23.97	22.63	20.42
R-2	BAKER-2	139.11	78.45	50.11	39.06	36.61	28.78	26.11	24.24	22.92	19.81
BJ1	BAKER	87.43	41.04	28.35	22.21	17.94	16.54	15.08	14.08	13.38	12.70
BJ2	CHARLIE	139.74	74.45	48.94	38.09	34.69	28.01	25.39	23.56	22.26	19.92
BJ3	DOG	146.25	79.06	51.46	40.09	36.62	29.52	26.77	24.85	23.49	20.79
BJ4	EASY	120.62	58.73	39.93	31.06	26.24	22.81	20.67	19.17	18.11	17.15
BJ5	SUGAR	98.00	50.30	34.46	27.10	24.06	20.22	18.44	17.19	16.32	15.03
BJ6	UNCLE	98.57	50.76	34.90	27.50	24.44	20.58	18.78	17.52	16.64	15.34
TS1	ABLE	206.71	123.33	76.71	59.72	58.84	43.95	39.85	36.99	34.96	28.95
TS2	BAKER	201.46	118.32	73.92	57.48	56.04	42.21	38.24	35.47	33.50	27.96
TS3	CHARLIE	153.26	82.73	53.15	41.17	37.71	30.07	27.19	25.18	23.74	20.83
TS4	DOG	124.16	62.88	41.80	32.45	28.32	23.77	21.51	19.94	18.82	17.28
TS5	EASY	81.95	38.05	26.97	21.13	17.31	15.68	14.26	13.27	12.57	12.35
TS6	FOX	95.99	48.98	32.52	25.17	23.09	18.42	16.69	15.49	14.64	14.39
TS7	GEORGE	95.31	48.47	33.04	25.90	22.69	19.23	17.50	16.29	15.44	14.30
TS8	HOW	77.88	34.92	25.12	19.79	15.79	14.85	13.58	12.71	12.11	11.80
UK1	ANNIE	94.36	47.02	32.31	25.32	21.33	18.81	17.11	15.93	15.10	14.50
UK2	NANCY	91.63	46.00	31.53	24.72	21.49	18.36	16.71	15.56	14.74	13.76
UK3	RUTH	123.56	68.43	44.71	35.01	32.65	25.97	23.62	21.99	20.83	18.25
UK4	DIXIE	125.38	63.86	42.54	33.07	28.95	24.29	22.00	20.41	19.27	17.67
UK5	RAY	120.81	66.01	43.42	34.00	31.49	25.22	22.94	21.35	20.22	17.83
UK6	BADGER	111.31	55.83	38.51	30.26	25.55	22.56	20.56	19.16	18.18	17.48
UK7	SIMON	98.02	46.34	31.36	24.39	19.91	17.97	16.32	15.17	14.37	13.54
UK8	ENCORE	150.94	77.34	50.31	38.81	34.16	28.19	25.43	23.51	22.13	20.00
UK9	HARRY	98.30	49.40	33.64	26.31	22.91	19.47	17.70	16.46	15.58	14.48
UK10	GRABLE	123.78	57.13	38.12	29.17	23.19	20.94	18.81	17.32	16.25	15.94
UK11	CLIMAX	120.90	56.68	39.07	30.34	24.94	22.25	20.15	18.68	17.64	17.08
TP1	WASP	97.37	45.60	31.35	24.34	19.81	17.83	16.14	14.96	14.12	13.74
TP2	MOTH	74.61	34.98	24.70	19.35	15.93	14.36	13.07	12.16	11.52	11.27
TP3	TESLA	78.93	40.24	27.47	21.56	18.89	16.05	14.61	13.61	12.91	11.98
TP4	TURK	79.23	39.38	27.02	21.15	17.80	15.68	14.26	13.27	12.57	12.07
TP5	HORNET	87.58	41.72	27.92	21.48	17.83	15.53	13.99	12.91	12.14	11.59
TP6	BEE	73.07	36.36	24.97	19.56	16.47	14.52	13.20	12.29	11.64	10.35
TP6	ESS	199.43	162.35	132.17	107.59	87.59	71.30	58.05	47.26	38.47	31.32
TP-6	BEE-ESS	136.25	99.36	78.57	63.58	52.03	42.91	35.63	29.77	25.06	20.83
TP7	APPLE-1	83.82	41.59	28.51	22.30	18.76	16.52	15.02	13.97	13.23	

APPENDIX D:

MULTIPLIERS: I-131 TO TOTAL FALLOUT
(nCi/sq meter)

SHOT	SHOT NAME	DAY 1	DAY 2	DAY 3	DAY 4	DAY 5	DAY 6	DAY 7	DAY 8	DAY 9	DAY 10
TP8	POST	72.17	35.95	24.69	19.35	16.30	14.37	13.07	12.17	11.53	11.08
		1	2	3	4	5	6	7	8	9	10
TP9	MET	87.56	43.40	29.73	23.25	19.55	17.22	15.64	14.55	13.78	13.23
TP10	APPLE-2	84.35	41.91	28.75	22.50	18.94	16.68	15.17	14.12	13.37	12.84
TP11	ZUCCHINI	84.83	38.63	27.59	21.59	17.58	16.00	14.54	13.53	12.81	12.67
PB1	BOLTZMANN	73.80	34.49	24.47	19.16	15.80	14.21	12.92	12.03	11.39	11.17
		1	2	3	4	5	6	7	8	9	10
PB2	WILSON	101.13	45.75	32.05	24.89	19.72	18.24	16.51	15.31	14.45	14.21
PB3	PRISCILLA	115.59	53.51	37.08	28.80	23.40	21.11	19.11	17.72	16.72	16.34
PB4	HOOD	110.15	50.27	35.21	27.38	22.01	20.11	18.22	16.90	15.96	15.77
		1	2	3	4	5	6	7	8	9	10
PB5	DIABLO	90.80	44.62	30.90	24.21	20.80	17.98	16.36	15.23	14.43	13.63
PB6(A) (K-O)	KEPLER	70.85	33.81	23.60	18.47	15.34	13.69	12.45	11.58	10.97	10.63
PB6(B)((K-O)	OWENS	85.90	40.25	27.74	21.55	17.50	15.81	14.32	13.27	12.53	12.24
AVERAGE K-O		78.38	37.03	25.67	20.01	16.42	14.75	13.38	12.43	11.75	11.43
		1	2	3	4	5	6	7	8	9	10
PB7	STOKES	107.84	48.15	33.70	26.11	20.60	19.07	17.25	15.97	15.06	15.05
PB8	SHASTA	91.83	45.10	31.19	24.43	20.95	18.13	16.50	15.36	14.55	13.74
PB9	DOPPLER	103.78	47.48	33.00	25.61	20.58	18.76	16.98	15.74	14.85	14.64
		1	2	3	4	5	6	7	8	9	10
PB10	FRANKLIN PRIME	108.01	48.07	33.68	26.08	20.60	19.05	17.22	15.95	15.04	15.02
PB11	SMOKY	86.99	40.72	28.76	22.52	18.62	16.71	15.20	14.14	13.40	13.06
PB12	GALILEO	91.88	50.06	32.99	25.85	23.74	19.18	17.45	16.24	15.39	13.65
		1	2	3	4	5	6	7	8	9	10
PB13(A) WCL	WHEELER	114.78	51.26	35.96	27.88	22.07	20.39	18.44	17.08	16.11	16.09
PB13(B) WCL	COULOUMB B	77.43	34.06	24.94	19.58	15.53	14.57	13.27	12.37	11.73	11.92
PB13(C) WCL	LAPLACE	118.66	54.98	38.01	29.49	24.11	21.58	19.53	18.10	17.08	16.59
PB13 AVERAGE WCL		103.62	46.77	32.97	25.65	20.57	18.85	17.08	15.85	14.97	14.87
		1	2	3	4	5	6	7	8	9	10
PB14	FIZEAU	75.16	34.18	24.42	19.11	15.45	14.17	12.88	11.99	11.35	11.29
PB15	NEWTON	97.16	49.14	32.95	25.67	22.10	18.91	17.15	15.93	15.06	14.01
PB16	WHITNEY	91.20	46.39	31.51	24.67	21.63	18.30	16.64	15.48	14.66	13.57
		1	2	3	4	5	6	7	8	9	10
PB17	CHARLESTON	105.93	48.52	33.70	26.16	21.02	19.16	17.34	16.08	15.17	14.95
PB18	MORGAN	102.79	46.27	32.32	25.06	19.87	18.33	16.59	15.37	14.50	14.43
SEDAN	SEDAN	236.59	133.24	93.87	76.97	57.50	60.91	56.76	53.96	52.08	55.68
		1	2	3	4	5	6	7	8	9	10

APPENDIX E

UNCERTAINTY ESTIMATE ASSOCIATED WITH THE REDUCTION OF THE ^{131}I GROUND DEPOSITION DATA AND TOTAL FALLOUT

Stelu Deaconu
Propulsion Research Center
University of Alabama at Huntsville

ABSTRACT

Uncertainty estimate of the NCI 131I and total fallout values are obtained by tracing the propagation of the original data uncertainties through the data reduction procedure. The emphasis of the error analysis is the validation of the mathematical manipulation of the data

MAIN

The sources of uncertainty in the NCI data total fallout estimate are: 1) measurement error in the NCI - ^{131}I data, 2) measurement error in the Hicks table values used as calibration data, and 3) error associated with the algebraic operations performed. While the first and the second sources are major contributors to the final uncertainty values the third provides a check of the validity of the mathematical treatment.

For each test (shot) the 131I – Hicks table data is curve-fitted and the best fit was found to be exponential

$$^{131}I = a\, e^{-bx} \qquad (1)$$

where a and b are coefficients, and x is the day number. This is consistent with the natural decay of radiation from particles. The scatter of the data about the regression curve is given by the standard error of estimate (SEE)[1] – The SEE statistics are:

$$SEE = \left(\frac{\sum_{i=1}^{N} \left({}^{131}I_i - a \cdot e^{-b \cdot x_i} \right)}{N-2} \right)^{\frac{1}{2}}$$

$$x_i = 1, 2, 5, 10 \qquad (2)$$

A ± 2SEE band will contain approximately 95% of the data points. The total uncertainty associated with this regression can be estimated as

$$U_{131I} = \left(\left(U_{131I}^{H} \right)^2 + (2SEE)^2 \right)^{\frac{1}{2}} \qquad (3)$$

where U_{131I}^{H} is the uncertainty in the initial Hicks data. Equation (3) takes into account a) the scattering of the original data about the mean for each data point, and b) the error introduced when fitting Equation (1) to the mean values. Calculation of SEE for random tests yielded the same uncertainty associated with the curve-fit for all tests

$$U_{131I} = \pm 2SEE = \pm 4\%$$

The small error found for the curve-fit suggests that indeed the behavior of the data is consistent and also verifies the natural decay law. Assuming an initial uncertainty of 30% in the Hicks data, the total uncertainty associated with the ^{131}I data regression becomes $U_{131I} = \pm 30.2\%$.

The total fallout (TF) from Hicks data was fitted with power curves.

$$TF = cx^{-d} \qquad (4)$$

The ± 2SEE band for each test is shown in percentages in Figure 1. The uncertainties are not uniform from test to test but vary from 2 – 43 %. The percentage uncertainties expressed so far are calculated with respect to the mean of all ^{131}I and TF daily values. The reason for this is that the SEE is indiscriminant with the x day value., i.e. it is an overall estimate of the uncertainty associated with the regression of all I-131 or Total Fallout day values. The percentage can be applied to the mean of each individual day resulting in slight underestimation on the low side of the data and overestimation on the high side of the data. However, a day-by-day analysis cannot be performed since the curve is obtained from only four day (1, 2, 5, 10) data. The advantage of the percentage approach is that it simplifies the next step of the data analysis.

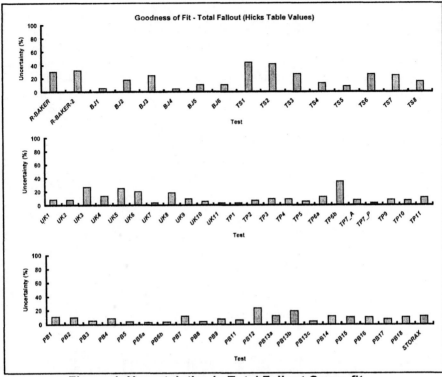

Figure 1. Uncertainties in Total Fallout Curve-fit

The magnification factor (MF) is defined as the ratio of the total fallout TF and the ^{131}I values. MF is a function of x_i-day value.

$$MF(x_i) = \frac{TF^H(x_i)}{^{131}_H I(x_i)} \qquad (5)$$

for simplicity the index H (Hicks table values) will be dropped. The uncertainty associated with the MF curve is given by

$$U^2_{MF} = \left(\frac{\partial MF}{\partial TF}\right)^2 U^2_{TF} + \left(\frac{\partial MF}{\partial^{131}I}\right)^2 U^2_{131I}$$
(6)

Again a percentage type uncertainty estimate is sought and therefore the TF and ^{131}I are treated as variables carrying associated uncertainty rather than replacing them with their regression formulae. Algebraic manipulation of (6) yields:

$$U^2_{MF} = \left(\frac{1}{^{131}I}\right)^2 U^2_{TF} + \left(-\frac{TF}{(^{131}I)^2}\right)^2 U^2_{131I}$$

$$= \left(\frac{TF}{^{131}I}\right)^2 \left(\frac{U^2_{TF}}{TF^2}\right) + \left(\frac{TF}{^{131}I}\right)^2 \left(\frac{U^2_{131I}}{(^{131}I)^2}\right)$$

$$= MF^2 \left[\left(\frac{U_{TF}}{TF}\right)^2 + \left(\frac{U_{131I}}{^{131}I}\right)^2\right]$$

or

$$\left(\frac{U_{MF}}{MF}\right)^2 = \left(\frac{U_{TF}}{TF}\right)^2 + \left(\frac{U_{131I}}{^{131}I}\right)^2$$

$$\left.\frac{U_{MF}}{MF}\right|_{\%} = \sqrt{\left(\frac{U_{TF}}{TF}\right)^2 + \left(\frac{U_{131I}}{^{131}I}\right)^2}$$
(7)

The uncertainties in MF associated with the errors made in using the curve-fit regressions Equations (1) and (4) and relationship (5) are presented in Table 1.

Shot	U_TF_Fit(%)	U_MF (%)	Shot	U_TF_Fit(%)	U_MF (%)	Shot	U_TF_Fit(%)	U_MF (%)
R-BAKER	30.21	30.48	UK5	24.78	25.10	PB2	9.69	10.48
R-BAKER-	32.23	32.48	UK6	20.26	20.65	PB3	5.42	6.74
BJ1	5.30	6.64	UK7	3.75	5.48	PB4	8.84	9.70
BJ2	17.92	18.36	UK8	18.26	18.69	PB5	4.16	5.77
BJ3	24.73	25.05	UK9	9.10	9.94	PB6a	2.91	4.95
BJ4	4.42	5.96	UK10	5.55	6.84	PB6b	3.51	5.32
BJ5	10.89	11.60	UK11	2.93	4.96	PB7	11.81	12.47
BJ6	10.82	11.54	TP1	2.99	4.99	PB8	4.19	5.79
TS1	43.91	44.09	TP2	6.52	7.65	PB9	7.59	8.58
TS2	41.67	41.86	TP3	9.23	10.06	PB11	6.30	7.46
TS3	26.60	26.90	TP4	8.67	9.55	PB12	23.43	23.76
TS4	13.10	13.70	TP5	5.31	6.64	PB13a	12.12	12.76
TS5	8.07	9.01	TP6a	12.15	12.79	PB13b	19.44	19.85
TS6	26.18	26.49	TP6b	34.75	34.98	PB13c	4.35	5.91
TS7	24.33	24.66	TP7_A	7.11	8.16	PB14	11.84	12.49
TS8	15.59	16.09	TP7_P	2.81	4.89	PB15	10.56	11.29
UK1	8.42	9.33	TP9	7.67	8.65	PB16	10.59	11.32
UK2	7.71	8.69	TP10	6.83	7.92	PB17	7.45	8.45
UK3	26.90	27.20	TP11	11.07	11.77	PB18	10.23	10.98
UK4	13.63	14.20	PB1	10.86	11.57	STORAX	11.75	12.42

Table 1. Regression Uncertainty Estimates of the TF and MF

Additional uncertainty will result when considering the uncertainty in the initial Hicks data *(see associated shot table data)*
The total test fallout (TTF) is obtained by multiplying the day value of the test MF by the NCI ^{131}I data value for each county. The data reduction equation is

$$TTF = \sum_{i=1}^{N=10} TF_{NCI}(x_i) = \sum_{i=1}^{N=10}\left[^{131}I_{NCI}(x_i) \cdot MF(x_i)\right]$$

(8)

The NCI data index will be dropped for simplicity keeping in mid that the ^{131}I values are from NCI data and the MF values are computed for each x_i day with Equation (5) from the Hicks table values. The uncertainty in the TTF is

$$U_{TTF}^2 = \sum_{i=1}^{N=10}\left(\frac{\partial TTF}{\partial^{131}I(x_i)}\right)^2 U_{131I_i}^2 + \sum_{i=1}^{N=10}\left(\frac{\partial TTF}{\partial MF(x_i)}\right)^2 U_{MF_i}^2 + \sum_{i=1}^{N=10}\sum_{\substack{j=1 \\ j \neq i}}^{N=10}\left(\frac{\partial TTF}{\partial MF(x_i)}\right)\left(\frac{\partial TTF}{\partial MF(x_j)}\right)\theta_{ij}U_{MF_i}^2$$

(9)

The first and second terms are contributions to uncertainty from the elemental variables in Equation (9), ^{131}I and MF. The third term is the contribution to uncertainty from the correlation that may exist between the individual uncertainties in $MF(x_i)$ and $MF(x_j)$ where $i, j = 1, 2, ...10$ and $i \neq j$. The correlation appears because the errors existent in the data that generated MF are essentially fossilized in the uncertainty associated with Equation (5). The degree of correlation is θ_{ij}. To simplify we make the argument that the use of Equation (5) will not necessarily result in the same type of consistent error for all days. In other words if (5) estimates MF(1) higher than the true value it does not necessarily estimates for instance MF(3) high. The reasons for this is that we can safely assume that the initial errors in the Hicks data were independent. Thus the problem can be simplified by considering $\theta_{ij} = 0$. Equation (9) becomes

$$U_{TTF}^2 \cong \sum_{i=1}^{N=10}\left(\frac{\partial TTF}{\partial^{131}I(x_i)}\right)^2 U_{131I_i}^2 + \sum_{i=1}^{N=10}\left(\frac{\partial TTF}{\partial MF(x_i)}\right)^2 U_{MF}^2$$

$$= \sum_{i=1}^{N=10}\left(\frac{\partial}{\partial^{131}I(x_i)}\sum_{i=1}^{N=10}[^{131}I(x_i) \cdot MF(x_i)]\right)^2 U_{131I}^2$$

$$+ \sum_{i=1}^{N=10}\left(\frac{\partial}{\partial MF(x_i)}\sum_{i=1}^{N=10}[^{131}I(x_i) \cdot MF(x_i)]\right)^2 U_{MF}^2$$

(10)

which can be further transformed to yield

$$U_{TTF}^2 \cong \sum_{i=1}^{N=10}\left(\sum_{j=1}^{N=10}[\delta_{ij} \cdot MF(x_i)]\right)^2 U_{131I_i}^2 + \sum_{i=1}^{N=10}\left(\sum_{j=1}^{N=10}[^{131}I(x_i) \cdot \delta_{ij}]\right)^2 U_{MF}^2$$

$$= \sum_{i=1}^{N=10}\left[(MF(x_i))^2 U_{131I}^2\right] + \sum_{i=1}^{N=10}\left[(^{131}I(x_i))^2 U_{MF}^2\right]$$

(11)

To find out the degree of uncertainty associated with the operation of addition and the use of the magnification factor MF we simply assign $U_{131I} = 0$ (i.e. no uncertainty in the NCI data) and obtain

$$U_{TTF} = U_{MF}\sqrt{\sum_{i=1}^{N=10}(^{131}I(x_i))^2}$$

(12)

It is evident from Equation (12) that a percentage expression is harder to obtain by algebraic manipulation because of the summation. One can input some county data and then divide by the TTF value to get a percentage estimate. The case study with just few of the counties showed that the Equation (8) is quite robust with respect to the uncertainties in MF. For test TS8 the uncertainty due to propagation of the MF errors is only about 5% (random selected counties). Even with an uncertainty in MF of 100% the TTF uncertainty is only about 32%. Obviously the uncertainties in the ^{131}I NCI data will change this since the neglected terms in Equation (11) are no longer negligible. However it can be concluded that the mathematical procedure for obtaining the total fallout/county is robust from the point of view of uncertainty propagation.

NOTES

[1] Coleman, H. W., and Steele, W. G., "Experimentation and Uncertainty Analysis For Engineers", 2nd edition, John Wiley and Sons, 1999

APPENDIX G
TABLES: NUCLEAR FALLOUT AND CANCER

1. **TABLE G-1-:** HIGHLY CORRELATED VARIABLES: RAINFALL DURING NUCLEAR TESTING ANDCANCER RATES 1951-1962

2. **TABLE G-2:** HIGHLY CORRELATED VARIABLES. TOTAL FALLOUT DURING NUCLEAR TESTING AND CANCER RATES 1951-1962

3. **TABLE G-3:** FALLOUT COMPONENTS HIGHLY CORRELATED WITH CANCER RATES TABLE

4. **TABLE G-4:** HIGHLY CORRELATED VARIABLES ALL PROBABILITY LEVELS: <0.0001

5. **TABLE G-5:** HIGHLY CORRELATED FALLOUT VARIABLES: WHITE FEMALE BREAST CANCER

6. **TABLE G-6**: HIGH TOTAL FALLOUT CORRELATIONS WITH COLON CANCER RATES

7. **TABLE G-7:** POISSON REGRESSION RELATIVE RATES: TOTAL FALLOUT AND CANCER

8. **TABLE G-8**: TOXICITY OF SELECTED RADIONUCLIDES
9. **TABLE G-9**: CANCER PARAMETER ABBREVIATIONS WITH ICD CODES

TABLE G-1
HIGHLY CORRELATED VARIABLES
RAINFALL DURING NUCLEAR TESTING AND
CANCER RATES 1951-1962
Spearman Correlations with T Score >5.0
Rainfall: ln(Rainfall Amt, Millimeters); N: 3094

RAINFALL DURING:	RANK	CANCER	R	T (N-2)	p-level
RANGER 1951	1	ALL CANCERS WF 1970-95	0.284	16.49	0
RANGER 1951	2	COLON WM 1970-95	0.279	16.14	0
RANGER 1951	3	RECTUM WM 1950-69	0.262	15.09	0
RANGER 1951	4	ALL CANCERS WF 1950-69	0.258	14.87	0
RANGER 1951	5	RECTUM WF 1950-69	0.255	14.68	0
RANGER 1951	6	COLON WF 1970-95	0.254	14.63	0
RANGER 1951	7	COLON WM 1950-69	0.244	14.01	0
RANGER 1951	8	RECTUM WM 1970-95	0.244	14.00	0
RANGER 1951	9	COLON WF 1950-69	0.243	13.91	0
RANGER 1951	10	RECTUM WF 1970-95	0.239	13.68	0
RANGER 1951	11	BLADDER WM 1970-95	0.236	13.52	0
RANGER 1951	12	BREAST WF 1970-95	0.232	13.27	0
RANGER 1951	13	ESOPHAGUS WM 1970-95	0.226	12.91	0
RANGER 1951	14	ALL CANCERS WM 1970-95	0.209	11.88	0
RANGER 1951	15	BLADDER WF 1970-95	0.201	11.41	0
RANGER 1951	16	BLADDER WM 1950-69	0.200	11.35	0
RANGER 1951	17	BREAST WF 1950-69	0.197	11.15	0
RANGER 1951	18	ALL CANCERS WM 1950-69	0.185	10.47	0
BUSTER-JANGLE 1951	1	LUNG WM 1970-95	0.324	19.04	0
BUSTER-JANGLE 1951	2	ALL CANCERS WM 1970-95	0.314	18.36	0
BUSTER-JANGLE 1951	3	ORAL WM 1970-95	0.222	12.64	0
BUSTER-JANGLE 1951	4	OTHER WM 1970-95	0.204	11.56	0
BUSTER-JANGLE 1951	5	CERVIX WF 1950-69	0.192	10.85	0
BUSTER-JANGLE 1951	6	LUNG WM 1950-69	0.184	10.41	0
BUSTER-JANGLE 1951	7	ORAL WM 1950-69	0.180	10.17	0
BUSTER-JANGLE 1951	8	ORAL WF 1950-69	0.177	10.00	0
BUSTER-JANGLE 1951	9	OTHER WM 1950-69	0.152	8.52	0
BUSTER-JANGLE 1951	10	OTHER WF 1970-95	0.147	8.25	0
BUSTER-JANGLE 1951	11	LUNG WF 1970-95	0.129	7.26	0
BUSTER-JANGLE 1951	12	ORAL WF 1970-95	0.128	7.16	0
BUSTER-JANGLE 1951	13	LIVER WM 1970-95	0.118	6.59	0

TABLE G-1
HIGHLY CORRELATED VARIABLES
RAINFALL DURING NUCLEAR TESTING AND
CANCER RATES 1951-1962
Spearman Correlations with T Score >5.0
Rainfall: ln(Rainfall Amt, Millimeters); N: 3094

RAINFALL DURING:	RANK	CANCER	R	T (N-2)	p-level
BUSTER-JANGLE 1951	14	CERVIX WF 1970-95	0.113	6.30	0
BUSTER-JANGLE 1951	15	ESOPHAGUS WF 1950-69	0.111	6.19	0
BUSTER-JANGLE 1951	16	BLADDER WF 1950-69	0.109	6.12	0
BUSTER-JANGLE 1951	17	OTHER WF 1950-69	0.100	5.57	0
BUSTER-JANGLE 1951	18	BLADDER WF 1970-95	0.091	5.07	0
TUMBLER-SNAPPER 1952	1	ALL CANCERS WM 1970-95	0.303	17.70	0
TUMBLER-SNAPPER 1952	2	LUNG WM 1970-95	0.264	15.20	0
TUMBLER-SNAPPER 1952	3	ORAL WM 1950-69	0.196	11.10	0
TUMBLER-SNAPPER 1952	4	ORAL WM 1970-95	0.185	10.49	0
TUMBLER-SNAPPER 1952	5	CERVIX WF 1950-69	0.182	10.32	0
TUMBLER-SNAPPER 1952	6	OTHER WM 1970-95	0.178	10.04	0
TUMBLER-SNAPPER 1952	7	ALL CANCERS WF 1970-95	0.174	9.83	0
TUMBLER-SNAPPER 1952	8	COLON WF 1970-95	0.164	9.22	0
TUMBLER-SNAPPER 1952	9	BLADDER WF 1950-69	0.155	8.72	0
TUMBLER-SNAPPER 1952	10	CERVIX WF 1970-95	0.149	8.39	0
TUMBLER-SNAPPER 1952	11	ESOPHAGUS WM 1970-95	0.149	8.37	0
TUMBLER-SNAPPER 1952	12	COLON WM 1970-95	0.148	8.29	0
TUMBLER-SNAPPER 1952	13	LUNG WF 1970-95	0.144	8.11	0
TUMBLER-SNAPPER 1952	14	BLADDER WF 1970-95	0.144	8.07	0
UPSHOT-KNOTHOLE 1953	1	COLON WM 1970-95	0.205	11.65	0
UPSHOT-KNOTHOLE 1953	2	COLON WM 1950-69	0.194	10.97	0
UPSHOT-KNOTHOLE 1953	3	COLON WF 1970-95	0.192	10.90	0
UPSHOT-KNOTHOLE 1953	4	ALL CANCERS WM 1950-69	0.178	10.06	0
UPSHOT-KNOTHOLE 1953	5	COLON WF 1950-69	0.178	10.05	0
UPSHOT-KNOTHOLE 1953	6	ALL CANCERS WF 1950-69	0.162	9.12	0
UPSHOT-KNOTHOLE 1953	7	RECTUM WM 1950-69	0.159	8.97	0
UPSHOT-KNOTHOLE 1953	8	ALL CANCERS WM 1970-95	0.141	7.91	0
UPSHOT-KNOTHOLE 1953	9	PROSTATE WM 1950-69	0.134	7.52	0
UPSHOT-KNOTHOLE 1953	10	BREAST WF 1950-69	0.128	7.17	0
UPSHOT-KNOTHOLE 1953	11	LEUKEMIA WM 1970-95	0.120	6.75	0
UPSHOT-KNOTHOLE 1953	12	RECTUM WM 1970-95	0.119	6.67	0

TABLE G-1
HIGHLY CORRELATED VARIABLES
RAINFALL DURING NUCLEAR TESTING AND
CANCER RATES 1951-1962
Spearman Correlations with T Score >5.0
Rainfall: ln(Rainfall Amt, Millimeters); N: 3094

RAINFALL DURING:	RANK	CANCER	R	T (N-2)	p-level
UPSHOT-KNOTHOLE 1953	13	NHL WM 1970-95	0.119	6.67	0
UPSHOT-KNOTHOLE 1953	14	LEUWM5	0.112	6.28	0
UPSHOT-KNOTHOLE 1953	15	STOWF5	0.112	6.24	0
UPSHOT-KNOTHOLE 1953	16	BLADDER WM 1950-69	0.107	6.00	0
UPSHOT-KNOTHOLE 1953	17	BRNWM5	0.100	5.56	0
UPSHOT-KNOTHOLE 1953	18	KIDWM7	0.096	5.37	0
UPSHOT-KNOTHOLE 1953	19	NHLWF7	0.096	5.35	0
UPSHOT-KNOTHOLE 1953	20	NHLWF5	0.095	5.32	0
UPSHOT-KNOTHOLE 1953	21	BREAST WF 1970-95	0.093	5.18	0
UPSHOT-KNOTHOLE 1953	22	OVWF5	0.093	5.17	0
TEAPOT 1955	1	LUNG WM 1970-95	0.499	32.05	0
TEAPOT 1955	2	ALL CANCERS WM 1970-95	0.356	21.18	0
TEAPOT 1955	3	CERVIX WF 1950-69	0.254	14.61	0
TEAPOT 1955	4	CERVIX WF 1970-95	0.230	13.13	0
TEAPOT 1955	5	ORAL WF 1950-69	0.206	11.72	0
TEAPOT 1955	6	OTHER WM 1970-95	0.201	11.40	0
TEAPOT 1955	7	LIVER WM 1950-69	0.153	8.61	0
TEAPOT 1955	8	LIVER WM 1970-95	0.147	8.28	0
TEAPOT 1955	9	OTHER WM 1950-69	0.136	7.63	0
TEAPOT 1955	10	ORAL WM 1970-95	0.124	6.95	0
TEAPOT 1955	11	LUNG WM 1950-69	0.116	6.51	0
TEAPOT 1955	12	ORAL WM 1950-69	0.115	6.43	0
TEAPOT 1955	13	OTHER WF 1950-69	0.109	6.12	0
TEAPOT 1955	14	LUNG WF 1970-95	0.108	6.05	0
TEAPOT 1955	15	BRAIN WM 1970-95	0.101	5.66	0
TEAPOT 1955	16	ESOPHAGUS WF 1950-69	0.099	5.52	0
TEAPOT 1955	17	OTHER WF 1970-95	0.096	5.34	0
PLUMBBOB 1957	1	LUNG WM 1970-95	0.424	26.02	0
PLUMBBOB 1957	2	ALL CANCERS WM 1970-95	0.294	17.09	0
PLUMBBOB 1957	3	ORAL WF 1950-69	0.189	10.70	0
PLUMBBOB 1957	4	ORAL WM 1970-95	0.162	9.14	0

TABLE G-1
HIGHLY CORRELATED VARIABLES
RAINFALL DURING NUCLEAR TESTING AND
CANCER RATES 1951-1962
Spearman Correlations with T Score >5.0
Rainfall: ln(Rainfall Amt, Millimeters); N: 3094

RAINFALL DURING:	RANK	CANCER	R	T (N-2)	p-level
PLUMBBOB 1957	5	CERVIX WF 1950-69	0.140	7.85	0
PLUMBBOB 1957	6	OTHER WM 1970-95	0.121	6.77	0
PLUMBBOB 1957	7	OTHER WM 1950-69	0.119	6.67	0
PLUMBBOB 1957	8	BRAIN WM 1970-95	0.117	6.57	0
PLUMBBOB 1957	9	ORAL WM 1950-69	0.115	6.45	0
PLUMBBOB 1957	10	BRAIN WF 1970-95	0.103	5.73	0
PLUMBBOB 1957	11	LIVER WM 1970-95	0.099	5.55	0
PLUMBBOB 1957	12	LUNG WM 1950-69	0.092	5.16	0
PLUMBBOB 1957	13	ORAL WF 1970-95	0.091	5.06	0
PLUMBBOB 1957	14	ESOPHAGUS WF 1950-69	0.090	5.05	0
SEDAN: JULY, 1962	1	COLON WF 1970-95	0.259	14.94	0
SEDAN: JULY, 1962	2	COLON WM 1970-95	0.209	11.85	0
SEDAN: JULY, 1962	3	COLON WF 1950-69	0.208	11.81	0
SEDAN: JULY, 1962	4	LIVER WF 1950-69	0.196	11.10	0
SEDAN: JULY, 1962	5	ALL CANCERS WF 1950-69	0.191	10.85	0
SEDAN: JULY, 1962	6	RECTUM WM 1950-69	0.181	10.21	0
SEDAN: JULY, 1962	7	RECTUM WM 1970-95	0.174	9.83	0
SEDAN: JULY, 1962	8	COLON WM 1950-69	0.162	9.12	0
SEDAN: JULY, 1962	9	RECTUM WF 1970-95	0.147	8.26	0
SEDAN: JULY, 1962	10	BREAST WF 1970-95	0.134	7.49	0
SEDAN: JULY, 1962	11	RECTUM WF 1950-69	0.127	7.11	0
SEDAN: JULY, 1962	12	BREAST WF 1950-69	0.124	6.96	0
SEDAN: JULY, 1962	13	CERVIX WF 1970-95	0.114	6.36	0
SEDAN: JULY, 1962	14	KIDNEY WF 1950-69	0.108	6.06	0
SEDAN: JULY, 1962	15	ALL CANCERS WF 1970-95	0.105	5.84	0
SEDAN: JULY, 1962	16	OVARIAN WF 1950-69	0.102	5.71	0
SEDAN: JULY, 1962	17	LIVER WF 1970-95	0.096	5.39	0

TABLE G-2
HIGHLY CORRELATED VARIABLES
TOTAL FALLOUT DURING NUCLEAR TESTING AND CANCER RATES 1951-1962
Spearman Correlations with T Score >5.0
TOTAL FALLOUT: Mean of Lognormal Distribution; N: 3094

TOTAL FALLOUT DURING: SHOT SERIES	RANK	VARIABLES	R	t(N-2)	p-level
RANGER 1951	1	ALL CANCERS WF1970-1995	0.298	17.35	0
RANGER 1951	2	COLON WM1970-1995	0.289	16.77	0
RANGER 1951	3	RECTUM WM 1950-1969	0.285	16.52	0
RANGER 1951	4	ALL CANCERS WF 1950-1969	0.276	15.99	0
RANGER 1951	5	COLON WF1970-1995	0.275	15.90	0
RANGER 1951	6	RECTUM WF 1950-1969	0.273	15.79	0
RANGER 1951	7	COLON WM 1950-1969	0.262	15.09	0
RANGER 1951	8	COLON WF 1950-1969	0.261	15.04	0
RANGER 1951	9	RECTUM WM1970-1995	0.259	14.89	0
RANGER 1951	10	RECTUM WF1970-1995	0.247	14.19	0
RANGER 1951	11	BLADDER WM1970-19950	0.246	14.11	0
RANGER 1951	12	BREAST WF1970-1995	0.242	13.84	0
RANGER 1951	13	ESOWM1970-1995	0.228	13.04	0
RANGER 1951	14	BLWF1970-19950	0.216	12.31	0
RANGER 1951	15	BREAST WF 1950-1969	0.209	11.89	0
RANGER 1951	16	ALL CANCERS WM1970-1995	0.209	11.85	0
RANGER 1951	17	BLADDER WM 1950-19690	0.208	11.85	0
RANGER 1951	18	ALL CANCERS _WM 1950-1969	0.191	10.82	0
RANGER 1951	19	LNGWF1970-1995	0.178	10.04	0
RANGER 1951	20	ESOWM 1950-1969	0.172	9.73	0
RANGER 1951	21	ORAWM 1950-1969	0.167	9.40	0
RANGER 1951	22	CRVWF 1950-1969	0.164	9.27	0
RANGER 1951	23	BLWF 1950-19690	0.162	9.12	0
RANGER 1951	24	ORAWM1970-1995	0.156	8.78	0
RANGER 1951	25	LNGWM 1950-1969	0.152	8.57	0
RANGER 1951	26	OVARIAN WF 1950-1969	0.146	8.20	0
RANGER 1951	27	ESOWF1970-1995	0.141	7.92	0
RANGER 1951	28	OVARIAN WF1970-1995	0.140	7.86	0
RANGER 1951	29	CRVWF1970-1995	0.129	7.25	0
RANGER 1951	30	LNGWF 1950-1969	0.128	7.19	0
RANGER 1951	31	ESOWF 1950-1969	0.099	5.52	0

TABLE G-2
HIGHLY CORRELATED VARIABLES
TOTAL FALLOUT DURING NUCLEAR TESTING AND CANCER RATES 1951-1962
Spearman Correlations with T Score >5.0
TOTAL FALLOUT: Mean of Lognormal Distribution; N: 3094

TOTAL FALLOUT DURING: SHOT SERIES	RANK	VARIABLES	R	t(N-2)	p-level
RANGER 1951	32	ORAWF1970-1995	0.095	5.31	0
BUSTER-JANGLE 1951	1	LNGWM1970-1995	0.324	19.01	0
BUSTER-JANGLE 1951	2	ALL CANCERS WM1970-1995	0.314	18.37	0
BUSTER-JANGLE 1951	3	ORAWM1970-1995	0.197	11.19	0
BUSTER-JANGLE 1951	4	CRVWF 1950-1969	0.192	10.86	0
BUSTER-JANGLE 1951	5	OTHWM1970-1995	0.191	10.79	0
BUSTER-JANGLE 1951	6	ORAWM 1950-1969	0.175	9.90	0
BUSTER-JANGLE 1951	7	ORAWF 1950-1969	0.167	9.41	0
BUSTER-JANGLE 1951	8	CRVWF1970-1995	0.155	8.70	0
BUSTER-JANGLE 1951	9	LNGWM 1950-1969	0.147	8.28	0
BUSTER-JANGLE 1951	10	ORAWF1970-1995	0.128	7.20	0
BUSTER-JANGLE 1951	11	OTHWM 1950-1969	0.124	6.93	0
BUSTER-JANGLE 1951	12	OTHWF1970-1995	0.124	6.93	0
BUSTER-JANGLE 1951	13	BLWF 1950-19690	0.118	6.59	0
BUSTER-JANGLE 1951	14	ESOWF 1950-1969	0.117	6.55	0
BUSTER-JANGLE 1951	15	LNGWF1970-1995	0.116	6.50	0
BUSTER-JANGLE 1951	16	OTHWF 1950-1969	0.094	5.22	0
BUSTER-JANGLE 1951	17	BLWF1970-19950	0.090	5.04	0
TUMBLER-SNAPPER 1952	1	BREAST WF 1950-1969	0.193	10.96	0
TUMBLER-SNAPPER 1952	2	COLON WM 1950-1969	0.184	10.42	0
TUMBLER-SNAPPER 1952	3	COLON WF 1950-1969	0.172	9.73	0
TUMBLER-SNAPPER 1952	4	COLON WF1970-1995	0.159	8.94	0
TUMBLER-SNAPPER 1952	5	OVARIAN WF1970-1995	0.158	8.88	0
TUMBLER-SNAPPER 1952	6	RECTUM WM 1950-1969	0.154	8.67	0
TUMBLER-SNAPPER 1952	7	OVARIAN WF 1950-1969	0.146	8.22	0
TUMBLER-SNAPPER 1952	8	COLON WM1970-1995	0.139	7.78	0
TUMBLER-SNAPPER 1952	9	KIDNEY WM 1950-1969	0.119	6.65	0
TUMBLER-SNAPPER 1952	10	NON-HODGKINS LYMPHOMA WF 1950-1969	0.118	6.62	
TUMBLER-SNAPPER 1952	11	BREAST WF1970-1995	0.114	6.39	0
TUMBLER-SNAPPER 1952	12	STOMACH WM 1950-1969	0.111	6.20	0

TABLE G-2
HIGHLY CORRELATED VARIABLES
TOTAL FALLOUT DURING NUCLEAR TESTING AND CANCER RATES 1951-1962
Spearman Correlations with T Score >5.0
TOTAL FALLOUT: Mean of Lognormal Distribution; N: 3094

TOTAL FALLOUT DURING: SHOT SERIES	RANK	VARIABLES	R	t(N-2)	p-level
TUMBLER-SNAPPER 1952	13	RECTUM WF 1950-1969	0.111	6.19	0
TUMBLER-SNAPPER 1952	14	BLADDER WM 1950-19690	0.107	6.00	0
TUMBLER-SNAPPER 1952	15	ALL CANCERS WF 1950-1969	0.101	5.63	0
TUMBLER-SNAPPER 1952	16	PROSTATE WM1970-1995	0.096	5.38	0
TUMBLER-SNAPPER 1952	17	NON-HODGKINS LYMPHOMA WM 1950-1969	0.094	5.24	0
TUMBLER-SNAPPER 1952	18	PROSTATE WM 1950-1969	0.093	5.18	0
UPSHOT-KNOTHOLE 1953	1	STOMACH WM 1950-1969	0.155	8.72	0
UPSHOT-KNOTHOLE 1953	2	LEUKEMIA WM 1950-1969	0.136	7.62	0
UPSHOT-KNOTHOLE 1953	3	STOMACH WF 1950-1969	0.126	7.06	0
UPSHOT-KNOTHOLE 1953	4	LIVER WF 1950-1969	0.105	5.88	0
UPSHOT-KNOTHOLE 1953	5	COLON WF 1950-1969	0.095	5.31	0
UPSHOT-KNOTHOLE 1953	6	COLON WF1970-1995	0.095	5.28	0
UPSHOT-KNOTHOLE 1953	7	COLON WM 1950-1969	0.093	5.17	0
TEAPOT 1955	1	LIVER WF 1950-1969	0.103	5.77	0
PLUMBBOB 1957	1	LEUKEMIA WM 1950-1969	0.076	4.22	0.000025
PLUMBBOB 1957	2	LIVER WF 1950-1969	0.003	0.18	0.857883
SEDAN July, 1962	1	COLON WF1970-1995	0.254	14.61	0
SEDAN July, 1962	2	COLON WF 1950-1969	0.217	12.33	0
SEDAN July, 1962	3	COLON WM1970-1995	0.201	11.43	0
SEDAN July, 1962	4	ALL CANCERS WF 1950-1969	0.192	10.89	0
SEDAN July, 1962	5	LIVER WF 1950-1969	0.190	10.73	0
SEDAN July, 1962	6	RECTUM WM 1950-1969	0.184	10.42	0
SEDAN July, 1962	7	COLON WM 1950-1969	0.172	9.71	0
SEDAN July, 1962	8	RECTUM WM1970-1995	0.165	9.29	0
SEDAN July, 1962	9	RECTUM WF1970-1995	0.144	8.07	0
SEDAN July, 1962	10	BREAST WF 1950-1969	0.139	7.81	0
SEDAN July, 1962	11	BREAST WF1970-1995	0.138	7.77	0
SEDAN July, 1962	12	RECTUM WF 1950-1969	0.126	7.06	0
SEDAN July, 1962	13	OVARIAN WF 1950-1969	0.114	6.39	0
SEDAN July, 1962	14	KIDNEY WF1970-1995	0.112	6.24	0

TABLE G-2
HIGHLY CORRELATED VARIABLES
TOTAL FALLOUT DURING NUCLEAR TESTING AND CANCER RATES 1951-1962
Spearman Correlations with T Score >5.0
TOTAL FALLOUT: Mean of Lognormal Distribution; N: 3094

TOTAL FALLOUT DURING: SHOT SERIES	RANK	VARIABLES	R	t(N-2)	p-level
TOTAL FALLOUT 1951-1962	1	COLON WF 1950-1969	0.166	9.36	0
TOTAL FALLOUT 1951-1962	2	COLON WF 1970-1995	0.164	9.26	0
TOTAL FALLOUT 1951-1962	3	LIVER WF 1950-1969	0.155	8.70	0
TOTAL FALLOUT 1951-1962	4	STOMACH WM 1950-1969	0.139	7.79	0
TOTAL FALLOUT 1951-1962	5	BREAST WF 1950-1969	0.132	7.42	0
TOTAL FALLOUT 1951-1962	6	COLON WM 1950-1969	0.128	7.16	0
TOTAL FALLOUT 1951-1962	7	COLON WM 1970-1995	0.122	6.83	0
TOTAL FALLOUT 1951-1962	8	ALL CANCERS WF 1950-1969	0.116	6.48	0
TOTAL FALLOUT 1951-1962	9	STOMACH WF 1950-1969	0.107	5.97	0
TOTAL FALLOUT 1951-1962	10	RECTUM WM 1950-1969	0.097	5.43	0

TABLE G-3
FALLOUT COMPONENTS
HIGHLY CORRELATED WITH CANCER RATES

RANK	FALLOUT COMPONENT	SUM RANK	AV Rank	SUM T SCORES	MAX T-SCORE
1	RANGER TOTAL FALLOUT	2513	84	194.28	16.77
2	W181	1599	53	189.81	13.12
3	Au198	1880	63	189.34	16.26
4	W185	1650	55	185.21	12.60
5	Au199	2005	67	181.71	15.78
6	RANGER In PRECIP	2746	92	177.99	16.14
7	UPSHOT-KNOTHOLE PRECIP	1385	46	176.94	11.65
8	Cu64	1863	62	152.06	12.13
9	SEDAN TOTAL FALLOUT	2222	74	151.27	14.61
10	TUMBLER-SNAPPER TOTAL FALLOUT	2115	71	148.26	10.96
11	SEDAN In PRECIP	2396	80	142.62	14.94
12	Be7	2399	80	134.97	14.02
13	Ag109	2427	81	127.59	15.18
14	Eu155	2567	86	123.26	11.42
15	Ag112	2427	81	120.87	13.64
16	I130	2547	85	120.19	15.65
17	I133	2653	88	113.23	14.30
18	Pd112	2660	89	111.43	13.16
19	Tb161	2932	98	110.17	7.95
20	Sm153	2644	88	108.28	11.08
21	Br82	2742	91	105.50	12.50
22	I135	2861	95	103.96	15.75
23	U240	2863	95	100.54	14.09
24	U237	2670	89	99.78	8.37
25	Nb95	2773	92	96.36	7.12
26	Na24	2927	98	95.71	9.66
27	UPSHOT KNOTHOLE TOTAL FALLOUT	2774	92	95.24	8.72
28	Tc99m	2980	99	92.96	9.56
29	Mo99	3026	101	92.17	9.66
30	I132	2994	100	91.84	9.84
31	IN115m	3179	106	87.08	9.30
32	Cd115	3262	109	84.94	9.15
33	I131	3168	106	83.99	7.79
34	Pd111m	3235	108	83.67	14.93

TABLE G-3
FALLOUT COMPONENTS
HIGHLY CORRELATED WITH CANCER RATES

RANK	FALLOUT COMPONENT	SUM RANK	AV Rank	SUM T SCORES	MAX T-SCORE
35	Sb126	3304	110	80.35	6.65
36	Cs137	3285	110	79.47	6.47
37	Ba137m	3291	110	79.46	6.47
38	Ru106, Rh106	3330	111	79.07	5.99
39	Te127m	3362	112	78.81	6.14
40	Np239	3370	112	77.89	6.14
41	Sr89	3411	114	76.11	6.79
42	Y91	3406	114	76.07	6.59
43	Sr90, Y90	3412	114	75.92	6.77
44	Zr95	3412	114	75.33	5.70
45	Ce144, Pr144	3444	115	74.87	5.66
46	NB95m	3856	129	63.62	5.57
47	La141	3682	123	54.20	13.27
48	TUMBLER-SNAPPER PRECIP	4135	138	49.38	9.22
49	Cm242	3982	133	32.92	6.53

TABLE G-4
HIGHLY CORRELATED VARIABLES
ALL PROBABILITY LEVELS: <0.0001

1	RANGER TOTAL FALLOUT	COLON CANCER WM 1970-94	0.289	16.77
2	RANGER TOTAL FALLOUT	RECTAL CANCER WM 1950-69	0.285	16.52
3	Au198	COLON CANCER WF 1970-94	0.281	16.26
4	RANGER PRECIP	COLON CANCER WM 1970-94	0.279	16.14
5	RANGER TOTAL FALLOUT	COLON CANCER WF 1970-94	0.275	15.90
6	RANGER TOTAL FALLOT	RECTAL CANCER WF 1950-69	0.273	15.79
7	Au199	COLON CANCER WF 1970-94	0.273	15.78
8	I135	STOMACH CANCER WM 1950-69	0.273	15.75
9	I130	STOMACH CANCER WM 1950-69	0.271	15.65
10	Ag109	STOMACH CANCER WM 1950-69	0.263	15.18
11	RANGER TOTAL FALLOUT	COLON CANCER WM 1950-69	0.262	15.09
12	RANGER PRECIP	RECTAL CANCER WM 1950-69	0.262	15.09
13	RANGER TOTAL FALLOUT	COLON CANCER WF 1950-69	0.261	15.04
14	SEDAN PRECIP	COLON CANCER WF 1970-94	0.259	14.94
15	Pd111m	STOMACH CANCER WM 1950-69	0.259	14.93
16	RANGER TOTAL FALLOUT	RECTAL CANCER WM 1970-94	0.259	14.89
17	RANGER PRECIP	RECTAL CANCER WF 1950-69	0.255	14.68
18	RANGER PRECIP	COLON CANCER WF 1970-94	0.254	14.63
19	SEDAN TOTAL FALLOUT	COLON CANCER WF 1970-94	0.254	14.61
20	Au198	COLON CANCER WF 1950-69	0.250	14.37
21	I133	STOMACH CANCER WM 1950-69	0.249	14.30
22	RANGER TOTAL FALLOUT	RECTAL CANCER WF 1970-94	0.247	14.19
23	U240	STOMACH CANCER WM 1950-69	0.246	14.09
24	Be7	COLON CANCER WF 1970-94	0.244	14.02
25	RANGER PRECIP	COLON CANCER WM 1950-69	0.244	14.01
26	RANGER PRECIP	RECTAL CANCER WM 1970-94	0.244	14.00
27	RANGER PRECIP	COLON CANCER WF 1950-69	0.243	13.91
28	RANGER TOTAL FALLOUT	BREASTL CANCER WF 1970-94	0.242	13.84
29	Au199	COLON CANCER WF 1950-69	0.241	13.83
30	RANGER PRECIP	RECTAL CANCER WF 1970-94	0.239	13.68
31	Ag112	STOMACH CANCER WM 1950-69	0.238	13.64
32	La141	STOMACH CANCER WM 1950-69	0.232	13.27
33	RANGER PRECIP	BREASTL CANCER WF 1970-94	0.232	13.27
34	Pd112	STOMACH CANCER WM 1950-69	0.230	13.16
35	W181	RECTAL CANCER WM 1950-69	0.230	13.12
36	Au198	COLON CANCER WM 1970-94	0.228	12.99

TABLE G-4
HIGHLY CORRELATED VARIABLES
ALL PROBABILITY LEVELS: <0.0001

37	Au198	RECTAL CANCER WM 1950-69	0.227	12.97
38	Au199	COLON CANCER WM 1970-94	0.221	12.60
39	W185	RECTAL CANCER WM 1950-69	0.221	12.60
40	Br82	STOMACH CANCER WM 1950-69	0.219	12.50
41	Au199	RECTAL CANCER WM 1950-69	0.219	12.45
42	SEDAN TOTAL FALLOUT	COLON CANCER WF 1950-69	0.217	12.33
43	Cu64	STOMACH CANCER WM 1950-69	0.213	12.13
44	RANGER TOTAL FALLOUT	BREAST CANCER WF 1950-69	0.209	11.89
45	SEDAN LN PRECIP	COLON CANCER WM 1970-94	0.209	11.85
46	W185	COLON CANCER WF 1970-94	0.208	11.82
47	SEDAN LN PRECIP	COLON CANCER WF 1950-69	0.208	11.81
48	W181	BREAST CANCER WF 1950-69	0.206	11.69
49	UPSHOT-KNOTHOLE PRECIPITATION	COLON CANCER WM 1970-94	0.205	11.65
50	Au198	RECTAL CANCER WM 1970-94	0.203	11.52
51	W181	COLON CANCER WF 1970-94	0.202	11.49
52	SEDAN TOTAL FALLOUT	COLON CANCER WM 1970-94	0.201	11.43
53	Eu155	COLON CANCER WF 1970-94	0.201	11.42
54	Au198	BREAST CANCER WF 1950-69	0.200	11.34
55	W185	COLON CANCER WF 1950-69	0.200	11.33
56	W181	COLON CANCER WF 1950-69	0.199	11.32
57	Au199	RECTAL CANCER WM 1970-94	0.198	11.22
58	W185	BREAST CANCER WF 1950-69	0.198	11.20
59	Be7	COLON CANCER WF 1950-69	0.197	11.17
60	RANGER PRECIP	BREAST CANCER WF 1950-69	0.197	11.15
61	Sm153	STOMACH CANCER WM 1950-69	0.195	11.08
62	UPSHOT-KNOTHOLE PRECIP	COLON CANCER WM 1950-69	0.194	10.97
63	TUMBLER-SNAPPER TOTAL FALLOUT	BREAST CANCER WF 1950-69	0.193	10.96
64	Au198	BREASTL CANCER WF 1970-94	0.193	10.92
65	Au199	BREAST CANCER WF 1950-69	0.193	10.92
66	UPSHOT-KNOTHOLE PRECIPITATION	COLON CANCER WF 1970-94	0.192	10.90
67	Be7	COLON CANCER WM 1970-94	0.190	10.77
68	In117	STOMACH CANCER WM 1950-69	0.189	10.67
69	Au199	BREASTL CANCER WF 1970-94	0.187	10.58
70	W181	COLON CANCER WM 1970-94	0.187	10.57
71	W181	RECTAL CANCER WM 1970-94	0.186	10.53

TABLE G-4
HIGHLY CORRELATED VARIABLES
ALL PROBABILITY LEVELS: <0.0001

72	W185	COLON CANCER WM 1970-94	0.186	10.50
73	W181	BREASTL CANCER WF 1970-94	0.184	10.44
74	SEDAN TOTAL FALLOUT	RECTAL CANCER WM 1950-69	0.184	10.42
75	TUMBLER-SNAPPER TOTAL FALLOUT	COLON CANCER WM 1950-69	0.184	10.42
76	W185	RECTAL CANCER WM 1970-94	0.181	10.24
77	SEDAN PRECIP	RECTAL CANCER WM 1950-69	0.181	10.21
78	BR83	STOMACH CANCER WM 1950-69	0.180	10.19
79	Au198	RECTAL CANCER WF 1950-69	0.179	10.10
80	W185	BREASTL CANCER WF 1970-94	0.178	10.05
81	UPSHOT-KNOTHOLE PRECIPITATION	COLON CANCER WF 1950-69	0.178	10.05

TABLE G-5
HIGHLY CORRELATED FALLOUT VARIABLES:
WHITE FEMALE BREAST CANCER

FALLOUT COMPONENT	TIME PERIOD	R	t(N-2)	p-level
RANGER TOTAL FALLOUT	1950-69	0.209	11.89	0
W181	1950-69	0.206	11.69	0
Au198	1950-69	0.200	11.34	0
W185	1950-69	0.198	11.20	0
RANGER PRECIP	1950-69	0.197	11.15	0
TUMBLER-SNAPPER TOTAL FALLOUT	1950-69	0.193	10.96	0
Au199	1950-69	0.193	10.92	0
Cu64	1950-69	0.159	8.98	0
Ag109	1950-69	0.142	7.97	0
SEDAN TOTAL FALLOUT	1950-69	0.139	7.81	0
I130	1950-69	0.138	7.75	0
Tb161	1950-69	0.134	7.52	0
Be7	1950-69	0.131	7.34	0
UPSHOT-KNOTHOLE PRECIPITATION	1950-69	0.128	7.17	0
I133	1950-69	0.128	7.17	0
Ag112	1950-69	0.128	7.15	0
SEDAN PRECIP	1950-69	0.124	6.96	0
I135	1950-69	0.124	6.95	0
Eu155	1950-69	0.120	6.72	0
Sm153	1950-69	0.119	6.65	0
Pd112	1950-69	0.118	6.58	0
Nb95	1950-69	0.114	6.39	0
Br82	1950-69	0.114	6.37	0
U240	1950-69	0.110	6.15	0
Tc99m	1950-69	0.103	5.77	0
Mo99	1950-69	0.102	5.72	0
Pd111m	1950-69	0.098	5.47	0
I132	1950-69	0.095	5.30	0
RANGER TOTAL FALLOUT	1970-94	0.242	13.84	0
RANGER PRECIP	1970-94	0.232	13.27	0
Au198	1970-94	0.193	10.92	0
Au199	1970-94	0.187	10.58	0
W181	1970-94	0.184	10.44	0

TABLE G-5
HIGHLY CORRELATED FALLOUT VARIABLES:
WHITE FEMALE BREAST CANCER

FALLOUT COMPONENT	TIME PERIOD	R	t(N-2)	p-level
W185	1970-94	0.178	10.05	0
SEDAN TOTAL FALLOUT	1970-94	0.138	7.77	0
SEDAN PRECIP	1970-94	0.134	7.49	0
Tb61	1970-94	0.125	6.98	0
Be7	1970-94	0.119	6.69	0
TUMBLER-SNAPPER TOTAL FALLOUT	1970-94	0.114	6.39	0
Eu55	1970-94	0.098	5.45	0
TUMBLER-SNAPPER PRECIP	1970-94	0.093	5.22	0
UPSHOT-KNOTHOLE PRECIP	1970-94	0.093	5.18	0

TABLE G-6
HIGH TOTAL FALLOUT CORRELATIONS WITH COLON CANCER RATES

RADIONUCLIDE	GENDER AND YEAR	R	T(N-2)	p level
Au198	WF 1970-94	0.281	16.26	0
Au198	WF 1950-69	0.250	14.37	0
Au198	WM 1970-94	0.228	12.99	0
Au198	WM 1950-69	0.177	9.99	0
Au199	WF 1970-94	0.273	15.78	0
Au199	WF 1950-69	0.241	13.83	0
Au199	WM 1970-94	0.221	12.60	0
Au199	WM 1950-69	0.169	9.52	0
Be7	WF 1970-94	0.244	14.02	0
Be7	WF 1950-69	0.197	11.17	0
Be7	WM 1970-94	0.190	10.77	0
Be7	WM 1950-69	0.121	6.80	0
Co60	WF 1970-94	0.137	7.68	0
Eu155	WF 1970-94	0.201	11.42	0
Eu155	WF 1950-69	0.172	9.74	0
Eu155	WM 1970-94	0.156	8.77	0
Eu155	WM 1950-69	0.135	7.56	0
SEDAN PRECIP	WF 1970-94	0.259	14.94	0
SEDAN PRECIP	WM 1970-94	0.209	11.85	0
SEDAN PRECIP	WF 1950-69	0.208	11.81	0
SEDAN PRECIP	WM 1950-69	0.162	9.12	0
SEDAN TOTAL FALLOUT	WF 1970-94	0.254	14.61	0
SEDAN TOTAL FALLOUT	WF 1950-69	0.217	12.33	0
SEDAN TOTAL FALLOUT	WM 1970-94	0.201	11.43	0
SEDAN TOTAL FALLOUT	WM 1950-69	0.172	9.71	0

TABLE G-7
POISSON REGRESSION: RELATIVE RATES
TOTAL FALLOUT AND CANCER

SEVEN EQUAL INTERVALS 513 COUNTIES IN IA, IL, KS, MO, NE 1951-1962
*SEE TABLE G-9 FOR KEY TO CANCER PARAMETER ABBREVIATIONS

RANK	CANCER* PARAMETER	COEFF	STD. ERROR	df	DEV VALUE	RATE RATIO	LCI	UCI	Z SCORE	PROB DUE TO CHANCE
1	FLS95	0.28080	0.01296	511	4.86E+03	1.32	1.29	1.36	7.03	0.00000
2	FCLN96	0.07805	0.00290	511	1.39E+04	1.08	1.08	1.09	5.16	0.00000
3	FBRAIN96	0.14680	0.00627	511	1.07E+04	1.16	1.14	1.17	5.12	0.00000
4	FBRST60	0.05684	0.00244	511	2.09E+04	1.06	1.05	1.06	3.65	0.00026
5	MEYE96	0.30350	0.08150	511	5.49E+02	1.36	1.16	1.59	3.59	0.00033
6	BMHST96	0.29070	0.06910	511	7.52E+02	1.34	1.17	1.53	3.47	0.00054
7	BNOD96	0.17630	0.03188	511	1.66E+03	1.19	1.12	1.27	3.07	0.00214
8	FCLN70	0.04183	0.00298	511	1.17E+04	1.04	1.04	1.05	2.94	0.00328
9	MMM60	0.09186	0.00793	511	9.54E+03	1.10	1.08	1.11	2.68	0.00736
10	MCONSAR96	0.21930	0.05654	511	1.15E+03	1.25	1.12	1.39	2.59	0.00960
11	MCLN60	0.03743	0.00304	511	1.18E+04	1.04	1.03	1.04	2.56	0.01047
12	MBONE96	0.11370	0.01497	511	4.84E+03	1.12	1.09	1.15	2.47	0.01351
13	MLIV96	0.07934	0.00975	511	5.56E+03	1.08	1.06	1.10	2.47	0.01351
14	FBRST96	0.02020	0.00227	511	6.78E+03	1.02	1.02	1.03	2.44	0.01469
15	FLS70	0.05841	0.00584	511	9.56E+03	1.06	1.05	1.07	2.31	0.02089
16	UT96	0.06237	0.00761	511	6.57E+03	1.06	1.05	1.08	2.29	0.02202
17	FBRST50	0.03051	0.00254	511	1.47E+04	1.03	1.03	1.04	2.24	0.02509
18	MSTOM60	0.04274	0.00365	511	1.46E+04	1.04	1.04	1.05	2.19	0.02852
19	MALL50	0.01308	0.00101	511	1.82E+04	1.01	1.01	1.02	2.17	0.03001
20	WFLIV96	0.07871	0.01175	511	5.05E+03	1.08	1.06	1.11	2.13	0.03317
21	BLCL96	0.16650	0.06323	511	8.05E+02	1.18	1.04	1.34	2.10	0.03573
22	FLS50	0.06464	0.00737	511	9.80E+03	1.07	1.05	1.08	2.00	0.04550
23	MCLN50	0.03032	0.00317	511	1.23E+04	1.03	1.02	1.04	1.95	0.05118
24	MLS60	0.04638	0.00521	511	1.08E+04	1.05	1.04	1.06	1.93	0.05361
25	MBRAIN50	0.05342	0.00625	511	1.03E+04	1.06	1.04	1.07	1.90	0.05743
26	MBRAIN50	0.05342	0.00625	511	1.03E+04	1.06	1.04	1.07	1.90	0.05743
27	MBRAIN96	0.03627	0.00526	511	7.13E+03	1.04	1.03	1.05	1.85	0.06431
28	FLS60	0.05060	0.00619	511	1.05E+04	1.05	1.04	1.07	1.81	0.07030
29	FLS60	0.05060	0.00619	511	1.05E+04	1.05	1.04	1.07	1.81	0.07030
30	FSTOM96	0.03978	0.00719	511	4.85E+03	1.04	1.03	1.06	1.80	0.07186

TABLE G-7
POISSON REGRESSION: RELATIVE RATES
TOTAL FALLOUT AND CANCER

SEVEN EQUAL INTERVALS 513 COUNTIES IN IA, IL, KS, MO, NE 1951-1962
*SEE TABLE G-9 FOR KEY TO CANCER PARAMETER ABBREVIATIONS

RANK	CANCER PARAMETER*	COEFF	STD. ERROR	df	DEV VALUE	RATE RATIO	LCI	UCI	Z SCORE	PROB DUE TO CHANCE
31	MLS70	0.03961	0.00495	511	1.02E+04	1.04	1.03	1.05	1.79	0.07345
32	FKDNY96	0.03763	0.00698	511	5.01E+03	1.04	1.02	1.05	1.72	0.08543
33	FCONSAR96	0.15150	0.07488	511	7.54E+02	1.16	1.01	1.35	1.67	0.09492
34	MCLN96	0.01661	0.00277	511	6.71E+03	1.02	1.01	1.02	1.65	0.09894
35	FSTOM60	0.03870	0.00502	511	1.12E+04	1.04	1.03	1.05	1.65	0.09894
36	FALL60	0.00957	0.00113	511	1.38E+04	1.01	1.01	1.01	1.63	0.10310
37	FMM50	0.07680	0.01269	511	7.19E+03	1.08	1.05	1.11	1.61	0.10740
38	FCLN60	0.02317	0.00302	511	1.19E+04	1.02	1.02	1.03	1.59	0.11184

TABLE G-8 TOXICITY OF SELECTED RADIONUCLIDES

Radionuclide	Half-Life	RADIATION (SEE NOTES)	TARGET ORGAN[1]	TARGET ORGANS AND MORTALITY AND MORBIDITY RISK		
				AIR	WATER	FOOD
Am241	432.2 years	α(5.637 MeV); e^-; γ		No single cancer type accounts for more than 40% of the total cancer mortality. Lung	No single cancer type accounts for more than 40% of the total cancer mortality.	No single cancer type accounts for more than 40% of the total cancer mortality.
Au198	2.694 days	β^-(0.812 MeV); e^-; γ	Soluble: GI (Lower Large Intestine) Insoluble: GI (Lower Large Intestine)	-------	-------	-------
Au199	3.14 days	β^-(0.453 MeV); e^-; γ	Soluble: GI (Lower Large Intestine) Insoluble: GI (Lower Large Intestine)	-------	-------	
Be7	53.82 days	EC (Electron Capture: 0.862 MeV); γ	Soluble: GI.: Lower Large Intestine, Total Body; Insoluble: Lung	-------	-------	
Br82	1.471 days	β^-(3.093 MeV); γ				
Cd115m	44.6 days	β^-(1.629 MeV); γ	Soluble: GI (Lower Large Intestine, Liver, Kidney; Insoluble: Lung, GI (Lower Large			

TABLE G-8 TOXICITY OF SELECTED RADIONUCLIDES

Radionuclide	Half-Life	RADIATION (SEE NOTES)	TARGET ORGAN[1]	TARGET ORGANS AND MORTALITY AND MORBIDITY RISK		
				AIR	WATER	FOOD
Ce144	284.6 days	β^-(0.310 MeV max), e-, γ	Soluble: GI (Lower Large Intestine), Bone, Liver; Insoluble: Lung, GI (Lower Large Intestine)	No single cancer type accounts for more than 40% of the total cancer mortality.; Lung	Colon	Colon
Cm242	162.8 days	α(6.126 MeV 74%; 6.07 MeV 25%)); e-; γ Pu X-rays. (NOTE: Parent: Pu238; daughters: Am242 and Cf246)	Soluble: GI (Lower Large Intestine), Liver Insoluble: Lung, GI (Lower Large Intestine)	Liver, Lung	Colon	Colon
Co60	5.27 years	β^-(1.322 MeV); γ	Soluble: GI (Lower Large Intestine); Total Body; Insoluble: Lung, GI (Lower Large Intestine).	Lung	No single cancer type accounts for more than 40% of the total cancer mortality.	No single cancer type accounts for more than 40% of the total cancer mortality.
Cs137	30.3 years	β^-(1.175 MeV); γ	Soluble: Total Body, Liver, Spleen, Muscle; Insoluble: Lung, GI (Lower Large Intestine)	Lung, No single cancer type accounts for more than 40% of the total cancer mortality.	———	No single cancer type accounts for more than 40% of the total cancer mortality.

TABLE G-8 TOXICITY OF SELECTED RADIONUCLIDES

Radionuclide	Half-Life	RADIATION (SEE NOTES)	TARGET ORGAN[1]	TARGET ORGANS AND MORTALITY AND MORBITITY RISK		
				AIR	WATER	FOOD
Eu155	4.71 years	β^-(1.969 MeV) e$^-$; γ	Soluble: GI (Lower Large Intestine), Kidney, bone; Insoluble: Lung: GI (Lower Large Intestine)			
Fe59	44.51 days	β^-(1.565 MeV); γ	Soluble: GI (Lower Large Intestine), Spleen; Insoluble: Lung, GI (Lower Large Intestine)	Lung	Colon	
I-131	8.04 days	β^-(0.606 Mev average); e$^-$; γ	Soluble: Thyroid; Insoluble: GI (Lower Large Intestine), Lung	Lung, Thyroid	Thyroid	No single cancer type accounts for more than 40% of the total cancer mortality.
I-133	20.8 hours	β^-(1.27 MeV max); γ	Soluble: Thyroid; Insoluble: GI (Lower Large Intestine), Lung	Thyroid, Lung	Thyroid	Thyroid
Mn54	312.2 days	EC (Electron Capture) (1.377 MeV); e$^-$; γ	Soluble: GI (Lower Large Intestine), Liver; Insol. Lung, GI	------	------	------
Mo99	2.748 days	β^-(1.357 MeV); γ	Soluble: Kidney, GI (Lower Large Intestine);		No single cancer type accounts for more than 40% of	No single cancer type accounts for more than 40% of

TABLE G-8 TOXICITY OF SELECTED RADIONUCLIDES

Radionuclide	Half-Life	RADIATION (SEE NOTES)	TARGET ORGAN[1]	TARGET ORGANS AND MORTALITY AND MORBITITY RISK		
				AIR	WATER	FOOD
			Insoluble: GI (Lower Large Intestine)		the total cancer mortality.	the total cancer mortality.
Na24	14.96 years	β^-(5.514 MeV); γ	Soluble: GI (Small Intestine); Insoluble: GI: (Lower Large Intestine)			
Nb95	34.97 days	β^-(0.926 MeV); γ	Soluble: GI (Lower Large Intestine), Total Body; Insoluble: Lung, GI (Lower Large Intestine)	----	Colon	Colon
Nb95m	90.0 hours	IT; e$^-$; γ	----	----	Colon	Colon
Np239	2,140,000 years	α (4.957 MeV); e$^-$; γ	Soluble: GI (Lower Large Intestine) Insoluble: GI (Lower Large Intestine)	Colon, Lung	Colon	Colon
Pb203	2.162 days	EC (Electron Capture: 0.97 MeV); e$^-$; γ	Soluble: GI (Lower Large Intestine) Insoluble: GI (Lower Large Intestine)	----	----	----
Ru106	1.02 years	β^-0.039 MeV); **no γ**	Soluble: GI (Lower Large Intestine);		Colon	Colon

TABLE G-8 TOXICITY OF SELECTED RADIONUCLIDES

Radionuclide	Half-Life	RADIATION (SEE NOTES)	TARGET ORGAN[1]	TARGET ORGANS AND MORTALITY AND MORBITITY RISK		
				AIR	WATER	FOOD
Sb126	12.4 days	β⁻(3.67 MeV); γ	Insoluble: Lung, GI, (Lower Large Intestine)		Colon	
Sr89	50.52 days	β⁻(1.492 MeV); γ	Soluble: Bone Insoluble: Lung, GI (Lower Large Intestine)	-------	Colon	Colon
Sr90	29.1 years	β⁻(0.546 Mev) no γ	Soluble: Bone Insoluble: GI (Lower Large Intestine)	-------	Leukemia	Leukemia
Tb161	6.91 days	β⁻(0.593 MeV); e⁻; γ	Soluble: GI (Lower Large Intestine), Bone, Kidney, Total Body; Insoluble: Lung, GI (Lower Large Intestine)	-------	-------	-------

TABLE G-8 TOXICITY OF SELECTED RADIONUCLIDES

Radionuclide	Half-Life	RADIATION (SEE NOTES)	TARGET ORGAN[1]	TARGET ORGANS AND MORTALITY AND MORBITITY RISK		
				AIR	WATER	FOOD
Te127m	109 days	IT(0.088 Mev 98%)β⁺(0.770 MeV 2%); e⁻; γ	Soluble: Kidney, Testis, GI (Lower Large Intestine); Insoluble: Lung, GI, (Lower Large Intestine)	Colon, Lungs	Colon	Colon
U237	6.75 days	β⁻(0.519 MeV); e⁻; γ		------	------	------
U240	14.1 hours	β⁻(0.360 MeV max) e⁻;γ		------	------	------
W185	74.8 days	β⁻(1.312 MeV); no γ"	Soluble: GI (Lower Large Intestine); Insoluble: GI (Lower Large Intestine)	------	------	------
Y90	2.67 days	β⁻(2.283 MeV) no γ	Soluble: GI (Lower Large Intestine) Insoluble: GI (Lower Large Intestine)		Colon	

U.S. FALLOUT ATLAS : TOTAL FALLOUT

TABLE G-8 TOXICITY OF SELECTED RADIONUCLIDES

Radionuclide	Half-Life	RADIATION (SEE NOTES)	TARGET ORGAN[1]	TARGET ORGANS AND MORTALITY AND MORBIDITY RISK		
				AIR	WATER	FOOD
Y91	58.5 days	β^-(1.545 MeV); γ	Soluble: GI (Lower Large Intestine), Bone Insoluble: Lung, GI (Lower Large Intestine)	------	------	------
Zr95	64.02 days	β^-(1.125 MeV); γ	------	------	Colon	Colon

REFERENCES:

1. Derived Concentration Limits of Radionuclides in Air and Water for Occupational Exposure (40 h/week). Table A1:lia. P. 59. Safe Handling of Radionuclides 1973 Edition. International Atomic Energy Agency, Vienna, 1973.
2. Safe Handling of Radionuclides 1973 Edition. Code of Practice : Sponsored by the International Atomic Energy Agency and the World Health Organization. International Atomic Energy Agency, Vienna, 1973. Table Al-IIa.
3. Federal Guidance Report No. 13, Part I Interim Version. Health Risks from Low-Level Environmental Exposure to Radionuclides. Radionuclide-Specific Lifetime Ragiogenic Cancer Risk Coefficients for the U.S. Population, Based on Age-Dependent Intake, Dosimetry, and Risk Models. Table2.1, 2.2, 2.3. Oak Ridge National Laboratory, Office of Radiation and Indoor Air, United States Environmental Protection Agency, Washington, DC. 20460. EPA 402-R-97-014. Jan, 1998.[1] Radiological Health Handbook. Ibid. p.336.
4. Radiological Health Handbook: US Department of Health, Education and Welfare. January, 1970.P.275

TABLE G-9
INTERNATIONAL CLASSIFICATION OF DISEASES CODES USED IN U.S. ATLAS OF NUCLEAR FALLOUT 1951-1962

ABBREV	VARIABLE	GENDER	DATES	ICD	ICD PUB. No.	DATA-BASE*	POPULATION EVALUATED**
ALLWF5	All Cancer	White Female	1950-69	140-208	6,7,8	NCI 2000	US
ALLWF7	All Cancers	White Female	1950-69	140-208 except 202.2-3, 202.5-.6	6,7,8	NCI 2000	US
ALLWM5	All Cancers	White Male	1970-95	140-208	8,9	NCI 2000	US
ALLWM7	All Cancers	White Male	1970-95	140-208 except 202.2-3, 202.5-.6	8,9	NCI 2000	US
BLCL96	Lymphosarcoma Cell Leukemia	White Female, Male	1979-96	207.8-207.8	9	CDC	MW
BLWF50	Bladder Cancer	White Female	1950-69	181	6,7,8	NCI 2000	US
BLWF70	Bladder Cancer	White Female	1970-95	188, 189.3-189.9	8,9	NCI2000	US
BLWM50	Bladder Cancer	White Male	1950-69	181	6,7,8	NCI 2000	US
BLWM70	Bladder Cancer	White Male	1970-95	188, 189.3-189.9	8,9	NCI 2000	US
BMHST96	Malignant Histiocytosis	White Female, Male	1979-96	202.3-202.3	9	CDC	MW
BNOD96	Nodular Lymphoma	White Female, Male	1979-96	202.0-202.0	9	CDC	MW
BOTHLR96	Leukemic Reticuloendotheliosis	White Female, Male	1979-96	202.4-202.4	9	CDC	MW
BRNWF5	Brain Cancer	White Female	1950-69	193	6,7,8	NCI 2000	US
BRNWF7	Brain Cancer	White Female	1970-95	191, 192	8,9	NCI 2000	US
BRNWM5	Brain Cancer	White Male	1950-69	193	6,7,8	NCI2000	US
BRNWM7	Brain Cancer	White Male	1970-95	191, 192	8,9	NCI2000	US
BSTWF5	Breast Cancer	White Female	1950-69	170	6,7,8	NCI2000	US
BSTWF7	Breast Cancer	White Female	1970-95	174, 175	8,9	NCI2000	US
COLWF5	Colon Cancer	White Female	1950-69	153	6,7,8	NCI2000	US
COLWF7	Colon Cancer	White Female	1970-95	153, 159.0	8,9	NCI2000	US

U.S. FALLOUT ATLAS : TOTAL FALLOUT

TABLE G-9
INTERNATIONAL CLASSIFICATION OF DISEASES CODES USED IN U.S. ATLAS OF NUCLEAR FALLOUT 1951-1962

ABBREV	VARIABLE	GENDER	DATES	ICD	ICD PUB. No.	DATA-BASE*	POPULATION EVALUATED**
COLWM5	Colon Cancer	White Male	1950-69	153	6,7,8	NCI2000	US
COLWM7	Colon Cancer	White Male	1970-95	153, 159.0	8,9	NCI2000	US
CONGEN	Congenital Anomalies	White Female, Male	1979-95	740.0-759.9	9	CDC	MW
CRVWF5	Cervical Cancer	White Female	1950-69	171, 180	6,7,8	NCI2000	US
CRVWF7	Cervical Cancer	White Female	1970-95	180	8,9	NCI2000	US
ESOWF5	Cancer of Esophagus	White Female	1950-69	150	6,7,8	NCI2000	US
ESOWF7	Cancer of Esophagus	White Female	1970-95	150	8,9	NCI2000	US
ESOWM5	Cancer of Esophagus	White Male	1950-69	150	6,7,8	NCI2000	US
ESOWM7	Cancer of Esophagus	White Male	1970-95	150	8,9	NCI2000	US
FALL50	All Cancer	White Female	1950-59	140-208	9	NCI1983	MW
FALL60	All Cancer	White Female	1960-69	140-208	9	NCI1983	MW
FALL70	All Cancer	White Female	1970-79	140-208	9	NCI1983	MW
FALL96	All Cancers	White Female	1979-96	140-208.	9	CDC	MW
FBONE50	Bone Cancer	White Female	1950-59	170	9	NCI1983	MW
FBONE60	Bone Cancer	White Female	1960-69	170	9	NCI1983	MW
FBONE70	Bone Cancer	White Female	1970-79	170	9	NCI1983	MW
FBONE96	Bone Cancer	White Female	1979-96	170	9	CDC	MW
FBRAIN50	Brain and other Parts of Nervous System Cancer	White Female	1950-59	191-192	9	NCI1983	MW
FBRAIN60	Brain and other Parts of Nervous System Cancer	White Female	1960-69	191-192	9	NCI1983	MW

U.S. FALLOUT ATLAS : TOTAL FALLOUT

TABLE G-9
INTERNATIONAL CLASSIFICATION OF DISEASES CODES USED IN U.S. ATLAS OF NUCLEAR FALLOUT 1951-1962

ABBREV	VARIABLE	GENDER	DATES	ICD	ICD PUB. No.	DATA-BASE*	POPULATION EVALUATED**
FBRAIN70	Brain and other Parts of Nervous System Cancer	White Female	1970-79	191-192	9	NCI1983	MW
FBRAIN96	Brain Cancer	White Female	1979-96	191.0-191.9	9	CDC	MW
FBRST50	Breast Cancer	White Female	1950-59	174-175	9	NCI1983	MW
FBRST60	Breast Cancer	White Female	1960-69	174-175	9	NCI1983	MW
FBRST70	Breast Cancer	White Female	1970-79	174-175	9	NCI1983	MW
FBRST96	Breast Cancer	White Female	1979-96	174.0-174.9	9	CDC	MW
FCLN50	Colon Cancer	White Female	1950-59	153, 159	9	NCI1983	MW
FCLN60	Colon Cancer	White Female	1960-69	153, 159	9	NCI1983	MW
FCLN70	Colon Cancer	White Female	1970-79	153, 159	9	NCI1983	MW
FCLN96	Colon Cancer	White Female	1979-96	153, 159	9	CDC	MW
FCONSAR9	Connective and Soft Tissue Neoplasms	White Female	1979-96	171.0-171.0	9	CDC	MW
FKDNY50	Kidney Cancer	White Female	1950-59	189 except 189.3	9	NCI1983	MW
FKDNY60	Kidney Cancer	White Female	1960-69	189 except 189.3	9	NCI1983	MW
FKDNY70	Kidney Cancer	White Female	1970-79	189 except 189.3	9	NCI1983	MW
FKDNY96	Kidney Cancer	White Female	1979-96	189 except 189.3	9	CDC	MW
FLEUK50	Leukemia	White Female	1950-59	204-208, 202.4, 203.1	9	NCI1983	MW
FLEUK60	Leukemia	White Female	1960-69	204-208, 202.4, 203.1	9	NCI1983	MW
FLEUK70	Leukemia	White Female	1970-79	204-208, 202.4, 203.1	9	NCI1983	MW
FLEUK96	Leukemia	White Female	1979-96	204.0-208.9	9	CDC	MW

U.S. FALLOUT ATLAS : TOTAL FALLOUT

TABLE G-9
INTERNATIONAL CLASSIFICATION OF DISEASES CODES USED IN U.S. ATLAS OF NUCLEAR FALLOUT 1951-1962

ABBREV	VARIABLE	GENDER	DATES	ICD	ICD PUB. No.	DATA-BASE*	POPULATION EVALUATED**
FLS50	Lymphosarcoma reticulum cell sarcoma and other lymphoma	White Female	1950-59	200, 202, 159.1, 202.0, 1, 8, 9	9	NCI1983	MW
FLS60	Lymphosarcoma reticulum cell sarcoma and other lymphoma	White Female	1960-69	200, 202, 159.1, 202.0, 1,8,9	9	NCI1983	MW
FLS70	Lymphosarcoma reticulum cell sarcoma and other lymphoma	White Female	1970-79	200, 202, 159.1, 202.0, 1,8,9	9	NCI1983	MW
FLS95	Lymphosarsoma	White Female	1979-95	200.1	9	CDC	MW
FMM50	Multiple Myeloma	White Female	1950-59	203 except 203.1	9	NCI1983	MW
FMM60	Multiple Myeloma	White Female	1960-69	203 except 203.1	9	NCI1983	MW
FMM70	Multiple Myeloma	White Female	1970-79	203 except 203.1	9	NCI1983	MW
FMM96	Multiple Myeloma	White Female	1979-96	203 except 203.1	9	CDC	MW
FPAN96	Pancreatic Cancer	White Female	1979-96	157	9	CDC	MW
FSTOM50	Stomach Cancer	White Female	1950-59	151	9	NCI1983	MW
FSTOM60	Stomach Cancer	White Female	1960-69	151	9	NCI1983	MW
FSTOM70	Stomach Cancer	White Female	1970-79	151	9	NCI1983	MW
FSTOM96	Stomach Cancer	White Female	1979-96	151.0-151.9	9	CDC	MW
FTHYM95	Thymus Cancer	White Female	1979-95	164.0-164.0	9	CDC	MW
FTHYR50	Thyroid Cancer	White Female	1950-59	193	9	NCI1983	MW
FTHYR60	Thyroid Cancer	White Female	1960-69	193	9	NCI1983	MW
FTHYR70	Thyroid Cancer	White Female	1970-79	193	9	NCI1983	MW
FTHYR96	Thyroid Cancer	White Female	1979-96	193.0-193.9	9	CDC	MW
KIDWF5	Kidney Cancer	White Female	1950-69	180, 189.0-189.2	6, 7, 8	NCI2000	US

U.S. FALLOUT ATLAS : TOTAL FALLOUT

TABLE G-9
INTERNATIONAL CLASSIFICATION OF DISEASES CODES USED IN U.S. ATLAS OF NUCLEAR FALLOUT 1951-1962

ABBREV	VARIABLE	GENDER	DATES	ICD	ICD PUB. No.	DATA-BASE*	POPULATION EVALUATED**
KIDWF7	Kidney Cancer	White Female	1970-95	189.0-189.2	8,9	NCI2000	US
KIDWM5	Kidney Cancer	White Male	1950-69	180, 189.0-189.2	6, 7, 8	NCI2000	US
KIDWM7	Kidney Cancer	White Male	1970-95	189.0-189.2	8, 9	NCI2000	US
LEUWF5	Leukemia	White Female	1950-69	204, 204-207	6, 7, 8	NCI2000	US
LEUWF7	Leukemia	White Female	1970-95	202.4, 203.1, 204-208	8,9	NCI2000	US
LEUWM5	Leukemia	White Male	1950-69	204, 204-207	6, 7, 8	NCI2000	US
LEUWM7	Leukemia	White Male	1970-95	202.4, 203.1, 204-208	8, 9	NCI200	US
LNGWF5	Lung Cancer	White Female	1950-69	162, 163	6, 7, 8	NCI2000	US
LNGWF7	Lung Cancer	White Female	1970-95	162, 163	8, 9	NCI2000	US
LNGWM5	Lung Cancer	White Male	1950-69	162, 163	6, 7, 8	NCI2000	US
LNGWM7	Lung Cancer	White Male	1970-95	162, 163	8,9	NCI2000	US
LVWF5	Liver Cancer	White Female	1950-69	155, 156, 197.8	6, 7, 8	NCI2000	US
LVWF7	Liver Cancer	White Female	1970-95	155, 156, 197.8	8,9	NCI2000	US
LVWM5	Liver Cancer	White Male	1950-69	155, 156, 197.8	6,7,8	NCI2000	US
LVWM7	Liver Cancer	White Male	1970-95	155, 156, 197.8	8,9	NCI2000	US
MALL50	All Cancer	White Male	1950-59	140.0-205.0	9	NCI1983	MW
MALL60	All Cancers	White Male	1960-69	140.0-208.0	9	NCI1983	MW
MALL70	All Cancer	White Male	1970-79	140.0-208.0	9	NCI1983	MW
MALL96	All Cancers	White Male	1979-96	140.0-208.0	9	CDC	MW
MBONE50	Bone Cancer	White Male	1950-59	170	9	NCI1983	MW

U.S. FALLOUT ATLAS : TOTAL FALLOUT

TABLE G-9
INTERNATIONAL CLASSIFICATION OF DISEASES CODES USED IN U.S. ATLAS OF NUCLEAR FALLOUT 1951-1962

ABBREV	VARIABLE	GENDER	DATES	ICD	ICD PUB. No.	DATA-BASE*	POPULATION EVALUATED**
MBONE60	Bone Cancer	White Male	1960-69	170	9	NCI1983	MW
MBONE70	Bone Cancer	White Male	1970-79	170	9	NCI1983	MW
MBONE96	Bone Cancer	White Male	1979-95	170	9	CDC	MW
MBRAIN50	Brain and Other Parts of Nervous System Cancer	White Male	1950-59	191-192	9	NCI1983	MW
MBRAIN60	Brain and Other Parts of Nervous System Cancer	White Male	1960-69	191-192	9	NCI1983	MW
MBRAIN70	Brain and Other Parts of Nervous System Cancer	White Male	1970-79	191-192	9	NCI1983	MW
MBRAIN96	Brain Cancer	White Male	1979-95	191.0-191.9	9	CDC	MW
MCLN50	Colon Cancer	White Male	1950-59	153, 159	9	NCI1983	MW
MCLN60	Colon Cancer	White Male	1960-69	153, 159	9	NCI1983	MW
MCLN70	Colon Cancer	White Male	1970-79	153, 159	9	NCI1983	MW
MCLN96	Colon Cancer	White Male	1979-96	153.0-153.9	9	CDC	MW
MCONSAR9	Connective and Soft Tissue Neoplasms	White Male	1979-96	171.0-171.0	9	CDC	MW
MEYE50	Eye Cancer	White Male	1950-59	190	9	NCI1983	MW
MEYE60	Eye Cancer	White Male	1960-69	190	9	NCI1983	MW
MEYE70	Eye Cancer	White Male	1970-79	190	9	NCI1983	MW
MEYE96	Eye Cancer	White Male	1979-95	190	9	CDC	MW
MKDNY50	Kidney Cancer	White Male	1950-59	189 except 189.3	9	NCI1983	MW
MKDNY60	Kidney Cancer	White Male	1960-69	189 except 189.3	9	NCI1983	MW
MKDNY70	Kidney Cancer	White Male	1970-79	189 except 189.3	9	NCI1983	MW

U.S. FALLOUT ATLAS : TOTAL FALLOUT

TABLE G-9
INTERNATIONAL CLASSIFICATION OF DISEASES CODES USED IN U.S. ATLAS OF NUCLEAR FALLOUT 1951-1962

ABBREV	VARIABLE	GENDER	DATES	ICD	ICD PUB. No.	DATA-BASE*	POPULATION EVALUATED**
MKDNY96	Kidney Cancer	White Male	1979-96	189.0-189.1	9	CDC	MW
MLEUK50	Leukemia	White Male	1950-59	204-208, 202.4, 203.1	9	NCI1983	MW
MLEUK60	Leukemia	White Male	1960-69	204-208, 202.4, 203.1	9	NCI1983	MW
MLEUK70	Leukemia	White Male	1970-79	204-208, 202.4, 203.1	9	NCI1983	MW
MLEUK96	Leukemia	White Male	1979-96	204.0-208.9	9	CDC	MW
MLIV96	Liver Cancer	White Male	1979-96	155.0-155.1	9	CDC	MW
MLS50	Lymphosarcoma, reticulum cell sarcoma and other lymphoma	White Male	1950-59	200, 202, 159.1, 202.0, 1,8,9	6,7	NCI1983	MW
MLS60	Lymphosarcoma reticulum cell sarcoma and other lymphoma	White Male	1960-69	200, 202, 159.1, 202.0, 1,8,9	7,8	NCI1983	MW
MLS70	Lymphosarcoma reticulum cell sarcoma and other lymphoma	White Male	1970-79	200, 202, 159.1, 202.0, 1,8,9	8,9	NCI1983	MW
MLS95	Lymphosarcoma	White Male	1979-95	200.1	9	CDC	MW
MMM50	Multiple Myeloma	White Male	1950-59	203 except 203.1	9	NCI1983	MW
MMM60	Multiple Myeloma	White Male	1960-69	203 except 203.1	9	NCI1983	MW
MMM70	Multiple Myeloma	White Male	1970-79	203 except 203.1	9	NCI1983	MW
MMM96	Multiple Myeloma	White Male	1979-96	203.0-203.9	9	CDC	MW
MPAN96	Pancreatic Cancer	White Male	1979-96	157.0-157.9	9	CDC	MW
MSTOM50	Stomach Cancer	White Male	1950-69	151	9	NCI1983	MW
MSTOM60	Stomach Cancer	White Male	1960-69	151	9	NCI1983	MW
MSTOM70	Stomach Cancer	White Male	1970-79	151	9	NCI1983	MW
MSTOM96	Stomach Cancer	White Male	1979-96	151.0-151.9	9	CDC	MW

TABLE G-9
INTERNATIONAL CLASSIFICATION OF DISEASES CODES USED IN U.S. ATLAS OF NUCLEAR FALLOUT 1951-1962

ABBREV	VARIABLE	GENDER	DATES	ICD	ICD PUB. No.	DATA-BASE*	POPULATION EVALUATED**
MTHYM95	Thymus	White Male	1979-95	164.0-164.0	9	CDC	MW
MTHYR50	Thyroid Cancer	White Male	1950-69	193	9	NCI1983	MW
MTHYR60	Thyroid Cancer	White Male	1960-69	193	9	NCI1983	MW
MTHYR70	Thyroid Cancer	White Male	1970-79	193	9	NCI1983	MW
MTHYR96	Thyroid Cancer	White Male	1979-96	193	9	CDC	MW
NHLWF5	Non-Hodgkins Lymphoma	White Female	1950-69	200, 202, 205	6,7,8	NCI2000	US
NHLWF7	Non-Hodgkins Lymphoma	White Female	1970-95	159.1, 200, 202 except 202.2-6, 205	8,9	NCI2000	US
NHLWM5	NonHodgkins Lymphoma	White Male	1950-69	200, 202, 205	6,7,8	NCI2000	US
NHLWM7	Non-Hodgkin's Lymphoma	White Male	1970-95	159.1, 200, 202 except 202.2-6, 205	8,9	NCI2000	US
ORAWF5	Oral Cancer	White Female	1950-69	141, 143-5, 147-8	6,7,8	NCI2000	US
ORAWF7	Oral Cancer	White Female	1970-95	141, 143-6, 148-9	8,9	NCI2000	US
ORAWM5	Oral Cancer	White Male	1950-69	141, 143-5, 147-8	6,7,8	NCI2000	US
ORAWM7	Oral Cancer	White Male	1970-95	141, 143-6, 148-9	8,9	NCI2000	US
OTHWF5	Other and Unspecified Cancer	White Female	1950-69	152, 156.2, 158-9, 163.1, 163.9,164, 165, 173.5, 176, 179, 184.8-9,195-9, 198-9 except 197.8	6,7,8	NCI2000	US
OTHWF7	Other and Unspecified Cancer	White Female	1970-95	152, 159-9, 163.1, 164.2-.9, 165, 173.5, 184.8-.9, 195-9 except 197.8	8,9	NCI2000	US

U.S. FALLOUT ATLAS : TOTAL FALLOUT

TABLE G-9
INTERNATIONAL CLASSIFICATION OF DISEASES CODES USED IN U.S. ATLAS OF NUCLEAR FALLOUT 1951-1962

ABBREV	VARIABLE	GENDER	DATES	ICD	ICD PUB. No.	DATA-BASE*	POPULATION EVALUATED**
OTHWM5	Other and Nonspecified Cancers	White Male	1950-69	152, 156.2, 158-9, 163.1, 163.9,164, 165, 173.5,176, 179, 184.8-9,195-9, 198-9 except 197.8	6,7,8	NCI2000	US
OTHWM7	Other and Nonspecified Cancers	White Male	1970-95	152, 159-9, 163.1, 164.2-9, 165, 173.5, 184.8-.9, 195-9 except 197.8	8,9	NCI2000	US
OVWF5	Ovarian Cancer	White Female	1950-69	175, 183	6,7,8	NCI2000	US
OVWF7	Ovarian Cancer	White Female	1970-95	183	8,9	NCI2000	US
PANWF5	Pancreatic Cancer	White Female	1950-69	157	6,7,8	NCI2000	US
PANWF7	Pancreatic Cancer	White Female	1970-95	157	8,9	NCI2000	US
PANWM5	Pancreatic Cancer	White Male	1950-69	157	6,7,8	NCI2000	US
PANWM7	Pancreatic Cancer	White Male	1970-95	157	8,9	NCI2000	US
PROWM5	Prostate Cancer	White Male	1950-69	177, 185	6,7,8	NCI2000	US
PROWM7	Prostate Cancer	White Male	1979-95	185	9	NCI2000	US
RECWF5	Cancer of Rectum	White Female	1950-69	154	6,7,8	NCI2000	US
RECWF7	Cancer of Rectum	White Female	1970-95	154 except 154.3	8,9	NCI2000	US
RECWM5	Cancer of Rectum	White Male	1950-69	154	6,7,8	NCI2000	US
RECWM7	Cancer of Rectum	White Male	1970-95	154 except 154.3	8,9	NCI2000	US
STOWF5	Stomach Cancer	White Female	1950-69	151	6, 7, 8	NCI2000	US
STOWF7	Stomach Cancer	White Female	1970-95	151	8, 9	NCI2000	US
STOWM5	Stomach Cancer	White Male	1950-69	151	6, 7, 8	NCI2000	US
STOWM7	Stomach Cancer	White Male	1970-95	151	8, 9	NCI2000	US

U.S. FALLOUT ATLAS : TOTAL FALLOUT

TABLE G-9
INTERNATIONAL CLASSIFICATION OF DISEASES CODES USED IN U.S. ATLAS OF NUCLEAR FALLOUT 1951-1962

ABBREV	VARIABLE	GENDER	DATES	ICD	ICD PUB. No.	DATA-BASE*	POPULATION EVALUATED**
UT50	Uterine Cancer	White Female	1950-59	179, 181, 182	9	NCI1983	US
UT60	Uterine Cancer	White Female	1960-69	179, 181, 182	9	NCI1983	US
UT70	Uterine Cancer	White Female	1970-79	179, 181, 182	9	NCI1983	US
UT96	Uterine Cancer	White Female	1979-96	182.0-182.0	9	CDC	MW
WFLIV96	Liver Cancer	White Female	1979-96	155.0-155.1	9	CDC	MW
WMLIV96	Liver Cancer	White Male	1979-96	155.0-155.1	9	CDC	MW

* Databases include:
CDC: Centers for Disease Control WONDER site: Disease rates per 100,000 population Age-Adjusted, Standard year: 1970.
NCI1983 Mason, Thomas J. and Wilson B. Riggan. U.S. Cancer Mortality Rates and Trends, 1970-1979. National Cancer Institute, 1983. Op cit.
NCI2000 U.S. Cancer Atlas 2000. National Cancer Institute, April, 2000.

**Populations analyzed:
US All U.S. counties including District of Columbia and 37 county subdivisions near the Nevada Test Site. Total N: 3094.
MW Midwest counties in the following states: Iowa, Illinois, Kansas, Missouri and Nebraska. Total N:

U.S. FALLOUT ATLAS : TOTAL FALLOUT

NOTES AND REFERENCES

RADIATION SYMBOLS

EC: Electron capture
IT: Isomeric Transition (decay from an excited metastable state to a lower state)
α: Alpha particle emission
β^- Negative beta particle emission
γ: Gamma ray emission
e^- electron production

REFERENCES TO APPENDIX G

1. Table A1:IIa. Derived Concentration Limits of Radionuclides in Air and Water for Occupational Exposure (40 h/week). P. 59. Safe Handling of Radionuclides 1973 Edition. International Atomic Energy Agency, Vienna, 1973.
2. Safe Handling of Radionuclides 1973 Edition. Code of Practice : Sponsored by the International Atomic Energy Agency and the World Health Organization. International Atomic Energy Agency, Vienna, 1973. Table AI-IIa.
3. Table 2.1, 2.2, 2.3. Federal Guidance Report No. 13, Part I Interim Version. Health Risks from Low-Level Environmental Exposure to Radionuclides. Radionuclide-Specific Lifetime Ragiogenic Cancer Risk Coefficients for the U.S. Population, Based on Age-Dependent Intake, Dosimetry, and Risk Models. Oak Ridge National Laboratory, Office of Radiation and Indoor Air, United States Environmental Protection Agency, Washington, DC. 20460. EPA 402-R-97-014. Jan, 1998.[1] Radiological Health Handbook. Ibid. p.336.
4. Radiological Health Handbook: US Department of Health, Education and Welfare. January, 1970. P.275

BIBLIOGRAPHY

1. Allen, Philip W. and Lester Machta, *Operation Buster: Transport of Radioactive Debris from Operations Buster and Jangle. Project 7.1.* Armed Forces Special Weapons Project, Washington, D.C. March 15, 1952, WT 308.

2. *Announced United States Nuclear Tests July 1945 Through December 1982.* NVO-209 (Rev 3) Office of Public Affairs, U.S. Department of Energy, Nevada Operations Office. in Cooperation with Los Alamos National Laboratory, Lawrence Livermore National Laboratory, Sandia National Laboratories. January, 1983.

3. Bogen, Kenneth T. and Abraham S. Goldin. *Population Exposure to External Natural Radiation Background in the United States*, ORP/SEPD-80-12. Surveillance and Emergency Preparedness Division, Office of Radiation Programs, U.S. Environmental Protection Agency, Washington, DC. 20460. April, 1981.

4. Breiman, Leo, Jerome Friedman, Richard Olshen, and Charles Stone. *Classification and Regression Trees.* Pacific Grove: Wadsworth, 1984.

5. Cotton, William R. and Richard A. Anthes, *Storm and Cloud Dynamics*, Vol 44 in International Geophysics Series. Academic Press, Inc. Harcourt Brace Jovanovich, Publishers. San Diego, CA. 1989.

6. Devesa, Susan S., Dan J. Grauman, William J. Blot, Gene A. Pennello, Robert N. Hoover, Joseph F. Fraumeni Jr; *Atlas of Cancer Mortality in the United States 1950-94.* National Institutes of Health, National Cancer Institute NIH Publication No. 99-4564, September, 1999.

7. Beck, Harold L., Irenke K. Helfer, Andre Bouville, and Mona Dreicer. "Estimates of Fallout in the Continental U.S. from Nevada Weapons Testing Based on Gummed-Film Monitoring Data" *Health Physics* Vol. 59. No. 5. (November) pp 565-576, 1990.

8. Bouville, Andre et al. *Estimated Exposures and Thyroid Doses Received by the American People from Iodine-I-131 in Fallout Following Nevada Atmospheric Nuclear Bomb Tests. A Report from the National Cancer Institute.* U.S. Department of Health and Human Services, National Institutes of Health, National Cancer Institute. October, 1997.

9. *Cancer Epidemiology and Prevention.*, Editors: Schottenfeld, David and Joseph F. Fraumeni, Jr... Second Edition. Oxford University Press. 1996.

10. Centers for Disease Control: http://wonder.cdc.gov/

11. Coleman, Hugh W. and W. Glenn Steele, *Experimentation and Uncertainty Analysis for Engineers.* 2nd Ed. John Wiley and Sons, Inc. New York. 1999.

12. Corcoran, Brent Coull and Aneesh Patel. *EGRET for Windows. User Manual.* Cytel Statistical Software, Cytel Software Corporation 765 Massachusetts Avenue, Cambridge, MA 02139. August 1999.

13. Defense Nuclear Agency 6001F-6008F.

14. *Defense Nuclear Agency Fact Sheet*: Subject: Teapot Series DNA 6010F-6013F.

15. *Operation Tumbler-Snapper.* Defense Nuclear Agency 6020F. "

16. Doviak, Richard J. and Dušan S. Zrnić. Doppler Radar and Weather Observations. 2nd Ed. Academic Press, Inc. Harcourt Brace Jovanovich, Publishers, San Diego., 1993.

17. Eckerman, Keith F. et al. *Federal Guidance Report No. 13: Health Risks from Low-Level Environmental Exposure to Radionuclides. Part I. Interim Version.* EPA 402-R-97-014. Oak Ridge Natonal Laboratory, Office of Radiation and Indoor Air, United States Environmental Protection Agency, Washington, DC. 20460. Jan 1998.

18. Eckerman, Keith F. and Jeffrey C. Ryman. *Federal Guidance Report No. 12: "External Exposure to Radionuclides in Air, Water and Soil" Exposure-to-Dose Coefficients for General Application*, Based on the 1987 Federal Radiation Protection Guidance. Oak Ridge National Laboratry, Office of Radiation and Indoor Air. U.S. Environmenal Protection Agency, Washington DC. 20460. 1993.

19. *Effects on Populations of Exposure to Low Levels of Ionizing Radiation:* 1980. National Research Council, Washington DC. November 1980

20. Elwood, Mark, *Critical Appraisal of Epidemiological Studies and Clinical Trials* 2nd Ed. Oxford University Press, Oxford, UK, 1998.

21. Emsley, John, *The Elements.* 3rd Ed. Clarendon Press. Oxford UK, 1998.

22. *Exposure of the American People To Iodine-131 From Nevada Nuclear-Bomb Tests. Review of the National Cancer Institute Report and Public Health Implications.* Institute of Medicine., National Research Council. National Academy Press. Washington DC 1999.

23. Friedman J.H. *Multivariate Adaptive Regression Splines,* Annals of Statistics, 19, 1-141 (March). Salford Systems. San Diego, CA. 1999.

24. "Fallout From Nuclear Weapons Tests": *Hearings Before the Special Subcommittee on Radiation of the Joint Committee on Atomic Energy, Congress of the United States, 86th Congress, First Session, On the Fallout From Nuclear Weapons Tests.* Washington, DC. 1959.

25. Ferber, Gilbert J. et al. *Long-Range Transport and Diffusion Experiments.* AD-764 893. Advanced Research Project Agency. May, 1973.

26. Francis, Magnus B. *Toxic Substances in the Environment.* Wiley-Interscience. New York, 1994.

27. *General and Applied Toxicology.* Ed: Bryan Ballantine, Timothy Marrs, Paul Turner, Stockton Press, New York, NY., 1995.

28. Glasstone, Samuel and Philip J. Dolan, *The Effects of Nuclear Weapons.,* U.S. Department of Energy and Department of Defense, 1977.

29. HASL-142, *Quarterly Summary Report,* Jan 1, 1964, United States Atomic Energy Commission, Health and Safety Laboratory, New York, NY. 1964.

30. *Health Impacts of Large Releases of Radionuclides.* Ciba Foundation Symposium 203. John Wiley & Sons, New York, NY, 1997.

31. Hicks, Harry G. *Results of Calculations of External Gamma-Radiation Exposure Rates from Fallout and the Related Radionuclide Compositions. Operation Ranger 1951..* Lawrence Livermore National Lab. CA. Jul 1981. UCRL-53152 Pt. 1.

32. Hicks, Harry G. *Results of Calculations of External Gamma-Radiation Exposure Rates from Fallout and the Related Radionuclide Compositions. Operation Buster-Jangle 1951.* Lawrence Livermore National Lab. CA. Jul 1981. UCRL-53152 Pt. 2

33. Hicks, Harry G. *Results of Calculations of External Gamma-Radiation Exposure Rates from Fallout and the Related Radionuclide Compositions. Operation*

Tumbler-Snapper 1952. Lawrence Livermore National Lab. CA. Jul 1981. UCRL-53152 Pt. 3

34. Hicks, Harry G. *Results of Calculations of External Gamma-Radiation Exposure Rates from Fallout and the Related Radionuclide Compositions. Operation Upshot-Knothole 1953* Lawrence Livermore National Lab. CA. Jul 1981. UCRL-53152 Pt. 4

35. Hicks, Harry G. *Results of Calculations of External Gamma-Radiation Exposure Rates from Fallout and the Related Radionuclide Compositions. Operation Teapot 1955.* Lawrence Livermore National Lab. CA. Jul 1981. UCRL-53152 Pt. 5.

36. Hicks, Harry G. *Results of Calculations of External Gamma-Radiation Exposure Rates from Fallout and the Related Radionuclide Compositions. Operation Plumbbob 1957.* Lawrence Livermore National Lab. CA. Jul 1981. UCRL-53152 Pt. 6.

37. Hicks, Harry G. *Results of Calculations of External Gamma-Radiation Exposure Rates from Fallout and the Related Radionuclide Compositions. Operations Nougat through Bowline. 1962-1968* Lawrence Livermore National Lab. CA. Jul 1981. UCRL-53152 Pt. 8.

38. Hicks, Harry G. "Calculation of the Concentration of Any Radionuclide Deposited on the Ground by Offsite Fallout from a Nuclear Detonation," *Health Physics. Vol 42. No 5. (May), pp 585-600. 1982.*

39. Hicks, Harry G. "Additional Calculations of Radionuclide Production Following Nuclear Explosions and Pu Isotopic Ratios for Nevada Test Site Events. " *Health Physics.* Vol 59. No. 5 (November) pp 515-523. 1990.

40. Hirsch, Robert P. and Richard K. Riegelman: *Statistical First Aid: Interpretation of Health Research Data.* Blackwell Science. Cambridge, MA, 1992.

41. Houze, Robert A. Jr. *Cloud Dynamics.* Volume 53 in the International Geophysics Series. Academic Press, Inc. Harcourt Brace and Company. San Diego, CA, 1993.

42. *ICD 9 CM Code Book.* Volumes 1,2,3. 1998. St. Anthony Publishing, Inc. Reston VA. September 1997.

43. Kachigan, Sam Kash *Multivariate Statistical Analysis: A Conceptual Introduction.* 2nd Ed. Radius Press New York. 1991.

44. Keller, Edward A. *Environmental Geology*, 2nd Ed. Charles E. Merrill Publishing Co. Columbus, OH, 1979.

45. Knapp, Rebecca G. and M. Clinton Miller III. *Clinical Epidemiology and Biostatistics*, Williams and Wilkins. Baltimore MD, Philadelphia PA, 1992.

46. Krey, Philip W. , Merrill Heit and Kevin M. Miller. "Radioactive Fallout Reconstruction from Contemporary Measurements of Reservoir Sediments." *Health Physics* Vol. 59. No. 5. (November) pp 541-554, 1990.

47. Larson, Kermit H., Howard A. Hawthorne and Keith R. Price. *Preliminary Report on Rain Out from Shots Diablo and Stokes in the Belle Fource-Rapid City, South Dakota Area. Program 37.* August 12-August 17, 1957. Date of Report: Aug 27, 1957.

48. Lippman, Morton and Richard B. Schlesinger, *Chemical Contamination in the Human Environment.* Oxford University Press. New York, 1979.

49. List, Robert J. *Radioactive Debris from Operations Tumbler and Snapper, " NYO-4512.* U.S. Weather Bureau, February 25, 1953.

50. List, Robert J. *The Transport of Atomic Debris from Operation Upshot-Knothole" NYO-4602.* United States Atomic Energy Commission, Technical Information Services, Oak Ridge, TN. Weather Bureau, Washington, DC. Jun 25, 1954.

51. List, Robert J. *Radioactive Fallout in North America from Operation Teapot, NYO-4696*, U.S. Atomic Energy Commission, Washington, DC. February, 1956.

52. Miller, Richard L. *Under The Cloud: The Decades of Nuclear Testing* Macmillan. New York, 1986; Two-Sixty Press 1998.

53. Mason, Thomas J., Frank W. McKay, Robert Hoover, William J. Blot, Joseph F. Fraumeni Jr. *Atlas of Cancer Mortality for U.S. Counties: 1950-1969.* DHEW Publication No. (NIH) 75-780. . Epidemiology Branch, National Cancer Institute. 1975.

54. Mason, Thomas J. , Wilson B. Riggan, and John F. Acquavella *U.S. Cancer Mortality Rates and Trends 1950-1979* Vol I, II, III. NCI/EPA Interagency Agreement on Environmental Carcinogenesis. EPA-600/ 1-83-015b, September, 1983.

55. *Microsoft Excel 97 Worksheet Function Reference.* Microsoft Press. 1997.

56. Morganthau, Manfred; Harvey Meieran, Richard Showers, Jeffrey Morse, Norman Dombeck, Arnoldo Garcia., *Local Fallout from Nuclear Test Detonations*, and *Compilation of Fallout Patterns and Related Test Data*. DASA 1251 Vols 1 and Vol 2.. U.S. Army Nuclear Defense Laboratory., Defense Atomic Support Agency. Undated.

57. NOAA Technical Memorandum ERL ARL-36: March 1972.

58. Norman, Geoffrey G. and David L. Streiner, *Biostatistics: The Bare Essentials*. B.C. Decker, Inc. Hamilton, Ontario, Canada. 1998.

59. *Occupational, Industrial, and Environmental Toxicology.* Hamilton, Richard J MD, Ed. Scott D Phillips MD ,Ed. Michael I Greenberg MD, MPH Ed. in Chief: Mosby Publishing Co. St. Louis. 1997.

60. Page, Randy M. ; Galen E. Cole and Thomas C. Timmreck. *Basic Epidemiological Methods and Biostatistics.,* Jones and Bartlett Publishers, Sudbury, MA. 1998.

61. Petrie, Aviva and Caroline Sabin, *Medical Statistics at a Glance*. Blackwell Science, Oxford, UK. 2000.

62. Proctor, Robert N. *Cancer Wars: How Politics Shapes What We Know and Don't Know About Cancer*. Perseus Books LLC. New York. 1995.

63. *Radioactive Debris from Operations Upshot and Knothole NYO-4552* . Health and Safety Laboratory, New York Operations Office. Jun 25, 1954.

64. *Radiological Effluents Released From U.S. Continental Tests 1961 Through 1992.* United States Department of Energy Nevada Operations Office.DOE/NV-3117 (Rev. 1) UC-702. August 1996.

65. *Radiation Protection Theory and Practice* Proceedings of the 25th Anniversary Symposium of the Society for Radiological Protection, Malvern 4-9. Edited by E.P. Goldfinch. IOP Publishing Ltd. Bristol, UK.1989.

66. Roman, Steven, *Writing Excel Macros*, O'Reilly Publications, Cambridge, MA. May, 1999.

67. *Safe Handling of Radionuclides*, 1973 Edition. International Atomic Energy Agency, Vienna, 1973.

68. Snedecor, George W. and William G. Cochran. *Statistical Methods*. Iowa State University Press, Ames IA 1989.

69. *Spatial Epidemiology Methods and Applications.*, Ed: P. Elliott, J.C. Wakefield, N.G. Best and D.J. Briggs. Oxford University Press. Oxford, UK. 2000.

70. Steinberg, Dan and Phillip Colla. *CART: A Tree-Structured Non-Parametric Data Analysis.* San Diego, CA. Salford Systems, 1995.

71. Szklo, Moyses and E. Javier Nieto. *Epidemiology: Beyond the Basics.* Aspen, Gaithersburg, MD 2000.

72. *Thunderstorm Morphology and Dynamics*: Vol II of *Thunderstorms: A Social, Scientific and Technological Documentary.* 2nd Ed. Edited by Edwin Kessler, University of Oklahoma Press, Norman, OK 1983.

73. *Toxicological Profile for Ionizing Radiation. Draft for Public Comment.* U.S. Department of health and Human Services, Public Health Service, Agency for Toxic Substances and Disease Registry,. September, 1997.

74. *Transport, Deposition and Meteorological Studies of Nuclear Debris in the Atmosphere.* R.J. List (Ed). T.W. Ashenfelter, G.J. Ferber, J.L. Heffter, L. Machta, A.D. Taylor and K. Telegadas. AEC, Division of Biology and Medicine Fallout Studies Branch, Washington, DC.

75. *United States Nuclear Tests July 1945 through September 1992.* Department of Energy, DOE/NV-209 (Rev 14) December 1994.

76. *U.S. Cancer Mortality Rates and Trends 1950-1969* U.S. Environmental Protection Agency and National Cancer Institute William B. Riggan and Thomas J. Mason Eds. EPA-600/1-83-015a. Sep 1983.

77. *U.S. Cancer Atlas.* National Cancer Institute. 2000. http://www2.nci.nih.gov

78. Yu C. et al. . *Manual for Implementing Residual Radioactive Material Guidelines Using RESRAD, Version 5.0.* Working Draft for Comment. Environmental Assessment Division, Argonne National Laboratory. Operated by The University of Chicago under Contract W-31-109-Eng-38 for the United States Department of Energy. September, 1993.

79. Webb, Jeff: *Using Excel Visual Basic* 2nd Edition. Que. Corporation. Indianapolis, IN. 1996.

80. *What Risk? Science, Politics and Public Health.* Ed. Roger Bate. Butterworth Heinemann. Oxford, UK. 1997.

81. Wilhelmson, Robert. et al. *Study of a Numerically Modeled Severe Storm.* PATHFINDER Sampler, COMMAS Images and Multimedia Teaching Modules. National Center for Supercomputing Applications, Champaign IL. 1988.

ABOUT THE AUTHOR

Richard L. Miller is an industrial health specialist with field experience in onsite coordination of more than 500 safety and health investigations primarily in petrochemical and energy-related industries. His major areas of focus have included retrospective event and exposure reconstruction, and event cluster evaluation. In 1979 Miller directed epidemiological field investigations on the relationship between rare brain cancers and chemical workplace exposures, and for this work received an OSHA Special Incentive Award and letter of commendation from the director of NIOSH for "findings of national impact and importance."

Miller has testified as an expert in federal cases involving toxic exposure and has performed onsite risk evaluations of industrial sites in such diverse places as Mexico, Soviet Latvia, Estonia, the Russian Federation (Tataria) and the Arabian Gulf. He has also performed extensive research on the nuclear test program and is the author of the 1986 book *Under The Cloud: the Decades of Nuclear Testing*, chosen by the *Library Journal* as one of the top science-technical books of 1986. Miller is a member of the American Statistical Association and is currently completing a technical book on calculating industrial exposures.

U.S. FALLOUT ATLAS : TOTAL FALLOUT